The Wild Mammals of Missouri

The Wild Mammals of Missouri

Third Revised Edition

Charles W. Schwartz and Elizabeth R. Schwartz

Edited by Debby K. Fantz and Victoria L. Jackson

Partial funding was made possible by the Missouri Department of Conservation

UNIVERSITY OF MISSOURI PRESS and MISSOURI DEPARTMENT OF CONSERVATION
Columbia

University of Missouri Press, Columbia, Missouri 65211
Printed and bound in the United States of America
 First Printing 2016

Cataloging-in-Publication data available from the Library of Congress

ISBN: 978-0-8262-2088-2

∞™ This paper meets the requirements of the American National Standard for Permanence of Paper for Printed Library Materials, Z39.48, 1984.

Typefaces: Minion and Frutiger

The University of Missouri Press and the Missouri Department of Conservation gratefully acknowledge the State Historical Society of Missouri for providing access to the original artwork for the plates by Charles Schwartz reproduced herein.

Contents

CHARLES W SCHWARTZ

Species Illustrations

Foreword

How many a man has dated a new era in his life from the reading of a book!

–Thoreau, "Reading," ***Walden*** **1854**

I remember holding my first copy of *The Wild Mammals of Missouri* and thinking what an incredible book it was. I still have the 1981 revised edition. Although it has lost its glossy newness from years of reading and rereading, penciling in personal notes, and tabbing pages to mark points of personal relevance, that copy of *The Wild Mammals of Missouri* remains a prized possession. It informed and enriched my life growing up in central Missouri. Since 2001, the second revised edition became my new go-to mammal guide while also marking the passage of my life as a Missourian, and this third revised edition will guide me into the future.

The Wild Mammals of Missouri was first published in 1959, and although it continues to be "widely acclaimed as the definitive work on its subject," as former Missouri Department of Conservation director Larry R. Gale put it, this book has proven to be more than an academic resource. Each edition is a chronicle of the reintroduction of extirpated mammals and the recovery of waning mammal populations in Missouri. It is a success story of conservation efforts in management and habitat work.

As you turn the pages of this book, I encourage you to appreciate not only the value of the words but also the priceless wonder of Charles Schwartz's art. Mark Raithel's illustrations, new in this edition, maintain the high standards of knowledge, quality, scientific accuracy, and artistic legacy set forth by Charles and Libby Schwartz. Mark's art commemorates the reintroduction of elk into Missouri and elevates this large mammal from a mere reference in the back of earlier editions to a fully illustrated field guide section of its own. Mark also introduces Missourians to the Seminole bat, a new but rare occurrence in Missouri.

Missourians can be proud of this third revised edition of a classic natural history reference. It demonstrates their continued commitment to conservation efforts and the value of partnerships with local communities, landowners, and organizations. It reflects successful layers of management decisions based on technical research to ensure the existence of mammals in future landscapes. It celebrates the economic value of Missouri's mammals and the important part they play in preserving Missouri's hunting heritage. It provides detailed natural history information in a way that enhances understanding of Missouri mammals and empowers Missourians to enjoy wildlife viewing and hunting opportunities.

This edition of *The Wild Mammals of Missouri* validates the support and confidence Missourians have in conservation efforts. It reinforces the Department's promise to continue efforts that will provide Missourians the excitement of seeing black bear and elk tracks alongside those of deer and bobcat.

I invite you to make this book your go-to guide on Missouri mammals. I know it will be my companion through coming decades of Missouri habitat and wildlife conservation efforts and success stories.

Robert L. Ziehmer, Director
Missouri Department of Conservation
June 23, 2015

Preface to the Third Revised Edition

Many students, biologists, and other conservation-minded lovers of nature have enjoyed *The Wild Mammals of Missouri* by Charles W. and Dr. Elizabeth R. Schwartz since its first publication in 1959. This successful book, which focused on amazing drawings by Charles and an enormous amount of natural history information, is the definitive guide to the identification of the wild mammals of Missouri and a reference to facts about their lives. Its enormous popularity as a college textbook and general reference led to two revised editions: the 1981 first revised edition, edited by both authors, and the 2001 second revised edition, edited by Elizabeth Schwartz.

The wildlife world lost two great conservationists when Charles Schwartz died in 1991 and Elizabeth died in 2013. These distinguished naturalists were a formidable team with over seventy years of combined work for the Missouri Department of Conservation. Their contributions, through this book and many other publications, wildlife research, wildlife photography, wildlife art, motion pictures, and numerous television and other programs, helped people to better understand mammals and other wildlife of Missouri. This third revised edition is dedicated to them.

As biologists with careers focused on mammalian research and wildlife education, it was a professional dream of ours to ensure that this portion of the Schwartzes' conservation legacy continued to educate, amaze, and inspire others through a third revised edition of *The Wild Mammals of Missouri*. We feel that we have captured the spirit of the book, which is to promote a better understanding of, greater appreciation for, and deeper commitment to the conservation of these interesting and important elements of our wild heritage.

The Wild Mammals of Missouri has evolved from 63 full species accounts in the 1959 edition to 72 in this third revised edition. The basic structure of the book was maintained, but some changes include recent revisions in both common and scientific names; the updating of distribution maps, including a new focus on Missouri county-level range information; the inclusion of metric measurements throughout the entire book; and incorporation of recent and ongoing research. Notable changes in species include the addition of nine new permanent resident mammals; successful reintroductions of the river otter and elk; range expansions of the black bear, the armadillo, and the meadow vole; and the unfortunate extirpations of the red wolf and white-tailed jackrabbit.

We are indebted to many persons for their assistance during the revision and publication of this book. For their support and encouragement throughout this process, we thank Robert L. Ziehmer, Director of the Missouri Department of Conservation (MDC); James T. Blair, IV, David W. Murphy, Marilynn J. Bradford, and Don C. Bedell of the Missouri Conservation Commission; Dr. Mike Hubbard, Dr. Rochelle Renken, Tom Kulowiec, and Dr. Charles Anderson of the MDC Resource Science Division; Mike Huffman, Mary Lyon, and Chris Weiberg of the MDC Outreach and Education Division; staff at the University of Central Missouri, especially Alice Greife, Dean of College of Health, Science, and Technology; and Clair Willcox, Sara Davis, and Jane Lago with the University of Missouri Press.

We especially thank Philip Marley, MDC GIS Specialist, for the many, many hours he spent designing, creating, and editing the range maps. Special thanks go to several additional MDC staff, including Dr. Lonnie Hansen, Dr. Barb Keller, and Dave Hasenbeck, for writing the elk account and reviewing the elk maps; Tony Elliott and Shelly Colatskie for extensive review of the bat chapter and range maps; and Jason Sumners for his thorough review of the white-tailed deer account and maps. In addition we are very grateful for the help of Dr. Lynn Robbins of Missouri State University for reviewing the armadillo and black-tailed jackrabbit chapters, and all range maps; and Dr. John Scheibe of Southeast Missouri State University for editing the swamp rabbit account and all range maps. We thank numerous other colleagues who reviewed range maps, provided data, and edited species accounts, including Rachel Wright, Cliff White, Krista Noel, Julie Fleming, Dorothy Butler, Justan Blair, Jeff Beringer, Susan Farrington, Clay Creech, Andrea Schumann, John Vogel, Larry Rizzo, Rhonda

Rimer, Beth Emmerich, Tim Russell, Marsha Jones, Mike Jones, Mark Reed, Jim Braithwait, Eric Abbott, Tom Meister, Daryl Damron, John George, and Chris Newbold with MDC; Waylon Hiller with Missouri Valley College; Dr. Jay McGhee with Northwest Missouri State University; Dr. Cody Thompson with the University of Michigan; and Dr. Russel Pfau with Tarleton State University. The University of California Press supplied the nine-banded armadillo plate, and MDC supplied the armadillo drawing. Mark Raithel, MDC Wildlife Artist, created the elk and Seminole bat drawings. Susan Farrington, MDC Natural History Biologist, supplied the Jacks Fork River photograph; Dr. Jeff Briggler, MDC Herpetologist, supplied the Big Cane Conservation Area slough photograph; and Noppadol Paothong, MDC Photographer, supplied the Golden Prairie and Dunn Ranch Prairie photographs. Kris Anstine, Gary Cox, and the Archives of the University of Missouri in Columbia allowed us access to historical mammal information and the use of photographs of the Schwartzes working on the 1959 edition of *The Wild Mammals of Missouri*. Staff from numerous vertebrate museums generously supplied us with information on Missouri mammals maintained in their collections. On a personal note, we thank our spouses, John R. Fantz, Jr., and Dr. Chad King, and our families, dear friends, and mentors, especially Dave Hamilton, Dr. Erik Fritzell, Vicki Heidy, Dr. Jerry Choate, and Dr. Earl Zimmerman, for their support of our conservation work.

Debby K. Fantz
Columbia, Missouri
Dr. Victoria L. Jackson
Edmond, Oklahoma
June 2015

Preface to the Second Revised Edition

Mammals have been the dominant animals on the earth for the last 65 million years, following the Age of Dinosaurs. During this time, many kinds of mammals have lived in the interior plains of North America where our state of Missouri is located. It is only recently—a mere 10,000 or so years ago—that most of the large mammals such as camels, giraffes, giant sloths, and elephants have disappeared from this part of the world, although some of these species continue to survive elsewhere on our planet.

The mammals living in the interior plains of North America today descended from species that had the ability to adapt to subsequent changes in their environment in one way or another. Their lives have been portrayed in earlier editions of *The Wild Mammals of Missouri,* published in 1959 and 1981. In this present revised edition, we want to recognize them as fellow travelers in space and time.

It came as a complete surprise: "Would you consider revising, again, *The Wild Mammals of Missouri?*" As I looked over the 1981 edition, I saw some things that could be brought up to date—twenty years had brought about changes in Missouri's mammalian fauna as well as new biological information about the species—so I said "Yes." The core text remains the same; the additions, deletions, and other adjustments make it useful to today's reader.

I want to thank Julie K. Miller, who helped me with the research and editing and stayed on through the publication. As a student of wildlife during the 1990s, she brought a fresh and contemporary viewpoint to my research. I am much indebted to her.

I also want to thank many others who helped me along the way: Jerry M. Conley, Director, and Anita B. Gorman, Randy Herzog, Ronald J. Stites, and Howard Wood of the Conservation Commission; Oliver Torgerson, Wildlife Division Administrator, Janet Sternburg, Wildlife Ecologist, and Julie Fleming, Natural History Biologist, of the Missouri Department of Conservation; Beverly Jarrett, Director and Editor in Chief, Dwight Browne, Production Manager, Julie Schroeder, Editor, and Kristie Lee, Designer, of the University of Missouri Press; and Dr. Erik Fritzell, Oregon State University, Dr. Scott Burt, Truman State University, Alix Fink, University of Missouri, Dr. Cheryl A. Schmidt, Central Missouri State University, Dr. Barbara S. Miller, Eastern Washington University, Craig Miller, University of Idaho, and Glen Chambers, retired from the Missouri Department of Conservation, all of whom provided helpful input and suggestions. I am indebted to the University of California Press for supplying the armadillo plate and to the Missouri Department of Conservation for the armadillo family. Unfortunately, my husband died in 1991.

Elizabeth R. Schwartz
Coeur d'Alene, Idaho
February 2001

Preface to the First Revised Edition

When the last copies of the Sixth Printing were nearly gone, the request came to us, "Do you have something to include in the next printing that will update the book?" We could think of several things we would like to add and started to fulfill this request. But in doing so, we found that one alteration here required another one there, until no end seemed in sight. Obviously, to do what was necessary required a revision.

As we considered this aspect, we found our basic structure was sound and the desired changes could be likened to the minor alterations in remodeling a good house. There was no need to tear up more than necessary, and compromises could be made that would provide a more up-to-date edition. This revision includes one basic change (the addition of eight new species to Missouri's fauna) and several alterations (recent revisions in accepted nomenclature, updating of distribution maps, addition of the metric system, and inclusion of current research, especially in the fields of social communication and behavior, age determination, and movements of individual animals). We hope we have successfully followed the web of references throughout the text and that this edition will be useful for as long as the first one was.

We wish to acknowledge the following persons who contributed to this effort and extend them our thanks: Larry R. Gale, Director, Missouri Department of Conservation, who granted us the time to devote to this effort without interruption from other assignments; W. Robert Aylward, Carl Disalvo, J. Ernest Dunn, Jr., and Jack Waller, Missouri Conservation Commissioners, who likewise supported this endeavor; and these other associates with the Missouri Department of Conservation, Donald Christisen, David Erickson, Margaret LaVal, Richard LaVal, Wayne Porath, Josephine Radmacher, Kenneth Sadler, and Oliver Torgerson. Various assistance was also received from David Easterla, Northwest Missouri State University; David Hamilton, University of Missouri; Robert Hoffman, University of Kansas; Karl Maslowski, Cincinnati, Ohio; Orin Mock, Kirksville College of Osteopathic Medicine; David Murphy, University of Missouri; George Pollak, University of Texas; Philip Wright, Montana State University; and the Museum of Natural History, University of Kansas. All of the artwork in this volume belongs to the State Historical Society of Missouri.

Charles W. Schwartz
Elizabeth R. Schwartz
1 September 1980
Jefferson City, Mo.

Charles and Elizabeth Schwartz at work on the first edition of *The Wild Mammals of Missouri*, courtesy of University Archives (Collection UW:4/151/4).

Preface to the First Edition

This book is designed as a guide to the identification of the wild mammals of Missouri and as a reference to facts about their lives. Aboriginal man has been excluded and domestic mammals have been included only in a few cases where their skulls or tracks might be confused with those of wild species. In order that persons of many ages and backgrounds can understand the text, technical terms have been used as little as possible.

The text has been written from many sources: from our original fieldwork and observations, from information obtained through the Missouri Conservation Commission and the University of Missouri, and from the existing literature on the subject, the most important of which is listed at the end of each species' account. For the most part, the drawings of the mammals were made from fresh specimens and from reference to our mammal photographs; the skull drawings were made largely from skulls in the mammal collection of the University of Missouri. All of this material is presented here as a contribution to a better understanding of the mammals of Missouri.

One will notice that the information available for some species is very meager and that large gaps exist in our knowledge of them. This should prove an incentive to investigate the lives of our mammals and to contribute to the facts accumulating about them.

The Missouri Conservation Commission has long recognized the need for such a reference as this, and it is because of the Commission's interest that this book was written. During its preparation, forty-seven species' accounts were abstracted and with their plates were published in the *Missouri Conservationist;* these appeared monthly from July 1953 through September 1957. The University of Missouri realized the value of such a book and has contributed both editorial and financial assistance to its publication.

We are indebted to many persons for help during the preparation of this book and for assistance during its publication. The following have read parts or all of the manuscript and have helped in its preparation: Rollin Baker, Michigan State University; Edmund Jaeger, Riverside Junior College, Riverside, California; James Layne, University of Florida; Remington Kellogg, U.S. National Museum; Oliver Pearson, University of California; Donald Progulske, South Dakota State College; Harold Reynolds, University of Tasmania; Glen Sanderson, Illinois Natural History Survey; Milton Weller, Iowa State College; Philip Wright, Montana State University; Anna Benjamin, Robert Campbell, Clinton Conaway, William Elder, Robert Morris, Richard Myers, William Peden, and Fred Robins, University of Missouri; and from the Missouri Conservation Commission, Allen Brohn, Donald Christisen, George Dellinger, Robert Dunkeson, Richard Grossenheider, James Keefe, Leroy Korschgen, Dean Murphy, Werner Nagel, William Nunn, Dunbar Robb, Dan Saults, Frank Sampson, Charles Shanks, Reed Twichell, and Howard Wight. We wish to acknowledge also the following persons, all of the Missouri Conservation Commission: Irwin Bode, formerly Director, and Melvin Steen, formerly Chief, Division of Fish and Game, under whose guidance this book was started; William Towell, Director, and Larry Gale, Chief, Division of Fish and Game, under whose guidance this book was completed; and Frank Briggs, Robert Brown, Jr., Ewart Burch, Ted Butler, Ben Cash, Roscoe Clark, the late Clifford Gaylord, the late Edward Love, Dru Pippin, and Joe Roberts, former and present Commissioners, who have provided continual encouragement. We are also grateful to Mr. Ed Wilson, Smith-Grieves Co., Printers-Lithographers, who was extremely helpful in guiding this book through the tortuous channels of its publication.

Charles W. Schwartz
Elizabeth R. Schwartz

The Wild Mammals of Missouri

1
Introducing the Mammals

Mammals are a large group of animals that are remarkably diverse in their size, appearance, form, and adaptations. The group includes both wild and domestic animals (our companion animals, working and sporting animals, and livestock), is of great importance to us, and affects our lives in both positive and negative ways. Globally, there are more than 5,400 species of mammals alive today, inhabiting all continents and our oceans.

Mammals exert many influences upon the land on which they live, upon one another, upon other animal and plant species, and both directly and indirectly upon humans. They are part of our diverse wildlife fauna, play key roles within the systems in which they live, and ultimately contribute to maintaining nature's balance. They can influence the composition of forest and fields by choosing to forage on selected plants and by their movements of seeds from one location to another. The tunnels of shrews and mice in the leafy litter of a healthy forest floor catch and help to slow down the runoff of water, help prevent flooding of streams and rivers, and ultimately affect aquatic life. These tunnels also provide refuge for other small animals. The physical and chemical contributions that these small mammals make to the soil often are not considered. By tunneling in search of food and homes, infinite numbers of small mammals have worked and aerated the soil. The green leafy foods and seeds they have stored underground, their daily waste products, and the ultimate decay of their bodies contribute rich organic matter to this great resource upon which human welfare depends.

Mammals are also important to our economy. The hunting of game animals supports the businesses that manufacture sporting arms, ammunition, and supplies; contributes tourist money to local economies; and provides meat for our consumption. The many generations of mice that serve as food for minks and other furbearers affect the income of trappers and fur buyers, and even the quality of fur garments. Mammals are responsible for economic loss when they cause agricultural damage and when they collide with our vehicles. Unfortunately, diseases can be carried or transmitted from wild mammals to our domestic

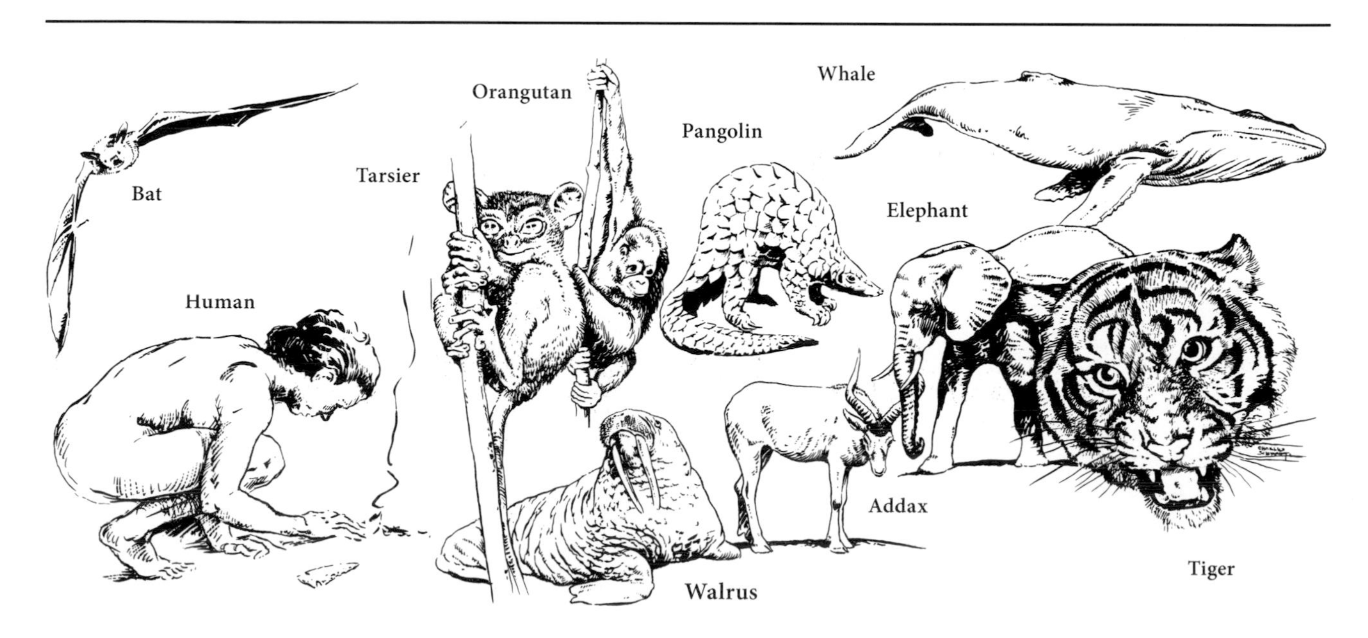

animals and ourselves. Coyotes and other predators are often condemned when they prey on small game or livestock, but are never praised for consuming countless rodents that compete with sheep for forage or that eat grain meant to feed chickens and hogs. In addition, the value of bats that consume vast amounts of insects, including mosquitoes and some of the most damaging agricultural pests, and pollinate and disperse seeds of countless plants is often overlooked because these mammals perform this activity at night and are not easily observed.

Mammals have another value that is extremely difficult to assess: their aesthetic value. Increasing numbers of people enjoy catching glimpses of wild mammals and knowing that wildlife exists. Some of Missouri's wild animals could be and have been kept as pets; this is generally problematic, and anyone interested should check the *Wildlife Code of Missouri*, ask their local conservation agent, or contact their local Missouri Department of Conservation office.

It is lamentable that humans do not always understand the complications of our world. Our interest is often directed toward one objective without considering all other aspects. Some of the major ways that we adversely influence mammals today are through habitat degradation, fragmentation, and destruction; and pollutants (pesticides, herbicides, and so on) released into the environment. Several mammals have declined due to overexploitation and habitat loss or degradation and are now considered extinct (no longer in existence) or extirpated (extinct in a certain area, but continuing to exist elsewhere), or are listed as species of conservation concern (comprising rare, vulnerable, imperiled, threatened, and endangered species). It is commendable that the decline of species has been recognized and official steps have been taken by conservationists to protect and manage for these rare animals. All of the varied contributions and values of mammals are worth considering in a big-picture conservation approach directed toward the perpetuation, utilization, and management of all species.

Some Characteristics of Mammals

Mammals, like all other animal groups, have their own distinguishing characteristics. Females have *mammary glands* that produce milk, and it was from the Latin word *mamma*, meaning "breast," that the name *mammal* was derived. Another obvious characteristic of this group is hair, present at least at some time in the individual's life. Mammals also have many distinctive internal structures, but discussing these is beyond the realm of this book. Instead, certain features are described here that are basic to the identification of mammals and to a better understanding of their way of life.

Hair. Hair usually covers the body of mammals, but in certain marine forms hair may be present only on the young or may consist of a few whiskers about the lips in adults.

Two main types of hair form the coat, or *pelage*: the *underhairs*, or *fur*, are thick and soft and lie next to the skin; the *guard hairs* are relatively long and coarse and lie over the underhairs. In addition, most mammals have special sensory hairs such as whiskers (*vibrissae*).

Hair is an outgrowth of the skin. When a mammal molts, the individual hairs are replaced by new ones. Except for differences resulting from fading and wear, changes in pelage color are due to new, differently pigmented hairs and not to color changes in existing hairs. In fur-trade language, a mammal's coat is best, or *prime*, when the new winter coat is complete.

One of the most important functions of hair is insulation. Air is retained within the covering of hair, preventing the loss of body heat and penetration of cold air.

Limbs. Because the feet reveal much about an animal's habits with respect to gathering food, building homes, and the ability to pursue prey or escape predators, the feet of most Missouri species are shown in detail in this book. Mammals typically have four legs, or *limbs*, with toes, or *digits*, that terminate in nails, claws, or hooves. However, in different mammals these are modified for various uses. They tend to be specialized for use in locomotion and in one particular type of locomotion more than any other.

Mammals that spend their entire lives in the water, for example cetaceans (whales, porpoises, and dolphins) and sirenians (manatees and dugong), have front limbs specialized into flippers used in swimming. Such mammals have no external hind limbs, but some of them possess small internal bones that are remnants of the hip girdle and hind limbs. Other mammals that are primarily aquatic but spend some time on land, for example pinnipeds (seals, sea lions, and walruses), have both front and hind limbs present externally. These are developed into finlike paddles with fully webbed toes. River otters, muskrats, and beavers are partially aquatic, and their hind feet are webbed as an adaptation for swimming.

Bats, the only mammals capable of sustained flight, have front limbs modified into wings. Such wings consist of paired membranes, or *patagia*, extending from the front limbs and sides of the body to the hind limbs. Another pair of membranes may connect the hind limbs and tail. The wings are strengthened by the elongated digits of the front limbs and by the hind limbs. The hind limbs support the animal while it hangs from a tree, cave wall, or other structure used for roosting. They also assist the front limbs when the bat crawls.

Burrowing mammals, like moles and pocket gophers that spend all of their lives in the ground, have short, powerful front limbs with strong claws that they use in digging passageways. Badgers, which dig a great deal either for homes or food, also have strong front limbs with long claws.

Tree squirrels and other mammals that live mainly in trees (that are *arboreal*) have sharp claws and a long tail. These adaptations aid these agile tree dwellers in jumping from one swaying branch to another and in maintaining precarious footholds. Flying squirrels (which do not truly fly but rather glide between trees) and certain other unrelated arboreal gliding mammals, for example the sugar glider of Australia, have developed an accessory membrane between front and hind limbs. This forms a winglike surface that these animals use in gliding from a high perch to a lower one.

Mammals that live mostly on the ground either walk on the entire foot, like bears, shrews, and humans, or on their toes. Mammals that walk on the soles of their feet are known as *plantigrade*, whereas *cursorial* (running) mammals can either run on their digits (*digitigrade*) or on the tips of their toes (*unguligrade*). Examples of digitigrade mammals include dogs and domestic cats, while our hooved animals like horses and white-tailed deer are unguligrade.

Teeth. In this book, the teeth of most Missouri mammals are shown on the plates with the skull and whole animal. In addition to often being important in identifying a species, the teeth tell a great deal about an animal's feeding habits.

Almost all adult mammals have teeth; notable exceptions include baleen whales and some anteaters, which are toothless (*edentate*). Many species have two sets of teeth (they are *diphyodont*): the *milk teeth*, which are usually lost sometime after birth, and the *permanent teeth*, which are present during most of their lives. In most mammals, four general types of teeth occur in order from front to rear in each jaw: *incisors*, *canines*, *premolars*, and *molars*. However, the number of the respective types varies with the species.

The bulk of each tooth is formed of *dentine*, which is chiefly calcium phosphate. Within this is a central soft pulp, well supplied with nerves and blood when the tooth is growing. Partially or completely covering the dentine is a thin layer of hard *enamel*. A series of thin layers of *cementum* normally covers the roots of the tooth, but in certain plant eaters cementum may cover the entire tooth. Teeth wear very rapidly and in old animals may be worn down to the gums. Some teeth, such as the incisors of rodents, rabbits, and hares, continue to grow throughout life. The number of teeth, the amount of wear on the teeth, and the number of layers of cementum are often used by mammalogists as indicators of the individual's age.

Mammals like rabbits, hares, and rodents that clip and gnaw stems and other plant fibers have large, chisel-shaped incisors. These are separated from the cheek teeth by a wide space, the *diastema*. The lips often pinch into this space preventing chips of whatever is being gnawed from getting into the mouth cavity. The cheek teeth are generally flattened or low crowned (*bunodont*) and well suited to grinding.

Meat-eating mammals, like carnivores, have lage, sharp canine teeth. These are used for piercing and tearing flesh. The last upper premolar and first lower molar of most carnivores are developed into specialized flesh-cutting teeth, the *carnassials*. However, some members of this order, like the American black bear, that eat large quantities of both plant and animal food, lack the extreme development for eating flesh and instead possess lower and more rounded cheek teeth.

Deer browse extensively on twigs and fibrous plant parts. These foods are cut by the sharp-cusped, high-crowned (*hypsodont*) cheek teeth before being swallowed and are later ground by the same teeth when the food is regurgitated as the "cud."

Mammals like shrews and bats, which feed principally on chitinous-bodied insects and other small invertebrates, have sharp-cusped molar teeth that cut up the hard parts of their prey. The incisors of shrews are *procumbent*, modified into grasping pincers that are well suited for holding the hard bodies of large struggling insects, which the tiny shrew feeds upon.

The types, arrangement, and numbers of teeth for any species are expressed as a *dental formula* that helps in identifying species (see accompanying illustration). The first letter of each type of tooth is used in this formula: *I* for incisor, *C* for canine, *P* for premolar, and *M* for molar. The number of teeth of each type in the upper and lower jaws on one half only is listed from front to rear; those of the upper jaw are placed above those of the lower jaw. The total number of teeth is derived by adding the numbers given in the dental formula and doubling this to include a similar number of teeth on the other half of the jaw. Thus the dental formula of the gray fox is written:

$$I\ \frac{3}{3}\ C\ \frac{1}{1}\ P\ \frac{4}{4}\ M\ \frac{2}{3} = 42$$

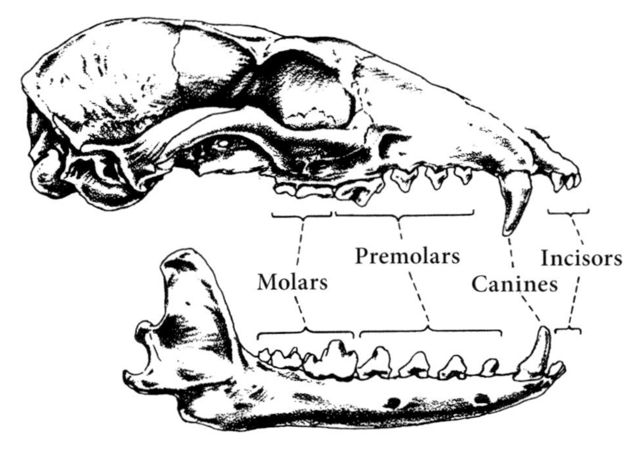

Gray fox skull showing types, arrangement, and numbers of teeth.

The dental formulas given in this book are for adult specimens; those of young animals may be different.

Reproduction. Mammals typically nourish their young while they are developing within their bodies and give birth to them alive. The exceptions are the egg-laying mammals: four echidnas (spiny anteaters) of Australia and New Guinea, and the duck-billed platypus of Australia and Tasmania.

The following is a generalized account of reproduction in mammals. The period of reproduction, or the *breeding season*, extends from the beginning of mating activities through the period of bearing and caring for the young.

The paired reproductive organs of males, the *testes*, form reproductive cells called *spermatozoa* or *sperm*, each of which carries one-half of the male parent's genetic makeup. These are released in fluid secreted by certain glands and pass to the outside of the body through the penis. Some species have a penis bone, or *baculum*; the size, shape, and weight of this bone are often used by mammalogists to determine the age of the individual.

Depending upon the species, the testes may lie within the body cavity permanently, descend to an external sac, or *scrotum*, during the breeding season, or remain in the scrotum throughout the year. The males of some species can mate at any time of year, but males of other species mate only during a restricted period. During the nonbreeding season in those species that do not breed continuously, the testes frequently decrease in size, cease sperm production, and usually return from the scrotum to the body cavity.

The paired reproductive organs of females, the *ovaries*, produce reproductive cells called *eggs*, or *ova*, each of which carries one-half of the genetic makeup of the mother. The eggs are shed from the ovaries and pass into the paired *Fallopian tubes* (*oviducts*) and on into the *uterus* (*uteri*, plural). Sperm *fertilizes* ova within the oviduct or upper portion of the uterus. Each fertilized egg starts to develop into an *embryo*, which becomes attached, or *implanted*, in the uterine wall. A structure known as the *placenta* then develops from both the embryo and uterus. Nourishment from the bloodstream of the mother passes into the bloodstream of the developing young through the network of blood vessels in the placenta. The period during which the young develop within the female's body is known as *pregnancy* or *gestation*. After this period, the young pass out of the female's body through the *vagina* during the process of birth, or *parturition*. Following birth, the mammary glands in females begin to *lactate*, or secrete milk, which the young, in most species, obtain by sucking nipples, or *teats*. The number and arrangement of teats vary with the species.

In some mammals, such as the badgers and weasels, there is an interruption in the usual sequence of development of the young. After the fertilized egg starts to develop there is a dormant period, sometimes of several months, before the embryo becomes implanted in the uterus and resumes development. This is known as *delayed implantation* or *embryonic diapause*.

Another modification of the typical pattern of mammalian reproduction occurs in bats. Some species mate in fall and winter as well as in spring, but the eggs are not shed from the female's ovaries until spring. The sperm from the early matings may remain viable and capable of fertilizing the eggs in spring.

Marsupial mammals, like the opossums, have a decreased gestation and thus offspring are born in a very early stage; these relatively very immature (*altricial*) young need to continue growth and development in the female's pouch (*marsupium*). By contrast, the young of other species are better developed, although varying in degree. The young of rabbits and most rodents and carnivores are blind, helpless, and either naked or lightly furred; those of certain other species, like deer and hares, are relatively well developed (*precocial*) at birth: they have their eyes open, are able to move about shortly after birth, and are well furred.

Stages in the development of the eastern cottontail embryo are shown on page 135.

Where Mammals Live

Although mammals live all over the world—on land, in water, in the ground, in trees above the ground, and in the air—each species has a definite geographic distribution, or *range*. Humans, the widest ranging of all mammals, live nearly everywhere, and domestic mammals, rats, and mice generally live where we do. Some whales, for example blue whales and fin whales, may swim into every ocean. However, most other mammals occupy more limited areas.

Within a range there are one or more types of natural settings, or *habitats*, suited to the species' livelihood (ecological role, or *niche*). These habitats result from the interaction of many factors, such as climate, soil, geology, topography, plant and animal life, and land use. Through the course of time, each species has developed certain physical features that adapt it to live best in a specific habitat. In primitive times the range of many animal species was definitely associated with particular native habitats. This is still true today in wilderness areas, but for the most part, humans have altered the original landscape so much that frequently it is difficult to associate present habitat conditions and the distribution of a species. Our many ways of altering land and resources, such as forest destruction, marsh reclamation, stream pollution, overgrazing, cultivation, abuse of caves, and fencerow cleaning, may reduce the range of some species and increase the range of others.

Some mammals live primarily in one type of habitat, while others live in many. Southern flying squirrels occupy, in general, heavy deciduous forests—in Missouri, oak-hickory forest. Texas mice live in the eastern red cedar and grass glade habitat, and in crevices in rocky bluffs in southwest Missouri. Eastern cottontails live in many varied conditions throughout their range but seemingly prefer the forest edge and the by-product of cultivation—open, brushy fields. Some species of bats occupy two entirely different types of habitats during the year; they spend the winter hibernating in caves and the summer foraging over forested areas or along timbered watercourses.

Within the range and habitat a species occupies, each individual has a certain area, or *home range*, where it has its home, gets its food, and rears its young. As a result of these activities, it becomes extremely familiar with this area and spends considerable time in restricted parts of it. Some species tend to have permanent home ranges that they seldom leave; others may occupy a particular area for a certain period, then move on to another. If this movement is seasonal and occurs regularly each year, it is called *migration*. Therefore, migratory species, such as bats, have separate seasonal home ranges.

There is another aspect of a home range that is important—the spatial dimension. Some species may combine vertical and horizontal distribution as in the case of tree squirrels; others, like bats, have a large range in the air above the ground and a restricted area on the ground or in caves under the ground; still others, like moles or pocket gophers, have a subterranean range. Our concepts of the use of an area by one or more species should consider the complexities of animal movements. The size of the home range is related to many factors, such as the size of the mammal, its sex, the general abundance of food and other essential habitat features, and the social pressures by members of the same or different species. A small mammal like a short-tailed shrew may live within 0.4 ha (1 ac.) of land, while a large mammal like a coyote may have a home range as much as 48 km (30 mi.) in diameter. In most species, males tend to have larger home ranges than females.

In general, plant-feeding, or *herbivorous*, mammals seldom need to travel far in search of food and thus have small home ranges. For example, eastern cottontails living in an area with good food and cover have home ranges varying from only 0.4 to 2 ha (1 to 5 ac.).

However, when seasonal changes make some types of food unavailable over vast areas, as in arctic regions or in parts of Africa, herds of herbivorous mammals may have to migrate many kilometers (miles) to reach new food supplies. Local weather conditions, too, may cause mammals to make small, temporary shifts in ranges. During periods of deep snow, eastern cottontails may move from their usual homes in brushy fields or fencerows to nearby woodlands. Here they gnaw the bark of trees and shrubs above the snow and subsist on this diet until the snow melts enough to permit them to find food in their former homes. In contrast to most plant feeders, *carnivorous* mammals, those that mainly or exclusively eat animals, must traverse relatively large areas because their food is less readily obtained.

When many individuals of a species live in the same general area, their home ranges may overlap. If an individual defends part of the home range from intruders, particularly of the same species, that part is known as a *territory.* Most territories are established in the vicinity of nests or dens containing young or around stores of food.

The movements of mammals within their home ranges and the behavior of mammals in their territories have traditionally been hard to study because of the difficulty of identifying individuals or following them for a long period of time. Various methods of studying individuals have been used, particularly the capture-mark-recapture method, but none has provided the exact information that has been obtained by the use of animal tracking systems; very high frequency (VHF) radio telemetry and, more recently, global positioning system (GPS) satellite telemetry. These biotelemetry techniques of attaching to a mammal an electronic device that gives off or captures radio signals allow an observer to locate the individual at any time and to monitor its movements continuously. Many of the species accounts here include information on the home range obtained by these techniques. Wildlife biologists are currently using satellite telemetry to track elk, white-tailed deer, and black bear movements in Missouri.

Ecological Sections of Missouri

In this book, species accounts include a *Distribution and Abundance* section in which the ecological sections where a species lives are listed and general information about species abundance is presented. Missouri has four ecological sections: the *Central Dissected Till Plains*, the *Osage Plains*, the *Ozark Highlands*, and the *Mississippi River Alluvial Basin* (see accompanying map and photographs). The following descriptions of each ecological section, summarized from the *Atlas of Missouri Ecoregions* (Nigh and Schroeder, 2002), provide a general picture of our state. Most Missouri mammals are not strictly confined to any one section, either because a species may be a habitat generalist and occupy different habitat types, or because the habitats occupied by a species may occur within more than one section. But the density of different species is frequently influenced by a given section, chiefly because of the abundance and quality of preferred habitats.

The *Central Dissected Till Plains* covers almost all of Missouri north of the Missouri River and extends into southern Iowa and portions of Kansas, Nebraska, Illinois, and South Dakota. This section is characterized by moderately dissected glaciated plains that slope toward the Missouri and Mississippi Rivers.

The Central Dissected Till Plains in Missouri is covered with Pleistocene loess (wind-blown sediment) over glacial till (sediment deposited directly by a glacier) that varies in thickness from 0 to over 91 m (0 to 300 ft.). The till was deposited over 400,000 years ago. Smooth plains and broad ridges occupy areas away from major drainages, with native prairie and claypan soils found on the flattest uplands. Closer to drainages, rolling hills with gentle slopes have prairie soils higher in the landscape and transitional savanna or woodland soils on the side slopes. More highly

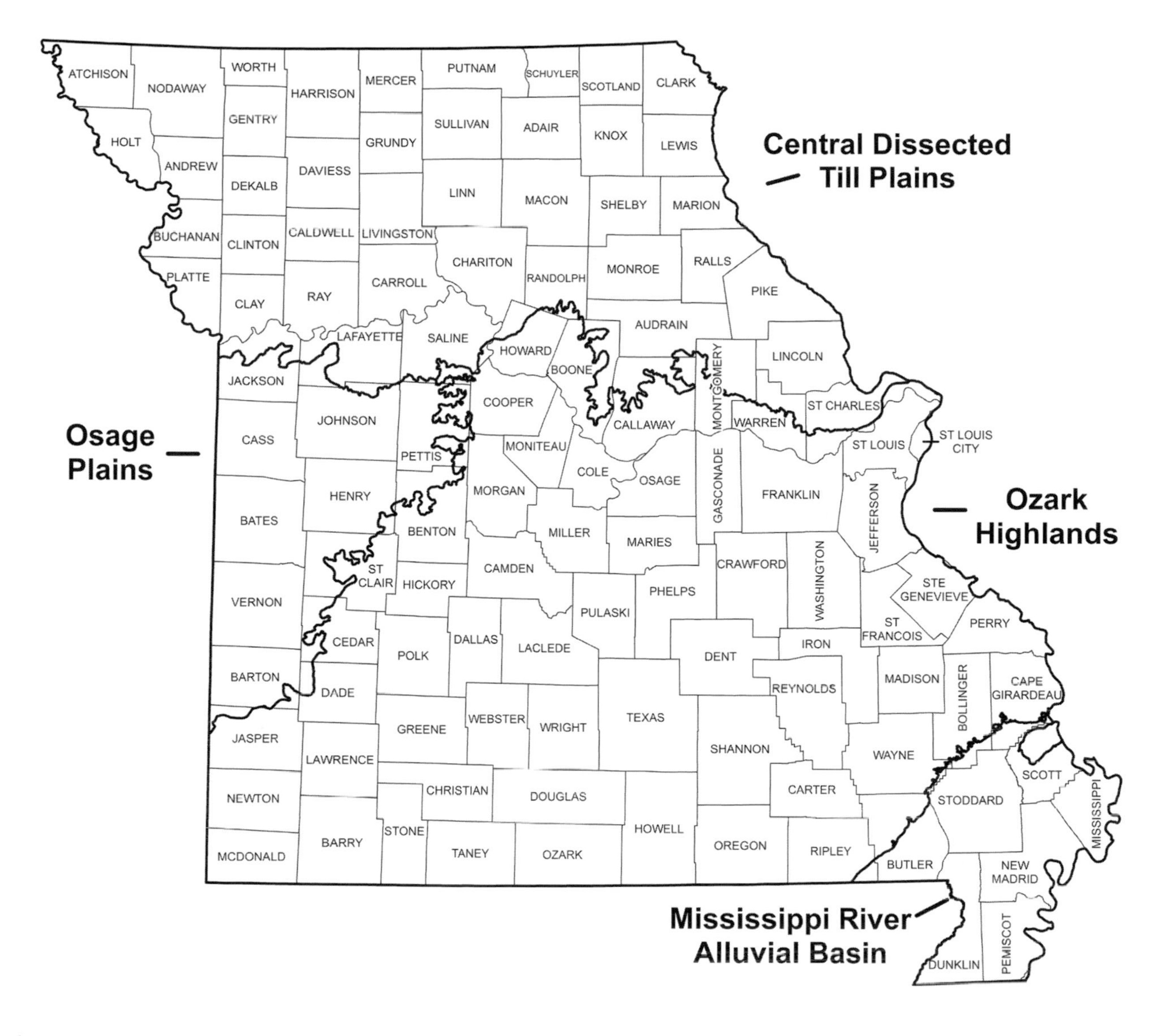

dissected lands, with narrow ridges and deep valleys, occur near the major rivers. These areas have transitional soils high in the landscape and true forest soils low in the landscape. In general, the complete dissection of the Central Dissected Till Plains resulted in a mosaic of uplands, side slopes, and bottomlands, with prairie, woodland, and forest ecosystems.

Most of the presettlement tallgrass prairie in this ecoregion is now pasture or cropland, and much of the original oak and hickory timber has been cleared. The mammals living here today are both open-land and forest species and include the Franklin's ground squirrel, thirteen-lined ground squirrel, meadow vole, least weasel, and cinereus shrew.

The *Osage Plains* is an unglaciated prairie that extends from west-central Missouri into eastern Kansas and northeastern Oklahoma. A flat to gently rolling landscape underlain mainly by Pennsylvanian-age shale, sandstone, and limestone, the Osage Plains was historically nearly pure tallgrass prairie. Today, most of the presettlement prairie is pasture or cropland, but some large prairie blocks still exist. There are also extensive wetlands associated with streams of the upper Osage River watershed. The mammals living here include the badger, cotton rat, meadow jumping mouse, least shrew, and western harvest mouse.

The *Ozark Highlands* is the western extension of a large eastern deciduous forest and includes most of southern Missouri, much of northern Arkansas, and small parts of Illinois, Oklahoma, and Kansas. This area has been undergoing erosion and weathering for a quarter billion years and is now a thoroughly dissected plateau and region of unique ecosystems.

The Ozark Highlands in Missouri is dominated by soluble carbonate bedrock and karst features including caves, sinkholes, and underground drainage systems. The highest and least dissected areas are flat to gently rolling plains that have very deep accumulations of

Mississippi River Alluvial Basin. A natural slough in Big Cane Conservation Area in Butler County, southeast Missouri.

Central Dissected Till Plains. The open, rolling hills around Dunn Ranch Prairie in Harrison County, northwest Missouri.

Osage Plains. Golden Prairie in Barton County, southwest Missouri.

Ozark Highlands. Flattop Bluff along the Jacks Fork River in Shannon County, south-central Missouri.

limestone, dolomite, and chert. These upland plains formerly supported prairies, savannas, and open woodlands. The plains give way to rolling hills closer to drainages, and then to the rugged, highly dissected hills and breaks bordering major streams. Soils in the hills are deep, rocky, and highly weathered and formerly supported oak and oak–shortleaf pine woodlands and forests. Areas of shallow soils and bedrock exposure are common, and rare and unique species are associated with shallow-soil glades. Some stream channels and valleys lose water to subterranean passageways, while others receive water through seepage and springs, sometimes from areas far beyond their surface watershed.

This area is suited more for forestry than for agriculture, but in many places the land has been cleared and now is pasture or cropland. Most of the original timber has been harvested or burned, resulting in a second-growth forest dominated by oaks and hickories and, to a lesser extent, eastern red cedar and shortleaf pine. The mammals living here include bobcat, golden mouse, woodrat, black bear, flying squirrel, and chipmunk. Most of Missouri's caves occur here, and this is where most of the cave-dwelling bats hibernate in winter.

The *Mississippi River Alluvial Basin* extends from southeastern Missouri down the Mississippi River to the Gulf of Mexico. It includes a complex arrangement of alluvial surfaces (made up of materials left by the waters of rivers and floods) formed during the long evolution of the system. In Missouri, it is covered to great depths by clay, sand, and gravel deposited by the Mississippi and Ohio Rivers as they meandered across the region. The topography is gently undulating to flat, with low ridges arising a few meters (feet) above the plains.

Formerly, most of this area was a forested swamp filled with bald cypress, sweet gum, and associated wetland plants. Most of the Missouri section was poorly drained naturally, and beginning in the late 1800s, extensive logging, land drainage, and flood protection projects converted virtually the entire area to cropland. This area now has proportionately more kilometers (miles) of streams and drainage ditches than any other area in the state. Mammals living here include the rice rat, swamp rabbit, southeastern myotis, and cotton mouse that typically occur in low, wet habitats, and some species that live in timber and open lands.

About the Range Maps

Maps for North American and Missouri ranges are included with each species account. The North American maps are derived from those published by the Smithsonian National Museum of Natural History with additional information from state and federal natural resource agencies, other conservation organizations, and published literature.

The Missouri range is presented as presence, probable presence, or probable absence within counties. Known county record information comes from several sources, including vertebrate museum records; collection records submitted by biologists; harvest records; research, surveys, and monitoring projects; project summary reports; published literature; and observations.

The Missouri maps show counties where a species is *known to occur* (indicated by dark gray shading; at least one recent record) or where a species *probably occurs* (indicated by light gray shading; based on knowledge of the species' natural history and habitats available in a county). Observations for a species may exist for the *probably occurs* counties but the information has not been passed on to the biologists who maintain the databases, or the species has not been discovered yet. Counties with no shading are where no records exist, and where the species is not expected to occur (*probably does not occur*).

For a few species that have been extensively monitored (for example, swamp rabbits), *historical* counties have been indicated (indicated with hatched lines; where a species once lived but is thought to no longer occur). In addition, armadillos, black bears, and badgers occasionally show up in counties with no known populations of those species. For example, dispersing male bears wander outside of their range while looking for new, unoccupied territories but do not establish there. A black dot in the middle of a county indicates there has been at least one of these *rare occurrences* within that county. Also, Missouri has no breeding population of mountain lions, but the Missouri Department of Conservation has confirmed 51 instances since 1994. A black triangle in the middle of a county indicates there has been at least one *confirmed sighting* within that county. The Missouri distributions are not exact but do give general pictures of where mammalian species can be found within the state.

Social Communication and Behavior

Mammals communicate with each other in various ways—by sound, visual display, scent, and touch. It is only in the last half of the twentieth century that emphasis has been placed on the importance of communication in mammalian biology and that acute

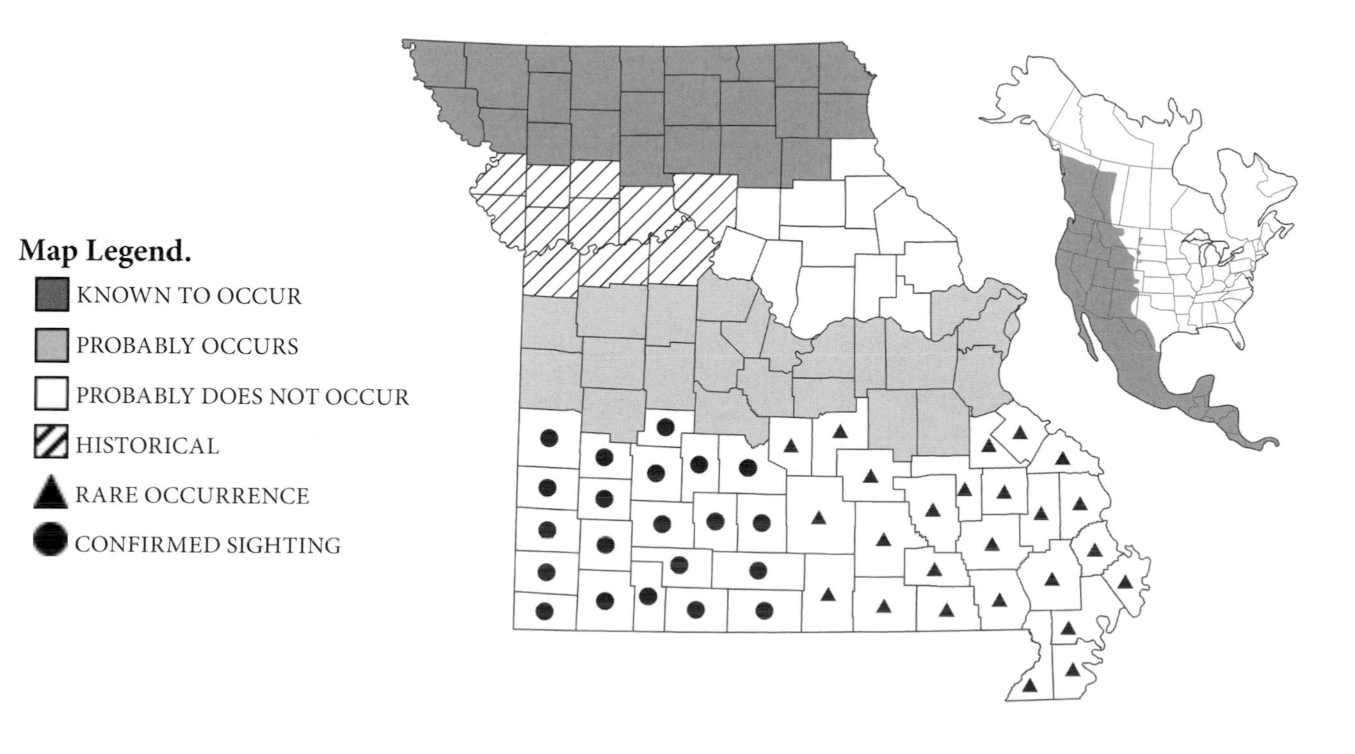

observations of different species have added to our understanding of their behavior.

Sound carries for great distances and does not require visual contact of the individuals. Most mammals have a large vocabulary that indicates alarm, distress, contentment, resentment, assembly, recognition, and many other messages. They also produce nonvocal sounds by various actions such as stamping their feet, clicking their teeth, snorting, and slapping their tails on the water.

Visual display is another type of communication, requiring that the individuals see each other. Various body stances or actions, different positions of the ears or tail or legs, the amount of opening the mouth (*gape*), staring or avoidance of eye contact, erection of certain hairs or patches of fur on the body—all send a message to the observer. These actions indicate the mood of the displayer in such behaviors as greeting, threatening, submission, or fright.

Certain glands produce chemicals that are diffused or carried in a current, such as water, air, or urine. Chemicals emitted and received by members of the same species are known as *pheromones*. Pheromones can indicate the sex and state of sexual condition of the sender, identity (age, dominance, or strangeness) of an individual, or mark a territory. Other chemicals, such

as the odor of a skunk, are a means of communication within, as well as between, species. Compared to other means of communication, chemical scents have the advantage of lasting after the animal has departed.

Tactile contact, such as grooming, snuggling, and nuzzling, is likewise a mode of communication. This is especially important during courtship and in parent-offspring relationships. Social mammals use tactile communication to reinforce bonds and establish social hierarchies.

All these means of communication may be used individually or in combination and in varying degrees of intensity, and they often accompany other behaviors, both sexual and social. One of these social behavior patterns is aggression, which takes many forms. Aggression may be demonstrated in a mild way, by physical indicators such as stance, stare, facial grimace, or raising of certain hairs. Threats, with and without accompanying vocalizations, are a slightly stronger means of displaying aggression, while the most active means involve a chase and physical combat. Each species has its own varieties of expression, and these are understood by all members of the species, and often by other species as well.

Social relationships between individuals of the same species are extremely varied. Some species, like the Virginia opossum or the mountain lion, live a solitary life, except at breeding times and when there are unweaned young. Adult foxes and coyotes form a pair bond, the male staying with the female for some period of time and assisting with the care of the young. This type of relationship is more permanent in gray wolves, which form packs that live and hunt together and exhibit a complex social structure.

Large groups of individuals are generally classified as to whether they move about or remain in a particular area. Thus, the American bison and the various hoofed mammals that wander over the prairies and tundra live in herds. Prairie dogs, beavers, and other groups like brown rats and voles that remain in one particular area live in colonies.

Winter colonies of bats and temporary aggregations of rodents during the winter exhibit little if any social organization. During the summer, many species of bats form colonies by gender—the females living together in a nursery colony and the males living in a separate part of the colony or elsewhere. Other species that live in colonies separate into units in which a social rank or hierarchy is developed by means of aggression. In these colonies, some individuals become dominant over others and each individual knows its place. The establishment of dominance reduces physical combat and stress among the members. Dominance provides priority in the necessities of life including the best place to live, the choicest food supply, and preference in mate selection.

Generally, an adult male is the dominant member of the unit; he may have one or more females that he mates and lives with plus their offspring. As the young mature, females are tolerated in the unit more than males, and most aggression and dispersal is among the young male segment. Upon dispersal, a young male may fill a vacancy left by the death of a dominant male; move to an uninhabited, usually peripheral, part of the range; or succumb to predation or some other form of death while struggling to establish his own territory and family unit. The members of one unit recognize each other and are aggressive to other individuals of the same species that come into their unit.

Host-Parasite Relationships

In addition to the obvious interrelations between mammals and various other animal species (such as competition for a food source or living space), there are not-so-obvious associations between mammals and their parasites. The complexities of the host-parasite relationship and the life histories of parasites are of tremendous importance in understanding the well-being of a mammal and its relation to its environment as well as the evolutionary significance of the parasitic forms. Because parasites obtain their nourishment at the expense of the host, they can have detrimental effects on the host's health. In the following species accounts only a brief indication is given of the many kinds of parasites, both internal and external, and their significance.

The complexities of host-parasite relationships are well exemplified by the eastern cottontail. The cottontail is a source of food for many carnivorous mammals like the coyote, foxes, domestic dogs, and bobcat.

When the cottontail is eaten, and if it is infected by the intermediate stage of one or more particular species of *tapeworm*, these carnivores consume the parasite in its *bladderworm* stage from the liver, body cavity, mesenteries, or less often from the muscles. The bladder housing of the larval worm is digested in the digestive tract of the new host, for example in the coyote, and the head, or *scolex*, of the parasite is freed. The scolex extends its moveable spines and attaches itself to the wall of the coyote's small intestine, where it develops into a mature, segmented tapeworm. These segments produce eggs sexually that develop into embryos covered by a shell. The embryos pass out of the coyote's body along with the host's feces. Another cottontail becomes infected by eating vegetation contaminated with the tapeworm embryos. Once in the digestive tract of the new cottontail host, the shell of the embryo is digested and the embryo is freed, penetrates the wall of the digestive tract, and gets into a blood vessel. Transported by the blood, it reaches a suitable place in the cottontail where it can grow into a bladderworm and awaits another host that will continue the life cycle by eating the cottontail someday.

But this is only one cycle. The eastern cottontail is part of others. In the intestine of the cottontail, there may be adult tapeworms of different species than the ones living in the intestine of the coyote (the cottontail is known to harbor at least ten species of tapeworms). A cottontail tapeworm produces eggs that grow into embryos and pass out of the cottontail with its droppings. These eggs are ingested by mites that live in the soil in places that cottontails frequent, like nests or resting spots. Once inside the mite, the tapeworm embryos emerge from their shells, penetrate the intestinal wall, and get into the body cavity. There they change into the bladderworm stage. When the mite is accidentally ingested by a cottontail that is grooming its coat, licking soil, or eating some of its own droppings contaminated by mites, the bladderworms are released, become attached to the intestinal wall of the cottontail, and the cycle is completed.

In this simple example, the cottontail is the host for the adult stage of one species of tapeworm and the larval stage of another. But it is also the host for other internal parasites such as *protozoa*, *roundworms*, and certain types of *flatworms*, as well as many external parasites like *ticks*, *parasitic mites*, *fleas*, and *certain insects* that have complicated life cycles, too.

Age Determination

It is often important to know the age of a wild mammal, particularly to understand the structure of the population. By determining the age of individuals of a given sample at critical periods, it is possible to assess the annual reproduction rate, longevity, and turnover (replacement of all individuals) in the population.

There are various techniques for doing this, some of which may seem complicated to the casual observer, but each of which has its merits. Some are useful for only a limited time, some are more reliable than others, and some are easy to use, while others require more specialized equipment and techniques. In the end, there may have to be a compromise as to which method offers acceptable reliability in terms of efficiency.

All mammals grow at a rapid rate until some point in life when growth slows down or is completed. During this period of rapid growth, changes in body weight or body measurements are useful in determining age. Sexual maturation and breeding condition are good indicators of age but have limitations in those species that first breed when still less than adult size. Changes in *pelage* (body hair) coloration and condition of the molt may indicate age at certain times. The closure of certain bones of the skull also occurs at particular ages.

There has been a search for some measurable character that grows throughout life, or that continues to grow after general body growth has ceased. Tooth eruption, development, and wear have proved especially useful; the growth of certain bones (penis bone and long bones of the legs) has afforded limited opportunities for determining age; and the lens of the eye has contributed a valuable means. These various methods of age determination are reported under the respective species' accounts, but three more technical ones are discussed here.

One of the most accurate methods of aging a mammal uses the lens of the eye. The lens grows throughout life by the increase of proteins, one fraction of which is soluble in water, the other fraction insoluble. The *wet weight* of the lens can be taken in the field in a short time and is comparable to the *dry weight*, which requires fixation and a drying procedure in the laboratory. Two other measurements, both biochemical, can be made of the protein fractions: that of the soluble proteins is not a reliable indicator of age after the initial period of growth of the animal, but the measure of insoluble proteins is very closely related to age, especially of older individuals. Unfortunately, in spite of this accuracy, the technique of measuring insoluble proteins is limited because of the time and specialized equipment required.

A useful and fairly accurate method of age determination, particularly in young and middle-aged animals, is a count of the cementum layers of the teeth. As a tooth grows, a layer of cementum is deposited on the roots. This appears to be laid down in annual rings, the first one often forming in the second year of life. The technique involves extraction of a tooth (different teeth being preferred in different species or for various reasons), which is then decalcified and cut longitudinally or horizontally into very thin sections and mounted on a slide. The material is stained and examined under a microscope. However, counting the rings is often difficult and the relation to years is sometimes hard to determine accurately.

A promising method for aging mammals that has been used in living ground squirrels is to measure the breaking strength of tendon fibers in the tail. The denaturation of tail-tendon collagen fibers over time is a measure of physiological age that can be correlated with absolute age. This is a technical procedure but has the advantage of providing repetitive measurements over time for a single individual. Tail biopsies can be taken in the field and age determined from the amount of collagen present.

Mammal Signs

Because most mammals are nocturnal or are active at twilight, we seldom observe them, but they do leave tracks and other evidence of their activities. It is by these signs that alert observers can learn much about the mammals around them.

Tracks are the most universal sign of a mammal's presence. These are best seen in soft mud along the shores of ponds, lakes, rivers, roadside ditches, or wherever water has evaporated and left a sticky surface. They can also be observed readily in fresh snow, in the fine powdery dust under overhanging rock ledges, on dusty paths, or in sand.

The tracks of larger mammals are usually easy to identify, but as the size of the animal diminishes, the difficulty of identifying the tracks increases. Unfortunately, too, complete prints are not always available, but with a bit of sleuthing some valuable information can often be obtained from very little evidence.

First, observe the size and shape of each print. Count the number and arrangement of toes, and note whether or not claw marks show. Are there webs? Look for the number and shape of the pads, and determine if the fur on the bottom of the foot left any impression. If possible, figure out which imprints are from the front and hind feet.

Next, study the trail. Did the tail leave a mark? What is the distance between the prints of front and hind feet? Was the mammal walking, running, hopping, trotting, bounding, or loping?

Finally, consider which mammals live in the habitat where the tracks are. By a process of elimination, rule out one mammal after another until you arrive at the name of the mammal or group of mammals most likely to have made the tracks. Then refer to the illustration of their tracks and study the feet shown on the full-page plates.

In addition to tracks, mammals leave other signs. In the limited space here, only suggestions are given about what to look for. Many of these signs are discussed and illustrated in the individual species accounts.

Most mammals have dens or nests, and some of these can be found and their owners identified. Tree squirrels build leafy structures that are readily discernible, or they use knotholes or old woodpecker holes in trees. Claw marks around these openings and bits of fur lodged in the rough bark indicate the occupant of the hole. Underground dens usually have a tunnel system and entrance. More often than not, these entrances are marked by freshly excavated soil or compressed, smoothly worn piles. Ground squirrels and woodchucks make many of these dens, but other animals such as foxes and skunks use those of the latter also. Sometimes bits of fur or droppings near the entrance will indicate the owner. Pocket gophers and moles leave piles of freshly excavated soil (*tailings*) from their underground tunnel systems; the manner of the excavation indicates which species made it. In addition, raised places in the ground show the location of a mole's tunnel underground. Look for shrew and mouse nests under fallen logs, under rocks, or in old birds' nests. These nests may be small, globular aggregations of shredded grass or bark. Mice frequently line their nests with fur or feathers from some dead animal they encounter in the field. They and flying squirrels pilfer woolen or cotton materials from clothing when available and incorporate these into their nests. Some animals such as cottontails make a simple nest for their young in a shallow depression in the ground and line it with grass and with fur plucked from their own bodies. The large, conical homes that beavers make of sticks and mud are easily identified. Muskrats make somewhat similar but smaller homes out of marsh vegetation they accumulate. These two species also make dens underground, and the entrances to these homes may be underwater in the banks of rivers or ponds. The disarray of small sticks and other forest debris that woodrats often assemble at the base of rock piles, rocky ledges, downed timber, or even in the attics of old cabins and barns tells of their presence.

Where mammals are very active, they habitually use the same paths, which become discernible runways. Beavers make large, worn runways from the water up the banks to fields where they feed. Voles create intricate runway systems in grass; these are readily identified. Small creatures like shrews and mice have runways beneath the forest leaf litter. Some runways are made by one species and used by others, but invariably each leaves evidence of its own presence.

Tiny caches of food in the form of seeds, snail shells, or insect parts mark the places where mice or shrews have provisioned their stores or fed. The remains of small animals that are caught and fed upon indicate the presence of a large predator. Piles of mussel shells on a partly submerged log show where a muskrat has fed. River otters leave catfish heads along the shoreline and large, partially eaten fish lying on the bank.

Peromyscus food cache of wild plum seeds. Each seed has been gnawed open on the side and the contents removed.

The different species of mammals all have identifiable *scats*, or feces. These frequently tell a great deal about the animal's food habits. Muskrats deposit their droppings on logs or rocks in a marsh or along a river. River otter latrines are found on high banks, dams, or points that extend into the water and contain crayfish remains and fish scales. Scats showing the remains of seeds, insects, crayfish, and other items invariably point to the omnivorous raccoon. The droppings of coyotes and foxes show the remains of rodents, rabbits, and even carrion they have fed upon. Runways in the grass containing bright green droppings indicate the presence of southern bog lemmings, while brownish droppings tell that the prairie vole lives there. The round, fibrous droppings of cottontails are common in their runways and forms, but when located on top of tree stumps or fallen timber, these droppings may be indicative of swamp rabbits. White-tailed deer leave similar round, fibrous droppings along their trails. The tiny droppings of house mice, deer mice, and white-footed mice indicate their presence in our homes; droppings of rats along their trails and at their feeding places tell on them. The evidence of bats is quite diagnostic because of the accumulation of their droppings in caves or at the base of rocky shelters or around buildings. Bats also leave stains from their bodies where they enter crevices or buildings or cluster on cave ceilings.

Trees and branches that have been gnawed or felled by beavers are easily discernible. Squirrels and raccoons damage standing corn in a characteristic way. White-tailed deer rub or polish their antlers against saplings. Searching for and interpreting mammal signs can be a rewarding experience for the outdoor enthusiast.

Size of Mammals

Mammals vary considerably in size. The smallest is the bumblebee bat of western Thailand and southeast Burma; it is about 3.3 cm (1.3 in.) long and weighs a mere 2 g (1⁄16 oz.). At the other extreme is the endangered blue whale, the world's largest animal, which measures up to 33 m (108 ft.) in length and can weigh 136 metric tons (150 tons). The cinereus shrew and the least shrew are the smallest mammals in Missouri; they measure about 7 cm (2¾ in.) in length and weigh the same as a penny—about 3 g (1⁄10 oz.). The largest mammal now living in Missouri is the recently reintroduced elk. It reaches 242 cm (9 ft.) in length and can weigh up to 376 kg (830 lb.).

Mammalogists customarily take certain measurements that are useful in identifying mammals. Not only is the overall size important, but various parts of the body that can be measured conveniently are also used for comparison between individuals and species. The following measurements are most generally taken and are given for those species discussed in detail in this book (see accompanying illustrations). All measurements presented throughout the book are given in both metric and English systems. For small mammals, lengths are usually reported in millimeters and weight in grams.

Measurements and weight of a mammal are recorded in the following order:

total length–tail length–hind foot length–ear length–weight

Total length. The distance from the tip of the nose to the tip of the fleshy part of the tail (but not including the hairs on the tail). This measurement is taken with the animal placed on its back.

Tail. The distance from the base of the tail to the tip of the fleshy part of the tail (not including the hairs on the tail). This measurement is taken on the upper

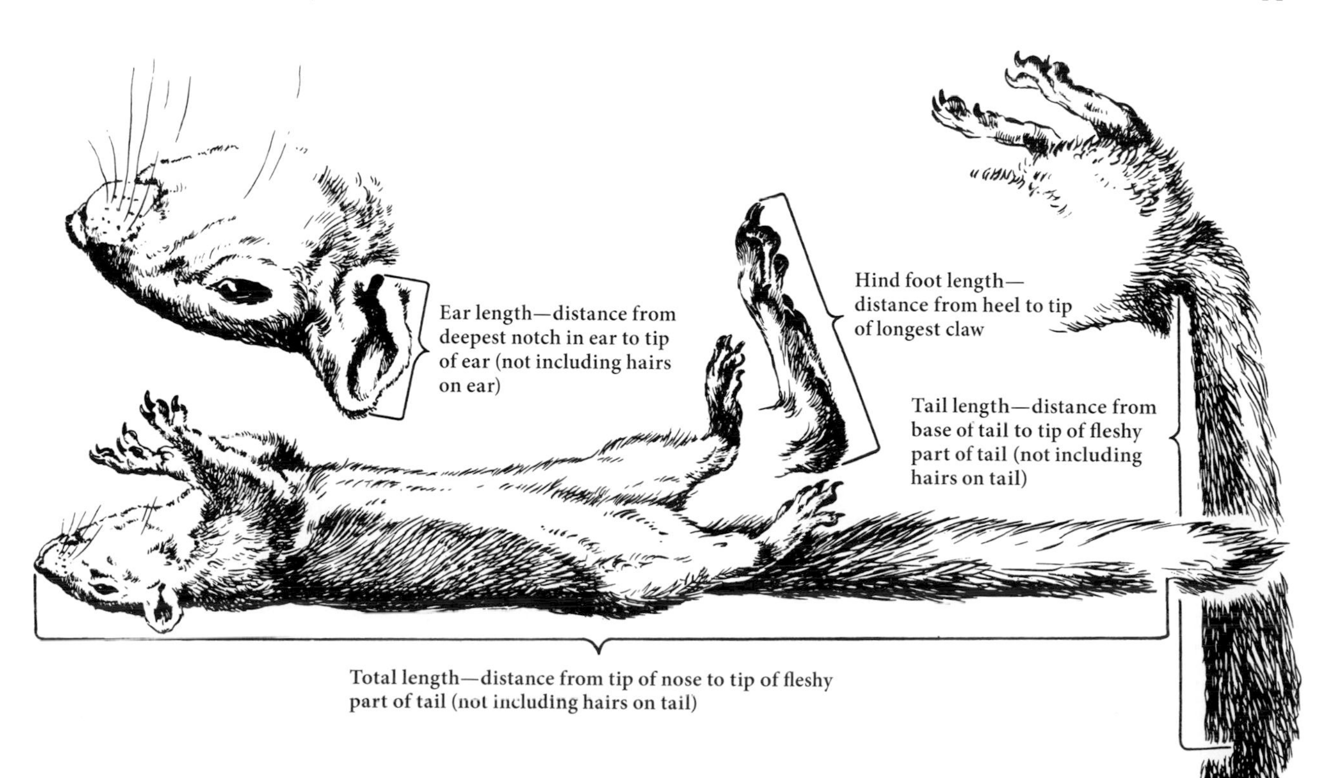

surface of the tail when the tail is bent back at a right angle to the body.

Hind foot. The distance from the heel to the tip of the claw on the longest toe.

Ear. The distance from the lowest or deepest notch in the ear to the tip of the ear (but not including the hairs on the ear).

Skull length. The distance between the foremost and hindmost points.

Skull width. The distance between the widest points.

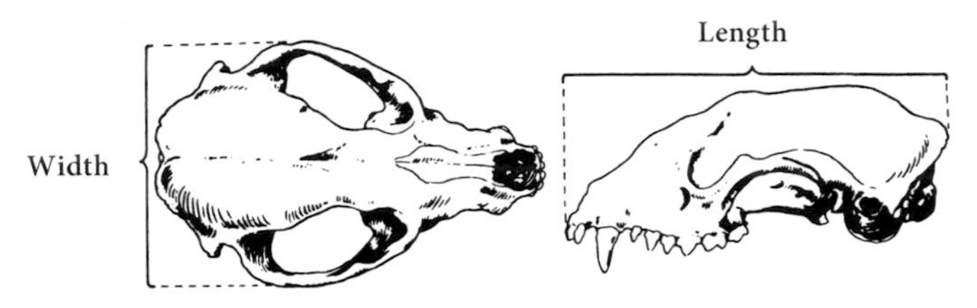

Measurements given for each species are from specimens collected in Missouri when these are available; otherwise, they are taken from some other source for the subspecies occurring in Missouri or for the species as a whole, whichever is available or more applicable. The measurements apply to both sexes, unless stated otherwise, and indicate the normal range in adults; young specimens will be smaller. Weight necessarily varies with such conditions as pregnancy, storage of fat before hibernation, and loss of weight during winter.

Human–Wild Mammal Conflicts

Mammals are an important part of a healthy ecosystem, but unfortunately they sometimes become a nuisance, damage property, or become a health or safety concern. Some minor problems are to be expected, but others are more serious and require some type of action to resolve the issue. This involves removing whatever is attracting the animal(s), or excluding, frightening away, or removing the animal(s).

It is important to first identify the problem animal, since knowing the species determines the best management approach. Look for signs and try to estimate the number of animals involved. Outdoor motion-activated trail cameras can be used to capture photos of the offending animal. After identifying the species and deciding if the problem is severe and persistent enough to require action, and before implementing any control measure, consult the *Wildlife Code of Missouri*, updated annually by the Missouri Department of Conservation (MDC), and for guidance contact a wildlife professional such as your county conservation agent; local MDC office; local, state, or federal animal control officers or wildlife damage biologists; private lands conservationists; or a private pest control business. Also check federal, state, and local city or county regulations governing the use of all pest control methods. Some methods that are effective for one species may be ineffective for another, and some methods may not be legal in certain areas. Special restrictions apply to endangered and other protected species.

Plan ahead. Advanced planning can help prevent some problems from occurring. Before building a structure, planting a garden, or investing considerable time and effort into attracting desirable animals, consider how your plans may attract undesirable ones. Changing your design to discourage unwanted animals is easier and probably more cost-effective than dealing with problems after they occur.

First, decide which species are desirable and undesirable to you. Next, ask wildlife professionals their opinions on how to attract the desirable species but not the undesirable ones. Then examine your proposed design and determine if you are providing the four basic habitat requirements for survival: sufficient space, food, water, and shelter. When features that will attract unwanted animals are recognized, changes can be made that will eliminate habitat requirements important to those species.

Habitat modification. Habitat modification in a problem area can be used to discourage unwanted animals by removing components important to that species. Available food sources (for example, pet food, garbage, or animal feed) are the most common attractants for problem animals. In most cases simply removing the food will discourage unwanted animals. Modification or elimination of shelter is another important control method. Common examples of shelter for some mammals include attics for squirrels and bats, brush piles for cottontails, woodpiles and thick vegetation for small rodents, and access under porches and houses for woodchucks, striped skunks, opossums, and small mammals. Removing or making these areas unavailable will encourage animals to find homes elsewhere. Water can also be a strong lure, especially during drought years, and a pool of water under a leaking outside faucet will attract animals.

Exclusion. Excluding animals involves creating a physical barrier around the problem area. Common examples of barriers used to exclude mammals include permanent net or woven-wire fences around gardens; permanent or temporary electric fences; hardware cloth over openings of structures; screening windows, vents, and other ground-level openings; hardware cloth cylinders around smaller trees and bushes; and aluminum flashing around larger trees.

Frightening. Frightening, by sight, sound, or smell, may deter some animals, but it may only be a short-term solution. Some examples of frightening devices and techniques used to deter mammals include pyrotechnics, guard dogs, and predator odors such as coyote or fox urine to deter their prey. Making an animal's visit unpleasant may encourage it to find someplace else to live.

Repellants. Few effective repellents to control damage-causing mammals are known, and legal restrictions prevent the use of most of them.

Poisons. Most poisons are not legal for mammals. Poisons can have detrimental effects on nontarget species.

Fumigants. Fumigants are poisonous gases that are most often used to kill mammals that live in burrows.

Trapping. The taking of game mammals during their prescribed open trapping seasons is an effective technique for controlling animal numbers and avoiding potential problems. Once an individual animal becomes a problem, trapping is a preferred tool for removing it. Knowledge of traps and trapping, as well as the life history of the species, are important for capturing and removing animals.

Traps are classified as either nonlethal or lethal. Nonlethal traps for mammals include cage or box-type traps, foothold traps, and snares. Lethal traps for mammals include body-gripping traps, mouse and rat-sized snap-traps, and harpoon and scissor-type traps. Rental companies may offer cage-type traps, or they may be purchased at hardware and garden-supply stores. Animal-control agencies may have cage-type traps to loan and may assist in setting and checking them. Some mammals are easier to trap with a foothold or body-gripping trap, but most people lack the skill and experience needed to use them. For assistance, contact a wildlife professional and ask for the name of a local trapper who can help you when these types of traps are needed.

Relocation not recommended. Before setting traps to remove problem mammals, you must have a plan for dealing with the captured animal. Although live-trapping followed by relocating the animal some distance from your property may seem like a good idea, it is usually not recommended. Moving an animal can confuse it and increase its vulnerability to vehicles or predators; spread disease to other animals living in the relocation area or to the relocated animal; make it difficult for the animal to find adequate food, water, or shelter in its new environment; upset the social order of an established population of the same species in the relocation area; and create a problem for another person at the new location. Also, federal, state, and local agencies may prohibit the release of wildlife on public lands. For these reasons, euthanizing the captured animal is usually recommended. Contact a wildlife professional for information on humane and legal methods to euthanize an animal.

Shooting. The taking of game mammals during their prescribed open hunting seasons is also an effective technique for controlling animal numbers and avoiding potential problems. Once an animal has become a problem, shooting is a preferred tool for removing it in some situations. Contact a wildlife professional, and check local government regulations, particularly if you are in an urban area, as most have ordinances against shooting within city limits.

How Mammals Are Named

The number and variety of living things are so great that it has been necessary to classify and name them, according to a standard international cataloguing system, for accurate reference. Mammals fall under the International Code of Zoological Nomenclature. Because the rules of this system are complex, only a general explanation is given here. The basis of the classification is similarity of structure and hereditary relationship.

All animals belong to the Animalia *kingdom*. This is separated into main divisions, known as *phyla* (singular *phylum*). Each phylum is further broken down into successively smaller units wherein the individuals possess closer and closer relationships. The major divisions of a phylum are, in order: *class*, *order*, *family*, *genus* (plural *genera*), and *species* (singular and plural the same). There are also intermediate divisions, such as subphyla, subfamily, and subspecies, but these are seldom used in this book.

The phylum Chordata contains the class Mammalia, or mammals. The wild representatives of this class established in Missouri belong to 8 orders, 20 families,

50 genera, and 72 species. Each species discussed in this book is listed in its respective order in this overall classification.

Under the biological species concept, species are simply defined as a group whose members breed only with each other in nature and are more like each other than members of other similar groups. Modern molecular methods have greatly altered our understanding of species relatedness and thus how species are defined. Each recognized species has been given a scientific name that is usually accepted as standard. However, for various reasons, some scientific names are changed and, until the revision becomes universally accepted, the species may be known by more than one scientific name. Several of these changes in names and classification have occurred since the earlier editions of this book. These have required the necessary alterations here to bring them into accordance with those accepted by most mammalogists.

The scientific name of a species consists of two words and is customarily written in italics. The first letter of the first word is capitalized. Most of these scientific names are derived from Latin or Greek, but some are Latinized forms of modern words or names. Frequently the scientific name refers to some physical characteristic or habit of the animal, or to a person associated with its collection. For example, the scientific name of the coyote is *Canis latrans.* The first part of this name is that of the genus to which the species belongs; the second part distinguishes this particular species from any others belonging to the same genus and is called a *specific epithet.* The name *Canis* is the Latin word for "dog" and denotes the close relationship between coyotes and dogs; *latrans* is the Latin word for "barker" and refers to this animal's habitual barking.

Different members of a species often vary considerably in minor characteristics, such as size or coloration, and these variations tend to be constant within a particular part of the species' range due to some isolating mechanism, frequently geographic, structural, or behavioral. Such variations are sometimes recognized,

and animals possessing them are designated as a *subspecies* by adding a third word to the species' name. Thus the two subspecies of the coyote found in Missouri would be classified like this:

Phylum: Chordata (animals with a notochord, or primitive backbone)
Subphylum: Vertebrata (animals with a backbone of vertebrae)
Class: Mammalia (mammals)
Order: Carnivora (flesh-eating mammals)
Family: Canidae (dogs, coyotes, foxes, and relatives)
Genus: *Canis* (dogs, wolves, coyotes, and jackals)
Species: *Canis latrans* (coyote)
Subspecies:
Canis latrans thamnos (northeastern coyote)
Canis latrans frustror (southeastern coyote)

Although variations in size or coloration are obvious in some individuals, it is often difficult to identify an animal to its subspecies with certainty unless a large series of specimens is available for comparison. Also, where the ranges of two subspecies meet, intergradation often occurs and individuals in such an area possess characteristics or intermediate characteristics of both subspecies. Because this distinction of subspecies is of more interest to those who classify animals than to persons who may wish merely to know the identity of a particular species, we have only infrequently given or discussed the subspecific names.

In addition to the names of the species and subspecies as given here, a complete catalog of scientific names includes the name of the person or persons who first described and named the species or subspecies. The describer's name is not italicized; if it is given in parentheses, this indicates that the species was originally placed in a different genus than in the one now used. The complete form of the coyote's name is written *Canis latrans* Say; this means that a person named Say gave the original scientific name to this species and that the name he gave to this animal still stands for it today. For simplicity, this system has been used only once in this book—in the following list of the wild mammals of Missouri.

Many different common names are often used for a species in various parts of its range. If it were not for the standard scientific names, the identity of a species might be hard to establish. In some Missouri localities the name *gopher* refers to a ground squirrel, either the thirteen-lined ground squirrel or the Franklin's ground squirrel; in other places in this state, *gopher* refers to an entirely different animal, the plains pocket gopher, while in the extreme western United States, it sometimes refers to a California ground squirrel. In this book we have given the meaning of the scientific name of each animal as an aid to understanding and remembering it and have used the most widely accepted common name. Other common names sometimes applied to the species are also listed.

List of the Wild Mammals of Missouri

Order Didelphimorphia—Pouched Mammals (Opossums)

Family Didelphidae—Opossums
Didelphis virginiana Kerr—Virginia Opossum

Order Soricomorpha—Shrew-form Mammals

Family Soricidae—Shrews
Sorex cinereus Kerr—Cinereus Shrew
Sorex longirostris Bachman—Southeastern Shrew
Blarina brevicauda (Say)—Northern Short-tailed Shrew
Blarina carolinensis (Bachman)—Southern Short-tailed Shrew
Blarina hylophaga Elliot—Elliot's Short-tailed Shrew
Cryptotis parva (Say)—North American Least Shrew
Family Talpidae—Moles and Relatives
Scalopus aquaticus (Linnaeus)—Eastern Mole

Order Chiroptera—Flying Mammals

Family Vespertilionidae—Evening Bats
Myotis lucifugus (Le Conte)—Little Brown Myotis
Myotis grisescens A. H. Howell—Gray Myotis
Myotis austroriparius (Rhoads)—Southeastern Myotis
Myotis septentrionalis (Trouessart)—Northern Long-eared Myotis
Myotis sodalis Miller and Allen—Indiana Myotis
Myotis leibii (Audubon and Bachman)—Eastern Small-footed Myotis
Lasionycteris noctivagans (Le Conte)—Silver-haired Bat
Perimyotis subflavus (F. Cuvier)—Tri-colored Bat
Eptesicus fuscus (Beauvois)—Big Brown Bat
Lasiurus borealis (Muller)—Eastern Red Bat
Lasiurus cinereus (Palisot de Beauvois)—Hoary Bat
Nycticeius humeralis (Rafinesque)—Evening Bat
Corynorhinus townsendii (Cooper)—Townsend's Big-eared Bat
Corynorhinus rafinesquii (Lesson)—Rafinesque's Big-eared Bat

Order Cingulata—Armored Mammals

Family Dasypodidae—Armadillos
Dasypus novemcinctus Linnaeus—Nine-banded Armadillo

Order Lagomorpha—Hare-shaped Mammals

Family Leporidae—Hares and Rabbits
Sylvilagus floridanus (J. A. Allen)—Eastern Cottontail
Sylvilagus aquaticus (Bachman)—Swamp Rabbit
Lepus californicus Gray—Black-tailed Jackrabbit

Order Rodentia—Gnawing Mammals

Family Sciuridae—Squirrels, Chipmunks, Marmots, and Relatives
Tamias striatus (Linnaeus)—Eastern Chipmunk
Marmota monax (Linnaeus)—Woodchuck
Ictidomys tridecemlineatus (Mitchill)—Thirteen-lined Ground Squirrel
Poliocitellus franklinii (Sabine)—Franklin's Ground Squirrel
Sciurus carolinensis Gmelin—Eastern Gray Squirrel
Sciurus niger Linnaeus—Eastern Fox Squirrel
Glaucomys volans (Linnaeus)—Southern Flying Squirrel
Family Geomyidae—Pocket Gophers
Geomys bursarius (Shaw)—Plains Pocket Gopher
Family Heteromyidae—Pocket Mice and Relatives
Perognathus flavescens Merriam—Plains Pocket Mouse
Family Castoridae—Beavers
Castor canadensis Kuhl—American Beaver
Family Cricetidae—New World Rats and Mice, Voles, and Relatives
Oryzomys palustris (Harlan)—Marsh Rice Rat
Reithrodontomys montanus (Baird)—Plains Harvest Mouse
Reithrodontomys megalotis (Baird)—Western Harvest Mouse
Reithrodontomys fulvescens J. A. Allen—Fulvous Harvest Mouse
Peromyscus maniculatus (Wagner)—Deer Mouse
Peromyscus leucopus (Rafinesque)—White-footed Mouse
Peromyscus gossypinus (Le Conte)—Cotton Mouse
Peromyscus attwateri J. A. Allen—Texas Mouse
Ochrotomys nuttalli (Harlan)—Golden Mouse
Sigmodon hispidus Say and Ord—Hispid Cotton Rat
Neotoma floridana (Ord)—Eastern Woodrat
Microtus pennsylvanicus (Ord)—Meadow Vole

Microtus ochrogaster (Wagner)—Prairie Vole
Microtus pinetorum (Le Conte)—Woodland Vole
Ondatra zibethicus (Linnaeus)—Common Muskrat
Synaptomys cooperi Baird—Southern Bog Lemming

Family Muridae—Old World Rats, Mice, and Relatives
Rattus rattus (Linnaeus)—Black Rat
Rattus norvegicus (Berkenhout)—Brown Rat
Mus musculus Linnaeus—House Mouse

Family Dipodidae—Jumping Mice and Relatives
Zapus hudsonius (Zimmermann)—Meadow Jumping Mouse

Order Carnivora—Flesh-eating Mammals

Family Canidae—Dogs, Coyotes, Foxes, and Relatives
Canis latrans Say—Coyote
Vulpes vulpes (Linnaeus)—Red Fox
Urocyon cinereoargenteus (Schreber)—Gray Fox

Family Ursidae—Bears
Ursus americanus Pallas—American Black Bear

Family Procyonidae—Raccoons and Relatives
Procyon lotor (Linnaeus)—Raccoon

Family Mustelidae—Weasels, Badgers, Otters, Minks, and Relatives
Mustela nivalis Linnaeus—Least Weasel
Mustela frenata Lichtenstein—Long-tailed Weasel
Neovison vison (Schreber)—American Mink
Taxidea taxus (Schreber)—American Badger
Lontra canadensis (Schreber)—North American River Otter

Family Mephitidae—Skunks and Relatives
Spilogale putorius (Linnaeus)—Eastern Spotted Skunk
Mephitis mephitis (Schreber)—Striped Skunk

Family Felidae—Cats
Puma concolor (Linnaeus)—Mountain Lion
Lynx rufus (Schreber)—Bobcat

Order Artiodactyla—Even-toed Hoofed Mammals

Family Cervidae—Deer, Elk, and Relatives
Odocoileus virginianus (Zimmermann)—White-tailed Deer
Cervus elaphus Linnaeus—Elk

How to Identify Mammals by the Keys

There are two keys given in this book to help identify the wild mammals of Missouri. One is based on whole adult animals; the other, on skulls of adults. The first key can be used properly only if the animal can be examined closely. The skull key requires a complete skull in most cases and a good hand lens or dissecting microscope for the proper observation of teeth in small rodents, shrews, and bats. Where measurements are called for in the key, follow the directions on pages 15–16. Where a dental formula is given in the key, refer to pages 3–4 to understand its application.

Very frequently bones are found and the skull or its fragments may be of help in identifying the animal's remains. Entire or partial skulls of small mammals are also found in the food pellets cast up by hawks and owls; when identified, they furnish information about the feeding habits of these avian predators. Skulls are sometimes necessary in the accurate identification of a species; their general structure and teeth also explain much about the animal's way of living.

To help identify skulls, three views are shown on most of the plates for the different species and a key has been prepared. Since the average person is not acquainted with the technical vocabulary used to describe skull structures, such terms have been kept to a minimum and label lines have been used freely.

To identify a specimen, first use either of the keys to the orders, starting by reading statements 1a and 1b. Where it has been necessary to use technical terms, they are explained in the illustration to which you are referred. Choose the statement that applies to your specimen and go to the next number as directed. Follow the chain of numbers until you come to the order to which your specimen belongs.

Then look up the pages to which you are referred. If there is more than one species in the order, you will find a key to the species. Follow this in the same manner as the key to the orders. When you come to the name of your specimen, look up the pages where this species is discussed and illustrated in more detail. Make sure everything agrees. If something disagrees, you may have misidentified your specimen. Work backward in the key by following the numbers marked "from" until you find the mistake. Then take the correct path through the key to your specimen.

Key to the Orders
by Whole Adult Animals

1a. Body covered with scutes (see plate 19). **Armored Mammals** (Cingulata) p. 121
1b. Body partially or entirely covered with fur. **Go to 2**

2a. (From 1b) Front limbs modified into wings with paired membranes stretched between the elongated fingers, front limbs, body, and hind legs (see plates 8–9); another pair of membranes between hind legs and tail; capable of true flight. **Flying Mammals** (Chiroptera) p. 63

2b. (From 1b) Front limbs not modified into wings; no membranes between hind legs and tail; not capable of true flight. **Go to 3**

3a. (From 2b) Feet provided with 2 hoofed toes that support the weight of the body (see plate 64); antlers sometimes present. **Even-toed Hoofed Mammals** (Artiodactyla) p. 393
3b. (From 2b) Feet provided with claws. **Go to 4**

4a. (From 3b) First toe on hind foot clawless, thumblike, and opposing other 4 toes (see plates 1–2); female with pouch on belly in which young are carried for a time after their birth. **Pouched Mammals (Opossums)** (Didelphimorphia) p. 27

4b. (From 3b) First toe on hind foot possessing a claw, not thumblike, and not opposing other toes; female without pouch on belly. **Go to 5**

5a. (From 4b) Incisor teeth chisel shaped and separated from cheek teeth (premolars and/or molars) by a very wide space (in upper jaw equal to or greater than length of entire row of grinding teeth); canine teeth absent. **Go to 6**
5b. (From 4b) Incisor teeth not chisel shaped; teeth in a continuous row and in upper jaw spaces between teeth smaller than width of largest tooth; canine teeth present. **Go to 7**

6a. (From 5a) Two incisors in each half of upper jaw, the smaller one directly behind the larger one (see plates 20–22); ears longer than tail; tail a cottony tuft; hind foot with 4 clawed toes; entire bottoms of feet densely furred. **Hare-shaped Mammals** (Lagomorpha) p. 127
6b. (From 5a) One incisor in each half of upper jaw (see plates 23–49); ears shorter than tail; tail not a cottony tuft; hind foot with 5 clawed toes; entire bottoms of feet not densely furred. **Gnawing Mammals** (Rodentia) p. 151

7a. (From 5b) No movable snout projecting over mouth (see plates 50–63); eyes large; total length more than 203 mm (8 in.); upper canine teeth enlarged and decidedly longer than adjacent teeth; first upper incisors not enlarged; fur not plushlike. **Flesh-eating Mammals** (Carnivora) p. 305
7b. (From 5b) Movable snout projecting well over mouth (see plates 3–7); eyes small or not visible; total length 203 mm (8 in.) or less; upper canine teeth not enlarged; first upper incisors enlarged; fur plushlike. **Shrew-form Mammals** (Soricomorpha) p. 37

Key to the Orders
by Skulls of Adults[1]

1a. Incisor teeth present in upper jaw. **Go to 3**

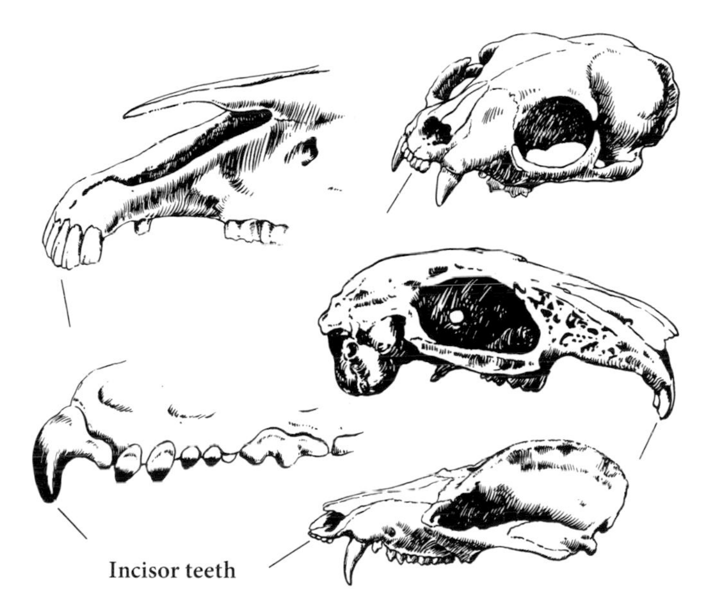

1b. Incisor teeth absent in upper jaw. **Go to 2**

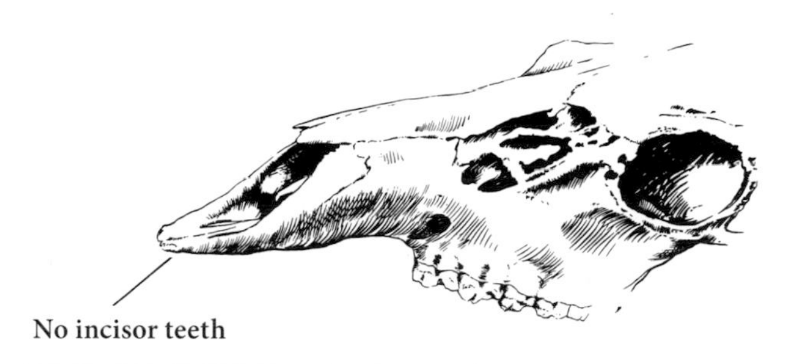

1. This key includes the horse, ass, and mule, which are members of the Odd-toed Hoofed Mammals (order Perissodactyla), because the skulls of these domestic animals might be found in the same general area where white-tailed deer live and might be confused with deer skulls.

2a. (From 1b) Teeth peg-shaped. **Armored Mammals** (Cingulata) p. 121

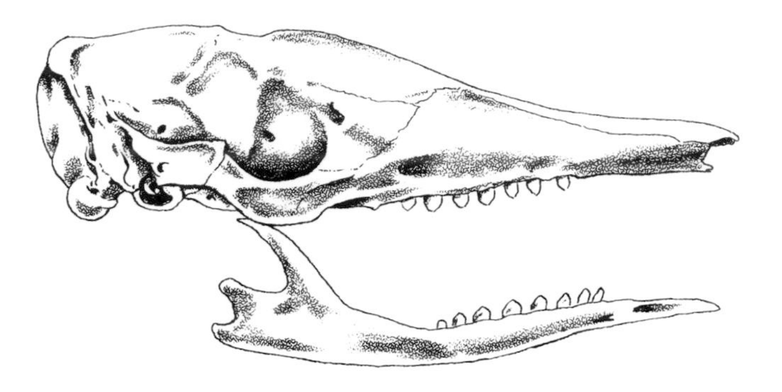

2b. (From 1b) Teeth not peg-shaped. **Even-toed Hoofed Mammals** (Artiodactyla) p. 393

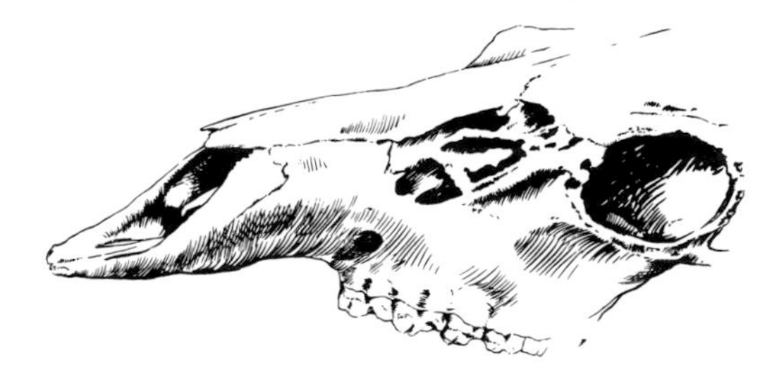

3a. (From 1a) Upper incisor teeth separated from cheek teeth by a space greater than width of largest tooth; upper canine teeth absent (except in some horses, asses, or mules). **Go to 4**

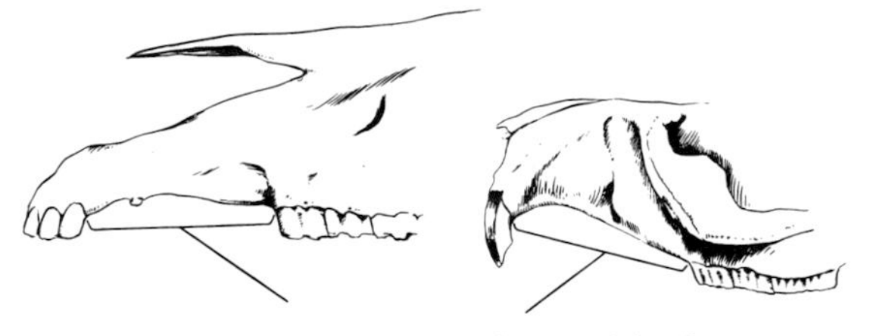

This space greater than width of largest tooth

3b. (From 1a) Teeth in a continuous row and spaces between teeth smaller than width of largest tooth; upper canine teeth present. **Go to 6**

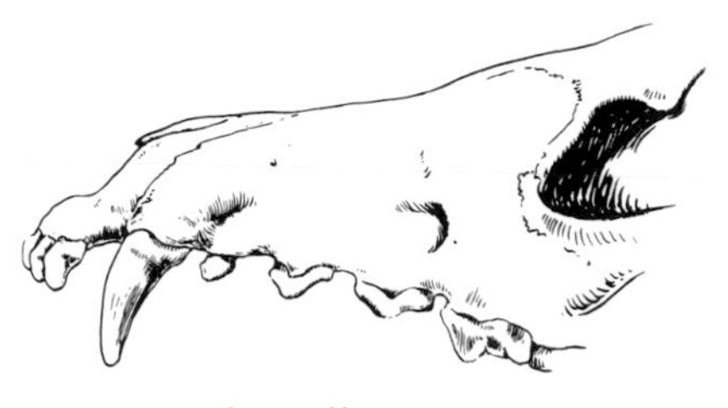

Spaces between teeth smaller than width of largest tooth

4a. (From 3a) Three incisors in each half of upper jaw; length of skull more than 228 mm (9 in.). **Odd-toed Hoofed Mammals** (Perissodactyla) p. 393

4b. (From 3a) One or 2 incisors in each half of upper jaw; length of skull less than 152 mm (6 in.). **Go to 5**

Three incisors in each half of upper jaw

5a. (From 4b) Two incisors in each half of upper jaw, the smaller one directly behind the larger one; bony network on side of face in front of eye socket; total teeth 28. **Hare-shaped Mammals** (Lagomorpha) p. 127

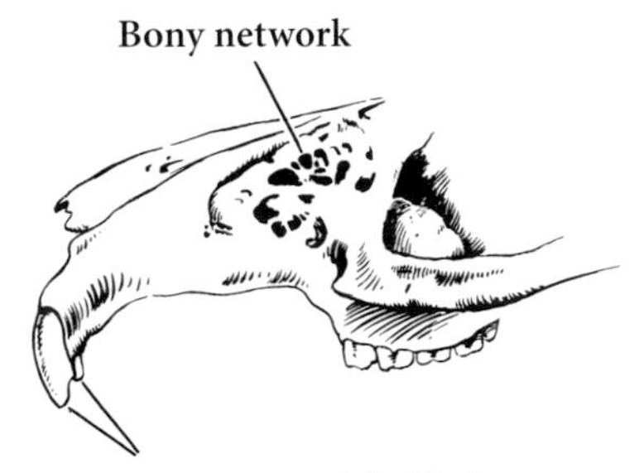

Two incisors in each half of upper jaw

5b. (From 4b) One incisor in each half of upper jaw; no bony network on side of face in front of eye socket; total teeth from 16 to 22. **Gnawing Mammals** (Rodentia) p. 151

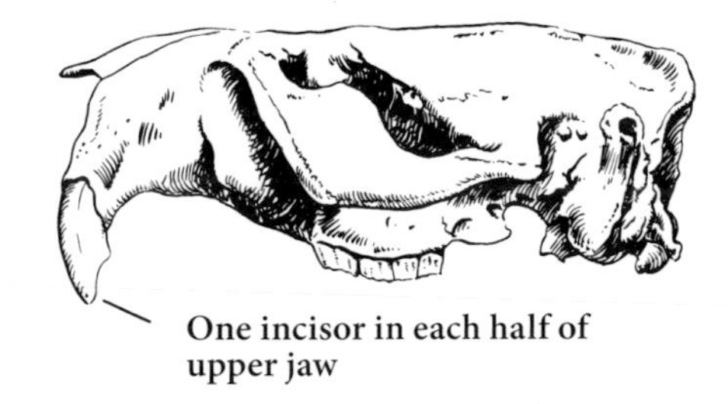

One incisor in each half of upper jaw

6a. (From 3b) First upper incisors greatly enlarged. **Shrew-form Mammals** (Soricomorpha) p. 37

6b. (From 3b) First upper incisors not greatly enlarged. **Go to 7**

7a. (From 6b) Front of skull with prominent U-shaped notch in hard palate; 1 or 2 incisors in each half of upper jaw. **Flying Mammals** (Chiroptera) p. 63

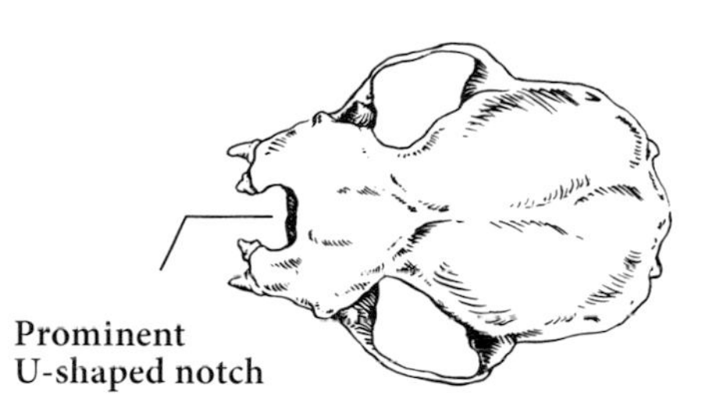

7b. (From 6b) Front of skull without prominent U-shaped notch in hard palate; 3 or 5 incisors in each half of upper jaw. **Go to 8**

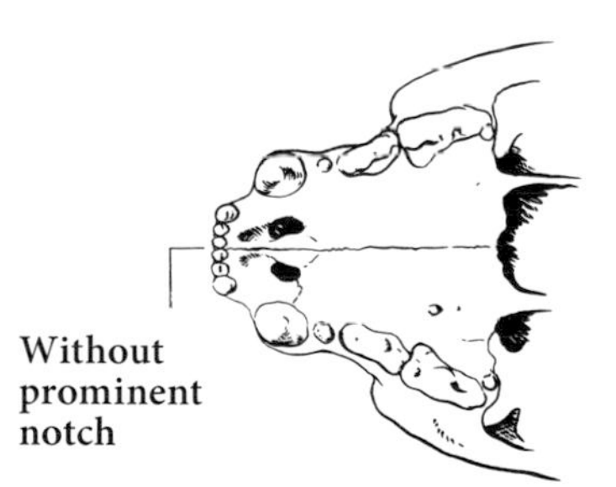

8a. (From 7b) Nasal bones broadly expanded on face in front of eye socket; 5 incisors in each half of upper jaw. **Pouched Mammals (Opossums)** (Didelphimorphia) p. 27

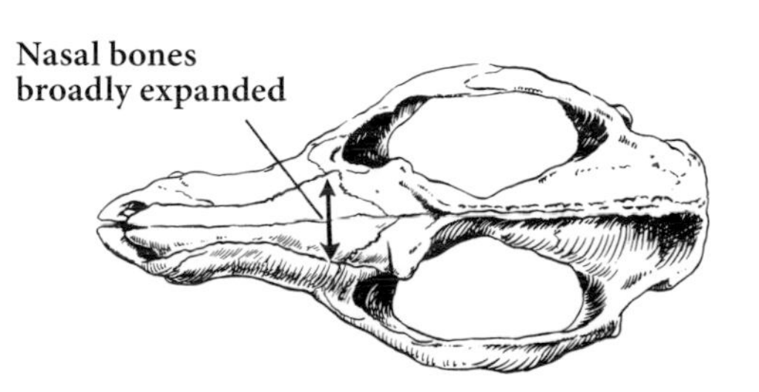

8b. (From 7b) Nasal bones not expanded on face in front of eye socket; 3 incisors in each half of upper jaw. **Go to 9**

9a. (From 8b) Upper canine teeth project outward when viewed from front, and often upward. **Even-toed Hoofed Mammals** (Artiodactyla) p. 393

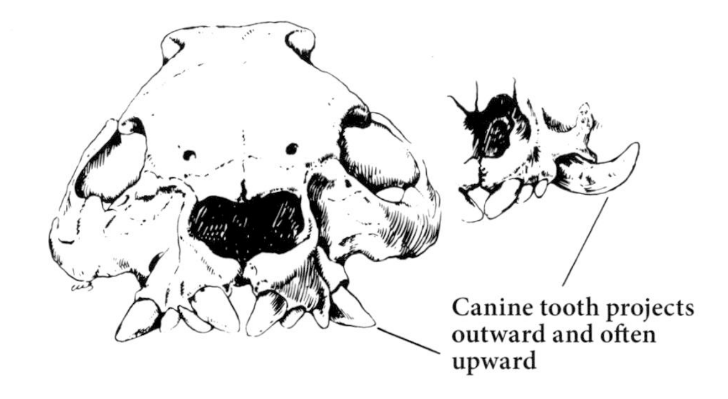

9b. (From 8b) Upper canine teeth project downward. **Flesh-eating Mammals** (Carnivora) p. 305

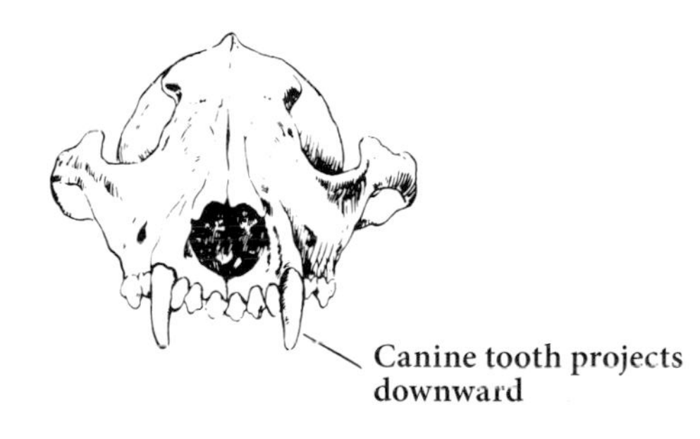

GENERAL REFERENCES

See also general references for some orders or families, and selected references for each individual species given at the end of each species' account.

Allen, T. B., ed. 1995. *Wild animals of North America.* 3rd ed. National Geographic Society, Washington, DC. 406 pp.

Barbour, R. W., and W. H. Davis. 1974. *Mammals of Kentucky.* University Press of Kentucky, Lexington. 368 pp.

Bauer, E. A. 1998. *Big game of North America.* Voyageur Press, Minneapolis, MN. 160 pp.

Bee, J. W. 1981. *Mammals in Kansas.* University Press of Kansas, Lawrence. 302 pp.

Bennitt, R., and W. O. Nagel. 1937. *A survey of the resident game and furbearers of Missouri.* University of Missouri Studies, Columbia, vol. 12. 215 pp.

Beringer, J., and J. Blair. 2013. *2013 furbearer program annual report.* Missouri Department of Conservation, Resource Science Division, Jefferson City. 53 pp.

Brown, G. H. 1952. *Illustrated skull key to the recent land mammals of Virginia.* Virginia Cooperative Wildlife Research Unit, Blacksburg, Release 52-2. 75 pp.

Brown, L. N. 1997. *A guide to the mammals of the southeastern United States.* University of Tennessee Press, Knoxville. 236 pp.

Caire, W., B. P. Glass, and M. A. Mares. 1989. *Mammals of Oklahoma.* University of Oklahoma Press, Norman. 544 pp.

Connior, M. B. 2010. Annotated checklist of the recent wild mammals of Arkansas. *Occasional Papers, Museum of Texas Tech University* 293. 12 pp.

Elbroch, M. 2003. *Mammal tracks and sign: A guide to North American species.* Stackpole Books, Mechanicsburg, PA. 768 pp.

———. 2006. *Animal skulls: A guide to North American species.* Stackpole Books, Mechanicsburg, PA. 716 pp.

Feldhamer, G. A., L. C. Drickamer, S. H. Vessey, and J. F. Merritt. 2007. *Mammalogy: Adaptation, diversity, and ecology.* 3rd ed. Johns Hopkins University Press, Baltimore, MD. 643 pp.

Feldhamer, G. A., B. C. Thompson, and J. A. Chapman. 2003. *Wild mammals of North America: Biology, management, and conservation.* 2nd ed. Johns Hopkins University Press, Baltimore, MD. 1232 pp.

Gotch, A. F. 1996. *Latin names explained: A guide to the scientific classification of reptiles, birds, and mammals.* 1st US ed. Facts on File, New York. 720 pp.

Hall, E. R. 1981. *The mammals of North America.* 2nd ed. John Wiley and Sons, New York. Vol. 1: 600 pp.; vol. 2: 580 pp.

Hall, E. R., and K. R. Kelson. 1959. *The mammals of North America.* Ronald Press, New York. Vol. 1: 546 pp.; vol. 2: 536 pp.

Harvey, M. J., J. S. Attenbach, and T. L. Best. 2011. *Bats of the United States and Canada.* Johns Hopkins University Press, Baltimore, MD. 224 pp.

Hayssen, V., A. Tienhoven, and A. Tienhoven. 1993. *Asdell's patterns of mammalian reproduction: A compendium of species-specific data.* 2nd ed. Cornell University Press, Ithaca, NY. 1023 pp.

Hoffmeister, D. F. 2002. *Mammals of Illinois.* University of Illinois Press, Urbana. 384 pp.

Jones, J. K., Jr. 1964. *Distribution and taxonomy of mammals of Nebraska.* University of Kansas Publications, Museum of Natural History, Lawrence. Vol. 16, no. 1. 356 pp.

Jones, J. K., Jr., D. M. Armstrong, and J. R. Choate. 1985. *Guide to the mammals of the plains states.* University of Nebraska Press, Lincoln. 368 pp.

Jones, J. K., Jr., and R. W. Manning. 1992. *Illustrated key to skulls of genera of North American land mammals.* Texas Tech University Press, Lubbock. 80 pp.

Kayr, R. W., and D. E. Wilson. 2009. *Mammals of North America.* 2nd ed. Princeton University Press, Princeton, NJ. 248 pp.

McKenna, M. C., and S. K. Bell. 2000. *Classification of mammals.* Columbia University Press, New York. 640 pp.

McKinley, D. 1960. *A chronology and bibliography of wildlife in Missouri.* University of Missouri Bulletin, Columbia. Vol. 61, no. 13. 128 pp.

McKinley, D., and P. Howell. 1976. Missouri's wildlife trail. Parts I and II. *Missouri Conservationist* 37. 61 pp.

Murray, D. K. 1991. Mammals of the Ozark National Scenic Riverways. M.S. thesis, University of Missouri, Columbia. 458 pp.

Nelson, P. W. 2005. *The terrestrial natural communities of Missouri.* Missouri Department of Natural Resources, Jefferson City. 550 pp.

Nigh, T. A., and W. A. Schroeder. 2002. *Atlas of Missouri ecoregions.* Missouri Department of Conservation, Jefferson City. 212 pp.

Novak, M., J. A. Baker, M. E. Obbard, and B. Malloch, eds. 1987. *Wild furbearer management and conservation in North America.* Ontario Trappers Association, Canada. 1150 pp.

Nowak, R. M. 1994. *Walker's bats of the world.* Johns Hopkins University Press, Baltimore, MD. 288 pp.

———. 1999. *Walker's mammals of the world.* 6th ed. Johns Hopkins University Press, Baltimore, MD. 2015 pp.

———. 2005. *Walker's carnivores of the world.* Johns Hopkins University Press, Baltimore, MD. 309 pp.

Polder, E. 2006. *Native and introduced mammals of Iowa.* Heritage Printing, Dyersville, IA. 218 pp.

Rezendes, P. 1999. *Tracking and the art of seeing: How to read animal tracks and sign.* 2nd rev. ed. HarperCollins, New York. 336 pp.

Sampson, F. W. 1980. *Missouri fur harvests.* Terrestrial Series no. 7. Missouri Department of Conservation, Jefferson City. 60 pp.

Schmidt, J. L., and D. L. Gilbert, eds. 1978. *Big Game of North America. Ecology and management.* Stackpole Books, Harrisburg, PA. 494 pp.

Schwartz, C. W., and E. R. Schwartz. 1959. *The wild mammals of Missouri.* University of Missouri Press, Columbia. 341 pp.

———. 1981. *The wild mammals of Missouri.* Rev. ed. University of Missouri Press, Columbia. 356 pp.

———. 1993. *About mammals and how they live.* Missouri Department of Conservation, Jefferson City. 191 pp.

———. 2001. *The wild mammals of Missouri.* 2nd rev. ed. University of Missouri Press, Columbia. 368 pp.

Sealander, J. A., and G. A. Heidt. 1990. *Arkansas mammals: Their natural history, classification, and distribution.* University of Arkansas Press, Fayetteville. 308 pp.

Vaughan, T. A., J. M. Ryan, and N. J. Czaplewski. 2000. *Mammalogy.* 4th ed. Saunders College Publishing, Fort Worth, TX. 565 pp.

Whitaker, J. O., Jr. 1996. *National Audubon Society field guide to North American mammals.* 2nd ed. Alfred A. Knopf, New York. 937 pp.

Whitaker, J. O., Jr., and W. J. Hamilton Jr. 1998. *Mammals of the eastern United States.* Cornell University Press, Ithaca, NY. 583 pp.

Wilson, D. E., and D. M. Reeder, eds. 2005. *Mammal species of the world: A taxonomic and geographic record.* 3rd edition. The Johns Hopkins University Press, Baltimore, MD. 2142 pp.

Wilson, D. E., and S. Ruff, eds. 1999. *The Smithsonian book of North American mammals.* Smithsonian Institution Press, Washington, DC. 750 pp.

2

Pouched Mammals (Opossums)
Order Didelphimorphia

In earlier editions of this book, pouched mammals belonged in the order Marsupialia. Marsupial species were reclassified and the opossums now belong in the order Didelphimorphia. Today, most taxonomists assign marsupials to the infraclass Metatheria, but others have elevated Marsupialia to infraclass with two superorders: Ameridelphia (the American marsupials) and Australidelphia (marsupials inhabiting Australia and nearby islands). Most marsupials, including the kangaroos, wombats, koala, and bandicoots, inhabit Australia and the neighboring islands. Fewer opossum and shrew opossum marsupials occur in North and South America.

The word "marsupial" comes from the Latin word meaning "pouch" and refers to the prominent marsupium, or pouch, on the belly of females in most species of this group. The young are born incompletely formed and are carried in this pouch, where they continue growth and development. The unique presence of the epipubic bones help support the pouch on the pelvis. The brain is relatively small and primitive.

Opossums (Family Didelphidae)

The characters of this family are exemplified by the Virginia opossum, the only marsupial found in North America north of Mexico.

Virginia Opossum (*Didelphis virginiana*)

Name

The first part of the scientific name, *Didelphis*, is of Greek origin and means "double womb" (*di*-, "double," and *delphys*, "womb"). This name is subject to two interpretations; it may refer to the opossum's paired uteri or to the pouch as an accessory "womb" where the young are carried until they are able to care for themselves. The second part, *virginiana*, means "of Virginia," for the location of its original collection.

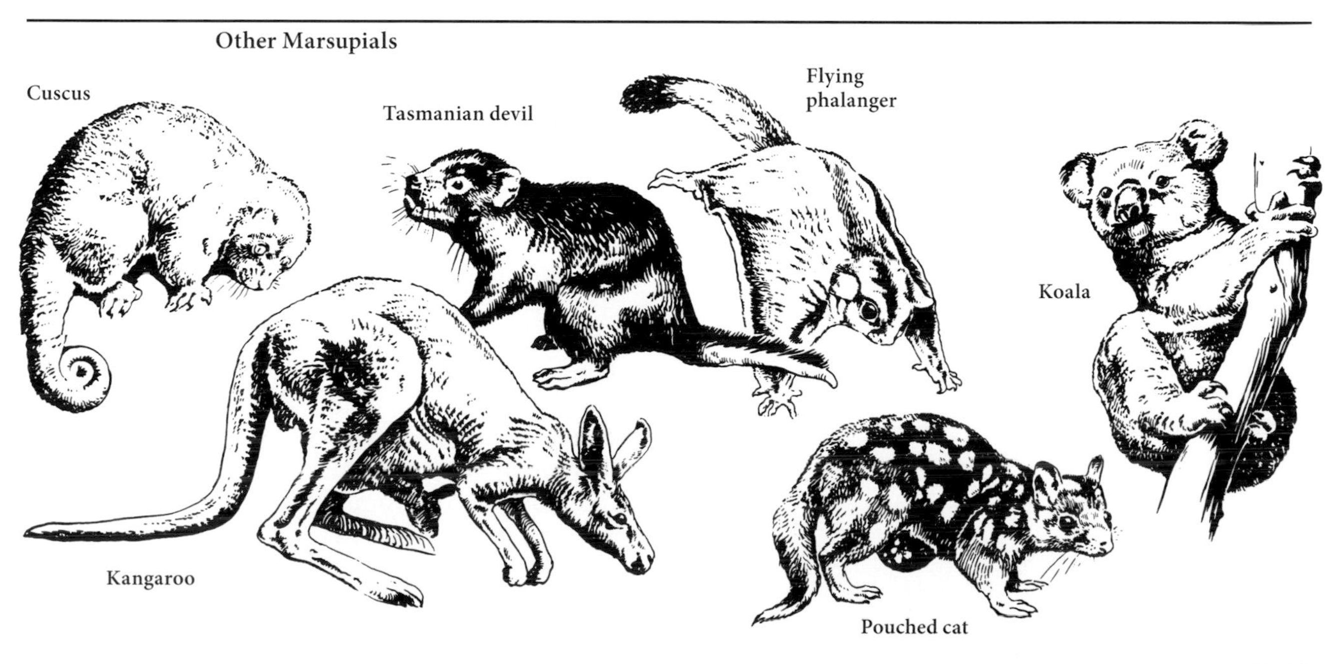

The common name is derived from the Algonquian word *apasum* meaning "white animal." This species was formerly known as *Didelphis marsupialis*.

Description (Plate 1)

The Virginia opossum is a medium-sized animal with long, rather coarse fur; a sharp, slender muzzle; prominent, thin, naked ears; short legs of about equal length; and a long, prehensile (able to grasp) tail covered with scales and scant hairs. Each of the front feet has 5 claw-bearing toes, while each of the hind feet has 4 claw-bearing toes and a large first toe that is thumblike, clawless, and opposable (capable of moving toward and touching the other digits on the same hand). The opossum walks on the entire sole of the foot (plantigrade), which is naked. The prominent, fur-lined pouch, or marsupium, which contains the teats, is present on the belly of mature females. In many adults, and particularly in males, the margins of the ears and the tip of the tail are often damaged by freezing.

Color. The body is predominantly grayish white with the front and hindquarters darker and the belly lighter. The guard hairs are all white, while the underfur is composed of both pure white hairs and white hairs tipped with black. The head is white to yellowish and sometimes has a dusky area encircling the eye. The nose is pink, the eyes are black, and the ears bluish black to black with a pink to white margin in some individuals. The tail is black at its base up to slightly more than half its length and yellowish white to pink the rest of its length. The feet and toes are pink to white.

Although gray is the most common color in Missouri, several other colorations, or color phases, occur. In the black phase, the white underfur is black tipped and the guard hairs are generally black; in the brown phase, the white underfur is brown tipped and the guard hairs are either white or brown. A very few white opossums lack dark pigment and are true albinos; most white ones possess some black pigment in the ears, eyes, skin of the feet, and base of the tail. Males and females are colored alike and do not change color with the seasons.

Measurements.

Total length	609–863 mm	24–34 in.
Tail	241–381 mm	9½–15 in.
Hind foot	50–76 mm	2–3 in.
Ear	50 mm	2 in.
Skull length	79–127 mm	3⅛–5 in.
Skull width	57–69 mm	2¼–2¾ in.
Weight	1.6–6.8 kg	3½–15 lb.

Males are usually larger than females and at the same age may be more than twice as large.

In 2012, a 6.1 kg (13.4 lb.) male became the official record-weight Virginia opossum confirmed by the Missouri Department of Conservation. The state furbearer record program was started in 2011.

Teeth and skull. An adult opossum has 50 teeth, the largest number present in any Missouri mammal. The dental formula is:

$$I\,\frac{5}{4}\;C\,\frac{1}{1}\;P\,\frac{3}{3}\;M\,\frac{4}{4} = 50$$

Individuals have been found with their front upper teeth worn down to the gums and other teeth abscessed or missing.

The opossum's skull is easily distinguished from skulls of other mammals of similar size by the number and arrangement of the teeth, the prominent sagittal crest, the shape of the nasal bone, which is broadly expanded toward the rear, and the angle of the lower jaw (turned inward). The braincase is extremely small for the size of the animal.

Sex criteria and sex ratio. Mature females can be identified by the pouch on the belly, and males by the scrotum that lies in front of the penis and by the forked penis. In cased pelts, sex can be recognized by the pouch or scrotum that usually remains on the skin. Males and females occur in approximately equal numbers among both young and adults.

Age criteria, age ratio, and longevity. The best criterion of age up to 10 months, when adult dentition is complete, is the number of teeth. At 3 months, opossums have a total of 30 teeth; at 4 months, 40; at 5 months, 44; and at 6 to 7 months, 48 teeth.

A female that has not reached sexual maturity is identified by her full-furred, white, shallow pouch with a marginal lip less than 6 mm (¼ in.) high and by tiny, white teats. On pelts, these pouches are not as flabby, dark, or as inclined to tear as those of females that have reared young. In a female that has experienced pregnancy, the pouch contains scant, rusty-colored fur and enlarged teats and has a marginal lip 1.6 to 2.2 cm (⅝ to ⅞ in.) high.

In skeletons of Virginia opossums under 10 months of age, the epipubic bones (which extend forward from each side of the pelvic, or hip, girdle in both sexes and support the pouch in females) generally measure less than 4.1 cm (1⅝ in.) in length while in opossums older than 10 months, the epipubic bones exceed 4.4 cm (1¾ in.) in length.

The age ratio in summer and early fall is 7 young per adult female. By winter this ratio has decreased to 6 young per adult female. Virginia opossums, for their size, are one of the shortest-lived mammals

Plate 1

Virginia Opossum (*Didelphis virginiana*)

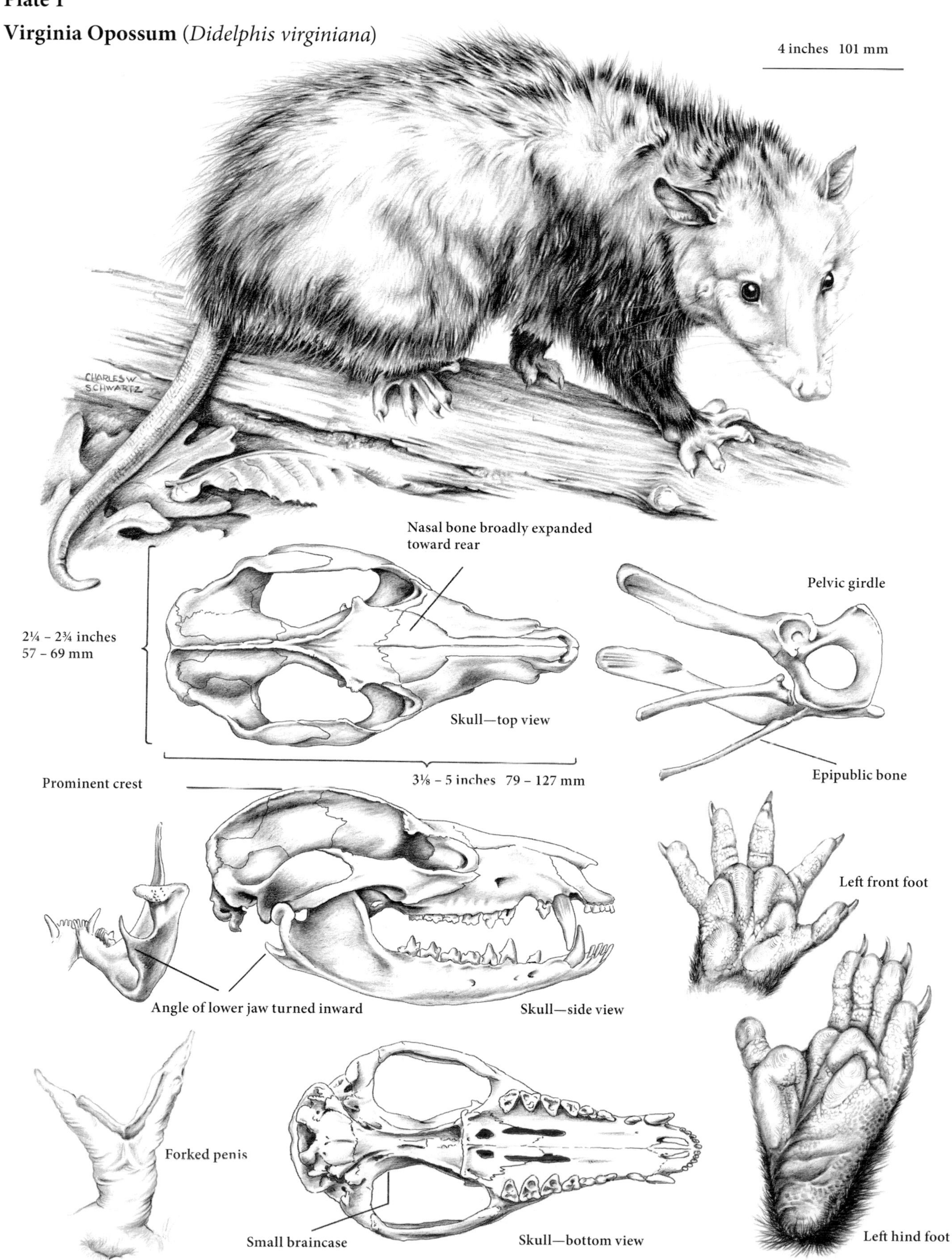

in the world with most living no more than 2 years in the wild. It is estimated that it takes 3½ years for the population of a given area to be replaced by new individuals.

Glands. Two glands on either side of the anus contain a foul-smelling greenish secretion that is often extruded in defensive behavior.

Voice and sounds. Although opossums are usually silent, they may give low growls or faint hissing sounds. A peculiar clicking sound made with the teeth or tongue is given commonly during the breeding period and less often at other times.

Distribution and Abundance

Prior to European settlement, the North American range of the Virginia opossum was limited in the north to Kentucky, Indiana, and Ohio. By 1917, their western boundary was central Kansas, and they continued to expand north and west. They now have reached into extreme southeastern Canada, but their range will eventually be limited by winter temperatures and snow depth to the north and by lack of water in the west. Their presence along the Pacific Coast and in Arizona, New Mexico, Colorado, and Idaho is the result of introductions.

Opossums are common throughout Missouri, and annual population surveys indicate statewide numbers continue to slowly increase. They are most common in the Central Dissected Till Plains and Osage Plains of northern and western Missouri. In favorable habitats, there may be 1 opossum per 0.8 to 1.2 ha (2 to 3 ac.). One study area of 725 ha (2.8 sq. mi.) in Illinois was estimated to have a population of 300 opossums per 259 ha (1 sq. mi.) in midsummer.

Habitat and Home

The Virginia opossum prefers to live in wooded areas mostly near streams. Densely forested sections are not inhabited as much as are farming areas interspersed with small, wooded streams. Timber near other sources of water, such as ponds, lakes, and swamps, provides additional habitat. The opossum is also now quite common in urban and suburban areas, which is likely due to their adaptability to various habitat types.

The opossum's home is any place that is dry, sheltered, and safe. Different homes may be used on successive nights and include dens or nests of other

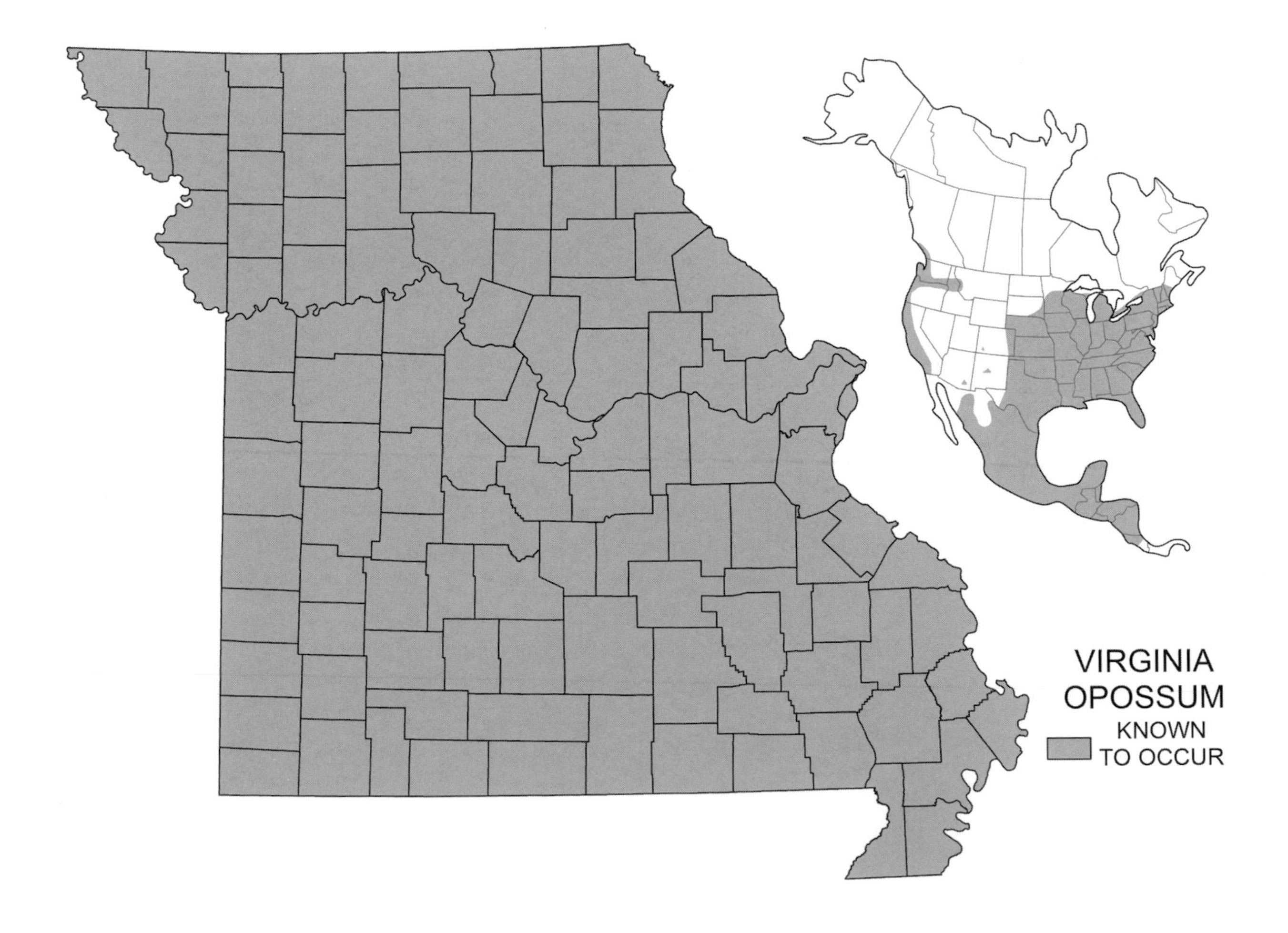

Opossum carrying nesting material

animals and cavities in rocks, brush piles, trash heaps, hollow trees, and fallen logs. They do not dig their own dens and are therefore dependent on others, such as woodchucks, for ground burrows. Inside the cavity, a nest is made of leaves and grass, or sometimes corn husks. In collecting this nesting material, the opossum picks up the leaves with its mouth, transfers them to the front legs, passes them under its body to the hind legs, and then carries the bundle in a loop made in the end of its tail. An opossum may transport as many as eight mouthfuls at once with its tail and carry additional material in its mouth.

Habits

The shy, secretive, and nocturnal opossum is seldom seen. It is most often observed along highways in the glare of automobile headlights as it feeds on animals killed by vehicles. While this mammal tends to be nomadic with shifting home ranges, the average home range estimate for females is about 61 ha (150 ac.) while males have an average home range of about 122 ha (300 ac.) outside of the mating season. Home ranges may exhibit moderate to extreme overlapping between individuals. Their activity is concentrated around the den they are using, and as they move from den to den, the area they use shifts correspondingly. Opossums have traveled as far as 302 m (990 ft.) between successive den sites, and a tracked opossum traveled as much

"Playing 'possum"

as 3.2 km (2 mi.) in one night. Adult males move farther than females, and their ranges may double during the breeding season.

Opossums are not aggressive and when pursued often climb trees or brush heaps in an attempt to escape. When frightened, they expose their teeth and drip saliva from the mouth. A common means of defense is feigning death, which is so characteristic it has given rise to the colloquial expression "playing possum." The animal rolls over on its side, becomes limp, shuts its eyes, and lets its tongue hang out of its open mouth; the heartbeat slows. This reaction is a brief nervous shock similar to fainting, but the animal recovers quickly and is able to take the first opportunity for escape.

In general, opossums show more activity in spring and summer than in fall and winter.

Although neither sex is particularly active when the temperature is below –7°C (20°F), females show a greater tendency than males to "hole up" for several days at a time during periods of very cold weather. Opossums generally lead solitary lives, but as many as five of the same sex have been found together in one den.

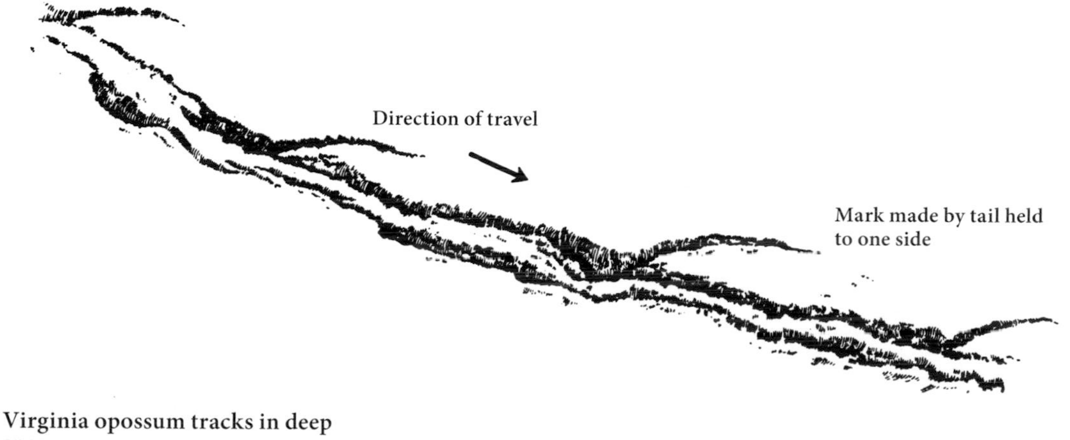

Virginia opossum tracks in deep snow

These mammals are well adapted for climbing, and the opposable toe on the hind foot assists them in holding onto small branches or similar structures. Opossums can support themselves for short periods entirely by the prehensile tail if at least half of it grasps a thick branch. In walking, they move with a slow, ambling gait and swing the tail from side to side. They sometimes dig in their quest for food. They are strong swimmers and may take to water voluntarily.

Foods

The omnivorous Virginia opossum eats a wide variety of foods but prefers animal matter during all seasons. Insects, such as grasshoppers, crickets, squash bugs, stink bugs, ground beetles, lamellicorn beetles, and ants, are the most common items. Although insects are not abundant in winter, they are still sought under fallen logs or rocks and in leaf litter or decaying animal matter. Since opossums are scavengers, most of the flesh food is carrion. The cottontail is the most important single kind of animal food, but dead opossums, domestic cats, squirrels, raccoons, skunks, moles, and mice are also eaten. Other animal foods such as reptiles, amphibians, birds and their eggs, land snails, crayfish, and earthworms compose smaller amounts of the diet. Domestic chickens may be eaten but are not an important food item.

Fruits form a substantial portion of the fare only during fall and early winter. These include pokeberries, mulberries, hackberries, greenbriers, groundcherries, grapes, blackberries, apples, pawpaws, haws, and persimmons. Although persimmons are a favorite food, the seeds are rarely eaten and do not show up commonly in the droppings. Corn, the only cultivated grain consumed to any extent, occurs most commonly in the diet during winter when other and perhaps more desirable foods are not available.

Reproduction

In Missouri the breeding season of opossums begins about the first of February. Mating occurs during the first three weeks of the month with the second week representing the peak. Gestation, or pregnancy, takes only 12½ to 13 days. Although the earliest recorded young in Missouri were born on 12 February, most litters are produced toward the end of February. This breeding period accounts for most of the year's production. When the young are ready to be weaned, sometime in May, the females mate again. The second litters are born toward the end of May or the early part of June, and these young are weaned around mid- or late September. Approximately three-fourths of the adult females have pouched young at one or both of these periods.

Contrary to some folklore, opossums do not mate through the nose. In mating, the penis enters the single opening of the female reproductive tract, the forked ends penetrating the paired vaginal canals leading to the two uteri. While the male is capable of fertile mating at any time of the year, the female has a restricted period of mating, or estrus, at approximately 28-day intervals during the breeding season. Females that do not become pregnant at their first mating period come into estrus again about four weeks later.

The average number of eggs shed from the ovaries at each estrous cycle is 22 and the maximum is 56, but not all are fertilized or develop. Many of the young that are born subsequently perish because the number of young that can survive depends upon the number of teats available for them. Most females possess 13 teats, 12 of which are typically arranged in a horseshoe shape with 1 in the center opposite the third pair from the rear. In rare cases some females have a few more or less than 13. The front 2 to 6 are smaller than the rest and often are not used. The average number of pouch young in Missouri is 9 with extremes from 5 to 13.

At birth the young are blind and very incompletely developed except for well-formed claws on the front limbs (see plate 2). Each young is less than 1.3 cm (½ in.) long and weighs about 0.2 g (less than 1⁄100 oz.). In preparation for birth of her young, the female assumes a position with the opening of the pouch directed upward and she licks her fur-lined pouch and lower abdomen. During birth the young pass from the

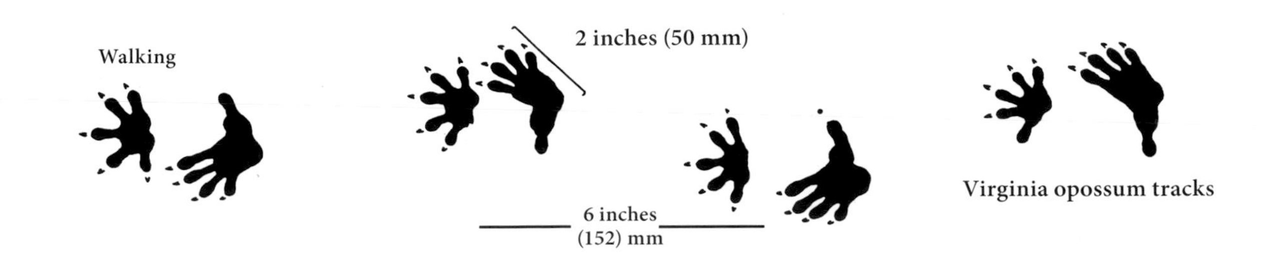

Virginia opossum tracks

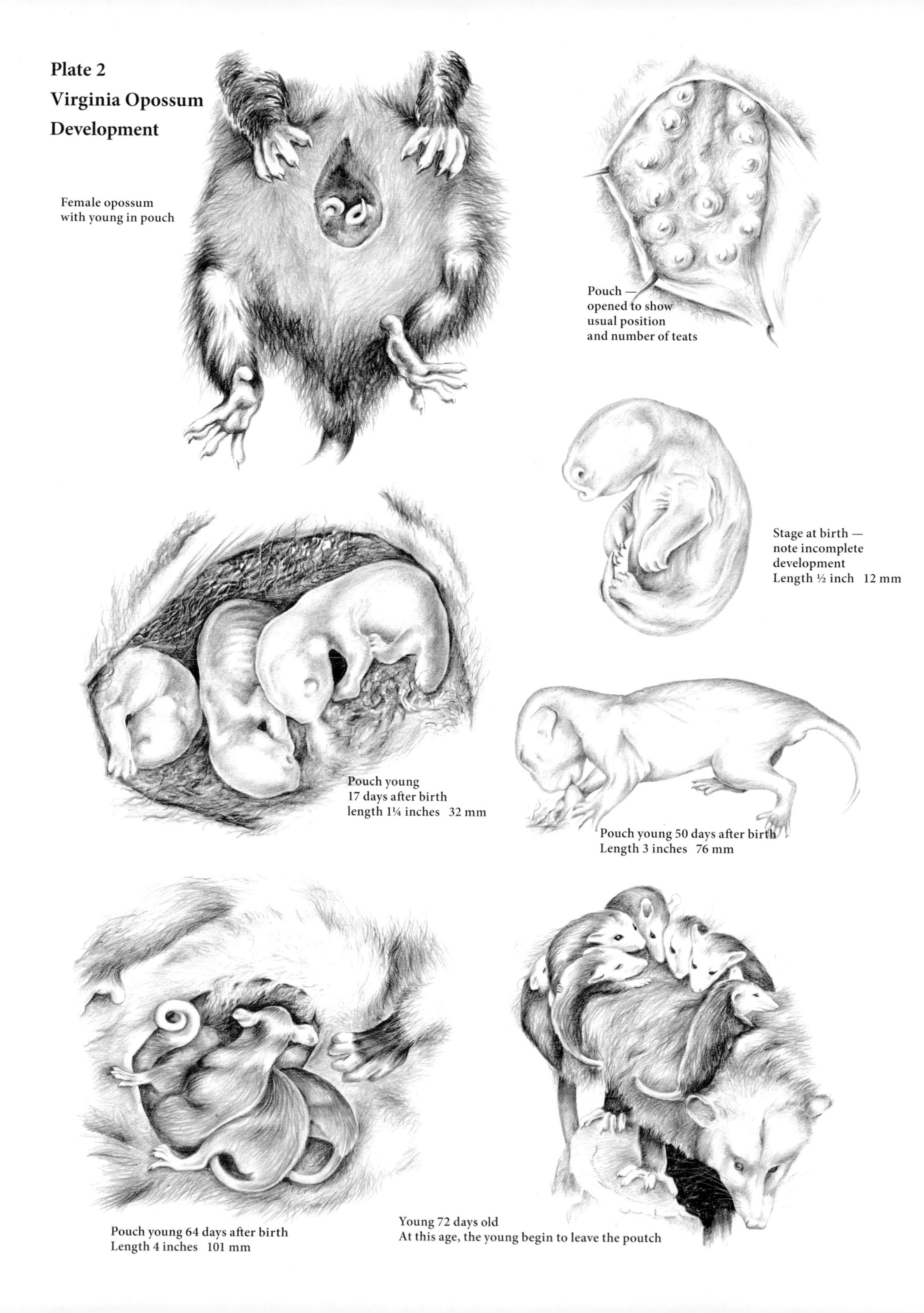
Plate 2
Virginia Opossum
Development
Female opossum
with young in pouch
Pouch —
opened to show
usual position
and number of teats
Stage at birth —
note incomplete
development
Length ½ inch 12 mm
Pouch young
17 days after birth
length 1¼ inches 32 mm
Pouch young 50 days after birth
Length 3 inches 76 mm
Pouch young 64 days after birth
Length 4 inches 101 mm
Young 72 days old
At this age, the young begin to leave the poutch

paired uteri to the outside of the body through a newly formed, temporary birth canal (the medial or pseudovaginal canal) rather than through the permanent, paired, vaginal canals. The young then climb up the female's fur with their front feet and claws to the opening of the pouch. Since the flap on the pouch is relaxed, the young have to climb only three or four times their own length. Once the young reach the pouch, those that find a teat become firmly attached to it and do not relinquish their hold for about 60 days.

As the young opossums continue to develop and grow larger, the teats elongate to 3.8 cm (1½ in.), enabling the young to nurse from outside the pouch. The female opens and closes the pouch at will and often permits her offspring to sun for extended periods. She keeps them dry by closing the pouch when she swims. About 2 weeks after birth females can be identified by the outline of the pouch and nipples, while males show the outline of a scrotum. At around 60 days of age, when the young are about the size of house mice, they begin to open their eyes and lips and let go of the nipples for the first time. At 80 days of age they are developing control of their body temperatures and can leave the pouch for short periods. After the young become too large for all to fit inside the pouch at once, some ride on the female's back, clinging to her fur with their feet, while others remain in the pouch. The young stay with the mother continuously for about 100 days. After weaning, the young become increasingly susceptible to predation as they begin to forage outside the protected weaning den. The juveniles typically will not disperse from this area until the winter mating season. Young of both sexes breed the first year following birth.

Some Adverse Factors

The predators of Virginia opossums are dogs, humans, foxes, coyotes, bobcats, and great horned owls. The parasites known to occur on or in opossums are mites (including four abundant host-specific mites), ticks, lice, fleas, roundworms, flukes, tapeworms, and numerous other endoparasites. The opossum has been reported to serve as the reservoir for the causative agents of many protozoan, fungal, bacterial, and viral diseases. Opossums are frequently killed by automobiles as they cross our highways.

Importance

The Virginia opossum is a common commercial furbearer in Missouri and throughout its range in the United States. The fur is used chiefly to trim coats, although choice skins may be made into whole coats.

In the 1940s the average number of pelts sold annually in Missouri was 288,344. This harvest dwindled in the 1950s to 58,751, was still lower (33,823) in the 1960s, increased somewhat in the 1970s to 62,082, and then declined again to 51,187 in the 1980s, 14,913 in the 1990s, and 8,445 in the first decade of the 2000s. During this seven-decade period, the highest annual take was 463,119 in 1940 and the lowest was 2,413 in 2000. The average annual price paid per pelt was $0.45 in the 1940s, $0.20 in the 1950s, $0.40 in the 1960s, $1.66 in the 1970s, $1.36 in the 1980s, $1.15 in the 1990s, and $1.57 in the first decade of the 2000s. The highest average annual price per pelt was $3.70 in 1979 and the lowest was $0.10 in 1954 and 1955. It is hard to measure the relative effects of the several factors that influence harvest numbers and pelt prices. Obviously, population abundance is involved, but to what extent has not been determined. Changes in demand for furs and the fluctuating values of pelts influence economic reasons for trapping opossums. In addition, the numbers of trappers working in Missouri influence the harvest.

Many people eat opossum, and baked opossum is a traditional dish in some localities. Although opossums occasionally raid poultry yards, their reputation as poultry destroyers is largely undeserved. However, in certain localities they can be notable predators of waterfowl and other ground-nesting birds.

Conservation and Management

Since this species is sufficiently abundant in Missouri, there is no immediate need for management other than to regulate the harvest. The proximity to water seems the most important factor in the opossum's environment.

In locations where opossum populations are high, predation upon birds and their nests can be a problem. If you are experiencing problems with opossums, contact a wildlife professional for advice, assistance, regulations, or special conditions for handling these animals.

SELECTED REFERENCES

See also discussion of this species in general references, page 23

Barr, T. R. B. 1963. Infectious diseases in the opossum: A review. *Journal of Wildlife Management* 27:53–71.

Fitch, H. S., and L. L. Sandidge. 1953. Ecology of the opossum on a natural area in northeastern Kansas. *University of Kansas Publications, Museum of Natural History* 7:305–338.

Fitch, H. S., and H. W. Shirer. 1970. A radiotelemetric study of spatial relationships in the opossum. *American Midland Naturalist* 84:170–186.

Gipson, P. S., and J. F. Kamler. 2001. Survival and home ranges of opossums in northeastern Kansas. *Southwestern Naturalist* 46:178–182.

Holmes, A. C. V., and G. C. Sanderson. 1965. Populations and movements of opossums in east-central Illinois. *Journal of Wildlife Management* 29:287–295.

Holmes, D. J. 1992. Sternal odors as clues for social discrimination by female Virginia opossums, *Didelphis virginiana*. *Journal of Mammalogy* 73:286–291.

Krause, W. J. and W. A. Krause. 2006. *The opossum: Its amazing story.* Department of Pathology and Anatomical Sciences, School of Medicine, University of Missouri, Columbia. 80 pp.

McManus, J. J. 1974. *Didelphis virginiana. Mammalian Species* 40. 6 pp.

Petrides, G. A. 1949. Sex and age determination in the opossum. *Journal of Mammalogy* 30:364–378.

Reynolds, H. C. 1945. Some aspects of the life history and ecology of the opossum in central Missouri. *Journal of Mammalogy* 26:361–379.

———. 1952. Studies on reproduction in the opossum (*Didelphis virginiana virginiana*). *University of California Publications in Zoology* 52:223–284.

Ryser, J. 1992. The mating system and male mating success of the Virginia opossum (*Didelphis virginiana*) in Florida. *Journal of Zoology* 228:127–139.

Sanderson, G. C. 1961. Estimating opossum populations by marking young. *Journal of Wildlife Management* 25:20–27.

Sandidge, L. L. 1953. Food and dens of the opossum (*Didelphis virginiana*) in northeastern Kansas. *Transactions of the Kansas Academy of Science* 56:97–106.

Shirer, H. W., and H. S. Fitch. 1970. Comparison from radiotracking of movements and denning habits of the raccoon, striped skunk, and opossum in northeastern Kansas. *Journal of Mammalogy* 51:491–503.

Wiseman, G. L., and G. O. Hendrickson. 1950. Notes on the life history and ecology of the opossum in southeast Iowa. *Journal of Mammalogy* 31:331–337.

Wright, D. D. 1989. Mortality and dispersal of juvenile opossum, *Didelphis virginiana*. M.S. thesis, University of Florida, Gainesville. 100 pp.

Wright, J. D., M. S. Burt, and V. L. Jackson. 2012. Influences of an urban environment on home range and body mass of Virginia opossums (*Didelphis virginiana*). *Northeastern Naturalist* 19:77–86.

3

Shrew-form Mammals
Order Soricomorpha

In previous editions of this book, the shrews and eastern mole belonged in the order Insectivora, which contained a variety of small insectivorous (insect-eating) mammals. Insectivora has been divided into several new orders and is no longer a biological grouping. Today, shrews and moles belong in the order Soricomorpha (from Latin meaning "shrew-form" or "shrewlike"), which contains three living families: true shrews (found on all continents except Australia and Antarctica, and other smaller areas); moles and their relatives (distributed throughout temperate Europe and Asia, except Ireland and Norway, in Southeast Asia, and in North America); and two solenodons (one found in Cuba and one on Hispaniola).

Soricomorphs are small mammals having an elongated head, often with a movable snout projecting well over the mouth; small to no external ears; very small eyes; teeth that are not well differentiated into incisors, canines, and premolars; 5 toes on each foot; and a primitive brain.

Key to the Species
by Whole Adult Animals

1a. Front feet more than twice as wide as hind feet, possessing well-developed claws, and adapted for digging (see plate 7); eyes completely hidden in fur and covered by fused eyelids. **Eastern Mole** (*Scalopus aquaticus*) p. 56

1b. Front feet less than twice as wide as hind feet; eyes very small but visible. **Go to 2**

2a. (From 1b) Tail short (less than one-half the length of head and body). **Go to 3**

2b. (From 1b) Tail long (more than one-half the length of head and body). **Go to 4**

3a. (From 2a) Total length 76 mm (3 in.) or more; dark gray above; 4 (and sometimes 5) small single-pointed teeth (unicuspids) visible from side between first upper incisor and first upper cheek tooth (see plate 5). **Short-tailed Shrews** (*Blarina* spp.) p. 45

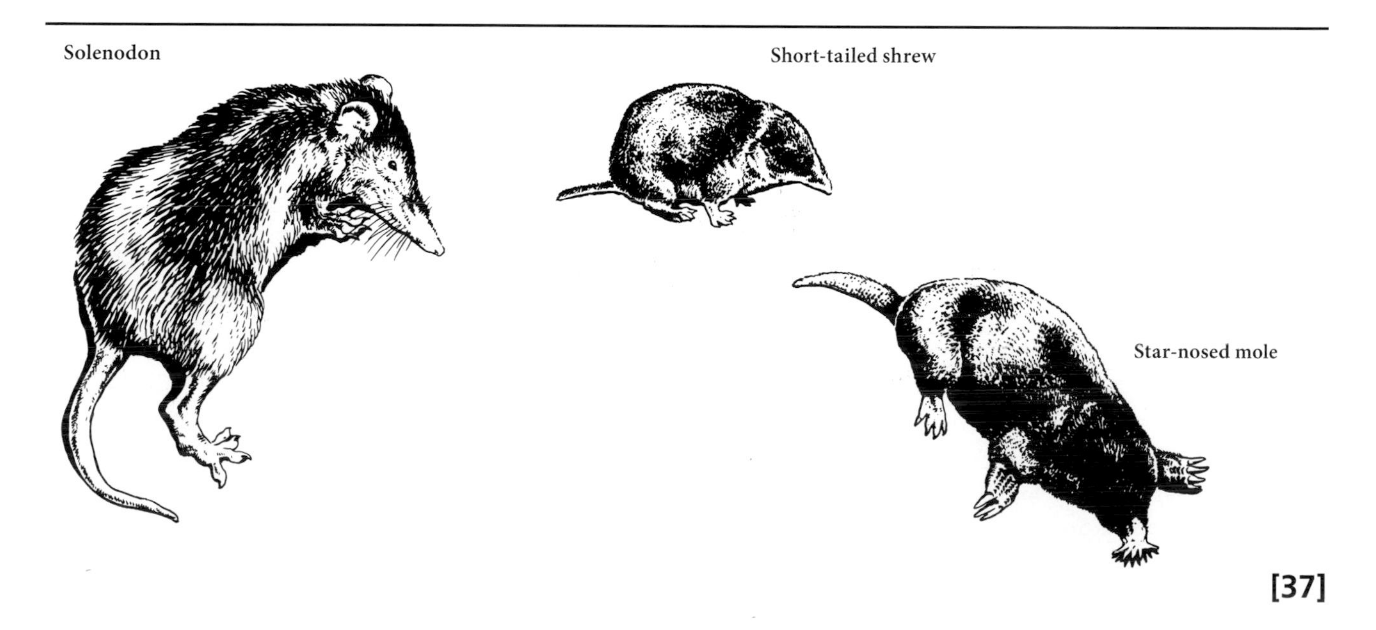

3b. (From 2a) Total length less than 82 mm (3¼ in.); dark brown to brownish gray above; 3 small single-pointed teeth (unicuspids) visible from side between first upper incisor and first upper cheek tooth (see plate 6). **North American Least Shrew** (*Cryptotis parva*) p. 52

4a. (From 2b) Grayish brown above; third upper single-pointed tooth (unicuspid) slightly larger than fourth upper single-pointed tooth; slightly longer tail with obvious constriction at base (see plate 3). **Cinereus Shrew** (*Sorex cinereus*) p. 38

4b. (From 2b) Reddish brown above; third upper single-pointed tooth (unicuspid) slightly smaller than fourth upper single-pointed tooth; slightly shorter tail without obvious constriction at base (see plate 4). **Southeastern Shrew** (*Sorex longirostris*) p. 42

Key to the Species
by Skulls of Adults

1a. Inflated bone surrounding inner ear (see plate 7); bony arch (zygomatic arch) present at side of eye socket. **Eastern Mole** (*Scalopus aquaticus*) p. 56

1b. Ringlike bone surrounding inner ear (see plates 3–6); no bony arch (zygomatic arch) at side of eye socket. **Go to 2**

2a. (From 1b) Total teeth 30; 3 small single-pointed teeth (unicuspids) visible from side between first upper incisor and first upper cheek tooth (see plate 6). **North American Least Shrew** (*Cryptotis parva*) p. 52

2b. (From 1b) Total teeth 32; 4 or 5 small, single-pointed teeth (unicuspids) visible from side between first upper incisor and first upper cheek tooth. **Go to 3**

3a. (From 2b) Prominent sharp projection on each side of braincase (seen in top view) (see plate 5); length of skull 19 mm (¾ in.) or more. **Short-tailed Shrews** (*Blarina* spp.) p. 45

3b. (From 2b) No prominent sharp projection on each side of braincase (seen in top view) (see plates 3–4); length of skull less than 19 mm (¾ in.). **Go to 4**

4a. (From 3b) Braincase high domed; rostrum narrow; third upper single-pointed tooth (unicuspid) slightly larger than fourth upper single-pointed tooth. **Cinereus Shrew** (*Sorex cinereus*) p. 38

4b. (From 3b) Braincase flattened; rostrum broad; third upper single-pointed tooth (unicuspid) slightly smaller than fourth upper single-pointed tooth. **Southeastern Shrew** (*Sorex longirostris*) p. 42

Shrews (Family Soricidae)

The family name, Soricidae, is based on the Latin word, *sorex*, for "shrew-mouse." Members of this family have the first or central incisors of both upper and lower jaws greatly enlarged and specialized into grasping pincers. The front and hind feet are about equal in size. Small external ears, or flaps, may be present. The eyes are tiny and probably capable of very limited vision. The fur has a plushlike quality and will lie either forward or backward.

Six species of shrews occur in Missouri. Their general characteristics are given in the accounts of the short-tailed and least shrews.

GENERAL REFERENCES

Churchfield, S. 1990. *The natural history of shrews*. Comstock Publishing, Ithaca, NY. 178 pp.

Whitaker, J. O., Jr., and R. E. Mumford. 1972. Food and ectoparasites of Indiana shrews. *Journal of Mammalogy* 53:329–335.

Cinereus Shrew (*Sorex cinereus*)

Name

The first part of the scientific name, *Sorex*, is the Latin word for "shrew-mouse." The last part, *cinereus*, is from the Latin word, *cinis*, for "ash-colored." This species is also known as the masked shrew, which describes the slightly darker (but not always apparent) coloration over the eyes, while "shrew" is from the Anglo-Saxon name *screawa*.

Description (Plate 3)

The cinereus shrew is distinguished as follows: from the short-tailed and least shrews by a longer tail (tail more than one-half the length of head and body) and from the southeastern shrew by the grayish brown color on the upperparts and slightly longer tail with an obvious constriction at the base.

Measurements

Total length	69–111 mm	2¾–4⅜ in.
Tail	25–50 mm	1–2 in.
Hind foot	9–14 mm	⅜–9/16 in.
Ear	3–7 mm	⅛–5/16 in.
Skull length	14–17 mm	9/16–11/16 in.
Skull width	7 mm	5/16 in.
Weight	2–5 g	1/16–1/5 oz.

Plate 3

Cinereus Shrew (*Sorex cinereus*)

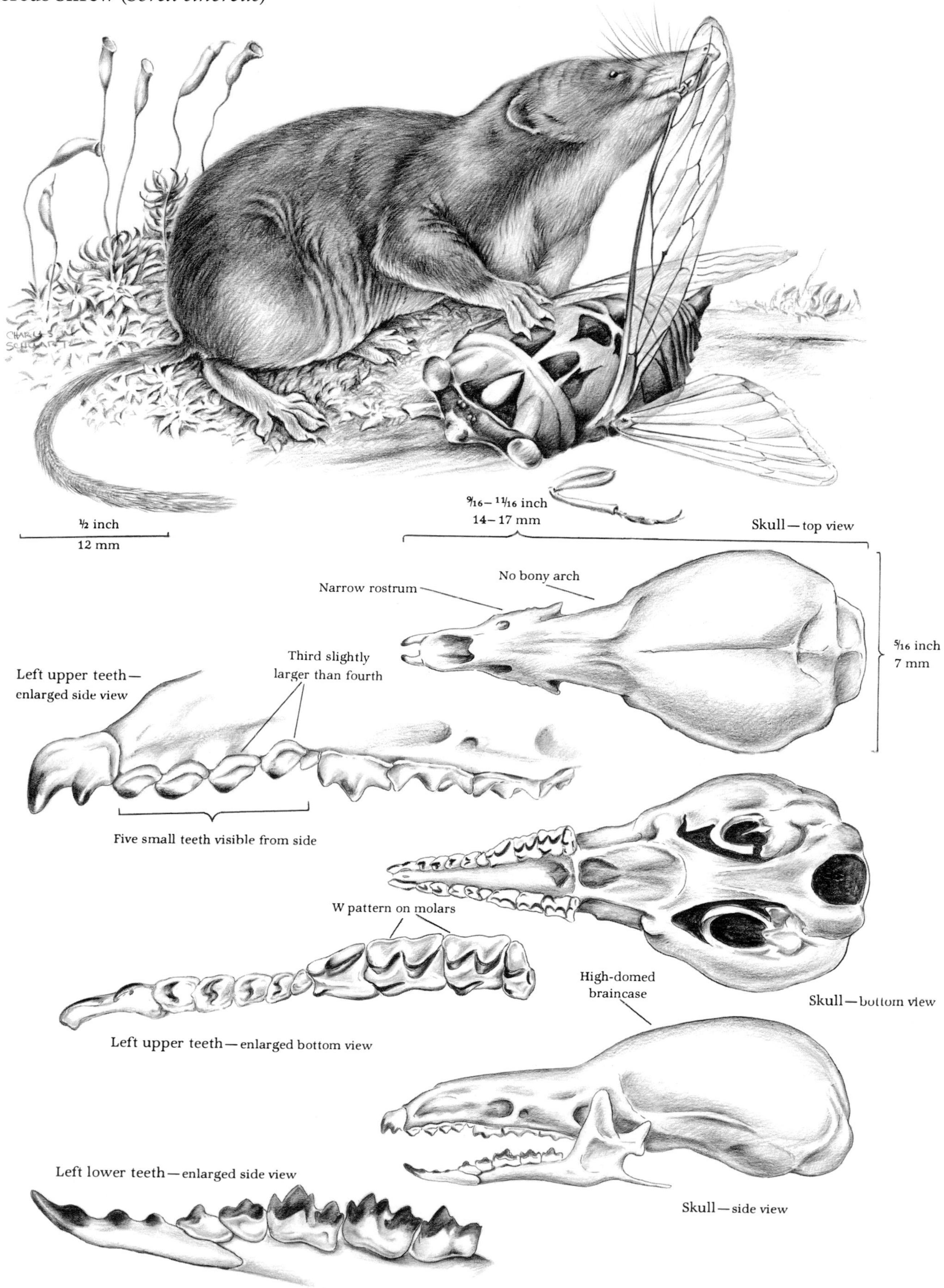

Teeth and skull. The dental formula of the cinereus shrew is:

$$I\ \frac{3}{1}\ C\ \frac{1}{1}\ P\ \frac{3}{1}\ M\ \frac{3}{3} = 32$$

There are 5 single-pointed teeth (unicuspids) visible from the side between the first upper incisor and the first upper cheek tooth. The third of these is slightly larger than the fourth.

The skull has a relatively narrow rostrum and high-domed braincase. The skull of the cinereus shrew is distinguished from that of the least shrew by having 32 teeth and 5 small single-pointed teeth (unicuspids) visible from the side between the first upper incisor and first upper cheek tooth; from that of the short-tailed shrews by lacking the prominent sharp projection on each side of the braincase (seen in top view) and by the smaller size; and from that of the southeastern shrew by the narrow rostrum and high-domed braincase, by having the third upper single-pointed tooth (unicuspid) slightly larger than the fourth, and by pigmented inner ridges in the upper unicuspids.

Longevity. The average life span is short, probably 12 to 18 months. In captivity, one lived for 23 months.

Voice and sounds. Cinereus shrews have several calls: staccato squeaks, especially when fighting or mad; faint twittering noises when searching for food; and a slow grinding of the teeth. Many shrew species echolocate by emitting ultrasonic sounds inaudible to humans. Given their poor eyesight, echolocation could be important for communication, detecting predators and food, and exploring new areas in the dark.

Glands. Small paired glands on the flanks of the body have a strong odor, particularly in males at the onset of the breeding season. The odor, possibly, is related to sexual attraction.

Distribution and Abundance

This little shrew ranges throughout Alaska and Canada, and south into the northern United States. While it is relatively scarce over the entire range, in some places it may be very abundant at times.

The cinereus shrew was first reported in Missouri in the 1970s and since then has been captured in 17 counties within the Central Dissected Till Plains of north Missouri. It is not known if this shrew has always lived in Missouri and was overlooked or if a recent southward range expansion occurred from Iowa. The latter explanation is favored because intensive collecting in the same area for many years did not find them prior to the 1970s. It probably occurs in a few additional northern Missouri counties in river-bottom habitats.

Habitat and Home

Cinereus shrews prefer low, damp areas in stream valleys or floodplains. They live in some of the same general areas as least shrews but utilize restricted microhabitats and are not generally found in grasslands like least shrews.

For their home they dig burrows 1.9 cm (¾ in.) in diameter, going down about 2.3 cm (9 in.). In this tunnel system they have several chambers; some for food storage, some for resting, and some for a nest. The nest is a woven sphere of dry or fresh grass and leaves, with one or more exits.

Habits

Cinereus shrews prefer a moist environment. While they are active all hours of the day and night, they are mostly nocturnal. They show increased activity during and after a rain, on cloudy nights, and in warmer weather. Their response to rain may indicate a preference for moist conditions or a change in available foods related to more moisture.

Many shrews often live in one general area and call to each other as if carrying on some sort of communication. Twenty shrews were seen and heard calling over 21 sq m (25 sq. yd.); in another location several were seen and heard in an area of 9.3 sq m (100 sq. ft.), while more were heard in the surrounding 929 sq m (10,000 sq. ft.). When two of these shrews that had been observed calling met in a small runway, they reared up on their hind legs, struck each other in the face until one or both fell over backward, then ran in the opposite direction.

Cinereus shrews have been reported to climb a sloping tree trunk, navigate a horizontal branch, and plunder a small bird's nest.

Cinereus shrews do not hibernate; they survive the winter by feeding on resources such as dormant insects. Favorable overwinter survivorship is attributable to their ability to generate heat without shivering and their conservation of energy through reduced winter body mass and foraging in thermally stable environments (that is, underground).

Foods

The most important foods of cinereus shrews include larvae of butterflies and moths, adult and larval beetles, harvestmen (daddy longlegs), centipedes,

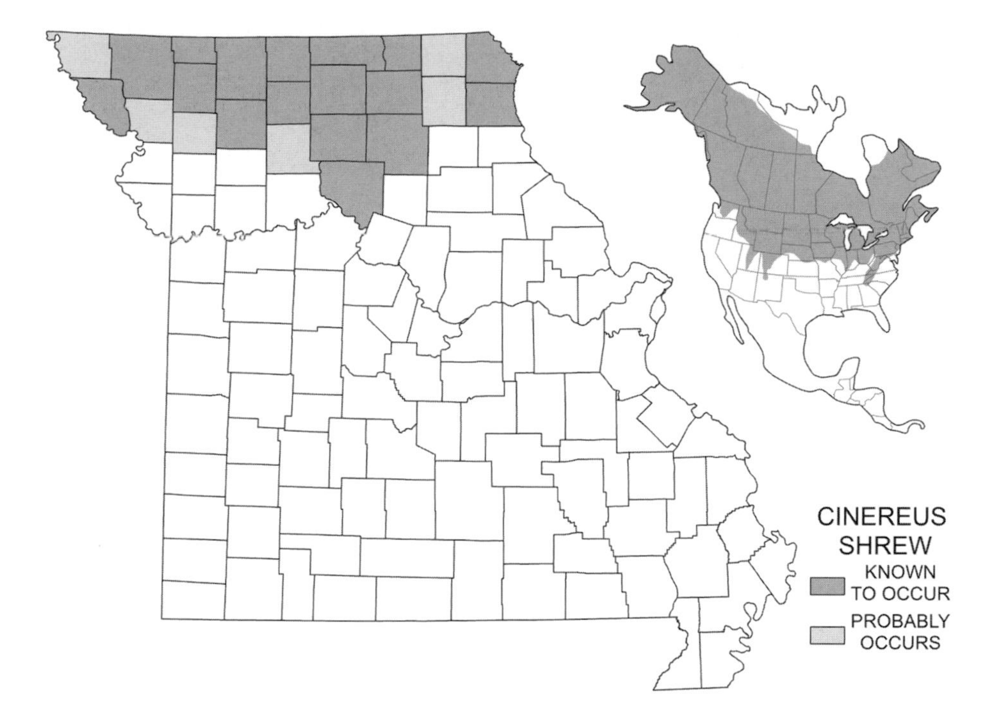

slugs, snails, spiders, and miscellaneous insects. Only small amounts of earthworms, vertebrates, and vegetable matter occur in the diet. Cinereus shrews are voracious feeders, eating three times their own weight every 24 hours.

Reproduction

There are several litters per season, and individual females can be both nursing and pregnant. Litters are recorded from March to September, with young of the year producing their first litter in their first fall. From 4 to 10 young constitute a litter.

At birth, the shrew is very tiny and weighs about 0.3 g (1⁄100 oz.). The skin is transparent, the eyelids are sealed, and the ears are closed and fused to the head. At about eight days of age, guard hairs show for the first time and in a few more days the underfur appears. In the next week the incisors erupt through the gums and the eyes and ears open. Weaning occurs after the twentieth day. The nestlings now have nearly four times as much subcutaneous fat as a grown shrew, and it is thought this fat may help the young survive while they are learning to hunt.

Four wild young captured from the same nest at nest-leaving time exhibited an interesting behavior. They ran about their new quarters in single file—a caravan formation—each shrew maintaining contact with the rump of the one in front by burying its nose in the fur near the tail. This was continued repeatedly until the fifth day after capture.

Some Adverse Factors

Mortality is greatest during the first two months after birth. Excessive rainfall and wetness are major causes of death in the nest. After the young leave the nest, mortality declines gradually. In the population as a whole, losses are lower in winter than in summer. These shrews vary greatly in abundance from place to place and year to year.

Hawks, owls, larger shrews, shrikes, herons, mergansers, snakes, frogs, foxes, weasels, and bobcats are predators on cinereus shrews. They are often killed but not eaten, probably because of their strong odor. Owing perhaps to their small size, these shrews exhibit few external parasites. They do, however, harbor a nematode in the urinary bladder.

Importance and Conservation and Management

These animals should be appreciated because of their rareness, their valuable role in the wildlife community, and their impact on insect populations.

SELECTED REFERENCES

See also discussion of this species in general references, page 23.

Chromanski-Norris, J. F., and E. K. Fritzell. 1983. Status and distribution of ten Missouri mammals. A report to the Missouri Department of Conservation, Jefferson City. 38 pp.

Doucet, G. J., and J. R. Bider. 1974. The effects of weather on the activity of the masked shrew. *Journal of Mammalogy* 55:348–363.
Easterla, D. A., and D. L. Damman. 1977. The masked shrew and meadow vole in Missouri. *Northwest Missouri State University Studies Quarterly* 37. 26 pp.
Forsyth, D. J. 1976. A field study of growth and development of nestling masked shrews (*Sorex cinereus*). *Journal of Mammalogy* 57:708–721.
French, T. W. 1984. Dietary overlap of *Sorex longirostris* and *S. cinereus* in hardwood floodplain habitats in Vigo County, Indiana. *American Midland Naturalist* 111:41–46.
Goodwin, M. K. 1979. Notes on caravan and play behavior in young captive *Sorex cinereus. Journal of Mammalogy* 60:411–413.
Hawes, M. L. 1976. Odor as a possible isolating mechanism in sympatric species of shrews (*Sorex vagrans* and *Sorex obscurus*). *Journal of Mammalogy* 57:404–406.
Horvath, O. 1965. Arboreal predation on bird's nest by masked shrew. *Journal of Mammalogy* 46:495.
Merritt, J. F. 1995. Seasonal thermogenesis and changes in body mass of masked shrews, *Sorex cinereus. Journal of Mammalogy* 76:1020–1025.
Pruitt, W. O., Jr. 1954. Aging in the masked shrew, *Sorex cinereus cinereus* Kerr. *Journal of Mammalogy* 35:34–39.
Short, H. L. 1961. Fall breeding activity of a young shrew. *Journal of Mammalogy* 42:95.
Tuttle, M. D. 1964. Observation of *Sorex cinereus. Journal of Mammalogy* 45:148.
Whitaker, J. O., Jr. 2004. *Sorex cinereus. Mammalian Species* 743. 9 pp.
Woolfenden, G. E. 1959. An unusual concentration of *Sorex cinereus. Journal of Mammalogy* 40:437.

Southeastern Shrew (*Sorex longirostris*)

Name

The first part of the scientific name, *Sorex*, is the Latin word for "shrew-mouse." The last part, *longirostris*, is from two Latin words, *longus* and *rostrum*, meaning "long snout." The common name, "southeastern," refers to the general area of North America where it lives, and "shrew" is from the Anglo-Saxon name *screawa*.

Description (Plate 4)

The general characteristics of the shrews in Missouri are given in the accounts of the short-tailed and least shrews. The southeastern shrew is distinguished as follows: from the short-tailed and least shrews by a longer tail (tail more than one-half the length of head and body) and from the cinereus shrew by the reddish

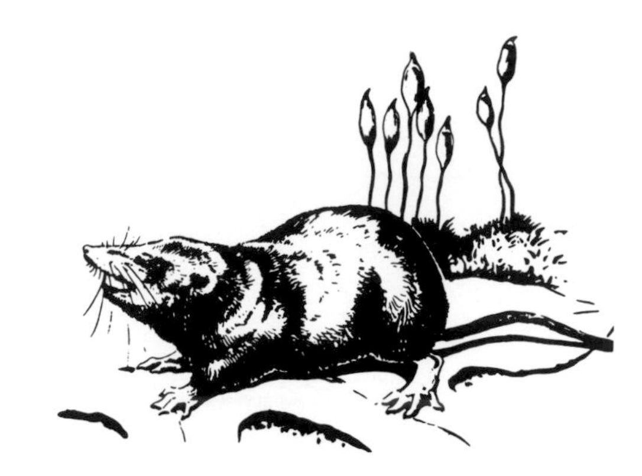

brown color on the upperparts and a slightly shorter tail without an obvious constriction at the base.

Measurements

Total length	76–107 mm	3–4¼ in.
Tail	26–41 mm	1¹⁄₁₆–1⅝ in.
Hind foot	9–12 mm	⅜–½ in.
Ear	6 mm	¼ in.
Skull length	14–17 mm	⁹⁄₁₆–¹¹⁄₁₆ in.
Skull width	7 mm	⁵⁄₁₆ in.
Weight	2–3.2 g	¹⁄₁₆-¹⁄₁₀ oz.

Teeth and skull. The dental formula of the southeastern shrew is:

$$I\ \frac{3}{1}\ C\ \frac{1}{1}\ P\ \frac{3}{1}\ M\ \frac{3}{3} = 32$$

There are 5 single-pointed teeth (unicuspids) visible from the side between the first upper incisor and the first upper cheek tooth. The third of these is slightly smaller than the fourth.

The skull of the southeastern shrew has a relatively broad rostrum and a flattened braincase. It is distinguished from that of the least shrew by having 32 teeth and 5 small single-pointed teeth (unicuspids) visible from the side between the first upper incisor and first upper cheek tooth; from that of the short-tailed shrews by lacking the prominent sharp projection on each side of the braincase (seen in top view) and by the smaller size; and from that of the cinereus shrew by the broader rostrum and flattened braincase, by having the third upper single-pointed tooth (unicuspid) slightly smaller than the fourth, and by the lack of pigment on the inner ridge of the unicuspids.

Longevity. These shrews may live up to 19 months, with most individuals probably living less than a year.

Distribution and Abundance

This species lives in the southeastern part of the United States. When this book was first written, the southeastern shrew was not known from Missouri. Today, they have been captured in at least 21 counties and probably occur in all but the northwest corner of the state. They are probably uncommon across Missouri.

Plate 4

Southeastern Shrew (*Sorex longirostris*)

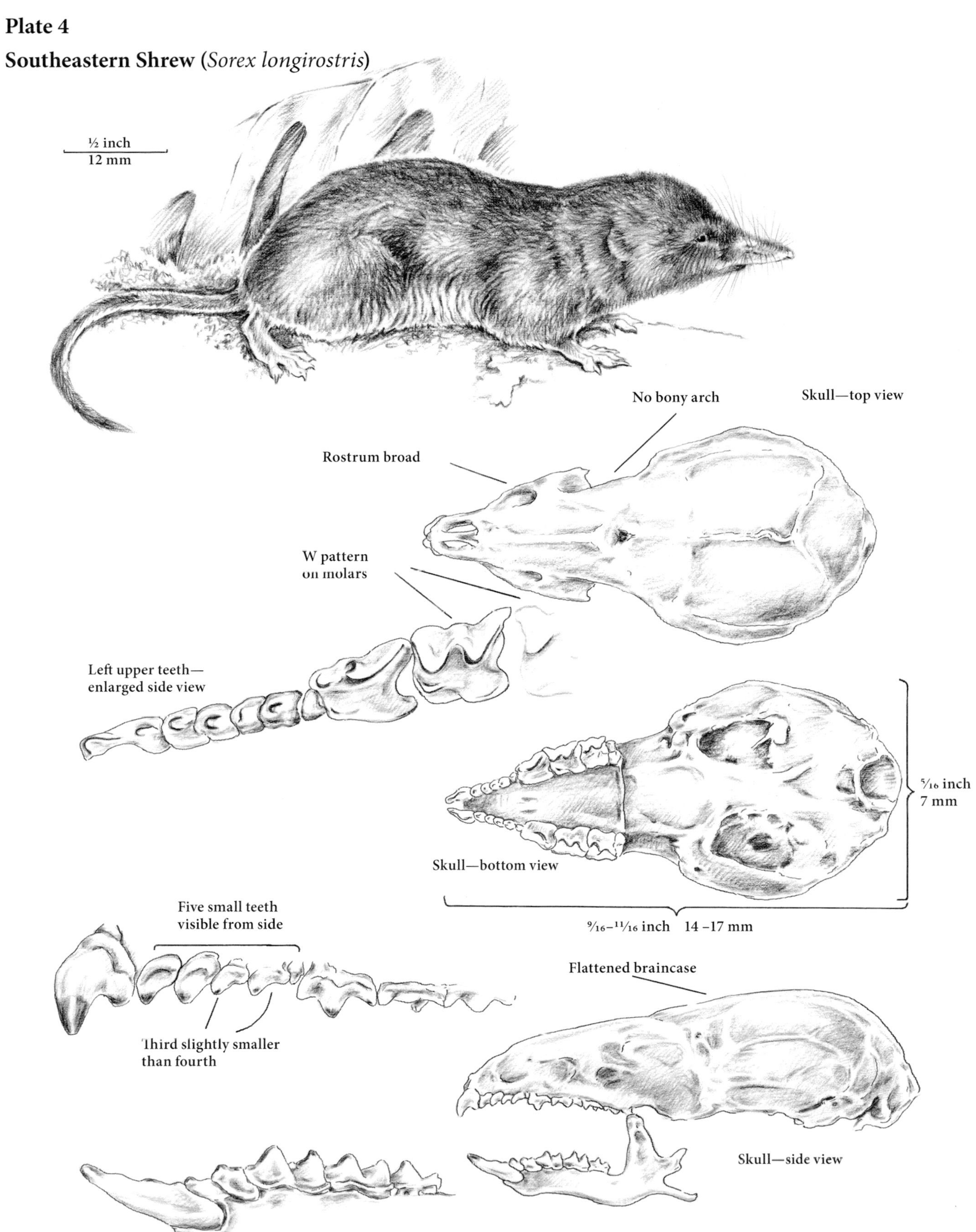

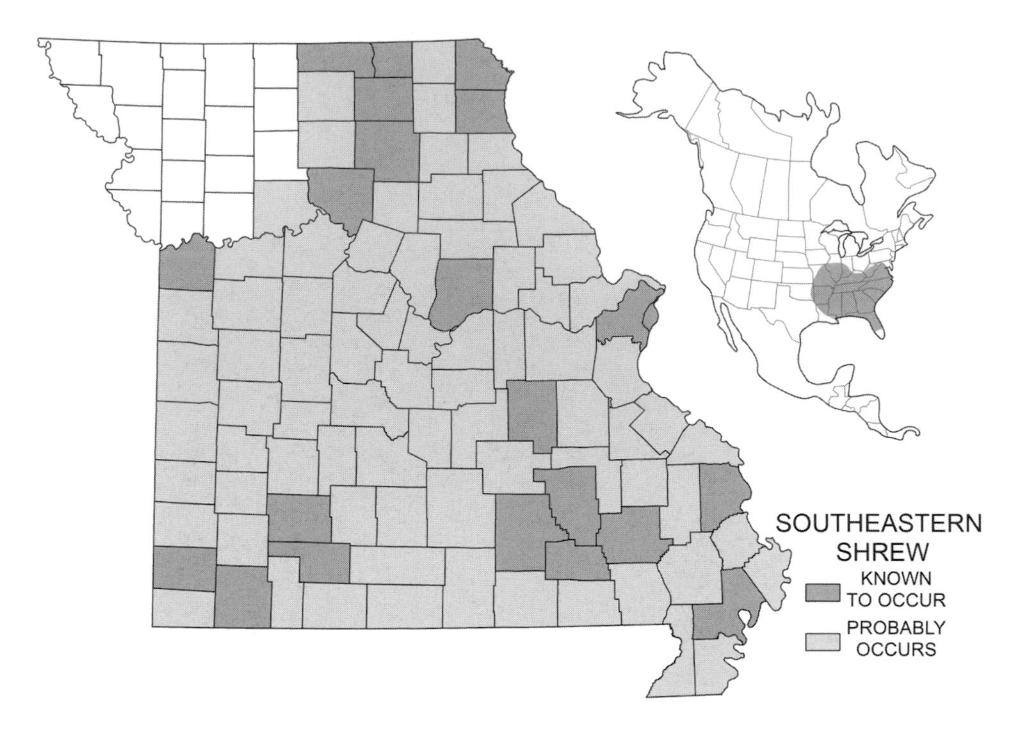

Shrews are best captured in sunken pitfall-type traps and are seldom taken in small mammal live box traps or snap traps.

Habitat and Home

Throughout its range, the southeastern shrew is found in a variety of habitats, but it is more often associated with bogs, marshy or swampy areas, or dense ground cover (briers or honeysuckle vines) in wooded areas. It has also been taken in dry upland fields, agricultural lands, and dry upland wooded areas. Its known Missouri habitat consists of generally dry upland sites with some woods.

Nests are shallow depressions lined with leaf litter. They are built under a decaying log or deep inside the log.

Habits

These shrews generally live underground, coming above ground mostly during or after a rain or on dewy nights.

Foods

Foods of this little-known shrew consist predominantly of spiders, followed by larvae of butterflies and moths, slugs, beetle larvae, snails, some vegetable matter, and centipedes.

Reproduction

Females have been found pregnant from March 31 to October 6, with some females breeding during their first season. They may give birth to 2 or more litters of 1–6 young per litter (average 4). The young stay in the nest for 3–4 weeks until fully grown.

Some Adverse Factors

Southeastern shrews are preyed upon by opossums, barn owls, barred owls, hooded mergansers, and domestic cats.

Importance and Conservation and Management

These subjects are discussed under accounts of the other shrews in Missouri.

SELECTED REFERENCES

See also discussion of this species in general references, page 23.

Brown, L. N. 1961. *Sorex longirostris* in southwestern Missouri. *Journal of Mammalogy* 42:527.

Chromanski-Norris, J. F., and E. K. Fritzell. 1983. Status and distribution of ten Missouri mammals. A report to the Missouri Department of Conservation, Jefferson City. 38 pp.

Dusi, J. L. 1959. *Sorex longirostris* in eastern Alabama. *Journal of Mammalogy* 40:438–439.

French, T. W. 1980a. *Sorex longirostris. Mammalian Species* 143. 3 pp.

———. 1980b. Natural history of the southeastern shrew, *Sorex longirostris* Bachman. *American Midland Naturalist* 104:13–31.

———.1984. Dietary overlap of *Sorex longirostris* and *S. cinereus* in hardwood floodplain habitats in Vigo County, Indiana. *American Midland Naturalist* 111:41–46.

Mock, O. B., and V. K. Kivett. 1980. The southeastern shrew, *Sorex longirostris*, in northeastern Missouri. *Transactions of the Missouri Academy of Science* 14:67–68.

Rose, R. K. 1980. The southeastern shrew, *Sorex longirostris*, in southern Indiana. *Journal of Mammalogy* 61:162–164.

Tuttle, M. D. 1964. Additional record of *Sorex longirostris* in Tennessee. *Journal of Mammalogy* 45:146–147.

Northern Short-tailed Shrew (*Blarina brevicauda*),

Southern Short-tailed Shrew (*Blarina carolinensis*), and

Elliot's Short-tailed Shrew (*Blarina hylophaga*)

Short-tailed shrews (*Blarina* spp.) have long been under taxonomic and geographic revision. Four species are currently recognized, and three occur in Missouri (*B. brevicauda*, *B. carolinensis*, and *B. hylophaga*). *B. brevicauda* is the largest of the three species, followed by *B. hylophaga* (intermediate in size), and *B. carolinensis* (smallest). Their natural histories are similar; therefore, they are covered here as a single account. Any differences in morphology are presented.

Name

The origin of *Blarina*, the first part of these three scientific names, is uncertain; it is considered to be a coined name. The second part for the northern short-tailed shrew, *brevicauda*, is from two Latin words and means "short tail" (*brevis*, "short," and *cauda*, "tail"). This appropriate term is also used for the common name. The second part for the southern short-tailed shrew, *carolinensis*, is a Latinized word meaning "of Carolina," presumably for the place from which the type specimen of this species comes. The second part for Elliot's short-tailed shrew, *hylophaga*, is Greek for "wood-eater," presumably related to the forested environment it inhabits. "Shrew" is from the Anglo-Saxon word *screawa*, which was applied to the common shrew of Europe as well as to persons with a harmful influence similar to the then-supposed, and rightly so, poisonous bite of the shrew.

Description (Plate 5)

Short-tailed shrews are very small mammals with an exceedingly pugnacious and energetic nature. They have a pointed head; distinct neck; cylindrical body; short, slender legs with 5 clawed toes on each foot; and a short, furred tail. The flexible, sensitive snout, which contains the nostrils, projects slightly over and beyond the mouth. The tiny black eyes are capable of light perception but probably see only about 2/20 on an optician's scale compared to the 20/20 that is normal for humans. The external ear opening is large but concealed in the fur. The senses of touch and hearing are well developed; that of smell is poorly developed. The velvety fur is short and brushes equally well in any direction. A microscopic examination shows whiplike tips on the hairs, similar to the hairs of moles but unlike those of any other mammal.

Short-tailed shrews are distinguished from the cinereus and southeastern shrews by their grayer color and shorter tails, and from the least shrew by their larger size and dark gray color on the upperparts.

Color. While short-tailed shrews are generally gray in color, they are darker on the back and lighter on the underparts, feet, undersurface of the tail, and around the mouth. Only rarely does a white shrew occur. Short-tailed shrews are typically darker in winter than in summer but show no sexual difference in coloration.

Feet of the short-tailed shrew

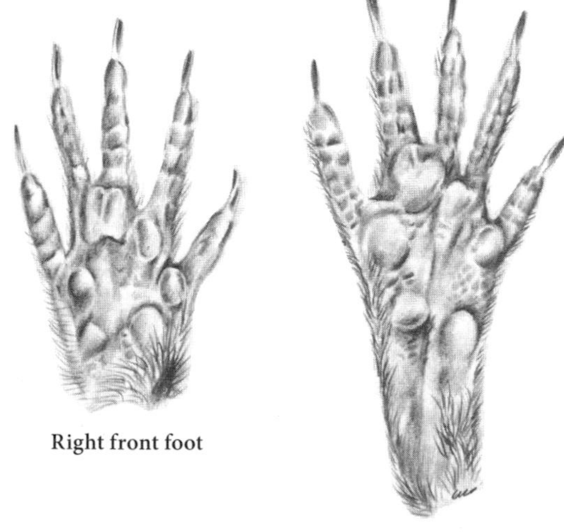

Plate 5
Northern Short-tailed Shrew (*Blarina brevicauda*)

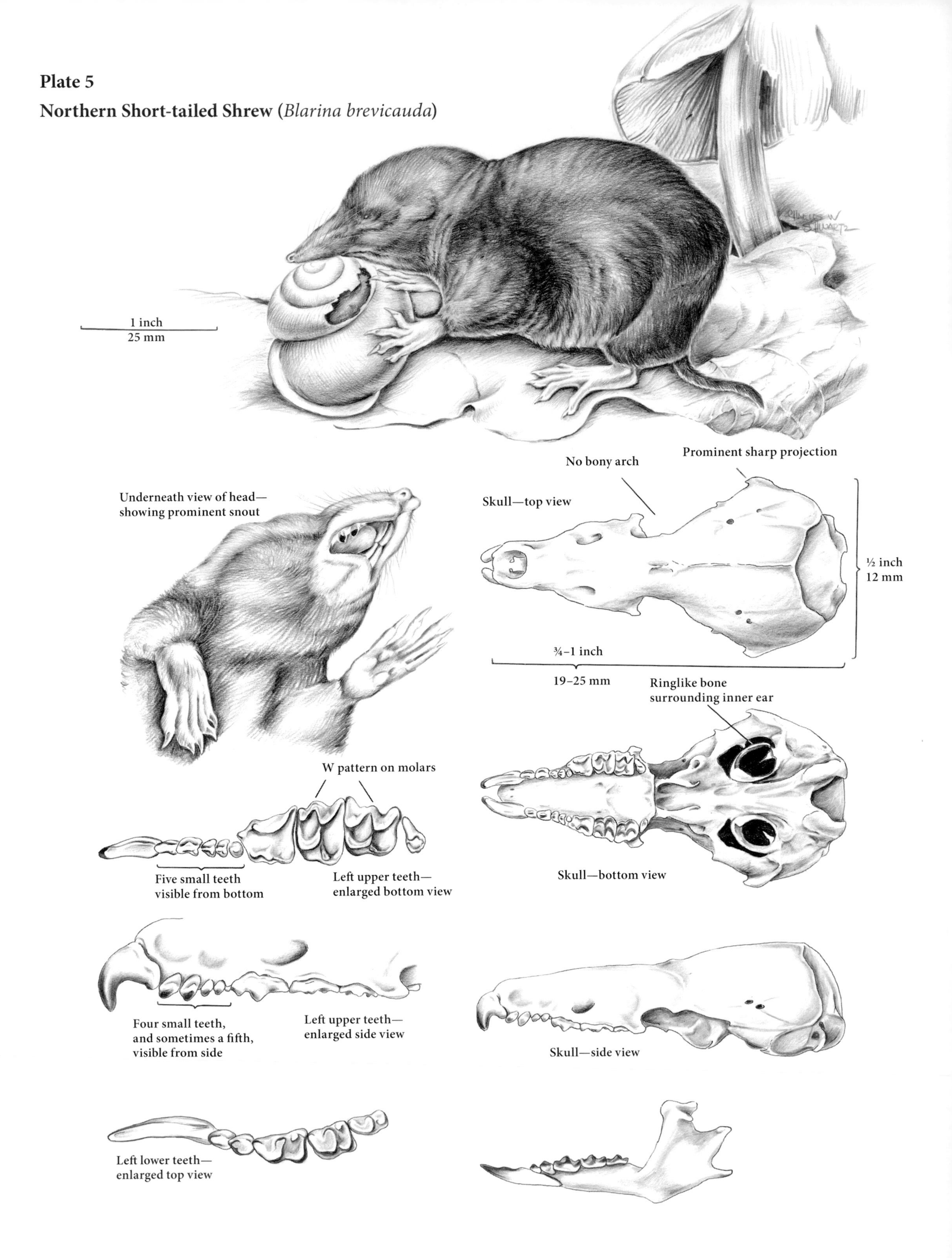

Molting occurs in spring and fall. In the spring molt, females follow a regular progression of hair replacement from the head toward the tail; males, however, have an irregular pattern. In the fall molt, second-year shrews and early-born, first-year animals start to molt at the rear of the body and show a gradual forward replacement of hairs. Late-born, first-year animals start to molt in the head region and show a gradual replacement of hairs toward the rear.

***Measurements**—Blarina brevicauda*

Total length	95–127 mm	3¾–5 in.
B. carolinensis	69–108 mm	2¾–4¼ in.
B. hylophaga	92–121 mm	3⅝–4¾ in.
Tail	19–25 mm	¾–1 in.
B. carolinensis	12–26 mm	½–1 in.
B. hylophaga	19–25 mm	¾–1 in.
Hind foot	12–19 mm	½–¾in.
Skull length	19–25 mm	¾–1 in.
Skull width	12 mm	½ in.
Weight	14–28 g	½–1 oz.
B. carolinensis	5.5–13 g	⅕–½ oz.
B. hylophaga	13–17 g	½–⅗ oz.

Males are slightly larger and heavier than females.

Teeth and skull. Shrews are readily recognized by their distinctive teeth, but they are not well differentiated as to kind. The dental formula of short-tailed shrews is given in two ways, depending upon the name assigned to specific teeth:

$$I\,\frac{3}{1}\;C\,\frac{1}{1}\;P\,\frac{3}{1}\;M\,\frac{3}{3}\quad\text{or}$$

$$I\,\frac{4}{2}\;C\,\frac{1}{0}\;P\,\frac{2}{1}\;M\,\frac{3}{3} = 32$$

The first or central incisors on both upper and lower jaws are greatly enlarged and specialized into grasping pincers. Between these large incisors and the first, large cheek teeth on each half of the upper jaw are 5 small, single-pointed teeth (unicuspids); usually only 4 of these are visible from the side and the very tiny fifth one, which may be seen from the side, is best viewed from below. The 2 large upper molars have a distinct W pattern on their surfaces. The teeth are usually tipped with reddish brown, being more so in younger individuals.

All shrew skulls have a ringlike bone surrounding the inner ear and lack a bony arch (zygomatic arch) at the side of the eye socket. The skulls of short-tailed shrews possess a prominent sharp projection on each side of the braincase when seen in top view. The skull of the southern short-tailed shrew is generally less massive than that of the northern short-tailed shrew.

The skull of a short-tailed shrew is distinguished from that of the least shrew by 32 teeth, by 4 or 5 small single-pointed teeth (unicuspids) visible from the side between the first upper incisor and first upper cheek tooth, and by its larger size. It is distinguished from cinereus and southeastern shrews by the prominent sharp projection on each side of the braincase (seen in top view) and length of 19 mm (¾ in.) or more.

Sex criteria and sex ratio. It is difficult to determine the sex of shrews externally. Males do not possess a scrotum, the testes being contained within the body cavity throughout life. Males are slightly larger and heavier than females and have better developed scent glands and thicker skin over the throat during the breeding season. Females have three pairs of teats in the groin region. Males tend to outnumber females in the shrew population.

Age criteria and longevity. Young shrews are distinguished from adults by the darker color of their fur, narrower heads, and thinner snouts. As shrews age, the brown color on the teeth gradually becomes less noticeable and the teeth show wear. In very old shrews the teeth may be worn down to the gums. While most shrews probably live no longer than one winter in the wild, some survive for two. In captivity a few have lived through three breeding seasons.

Glands. Short-tailed shrews possess a pair of small scent glands, about 3 mm (⅛ in.) long, on the flanks. These are active in males at all times and in females at times other than the breeding season. The rank, musky odor from these glands, when rubbed on the tunnel, serves as a sign of ownership and of the occupant's sex; during the breeding season males probably avoid tunnels with a strong scent (made by a male occupant) and look elsewhere for a female.

A single, large gland (32 mm long by 9 mm wide; 1 ¼ in. long by ⅜ in. wide) occurs on the belly of both sexes. The odor from this gland possibly discourages would-be predators but, in spite of this, many shrews are killed and eaten.

Certain salivary glands in the mouth, opening between the bases of the large lower incisors, contain a poison that flows with the saliva and enters an animal when it is bitten by the shrew. This poison slows the heart and breathing of the victim and may cause disintegration of muscle. The submaxillary glands of one adult short-tailed shrew contain enough poison to kill 200 mice. This poison also immobilizes insects: acting slowly, it extends the time fresh food is available, and it may facilitate food hoarding.

Voice and sounds. The vocalizations of short-tailed shrews extend from low to high ranges (audible to humans) into the ultrasonic range inaudible to humans. The sounds we hear are loud squeaks, a long shrill chatter, a low note of contentment that often accompanies feeding, a slow grinding of the teeth, and a series of clicks given by males when courting or as a warning or challenge. Ultrasonic calls are used to detect objects in the path by means of echolocation. They are given when exploring the tunnel system and may be used in identifying predators and food, especially in low-light conditions.

Distribution and Abundance

Prior to 1972, short-tailed shrews were considered as one widespread species, *B. brevicauda*, with three or more subspecies, and *B. telmalestes* (now *B. brevicauda telmalestes*), found only in the Dismal Swamp of Virginia and North Carolina. By the mid-1970s, biologists suggested that *B. brevicauda* was two species: a larger form in the north (*B. brevicauda*) and a smaller southern species (*B. carolinensis*). Additional research in the early 1980s suggested that the southern populations represented two, not one, distinct species: *B. carolinensis* in the southeast and *B. hylophaga* in the southwest. It was also suggested that a zone of contact where both *B. brevicauda* and *B. hylophaga* occurred existed in southern Iowa and northern Missouri. These conclusions were based on both morphometric (comparing a series of cranial measurements) and karyologic (comparing numbers and appearance of chromosomes) analyses. The North American range map based on this information for these three species is presented with the Missouri state range map.

Ongoing genetics work using mitochondrial DNA sequencing suggests that the distributions of these three species are different than previously thought, especially in Missouri, Arkansas, Oklahoma, Kansas, and Nebraska. Biologists currently conducting this genetic analysis suggest that their distributions are as presented in the second, stand-alone North American range map.

Short-tailed shrews range from southeastern Canada through much of the eastern Great Plains and throughout the eastern United States. They occur throughout the state; the Missouri range map shows counties where any of the three short-tailed shrew species have been captured or are expected to occur. An exhaustive search for and genetic analysis of these elusive shrews is needed to accurately depict their distributions and contact zones in Missouri and adjoining states.

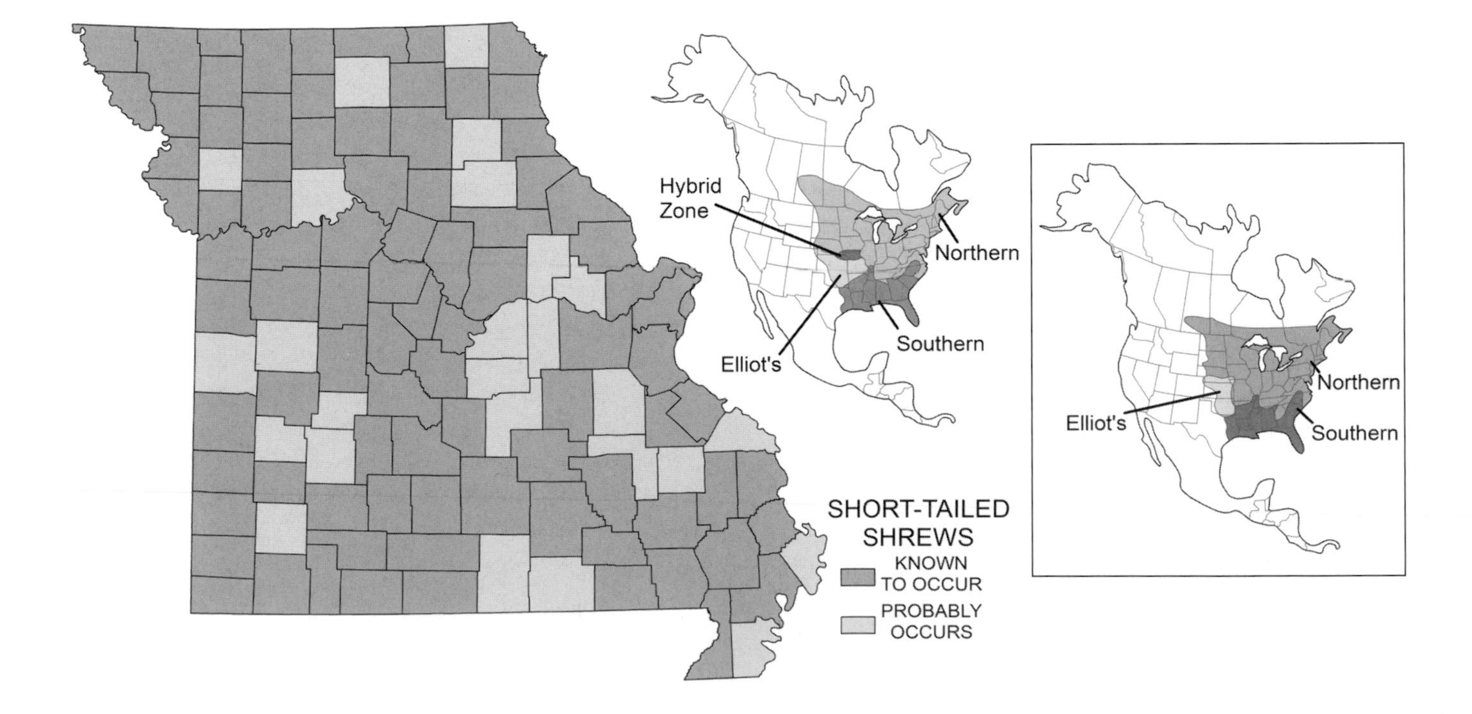

Nest and connecting runway of short-tailed shrew

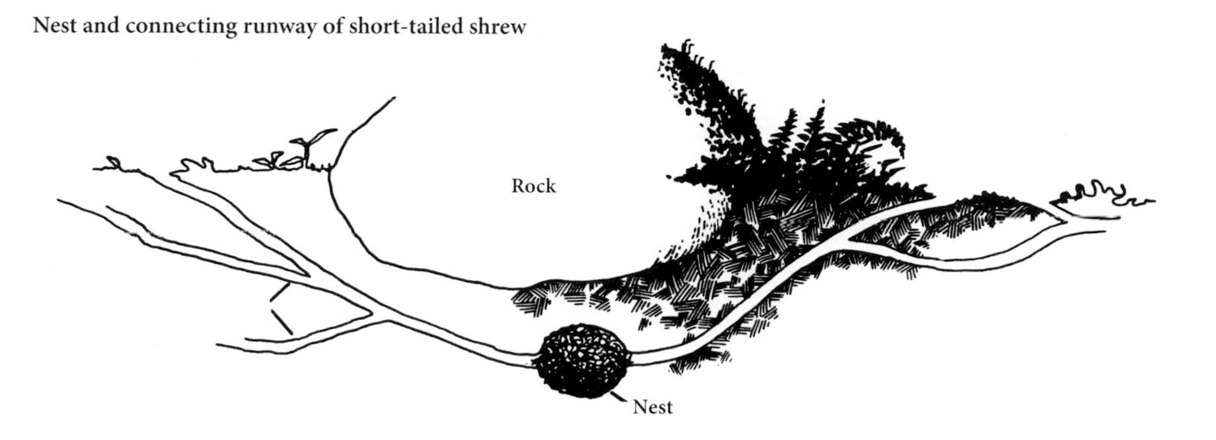

Shrew populations fluctuate violently from year to year. In peak years, there may be 50 or more short-tailed shrews per 0.4 ha (1 ac.), making them the most abundant mammal in a region. More generally, however, there are fewer than 1 to 4 shrews per 0.4 ha (1 ac.).

Habitat and Home

Short-tailed shrews live in dark, damp, or wet localities in flooded areas or fields covered with heavy, weedy growth. They occur less often in grassy cover. Shrews run about the surface of the ground or tunnel in mossy banks, under old logs, and in the leafy cover and humus of the forest floor. Runways and tunnels of other mammals, especially those abandoned by moles or mice, are commonly used, but on occasion a shrew digs its own system of tunnels. Such tunnels are from 2.5 to 6.4 cm (1 to 2½ in.) in diameter and are located just below the surface of the ground or deeper, some times as much as 56 cm (22 in.) underground.

Surface tunnels are constructed by merely raising and separating the leaf mold or sod from the soil beneath. Deeper tunnels are dug mainly with the front feet, although sometimes the snout and head are used as supplementary aids. In loose soil a shrew burrows at the rate of 30 cm (1 ft.) a minute, but beneath the mossy carpet in the forest one may burrow about 1 m (1 yd.) in 15 seconds. When there is snow, they may travel on the surface of the ground but just beneath the snow.

Somewhere in the tunnel system, usually under a rock, fallen log, or stump, or as much as 30 cm (12 in.) underground, the female builds a bulky globe-shaped nest for her young. The nest is made of partly shredded leaves and dried grass and has from one to three openings. It is from 15 to 20 cm (6 to 8 in.) in diameter with an inner cavity about 6.4 cm (2½ in.) across. Similar nests, used for resting or winter homes, are also constructed in the tunnel system.

Habits

Shrews are extremely active, very nervous, and highly belligerent. Their almost constant activity requires a tremendous amount of food and oxygen. They spend little time resting and, although they tend to be out mostly at night, search for food at any hour of the day or night all through the year.

Individual short-tailed shrews exhibit various behavioral patterns and threats, often accompanied by calls that initiate a response in another shrew. Such behavior allows this species to compete socially with reduced physical contact or fighting.

If properly cared for in captivity, short-tailed shrews get along well together. However, males quarrel more than females, and adults more than young. Sometimes, in confinement, one individual may kill and devour the other.

Although shrews do not defend any territory, they tend to stay in the same locality from month to month. An individual's home range is usually between 0.2

Short-tailed shrew tracks in thin snow

Short-tailed shrew tail in soft snow

and 0.4 ha (½ and 1 ac.) in extent and some have been known to occupy areas up to 1.8 ha (4½ ac.).

Shrews normally walk on the entire soles of their feet. They frequently run but rarely jump. They are able to make their way over fallen trees or to ascend inclined ones, but they are not good climbers. They can swim if necessary.

These mammals are very fastidious about their appearance and spend considerable time washing their fur with the tongue or combing it with their feet or toes. Droppings are deposited in unused portions of the tunnels.

Foods

Nearly the entire diet of short-tailed shrews consists of animal foods. Insects of all kinds predominate in the fare, but earthworms, snails, slugs, centipedes, millipedes, spiders, salamanders, snakes, birds, mice, and other shrews are also included. However, the vertebrate portion of this general diet has not been confirmed for the southern short-tailed shrew. Roots, nuts, fruits, berries, and fungi compose the vegetable foods; these are taken more in winter than in summer.

Short-tailed shrews may prey on mice and least shrews but are probably overrated as predators on these animals. Land snails are collected and hoarded in the tunnel system and, in order to be kept immobile, are moved to the surface of the ground as the outside air temperature falls and back into the cooler tunnel as the outside temperature rises. Mutilated insects and even dead mice may be cached, but some of the food probably spoils without being eaten. Prey is located by the shrew's keen sense of touch and echolocation, which compensate for its poor eyesight. The prey is then paralyzed by the powerful poison in the saliva.

In captivity, short-tailed shrews eat from one-half to more than their own weight in food daily. However, less is probably eaten in the wild. Although they are ravenous eaters, shrews have been known to fast from 24 to 36 hours without starving to death. Captive shrews take water greedily.

Reproduction

The breeding season of short-tailed shrews extends from early spring until late fall with peaks of production at the beginning and end of this season and only scattered breeding in winter. Most adult females have 1–3 and perhaps 4 litters annually. Many and prolonged matings are required by the female before the eggs are shed from the ovaries. The gestation period of short-tailed shrews is 21 or 22 days. From 3 to 10 young may be born per litter, but numbers between 5 and 7 are most common. Although mating is promiscuous, males and females may pair for a short time before the young are born. The female takes entire care of the family.

At birth the young are naked, wrinkled, and dark pink. Their eyes and ears are closed and no teeth are visible. They are only 2.2 cm (⅞ in.) long and weigh 0.3 g (1/100 oz.)—about the size of a honeybee. At 1 week, the young shrews have doubled their birth size and short hair is evident on the body. In another week, females can be identified by the presence of teats. The young are well furred at this age, their ears are open, and they begin to crawl. About 18 days of age, the upper incisors first show through the gums and, shortly afterward, the eyes open. The young shrews are well grown at 1 month and are soon on their own. Both sexes breed the year of their birth.

Some Adverse Factors

Snakes, hawks, owls, shrikes, weasels, skunks, foxes, coyotes, bobcats, domestic cats, domestic dogs, skunks,

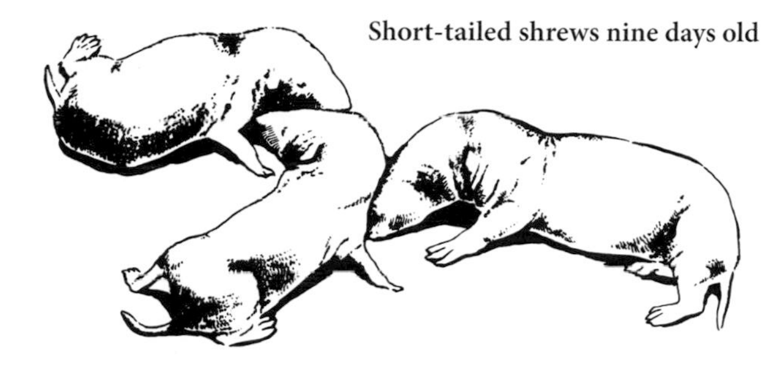
Short-tailed shrews nine days old

raccoons, and opossums all prey on shrews but sometimes do not eat them because of the disagreeable, musky odor.

Mites, ticks, fleas, botfly larvae, flukes, roundworms, and tapeworms are the parasites known to occur on or in short-tailed shrews, and a nest beetle, similar to the kind found on moles, has been reported from short-tailed shrews. This beetle feeds on the eggs and young of mites and other small animals in the nest.

Mortality is greatest the first two months after birth. Shrews have an unusually high mortality rate from accidents and may even die of shock. Cold-related stress has been suspected as the cause of up to 90 percent of winter mortality in populations of northern short-tailed shrews. Many individuals are scarred as a result of fighting with other shrews or their prey.

Importance

These numerous and widespread tiny mammals are extremely important and contribute in many ways to their environment. Through their predatory nature they help control insects and rodents; through their tunneling beneath the litter and debris of the forest floor, they aerate and mix the soil; and through their own bodies they furnish food for other animals and add to the organic content of the soil. These tiny animals help make the outdoors an interesting place.

Conservation and Management

Populations may decline precipitously in timbered areas that are cleared, and concentrations of pesticides could be a concern due to the predatory nature and high consumption rates of shrews.

SELECTED REFERENCES

See also discussion of this species in general references, page 23.

Benedict, R. A. 1999a. Morphological and mitochondrial DNA variation in a hybrid zone between short-tailed shrews (*Blarina*) in Nebraska. *Journal of Mammalogy* 80:112–134.

———. 1999b. Characteristics of a hybrid zone between two species of short-tailed shrews (*Blarina*). *Journal of Mammalogy* 80:135–141.

Dapson, R. W. 1968. Reproduction and age structure in a population of short-tailed shrews, *Blarina brevicauda*. *Journal of Mammalogy* 49:205–214.

Doremus, H. M. 1965. Heart rate, temperature, and respiration rate of the short-tailed shrew in captivity. *Journal of Mammalogy* 46:424–425.

Easterla, D. A. 1968. First records of *Blarina brevicauda minima* in Missouri and Arkansas. *Southwestern Naturalist* 13:448–449.

Ellis, L. S., V. E. Diersing, and D. F. Hoffmeister. 1978. Taxonomic status of short-tailed shrews (*Blarina*) in Illinois. *Journal of Mammalogy* 59:305–311.

Genoways, H. H., and J. R. Choate. 1972. A multivariate analysis of systematic relationships among populations of the short-tailed shrew (genus *Blarina*) in Nebraska. *Systematic Zoology* 21:106–116.

George, S. B., J. R. Choate, and H. H. Genoways. 1986. *Blarina brevicauda. Mammalian Species* 261. 9 pp.

George, S. B., H. H. Genoways, J. R. Choate, and R. J. Baker. 1982. Karyotypic relationships within the short-tailed shrews, genus *Blarina*. *Journal of Mammalogy* 63:639–645.

Kirkland, G. L., Jr. 1977. Response of small mammals to the clear cutting of northern Appalachian forests. *Journal of Mammalogy* 58:600–609.

Martin, I. G. 1981. Venom of the short-tailed shrew (*Blarina brevicauda*) as an insect immobilizing agent. *Journal of Mammalogy* 62:189–192.

McCay, T. S. 2001. *Blarina carolinensis. Mammalian Species* 673. 7 pp.

Moncrief, N. D., J. R. Choate, and H. H. Genoways. 1982. Morphometric and geographic relationships of short-tailed shrews (genus *Blarina*) in Kansas, Iowa, and Missouri. *Annals of the Carnegie Museum* 51:157–180.

Olsen, R. W. 1969. Agonistic behavior of the short-tailed shrew (*Blarina brevicauda*). *Journal of Mammalogy* 50:494–500.

Pearson, O. P. 1944. Reproduction in the shrew (*Blarina brevicauda* Say). *American Journal of Anatomy* 75:39–93.

Pfau, R. S. 2014. *Blaria brevicauda*, *B. hylophaga*, and *B. carolinensis* range in Missouri based on DNA sequencing. Personal communication.

Pfau, R. S., and J. K. Braun. 2013. Occurrence of the northern short-tailed shrew (Mammalia: Soricomorpha: Soricidae: *Blarina brevicauda*) in Oklahoma. *Proceedings of the Oklahoma Academy of Science* 93:1–6.

Pfau, R. S., D. B. Sasse, M. B. Connior, and I. F. Guenther. 2011. Occurrence of *Blarina brevicauda* in Arkansas and notes on the distribution of *Blarina carolinensis* and *Cryptotis parva*. *Journal of the Arkansas Academy of Science* 65:61–66.

Ritzi, C. M., B. C. Bartels, and D. W. Spraks. 2005. Ectoparasites and food habits of Elliot's short-tailed shrew, *Blarina hylophaga*. *Southwestern Naturalist* 50:88–93.

Rood, J. P. 1958. Habits of the short-tailed shrew in captivity. *Journal of Mammalogy* 39:499–507.

Sylvester, T. L., J. D. Hoffman, and E. K. Lyons. 2012. Diet and ectoparasites of the southern short-tailed shrew (*Blarina carolinensis*) in Louisiana. *Western North American Naturalist* 72:586–590.

Thompson, C. W., J. R. Choate, H. H. Genoways, and E. J. Fink. 2011. *Blarina hylophaga. Mammalian Species 43(*878). 10 pp.

Thompson, C. W., R. S. Pfau, J. R. Choate, H. H. Genoways, and E. J. Fink. 2011. Identification and characterization of the contact zone between short-tailed shrews (*Blarina*) in Iowa and Missouri. *Canadian Journal of Zoology* 89:278–288.

Tomasi, T. E. 1979. Echolocation by the short-tailed shrew, *Blarina brevicauda. Journal of Mammalogy* 60:751–759.

North American Least Shrew
(*Cryptotis parva*)

Name

The first part of the scientific name, *Cryptotis*, is from two Greek words and means "hidden ear" (*kryptos*, "hidden," and *ous*, "ear"). It refers to the extremely large ear opening that is well concealed in the fur. The second part, *parva*, is the Latin word for "small" and emphasizes the tiny size of this animal. "North American" refers to its continental range, "least" also refers to the size, and "shrew" is from the Anglo-Saxon name *screawa*.

Description (Plate 6)

The least shrew is one of the smallest mammals in Missouri, measuring only about 76 mm (3 in.) in total length. It has a long, pointed snout that extends considerably beyond the mouth and contains the nostrils at the end; tiny, black eyes; large ear openings hidden in the velvety fur; a distinct neck; moderately slender body; small front and hind limbs with 5 claw-bearing toes each; and an extremely short tail. The fur is soft and short and is not differentiated into underfur and overhair.

The least shrew is distinguished from cinereus and southeastern shrews by the shorter tail and from short-tailed shrews by its smaller size, brown to brownish gray color on the upperparts, and fewer number of teeth.

Color. The back is dark brown to brownish gray while the belly is gray. The upper surface of the tail is colored like the back and the undersurface like the belly. There is no difference in coloration between the sexes, and all individuals are typically darker in winter than in summer. Molting occurs at any season but most commonly in late spring and late fall. Albinos are rare.

Measurements

Total length	63–82 mm	2½–3¼ in.
Tail	12–19 mm	½–¾ in.
Hind foot	8–12 mm	⅜–½ in.
Skull length	12–15 mm	½–⅝ in.
Skull width	6 mm	¼ in.
Weight	2–10 g	⅟16–⅓ oz.

Teeth and skull. The least shrew typically has 30 teeth, but 2 of these may not always be present. Because of little differentiation into incisors, canines, and premolars, two dental formulas are given for this species depending upon the classification of certain teeth:

$$I\ \frac{3}{1}\ C\ \frac{1}{1}\ P\ \frac{2}{1}\ M\ \frac{3}{3}\quad \text{or}$$

$$I\ \frac{3}{1}\ C\ \frac{1}{1}\ P\ \frac{2}{1}\ M\ \frac{3}{3} = 32$$

The first or central incisors on both upper and lower jaws are greatly enlarged and specialized into grasping pincers. Between these large incisors and the first, large cheek teeth on each half of the upper jaw are 4 small, single-pointed teeth (unicuspids); 3 of these are visible from the side, but the last and smallest, which may not always be present, can be seen only from below. The 2 large upper molars show a distinct W pattern on their surfaces. All the teeth are tipped with brown, but this color diminishes with wear and age.

Feet of North American least shrew

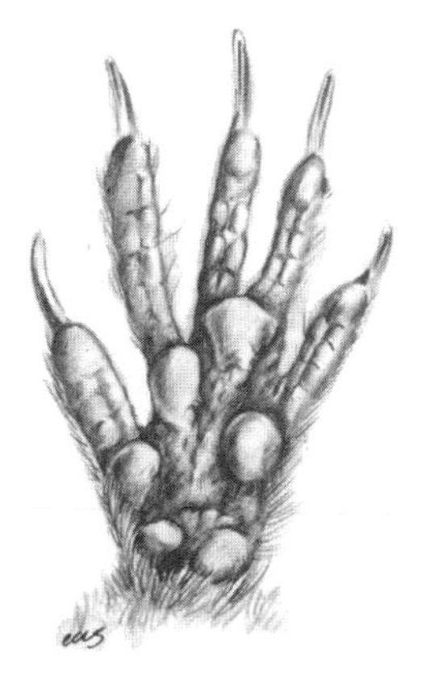

Left front foot

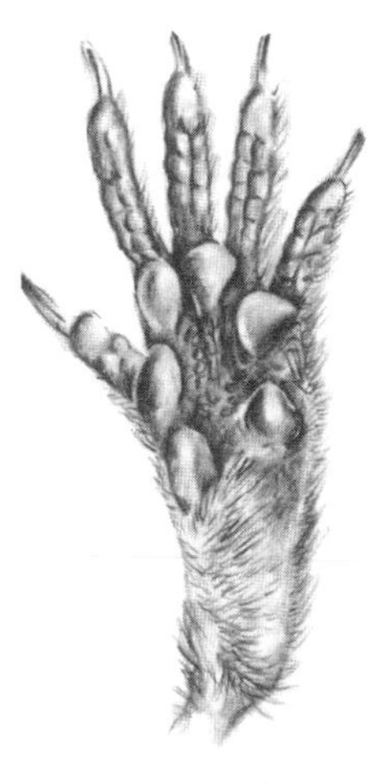

Left hind foot

Plate 6

North American Least Shrew (*Cyptotis parva*)

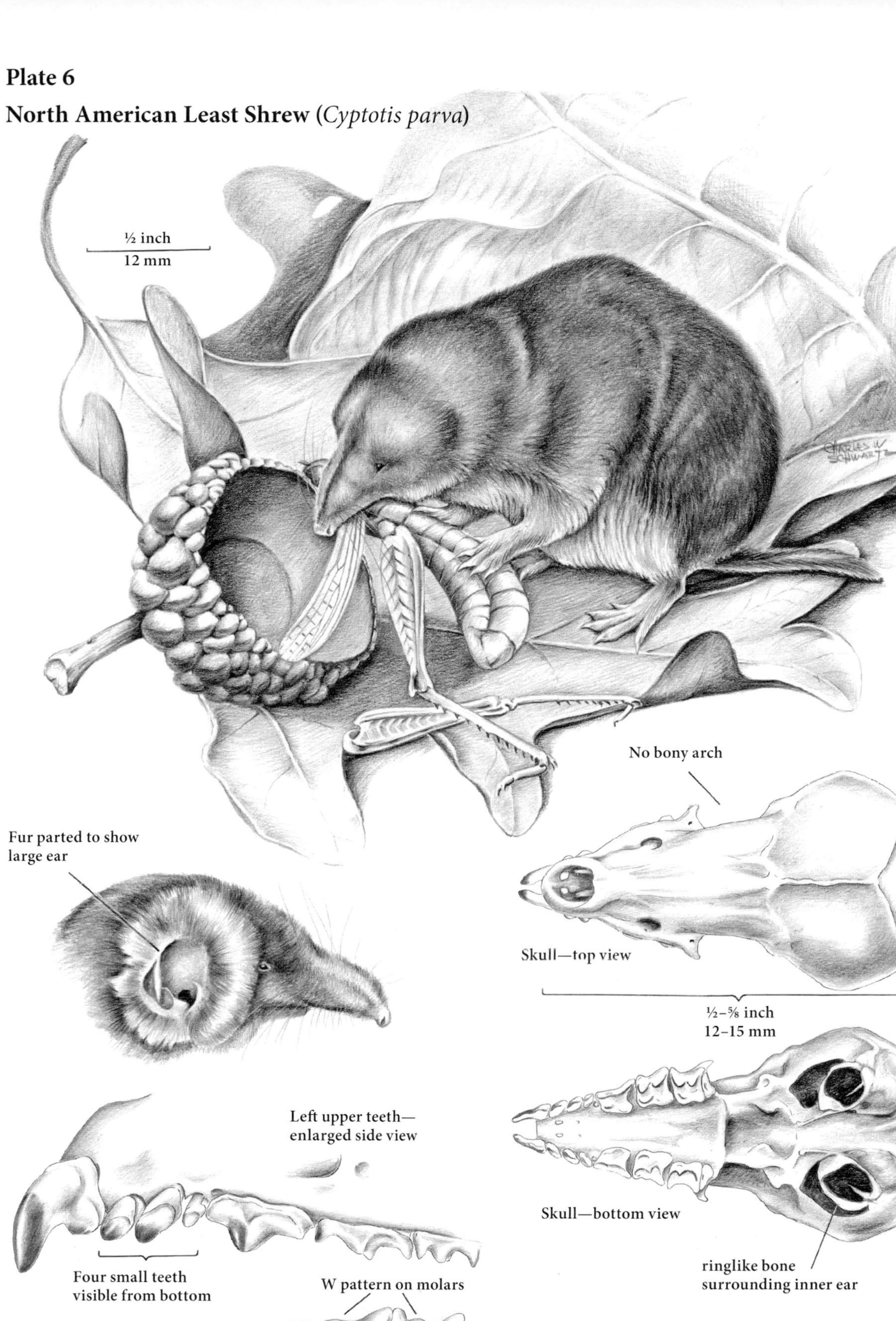

The skull of the least shrew is distinguished from those of other shrews in Missouri by having 30 teeth and by 3 small single-pointed teeth (unicuspids) visible from the side between the first upper incisor and first upper cheek tooth. The very small fourth single-pointed tooth is visible only from the bottom view.

Sex criteria. See account of short-tailed shrews for sex criteria.

Age criteria and longevity. The young are dark slate gray. In captivity one died of "old age" at 21 months.

Glands. Least shrews have a pungent odor. This comes from the small paired glands on the flanks of both sexes. Females also have scent glands in front of the lower part of the ear. Apparently, the absence of an emission of scent, or pheromone, from these glands indicates to a male that the female is sexually receptive, while the presence of a scent means she is pregnant and not receptive.

Voice and sounds. Least shrews have a vocabulary of sounds that includes clicks, *puts*, twitters, and chirps. Clicks are the most predominant and vary from low intensity (audible to humans) to high intensity (in the ultrasonic range and inaudible to humans). These vary from small bursts of 5 per minute to 13 per 5 seconds to a continuous buzz in the highest frequencies. Ultrasonic calls are used during intense exploration of the tunnel system and possibly in the identification of predators and prey, and may be interpreted as echolocation.

Distribution and Abundance

The least shrew ranges throughout most of the eastern United States and into portions of Mexico and Central America. They are generally abundant. They have been captured in several Missouri counties and probably occur statewide.

Habitat and Home

Open grass, brush, and dry, fallow fields are the sites preferred by least shrews, with marshy or timbered areas selected to a lesser extent. They dwell both on the surface of the ground and underground. These shrews use the runways and burrows of mice and the tunnels of moles or construct their own tunnels, which are about 2.5 cm (1 in.) in diameter and from 25 cm to 1.5 m (10 in. to 5 ft.) in length. Some least shrew tunnels are shallow and are made by merely separating the sod from the underlying soil; others are dug as much as 20 cm (8 in.) beneath the sod.

Sometimes several least shrews cooperate in building their tunnels. In captivity two shrews built a tunnel 61 cm (2 ft.) long with 4 openings, in a two-hour period. One of the shrews dug with its front feet, passed the soil beneath the body to the hind feet, and kicked it on behind. The other pushed the soil with its front feet and chest while using the hind feet for propulsion. Most of the soil was packed into the sides of the tunnels and only a little was scattered near the entrances.

Somewhere in the tunnel system the female constructs a compact ball-shaped nest of dried leaves and

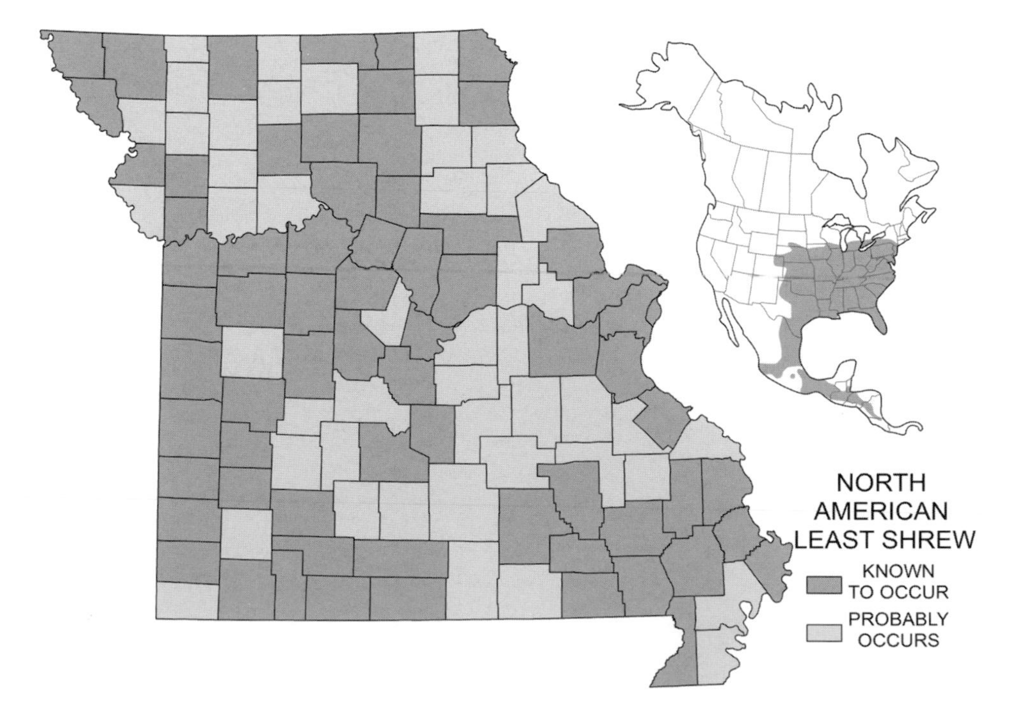

grass for her young and lines it with similar material that is partly shredded. The nest is usually built under the shelter of a rock, log, or stump or in a shallow tunnel or deep depression. The nest is from 7.6 to 17.8 cm (3 to 7 in.) in diameter and has 1 or 2 openings. Larger nests may be used as a winter home for several shrews—up to 31 have been reported in one nest, probably sharing body heat.

Habits

Like other shrews, least shrews are active at all times, resting only periodically. They have a very belligerent and nervous temperament but are somewhat more sociable than other shrew species. When food is scarce, however, they may fight, kill, and devour members of their own kind. They are least active during the hottest and coldest parts of the year.

These shrews frequently comb their fur with their feet or toes and cleanse it with their tongues. In captivity they use a special place in the cage for their droppings.

Captive least shrews do not sleep soundly but turn frequently and wake up often and stretch, yawn, and even wash themselves before going back to sleep. If several live in the same cage, they usually sleep in one pile, each one trying to burrow beneath the others and causing a constant shuffle.

Least shrews swim well but ride high in the water.

Foods

Least shrews eat mostly small insects and their larvae, snails, slugs, caterpillars, earthworms, centipedes, spiders, and the dead bodies of small animals. They take little vegetable matter.

Because of very poor eyesight, these animals locate their food mostly by their well-developed sense of touch. When catching prey, least shrews have been observed to hamstring the knee tendons in the hind legs of frogs and attack the joints of the jumping legs of crickets and grasshoppers. Captive least shrews appeared afraid of live mice. They do not possess poison and hence do not subdue their prey as do short-tailed shrews.

Least shrews have tremendous appetites and consume from three-fourths of their weight to more than their weight in food daily. This large intake is necessary to supply the energy for their nearly continuous activity. Digestion is very rapid, and when the diet is altered, fecal pellets change in consistency within two hours. These shrews will hoard in discrete locations near the nest when they are not hungry. Females will hoard more prey than males. Water is taken frequently.

Reproduction

Reproduction occurs from February until November, and several litters are produced annually. From 3 to 7 young compose a litter, with most litters containing 4 or 5. The duration of pregnancy is between 21 and 23 days.

At birth the naked young measure 2.2 cm (7/8 in.) in total length and weigh 0.3 g (1/100 oz.). Their eyes and ears are closed and no teeth are visible. At 1 week they are well furred. Shortly after this the ears open, and at 2 weeks the eyes open. By 3 weeks of age, the young shrews weigh 3 g (1/10 oz.) and their incisors are well developed. They are weaned at this time. When 1 month old, the young appear fully grown. Presumably both parents care for their offspring, as pairs of adults are found in nests with young.

Some Adverse Factors

One way the distribution and abundance of this seldom-seen creature is known is through recovery of parts of this mammal in the stomachs, droppings, or pellets of predators. Least shrews are eaten by owls, hawks, snakes, weasels, cats, and short-tailed shrews.

Fleas and mites are the only parasites recorded for the least shrew, but doubtless many of the parasites found on short-tailed shrews use this species as a host, too. Least shrews suffer many accidents and sometimes die of shock.

Importance

Shrews are valuable mammals. Although they are very small creatures, they contribute greatly to the wildlife community. They dig and work the soil, feed on abundant insect life, aid in decomposition of dead animals by their feeding activities, and serve as food for predators. Their presence enriches our environment.

Conservation and Management

No management measures are necessary at this time.

SELECTED REFERENCES

See also discussion of this species in general references, page 23.

Conaway, C. H. 1958. Maintenance, reproduction, and growth of the least shrew in captivity. *Journal of Mammalogy* 39:507–512.

Davis, W. B., and L. Joeris. 1945. Notes on the life-history of the little short-tailed shrew. *Journal of Mammalogy* 26:136–138.

Elder, W. H. 1960. An albino *Cryptotis* from Missouri. *Journal of Mammalogy* 41:506–507.

Formanowicz, D. R., Jr., P. J. Bradley, and E. D. Brodie Jr. 1989. Food hoarding by the least shrew (*Cryptotis parva*): Intersexual and prey type effects. *American Midland Naturalist* 122:26–33.

Hamilton, W. J., Jr. 1934. Habits of *Cryptotis parva* in New York. *Journal of Mammalogy* 15:154–155.

———. 1944. The biology of the little short-tailed shrew, *Cryptotis parva*. *Journal of Mammalogy* 25:1–7.

Kivett, V. K., and O. B. Mock. 1980. Reproductive behavior in the least shrew (*Cryptotis parva*) with special reference to the aural glandular region of the female. *American Midland Naturalist* 103:339–45.

Whitaker, J. O., Jr. 1974. *Cryptotis parva*. *Mammalian Species* 43. 8 pp.

Moles and Relatives (Family Talpidae)

The family name, Talpidae, is derived from the Latin word *talpa* meaning "mole." Characteristically, the front feet, legs, and shoulders of moles are greatly enlarged and modified, an adaptation for digging their underground tunnel systems. The ears have no external flaps, and the openings are hidden in the fur. The eyes are extremely minute and in some forms are hidden in the fur and even concealed beneath the skin. The body is stout and cylindrical. As in the shrews, the body fur is plushlike. Members of this family are found in North America, Europe, and Asia and include moles, shrew moles, and desmans. The only representative in Missouri is the eastern mole.

Eastern Mole (*Scalopus aquaticus*)

Name

The first part of the scientific name, *Scalopus*, is from two Greek words and means "digging" and "foot" (*skalops*, "mole," derived from the word "to dig," and *pous*, "foot"). This name refers to the large front feet of the mole, which are used in digging. The last part, *aquaticus*, is the Latin word meaning "found in water"; this misleading term was given to the mole because the webbed structure of the feet suggested that they were used in an aquatic habitat, and the original specimen was found dead in water.

The common name, "eastern," refers to the range of the species, and "mole" is from the Middle English *molle*, which is related to another Middle English word, *moldwarpe*, meaning "earth-thrower."

Description (Plate 7)

Moles live most of their lives underground and are highly specialized for their subterranean way of life. The eastern mole is a small, sturdy animal with a somewhat cylindrical body and an elongated head. The fleshy, movable snout, which serves as an organ of touch, projects over the mouth and contains the nostrils on the upper surface. The tiny, degenerate eyes are concealed in the fur and covered by fused eyelids; sight is limited merely to distinguishing light from dark. Although the ear opening is small and hidden in the fur, hearing is fairly acute.

The greatly enlarged front feet, which are broader than they are long, are normally held with the soles vertical and pointing outward. They possess well-developed claws and have a specialized sesamoid bone attached to the wrist that aids in digging. The hind feet are small. Both front and hind feet have a fringe of sensory hairs that helps in the excavating operations. The short tail is thick and scantily furred; it functions as an organ of touch and guides the mole when it moves backward in the tunnel. The short, velvety fur offers little resistance to travel through tunnels.

The bones of front limbs and breast are greatly enlarged, providing strong attachments for the powerful muscles used in digging. The extremely narrow hip girdle permits the mole to turn around in its tunnel; this it does by slowly performing a partial somersault or doubling back on itself.

Color. The eastern mole is grayish brown, being darker above and paler to browner beneath. The face, feet, and tail are whitish to pink. When viewed from different angles, the velvety fur often has a silvery sheen. The winter pelage is darker than the summer pelage. Compared to females, males tend to have a brighter orange strip on the belly caused by the secretion of skin glands in this region. Occasionally bright orange, cinnamon yellow, or white individuals occur. These variously colored moles are albinos with their white fur stained by the secretion of skin glands.

Moles molt in spring and fall. The fresh pelage comes in first on the chest and belly and then covers the underparts. On the back the new fur is first apparent near the tail and then gradually works forward. Usually there is a sharp line demarking old and new fur. There is no distinct underfur, all hairs being the same length. When viewed microscopically, the hairs can be seen to possess whiplike tips similar

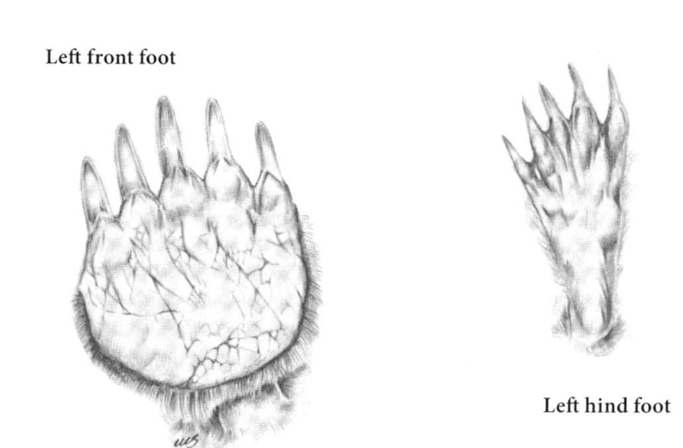

Feet of eastern mole

to the hairs of shrews but unlike those of any other mammal.

Measurements

Total length	139–203 mm	5½–8 in.
Tail	19–38 mm	¾–1½ in.
Hind foot	15–25 mm	⅝–1 in.
Skull length	28–38 mm	1⅛–1½ in.
Skull width	19 mm	¾ in.
Weight	28–141 g	1–5 oz.

Males are usually larger than females.

Teeth and skull. The dental formula of the eastern mole is:

$$I\,\frac{3}{2}\;C\,\frac{1}{0}\;P\,\frac{3}{3}\;M\,\frac{3}{3} = 36$$

However, the teeth of moles show little specialization into incisors, canines, premolars, and molars. The first upper incisors are large and curved downward and backward, but the second and third incisors are very small.

The eastern mole's skull is distinguished from the skulls of other adult Missouri mammals of similar size by the following combination of characteristics: number and shape of teeth as described above; weak zygomatic arch (the bony arch at the side of the eye socket); inflated bone surrounding the inner ear; and hard palate extending beyond the molars.

Sex criteria. The sexes are determined externally by the number of openings in the groin region. In females there are three openings. The forward one is the urinary opening, which is in the small urinary papilla, or projection; the middle one is the vagina; and the opening nearest the tail is the anus. In males only two openings occur, the combined urinary-reproductive

Plate 7

Eastern Mole (*Scalopus aquaticus*)

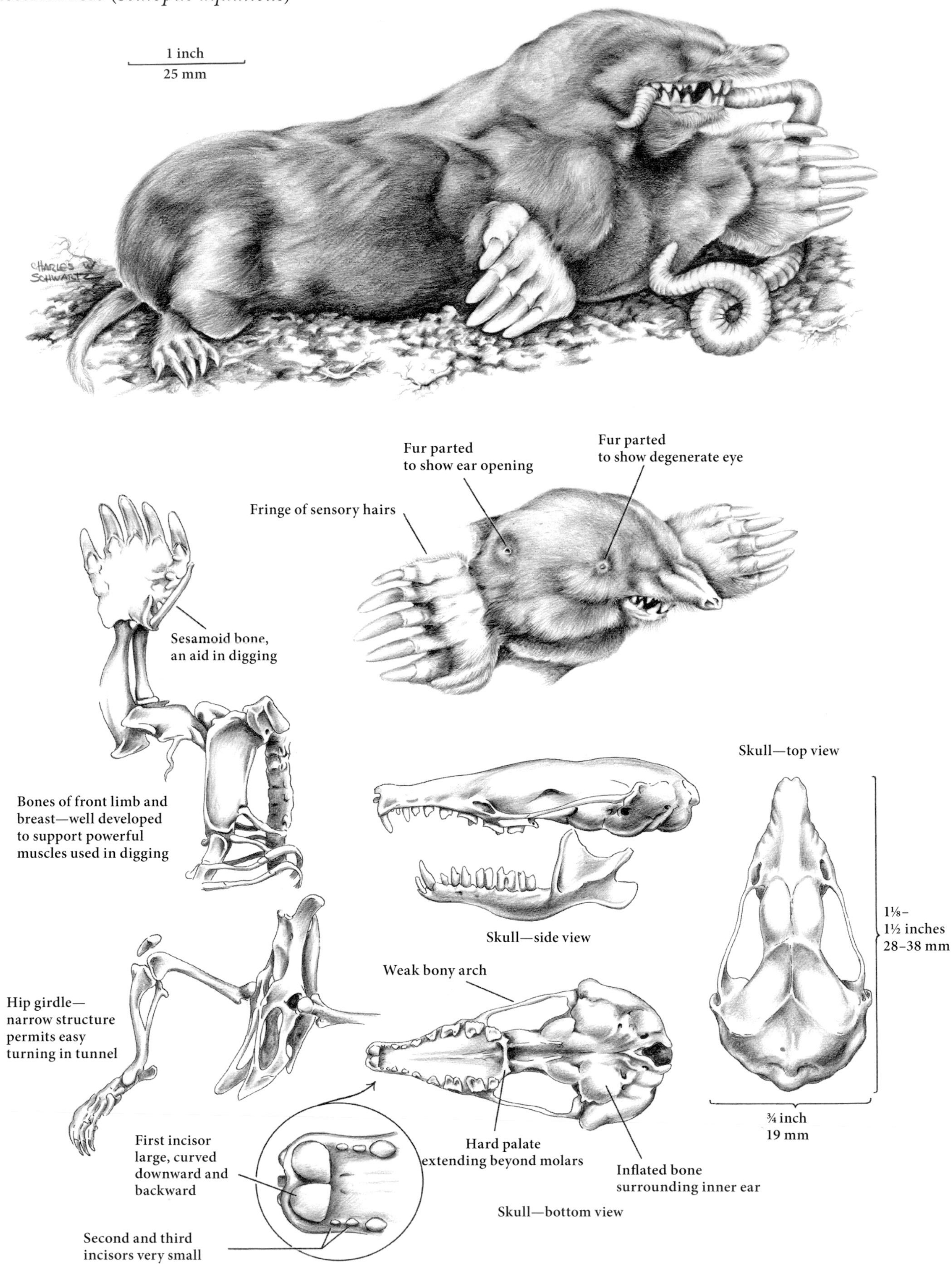

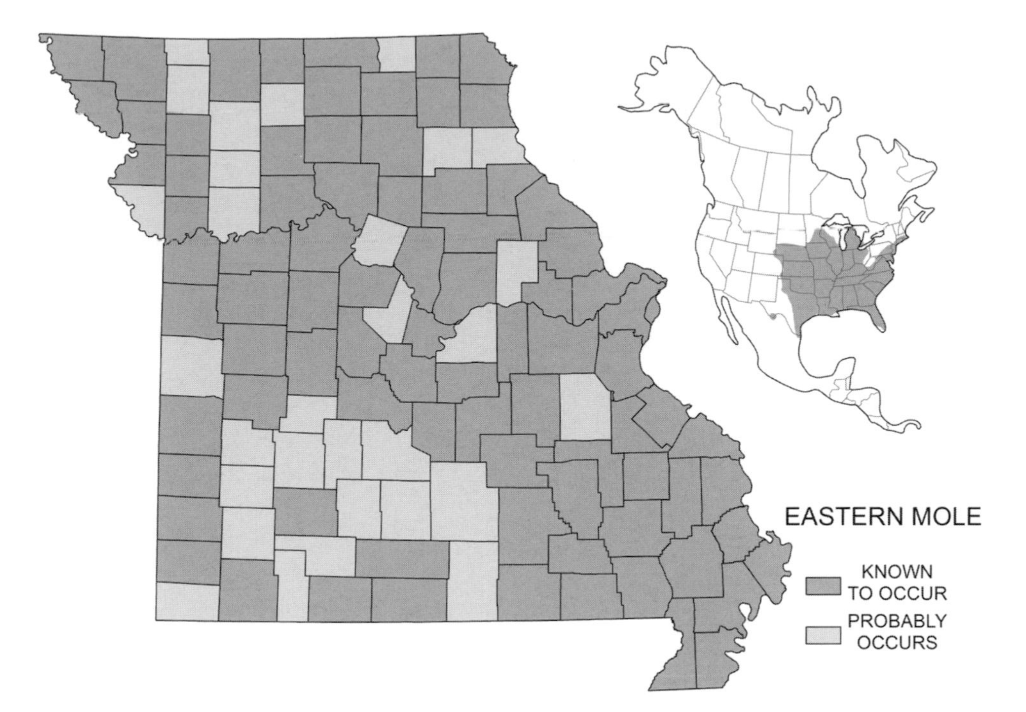

opening in the penis, and the anus. The testes are located within the body cavity of males and never descend into a scrotum. There are 6 teats on the belly, of which 1 pair is in the chest region, 1 pair in the groin region, and 1 pair between the others.

Age criteria and longevity. Young moles are distinguished from adults by their grayer coloration. With age, the skull becomes flattened and the teeth show wear.

Moles probably have a relatively long life span, as inferred by the low reproductive rate and paucity of predators. Estimates of greater than 3½ years (and perhaps greater than 6 years) exist. Females live longer than males.

Glands. A rank, musky odor is given off by a scent gland on the belly. This scent is left on the floor of the tunnel as the mole passes by and may serve as a means of communication between the sexes during the breeding season. It also renders the mole undesirable to many predators. Other large scent glands occur near the anus.

Voice and sounds. The various sounds made by moles are high-pitched squeals; harsh, guttural squeaks; short, snorting sounds; and grating of the teeth.

Distribution and Abundance

The eastern mole is common throughout most of the eastern United States. It is largely absent from the Appalachian Mountain region where soils are less favorable. The mole is common in Missouri and is found throughout the state. A high population is 3 to 5 moles per 0.4 ha (1 ac.).

Habitat and Home

Moles live underground in meadows, pastures, lawns, open woodlands, gardens, and stream banks where the soil is loose, contains humus, and is well drained but moist. In such soil conditions, digging is easy and foods the mole seeks usually abound. Soils too loose in structure to support tunnels are avoided unless they have abundant food.

Moles construct a series of tunnels in the ground. Temporary ones, barely under the surface but at a uniform depth, are made by raising the sod or ground cover with the front feet. Many of these are built after rains when the mole is in search of new sources of food and are seldom reused. These are the tunnels most commonly observed by humans. Permanent tunnels occur from 25 to 46 cm (10 to 18 in.) or more beneath the ground surface, which in winter is below the frost line. All are just large enough in diameter to allow the mole easy passage. The tunnels of males are slightly larger than those of females because of their respective sizes.

A large chamber in the deeper runways is used as a retreat during drought, heat, or cold and as a nest for the young. This chamber, between 13 and 20 cm (5 and 8 in.) in diameter, is customarily located under a stump, stone wall, bush, or other feature affording some permanent protection. There are usually several

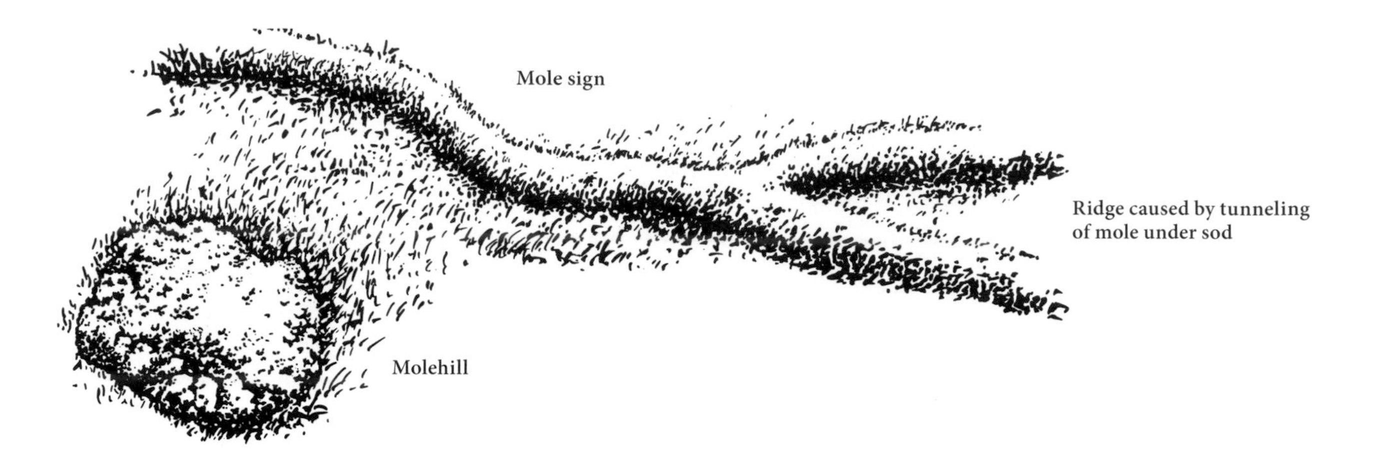

openings into this home, including one from below. When this chamber is used as a nest, a lining may be added of roots, grass, or leaves.

Habits

The digging of surface tunnels normally proceeds at the rate of about 30 cm (1 ft.) per minute, while deeper excavation goes on at approximately 4 to 5 m (12 to 15 ft.) per hour, including stops for rest and food. When working near the surface, a mole's movements may be detected by the upheaval of soil as it progresses with its excavations. During winter when the ground is frozen, moles dig less than at other seasons. However, they may still be active underground in their deeper tunnels.

In digging, the mole braces itself against the sides of the tunnel with the hind feet and loosens the soil with the front feet. The soil is passed under the body to the hind feet, which kick it further behind. When a pile has accumulated, the mole turns around and pushes the soil ahead by using one of its front feet. The soil is either packed into an unused tunnel or pushed out onto the surface of the ground through a vertical tunnel. At the sites of such excavations, piles of earth or "molehills" are built up. The tunnel is usually plugged from below, and no surface opening is present. However, openings sometimes occur that the mole uses for a rare visit to the surface. Molehills are not made as often by the eastern mole as by other kinds of moles. Molehills are occasionally confused with mounds made by pocket gophers, but the latter have soil piled fanwise around a plugged opening.

Each mole has its own system of tunnels and lives a solitary life. Other moles may be tolerated where the tunnels are next to one another or during the breeding season when some traveling probably occurs in search of mates. Often mice, shrews, and other animals use the tunnels more than do moles. There is a special place in the tunnel system for sanitation.

The home range varies from 0.2 to 1.8 ha (½ to 4½ ac.). Because males have a larger home range than females, a trap set in a runway has a better chance of catching a male than a female.

Moles are active during all hours of the day and night. They are most active on damp and cloudy days during spring and fall. By the use of telemetry, it has been determined that they rest about 3 hours, then are active about 5 hours. The resting period is spent in the

Sign in thin snow of mole tunnel underground

Mole pushing dirt through vertical tunnel onto suface of ground

nest during cold weather but in no specific location in the tunnel during warm weather.

Moles breathe very rapidly, respiring between 40 and 42 times a minute while at rest. This high metabolism seems to be associated with their general nervous disposition and high consumption of food. Their sleep is very profound. During sleep the head is curled under the body and the front feet are directed backward.

The eastern mole can swim but apparently does so only in an emergency.

Foods

Animal foods constitute almost all of the eastern mole's diet while plant matter, such as the seeds of oats, wheat, corn, and grass, forms the remaining amount. Earthworms, beetle grubs, ants, centipedes, millipedes, slugs, snails, spiders, sowbugs, and other larval and adult insects are favorite items. Occasionally a meadow vole or gartersnake may be added to the fare.

The mole kills its prey by crushing the animal against the side of the tunnel with its strong front feet. Sometimes the mole piles soil on the victim and bites off its head when it attempts to escape. If the prey offers no resistance, however, the mole may start to devour the animal without killing it.

Moles are insatiable eaters and in captivity consume, on the average, the equivalent of one-half of their body weight daily. Captive moles take water readily.

Reproduction

Mating occurs in the spring and, after a pregnancy of 4 to 6 weeks, the single annual litter is born in March, April, or the first week of May. The most usual number of young is 4, but from 2 to 5 may compose a litter. The female takes full care of her offspring.

The young are blind and naked at birth. Growth is rapid, and when 10 days old the young moles have a fine covering of gray, velvety fur that remains for several weeks. At 1 month of age, they are able to care for themselves and probably leave the nest about this time. They breed first in the spring following their birth.

Some Adverse Factors

Few animals prey upon moles because of their subterranean habits and musky odor. Snakes may overpower some moles in their tunnels, and shrews using mole runways may take some young in the nest. Hawks, owls, skunks, coyotes, foxes, domestic dogs, domestic cats, and humans are additional predators.

Mites, lice, fleas, and roundworms parasitize eastern moles. A mammal-nest beetle has been found on moles. This kind of beetle is known to live in the nests of rodents, soricomorphs, and bumblebees, where it probably feeds on eggs and young of mites.

Mole few days old

4

Flying Ma
Order Chir

The flying mammals, or order Chiroptera, are the sustained flight. The name two Greek words and mean to the winged limbs comm this large order (divided int roptera and Microchiropte bution, the greatest numbe tropics. Some representati flying "foxes" and other fru bats. Worldwide, there are family of bats (in suborder nates in Missouri, the eveni

The bats in Missouri are tivorous bats. Their ears a each generally has a well process, the *tragus* (plural enhance sound definition.

Vampire bat

Red bat

How bats find their way has been a subject of much research. Echolocation is useful for fairly close navigation, but vision may assist. Although their eyesight is poor (with some exceptions), bats can detect light and dark, large objects, and physiographic features.

Bats have been maligned as disease-causing nuisances. In Mexico and in Central and South America, rabies is known to be transmitted to humans and livestock by the bite of the vampire bat. There the incidence of rabies in livestock constitutes a serious economic loss. Bats infected with rabies have been found in Missouri, but their numbers are very low. Bats suffering from rabies exhibit behavioral changes; for example, normally nocturnal bats may be more active during the day–but it is impossible to determine if a bat has rabies without laboratory tests. The potential for contracting rabies is slight, but handling bats (or any other wild mammal) is not recommended. Anyone who has been bitten by a bat should try to capture it, wash and disinfect the wound, and contact a physician or local health official immediately.

Bats are important as predators of agricultural pests, as pollinators, and as seed dispersers, and their benefits far outweigh their potential for damage.

Current threats to bats in North America include habitat loss and degradation, cave disturbance, and the use of pesticides, all of which have been threats to our bat populations for many years; however, two new threats are causing noticeable declines: wind power and white-nose syndrome.

As we explore alternatives to fossil fuels, wind energy is becoming a viable alternative to more traditional fuel sources. Unfortunately, wind turbines cause mortality to bats and birds. The prominent causes for bat mortality have been identified as direct collision resulting in bone fractures and barotrauma, the damage to body tissue due to the abrupt change in pressure close to wind turbines.

First recognized in New York in 2006, white-nose syndrome is a fungus that appears to be of Eurasian origin. It infects the skin of cave-dwelling bats, disrupting hibernation and often resulting in mortality. The first fully developed case of white-nose syndrome was confirmed in Missouri in March 2012. Seven of our 12 species of bats have been confirmed to have white-nose syndrome, including the federally endangered Indiana and gray bats. By August 2014, white-nose syndrome had been detected in 17 counties in Missouri but was suspected in at least 3 more.

A multiagency white-nose syndrome decontamination team has created and periodically updates protocols to limit spread of the white-nose fungus. Field biologists are required to follow these guidelines when working with bats or working in caves. All people visiting caves are strongly urged to reduce the human-assisted movement of the fungus by not wearing the same clothing or footwear or using gear that has been in any other cave without first being properly disinfected; follow a "clean caving" strategy. For more information, including decontamination guidelines, contact the United States Fish and Wildlife Service or the Missouri Department of Conservation, or visit their websites.

Key to the Species
by Whole Adult Animals

1a. Ears 25 mm (1 in.) or more in length. **Go to 2**
1b. Ears less than 25 mm (1 in.) in length. **Go to 3**

2a. (From 1a) Hairs on belly with pinkish buff tips; little contrast in color between basal portions and tips of hairs on both back and belly (see plate 18); first upper incisor usually with 1 cusp; absence of long hairs projecting beyond the toes. **Townsend's Big-eared Bat** (*Corynorhinus townsendii*) p. 113
2b. (From 1a) Hairs on belly with white tips; strong contrast in color between the basal portions and tips of hairs on both back and belly (see illustration, p. 117); first upper incisor usually with 2 cusps; presence of long hairs projecting beyond the toes. **Rafinesque's Big-eared Bat** (*Corynorhinus rafinesquii*) p. 117

3a. (From 1b) Upper surface of tail membrane completely furred to tip of tail; undersurface of wing with patch of fur along forearm, wrist, and up to one-third of the way down the fingers (see plates 15–16). **Go to 4**

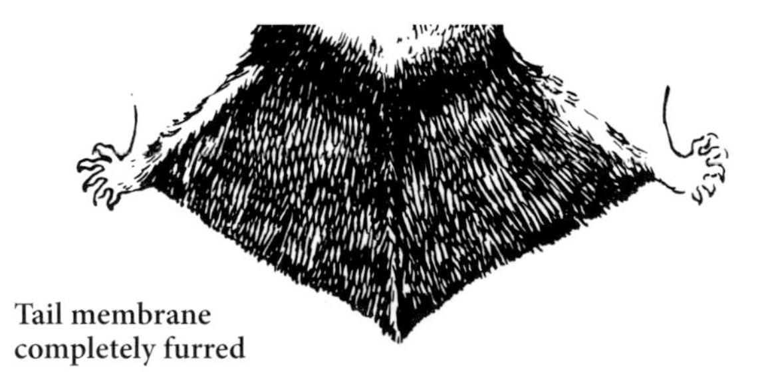

Tail membrane completely furred

3b. (From 1b) Upper surface of tail membrane naked or furred outward from body up to slightly more than one-half of its length; undersurface of wing without patch of fur along forearm, wrist, or fingers. **Go to 5**

4a. (From 3a) Body orange-red to buff and tips of hairs often frosted; yellowish white patch in front of each shoulder; ears without black rims (see plate 15). **Eastern Red Bat** (*Lasiurus borealis*) p. 102

4b. (From 3a) Body yellowish to dark brown with a frosted appearance; tan throat patch; ears with black rims (see plate 16). **Hoary Bat** (*Lasiurus cinereus*) p. 106

5a. (From 3b) Upper surface of tail membrane furred outward from body up to slightly more than half its length. **Go to 6**

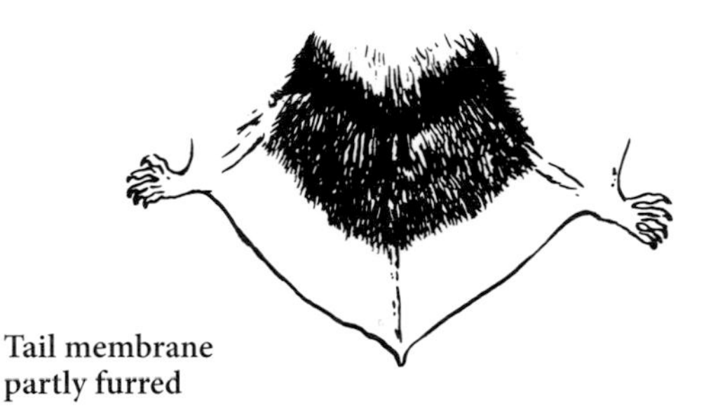

Tail membrane partly furred

5b. (From 3b) Upper surface of tail membrane naked or only furred close to body. **Go to 7**

6a. (From 5a) Tail membrane heavily furred outward from body for slightly more than one-half its length; body blackish brown with silvery white tips to the hairs (see plate 12). **Silver-haired Bat** (*Lasionycteris noctivagans*) p. 91

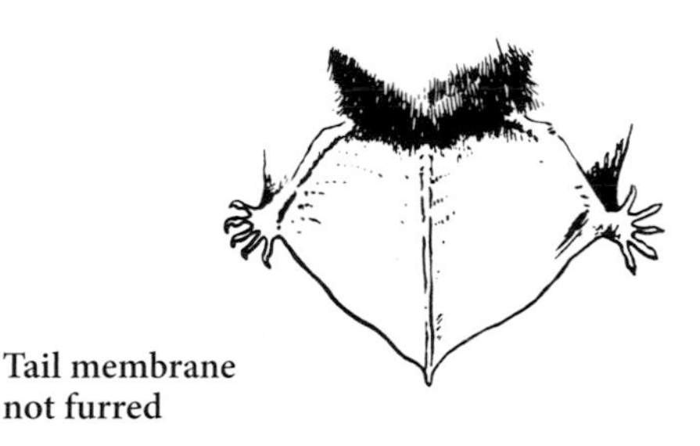

Tail membrane not furred

6b. (From 5a) Tail membrane sparsely furred outward from body for one-third of its length; body dark reddish brown to light yellowish brown or gray (see plate 13). **Tri-colored Bat** (*Perimyotis subflavus*) p. 94

7a. (From 5b) Tragus less than one-half length of ear, slightly curved forward, and with broadly rounded or blunt tip (see plates 14, 17). **Go to 8**

7b. (From 5b) Tragus about one-half or more length of ear, nearly straight, and with narrowly rounded or pointed tip (see plates 8, 10, 11). **Go to 9**

8a. (From 7a) Total length up to 95 mm (3¾ in.) (see plate 17). **Evening Bat** (*Nycticeius humeralis*) p. 109

8b. (From 7a) Total length more than 95 mm (3¾ in.) (see plate 14). **Big Brown Bat** (*Eptesicus fuscus*) p. 98

9a. (From 7b) Wing membrane attached at ankle (not always visible in dried skins); fur of back not dark at base (see plate 10). **Gray Myotis** (*Myotis grisescens*) p. 77

9b. (From 7b) Wing membrane attached along side of foot to base of toes; fur of back dark at base (see plates 8, 11). **Go to 10**

10a. (From 9b) When laid forward, ears project up to 4 mm (3/16 in.) beyond nostrils (see plate 11). **Northern Long-eared Myotis** (*Myotis septentrionalis*) p. 83

10b. (From 9b) When laid forward, ears do not project more than 1 mm (1/16 in.) beyond nostrils. **Go to 11**

11a. (From 10b) Calcar not keeled (see plate 8). **Go to 12**

11b. (From 10b) Calcar keeled (see plate 11). **Go to 13**

12a. (From 11a) Body yellowish to olive brown; tips of long and glossy back fur contrast with darker bases; hairs on toes long and dense. **Little Brown Myotis** (*Myotis lucifugus*) p. 67

12b. (From 11a) Body gray or bright orange-brown; tips of dense woolly back fur do not contrast with bases; hairs on toes extend beyond tips of claws. **Southeastern Myotis** (*Myotis austroriparius*) p. 81

13a. (From 11b) Body yellowish brown, usually with distinct black mask across face to the black ears (see plate 11); fur of back, when parted, shows blackish bases and yellowish brown outer portions; hind foot 6 mm (¼ in.) or less in length. **Eastern Small-footed Myotis** (*Myotis leibii*) p. 89

13b. (From 11b) Body dull grayish to purplish brown; fur of back, when parted, shows faint three-colored pattern, with basal two-thirds of each hair blackish brown, followed by narrow grayish band and cinnamon brown tip (see plate 11); hind foot more than 6 mm (¼ in.) in length. **Indiana Myotis** (*Myotis sodalis*) p. 86

Key to the Species
by Skulls of Adults

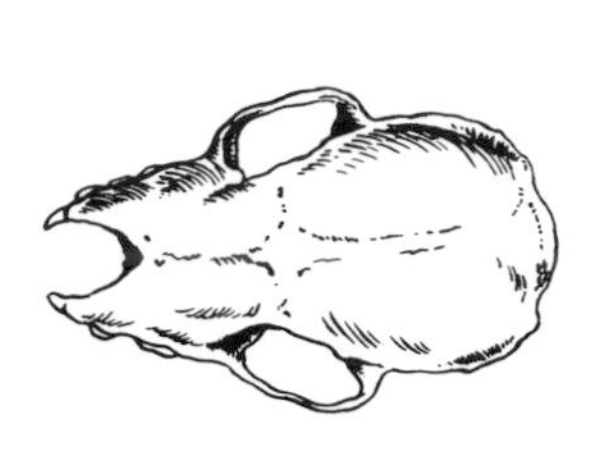

1a. Total teeth 38; in each half of upper jaw, incisors 2, premolars 3 (see plates 8, 10, 11). *Myotis* spp. **Little Brown Myotis** (*Myotis lucifugus*) p. 67; **Gray Myotis** (*Myotis grisescens*) p. 77; **Southeastern Myotis** (*Myotis austroriparius*) p. 81; **Northern Long-eared Myotis** (*Myotis septentrionalis*) p. 83; **Indiana Myotis** (*Myotis sodalis*) p. 86; **Eastern Small-footed Myotis** (*Myotis leibii*) p. 89
1b. Total teeth less than 38. **Go to 2**

2a. (From 1b) Total teeth 36; in each half of upper jaw, incisors 2, premolars 2 (see plates 12, 18, and illustration, p. 117). **Go to 3**
2b. (From 1b) Total teeth less than 36. **Go to 5**

3a. (From 2a) When viewed from side, upper profile of skull appears flat (see plate 12); rostrum strongly concave on each side at back of nasal opening. **Silver-haired Bat** (*Lasionycteris noctivagans*) p. 91
3b. (From 2a) Skull highly arched (see plate 18); rostrum convex on each side at back of nasal opening. **Go to 4**

4a. (From 3b) First upper incisor usually with 1 cusp (see plate 18). **Townsend's Big-eared Bat** (*Corynorhinus townsendii*) p. 113
4b. (From 3b) First upper incisor usually with 2 cusps (see illustration, p. 117). **Rafinesque's Big-eared Bat** (*Corynorhinus rafinesquii*) p. 117

5a. (From 2b) Total teeth 34; in each half of upper jaw, incisors 2, premolars 2 (see plate 13). **Tri-colored Bat** (*Perimyotis subflavus*) p. 94
5b. (From 2b) Total teeth less than 34. **Go to 6**

6a. (From 5b) Total teeth 30; in each half of upper jaw, incisor 1, premolar 1 (see plate 17). **Evening Bat** (*Nycticeius humeralis*) p. 109
6b. (From 5b) Total teeth 32 (see plates 14, 15, 16). **Go to 7**

7a. (From 6b) In each half of upper jaw, incisors 2, premolar 1. **Big Brown Bat** (*Eptesicus fuscus*) p. 98
7b. (From 6b) In each half of upper jaw, incisor 1, premolars 2 (first very small, at base of canine on tongue side). **Go to 8**

8a. (From 7b) Length of skull 15 mm (9/16 in.) or less. **Eastern Red Bat** (*Lasiurus borealis*) p. 102
8b. (From 7b) Length of skull 16 mm (⅝ in.) or more. **Hoary Bat** (*Lasiurus cinereus*) p. 106

GENERAL REFERENCES

Baer, G. M., and D. B. Adams. 1970. Rabies in insectivorous bats in the United States, 1953–1965. *Public Health Report* 85:637–645.

Barbour, R. W., and W. H. Davis. 1969. *Bats of America*. University Press of Kentucky, Lexington. 285 pp.

Boyles, J., J. Timpone, and L. Robbins. 2009. *Bats of Missouri*. Indiana State University Center for North American Bat Research and Conservation, Terre Haute, publication 3. 60 pp.

Cohn, J. P. 2008. White-nose syndrome threatens bats. *BioScience* 58:1098.

———. 2012. Bats and white-nose syndrome still a conundrum. *BioScience* 64:444.

Dunbar, M. B., J. O. Whitaker Jr., and L. W. Robbins. 2007. Winter feeding by bats in Missouri. *Acta Chiropterologica* 9:305–322.

Ehlman, S. M., J. J. Cox, and P. H. Crowley. 2013. Evaporative water loss, spatial distributions, and survival in white-nose-syndrome–affected little brown myotis: A model. *Journal of Mammalogy* 94:572–583.

Feldhamer, G. A., T. C. Carter, and J. O. Whitaker Jr. 2009. Prey consumed by eight species of insectivorous bats from southern Illinois. *American Midland Naturalist* 162:43–51.

Folk, G. E. 1940. Shift of population among hibernating bats. *Journal of Mammalogy* 21:306–315.

Galambos, R., and D. R. Griffin. 1942. Obstacle avoidance by flying bats: The cries of bats. *Journal of Experimental Zoology* 89:475–490.

Gould, E. 1955. The feeding efficiency of insectivorous bats. *Journal of Mammalogy* 36:399–407.

Griffin, D. R. 1940a. Notes on the life histories of New England bats. *Journal of Mammalogy* 21:181–187.

———. 1940b. Migrations of New England bats. *Harvard University, Bulletin of Museum of Comparative Zoology* 86:217–246.

———. 1945. Travels of banded cave bats. *Journal of Mammalogy* 26:15–23.
———. 1946. Mystery mammals of the twilight. *National Geographic* 90:117–134.
Grodsky, S. M., M. J. Behr, A. Gendler, D. Drake, B. C. Dieterle, R. J. Rudd, and N. L. Walrath. 2011. Investigating the causes of death for wind turbine–associated bat fatalities. *Journal of Mammalogy* 92:917–925.
Kunz, T. H., and M. B. Fenton, eds. 2003. *Bat ecology.* University of Chicago Press, Chicago, IL. 779 pp.
LaVal, R. K., R. L. Clawson, W. Caire, L. R. Wingate, and M. L. LaVal. 1977. *An evaluation of the status of Myotine bats in the proposed Meramec Park Lake and Union Lake project areas, Missouri.* U.S. Army Corps of Engineers, St. Louis District. 136 pp.
LaVal, R. K., and M. L. LaVal. 1980. *Ecological studies and management of Missouri bats, with emphasis on cave-dwelling species.* Missouri Department of Conservation, Jefferson City, Terrestrial Series No. 8. 53 pp.
Neuweiler, G. 2000. *The biology of bats.* Oxford University Press, New York. 310 pp.
O'Shea, T. J., P. M. Cryan, D. T. S. Hayman, R. K. Plowright, and D. G. Streicker. 2016. Multiple mortality events in bats: a global review. Mammal Review doi: 10.1111/mam.12064.
Pierce, G. W., and D. R. Griffin. 1938. Experimental determination of supersonic notes emitted by bats. *Journal of Mammalogy* 19:454–455.
Sales, G., and D. Pye. 1974. *Ultrasonic communication.* Chapman and Hall, London. 281 pp.
Thomas, D. W. 1995. Hibernating bats are sensitive to nontactile human disturbance. *Journal of Mammalogy* 76:940–946.
Turabelidze, G., H. Pue, A. Grim, and S. Patrick. 2009. First human rabies case in Missouri in 50 years causes death in outdoorsman. *Missouri Medicine* 106:417–419.
Twente, J. W., Jr. 1955. Aspects of a population study of cavern-dwelling bats. *Journal of Mammalogy* 36:379–390.
Wimsatt, W. A. 1944. Further studies on the survival of spermatozoa in the female reproductive tract of the bat. *Anatomical Record* 88:193–204.
———. 1945. Notes on breeding behavior, pregnancy, and parturition in some Vespertilionid bats of the eastern United States. *Journal of Mammalogy* 26:23–33.
Wimsatt, W. A., ed. 1970. *Biology of bats.* Academic Press, New York. Vol. 1: 406 pp.; vol. 2: 477 pp.; vol. 3: 651 pp.

Myotis scooping up moth with wing

Evening Bats (Family Vespertilionidae)

This family occurs worldwide with 14 species regularly occurring in Missouri. The family name, Vespertilionidae, is based on a Latin word meaning "bat." The name probably came from the Latin word for "evening"—the time when these animals are seen flying about. Members of this family are small (averaging 12 g and 65 mm long, ½ oz. and 2½ in. long). Their muzzles are simple and lack leaflike outgrowths; their ears have well-developed, straight or slightly curved tragi; and their long tails extend to the edge of the wide tail membrane but never much beyond.

Mouse-eared Bats (*Myotis* spp.)

There are six species of closely related bats in Missouri that are referred to commonly as the "mouse-eared bats" or technically as the *Myotis* bats. These are the little brown myotis, *Myotis lucifugus*; the gray myotis, *Myotis grisescens*; the southeastern myotis, *Myotis austroriparius*; the northern long-eared myotis, *Myotis septentrionalis*; the Indiana myotis, *Myotis sodalis*; and the eastern small-footed myotis, *Myotis leibii*. These bats are very similar in appearance and may be difficult to distinguish from each other unless examined closely (see plates 8–11). They are discussed separately where they warrant it but together where they are similar. The description for the little brown myotis contains information relevant for many other bat species.

Little Brown Myotis (*Myotis lucifugus*)

Name

The first part of the scientific name, *Myotis*, is from two Greek words and means "mouse ear" (*mys*, "mouse," and *otis*, "ear"). This refers to the general resemblance of this bat's ears to those of certain mice. The last part, *lucifugus*, is from two Latin words and means "light fleer" (*lux*, "light," and *fugio*, "to flee"). This alludes to the nocturnal habits of the bat.

The common names, "little" and "brown," describe the size and predominant color. "Bat" is from the Middle English word *bakke*, which is apparently of Scandinavian origin and related to an Old Norse word, *blaka*, meaning "to flutter." The little brown myotis is also known as the little brown bat.

Description (Plates 8 and 9)

The little brown myotis has relatively narrow, naked ears with bluntly rounded tips; when laid forward, they do not extend more than 1.6 mm (1/16 in.) beyond the

Plate 8

Little Brown Myotis (*Myotis lucifugus*)

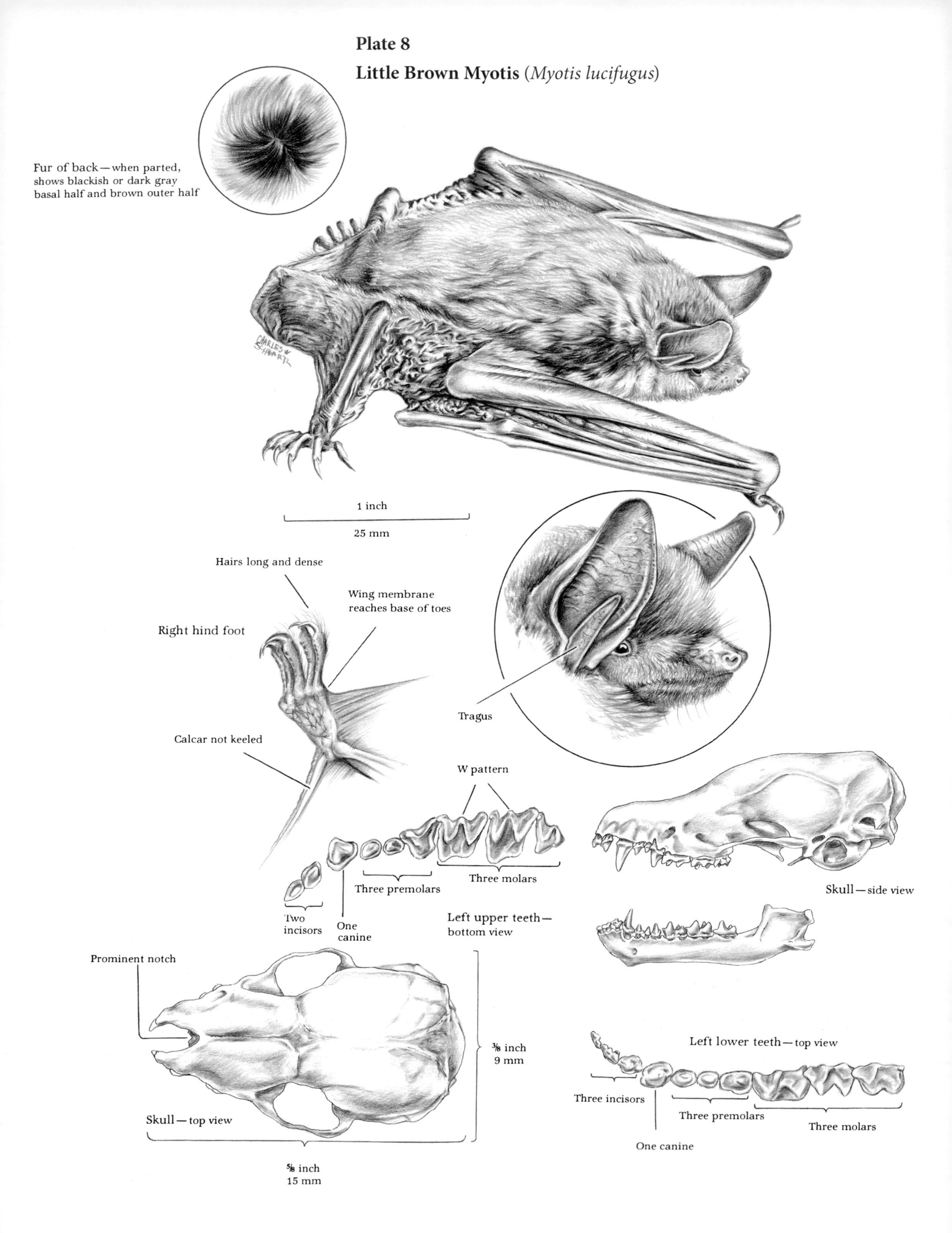

nostrils. The tragus is about one-half the length of the ear, is nearly straight, and tapers to a narrowly rounded or somewhat pointed tip. The wing membrane is naked, except for sparse furring next to the body, and is attached along the side of the foot, reaching to the base of the toes. The fifth finger is shorter than the fourth. The tail membrane is naked except for a few scattered hairs on the upper surface near the body. The calcar is not keeled. The tail is about as long as the outstretched leg. The body fur is long and soft and often wavy on the back. Hairs on the toes are long and dense.

The little brown myotis and other mouse-eared bats are most likely to be confused with the big brown bat, tri-colored bat, and evening bat. However, these species can be distinguished by some or all of the following characteristics: size, dental formula, color and character of the fur, shape and length of the ear and tragus, and hairs on the toes.

Color. The upper surface of the body varies from yellowish brown to olive brown. The basal half of each hair is blackish or dark gray while the outer half is brown. This two-toned coloration is easily seen when the fur is parted by blowing into it. There is a metallic sheen to the long, glossy tips of the fur on the back. The underparts are gray washed with buff, the individual hairs being black at the base and tipped with buff for the outer third. The wing and tail membranes and ears are glossy dark brown. In regions of high humidity, the color tends to be darker; under dry conditions, it is paler. Only rarely are individuals nearly black or blotched with white. The sexes are colored alike and show no difference in coloration with the seasons.

Measurements

Total length	76–95 mm	3–3¾ in.
Tail	31–44 mm	1¼–1¾ in.
Hind foot	6–11 mm	¼–7⁄16 in.
Ear	12–15 mm	½–⅝ in
Skull length	15 mm	⅝ in.
Skull width	9 mm	⅜ in.
Weight	7–9 g	¼ oz.

Weight is greatest in the fall before hibernation.

Teeth and skull. The dental formula of the little brown myotis is:

$$I\ \frac{2}{3}\ C\ \frac{1}{1}\ P\ \frac{3}{3}\ M\ \frac{3}{3} = 38$$

Occasionally skulls lack the second upper premolar teeth. As in all Missouri bats, the first two upper molars have a distinct W pattern on their surfaces.

The skulls of mouse-eared bats can be distinguished from all other bats in Missouri by the total number of teeth. However, the skull of the little brown myotis is very difficult to distinguish from that of other mouse-eared bats and is best identified by comparison with skulls of known identity or by some authority. The crest on the top of the skull is usually not as distinct as that of the gray myotis (see plate 10). As is typical of the skulls of all Missouri bats, there is a prominent notch in the front of the hard palate.

Sex criteria and sex ratio. Males are identified by the penis. Females are distinguished by the opening of the vagina, which is visible between the large urethral papilla (often misidentified as a penis) and the anus. There is one pair of teats in the chest region of both males and females.

The sex ratio in young little brown myotis in summer is approximately even. However, in summer, bats collected in foraging areas near nursery colonies show 2 females to 1 male. Away from nursery colonies, collections report all males. The sex ratio of bats hibernating in caves varies with the time of year in which the collections are made because of slightly different arrival and departure dates of males and females. When all bats are hibernating, the sexes are usually found in equal numbers.

Age criteria, age ratio, and longevity. Fully furred young are much darker and duller than adults. Until midsummer or early fall, young bats can be distinguished from adults by an examination of the joints of the elongated finger bones (see page 70). If the bones gradually enlarge to the joint and show clear areas of cartilage between them, the bat is young; if the ends of the bones meet in a small, round, dark knob, the bat is an adult. Although there are several methods used to determine the age of bats by wear on the canine and molar teeth, recent recoveries of very old banded bats show these to be unreliable.

In certain European species of *Myotis*, the condition of the reproductive organs has been correlated with the age of females. In young females that have not produced offspring, the two branches of the uterus are small and nearly equal in size. In older females that have produced offspring, the two uterine branches show a marked difference in size, the right one being larger. Presumably this aging method may apply to the species of *Myotis* in Missouri. (For additional aging methods, see Townsend's big-eared bat, page 113.)

The age ratio in a nursery colony of little brown myotis showed 235 young to 297 adult females.

Recovery of banded individuals shows that little brown myotis commonly live and reproduce up to 12

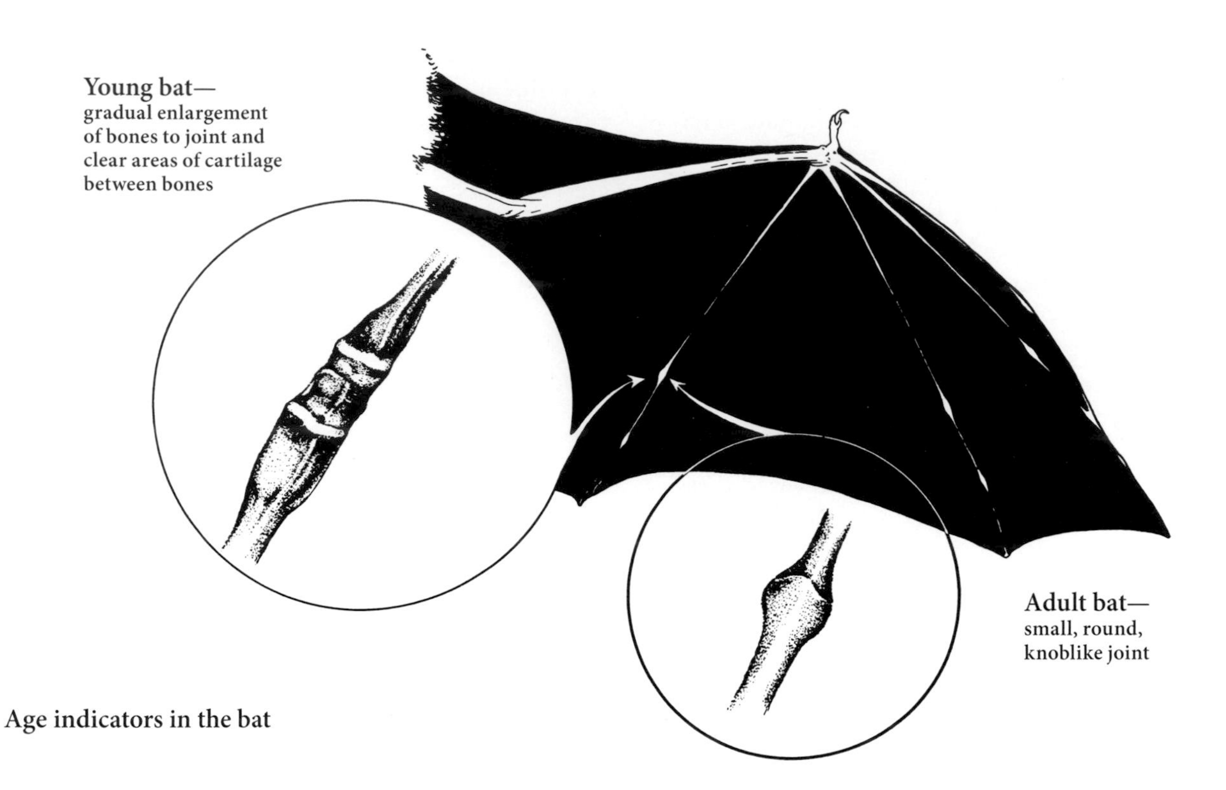

Age indicators in the bat

years of age. There are records of them reaching 29, 30, and even 34 years of age, based on a banding study. This longevity is reflected in the very low reproductive rate and low predation pressure.

Voice and sounds. As a means of guiding their flight and locating prey, most bats have developed a special technique of echolocation using ultrasonic sounds. Ultrasonic sounds, uttered at frequencies above the range of human hearing, are made with the bat's mouth and larynx (voice box). When a cry is given, a minute muscle in the bat's ear contracts momentarily; this prevents the bat from hearing its own cry, which might interfere with the reception of the echo. The ultrasonic sound strikes an object, and the echo is received by the bat's inner ear. By analyzing the echo, bats measure the distance to a given object; the direction it is from the bat; whether or not it is moving; the size, shape, and surface characteristics of the object; and the bat's own flight speed. In this way bats are adept at avoiding obstacles in their path and can pursue and intercept prey. In addition, they can utilize the calls of other bats to aid in the location of food and new roost sites.

Bat detectors can be used to pick up and record the high-frequency calls, allowing researchers to identify many species of bats. Unfortunately, it is difficult to distinguish between certain *Myotis* species using auditory means. Bats may call at any frequency in the ultrasonic range, may vary the frequency by "sweeping" from a high to a low range, and may change the duration of the individual calls. A "cruising" bat may emit cries at a rate of 8 to 15 per second, but when making a sudden maneuver it may increase the cries to around 150 to 200 per second. When the bat is closing in on prey, the cries become so fast they resemble a continuous buzz.

In addition to ultrasonic sounds, bats make sounds at low frequencies that are audible to humans. Squeaks given by infants are heard by mothers and probably aid in recognition of individual young. This may be especially useful when mothers return to a group of infants left in a large nursery colony. Squeaks by older bats, also audible to us, probably serve in attracting bats to good feeding sites, in helping bats keep contact with each other, and in aggression and mating. There is also a discrete copulation call given by males.

It is interesting that certain species of moths preyed upon by bats have evolved an anatomical structure that senses the echolocating sounds of bats. This warning mechanism allows the moths to take evasive action by increasing their flight speed or flying erratically.

Distribution and Abundance

The little brown myotis is one of the most common North American bats and ranges from southern Alaska across most of Canada and the United States. In the northeastern United States, this is one of the bat species most heavily affected by white-nose syndrome,

Plate 9

How to Distinguish Certain Species of *Myotis* by the Hind Foot and Fur

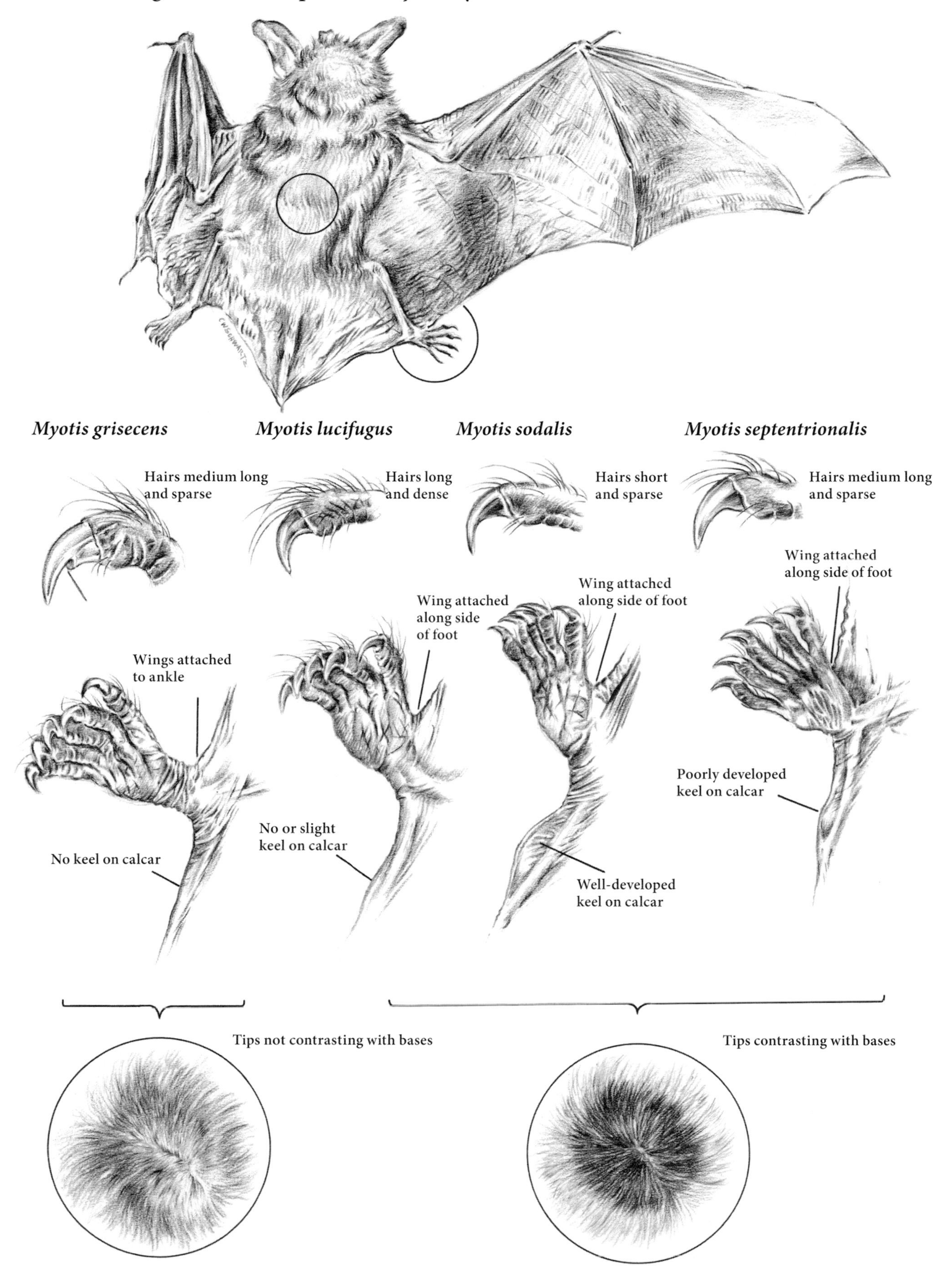

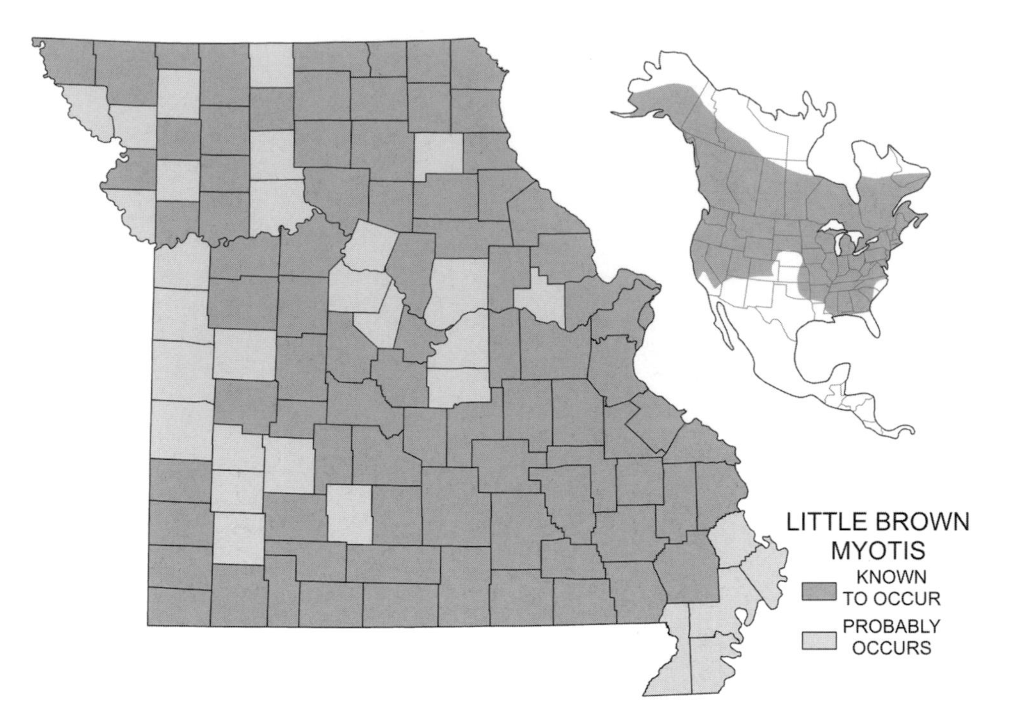

and it was recently ranked as vulnerable across its range.

The little brown myotis is widely distributed throughout Missouri all year but is common only in the immediate vicinity of large nursery colonies. The population is probably decreasing, possibly resulting from the destruction of bats roosting in buildings, the destruction of roosting sites in buildings, insecticide poisoning, and disturbance of hibernating colonies.

Habitat and Home

In Missouri little brown myotis hibernate during the winter in limestone caves and mines. Following hibernation they disperse, sometimes up to 998 km (620 mi.). In spring and summer the females live in nursery colonies, which may be in crevices of cliffs, in hollow trees, under loose bark of trees, in attics and other undisturbed parts of buildings, in towers, and in various other retreats (including bat houses). They seldom occur in caves during the summer.

The males are solitary or live in small groups in summer, using rocky crevices, hollow trees, or protected sites under siding, shingles, or loose bark. One colony was found in a cave. Rarely do they live in a segregated part of the nursery colony and then only toward the end of the rearing season. After the nursery colonies disperse in late summer, both sexes roost in trees or other locations.

Habits

In late summer, there is a tendency for little brown myotis to disperse rather widely from their summer areas. From early August to late October, they often gather at the mouths of certain large caves that they will later enter for hibernation. Here they fly about the entrance with bats of their own kind as well as other species. This activity is referred to as "swarming." It serves to acquaint young bats with places where they can survive the winter and makes the females available for mating.

They first enter the caves for their winter hibernation during October, and by November hibernation is well under way. In general, females start to hibernate first, followed by the males and young later.

During hibernation, little brown myotis hang upside down from the rough walls and ceilings of the caves, suspended by the strong hooked claws of their hind feet. They may hang alone or in clusters of usually 5 to 50 or more individuals. For the most part, the clusters are composed of one species, but occasionally a bat of another species may be found in the cluster. In Missouri, the Indiana myotis and the northern long-eared myotis are the species most frequently encountered with the little brown myotis.

Some caves may harbor several species of bats at one time, each selecting specific conditions. Little brown myotis prefer cooler and more humid parts of caves and are often covered with droplets of moisture that condense on their fur.

The body temperature of a hibernating bat becomes low like that of the adjacent rock, and the rates of heartbeat and breathing slow down. The bat enters a deep torpor (sleep) that may be broken by occasional brief periods of activity. The store of body fat, accumulated gradually during late summer, is the source of energy during this time. Older bats store more fat than young ones and are better able to survive hibernation in good condition. Temperatures close to freezing permit the most efficient utilization of the stored fat reserves.

Periodic activity may be stimulated by the decrease in body fluids, the accumulation of urine in the bladder, temperature changes above or below the optimum range for hibernation, or by an internal "clock." During these periods bats may fly around in their hibernating quarters, lap some moisture from the walls of the cave or stream within, groom, and sometimes mate; seldom do they feed. Arousal periods can include warming to normal body temperature and returning to torpor without perceptible physical movement. If the bats move because of some disturbance, they show a strong tendency to return to the same places.

Little brown myotis and other bat species are familiar with an area they frequent during feeding or when moving from their wintering to their summering area. The size of this area varies with the habits of the species. Little brown myotis taken as far as 290 km (180 mi.) from the cave where they were collected have been recaptured in the original cave. Marked bats have been collected in the same cave for three different years, and about 40 percent of the bats banded in one cave were retaken there the following winter. Marked bats are known to have moved as much as 250 km (156 mi.) from one cave to another, but most tend to stay in the same location all winter. However, it does appear that the farther a bat is released from "home," the less are its chances of returning.

With the coming of spring, females begin to leave the caves, the males remaining somewhat longer. The spring dispersal takes place over a few weeks, and females are found in their nursery colonies from the first of April on, reaching a peak in mid-May. The summer colonies may be in the same vicinity as the winter homes or as much as 455 km (283 mi.) away. There is a tendency for females to return to the same summer colony year after year, but some have been recovered in a summer colony as far as 250 km (158 mi.) from the one where they were first banded. There is seldom an exchange of bats from one summer colony to another. Bats removed from a summer colony and transported as far as 435 km (270 mi.) away have returned.

During the summer, females are generally segregated in the nursery colony; the pregnant females tend to hang together, while those that have already borne their young live in a different part of the colony. Little brown myotis seemingly like high temperatures but rarely over 38°C (100°F).

When the bats first enter the nursery colony in spring they are in good pelage. The fur is licked with the tongue, the ears are cleansed with the thumb, and the back is combed by rubbing it with a moistened leg. As the air temperature increases, external parasites become more common. Considerable restlessness occurs in the colonies because of the increasing temperature, the numerous young, and the occasional intrusion of males.

Young bats on their first flights outside of their maternity colony may be guided by scent posts along the main route of travel. Observations show that adult females, and perhaps young of the year, briefly rest on a selected tree and mark it by urine or a glandular secretion. Such scent posts serve as orientation cues as the young increase their range.

In flying, most bats do not soar like some birds but continually beat their wings. However, they may take short soaring sweeps when banking during pursuit of an insect. Little brown myotis take 15 strokes a second when flying at their usual rate of about 16 km (10 mi.) per hour. The flight is often erratic as they rapidly change their course to follow insects or to dart in and out of foliage or around obstacles.

When a bat starts to land, it flies toward the selected site, banks slightly, and then reaches out with its hind feet to catch a hold. This action is very rapid, and sometimes the bat seems to turn a "cartwheel" between flight and hanging by the hind feet.

Bats start to feed first at twilight. Where there is a colony, it is common for all adults to leave about the same time. They often disperse to different feeding areas. They may feed for an hour or two and then return to roost, either in their daytime retreat or in a temporary location. Nursing females return to feed their young. Some intermittent foraging occurs throughout the night, but in contrast to the early evening feeding, not all individuals participate in this activity at one time. There is another feeding period early in the morning, and by sunrise all bats have returned to their daytime roosts. Bats are rarely abroad in the day. They spend this period in their retreats, where they hang upside down by their hind feet or crawl into a crevice and become quiet in a horizontal position. Bats will use short-term torpor during the day to reduce energy expenditures.

When feeding, little brown myotis primarily select the border between more open areas and denser cover. They feed over meadows, in clearings in the forest, in shaded groves, around city streetlights, near open barns or sheds, over granaries, under bridges, and in other places where insect food is plentiful. They also fly along watercourses and over ponds and lakes where they occasionally dip down to lap up water or catch

insects attracted to the water. They are capable of swimming if they fall into water, but about 60 m (200 ft.) is all they can swim without exhaustion. Like other bats, little brown myotis crawl by supporting their bodies with the wrists and hind feet and climb by using the clawed thumb and the claws of the hind feet.

Foods

Little brown myotis are entirely insectivorous, taking mostly winged adults. When mayflies are available, they are fed upon heavily. Mosquitoes may be important before mayflies emerge. They commonly catch beetles, flies, caddisflies, lacewings, stoneflies, and moths, ranging in size from a wingspan of 3 mm to 13 mm (⅛ in. to ½ in.). For the bat to locate and catch the prey, the insect must be nearly the same size or larger than one wavelength of sound emitted by the bat. Bats can distinguish prey by comparing the echo spectra and their intensity.

Little brown myotis are very efficient feeders and fill their stomachs in an hour or two. In one night, an individual can consume one-half its body weight; lactating females can consume even more. Little brown myotis eat at the rate of one insect every few seconds. They must eat a great deal of food to maintain the high metabolic rate necessary for flight.

Bats capture small insects in the mouth or cupped tail membrane and feed on them while flying. Little brown myotis often feed by flying repeatedly through swarms of insects. If a large insect is not in the direct line of flight, the bat may use its wingtip to push the insect into the tail membrane. It is then grasped with the teeth and either eaten in flight ("eaten on the wing") or after the bat has landed. Rarely does a little brown myotis alight on a tree or telephone pole to capture a resting insect.

Reproduction

Mating occurs in the fall before hibernation, during winter if the bats become active (and even when the females are still torpid), and again in spring after hibernation.

The formation of spermatozoa in the male reproductive tract occurs only during the summer, but great masses of mature sperm are retained in the male during fall and winter. Some of the sperm received by the female in the fall and winter matings remain dormant while she is hibernating. These sperm and those received in spring matings are capable of fertilizing her ovum, or egg. When the female goes into hibernation, an egg has started to develop in one ovary. This egg undergoes no change until spring, but about the time the female leaves hibernation it is shed from the ovary and fertilization follows.

In the little brown myotis only one egg matures, so consequently only a single young can be produced annually. The gestation period is between 50 and 60 days. Although some young may be born in May, most are born by mid-June.

While giving birth, the female usually hangs head up with the tail membrane cupped below to form a pouch. The mother licks the young and helps free the wings and legs, which are stuck tightly to the infant's body. The female may or may not sever the umbilical cord or eat the placenta.

Compared to other mammals, the weight of the young at birth is large in proportion to the weight of the female. The young of the little brown myotis weighs 1.4 g (1/20 oz.), which is about one-fourth to one-fifth as much as the mother. At birth the eyes are closed and the body naked. Wings, ears, legs, tail, and tail membrane are dark brown to black, but the body and head are pinkish. The eyes and ears may open within hours. The young may soon acquire an infestation of mites from the mother or roost site. After 2 days young flutter their wings; they are capable of flight after 21 days.

When the mother leaves the nursery colony to feed, she leaves her young hanging there. Upon returning, she is seemingly particular about which young she nurses and searches through the clusters of young for her own. Infant isolation calls have been shown to exhibit considerable individual variation, and it is believed that such "signatures" aid in female recognition and retrieval of their young. If disturbed, the mother may carry her young in flight as long as it is small enough to transport.

The young are able to fly alone by 3 to 4 weeks of age but are not weaned until about 6 weeks of age. By 2 months, they have attained adult size. In little brown myotis, many young females mate their first fall and produce offspring when a year old, but most adult males do not mature sexually until their second summer.

During periods of high population density, adults have been observed to shift foraging areas during the onset of juvenile flight. Such segregation of resources may serve to enhance juvenile survival.

Some Adverse Factors

The wild predators of bats are minks, raccoons, skunks, domestic cats, rats, hawks, owls, and snakes. Some of this predation occurs within the roost: owls and snakes have been observed preying upon bats as they exit from their roosts in the evening. Humans destroy them in buildings and caves. Robins and blue jays often harass bats roosting in trees and may be destructive to young hanging there.

In general, infestations with external parasites are not heavy but are more so in the heat of summer and on females whose health is run down toward the end of the rearing season. Mites, ticks, bat bugs, chiggers, and parasitic flies are the common external parasites. Flukes, tapeworms, and roundworms occur internally.

The greatest mortality probably occurs during the first few weeks of life when young fall from their roosting locations and are not recovered by the mother. In summer, several successive nights of rain may prevent bats from foraging by causing insects, their food, to hide in sheltered spots. Strong winds, storms, and hail may beat bats onto the ground. Their wings may occasionally become caught on barbed-wire fences or burdock burrs and, thus trapped, the bats die. In winter when the temperature becomes very low in small caves, bats that do not leave may be entombed in huge icicles. Mummified bodies are found occasionally in dry caves. Caves usually buffer hibernating bats from the effects of weather, but extreme changes can cause flooding or temperature decreases leading to mortality. A heavy loss of young bats may occur during their first winter or spring because they do not store fat as well as adults.

Importance

Bats are one of the few wild mammals people routinely have an opportunity to watch. During the spring and summer, these agile little fliers offer an interesting spectacle as they forage at twilight over the countryside or in cities. Here is a creature that may spend a large part of its life in underground caverns or in human dwellings. Here is a mammal different from all others, which flies and even may migrate from one

part of the country to another as the seasons change. The bat is an animal associated with fables, superstition, and other folklore. Its ultrasonic cries and the manner in which it guides its flight by echolocation are intriguing to modern thought about aerodynamics and communication. Bats are remarkable members of our wildlife community and should be appreciated instead of destroyed.

From an economic standpoint bats have a definite value. It has been estimated that the gray myotis population in Missouri eats approximately 363 metric tons (400 tons) of insects per year. Since gray myotis are only a part of the total bat population of the state, the number of insects consumed by all the bat species in the state is tremendous. It is significant that many of the insects fed upon by bats are crop pests. In certain roosting places, such as caverns that have housed nursery colonies of gray myotis for many years, bat droppings may accumulate to a depth of several feet. Such deposits, known as guano, have been used for commercial fertilizer.

In localities where bats contract rabies, their bites can be dangerous to humans, mammalian pets, and cattle. At present, bats are not considered important vectors of rabies in Missouri. However, people sometimes do not recognize the rabies potential in these animals, possibly because of their small size, and do not take proper precautions. In 2014, the Missouri Department of Health and Senior Services tested 1969 animals for rabies; 27 tested positive. Of the 720 bats tested for rabies, 16 were positive. The 16 positive bats were found in Jackson County in west-central Missouri, Cole County in central Missouri, McDonald County in southwest Missouri, St. Francois County in southeast Missouri, and St. Louis and Franklin Counties in east-central Missouri. There are only one or two human rabies cases reported each year in the United States, and most of those cases are linked to contact with a bat.

Bat oil has been recommended for rheumatism, but actually it is the massage and not the oil that helps. The flesh of bats is edible. A common superstition that bats deliberately get into women's hair has no basis in fact. It is interesting to note that the Chinese culture considers bats a good omen—a sign of happiness.

Most bats in Missouri do not harm humans but, if handled, any Missouri species is capable of inflicting a sharp bite. Almost all occurrences in the United States of bats biting humans happen when a person handles a live bat without proper precautions (thick leather or similar gloves are necessary) and training.

Conservation and Management

Bats are protected by both state and federal laws. Those that are in a precarious status are listed under both the Federal and State Endangered Species Acts.

Where bats roost in and around inhabited buildings, they are often considered a nuisance. This is partly because of the odor or untidiness of their droppings, but mostly because people fear them. Where they are not desired in buildings, there are exclusion methods that do not require a person to come into contact with or harm the bats. Sometimes it is possible to build structures that bats will use for roosting sites instead of human habitations. However, little brown myotis are not likely to switch to such alternative structures unless they are excluded from their present roost. If you are experiencing problems with bats, contact a wildlife professional for advice, assistance, regulations, or special conditions for handling these animals.

The deliberate use of insecticides to destroy colonies or to make roosting sites undesirable is not recommended or humane. Likewise, the indiscriminate use of many questionable pesticides should be curtailed because the chain of events created by these pesticides plays havoc with many kinds of wildlife, including bats. Insects living in sprayed soil may acquire small amounts of the chemicals in their bodies, and when they transform to adults and are eaten by bats, the chemicals are released in the bats' bodies. Most of these chemicals are fat soluble and are deposited in the body fat of bats. As the fat is used up during hibernation, the chemicals are left to poison the bats when they emerge in the spring. These chemicals are also concentrated in the milk of mothers and affect their nursing young. Juvenile bats are one and one-half times more sensitive to insecticides than adults.

Caves are the only naturally occurring phenomenon that approximate constant temperature and humidity conditions and are sufficiently large to support immense aggregations of these highly selective animals. Because there are so few caves, their protection is important for bat survival. Commercialization of caves must be restricted; abuse of caves by spelunkers must be discouraged; harassment of hibernating or nursery colonies whether by casual intruders or biologists must be controlled. Hibernating bats are also sensitive to nontactile stimuli and will arouse and fly even following human visits. Such arousal can increase mortality due to premature depletion of fat reserves. It is recommended that caves that are crucial to the survival of a species be gated so access can be regulated by a responsible agency and limited to times of the year when disturbance to the bats is least.

Other changes in the environment that affect their food supply and homes, such as pollution of the streams over which bats forage, siltation of streams from mine effluents, and deforestation of their foraging habitats, specifically near cave entrances, should all be reduced as much as possible. In the case of some species, we probably have gone too far to save them; we must use foresight to save the remaining species of these intriguing animals.

SELECTED REFERENCES

See also discussion of this species in general references, page 23.

Adams, R. A. 1997. Onset of volancy and foraging patterns of juvenile little brown bats, *Myotis lucifugus*. *Journal of Mammalogy* 78:239–246.

Anthony, E. L. P., and T. H. Kunz. 1977. Feeding strategies of the little brown bat, *Myotis lucifugus*, in southern New Hampshire. *Ecology* 58:775–786.

Buchler, E. R. 1976. Prey selection by *Myotis lucifugus* (Chiroptera: Vespertilionidae). *American Naturalist* 110:619–628.

———. 1980. Evidence for the use of a scent post by *Myotis lucifugus*. *Journal of Mammalogy* 61:525–528.

Cagle, F. R., and E. L. Cockrum. 1943. Notes on a summer colony of *Myotis lucifugus lucifugus*. *Journal of Mammalogy* 24:474–492.

Cockrum, E. L. 1956. Homing, movements, and longevity of bats. *Journal of Mammalogy* 37:48–57.

Cope, J. B., and S. R. Humphrey. 1977. Spring and autumn swarming behavior in the Indiana bat, *Myotis sodalis*. *Journal of Mammalogy* 58:93–95.

Davis, W. B., and H. B. Hitchcock. 1995. A new longevity record for the bat *Myotis lucifugus*. *Bat Research News* 36:6.

Fenton, M. B., and R. M. R. Barclay. 1980. *Myotis lucifugus. Mammalian Species* 142. 8 pp.
Humphrey, S. R., and J. B. Cope. 1976. Population ecology of the little brown bat, *Myotis lucifugus*, in Indiana and north central Kentucky. Special Publication, American Society of Mammalogists, 4. 81 pp.
Kunz, T. H., and J. D. Reichard. 2010. *Status review of the little brown myotis* (Myotis lucifugus) *and determination that immediate listing under the Endangered Species Act is scientifically and legally warranted.* Report in collaboration with Friends of Blackwater Canyon, Wildlife Advocacy Project, Bat Conservation International, Center for Biological Diversity, and Meyer Glitzenstein and Crystal. 30 pp.
Miller, R. E. 1939. The reproductive cycle in male bats of the species *Myotis lucifugus lucifugus* and *Myotis grisescens. Journal of Morphology* 64:267–295.
Missouri Department of Health and Senior Services. 2014. Description of the incidence of confirmed rabies in laboratory exams: December 2014. Office of Surveillance Diseases/Conditions Annual Reports, Jefferson City. 2 pp.
Mohr, C. E. 1933. Observations on the young of cave-dwelling bats. *Journal of Mammalogy* 14:49–53.
Neilson, A. L., and M. B. Fenton. 1994. Responses of little brown *Myotis* to exclusion and to bat houses. *Wildlife Society Bulletin* 22:8–14.
Norquay, K. J. O., F. Martinez-Nuñez, J. E. Dubois, K. M. Monson, and C. K. R. Willis. 2013. Long-distance movements of little brown bats (*Myotis lucifugus*). *Journal of Mammalogy* 94:506–515.
Stegeman, L. C. 1954a. Notes on the development of the little brown bat, *Myotis lucifugus lucifugus. Journal of Mammalogy* 35:432–433.
———. 1954b. Variation in a colony of little brown bats. *Journal of Mammalogy* 35:111–113.
———. 1956. Tooth development and wear in *Myotis. Journal of Mammalogy* 37:58–63.
Thomson, C. E., M. B. Fenton, and R. M. R. Barclay. 1985. The role of infant isolation calls in mother-infant reunions in the little brown bat, *Myotis lucifugus* (Chiroptera: Vespertilionidae). *Canadian Journal of Zoology* 63:1982–1988.
Wimsatt, W. A. 1944. Growth of the ovarian follicle and ovulation in *Myotis lucifugus lucifugus. American Journal of Anatomy* 74:129–173.

Gray Myotis (*Myotis grisescens*)

Name

The first part of the scientific name, *Myotis*, is from two Greek words and means "mouse ear." The last part, *grisescens*, is from the Latin word *griseus* and means "becoming gray"; this and the common name describe the grayish brown color. This species is also known as the gray bat.

Description (Plates 9 and 10)

The gray myotis is very similar to the other mouse-eared bats in Missouri but can be identified by the following characteristics. The color is usually grayer than that of the other mouse-eared bats and the bases of the hairs of the back are not dark. It is usually the largest of these species. This is the only mouse-eared bat in which the wing is attached at the ankle joint instead of at the base of the toes. The calcar is not keeled. The hind feet are large and, because of the wing attachment at the ankle, appear conspicuous. The hairs on the toes are medium long and sparse. Each claw of the hind foot has a prominent notch.

Color. The upperparts range in color from a uniform dark grayish brown to russet, and the hairs, when parted, are seen to be nearly uniform in color. The underparts are paler than the back, and the hairs are dark gray at the base with whitish tips. Typically gray myotis also have small patches of orange under their chin. The fur has a velvety texture and is rather sparse and short. Rarely partial albino gray myotis occur.

Measurements

Total length	79–95 mm	3⅛–3¾ in.
Tail	31–44 mm	1¼–1¾ in.
Hind foot	7–11 mm	5/16–7/16 in.
Ear	12–15 mm	½–⅝ in.
Skull length	15 mm	⅝ in.
Skull width	9 mm	⅜ in.
Weight	5–14 g	⅕–½ oz.

Teeth and skull. The teeth are similar to those of the little brown myotis. Because the differences between the skulls of mouse-eared bats are so slight, all skulls of this group should be compared with specimens of known identity before attempting identification. The skull of the gray myotis can usually be distinguished from that of the little brown myotis by the more obvious crest on the top of the skull.

Longevity. Resulting from high juvenile mortality, the average life span of the gray myotis is only 5 years. However, individuals have been known to live up to 18 years.

Distribution and Abundance

The gray myotis once flourished in limestone caves across the southeastern United States, but due to human disturbance, populations plummeted during the first half of the 20th century. Today they have a much reduced range, occurring in parts of 14 states. Because of a rapid decline in numbers and habitat, this species

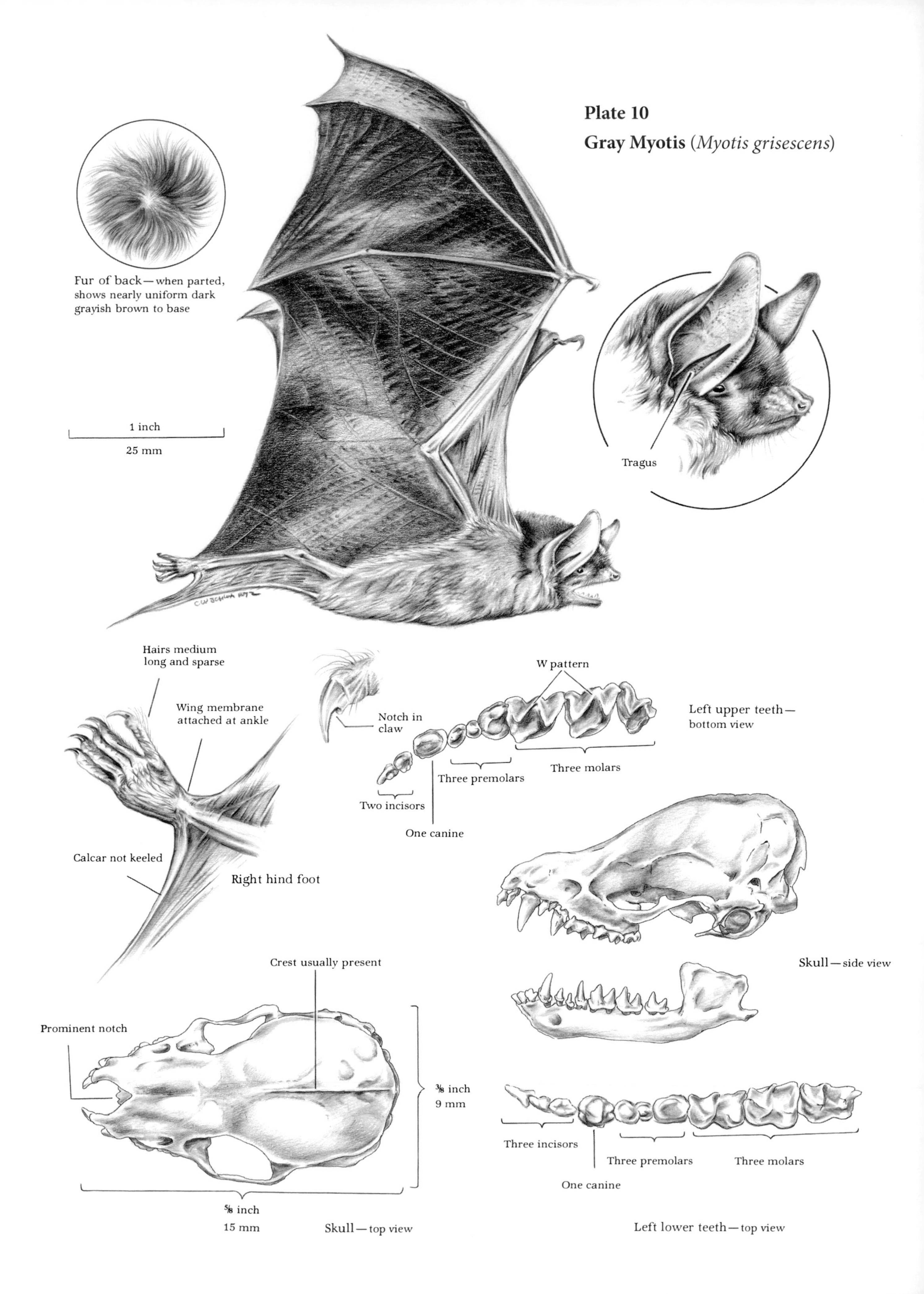
Plate 10
Gray Myotis (Myotis grisescens)
Fur of back—when parted, shows nearly uniform dark grayish brown to base
1 inch
25 mm
Tragus
Hairs medium long and sparse
Wing membrane attached at ankle
Notch in claw
W pattern
Left upper teeth—bottom view
Three premolars
Three molars
Two incisors
One canine
Calcar not keeled
Right hind foot
Skull—side view
Crest usually present
Prominent notch
⅜ inch
9 mm
Three incisors
Three premolars
Three molars
One canine
⅝ inch
15 mm
Skull—top view
Left lower teeth—top view

was placed on the Endangered Species list by the U.S. Fish and Wildlife Service in 1976. It is likewise listed as endangered in Missouri, and thus protected by federal and Missouri state laws. Due in part to protective measures enforced since the listing of this species, the population appears to be recovering.

In Missouri, the gray myotis lives in the Ozark Highlands, the Central Dissected Till Plains, and northeastern Missouri along the Mississippi River. In 1980 Missouri was estimated to have 515,000 gray myotis; however, a more recent estimate in 2006 was 745,000. Recently, they have been discovered utilizing abandoned limestone quarries throughout the state. The gray myotis is listed as a vulnerable species of conservation concern in Missouri and across its North American range.

Habitat and Home

The gray myotis is the only species currently prevalent in Missouri that inhabits caves all year. While some individuals of this species may use the same cave year-round, the bulk of the Missouri population spends the winter in a few southern Missouri caves and in spring moves to caves as much as 322 km (200 mi.) or more away. In these latter caves, females use areas that trap warm air, such as small chambers, high domes, or pockets, for their nursery sites.

In general, caves used for wintering quarters usually have a vertical opening or shaft; this may reduce predation and human disturbance. These vertical shafts also create lower temperatures. Those used for nursery colonies have large openings that may make flying easier if the adults need to carry the young at any time.

Sometimes gray myotis use large, man-made tunnels, such as huge storm sewers, as their summer quarters. There is a historic record of a large nursery colony in an old stone barn.

Habits

In fall, females are the first to migrate to their winter quarters. They are followed by yearlings and, later, adult males. They may travel hundreds of miles to reach the few caves suitable for hibernation. No significant population movement occurs after hibernation begins.

Gray myotis hang from cave walls in large masses composed of both sexes. These are sometimes several bats deep and may cover an area up to 186 sq m (2,000

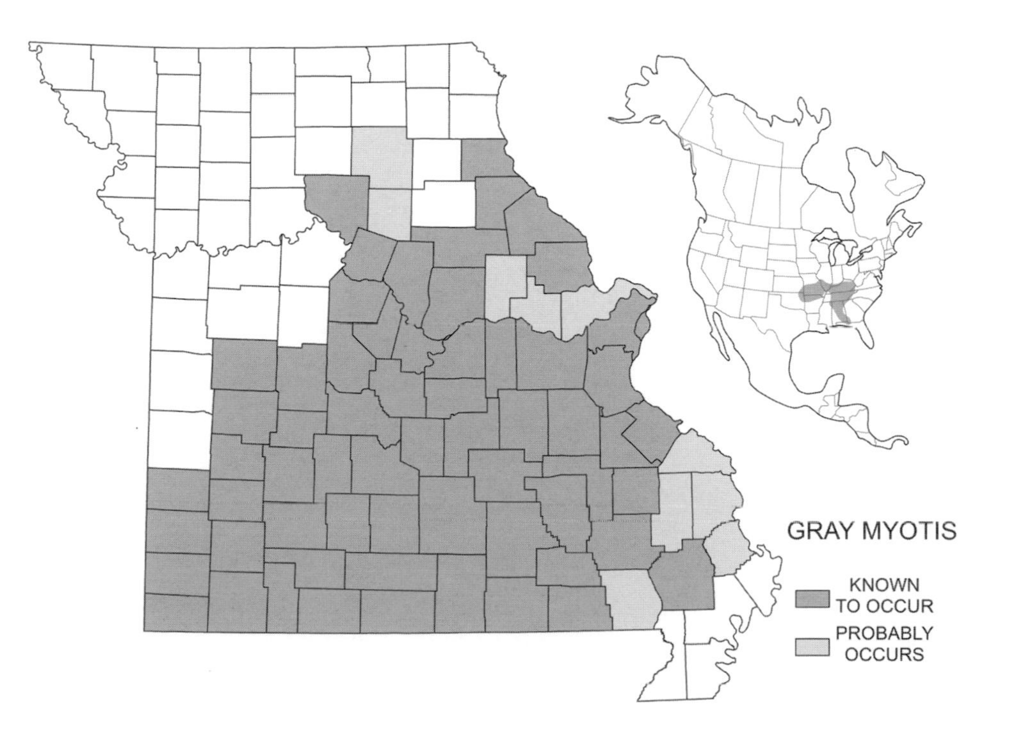

sq. ft.) at a density of 1,830 bats per sq m (170 bats per sq. ft.).

Emergence in spring is by females first. Yearlings are next, followed by males. The movement is progressive, groups of bats spending a few days to a week in one cave, then going on to another. Adult females select one cave as a maternity site while adult males and yearlings use other caves. If the two sexes live in the same cave they occupy different portions.

The young are left to cluster compactly in the nursery while the females forage for food. In general, gray myotis prefer warm maternity caves close to large rivers and reservoirs over which they feed. The close proximity of good caves and good feeding areas is advantageous to new-flying young.

Foods

In early spring, female gray myotis feed primarily on caddisflies and moths. As aquatic insects become available, like stoneflies and mayflies, they are predominant sources of food. Beetles are important during summer—mainly after maternity colonies disperse, when flies and moths are also taken in appreciable numbers. Juvenile gray myotis show little preference for specific insects at first, apparently taking what is most available. Later in summer they feed largely on beetles, especially the Asiatic oak weevil, which is most abundant in forested cliffs along rivers.

Reproduction

Breeding takes place in the fall and probably in winter and spring as in the little brown myotis. One young per female per year is born in May or June. The young are raised in large maternity colonies, and these colonies disperse in July when the young become volant.

Although young females may mate in their first fall, they are not mature enough to produce young their first spring; by contrast, young female little brown myotis produce offspring their first spring. Young male gray myotis become sexually mature in their second year, as do young male little brown myotis.

Some Adverse Factors; Importance; and Conservation and Management

The gray myotis, being a cave dweller during both summer and winter, has been subjected to increased disturbance due to the popularity of cave exploration and cave commercialization. If disturbed, pregnant females may abort their young, and very young bats may drop off the cave wall to the floor or into the stream. Some may be retrieved by the mother, but others may die. Because females have only one young a year, severe disturbance to a nursery colony could result in failure of the entire colony and affect not only one year's production but that of subsequent years as well.

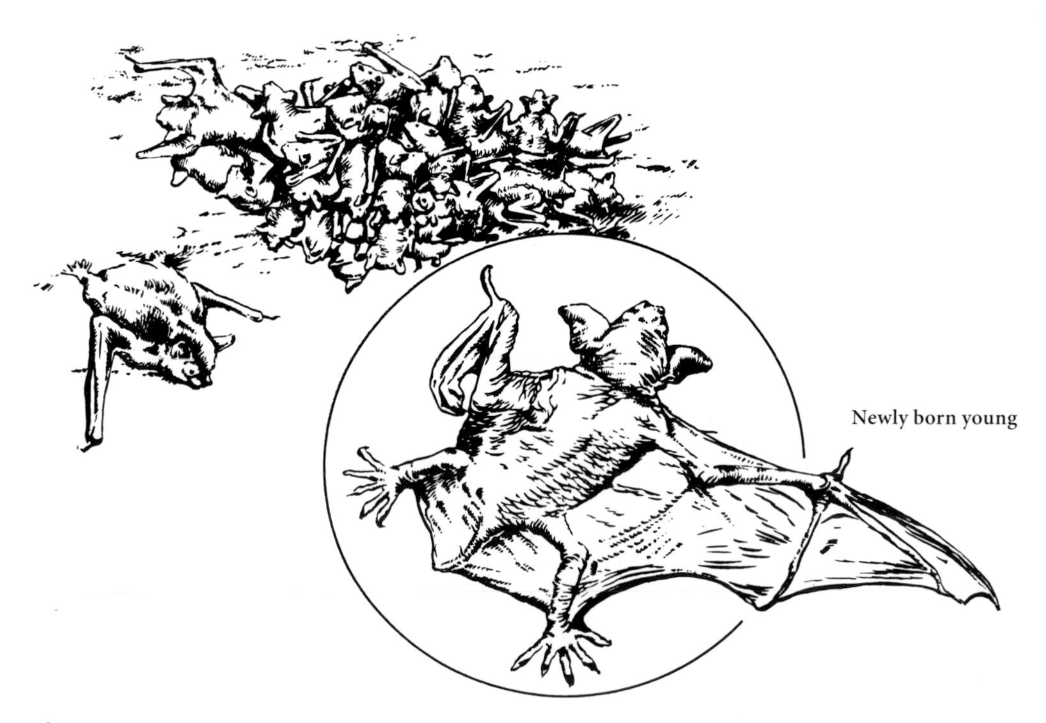

Nursery colony of young gray myotis hanging from roof of cavern

During hibernation, harassment may cause undue activity of the bats, using up stored fat the animals need for survival. This may lead to death during winter. Also, since the gray myotis lives in so few caves over a restricted range, interference in any one of these caves has a profound effect on this species' total population.

Many important caves throughout the gray myotis's range have been protected to halt this harassment and to protect its habitat. There have been increased efforts to gate cave entrances and to purchase land where critical nursery and wintering caves exist. In the face of such threats as white-nose syndrome, this species is still threatened, especially during winter. These protections must be maintained and expanded where needed, allowing this species to continue its recovery.

Pesticide contamination has also been implicated in the deaths of numerous bats in multiple summering caves in Missouri. It appears that pesticide residue contaminated the insect food base within the foraging habitat of the gray myotis. Although the use of the specific pesticides implicated in these mortalities has been curtailed, it is critical that future contamination of the gray myotis's food resources be avoided.

SELECTED REFERENCES

See also discussion of this species in general references, page 23.

Best, T. L., B. A. Milam, T. D. Haas, W. S. Cvilikas, and L. R. Saidak. 1997. Variation in diet of the gray bat (*Myotis grisescens*). *Journal of Mammalogy* 78:569–583.

Brack, V., Jr., and R. K. LaVal. 2006. Diet of the gray myotis (*Myotis grisescens*): Variability and consistency, opportunism, and selectivity. *Journal of Mammalogy* 87:7–18.

Clawson, R. L., and D. R. Clark Jr. 1989. Pesticide contamination of endangered gray bats and their food base in Boone County, Missouri, 1982. *Bulletin of Environmental Contamination and Toxicology* 42:431–437.

Decher, J., and J. R. Choate. 1995. *Myotis grisescens*. *Mammalian Species* 510. 7 pp.

Elder, W. H., and W. J. Gunier. 1978. Sex ratios and seasonal movements of gray bats (*Myotis grisescens*) in southwestern Missouri and adjacent states. *American Midland Naturalist* 99:463–472.

———. 1981. Dynamics of a gray bat population (*Myotis grisescens*) in Missouri. *American Midland Naturalist* 105:193–195.

Elliott, W. R., 2007. *Gray and Indiana bat population trends in Missouri*. National Cave and Karst Management Symposium, pp. 46–61.

Gunier, W. J., and W. H. Elder. 1971. Experimental homing of gray bats to a maternity colony in a Missouri barn. *American Midland Naturalist* 86:502–506.

Lacki, M. J., L. S. Burford, and J. O. Whitaker Jr. 1995. Food habits of gray bats in Kentucky. *Journal of Mammalogy* 76:1256–1259.

LaVal, R. K., R. L. Clawson, M. L. LaVal, and W. Caire. 1977. Foraging behavior and nocturnal patterns of Missouri bats, with emphasis on the endangered species *Myotis grisescens* and *Myotis sodalis*. *Journal of Mammalogy* 58:592–599.

Miller, R. E. 1939. The reproductive cycle in male bats of the species *Myotis lucifugus lucifugus* and *Myotis grisescens*. *Journal of Morphology* 64:267–295.

Tuttle, M. D. 1976. Population ecology of the gray bat (*Myotis grisescens*): Philopatry, timing and patterns of movement, weight loss during migration, and seasonal adaptive strategies. *Occasional Papers of the Museum of Natural History, University of Kansas, Lawrence* 54:1–38.

———. 1979. Status, causes of decline, and management of endangered gray bats. *Journal of Wildlife Management* 43:1–17.

Southeastern Myotis
(*Myotis austroriparius*)

Name

The first part of the scientific name, *Myotis*, is from two Greek words and means "mouse ear." The second part, *austroriparius*, is from the Latin words *austro*, meaning "southern," and *riparius*, which means "frequenting edges of streams"; these and the common name describe the location and habitat of this bat. This species is also known as the southeastern bat.

Description

The southeastern myotis closely resembles the other mouse-eared bats found in Missouri, especially the little brown myotis and the gray myotis. Like the little brown myotis, it lacks a keel on the calcar and the hair on the toes is long, extending beyond the tips of the claws. The southeastern myotis can be distinguished from the little brown myotis by its shorter and more woolly hair with tips that do not contrast with the base on the dorsal surface, and from the gray myotis by the absence of a notch in the claw and a wing attachment to the foot. The presence of a low sagittal crest can be felt under the skin in this bat. The baculum is smaller than in closely related forms.

Color. The most common color on the upperparts ranges from dull gray to gray-brown, though some

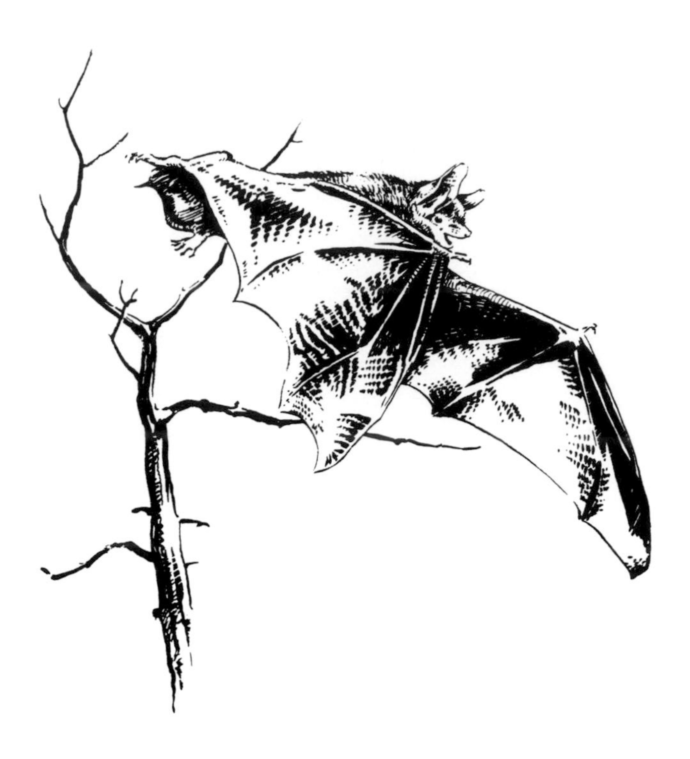

individuals may even be of a bright orange-brown to orange-red coloration. When parted, the hairs appear to be nearly uniform in color from tip to base (similar to that of the gray myotis; see plate 9). However, the fur on the ventral side has whitish tips with highly contrasting dark brown coloration at the base.

Measurements

Total length	77–97 mm	3–3⅞ in.
Tail	26–44 mm	1–1¾ in.
Hind foot	7–12.5 mm	¼–½ in.
Ear	9–16 mm	5⁄16–⅝ in.
Skull length	13.5–15 mm	½–⅝ in.
Weight	5–12 g	3⁄16–7⁄16 oz.

Females are slightly larger and heavier than males.

Teeth and skull. The teeth are similar to those of the little brown myotis. Because the differences between skulls of mouse-eared bats are so slight, all skulls of this group should be compared with specimens of known identity before attempting identification. The skull of the southeastern myotis is most similar to that of the gray myotis, and it can usually be distinguished from that of the little brown myotis by the presence of a more obvious sagittal crest on the top of the skull.

Distribution and Abundance

The southeastern myotis occurs in 15 states in the southeastern United States. The largest populations have been found in the northern half of Florida, but like so many of the other cave-dwelling bats, its numbers are declining.

In Missouri, the southeastern myotis is listed as a critically imperiled species of conservation concern and they are classified as vulnerable across their North American range. It is not known to hibernate within the state but is found in small numbers in the Mississippi River Alluvial Basin in late spring, summer, and early fall.

Habitat and Home

Caves are utilized as the primary roosting location for southeastern myotis. However, these bats will generally use other well-protected sites, such as manmade structures (bridges, mines, buildings, and storm sewers) or tree hollows, where caves are not available. These bats are often associated with bottomland hardwood forests near or in floodplains of major rivers (such as the Mississippi River in Missouri) as well as cypress and tupelo swamps. Many of their roosting sites are located in hollow trees over water, and this may lead to high juvenile mortality.

Habits

In the fall, maternity sites are abandoned around October. Southeastern myotis migrate minimal distances, from 8 to 24 km (5 to 15 mi.), moving locally to well-protected wintering sites. Within the southern reaches of the range, they do not really hibernate at all. In the north, including possibly Missouri, they hibernate in compact clusters in caves and mines. These wintering sites are then gradually abandoned from February through March.

Foods

The southeastern myotis typically feeds by flying low over water. They appear to feed predominately on beetles, moths, flies, and some mosquitoes.

Reproduction

Aside from a couple of locations of maternity roosts, nothing is known about the reproduction of the southeastern myotis in the northern reaches of their range.

In southern populations, breeding occurs from February through April, and maternity colonies begin forming in March. Usually two young are born into a pocket formed by the curling of the mother's tail, and the mother tears the amnion and bites the umbilical cord in two.

The eyes and ears of the young are closed at birth, yet at this time the milk teeth have erupted. The young become volant at approximately 5–6 weeks of age.

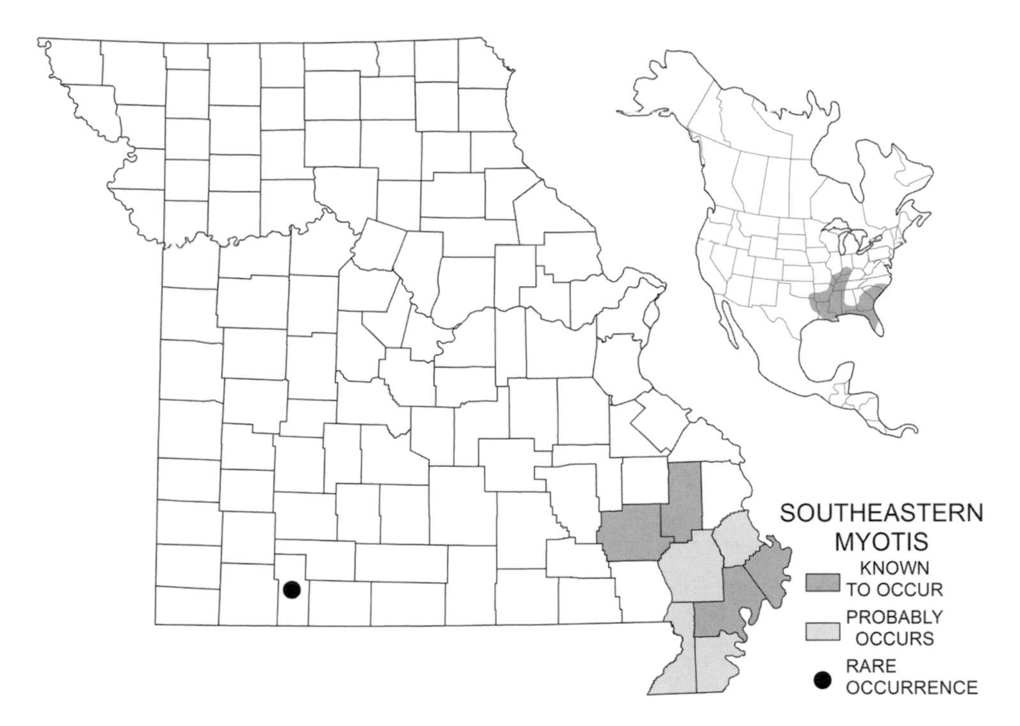

Some Adverse Factors; Importance; and Conservation and Management

These subjects have been discussed for bats in general at the beginning of the chapter and in the account of the little brown myotis.

As a result of high roost fidelity, southeastern myotis are particularly vulnerable to human destruction of roosting sites and increased exposure to white-nose syndrome. The disturbance of habitat surrounding roosts and the closure of cave entrances have previously resulted in the abandonment of certain large roost sites. Conservation of this species depends on the identification and protection of both maternity roosts and hibernacula. This is especially critical in the northern portions of the range where the status and life history of these bats are so poorly understood.

SELECTED REFERENCES

See also discussion of this species in general references, page 23.

Davis, W. H., and C. L. Rippy. 1968. Distribution of *Myotis lucifugus* and *Myotis austroriparius* in the southeastern United States. *Journal of Mammalogy* 49:113–117.

Foster, G. W., S. R. Humphrey, and P. P. Humphrey. 1978. Survival rate of young southeastern brown bats, *Myotis austroriparius*, in Florida. *Journal of Mammalogy* 59:299–304.

Gore, J. A., and J. A. Hovis. 1994. *Southeastern* Myotis *maternity cave survey. Final performance report January 1, 1991, to July 31, 1992.* Nongame Wildlife Program. Florida Game and Freshwater Fish Commission, Tallahassee. 33 pp.

Jones, C., and R. W. Manning. 1989. *Myotis austroriparius. Mammalian Species* 332. 3 pp.

Lowery, G. H., Jr. 1974. *The mammals of Louisiana and its adjacent waters.* Louisiana State University Press, Baton Rouge. 565 pp.

Mumford, R. E., and J. O. Whitaker Jr. 1982. *Mammals of Indiana.* Indiana University Press, Bloomington. 537 pp.

Rice, D. W. 1957. Life history and ecology of *Myotis austroriparius* in Florida. *Journal of Mammalogy* 38:15–32.

Rippy, C. L. 1965. The baculum in *Myotis sodalis* and *Myotis austroriparius austroriparius. Transactions of the Kentucky Academy of Science* 26:19–21.

Sherman, H. B. 1930. Birth of the young of *Myotis austroriparius. Journal of Mammalogy* 11:495–503.

Zinn, T. L., and S. R. Humphrey. 1981. Seasonal food resources and prey selection of the southeastern brown bat (*Myotis austroriparius*) in Florida. *Florida Scientist* 44:81–90.

Northern Long-eared Myotis (*Myotis septentrionalis*)

Name

The first part of the scientific name, *Myotis*, is from two Greek words and means "mouse ear." The last part, *septentrionalis*, is Latin for "northern." Formerly this species was considered a subspecies of *Myotis keenii*. In 1979 these two taxa were recognized as separate species based on distribution (Keen's bat being restricted to the northwest) and morphology. This species is also known as the northern long-eared bat, northern myotis, and northern bat (which is more commonly used to refer to *Eptesicus nilssonii*, which lives in Eurasia).

Description (Plates 9 and 11)

The northern long-eared myotis is very similar to the other mouse-eared bats in Missouri. It is identified by its general coloration and the following combination of characteristics. The wing is attached along the side of the foot and reaches the base of the toes. When laid forward, the ears extend up to 4.8 mm (3/16 in.) beyond the nostrils. The tragus is long (more than one-half the length of ear), narrow, and pointed. The fourth and fifth fingers are nearly equal in length. The calcar can bear a slight keel. The hairs on the toes are medium long and sparse.

Color. The northern long-eared myotis is light reddish brown above and buffy gray below. It is slightly lighter in color than the little brown myotis, and the hairs are shorter, less glossy, and more crinkled. When the hairs are parted, the bases are seen to be blackish while the outer portions are reddish to yellowish brown.

Measurements

Total length	76–95 mm	3–3¾ in.
Tail	34–41 mm	1⅜–1⅝ in.
Hind foot	7–11 mm	5/16–7/16 in.
Ear	15–19 mm	⅝–¾ in.
Skull length	15 mm	⅝ in.
Skull width	9 mm	⅜ in.
Weight	2–8 g	1/10–2/7 oz.

Females are generally larger and heavier than males.

Teeth and skull. The teeth are similar to those of the little brown myotis. Because the differences between the skulls of mouse-eared bats are so slight, all skulls of this group should be compared with specimens of known identity before attempting identification.

Distribution and Abundance

The North American range of the northern long-eared myotis is widespread, extending from central Canada throughout most of the eastern half of the United States, but over this range it is irregularly distributed. It is more common in its eastern range than in the western extreme.

This species is common and has been found in most Missouri counties. However, due to white-nose syndrome, it has become rarer. It is listed as a vulnerable species of conservation concern in Missouri but it is imperiled across its North American range. In 2015, it was listed as a threatened species under the federal Endangered Species Act.

Habitat and Home

The northern long-eared myotis is primarily a cave dweller in winter; in summer females seemingly prefer crevices and hollows of trees but may also be found in man-made structures such as barns. Females form

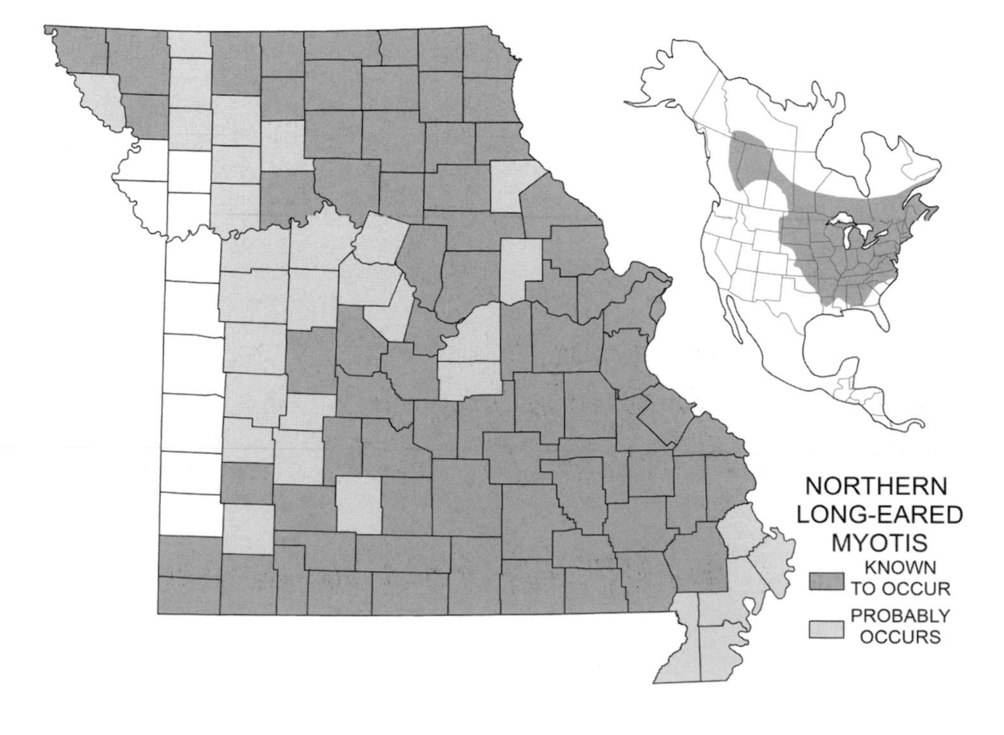

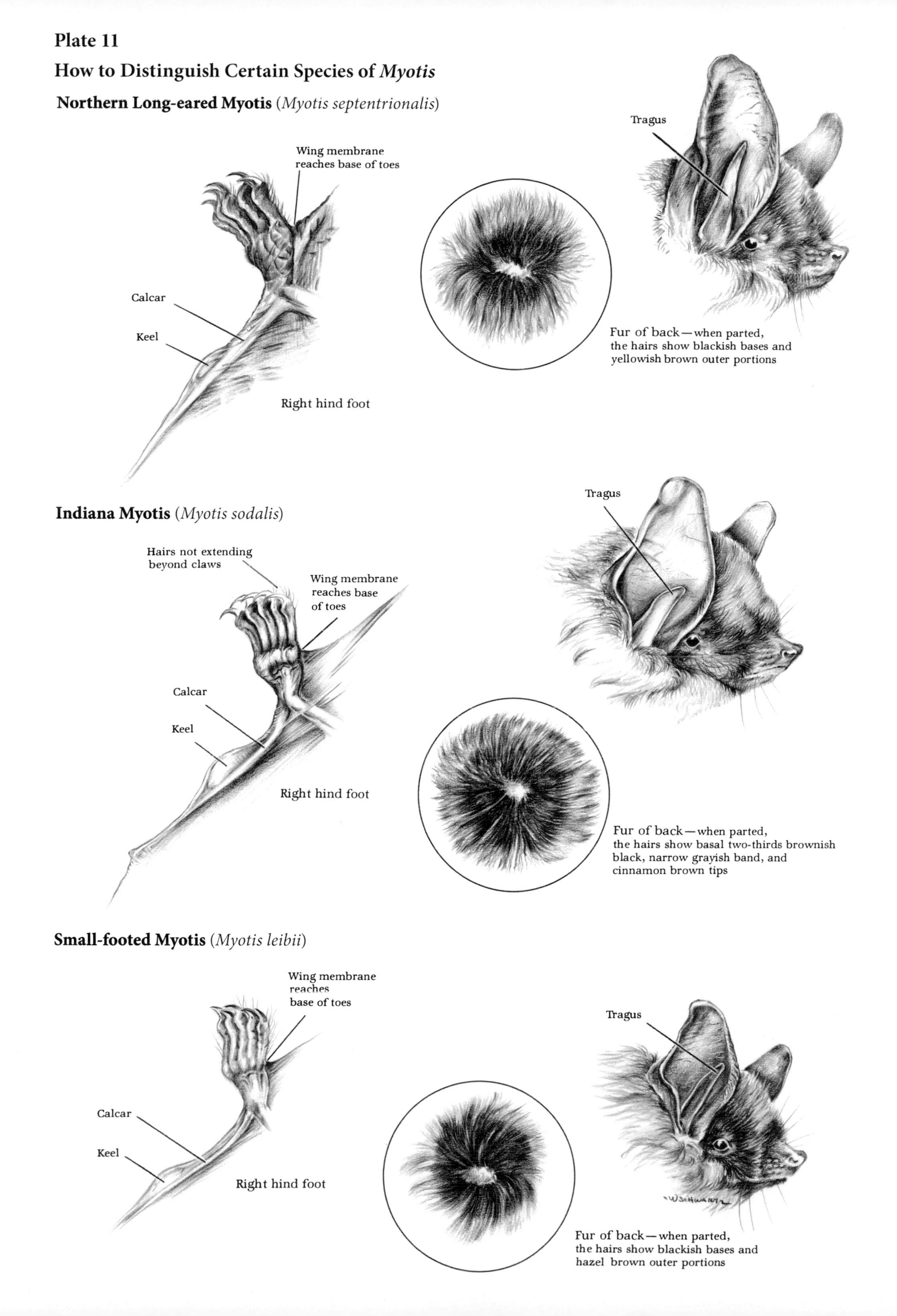
Plate 11
How to Distinguish Certain Species of Myotis
Northern Long-eared Myotis (Myotis septentrionalis)
Wing membrane reaches base of toes
Calcar
Keel
Right hind foot
Tragus
Fur of back—when parted, the hairs show blackish bases and yellowish brown outer portions
Indiana Myotis (Myotis sodalis)
Hairs not extending beyond claws
Wing membrane reaches base of toes
Calcar
Keel
Right hind foot
Tragus
Fur of back—when parted, the hairs show basal two-thirds brownish black, narrow grayish band, and cinnamon brown tips
Small-footed Myotis (Myotis leibii)
Wing membrane reaches base of toes
Calcar
Keel
Right hind foot
Tragus
Fur of back—when parted, the hairs show blackish bases and hazel brown outer portions

small summer maternity colonies involving up to several dozen individuals. Switching of summer roosts occurs frequently within a local area. Males sometimes use caves in late spring and summer.

Life History

Prior to hibernation, this species accumulates body fat. This is also the period of copulation, as mating presumably occurs while the bats swarm outside the wintering caves from August through September. In general, these bats hibernate from October to March. They show a tendency to return to the same caves each year. Individuals roost in deep crevices, alone or with another bat or two, and are thus often overlooked. Here temperatures are higher than in exposed parts of caves. They are often found hibernating in caves with large numbers of other bat species.

The northern long-eared myotis forages in summer among trees on hillsides and ridges, feeding below the crowns of trees. The diet includes moths, beetles, flies, caddisflies, and spiders. The longer tail and larger wing membrane of the northern long-eared myotis are indicators that this bat catches insects while on the wing. Feeding on flying insects as well as gleaning allows for a broader diet.

There is little information on breeding of this bat. As with the other *Myotis* species, the northern long-eared myotis uses delayed fertilization. One young per female is born in late May or June in Missouri. The nursery colonies are small.

Some Adverse Factors; Importance; and Conservation and Management

These subjects have been discussed for bats in general at the beginning of the chapter and in the account of the little brown myotis.

Today, white-nose syndrome is the predominant threat to this bat, especially in the northeastern portion of its range, where at many hibernation sites the species has declined by up to 99 percent from levels prior to the advent of white-nose syndrome.

SELECTED REFERENCES

See also discussion of this species in general references, page 23.

Caceres, M. C., and R. M. R. Barclay. 2000. *Myotis septentrionalis. Mammalian Species* 634. 4 pp.

Caire, W., R. K. LaVal, M. L. LaVal, and R. Clawson. 1979. Notes on the ecology of *Myotis keenii* (Chiroptera, Vespertilionidae) in eastern Missouri. *American Midland Naturalist* 102:404–407.

Carter, T. C., and G. A. Feldhamer. 2005. Roost tree use by maternity colonies of Indiana bats and northern long-eared bats in southern Illinois. *Forest Ecology and Management*, 219:259–268.

Faure, P. A., J. H. Fullard, and J. W. Dawson. 1993. The gleaning attacks of the northern long-eared bat, *Myotis septentrionalis*, are relatively inaudible to moths. *Journal of Experimental Biology* 178:173–189.

Foster, R. W., and A. Kurta. 1999. Roosting ecology of the northern bat (*Myotis septentrionalis*) and comparisons with the endangered Indiana bat (*Myotis sodalis*). *Journal of Mammalogy* 80:659–672.

Johnson, J. B., W. M. Ford, and J. W. Edwards. 2012. Roost networks of northern myotis in a managed landscape. *Forest Ecology and Management* 266:223–231.

Miller, L. A., and A. E. Treat. 1993. Field recordings of echolocation and social signals from the gleaning bat, *Myotis septentrionalis. Bioacoustics* 5:67–87.

Perry, R. W., and R. E. Thill. 2007. Roost selection by male and female northern long-eared bats in a pine-dominated landscape. *Forest Ecology and Management* 247:220–226.

Timpone, J. C., J. G. Boyles, K. L. Murray, D. P. Aubrey, and L. W. Robbins. 2010. Overlap in roosting habits of Indiana bats and northern bats. *American Midland Naturalist* 163:115–123.

U.S. Fish and Wildlife Service. 2015. Endangered and threatened wildlife and plants: Threatened species status for the northern long-eared bat with 4(d) rule. *Federal Register* 80:17974–18033.

van Zyll de Jong, C. G. 1979. Distribution and systematic relationships of long-eared *Myotis* in western Canada. *Canadian Journal of Zoology* 57:987–994.

Williams, D. F., and J. S. Findley. 1979. Sexual size dimorphism in vespertilionid bats. *American Midland Naturalist* 102:113–126.

Indiana Myotis (*Myotis sodalis*)

Name

The first part of the scientific name, *Myotis*, is from two Greek words and means "mouse ear." The second part, *sodalis*, is the Latin word for "companion" and refers to this bat's habit of hibernating in great masses. The common name is for the state of Indiana, from which the first specimen was described. This species is also known as the Indiana bat.

Description (Plates 9 and 11)

The Indiana myotis is very similar to the other mouse-eared bats in Missouri. It is identified by the general coloration and the following combination of characteristics. The wing is attached along the side of the foot and reaches the base of the toes. When laid forward, the ears do not project more than 1.6 mm (1/16 in.) beyond the nostrils. The calcar normally has a small keel. The hind foot is more than 6.3 mm (¼ in.) in length. Hairs on the toes are short and sparse, especially when compared to the little brown myotis.

Color. Above, the fur is a dull grayish brown to nearly black; when parted, it exhibits a faint three-colored pattern. The basal two-thirds of each hair is brownish black, followed by a narrow grayish band and a cinnamon brown tip. Below, the fur has a pinkish white cast coming from the gray bases, the grayish white tips, and the faint wash of cinnamon brown. The fur is fine and fluffy and the hairs tend to stand out from each other.

Measurements

Total length	69–98 mm	2¾–3⅞ in.
Tail	28–50 mm	1⅛–2 in.
Hind foot	7 mm	5/16 in.
Ear	9–15 mm	⅜–⅝ in.
Skull length	15 mm	⅝ in.
Skull width	9 mm	⅜ in.
Weight	2–10 g	1/10–⅓ oz

Teeth and skull. The teeth are similar to those of the little brown myotis. Because the differences between the skulls of mouse-eared bats are so slight, all skulls of this group should be compared with specimens of known identity before attempting identification.

Longevity. The maximum reported longevity of the Indiana myotis was for a bat banded as an adult and recovered 20 years later. Several bats aged 13–14 have also been recovered.

Distribution and Abundance

The Indiana myotis lives primarily in the eastern and midwestern United States, but it also occurs in parts of the southeast. Because the population has decreased drastically—as much as 58 percent during the period 1960–1991—and occupied so few caves, the Indiana myotis was placed on the Endangered Species List by the U.S. Fish and Wildlife Service in 1973. Attention is being directed toward saving both the bat and its cave habitat. Today, the spread of white-nose syndrome is impacting nearly all populations of Indiana myotis and has resulted in a 10.3 percent annual decline between 2006 and 2009.

The Indiana myotis occurs throughout most of Missouri. In this range it is widely dispersed in summer but concentrated in winter. In 2005 approximately 82 percent of the known population hibernated in 22 sites across their Missouri range. White-nose syndrome has been found in Missouri and its spread may result in drastic decreases of this species. The Indiana myotis is listed as a critically imperiled and state endangered species of conservation concern in Missouri, and it is listed as imperiled and federally endangered across its North American range.

Habitat and Home

Indiana myotis spend the winter in limestone caves, selecting the larger and cooler ones with temperatures between 3 and 6°C (37 and 43°F), and occasionally in an abandoned mine tunnel. They prefer places where the humidity is very high (66–90 percent).

There are breeding colonies formed in summer, but these are relatively small, containing up to several hundred individuals. Most females give birth to their young beneath the bark of live and dead trees or in tree cavities. Rarely do they live in buildings or under bridges. There is evidence in Missouri that suggests maternity colonies have both primary and alternative roost sites, most occurring in slightly different microhabitats. As temperatures and precipitation increase, the use of alternative roost trees increases. Maternity colonies are often associated with landscapes dominated by agricultural use. Some males summer in caves; most aggregate in small groups away from caves.

Habits

From late August to October, prior to hibernation, Indiana myotis start to swarm around cave entrances. During this period, mating occurs, and most of the fat used to sustain the bats during hibernation is accumulated.

By mid-October Indiana myotis begin to hibernate, females preceding males. More than other Missouri bats, this species shows a tendency for mass movement within the cave in response to temperature changes. At first, they select the warmer, deeper, and higher parts of a cave, but as outside temperatures decrease, they move and congregate in the colder parts of the cave, mostly near the entrance. Here they gather in tightly packed clusters of hundreds or thousands of individuals. It is not uncommon to find a few gray myotis or little brown myotis in some of these clusters.

When temperatures drop below freezing at the hibernating sites near cave entrances, the bats increase their metabolism sufficiently to maintain a body temperature a few degrees above that of the environment, or they awaken and move to warmer sites in the cave. At present few caves fill their requirements for hibernating sites, which may account for their return to the same caves year after year.

Some bats start to leave the caves in late March, but most leave during April.

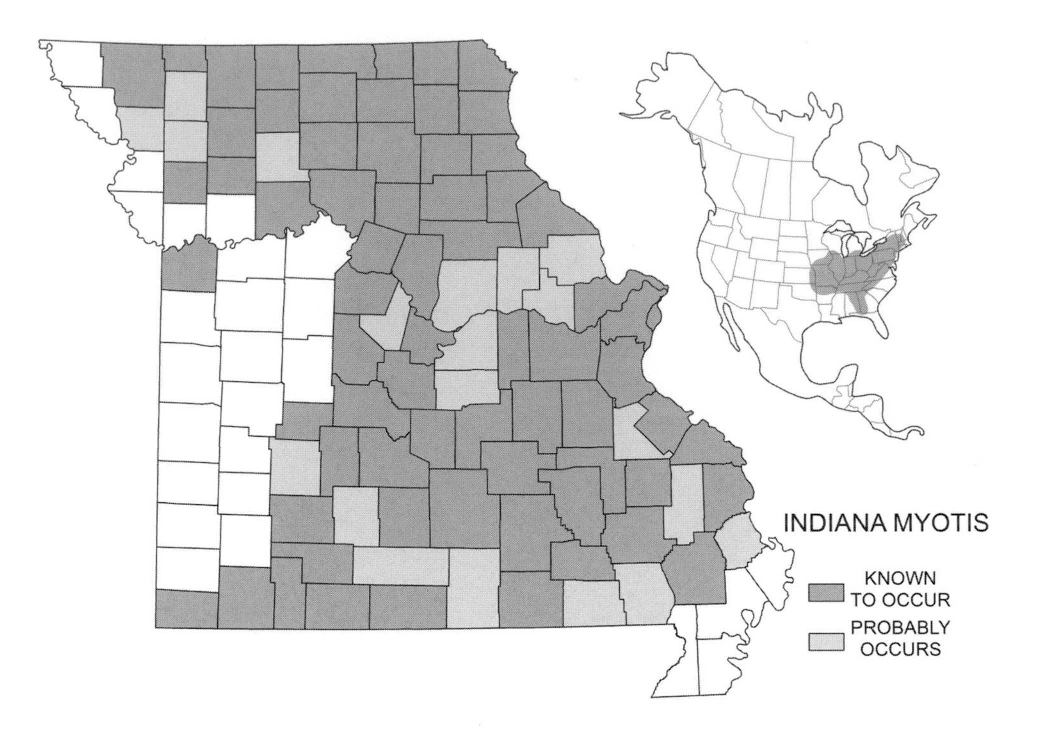

Indiana myotis forage mostly over forested areas, hunting more among trees than over water. They may fly among trees along streams or in the floodplain of rivers, or in the dense forests on hillsides or ridges. They are also known to forage along edges of agricultural fields.

Foods

The diet of the Indiana myotis varies with season and insect availability. The diet is predominantly moths, beetles, and true flies, but it may include caddisflies as well as other insects.

Reproduction

Most mating occurs in early October during the swarming period. At night the bats mate inside the cave near the entrance. During the day the sexes segregate in large clusters in different parts of the cave. Possibly there is less mating in winter and spring than in the little brown myotis. A single young is born in late June or early July. The young are reared in small maternity colonies of around 60 to 80 adults; colonies can reach as high as 200 individuals or more. The young bats are weaned at 25–37 days, and some have been found volant by mid-July.

Some Adverse Factors; Importance; and Conservation and Management

These subjects have been discussed for bats in general at the beginning of the chapter and in the account of the little brown myotis.

Because this bat is listed as endangered, it has received considerable attention. A 1971 proposal by the U.S. Army Corps of Engineers to dam the Meramec River and flood some of the remaining caves that harbored wintering colonies of this species was blocked, to a certain extent, by the plea to save this bat from extinction. In addition, other efforts have been directed toward protecting the winter homes of this species. Certain caves have been gated to discourage disturbance. However, research on these bats has shown that modified entrances can alter the thermal environment of the caves, which degrades the winter habitat. These findings have led to improved design of bat-friendly cave gates. Both the sites and the thermal environment must be conserved to protect hibernating Indiana myotis. Educational programs have called attention to the need to protect all colonies of bats in caves, mines, and other localities. For example, in 2012 a caver discovered the largest known hibernation site for the Indiana bat in a mine in Hannibal, Missouri. Mine portals were gated to protect the hibernating and summering bat populations. The mine and adjacent forested property became a city park, Sodalis Nature Preserve, that is protected in perpetuity through a conservation easement.

Despite many efforts to protect the caves used during hibernation, populations in Missouri continued to decline. Prior to the arrival of white-nose syndrome, the population estimate for Indiana myotis across their entire range had stabilized, but there is general concern that the disease will cause widespread decreases in the population. Additionally, some suspect that changes in habitat may be influencing bats during the active summer months. For summer roosts, these bats rely on the continual renewal of large-diameter standing dead trees; these are not a static resource, and specific forest management actions are necessary for their perpetuation.

SELECTED REFERENCES

See also discussion of this species in general references, page 23.

Brack, V., Jr., and R. K. LaVal Jr. 1985. Food habits of the Indiana bat in Missouri. *Journal of Mammalogy* 66:308–315.

Callahan, E. V., R. P. Drobrey, and R. L. Clawson. 1997. Selection of summer roosting sites by Indiana bats (*Myotis sodalis*) in Missouri. *Journal of Mammalogy* 78:818–825.

Clawson, R. L. 1991. Report on the status of priority 1 Indiana bat hibernacula, 1991. Unpublished report, U.S. Fish and Wildlife Service, Minneapolis, MN. 11 pp.

Clawson, R. L., R. K. LaVal, M. L. LaVal, and W. Caire. 1980. Clustering behavior of hibernating *Myotis sodalis* in Missouri. *Journal of Mammalogy* 61:245–253.

Cope, J. B., and S. R. Humphrey, 1977. Spring and autumn swarming behavior in the Indiana bat, *Myotis sodalis. Journal of Mammalogy* 58:93–95.

Elliott, W. R., 2007. Gray and Indiana bat population trends in Missouri. National Cave and Karst Management Symposium, pp. 46–61.

Garner, J., and J. E. Gardner. 1992. Determination of summer distribution and habitat utilization of the Indiana bat (*Myotis sodalis*) in Illinois. Unpublished report, Illinois Natural History Survey, Champaign. 25 pp.

Humphrey, S. R. 1978. Status, winter habitat, and management of the endangered Indiana bat, *Myotis sodalis. Florida Science* 41:65–76.

Humphrey, S. R., A. R. Richter, and J. B. Cope. 1977. Summer habitat and ecology of the endangered Indiana bat, *Myotis sodalis. Journal of Mammalogy* 58:334–346.

Kurta, A., D. King, J. A. Teramino, J. M. Stribley, and K. F. Williams. 1993. Summer roosts of the endangered Indiana bat (*Myotis sodalis*) on the northern edge of its range. *American Midland Naturalist* 129:132–138.

LaVal, R. K., R. L. Clawson, M. L. LaVal, and W. Caire. 1977. Foraging behavior and nocturnal activity patterns of Missouri bats, with emphasis on the endangered species *Myotis grisescens* and *Myotis sodalis. Journal of Mammalogy* 58:592–599.

Richter, A. R., S. R. Humphrey, J. B. Lope, V. Brack Jr. 1993. Modified cave entrances: Thermal effect on body mass and resulting decline of endangered Indiana bats (*Myotis sodalis*). *Conservation Biology* 7:407–415.

Sparks, D. W., C. M. Ritzi, J. E. Duchamp, and J. O. Whitaker Jr. 2005. Foraging habitat of the Indiana bat (*Myotis sodalis*) at an urban-rural interface. *Journal of Mammalogy* 86:713–718.

Thomson, C. E. 1982. *Myotis sodalis. Mammalian Species* 163. 5 pp.

Thorgmartin, W. E., R. A. King, P. C. McKann, J. A. Szymanski, and L. Pruitt. 2012. Population-level impact of white-nose syndrome on the endangered Indiana bat. *Journal of Mammalogy* 93:1086–1098.

Tuttle, N. M., D. P. Benson, and D. W. Sparks. 2006. Diet of the *Myotis sodalis* (Indiana bat) at an urban/rural interface. *Northeastern Naturalist* 13:435–442.

U.S. Fish and Wildlife Service (USFWS). 2007. *Indiana bat (*Myotis sodalis*) draft recovery plan: First revision.* U.S. Fish and Wildlife Service, Fort Snelling, MN. 258 pp.

Womack, K. M., S. K. Amelon, and F. R. Thompson III. 2013. Summer home range size of female Indiana bats (*Myotis sodalis*) in Missouri, USA. *Acta Chiropterologica* 15:423–429.

Eastern Small-footed Myotis (*Myotis leibii*)

Name

The first part of the scientific name, *Myotis*, is from two Greek words and means "mouse ear." The second part, *leibii*, is a Latinized name meaning "of Leib"; it was named for its discoverer, George C. Leib. The common name, "small-footed," is self-explanatory. This species was formerly known as *Myotis subulatus*. The western form is now considered a separate species, *Myotis ciliolabrum*. This species is also known as the eastern small-footed bat.

Description (Plate 11)

The eastern small-footed myotis is very similar to the other mouse-eared bats in Missouri. It is identified by the general coloration and the following combination of characteristics. It is the smallest of these species. The wing is attached along the side of the foot and reaches the base of the toes. When laid forward, the ears usually extend about 1.6 mm (1⁄16 in.) beyond the nostrils. The calcar is keeled. The hind foot is 6.3 mm (¼ in.) or less in length.

Color. The small-footed myotis is yellowish brown, usually with a distinct black mask across the face to the black ears. When the hairs of the back are parted, the bases are seen to be blackish. The wing and tail membranes are very dark brown.

Measurements

Total length	69–88 mm	2¾–3½ in.
Tail	28–34 mm	1⅛–1⅜ in.
Hind foot	6 mm	¼ in.
Ear	12 mm	½ in.
Skull length	12 mm	½ in.
Skull width	7 mm	5⁄16 in.
Weight	2–7 g	1⁄10–¼ oz.

Teeth and skull. The teeth are similar to those of the little brown myotis. Because the differences between the skulls of mouse-eared bats are so slight, all skulls of this group should be compared with specimens of known identity before attempting identification.

Distribution and Abundance

Throughout its range, extending from southern Canada through the northeastern United States and into the Midwest, the distribution of the eastern small-footed myotis is spotty and the species is considered to be very uncommon. They have only been found in a few scattered locations in Missouri. They are an imperiled species of conservation concern in Missouri and are listed as imperiled across their North American range.

Habitat and Home

During winter this bat hibernates in narrow cracks and crevices of caves or mine tunnels. In summer, caves, buildings, and cavities in the ground or beneath rocks are used for roosting and for nursery sites. Little is known about the habitat requirements of this bat except that it associates with forested landscapes.

Life History

In winter, the eastern small-footed myotis frequently hibernates under large rocks on the floor of the cave or tunnel. When it hangs on the cave wall, the forearms are somewhat extended, similar to those of gray myotis; whereas other bats usually hang with their wings parallel to the body's axis. It hibernates in small groups or often singly.

In Missouri, small-footed myotis have been found far from the cave's entrance, where neither light nor freezing temperatures penetrated. In other localities, they have been taken near the entrances of open, drafty mines and caves where temperatures drop below freezing.

This species is the latest to enter caves in winter and the earliest to leave in spring.

Moths, true flies, and beetles compose the majority of the diet.

Not much is known about the breeding of the eastern small-footed myotis. Mating occurs in the fall before hibernation, and fertilization is delayed until the spring. One young is born from May to July. There are small nursery colonies.

Some Adverse Factors; Importance; and Conservation and Management

These subjects have been discussed for bats in general at the beginning of the chapter and in the account of the little brown myotis.

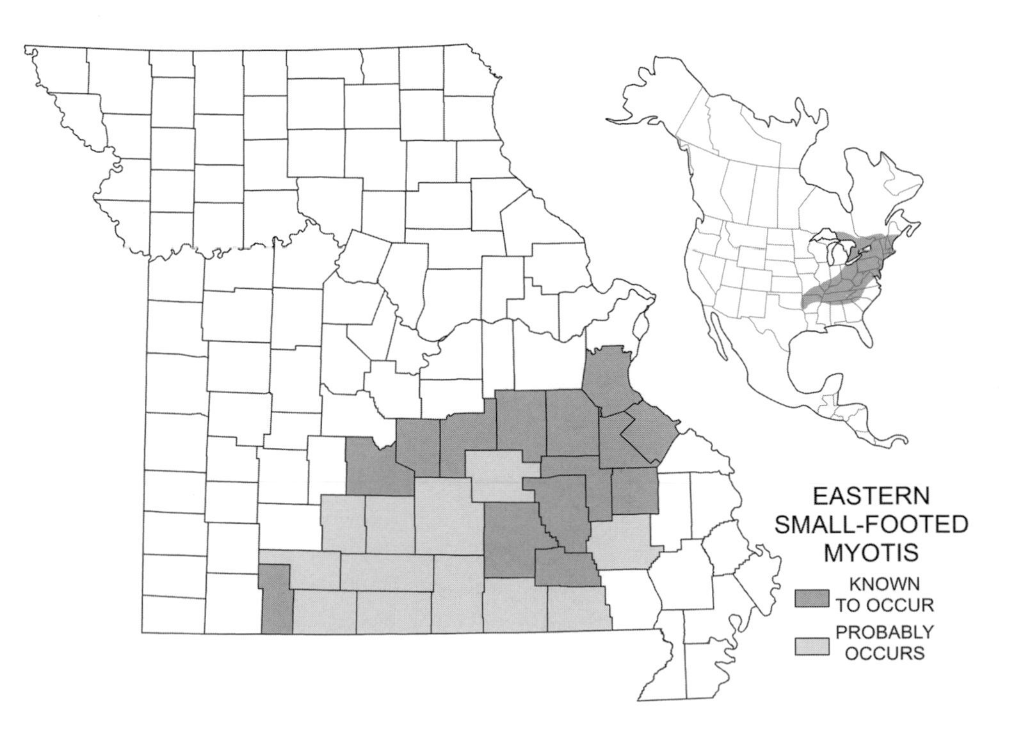

SELECTED REFERENCES

See also discussion of this species in general references, page 23.

Best, T. L., and J. B. Jennings. 1997. *Myotis leibii. Mammalian Species* 547. 6 pp.

Glass, B. P., and R. J. Baker. 1968. The status of the name *Myotis subulatus* Say. *Proceedings of the Biological Society of Washington* 81:257–260.

Gunier, W. J., and W. H. Elder. 1972. New records of *Myotis leibii* from Missouri. *American Midland Naturalist* 89:489–490.

Hitchcock, H. B., R. Keen, and A. Kurta. 1984. Survival rates of *Myotis leibii* and *Eptesicus fuscus* in southeastern Ontario. *Journal of Mammalogy* 65:125–130.

Moosman, P. R., H. H. Thomas, and J. P. Veilleux. 2007. Food habits of eastern small-footed bats (*Myotis leibii*) in New Hampshire. *American Midland Naturalist* 158:354–360.

Tuttle, M. D., and L. R. Heaney. 1974. Maternity habits of *Myotis leibii* in South Dakota. *Bulletin Southern California Academy of Science* 73:80–83.

van Zyll de Jong, C. G. 1984. Taxonomic relationships of Nearctic small-footed bats of the *Myotis leibii* group (Chiroptera: Vespertilionidae). *Canadian Journal of Zoology* 62:2519–2526.

Silver-haired Bat
(*Lasionycteris noctivagans*)

Name

The first part of the scientific name, *Lasionycteris,* is from two Greek words and means "hairy bat" (*lasios,* "hairy," and *nycteris,* "bat"). This refers to the heavy furring on the upper surface of the tail membrane near the body. The last part, *noctivagans,* is from two Latin words and means "night wanderer" (*nox,* "night," and *vagans,* "wanderer"). The common name comes from the silvery tips of the fur.

Description (Plate 12)

The naked ears have rounded tips and are nearly as broad as they are long; they barely reach the nostrils when laid forward. The tragus is short (less than one-half the length of the ear), broad, and bluntly rounded. The wing membrane is naked, but the tail membrane is heavily furred on the upper surface outward from the body for slightly more than one-half its length. The wing membrane is attached along the side of the foot to the base of the toes. The calcar is not keeled. The body fur is long.

The silver-haired bat is distinguished from all other bats in Missouri by its coloration.

Color. The fur of both back and belly is blackish brown tipped with silvery white. This frosted appearance is more conspicuous on the middle of the back and less so on the head and underparts. The wing and tail membranes and the ears are dark brown to black. The sexes are colored alike and there is no seasonal variation in color.

Measurements

Total length	90–113 mm	3½–4⅜ in.
Tail	27–50 mm	1¹⁄₁₆–2¹⁵⁄₁₆ in.
Hind foot	7–11 mm	¼–⁷⁄₁₆ in.
Ear	15 mm	⅝ in.
Skull length	15 mm	⅝ in.
Skull width	9 mm	⅜ in.
Weight	5–16 g	³⁄₁₆–⁹⁄₁₆ oz.

Teeth and skull. The dental formula of the silver-haired bat is:

$$I\,\frac{2}{3}\;C\,\frac{1}{1}\;P\,\frac{2}{3}\;M\,\frac{3}{3} = 36$$

Like all Missouri bats the first two upper molars have a distinct W pattern on their surfaces.

There is a prominent notch in the front end of the hard palate, as is typical of the skulls of all bats in Missouri. The skull is flat in its upper profile. The rostrum, the part of the skull in front of the eye sockets, is broad and almost as wide as the braincase when viewed from the top. The rostrum is strongly concave on each side at the back of the nasal opening. The skull of the silver-haired bat might be confused with skulls of big-eared bats because they have the same number and arrangement of teeth, but the aforementioned characteristics identify the silver-haired bat.

Sex criteria. The sexes are identified as in the little brown myotis. There is one pair of teats.

Age criteria. The young have long, silky, black hair with more pronounced silvery tips than those of adults.

Voice and sounds. Ultrasonic cries guide these bats in flight. (See the discussion of this subject in the account of the little brown myotis.)

Distribution and Abundance

Silver-haired bats range from southeastern Alaska, across southern Canada, throughout most of the United States, and into extreme northeastern Mexico. In general, this bat winters in the southern part of its range and summers in the northern. It occurs irregularly over this area but is most abundant in forested sections of the western United States.

In Missouri, silver-haired bats can be found statewide during spring and fall migration, but they are

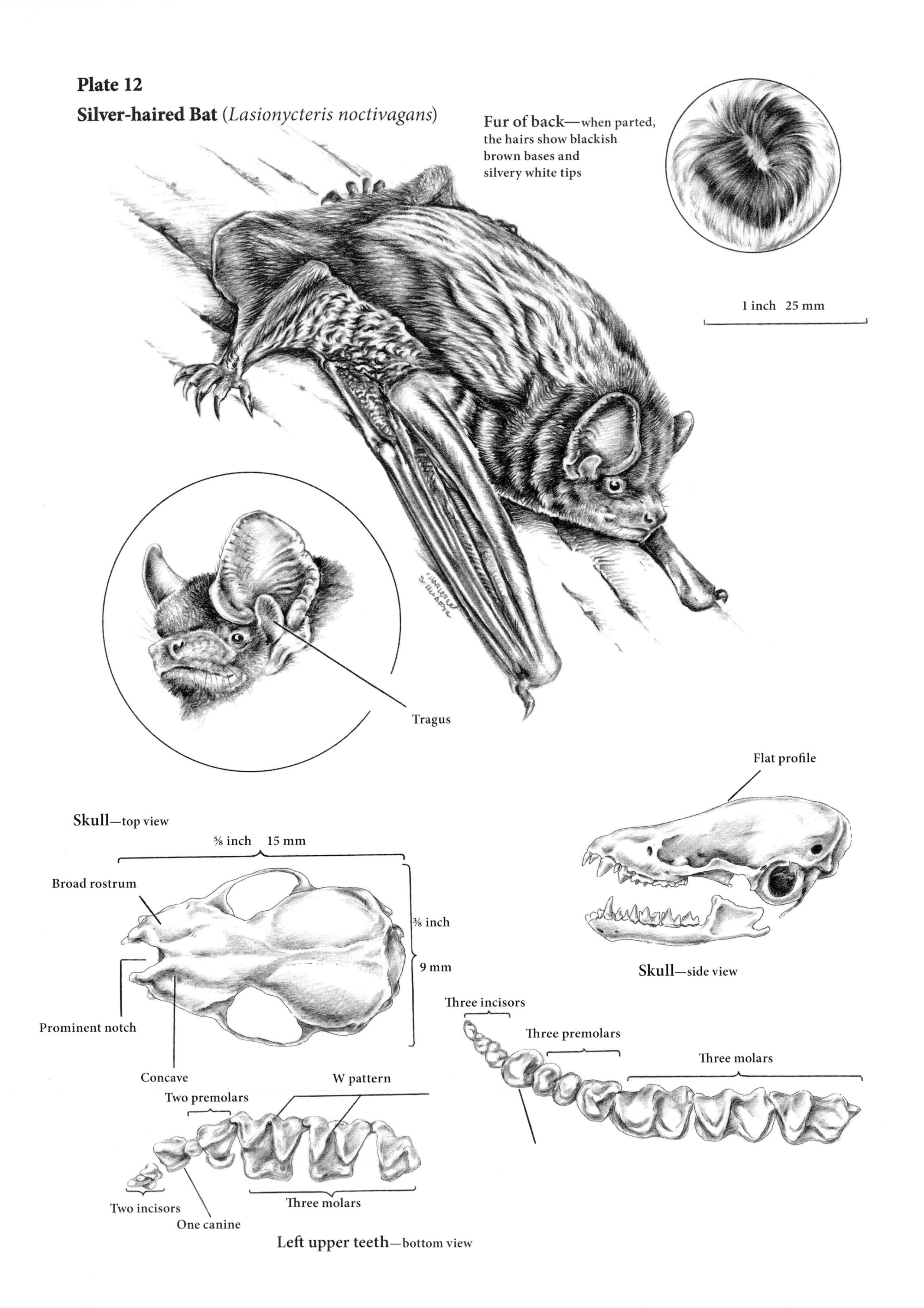

Plate 12
Silver-haired Bat (Lasionycteris noctivagans)
Fur of back—when parted, the hairs show blackish brown bases and silvery white tips
1 inch 25 mm
Tragus
Flat profile
Skull—top view
⅝ inch 15 mm
Broad rostrum
⅜ inch
9 mm
Prominent notch
Concave
Skull—side view
Three incisors
Three premolars
Three molars
W pattern
Two premolars
Two incisors
One canine
Three molars
Left upper teeth—bottom view

rare during winter and summer. Reproduction has been documented in only two counties, Putnam and Nodaway, during the summer. They are listed as a vulnerable species of conservation concern in Missouri but they are secure across their North American range.

Habitat and Home

This bat is primarily a tree-inhabiting species, living in forests, along wooded watercourses, and in semiwooded areas. It usually roosts in tree cavities and crevices under the bark of trees or in caves, rock crevices, or buildings. Location of roosts is often influenced by temperatures and incident solar radiation.

Habits

The silver-haired bat is generally migratory, moving south in the fall and north in the spring. It apparently flies great distances and migrates in rather large numbers; many silver-haired bats were observed to land on a boat far out at sea, and some individuals have reached Bermuda. During its migration it may fly in the daytime as well as at night.

Although most individuals of this species migrate, some apparently remain in northern parts of the range. The extent of hibernation is not well understood; some individuals rouse to forage during warmer days and utilize torpor to limit energy expenditures during colder times. Likewise, many males may stay south in their hibernation range throughout the year.

During summer, the females are gregarious, but the males live singly and apart from the females.

Silver-haired bats often come out to feed just before sunset. They forage along watercourses or the borders of hardwood and coniferous forests, darting in and out of the foliage or going up 12 m or more (40 ft. or more) above the ground. They generally fly slow, erratic courses with many twists and frequent, short glides. The flight itself is slow and has a fluttering quality. Although several bats may fly in the same general area, each bat has its own hunting route that is a sweeping circle 45 to 90 m (50 to 100 yds.) in diameter. Silver-haired bats can swim strongly for short distances.

Foods

Moths and true flies predominate in the diet of silver-haired bats. They also eat crane flies and wasps.

Reproduction

Mating occurs in late August or September. During spring the eggs are shed from the ovaries, as is true with other Missouri species of bats. Maternity colonies are typically found in tree cavities of declining or newly dead trees. Gestation occurs over a period of 50–60 days and twin pups (occasionally one) are born from mid-June to early July. At birth the young are black and wrinkled. They learn to fly and are weaned at 3 to 4 weeks of age.

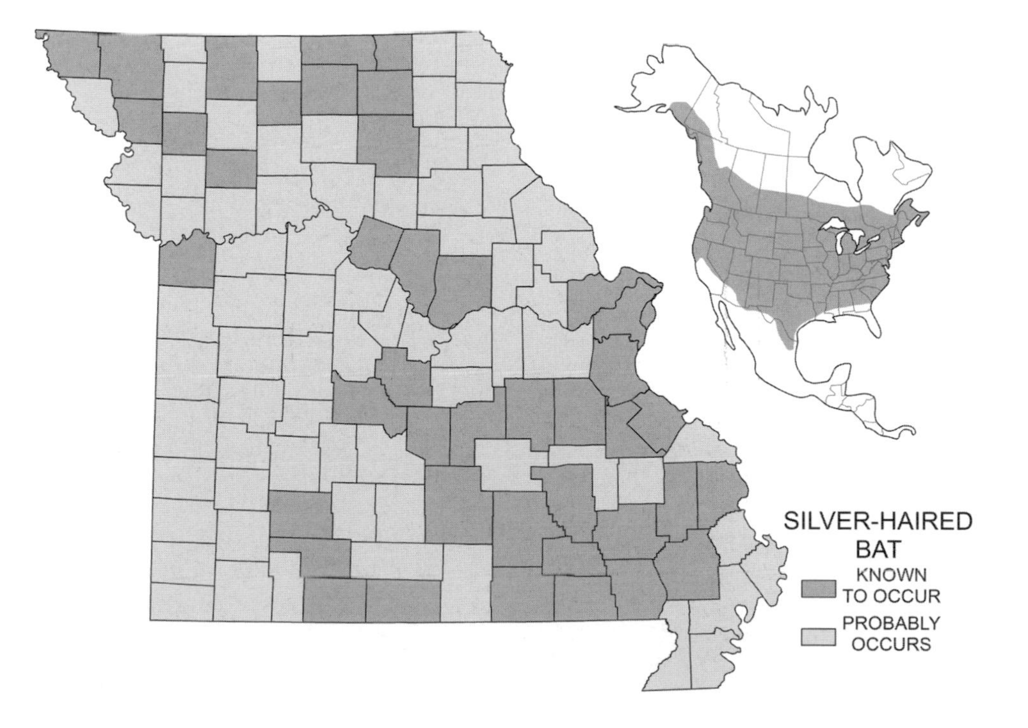

Some Adverse Factors; Importance; and Conservation and Management

These subjects have been discussed for bats in general at the beginning of the chapter and in the account of the little brown myotis. In addition, recruitment of snags and forest structural complexity are important for the conservation of the roost sites upon which these bats depend. This species constitutes a significant proportion of mortalities documented at wind turbine farms across North America.

SELECTED REFERENCES

See also discussion of this species in general references, page 23.

Barclay, R. M. R. 1985. Long- versus short-range foraging strategies of hoary (*Lasiurus cinereus*) and silver-haired (*Lasionycteris noctivagans*) bats and consequences for prey selection. *Canadian Journal of Zoology* 63:2507–2515.

Barclay, R. M. R., P. A. Faure, and D. R. Farr. 1988. Roosting behavior and roost selection by migrating silver-haired bats (*Lasionycteris noctivagans*). *Journal of Mammalogy* 69:821–825.

Betts, B. J. 1998. Roosts used by maternity colonies of silver-haired bats in northeastern Oregon. *Journal of Mammalogy* 79:643–650.

Campbell, L. A., J. G. Hallett, and M. A. O'Connell. 1996. Conservation of bats in managed forests: Use of roosts by *Lasionycteris noctivagans*. *Journal of Mammalogy* 77:976–984.

Easterla, D. A., and L. C. Watkins. 1970. Breeding of *Lasionycteris noctivagans* and *Nycticeius humeralis* in southwestern Iowa. *American Midland Naturalist* 84:254–255.

Frum, W. G. 1953. Silver-haired bat, *Lasionycteris noctivagans*, in West Virginia. *Journal of Mammalogy* 34:499–500.

Izor, R. J. 1979. Winter range of the silver-haired bat. *Journal of Mammalogy* 60:641–643.

Kunz, J. W. 1982. *Lasionycteris noctivagans*. *Mammalian Species* 172. 5 pp.

Parsons, H. I., D. A. Smith, and R. F. Whittam. 1986. Maternity colony of silver-haired bats, *Lasionycteris noctivagans*, in Ontario and Saskatchewan. *Journal of Mammalogy* 67:598–600.

Perry, R. W., D. A. Saugey, and B. G. Crump. 2010. Winter roosting ecology of silver-haired bats in an Arkansas forest. *Southeastern Naturalist* 9:563–572.

Reimer, J. P., E. F. Baerwald, and R. M. R. Barclay. 2010. Diet of hoary (*Lasiurus cinereus*) and silver-haired (*Lasionycteris noctivagans*) bats while migrating through southwestern Alberta in late summer and autumn. *American Midland Naturalist* 164:230–237.

Tri-colored Bat (*Perimyotis subflavus*)

Name

The first part of the scientific name, *Perimyotis*, is from three Greek words meaning "nearly mouse ear," referring to their similarities in tragi and dentition to that of *Myotis* bats. The second part, *subflavus*, is of Latin origin and means "yellowish belly" (*sub*, "below," and *flavus*, "yellowish"). The common name, tri-colored bat, refers to the presence of three distinct colors (black, yellow, and brown) on each hair. In previous editions of this book, this species was known as *Pipistrellus subflavus*, the eastern pipistrelle.

Description (Plate 13)

The thin, naked ears are longer than they are broad and have rounded tips; they extend slightly beyond the nostrils when laid forward. The tragus is less than one-half the length of the ear and has a bluntly rounded tip. The thin wing membrane, attached along the side of the foot to the base of the toes, is naked; the thin tail membrane is naked except on the upper surface for the third nearest the body, which is sparsely furred. The calcar is not keeled. The body fur is fluffy.

The tri-colored bat is most likely to be confused with the mouse-eared bats, the big brown bat, and the evening bat. However, these species can be distinguished by some or all of the following characteristics: dental formula, size, color and character of the fur, and shape of the ear and tragus.

Color. The color of the back varies from dark reddish brown to light yellowish brown or gray. The main hairs, when separated by blowing into the fur, are seen to be dark gray at the base, then broadly banded with yellowish brown and tipped with dark brown. Intermixed with these main hairs (6.3 mm, ¼ in. long) are some longer ones (9.5 mm, ⅜ in. long), which are entirely yellowish brown. The underparts of the body are yellowish brown, although the hairs have dark gray bases. The pale red skin over the forearm bones contrasts with the blackish wing membrane, but some pale splotches may occur on the membrane next to the bones. The tail membrane is blackish away from the body but tan near the body. The ears are tan. Only rarely are tri-colored bats blackish or do they have a white body with other parts normally colored. The sexes are colored alike and show no difference in coloration with the seasons.

Plate 13

Tri-colored Bat (*Perimyotis subflavus*)

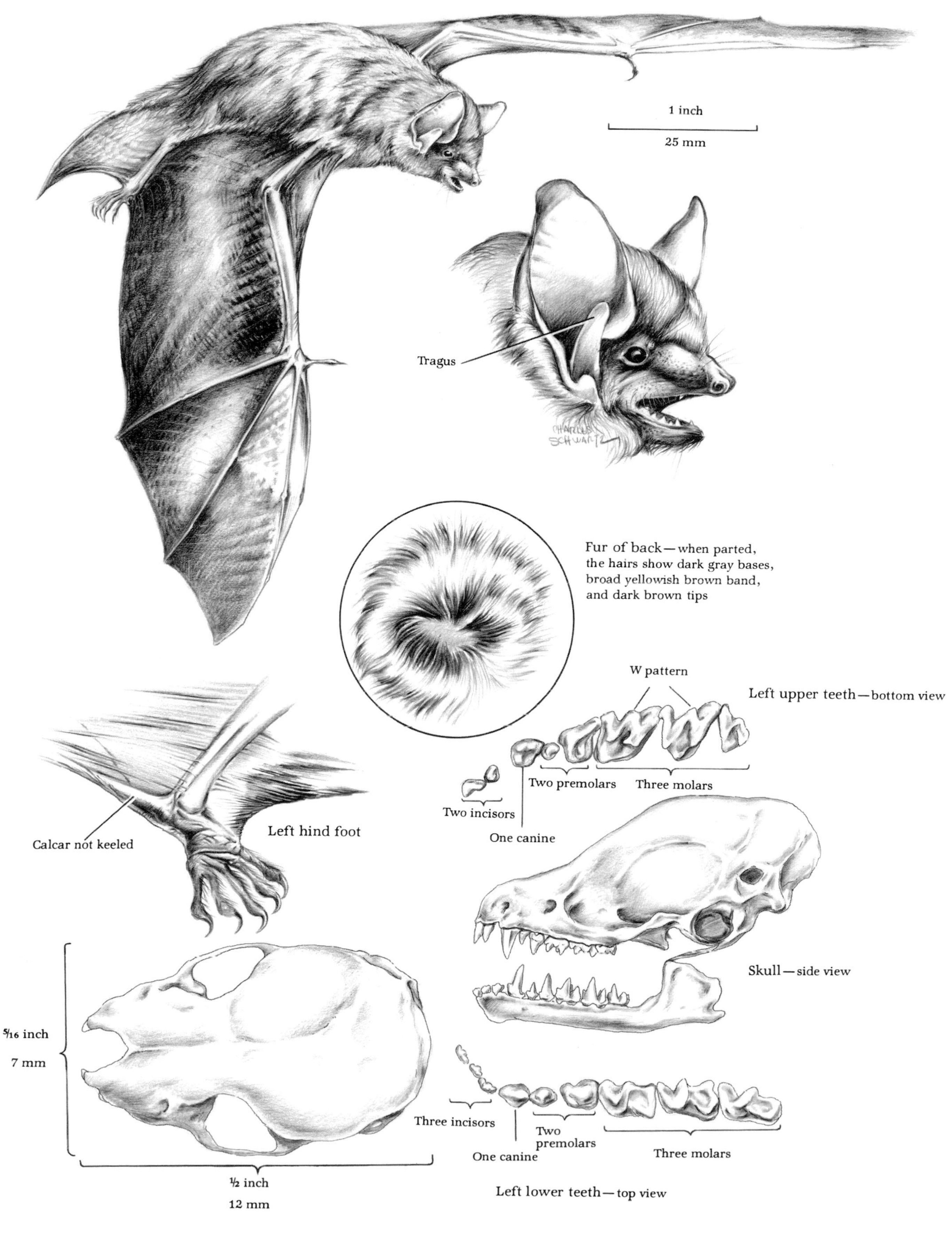

Measurements

Total length	79–88 mm	3⅛–3½ in.
Tail	34–44 mm	1⅜–1¾ in.
Hind foot	7–9 mm	5/16–⅜ in.
Ear	12–14 mm	7/16–9/16 in.
Skull length	12–13 mm	½ in.
Skull width	7 mm	5/16 in.
Weight	2–8 g	1/16–3/10 oz.

Females are generally larger than males.

Teeth and skull. This is the only bat in Missouri with 34 teeth. The dental formula of the tri-colored bat is:

$$I\,\frac{2}{3}\;C\,\frac{1}{1}\;P\,\frac{2}{2}\;M\,\frac{3}{3} = 34$$

As is typical of the skulls of all bats in Missouri, the first two upper molars have a distinct W pattern on their surfaces, and there is a prominent notch in the front end of the hard palate.

Sex criteria. The sexes are identified as in the little brown myotis. There is one pair of teats.

Age criteria and longevity. The young are darker than adults. Banded tri-colored bats have been documented to live up to 10 and even 15 years of age in the wild.

Glands. Oil glands occur as swellings on the upper lip and in the midline of the throat.

Voice and sounds. In addition to faint squeaks that are audible to humans, tri-colored bats have ultrasonic cries like other insect-eating bats that guide them in flight. (See account of the little brown myotis.)

Distribution and Abundance

The tri-colored bat is widely distributed from extreme southeastern Canada throughout the eastern half of the United States and into eastern Mexico and Central America.

Tri-colored bats occur throughout Missouri all year, spending winter in areas with caves. They are most common in the southern part of the range.

Habitat and Home

In Missouri, most tri-colored bats hibernate during the winter in caves, mines, and other cavelike structures, where they select the warmer and more humid parts. In summer, they roost in trees, in cracks and crevices of cliffs or buildings, in barns, or sometimes in high domes of caves. In the Ouachita Mountains of Arkansas, male and female bats are known to roost in mature hardwood forests with complex vertical structure, often using oak or pine trees.

Habits

In August, tri-colored bats return to certain caves where they swarm about the entrance, yet they may go to other caves for hibernation.

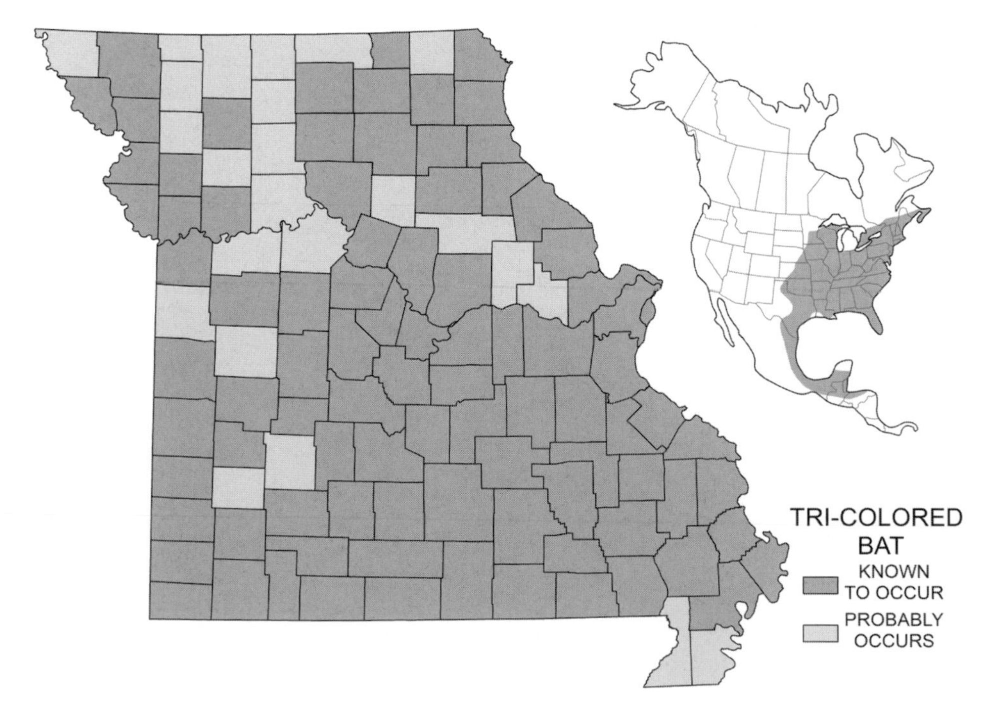

In general tri-colored bats enter their hibernating quarters in mid-October and stay until mid-April, with the females leaving before the males. However, a few individuals may start to hibernate even earlier and stay longer, making their inactive period as long as seven months. Tri-colored bats are profound hibernators and do not awaken and fly about as much during the winter as do other bats. In caves it is common to see hibernating tri-colored bats with droplets of moisture from condensation suspended in their fur. In the beam of a flashlight, they appear covered with frost.

When hibernating, tri-colored bats hang by their feet, usually alone, but sometimes near a few individuals of their own kind and will occasionally be mixed with other species. There is no segregation of sexes during hibernation.

In summer the females establish nursery colonies, ranging from a few dozen up to 50 individuals, in high domes of caves, under eaves of barns or other buildings, or in protected parts of cliffs, where they rear their young. Females from Indiana have been known to form small groups of two or three bats and use clumps of dead foliage in trees for maternity roosts. Each colony may use more than one roost during a given season. Presumably the summer colonies are in the same general area as the wintering sites, although banded tri-colored bats have been taken in summer as far as 137 km (85 mi.) from the caves where they were banded in winter. Tri-colored bats show a preference for the same summer roosting spots; three out of four bats banded in their roosting spot on a porch were found in the same place three years later. Males are presumably solitary.

In comparison to other bats, tri-colored bats appear weak fliers. They fly with a fluttering motion in an undulating course. During the season of activity, they leave their daytime retreats about sundown to feed. There is an irregular feeding period toward midnight and another one about daylight. They generally forage high over watercourses at the forest edge. This species is commonly visible foraging over ponds between sundown and dark.

Foods

This small bat feeds on tiny insects, particularly flies, moths, wasps, leafhoppers, and beetles, many of which are aquatic forms. Diets of tri-colored bats in Indiana were primarily composed of caddisflies with some beetles and ants secondarily included.

Reproduction

Mating occurs in fall, intermittently throughout winter, and again in spring. Relatively small maternity colonies start forming in mid-April. The one or two (rarely three) young are born from late May to mid-July, after a gestation period of at least 44–60 days.

The young are able to fly at about four weeks of age. They probably do not mate in the year of their birth.

Some Adverse Factors; Importance; and Conservation and Management

These subjects have been discussed for bats in general at the beginning of the chapter and in the account of the little brown myotis. This species is heavily impacted by white-nose syndrome.

SELECTED REFERENCES

See also discussion of this species in general references, page 23.

Briggler, J. T., and J. W. Prather. 2003. Seasonal use and selection of caves by the eastern pipistrelle bat (*Pipistrellus subflavus*). *American Midland Naturalist* 149:406–412.

Fujita, M. S., and T. H. Kunz. 1984. *Pipistrellus subflavus*. *Mammalian Species* 228. 6 pp.

Hoofer, S. R., R. A. Van Den Bussche, and I. Horáček. 2006. Generic status of the American pipistrelles (Vespertilionidae) with description of a new genus. *Journal of Mammalogy* 87:981–992

Humphrey, S. R., R. K. LaVal, and R. L. Clawson. 1976. Nursery populations of *Pipistrellus subflavus* (Chiroptera, Vespertilionidae) in Missouri. *Transactions of the Illinois State Academy of Science* 69:367.

Layne, J. N. 1993. Status of the eastern pipistrelle *Pipistrellus subflavus* at its southern range limit in eastern United States. *Bat Research News* 33:43–46.
MacDonald, K., E. Matsui, R. Stevens, and M. B. Fenton. 1994. Echolocation calls and field identification of the eastern pipistrelle (*Pipistrellus subflavus*: Chiroptera: Vespertilionidae), using ultrasonic bat detectors. *Journal of Mammalogy* 75:460–465.
Perry, R. W., and R. E. Thill. 2007. Tree roosting by male and female eastern pipistrelles in a forested landscape. *Journal of Mammalogy* 88:974–981.
Veilleux, J. P., J. O. Whitaker Jr., and S. L. Veilleux. 2003. Tree-roosting ecology of reproductive female eastern pipistrelles, *Pipistrellus subflavus*, in Indiana. *Journal of Mammalogy* 84:1068–1075.
Whitaker, J. O., Jr. 1998. Life history and roost switching in six summer colonies of eastern pipistrelles in buildings. *Journal of Mammalogy* 79:651–659.

Big Brown Bat (*Eptesicus fuscus*)

Name

The genus name, *Eptesicus*, is based on *petomai*, Latin for "house flier," and refers to this species' habit of living in houses. The second part, *fuscus*, is the Latin word for "brown" and describes the color. The common names, "big" and "brown," refer to the size and color.

Description (Plate 14)

The thick, naked ears are broad with rounded tips. The tragus is less than one-half the length of the ear, is slightly curved forward, and has a broadly rounded or blunt tip. The wing membrane is naked and is attached along the side of the foot, reaching to the base of the toes. The tail membrane is naked. The calcar is keeled. The body fur is long and soft.

The big brown bat is most likely to be confused with the mouse-eared bats, the tri-colored bat, and the evening bat. However, these species can be distinguished by some or all of the following characteristics: dental formula, size, color and character of the fur, and shape of the ear and tragus.

Color. The general coloration of the big brown bat is a uniform brown on the upperparts, varying in individuals from dark brown to cinnamon brown. The basal half of each hair is blackish while the outer half is brown. This two-toned coloration is easily seen when the fur is parted by blowing into it. On the underparts, the color is lighter than that of the back. The wing and tail membranes, face, and ears are blackish. Only rarely is an individual grizzled or blotched with white. The sexes are colored alike and show no seasonal difference in coloration.

Measurements

Total length	95–127 mm	3¾–5 in.
Tail	34–50 mm	1⅜–2 in.
Hind foot	9–12 mm	⅜–½ in.
Ear	15–19 mm	⅝–¾ in.
Skull length	19 mm	¾ in.
Skull width	12 mm	½ in.
Weight	14–21 g	½–¾ oz.

Females are larger than males.

Teeth and skull. The dental formula of the big brown bat is:

$$I\,\frac{2}{3}\;C\,\frac{1}{1}\;P\,\frac{1}{2}\;M\,\frac{3}{3} = 32$$

As in all Missouri bats the first two upper molars have a distinct W pattern on their surfaces and there is a prominent notch in the front end of the hard palate. When viewed from above, the entire skull has a very angular appearance. The rostrum is about as broad as long.

In the big brown bat the second upper incisor does not touch the canine tooth.

The eastern red bat, hoary bat, and the very rare Brazilian free-tailed bat also have 32 teeth, but these all have 1 incisor and 2 premolars on each side of the upper jaw.

Sex criteria and sex ratio. The sexes are identified as in the little brown myotis. There is one pair of teats. The sex ratio at birth is nearly even.

Age criteria and longevity. The young are darker and duller than adults.

Big brown bats in Colorado are estimated to live on average 5–7 years; however, the maximum recorded lifespan of a big brown bat was estimated to be 19 years.

Glands. Facial oil glands occur as swellings on the upper lip. These emit a pungent, musky odor under excitement. This odor may serve to attract the sexes, to aid in locating the daytime retreats, or to repel enemies.

Voice and sounds. The sounds of big brown bats that are audible to humans vary from long, drawn-out, raspy and guttural notes to staccato squeaks and clicks. Two bats flying close together probably communicate with each other by chattering. Young bats squeak, apparently calling to their mothers.

Big brown bats are guided in flight by means of their ultrasonic cries (inaudible to humans), which are

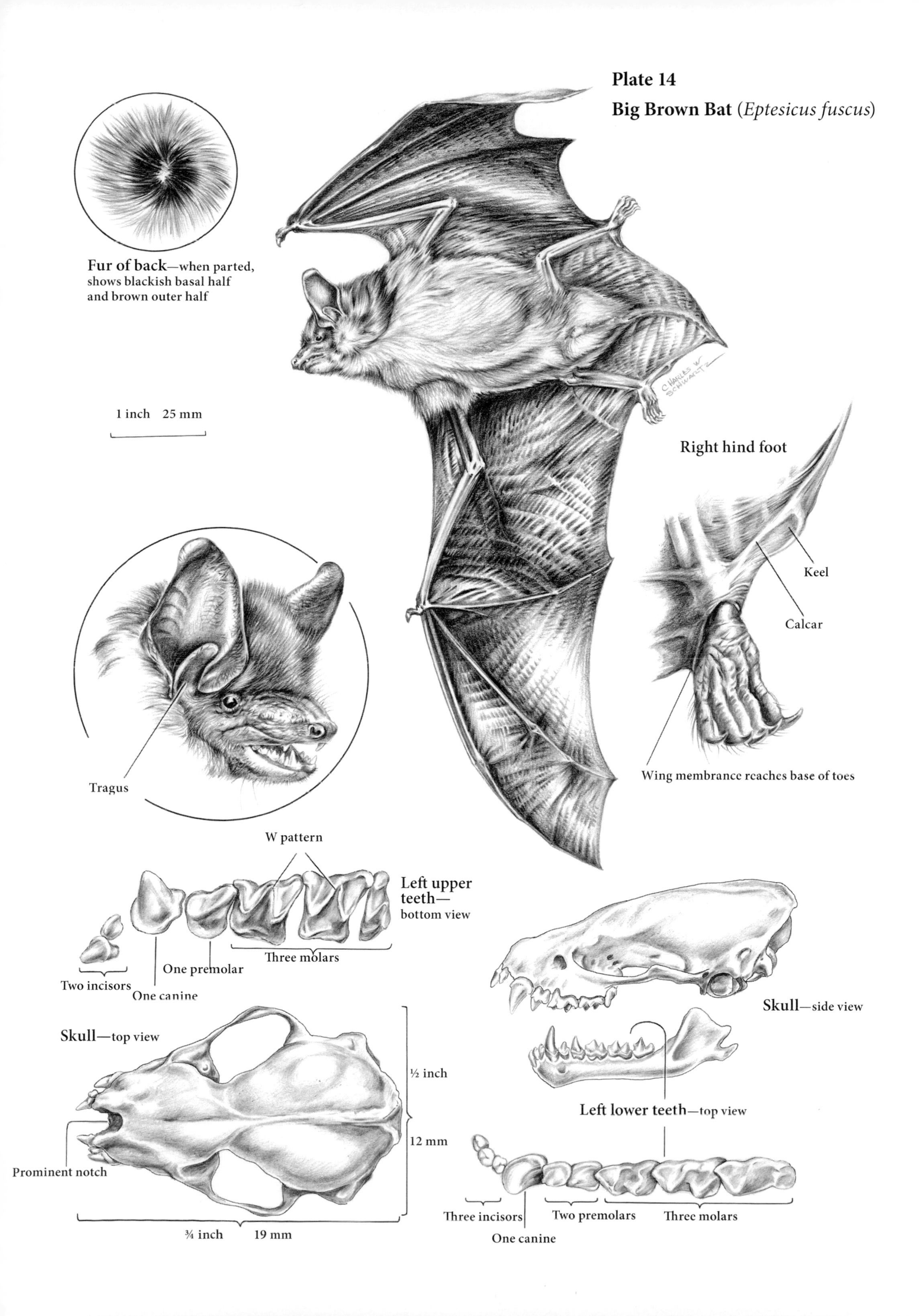
Plate 14
Big Brown Bat (*Eptesicus fuscus*)
Fur of back—when parted, shows blackish basal half and brown outer half
1 inch 25 mm
C HARLES W SCHWARTZ
Right hind foot
Keel
Calcar
Wing membrance reaches base of toes
Tragus
W pattern
Left upper teeth—bottom view
Two incisors
One canine
One premolar
Three molars
Skull—side view
Skull—top view
½ inch
12 mm
Prominent notch
¾ inch 19 mm
Left lower teeth—top view
Three incisors
One canine
Two premolars
Three molars

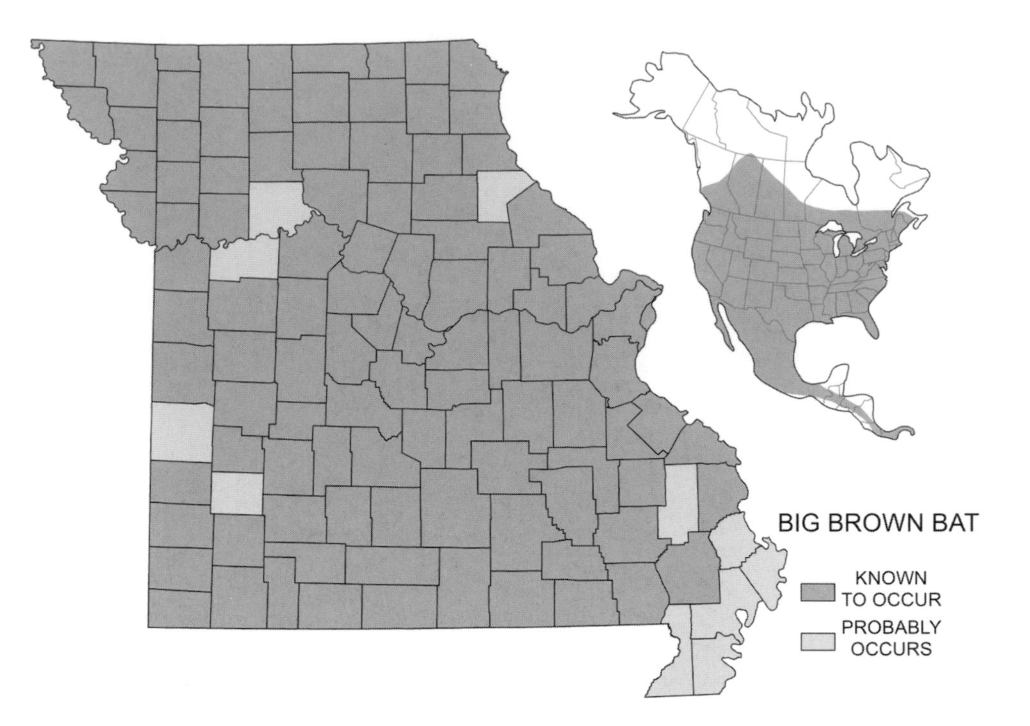

reflected back from obstacles and received by the inner ear (see account of the little brown myotis). Like all insect-eating bats, they hunt and pursue their prey by this same means.

Distribution and Abundance

The big brown bat ranges widely from southern Canada throughout the United States and Mexico and into Central America. It also occurs in extreme northern South America and the Caribbean. They are most common in the central part of their range.

Big brown bats dwell throughout Missouri all year.

Habitat and Home

During winter, big brown bats hibernate in many places: hollow trees, deep crevices in rocky cliffs, buildings, caves and mines, storm sewers, road culverts, and burial vaults. In summer the females establish nursery colonies of up to 2,000 individuals in hollow trees, attics, chimneys, lofts, or caves. Males generally roost solitarily in sheltered spots in man-made structures or in trees.

Habits

Most big brown bats migrate a relatively short distance (less than 80 km; 50 mi.) to winter hibernating spots. This species is very hardy and is one of the last to fly about in fall. But when the nightly temperatures approach freezing, big brown bats start to hibernate in well-protected sites such as mines, caves, and cooler human structures. They either hang by their hind feet or become dormant in a horizontal position depending upon the nature of the resting place. They usually occur singly or in groups of 2 to 20, and rarely up to a couple hundred. When they use caves for hibernation, they select the coldest portions and are often found near the entrance in total daylight.

The reduction in metabolism that accompanies hibernation is indicated by a lowered body temperature and rate of breathing. In dormant big brown bats the body temperature is approximately that of the surroundings, from –1.1 to 17.8°C (30 to 64°F), but when the bats are aroused their temperatures increase to over 36.7°C (98°F). In undisturbed hibernating big brown bats, breathing is intermittent; periods of 3 minutes or less of breathing at a rate of 25 to 50 inhalations per minute alternate with periods of 3 to 8 minutes without breathing. After being thoroughly aroused, breathing occurs at the rate of about 200 inhalations per minute.

In dormant bats, the spleen serves as a reservoir for red blood cells. A disturbance, such as awakening, induces a reflex action of the spleen that releases

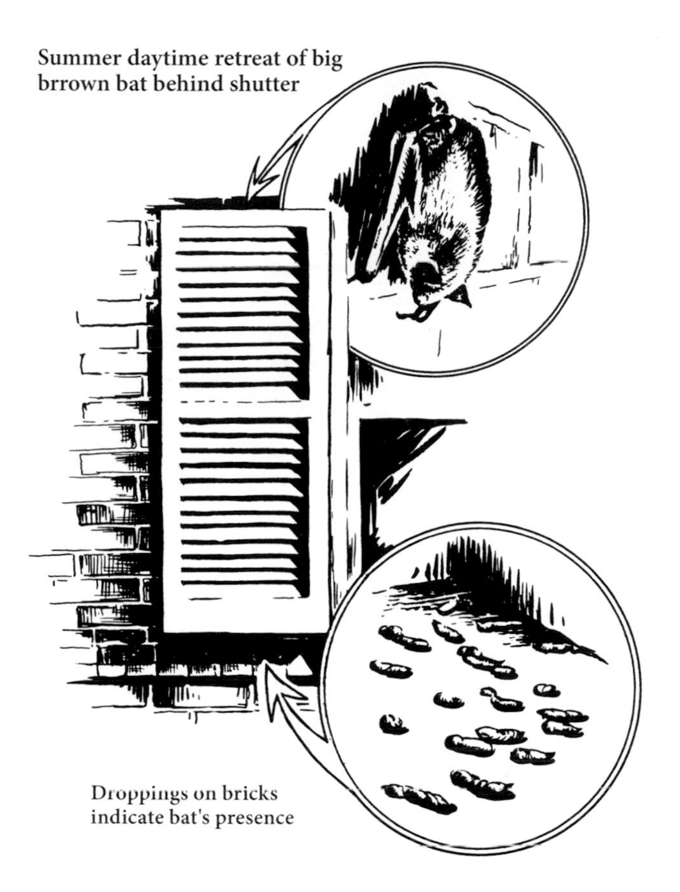

red blood cells into the bloodstream. Since red blood cells carry oxygen, this supplies more oxygen to the body with a resulting increase in activity. The body fat, which was accumulated gradually during the late summer months, serves as the source of energy during hibernation. At the beginning of hibernation, 30 percent of the bat's total weight is fat, but this decreases to 10 percent by the end of hibernation.

During winter there are brief periods when big brown bats become active, fly about, lap up moisture, and occasionally mate. If they live in caves or buildings, they fly about their hibernating quarters and may go outside; occasionally they even move to a different site.

Compared to some other bat species, big brown bats have only slight seasonal movements. Most marked individuals are recovered within a 16 km (10 mi.) radius of the banding site, but some range farther. The maximum known distance traveled by marked bats from winter to summer roosting sites is 53 km (33 mi.), while that between two winter sites is 98 km (61 mi.). Some big brown bats have returned to the site from which they were taken after being released 724 km (450 mi.) away.

Early in April the females leave their winter quarters and form nursery colonies of from 30 to 2,000 or so individuals. These may be in very protected places, but after a few weeks, when the young begin to fly, the whole colony may move to some less sheltered quarters. They probably return to the same site each year; two females banded in a nursery colony were recovered in the same locality two years after banding.

The males stay in their hibernating locations for some time after the females leave. During spring and early summer they are solitary, often roosting in a crevice. But after the nursery colonies have dispersed, they may be found in roosting sites with the females. Rarely do nursery colonies remain intact all summer.

Like most bats, big brown bats are nocturnal, leaving their daytime roosts about sunset to feed. Occasionally one is seen flying in midday. Compared to other bats, big brown bats are relatively slow fliers and follow a fairly steady course. However, they are highly maneuverable and can change directions abruptly. In addition to flying, they sometimes crawl on all four limbs and use the mouth to help grasp and climb up rough surfaces. In foraging, they fly along watercourses at the forest edge and above cropland and roadways. They generally fly below or within the canopy of trees.

Foods

Big brown bats feed on insects, beetles, and true bugs predominately. An examination of 2,200 droppings of this species showed the remains of 10 different orders of insects. The most common representatives belonged to the orders containing May beetles and click beetles, flying ants, and houseflies. It is evident that this efficient predator could capture moths, but it takes only a few. In captivity, big brown bats have been reported to eat other species of bats.

These bats catch and eat most of their food on the wing. Sometimes, however, they bring large insects back to the roost to devour. In captivity they consume one-third of their weight in food each day. In the wild, their stomachs are filled after one hour of feeding, and lactating females may consume a mass of insects equal to their own body mass.

Reproduction

Mating occurs regularly in the fall prior to hibernation, occasionally during winter when the bats become active, and again in spring. It is possible for the sperm received in the fall matings to fertilize the eggs in spring because the sperm remain dormant in the female during hibernation; however, the supply of stored sperm may be replenished by winter and spring matings.

In big brown bats, from 2 to 7 eggs are shed from the ovaries about the first week in April and become fertilized. Most or all embryos become implanted in the uterus, but usually only 1 or 2 develop fully. After a gestation of about 60 days, 1 or 2 young are born sometime from the end of May to late June.

At birth the young are naked and have their eyes closed. They weigh about 2.8 g (1⁄10 oz.). The eyes open about the second day. The female leaves her young in the nursery colony when she goes out to feed. When she returns, she carefully selects her own young to nurse. If the colony moves to an alternate roost, the female carries her young in flight.

The young are able to fly by themselves at about 3 to 4 weeks of age. When 2½ months old, they are the same size as adults. All young males become sexually mature in their first fall, but only about three-fourths of the young females do so at this age.

Big brown bat licking wing

Some Adverse Factors; Importance; and Conservation and Management

These subjects have been discussed for bats in general at the beginning of the chapter and in the account of the little brown myotis.

SELECTED REFERENCES

See also discussion of this species in general references, page 23.

Beer, J. R. 1955. Survival and movements of banded big brown bats. *Journal of Mammalogy* 36:242–248.

Beer, J. R., and A. G. Richards. 1956. Hibernation of the big brown bat. *Journal of Mammalogy* 37:31–41.

Christian, J. J. 1956. The natural history of a summer aggregation of the big brown bat, *Eptesicus fuscus fuscus. American Midland Naturalist* 55:66–95.

Davis, W. H., R. W. Barbour, and H. D. Hassell. 1968. Colonial behavior of *Eptesicus fuscus. Journal of Mammalogy* 49:44–50.

Hamilton, W. J., Jr. 1933. The insect food of the big brown bat. *Journal of Mammalogy* 14:155–156.

Kunz, T. H. 1974. Reproduction, growth, and mortality of the vespertilionid bat, *Eptesicus fuscus*, in Kansas. *Journal of Mammalogy* 55:1–13.

Kurta, A., and R. H. Baker. 1990. *Eptesicus fuscus. Mammalian Species* 356. 10 pp.

O'Shea, T. J., L. E. Ellison, and T. R. Stanley. 2011. Adult survival and population growth rate in Colorado big brown bats (*Eptesicus fuscus*). *Journal of Mammalogy* 92:433–443.

Rysgaard, G. N. 1942. A study of the cave bats of Minnesota with especial reference to the large brown bat, Eptesicus fuscus fuscus (Beauvois). *American Midland Naturalist* 28:245–267.

Wimsatt, W. A. 1944. Growth of the ovarian follicle and ovulation in *Myotis lucifugus lucifugus. American Journal of Anatomy* 74:129–173.

Eastern Red Bat (*Lasiurus borealis*)

Name

The first part of the scientific name, *Lasiurus*, is from two Greek words and means "hairy tail" (*lasios*, "hairy," and *oura*, "tail"). This refers to the furred upper surface of the tail membrane that is characteristic of this genus. The last part, *borealis*, is the Latinized word for "northern." The common name refers to this bat's range within the eastern United States and describes the predominant color.

Description (Plate 15)

The ear is short, broad, and rounded; it is sparsely furred inside but densely furred outside for the half nearest the head. The tragus is less than one-half the length of the ear; it has a decided forward curve and is broad at the base, tapering to the slightly rounded tip.

Both upper surfaces and undersurfaces of the wing membrane are furred outward from the body toward the elbow; on the undersurface the furring extends along the forearm to the wrist and up to one-third of the way down the fingers. The wing is attached along the side of the hind foot to the base of the toes. The large, wide tail membrane is thickly furred on its entire upper surface but thinly furred on the undersurface for only the half nearest the body. The calcar is keeled. The hind foot is well furred above. The body fur is soft and fluffy; it is longer on the neck than on the back, making a slight ruff.

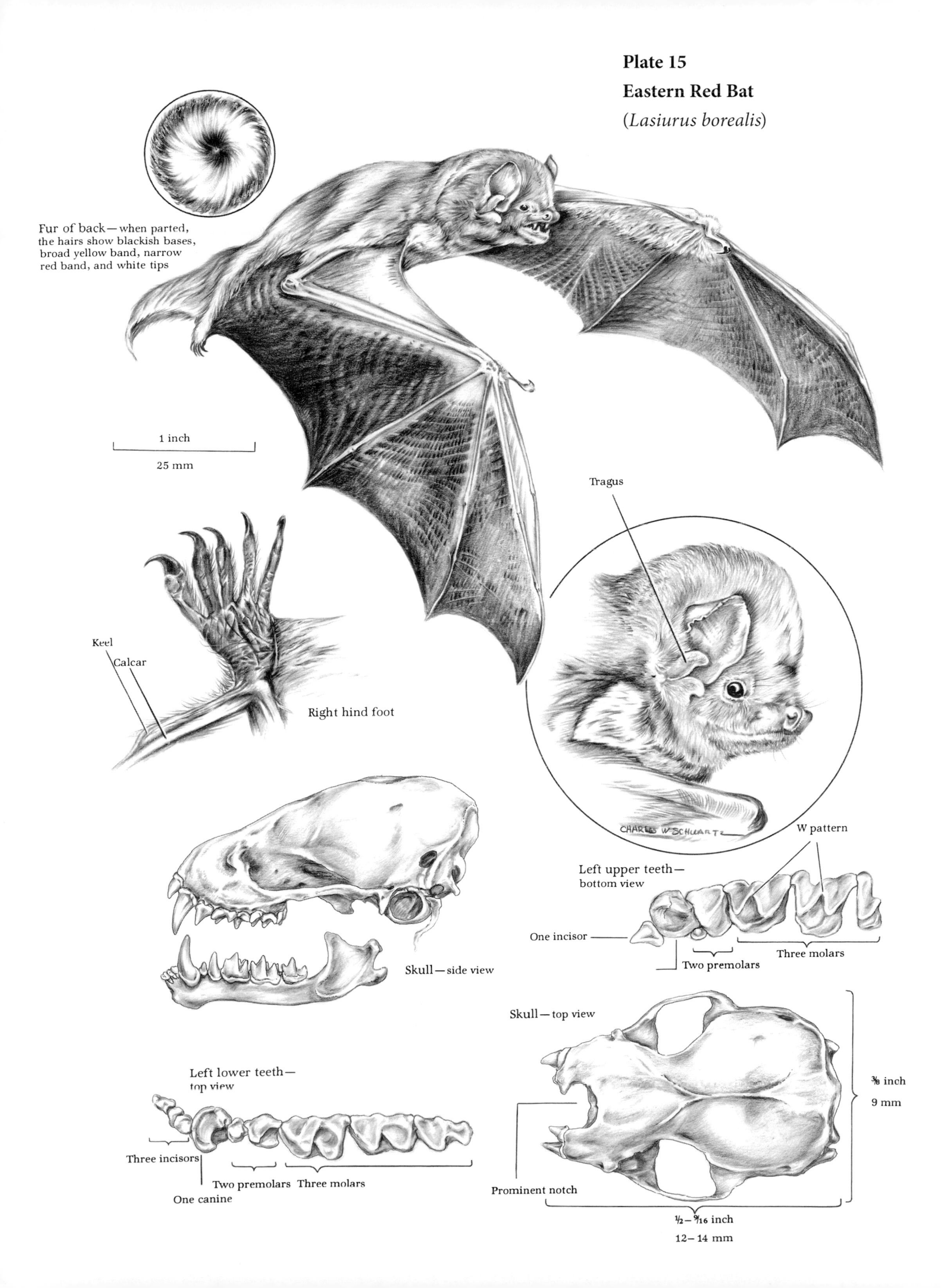

Plate 15
Eastern Red Bat
(Lasiurus borealis)
Fur of back—when parted, the hairs show blackish bases, broad yellow band, narrow red band, and white tips
1 inch
25 mm
Tragus
Keel
Calcar
Right hind foot
CHARLES W. SCHWARTZ
W pattern
Left upper teeth—bottom view
One incisor
Two premolars
Three molars
Skull—side view
Skull—top view
Left lower teeth—top view
Three incisors
One canine
Two premolars
Three molars
Prominent notch
3/8 inch
9 mm
1/2–9/16 inch
12–14 mm

The eastern red bat is easily distinguished from all other bats in Missouri, except for the very rarely encountered Seminole bat (see 414), by its color and from all except the closely related hoary bat by the well-furred tail membrane.

Color. Red bats exhibit a wide range in coloration. The upperparts vary from bright orange red to buff, with the tips of the hairs often frosted. The males are redder and less frosted than the females. When parted, the hairs on the back show blackish bases followed by a broad band of golden yellow, then a narrow band of red, and white tips. The underparts are paler than the back and lack much of the white tipping. Both sexes have a yellowish white patch in front of each shoulder that may continue in a broken but noticeable band across the chest. The wing membrane is brownish black, contrasting with the pale forearm bone and fingers. The unfurred portion of the tail membrane is brownish. The ears are similar in color to the back. There is no seasonal difference in coloration.

Measurements

Total length	95–126 mm	3¾–4¹⁵⁄₁₆ in.
Tail	41–63 mm	1⅝–2½ in.
Hind foot	7–11 mm	⁵⁄₁₆–⁷⁄₁₆ in.
Ear	9–12 mm	⅜–½ in.
Skull length	12–14 mm	½–⁹⁄₁₆ in.
Skull width	9 mm	⅜ in.
Weight	5–14 g	⅕–½ oz.

Teeth and skull. The dental formula of the eastern red bat is:

$$I\ \frac{1}{3}\ C\ \frac{1}{1}\ P\ \frac{2}{2}\ M\ \frac{3}{3} = 32$$

The upper incisor touches the canine. The first upper premolar is located at the base of the canine on the tongue side; it is so small that it might be overlooked without the aid of a hand lens. As in all Missouri bats, the first two upper molars have a distinct W pattern on their surfaces and there is a prominent notch in the front of the hard palate. The skull is small, broad, and short. It is different in appearance from any other bat in Missouri except the hoary bat.

There are two other bats in Missouri (hoary bat and big brown bat) with 32 teeth: the hoary bat's skull is longer, and the big brown bat has 2 incisors and 1 premolar on each side of the upper jaw.

Sex criteria. The sexes are identified by their external sex organs, described in the account of the little brown myotis. There are two pairs of functional teats.

Age criteria. The young have shorter and grayer fur than adults.

Glands. Red bats have facial oil glands.

Voice and sounds. Like all insect-eating bats, red bats are guided in flight by means of their ultrasonic cries whose echoes are received by the inner ear. (See account of little brown myotis.)

Distribution and Abundance

Eastern red bats are widespread across the eastern United States and range into extreme southeastern Canada and northeastern Mexico. Most individuals breed in the north and winter in the south.

The eastern red bat occurs throughout Missouri, mostly from May through September, but some are also present during winter, especially in the southern part of the state. These latter may be migrants from farther north or local overwintering bats. An estimate of the summer population in one locality was 1 red bat per 0.4 ha (1 ac.), or 1 family per 1.8 ha (4.5 ac.).

Habitat and Home

The eastern red bat is primarily a solitary tree-inhabiting species. During summer, it frequents forested edge habitats around cities or farms and roosts in trees, shrubs, or even on the ground. They have been found to frequent the outer foliage of the canopy of relatively large dominant trees. In spring and fall these bats often hang on shrubs or tall weeds during the day. Less often do they use undisturbed parts of buildings or caves. Red bats that spend the winter in Missouri mostly hide in clusters of dead leaves, hollows of trees, or under bark, but a very few use caves. Red bats have been found overwintering in eastern red cedars and hardwoods located on south-facing slopes in southwest Missouri. When ambient temperatures approach freezing, red bats switch to roosting under leaf litter.

Habits

In summer, the adults are solitary, although sometimes several may roost in the same tree. They spend the day among the foliage, hanging by one or both hind feet from the branches. They look much like dead leaves. The females do not form nursery colonies.

In fall, a few red bats swarm at the mouths of certain caves, mingling with other species of bats. Many eastern red bats may migrate southward, beginning in September and continuing until November. Although most of the migration occurs at night, sometimes small numbers may be seen traveling together in the daytime. Little is known of the pattern of migration, but they probably migrate several hundred kilometers (miles) and sometimes fly great distances over water.

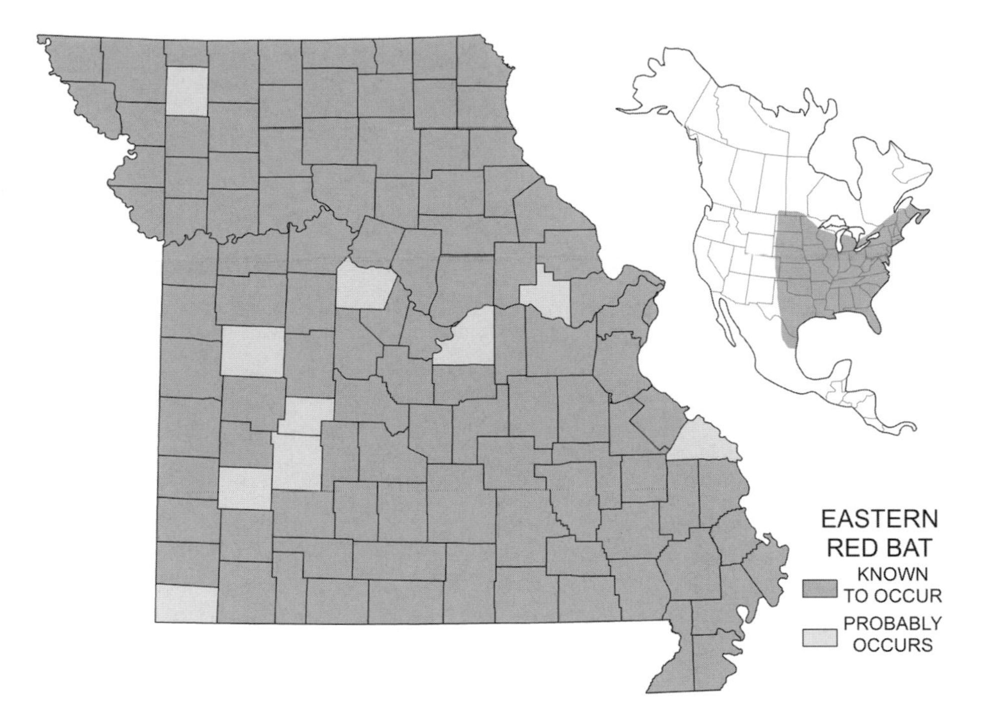

Prior to migration, these bats become very fat. This source of energy helps sustain them on their long flight and possibly supplements their meager winter fare.

Eastern red bats are adapted to survive drastic temperature fluctuations. Those that do not migrate normally spend the winter in sheltered spots such as tree cavities and respond to changes in temperature. When the environmental temperature rises, they may arouse and feed on what flying insects are available. In some caves dead red bats and their skeletons have been found. It is speculated that they spend the winter there under a generally constant temperature and are apparently unable to arouse spontaneously. Their stored energy is used to maintain body temperatures above freezing and, without additional food, they succumb.

Northward migration occurs late in spring.

Red bats start to feed 1–2 hours after dusk and continue into the night; however, some have been found active during the day, especially during winter. They forage, sometimes in great concentrations, along the edge of hill forests, floodplain timber, and fencerows. They are also seen over ponds and meadows and around streetlights in cities. Each bat has its own particular hunting area. In foraging they usually fly fairly high, erratic courses and occasionally descend spirally toward the ground. With their narrow, pointed wings they can maneuver with agility in a small area. On a level flight, one was timed at 64 km (40 mi.) per hour.

Foods

Red bats consume many types of flying insects: moths, beetles, crickets, flies, and cicadas. There is evidence suggesting that eastern red bats eavesdrop on the echolocation calls of other bats to identify the presence of prey.

Reproduction

Breeding occurs in August and September. The gestation period is between 80 and 90 days, but probably the eggs are not shed from the ovaries until spring, as occurs in other Missouri bats. The 1 to 4 young are born from late May through June.

At birth the young are blind and hairless, but the ears, tragus, and membranes are well developed. The eyes probably open in the first few days after birth, as occurs in most other Missouri bats. The female is very solicitous of her young, licking them and chewing droppings stuck to their fur.

The young help support themselves by holding onto a twig or leaf with a hind foot. They can fly when 4 to 6 weeks of age.

Some Adverse Factors; Importance; and Conservation and Management

These subjects have been discussed for bats in general at the beginning of the chapter and in the account of the little brown myotis. However, it is likely that red bats are subject to greater mortality than other bat species, as indicated by the higher number of young

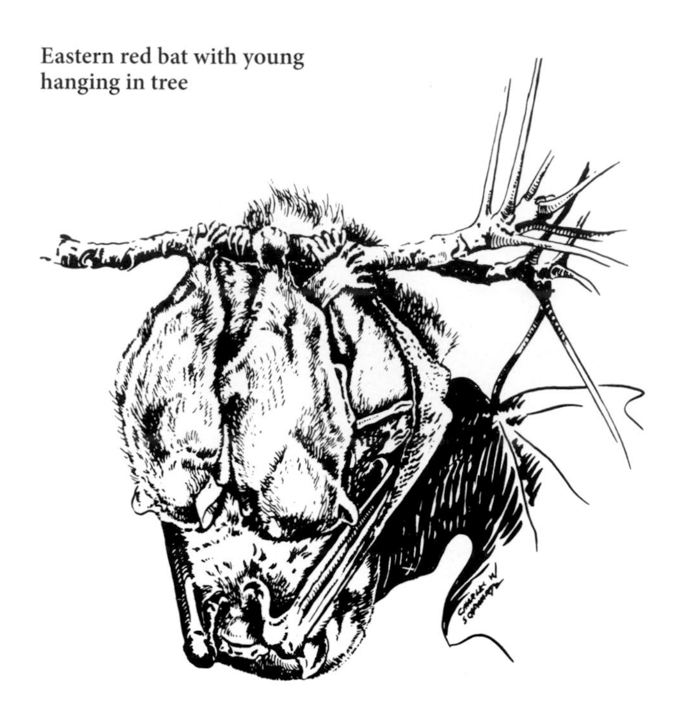

Eastern red bat with young hanging in tree

per female. Their relatively exposed roosts make them vulnerable to disturbances such as strong winds, rain, and predators. This species constitutes a large proportion of fatalities documented beneath wind turbines.

SELECTED REFERENCES

See also discussion of this species in general references, page 23.

Davis, W. H., and W. Z. Lidicker Jr. 1956. Winter range of the red bat, *Lasiurus borealis*. *Journal of Mammalogy* 37:280–281.

Hickey, M. B. C., and M. B. Fenton. 1990. Foraging by red bats (*Lasiurus borealis*): Do intraspecific chases mean territoriality? *Canadian Journal of Zoology* 68:2477–2482.

Hutchinson, J. T., and M. J. Lacki. 2000. Selection of day roosts by red bats in mixed mesophytic forests. *Journal of Wildlife Management* 64:87–94.

LaVal, R. K., and M. L. LaVal. 1942. Summer activities of bats (genus *Lasiurus*) in Iowa. *Journal of Mammalogy* 23:430–434.

———. 1979. Notes on reproduction, behavior, and abundance of the red bat, *Lasiurus borealis*. *Journal of Mammalogy* 60:209–212.

Mormann, B. M., and L. W. Robbins. 2007. Winter roosting ecology of eastern red bats in southwest Missouri. *Journal of Wildlife Management* 71:213–217.

Myers, R. F. 1960. *Lasiurus* from Missouri caves. *Journal of Mammalogy* 41:114–117.

Shump, K. A., Jr., and A. U. Shump. 1982. *Lasiurus borealis*. *Mammalian Species* 183. 6 pp.

Hoary Bat (*Lasiurus cinereus*)

Name

The first part of the scientific name, *Lasiurus*, is from two Greek words and means "hairy tail" (*lasios*, "hairy," and *oura*, "tail"). This refers to the furred upper surface of the tail membrane that is characteristic of this genus. The last part, *cinereus*, is Latin for "ash colored" and indicates the grayish color. "Hoary" refers to the grayish white or silvery, frosty look of the fur.

Description (Plate 16)

The ear is short, broad, and rounded and is furred both inside and out to the naked rim. The tragus is broad at the base, has a decided forward curve, and tapers to a rounded tip; it is less than one-half the length of the ear. The wing membrane is furred both above and below from the body outward toward the elbow; on the undersurface a strip of fur lies along the forearm to the wrist and about one-third of the way down the fingers. The wing membrane is attached along the side of the hind foot to the base of the toes. The large, wide tail membrane is well furred on its entire upper surface, but it is thinly furred below on only the half nearest the body. The calcar is keeled. The hind feet are well furred above. The body fur is long and soft; the hairs of the neck are longer than those of the back, forming a slight ruff.

The hoary bat is easily distinguished from all other bats in Missouri by its large size and coloration, and from all other bats except its close relative, the eastern red bat, by the well-furred tail membrane.

Color. The upperparts of the body are yellowish to dark brown with a frosty or hoary appearance. When the fur is blown into, each hair of the upperparts shows brownish black at the base, then cream colored grading to tan, followed by a narrow band of dark brown, and a whitish tip. The underparts are yellower and without so much white tipping. There is a conspicuous throat patch of long tan-tipped hairs with brownish black bases and a whitish shoulder patch that is usually continuous across the chest. The wing and tail membranes are brownish black; on the upper surface, the wing membrane is pale yellowish brown along the forearm and partway down the fingers, indicating the location of the furring below. The ears are tan and have black rims. A cream-colored spot is visible on the forearm just behind the thumb. The

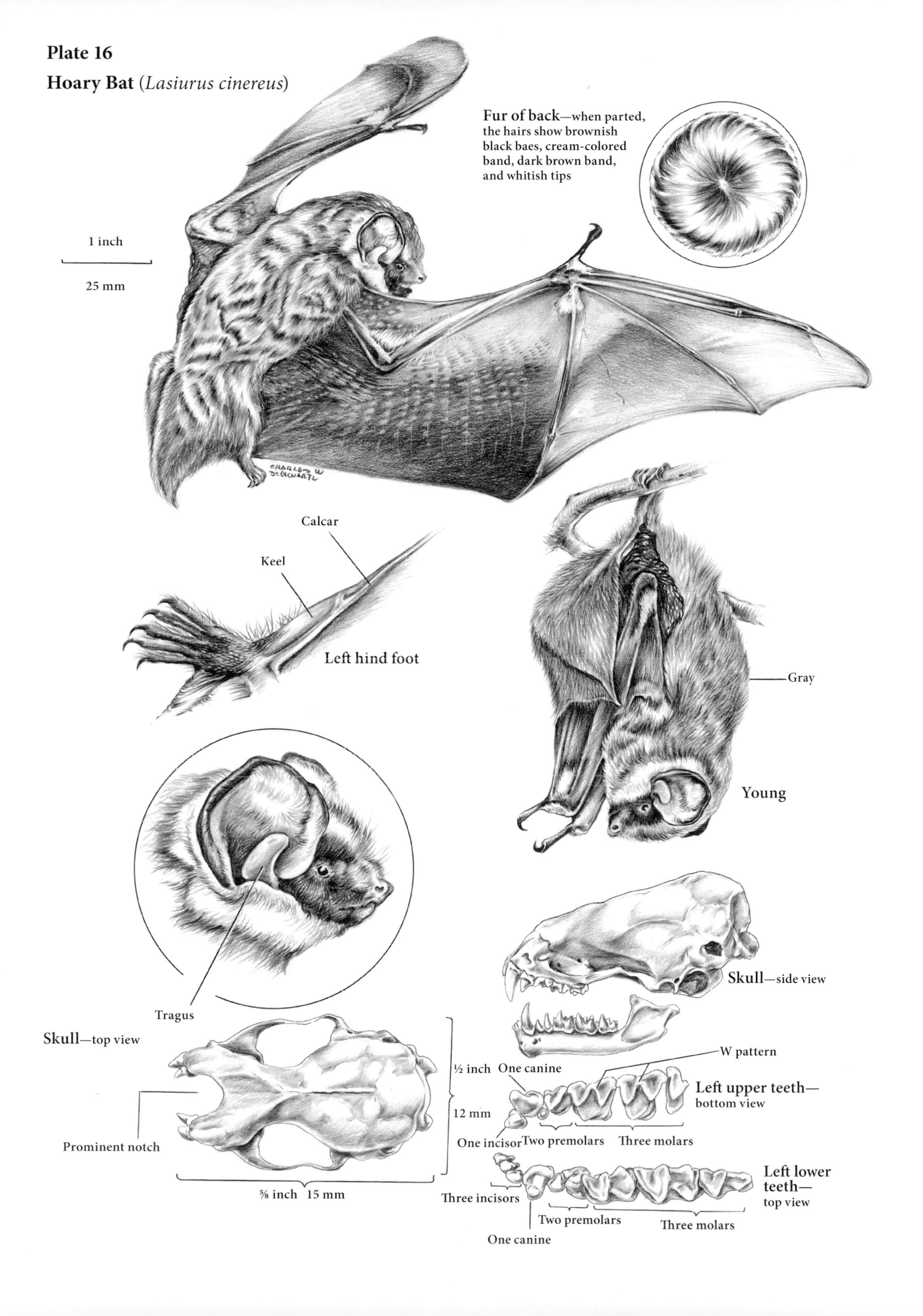

Plate 16

Hoary Bat (*Lasiurus cinereus*)

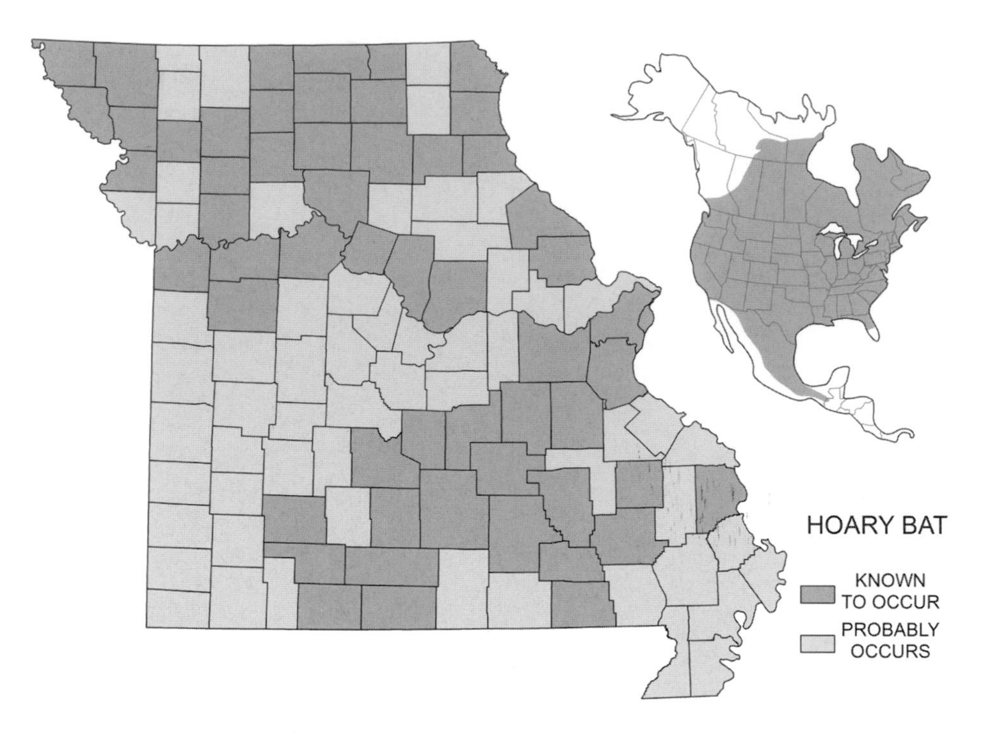

sexes are colored alike and show no seasonal change in coloration.

Measurements

Total length	102–149 mm	4–5⅞ in.
Tail	50–65 mm	2–2⁹⁄₁₆ in.
Hind foot	7–12 mm	¼–½ in.
Ear	19 mm	¾ in.
Skull length	15 mm	⅝ in.
Skull width	12 mm	½ in.
Weight	21–42 g	¾–1½ oz.

Females are larger than males.

Teeth and skull. The dental formula of the hoary bat is:

$$I\,\frac{1}{3}\;C\,\frac{1}{1}\;P\,\frac{2}{2}\;M\,\frac{3}{3} = 32$$

There are two other bats in Missouri (eastern red bat and big brown bat) with 32 teeth: the red bat's skull is shorter, and the big brown bat has 2 incisors and 1 premolar on each side of the upper jaw.

Sex criteria. The sexes are identified as in the little brown myotis. There are two pairs of teats.

Age criteria. The young are pale gray.

Voice and sounds. Like all insect-eating bats, hoary bats are guided in flight by means of their ultrasonic cries. (See account of the little brown myotis.) They also have a chatter that is audible to humans.

Distribution and Abundance

Hoary bats are the most widespread of all bats in North America. They range from the tree line in Canada south into Central America. They occur in all of the lower 48 U.S. states but have not yet been recorded in Alaska. The Hawaiian hoary bat, the ʻopeʻapeʻa, an endangered subspecies, is endemic to Hawaii and is the only bat that occurs there. Hoary bats also occur in South America, Iceland, Bermuda, and a few other locations.

Hoary bats are highly migratory with the northern part of their range used primarily in summer and the southern in winter. Hoary bats are generally uncommon in all of their range; they are most common in the southwestern United States.

Hoary bats occur in Missouri during spring, summer, and fall and are most commonly seen here in May. Some young are reared here in summer. This species is considered relatively rare throughout the state, but acoustic detectors have suggested that they may be more common than previously thought; hoary bats tend to fly high and many may be missed during mist-netting surveys.

Habitat and Home

This is a tree-inhabiting bat, usually roosting 3–5 m (9–17 ft.) above the ground in the foliage. Only rarely does it roost in caves in late summer.

Habits

Although a few hoary bats may stay in the north all year, most migrate in October and November to southern latitudes. The return migration occurs in spring. Although they typically travel individually or in small groups, large flocks have been reported.

In summer no nursery colonies are formed, but several individuals of the same sex may occur together. Males and females are geographically segregated during the summer, with females largely occupying the northern regions of the range, and males largely in the north and southwest. During the day hoary bats hang from tree branches by their hind feet. They choose a leafy site that is well covered above but generally open beneath. These sites are usually 3 to 5 m (10 to 15 ft.) above the ground and at the edge of a clearing.

Late in the evening they leave their roosting sites to forage. They fly above watercourses and meadows, following a very direct route. Their flight is fairly swift, strong, and high. Individuals do not share foraging areas.

Foods

Hoary bats feed extensively on moths and water boatmen. They also consume many wasps, dragonflies, beetles, and some mosquitoes.

Reproduction

Breeding occurs in late summer or early fall during migration. The length of gestation is unknown, but probably the eggs are not shed from the ovaries until spring. The young are born from late May to early July. There are usually 1 to 4 (average 2) young in a litter.

At birth the young are blind, covered with a fine fur, and weigh about 3.5 g (⅛ oz.) each. Females roost solitarily in foliage during pup-rearing. Growth of young is highly influenced by ambient temperature. They are free-flying around 1 month of age.

Some Adverse Factors; Importance; and Conservation and Management

These subjects have been discussed for bats in general at the beginning of the chapter and in the account of the little brown myotis. Their flight behavior might increase their risks of mortality at wind turbines.

SELECTED REFERENCES

See also discussion of this species in general references, page 23.

Bogan, M. A. 1972. Observation on parturition and development in the hoary bat, *Lasiurus cinereus. Journal of Mammalogy* 53:611–614.

Chromanski-Norris, J. F., and E. K. Fritzell. 1983. Status and distribution of ten Missouri mammals. A report to the Missouri Department of Conservation, Jefferson City. 38 pp.

Findley, J. S., and C. Jones. 1964. Seasonal distribution of the hoary bat. *Journal of Mammalogy* 45:461–470.

Koehler, C. E., and R. M. R. Barclay. 2000. Post-natal growth and breeding biology of the hoary bat (*Lasiurus cinereus*). *Journal of Mammalogy* 81:234–244.

Myers, R. F. 1960. *Lasiurus* from Missouri caves. *Journal of Mammalogy* 41:114–117.

Poole, E. L. 1932. Breeding of the hoary bat in Pennsylvania. *Journal of Mammalogy* 13:365–367.

Provost, E. E., and C. M. Kirkpatrick. 1952. Observations on the hoary bat in Indiana and Illinois. *Journal of Mammalogy* 33:110–113.

Reimer, J. P., E. F. Baerwald, and R. M. R. Barclay. 2010. Diet of hoary (*Lasiurus cinereus*) and silver-haired (*Lasionycteris noctivagans*) bats while migrating through southwestern Alberta in late summer and autumn. *American Midland Naturalist* 164:230–237.

Rolseth, S. L., C. E. Kochler, and R. M. R. Barclay. 1994. Differences in the diets of juvenile and adult hoary bats, *Lasiurus cinereus. Journal of Mammalogy* 75:394–398.

Shump, K. A., and A. U. Shump. 1982. *Lasiurus cinereus. Mammalian Species* 185. 5 pp.

Zinn, T. L., and W. W. Baker. 1979. Seasonal migration of the hoary bat, *Lasiurus cinereus*, through Florida. *Journal of Mammalogy* 60:634–635.

Evening Bat (*Nycticeius humeralis*)

Name

The first part of the scientific name, *Nycticeius*, is of Greek and Latin origin and means "belonging to the night" (*nyx*, "night," and *-eius*, "belonging to"). The second part, *humeralis*, is of Latin origin and means "pertaining to the humerus," which is the bone of the upper arm (*humerus*, "humerus," and *-alis*, "pertaining to"). The common name denotes the time when these bats are observed to fly about and feed.

Description (Plate 17)

The ears are thick and almost naked; their tips are rounded. The broad tragus is curved slightly forward and has a rounded or blunt tip; it is short, less than one-half the length of the ear. The tip of the wing is broadly rounded. The wing and tail membranes are thick and almost naked. The wing membrane is attached along the side of the foot to the base of the toes.

Plate 17
Evening Bat (*Nycticeius humeralis*)

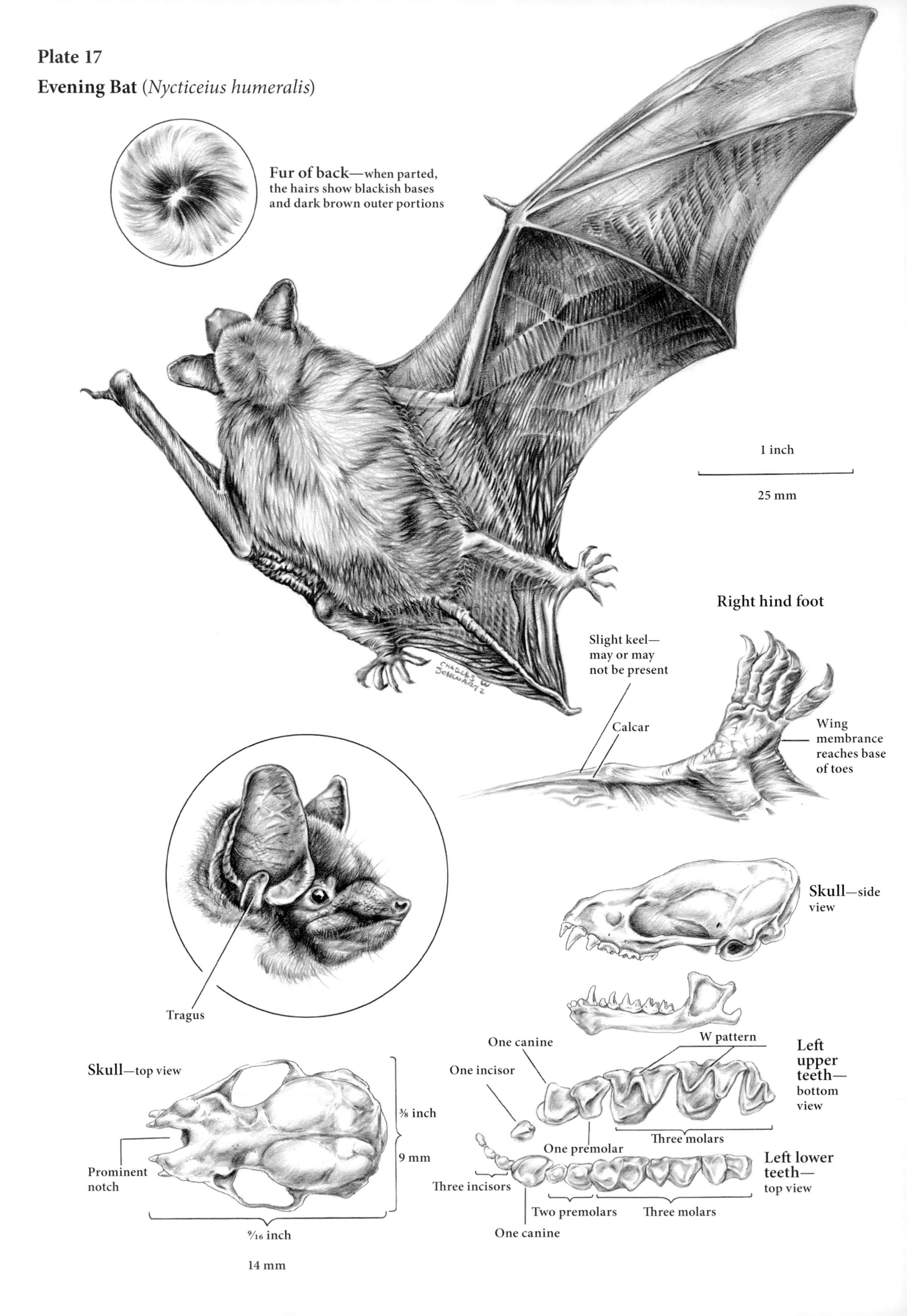

The calcar may or may not be keeled. The bones of the tail extend obviously (1.6 mm; 1⁄16 in.) beyond the tail membrane. The body fur is thick and moderately long.

The evening bat is most likely to be confused with the big brown bat, the mouse-eared bats, and the tricolored bat. However, these species can be distinguished by some or all of the following characteristics: dental formula, size, color and character of the fur, and shape of the ear and tragus.

Color. The general color of the upperparts is dark brown; when the hairs are parted by blowing into the fur, the bases can be seen to be blackish and the outer portions dark brown. The underparts are paler and more buffy. The wing and tail membranes, the ears, and the muzzle are blackish. The sexes are colored alike and show no seasonal variation.

Measurements

Total length	82–95 mm	3 3⁄16–3¾ in.
Tail	34–41 mm	1⅜–1⅝ in.
Hind foot	6–9 mm	¼–⅜ in.
Ear	12–14 mm	½–9⁄16 in.
Skull length	14 mm	9⁄16 in.
Skull width	9 mm	⅜ in.
Weight	2–14 g	1⁄10–½ oz.

Teeth and skull. The evening bat is the only bat in Missouri with 30 teeth. The dental formula is:

$$I\ \frac{1}{3}\ C\ \frac{1}{1}\ P\ \frac{1}{2}\ M\ \frac{3}{3} = 30$$

The upper incisor is distinctly separate from the canine. Like all Missouri bats, the first two upper molars have a distinct W pattern on their surfaces and there is a prominent notch in the front end of the hard palate. The skull is short and broad and has a low profile.

Sex criteria. The sexes are identified as in the little brown myotis. There is one pair of teats.

Age criteria. The young are darker than adults.

Glands. Oil glands are prominent on the muzzle.

Voice and sounds. Like other bats, this bat has ultrasonic cries that guide it in flight. (See account of the little brown myotis.)

Distribution and Abundance

The evening bat is found throughout much of the midwestern and eastern United States as well as northeastern Mexico. Within the Gulf Coast states and other areas, they are very common, but they are uncommon throughout the Appalachians and the northern part of the range and may be declining in some states.

Evening bats probably occur statewide. Male evening bats have been captured throughout the year in Missouri, and there is evidence to support year-long habitation by females as well.

Habitat and Home

The evening bat is mainly a tree-inhabiting species, but it frequently uses attics or other undisturbed places in buildings. Evening bats tend to select similar roosting trees and habitats during the winter months.

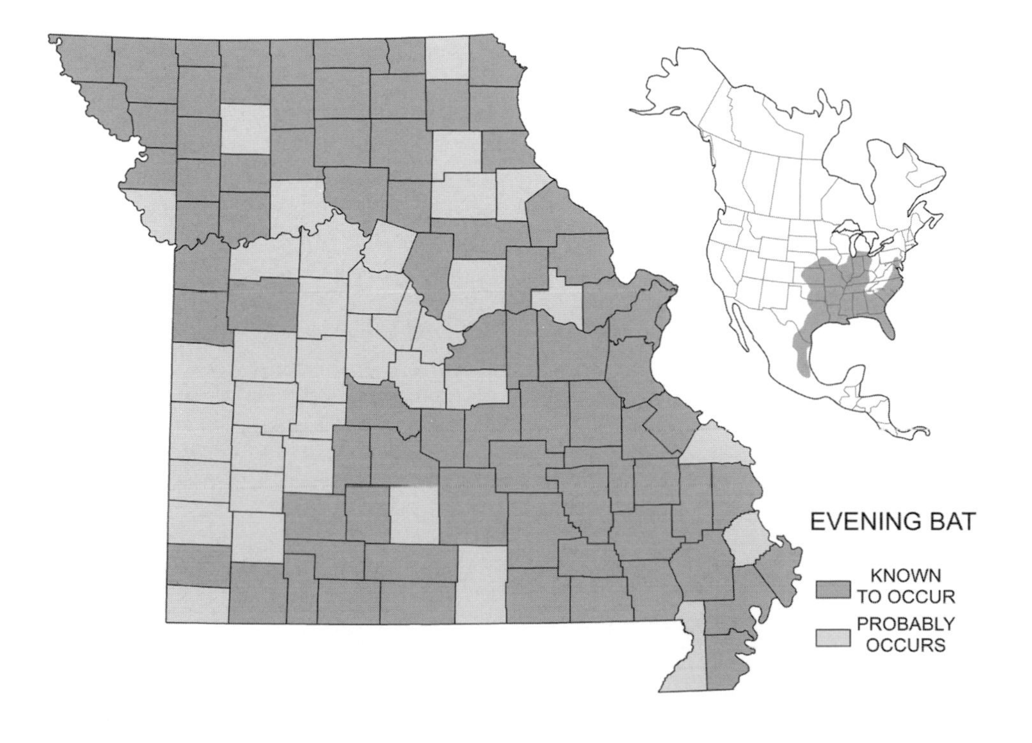

Habits

During the summer, nursery colonies are formed in cavities or under the bark of dead trees or in houses. Here there may be hundreds of young.

Females show an attachment to the nursery colony and have returned when released 155 km (96 mi.) away. The males do not migrate north in the spring with the females; instead, they remain in the southern latitudes.

When foraging during the last half hour of daylight, the evening bat flies from 12 to 23 m (40 to 75 ft.) high. But as darkness falls the bat comes closer to the ground, sometimes spiraling downward from these heights. In general, it is a slow, steady flier. Usually there is a peak of feeding activity after the bats leave the roost and a lesser peak before reentry in the morning. If young are in the roost, females return midway through the night to feed them.

When the nights become cool in fall, the bats spend less time away from the roost. They accumulate fat in autumn that is sufficient for a migration of possibly hundreds of miles, hibernation, or both. In the more southern reaches of their range evening bats may be active year-round.

Foods

Because this species generally forages in open habitat, flying insects, including mostly beetles, cicadas, and true bugs, form the primary foods.

Reproduction

Mating occurs in late summer and probably again in spring, during migration and with males occasionally visiting colonies. The length of the gestation period is unknown, but presumably the eggs are not shed from the ovaries until spring. The 1 to 3 young (usually twins) are born from the end of May through June. Each weighs about 1.4 g (1/20 oz.) at birth.

The newborn young are pink except for dark pigmentation on the feet, membranes, ears, and lips. Within 24 hours after birth, heavy pigmentation occurs on the skin and the eyes open. During the third week, they begin to fly about but continue to nurse, sometimes communally, until 4 weeks of age. In captivity, as the young grow older, the females are not particular about whose young they nurse.

Some Adverse Factors; Importance; and Conservation and Management

These subjects have been discussed for bats in general at the beginning of the chapter and in the account of the little brown myotis. It is important to point out that most evening bat maternity roosts are established in human structures, and this puts these bats at particular risk. As a result of human intolerance and of the removal of aging structures, most of these roosts do not last very long. The degree of impact of these habitat interruptions is unknown.

SELECTED REFERENCES

See also discussion of this species in general references, page 23.

Boyles, J. G., and L. W. Robbins. 2006. Characteristics of summer and winter roost trees used by evening bats (*Nycticeius humeralis*) in southwestern Missouri. *American Midland Naturalist* 155:210–220.

Easterla, D. A. 1965. A nursery colony of evening bats in southern Missouri. *Journal of Mammalogy* 46:498.

Gates, W. H. 1941. A few notes on the evening bat, *Nycticeius humeralis* (Rafinesque). *Journal of Mammalogy* 22:53–56.

Hooper, E. T. 1939. Notes on the sex ratio in *Nycticeius humeralis*. *Journal of Mammalogy* 20:369–370.

Mumford, R. E. 1953. Status of *Nycticeius humeralis* in Indiana. *Journal of Mammalogy* 34:121–122.

Watkins, L. C. 1970. Observations on the distribution and natural history of the evening bat (*Nycticeius humeralis*) in northwestern Missouri and adjacent Iowa. *Transactions of the Kansas Academy of Science* 72:330–336.

———. 1972a. A technique for monitoring the nocturnal activity of bats, with comments on the activity patterns of the evening bat, *Nycticeius humeralis. Transactions of the Kansas Academy of Science* 74:262–268.
———. 1972b. *Nycticeius humeralis. Mammalian Species* 23. 4 pp.
Whitaker, J. O., Jr., and P. D. Clem. 1992. Food of the evening bat, *Nycticeius humeralis*, from Indiana. *American Midland Naturalist* 127:211–214.
Whitaker, J. O., Jr., and S. L. Gummer. 1993. The status of the evening bat, *Nycticeius humeralis*, in Indiana. *Proceedings of the Indiana Academy of Science* 102:283–291.

Townsend's Big-eared Bat (*Corynorhinus townsendii*)

Name

The first part of the scientific name, *Corynorhinus*, is from a Latinized form of Greek roots meaning "club-nosed." The second part, *townsendii*, is a Latinized name meaning "of Townsend" and this latter was given in honor of John K. Townsend, M.D.

The common name, "big-eared," refers to the size of the ears. This bat is sometimes called the lump-nosed bat because of the conspicuous growths on the nose. Another common name, for the subspecies historically occupying Missouri, is the Ozark big-eared bat (*C. townsendii ingens*), referring to the geographical region with which this bat is associated in southwestern Missouri.

This species was formerly known as *Plecotus townsendii*.

Description (Plate 18)

The most conspicuous feature of this bat is its very large ears. The ears reach the middle of the body when laid back and are joined across the forehead at their bases. They are thin, naked except for scant furring along the edges, and narrowly rounded at the tips. The tragus is slender, slightly less than half the length of the ear, and rounded at the tip. Prominent, paired, glandular masses occur on the face between the nostril and eye and project upward; their function is at present unknown. The wing and tail membranes are very thin and naked. The wing membrane is attached along the side of the foot to the base of the lightly furred toes. The calcar is not keeled. The body fur is soft and long.

Townsend's big-eared bat is distinguished from its close Missouri relative, Rafinesque's big-eared bat, by the pinkish buff tips on the hairs of the belly, by the little contrast in color between the basal portions and tips of the hairs on the back and belly, and by the absence of long hairs that project beyond the toes. A difference in the upper incisor teeth is discussed under *Teeth and Skull* below. The large ears and lumps on the nose distinguish big-eared bats from all other bats in Missouri.

Color. The general color above is buffy tan to brown, while below it is pinkish buff. On both back and belly, when the fur is parted, little contrast shows between the basal portions of the hairs and the tips. The wing and tail membranes and the ears are dark brown. The sexes are colored alike and there is no seasonal color variation.

Measurements

Total length	90–112 mm	3½–4⅜ in.
Tail	41–63 mm	1⅝–2½ in.
Hind foot	9–12 mm	⅜–½ in.
Ear	30–39 mm	1³⁄₁₆–1⁹⁄₁₆ in.
Skull length	15 mm	⅝ in.
Skull width	9 mm	⅜ in.
Weight	5–14 g	⅕–½ oz.

Teeth and skull. The dental formula of Townsend's big-eared bat is:

$$I\ \frac{2}{3}\ C\ \frac{1}{1}\ P\ \frac{2}{3}\ M\ \frac{3}{3} = 36$$

The first upper incisor usually has 1 cusp. As in all Missouri bats, the first 2 upper molars have a distinct W pattern on their surfaces and a prominent notch occurs in the front end of the hard palate. The skull is highly arched. The rostrum, or part of the skull in front of the eye sockets, is narrow and about half as wide as the braincase when viewed from the top. The rostrum is convex on each side at the back of the nasal opening. The skull of Townsend's big-eared bat might be confused with that of the silver-haired bat because it has the same number and arrangement of teeth, but the aforementioned characteristics identify Townsend's big-eared bat. The skulls of the two species of big-eared bats are not easy to distinguish, but the first upper incisor usually has 1 cusp in Townsend's big-eared bat and 2 cusps in Rafinesque's big-eared bat.

Sex criteria and sex ratio. The sexes are identified as in the little brown myotis. There is one pair of teats. The sex ratio is equal at birth.

Age criteria and longevity. The age of big-eared bats can be determined by the condition of the reproductive organs and accessory structures. Adult (bred) females can be identified by the right branch of the uterus, which tends to be broad (1.6 mm; ¹⁄₁₆ in. in

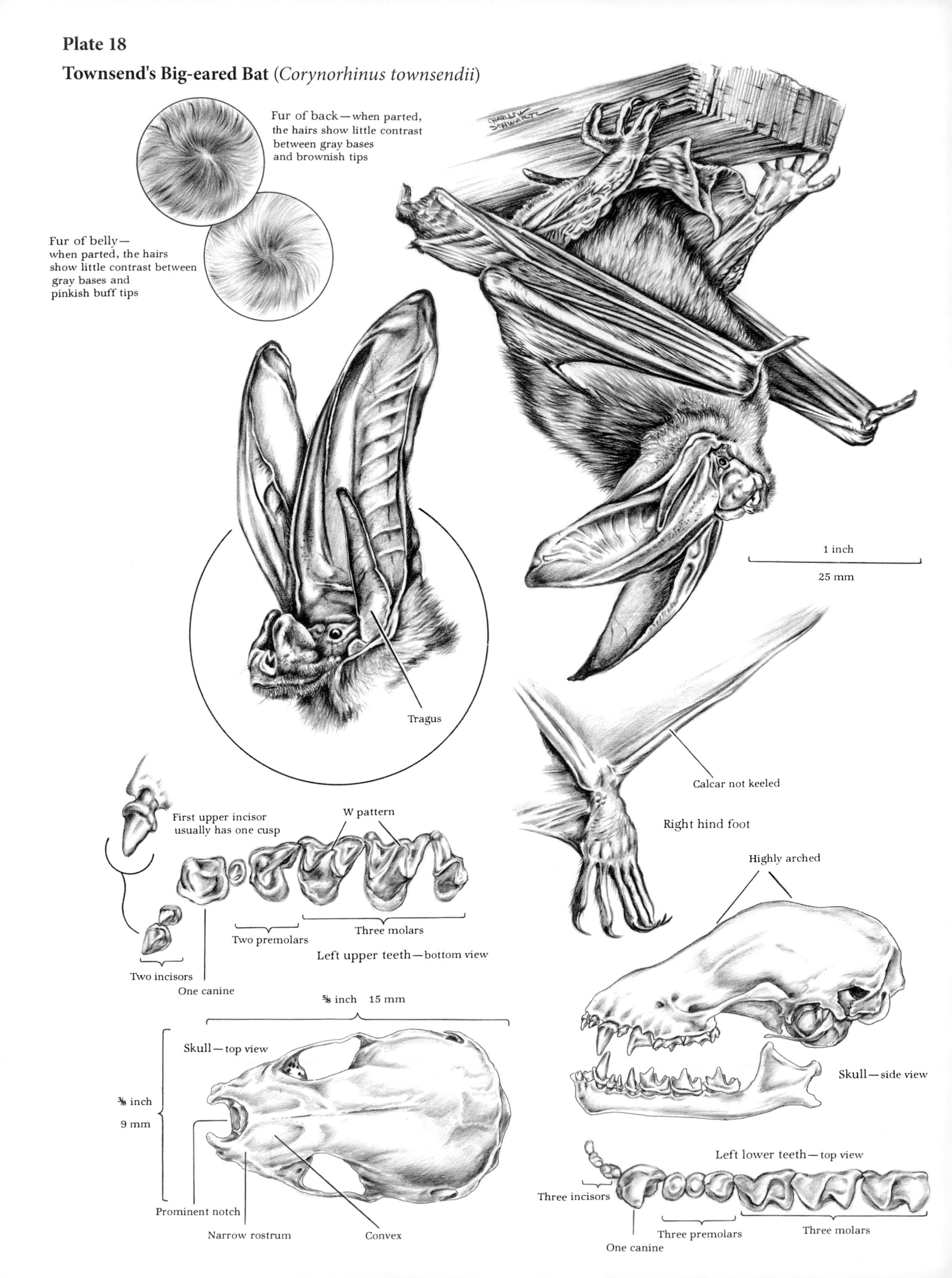

Plate 18

Townsend's Big-eared Bat (*Corynorhinus townsendii*)

diameter), resulting from a pregnancy; by the artery on the front of the right branch of the uterus, which is winding and surrounded by a conspicuous sheath; by the nipples, which are large; and by the skin surrounding the nipples, which, when the fur is blown aside, is usually pale. In contrast, young females under 9 months of age are identified by the right branch of the uterus not being broad (only 0.8 mm; 1⁄32 in. in diameter); by the artery on the front of the right branch of the uterus, which is straight and thin walled; by the nipples, which are tiny; and by the skin surrounding the nipples, which is frequently rosy or purple.

Adult males are recognized from November through March by their large epididymides (tubes containing sperm), which are conspicuous in the tail membrane behind the body; young males in their first winter do not show this sexual development. By April and May, it is difficult to distinguish age by the size of the epididymides because those of adults have dwindled to the size found in young males. (See discussion of other aging methods for bats given in the account of the little brown myotis.)

Big-eared bats are known to have reached 16 years of age in the wild.

Glands. It is believed that the knobby scent glands on the face may be related in some way to reproduction.

Voice and sounds. This bat directs its flight by means of ultrasonic cries, as do other bats (see the account of the little brown myotis). The echolocation calls of big-eared bats are of relatively low volume, leading them to be referred to as "whisper bats."

Distribution and Abundance

The Townsend's big-eared bat has a wide range in the western United States and Mexico, where it is abundant but listed as vulnerable. Isolated populations occur across the gypsum cave region of Texas, Oklahoma, and Kansas and in the limestone cave region of the Ozarks and Kentucky, and eastward. There are two federally endangered subspecies: the Ozark big-eared bat, which currently occurs in Missouri, Oklahoma, and Arkansas; and the Virginia big-eared bat (*C. townsendii virginianus*), which occurs in portions of Kentucky, North Carolina, Virginia, and West Virginia.

The Ozark big-eared bat subspecies was historically reported from Stone and Barry Counties in the Ozark Highlands; however, it has not been sighted in Missouri since the 1970s, and if present it is very rare. This subspecies is listed as extirpated among Missouri's species of conservation concern, and it is both state and federally endangered, and critically imperiled, across its subspecies range.

Habitat and Home

In winter Townsend's big-eared bats hibernate in caves and rocky crevices, in tunnels and old mine shafts, in hollow trees or beneath the loose bark of trees, and even in buildings. In summer they live in caves, old

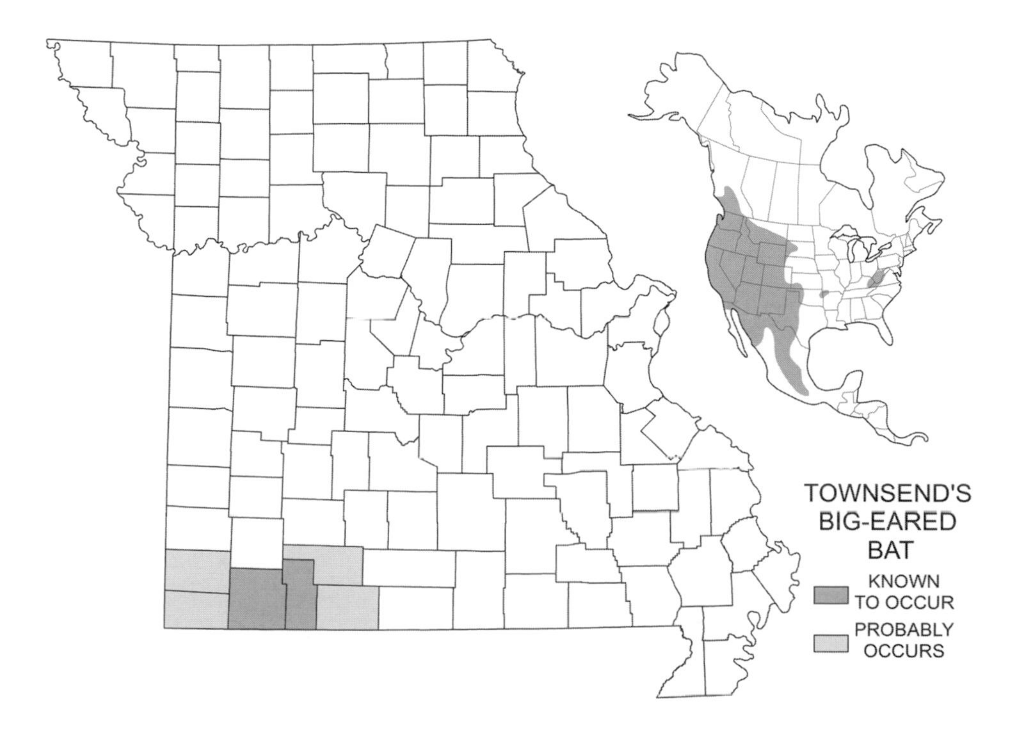

buildings, or rocky crevices. The Ozark subspecies is considered a year-round cave- or crevice-roosting bat.

Habits

Townsend's big-eared bats hibernate from October to April. During hibernation these bats usually hang singly or in clusters from a few to as many as 50 individuals of mixed sexes and ages. They generally select rather exposed and cold places; in caves they occur near the entrance in the "twilight zone." However, if temperatures become too severe, the bats move to more thermally stable areas in the cave. Although females are more profound hibernators than males, both sexes awaken periodically, fly about their hibernating quarters briefly, then return to their torpor. Most of this activity occurs at night. Sometimes they even change their hibernating site and fly from one cave to another.

When big-eared bats hibernate, they hang by their hind feet with the head downward and their wings slightly open, covering the body. Individuals that hang singly or in small groups coil their large ears spirally and flatten them against the neck.

When the bats awaken, the ears are straightened out and then held erect. There is a tendency for bats in larger clusters to keep their ears erect.

Movement from winter to summer quarters probably involves fairly short distances, and they may occupy the same location. In the western United States, almost all banded big-eared bats were recaptured at the same place where they were banded or within 2.4 km (1½ mi.) of the banding site. The greatest known distance traveled was 64.4 km (40 mi.). Three big-eared bats taken from a cave in May and released 45 km (28 mi.) away were recovered two days later in the original banding location.

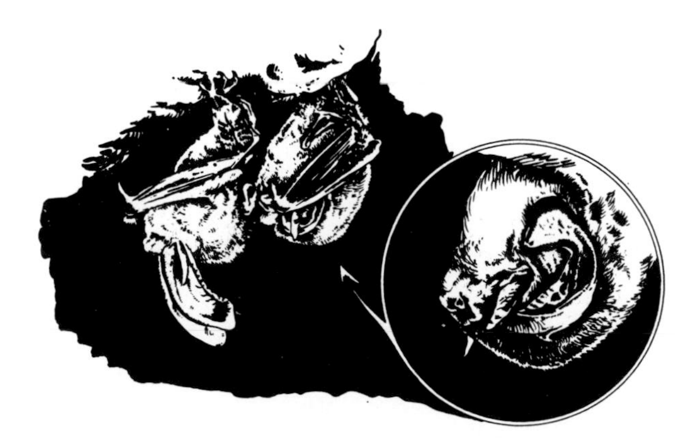

In summer, the females establish nursery colonies that may contain several hundred individuals. They ordinarily use the same site year after year. If disturbed, the entire female colony has been known to move 2.4 km (1½ mi.) to another location. In these colonies the females cluster close together, which is a means of conserving heat and keeping temperatures high enough for rapid development of the young. During summer the males are solitary.

Big-eared bats are swift fliers and agile dodgers. They usually do not come out until dark and, after a feeding period, rest in a different place from the daytime roost. They have another 1 or 2 (depending on reproductive condition) feeding period(s) before returning to their daytime retreat about dawn.

Foods

The edge between open and forested areas is the preferred foraging habitat. Moths constitute the main foods of Townsend's big-eared bats. Also taken are beetles, leafhoppers, and small flies.

Reproduction

Mating takes place in fall, during winter (even at times when the females are dormant), and again in spring. Gestation occurs in spring and lasts from 56 to 100 days, the shorter time occurring in females that maintain a higher body temperature by clustering together.

The single (rarely 2) young is born in May or June. While giving birth the female hangs by both her hind feet and thumbs with the wing and tail membranes spread to catch the young. The mother eats the placenta and licks all the fluids. A few minutes after birth the young attaches to a nipple.

At birth the young bat weighs 2.8 g (1/10 oz.). The body is pinkish gray and naked, the dark gray ears are long and pointed, the eyes are closed, the dark gray nose lumps are conspicuous, the tail membrane is pale gray, the wings are tiny, the thumbs are relatively large, the legs are long and stout, and the tail is short.

The young is left hanging in the nursery colony at night while the mother forages; upon returning, she finds her offspring in the large cluster of young and nurses it during the day. The eyes open when the young bat is little more than 1 week old. Although the young flies when 3 weeks old, it does not forage abroad at night until about 6 weeks of age. Young males do not mate in their first fall, but young females do.

Some Adverse Factors; Importance; and Conservation and Management

These subjects have been discussed for bats in general at the beginning of the chapter and in the account of the little brown myotis. However, Townsend's big-eared bats are particularly vulnerable to human disturbance and vandalism at maternity and hibernating caves and mines. Efforts to protect these bats include gating cave and mine entrances that have a high risk of human disturbance. Because of preferences for colder hibernating caves, it is imperative that such gates permit enough air flow to maintain the appropriate range of temperatures within the cave.

SELECTED REFERENCES

See also discussion of this species in general references, page 23.

Clark, B. K., B. S. Clark, D. M. Leslie Jr., and M. S. Gregory. 1996. Characteristics of caves used by the endangered Ozark big-eared bat. *Wildlife Society Bulletin* 24:8–14.

Clark, B. S., D. M. Leslie Jr., and T. S. Carter. 1993. Foraging activity of adult female Ozark big-eared bats (*Plecotus townsendii ingens*) in summer. *Journal of Mammalogy* 74:422–427.

Clark, B. S., B. K. Clark, and D. M. Leslie Jr. 2002. Seasonal variation in activity patterns of the endangered Ozark big-eared bat (*Corynorhinus townsendii ingens*). *Journal of Mammalogy* 83:590–598.

Dalquest, W. W. 1947. Notes on the natural history of the bat *Corynorhinus rafinesquii* in California. *Journal of Mammalogy* 28:17–30.

Handley, C. O., Jr. 1959. A revision of American bats of the genera *Euderma* and *Plecotus*. *Proceedings of the U.S. National Museum* 110:95–246.

Humphrey, S. R., and T. H. Kunz. 1976. Ecology of a Pleistocene relict, the western big-eared bat (*Plecotus townsendii*) in the southern Great Plains. *Journal of Mammalogy* 57:470–494.

Kunz, T. H., and R. A. Martin. 1982. *Plecotus townsendii*. *Mammalian Species* 175. 6 pp.

Pearson, O. P., M. R. Koford, and A. K. Pearson. 1952. Reproduction of the lump-nosed bat (*Corynorhinus rafinesquii*) in California. *Journal of Mammalogy* 33:273–320.

Sample, B. E., and R. C. Whitmore. 1993. Food habits of the endangered Virginia big-eared bat in West Virginia. *Journal of Mammalogy* 74:428–435.

Wethington, J. A., D. M. Leslie Jr., M. S. Gregory, and M. S. Wethington. 1996. Prehibernation habitat use and foraging activity by endangered Ozark big-eared bats (*Plecotus townsendii ingens*). *American Midland Naturalist* 135:218–230.

White, D. H., and J. T. Seginak. 1987. Cave gate designs for use in protecting endangered bats. *Wildlife Society Bulletin* 15:445–449.

Rafinesque's Big-eared Bat (*Corynorhinus rafinesquii*)

Name

The first part of the scientific name, *Corynorhinus*, is a Latinized form of Greek roots that mean "club-nosed." The second part, *rafinesquii*, is a Latinized name meaning "of Rafinesque" and was given in honor of Constantine Samuel Rafinesque, a famous early American naturalist.

The common name, "big-eared," refers to the size of the ears. This bat is sometimes called the lump-nosed bat because of the conspicuous growths on the nose.

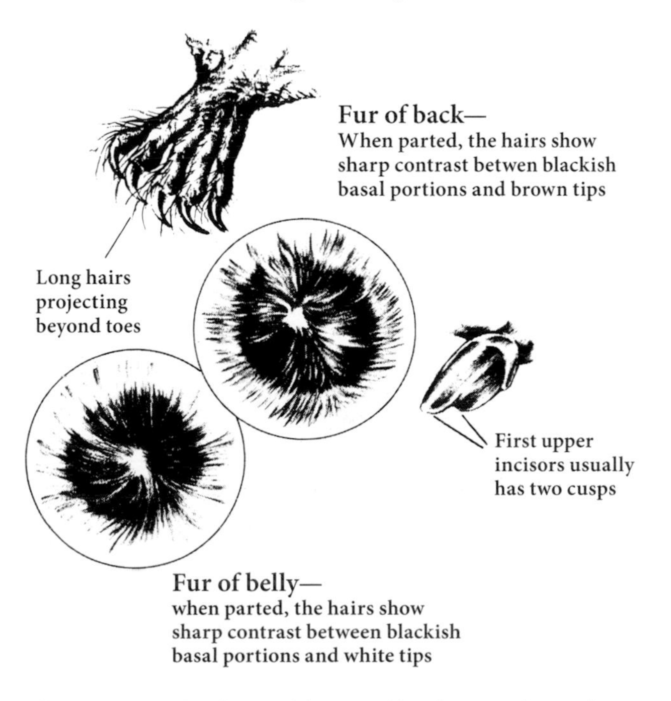

Characteristics of Rafinesque's big-eared bat distinguishing it from Townsend's big-eared bat

This species was formerly known as *Corynorhinus macrotis* and most recently as *Plecotus rafinesquii.*

Description

Rafinesque's big-eared bat closely resembles its relative, Townsend's big-eared bat, but is distinguished by the white-tipped hairs on the belly, by the strong contrast in color between the basal portions and tips of the hairs on the back and belly, and by the presence of long hairs that project beyond the toes. A difference in the upper incisor teeth is discussed under *Teeth and Skull* below.

Color. The back is pale brown and the belly whitish. When parted, the individual hairs of both back and belly show a sharp contrast between their blackish basal portions and brown (back) or white (belly) tips (see page 117).

Measurements

Total length	80–107 mm	3⅛–4¼ in.
Tail	47–50 mm	1⅞–2 in.
Hind foot	9 mm	⅜ in.
Ear	31 mm	1¼ in.
Skull length	15 mm	⅝ in.
Skull width	9 mm	⅜ in.
Weight	5–14 g	⅕–½ oz.

Teeth and skull. The dental formula is the same as that of the Townsend's big-eared bat. There are usually 2 cusps on the first upper incisor. In contrast, the first upper incisor of the Townsend's big-eared bat usually has 1 cusp.

Distribution and Abundance

Rafinesque's big-eared bat is uncommon throughout its range in the southeastern United States and is reported to be declining in some areas.

This species has been collected in five counties in southeastern Missouri, and it probably occurs throughout the Mississippi River Alluvial Basin. Nowhere is it abundant. It is classified as a critically imperiled species of conservation concern in Missouri but is listed as vulnerable to apparently secure across its North American range.

Habitat and Home

Rafinesque's big-eared bat is known to hibernate in caves in other parts of its range and may use caves or mines in Missouri during the winter. However, in some areas of its range where it is common, there are no caves. Distinguishing these from Townsend's big-eared bat, nursery colonies have been reported in tree cavities and in buildings but only rarely in caves. In summer, males tend to roost in buildings or behind the loose bark of trees and in hollow trees. They are associated with forested areas and riparian habitats throughout their range.

Life History

Due to its rarity there is a lack of information pertaining to this species in Missouri and throughout its North American range. It is presumed that this bat's way of life is very similar to that of Townsend's big-eared bat.

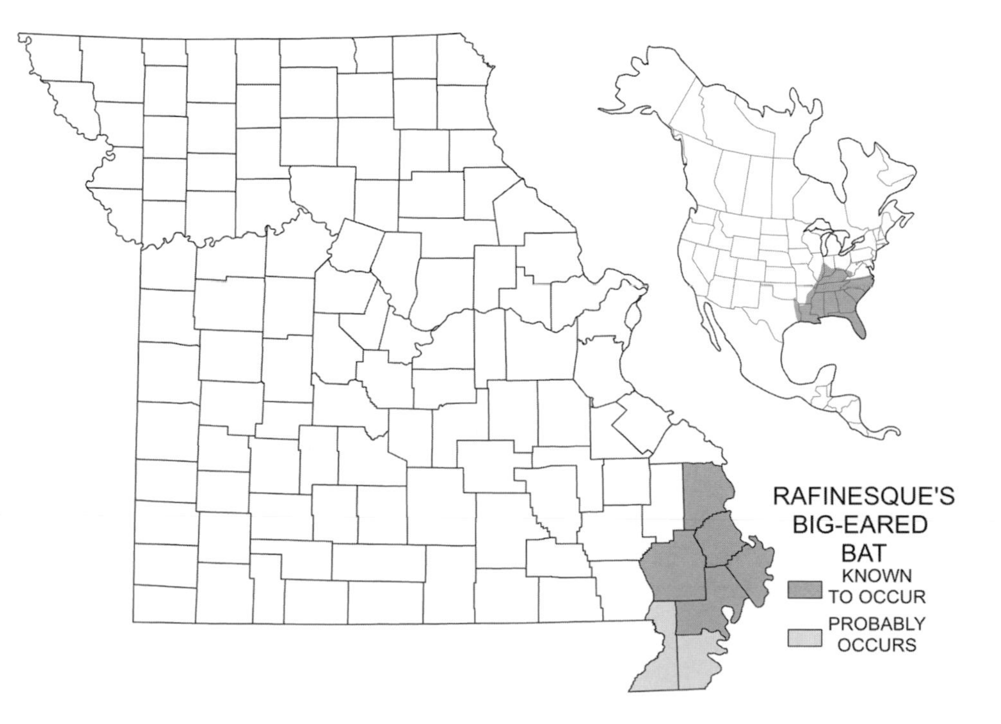

Some Adverse Factors; Importance; and Conservation and Management

These subjects have been discussed for bats in general at the beginning of the chapter and in the account of the little brown myotis. Like Townsend's big-eared bat, the locations and types of roosts expose these bats to a high susceptibility of disturbance.

SELECTED REFERENCES

See also discussion of this species in general references, page 23.

Handley, C. O., Jr. 1959. A revision of American bats of the genera *Euderma* and *Plecotus*. *Proceedings of the U.S. National Museum* 110:95–246.

Hurst, T. E., and M. J. Lacki. 1997. Food habits of Rafinesque's big-eared bat in southeastern Kentucky. *Journal of Mammalogy* 78:525–528.

Jones, C. 1977. *Plecotus rafinesquii*. *Mammalian Species* 69. 4 pp.

Jones, C., and R. D. Suttkus. 1975. Notes on the natural history of *Plecotus rafinesquii*. *Occasional Papers of the Museum of Zoology, Louisiana State University* 47:1–14.

5

Armored Mammals
Order Cingulata

In the previous edition of this book, the armadillo was listed in the order Xenarthra. The order Xenarthra has now been elevated to a superorder and contains a group of placental mammals living today only in the Americas and includes anteaters, tree sloths, and armadillos. The order Cingulata is within that superorder and contains the armored dasypodids, the armadillos, the only surviving family within that order. The name, Cingulata, comes from the Latin word *cingulum*, meaning "girdle," referring to the girdlelike shell.

Twenty-one armadillo species within 9 genera are recognized today in South, Central, and North America with the nine-banded armadillo being one of only two species found in North America, and the only species found in the United States. The other species found in North America is the northern naked-tailed armadillo, which ranges from southern Mexico to the extreme southwest corner of South America. The anteaters and tree sloths now belong in the order Pilosa.

Armadillos (Family Dasypodidae)

The characters of this family are exemplified by the nine-banded armadillo, the only species occurring in Missouri.

Nine-banded Armadillo
(*Dasypus novemcinctus*)

Name

The family name, Dasypodidae, and the generic name, *Dasypus*, mean "rough footed." The second part, *novemcinctus*, is Latin for "nine-banded" and refers to the typical 9 bands (ranging from 8 to 11) of shell-like skin that cover the back. The common name armadillo means "little armored one" in Spanish. This species is also known as the common long-nosed armadillo.

Description (Plate 19)

This unusual-looking animal cannot be confused with any other mammal in Missouri. The sides, back,

Armadillo

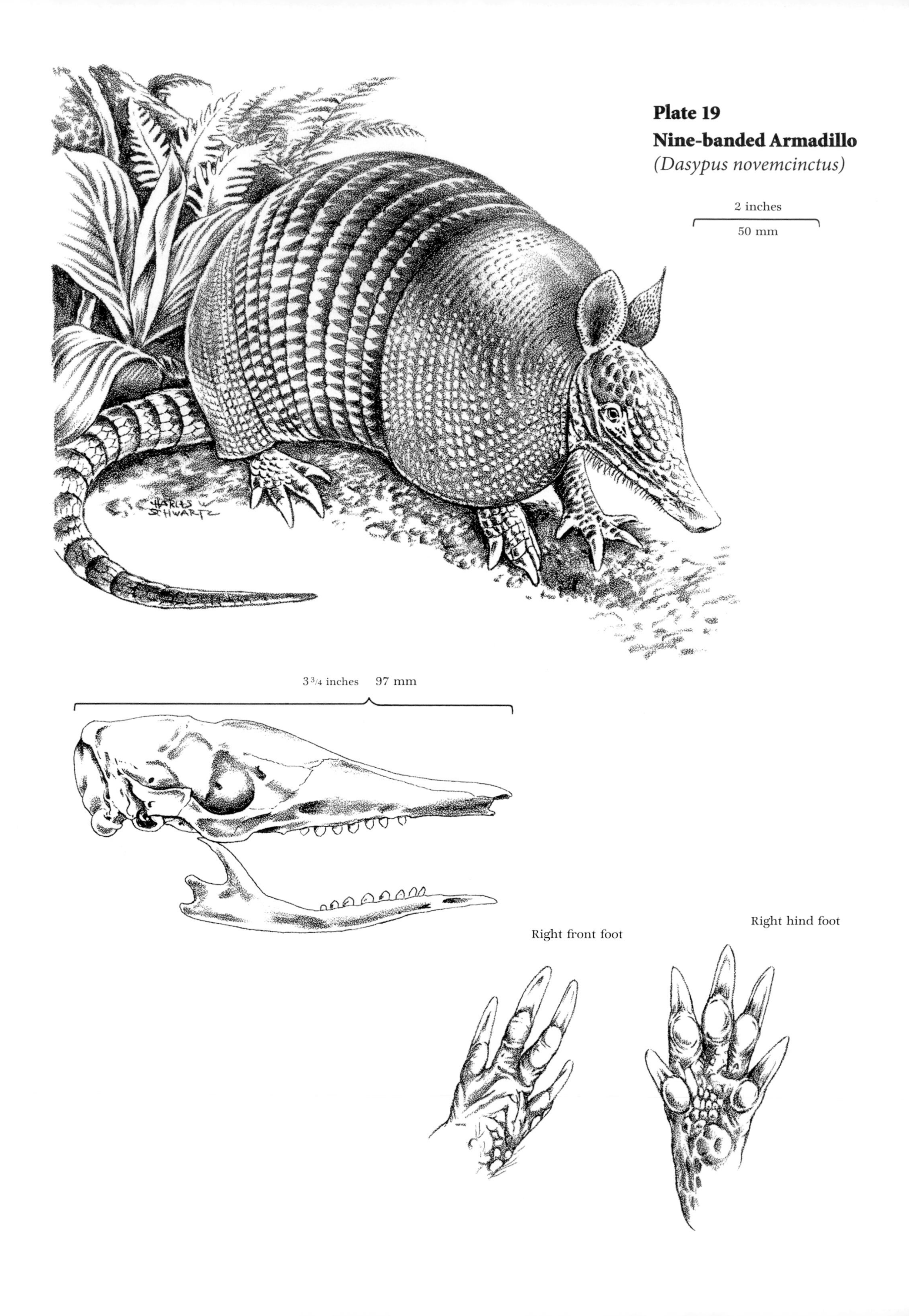
Plate 19
Nine-banded Armadillo
(Dasypus novemcinctus)
2 inches
50 mm
3 3/4 inches 97 mm
Right front foot
Right hind foot

tail, and top of the head are covered with bony dermal plates that are covered by a leathery skin. The head has 1 large plate from which the ears protrude; the body has 2 large plates with a series of 9 smaller, moveable "girdles" or "bands" around the midsection; the short legs and tail have other series of plates. It does not have furry skin like other Missouri mammals; it has hair only between the hardened plates and on the soft skin on the underside of the animal. The legs are short, and the toes (4 on the front feet and 5 on the hind) have well-developed claws. Sight and hearing are poor.

Color. The overall coloration is mottled dark brown to yellowish white.

Measurements

Total length	615–800 mm	23–31 in.
Tail	245–370 mm	9½–14⅜ in.
Hind foot	75–107 mm	2⅞–4⅛ in.
Ear	28–40 mm	1–1½ in.
Skull length	85–100 mm	3⁵⁄₁₆–3⅞ in.
Weight	2.4–7.9 kg	5¼–17¼ lb.
Usual	5.4–6.8 kg	11¾–14 lb.

Teeth and Skull. There are no teeth in the front of the mouth and only 28–32 simple teeth (technically considered premolars and molars) in the rear. These are rootless pegs and have no enamel covering.

Sex criteria and sex ratio. Males lack a scrotum and external testes, but the sexes can be distinguished by the presence of 4 teats on the females and an obvious penis on the male.

Glands. A pair of anal glands secretes a substance with a disagreeable odor that is used to ward off predators.

Distribution and Abundance

The nine-banded armadillo ranges from South America, through Central America and much of Mexico, and into the southeastern United States. Armadillos were first reported in extreme southern Texas in 1849, and since then human-caused landscape changes allowed them to expand their range northward and eastward. They prefer to use river valleys as dispersal corridors, so their invasion was especially rapid parallel to rivers.

The armadillo is a newcomer to Missouri's mammalian fauna. When this book was first published in 1959, the armadillo was not considered a resident but was included in the book as "possibly occurring within Missouri." Between 1947 and 1959, a few armadillo sightings had been reported from 10 counties (Barry, Camden, Cape Girardeau, Greene, Henry, Hickory, Jasper, Madison, Newton, and Randolph). These animals could have moved into the state from their range in Kansas and Arkansas, but they may also have been transported and liberated here. When the revised edition was published in 1981, several more sightings had been reported, but it was still listed as "possibly occurring in Missouri." In the 2001 second revised edition, it was given full resident status, having been observed in 66 counties. The current distribution is primarily the southern half of the state, but a few individuals have

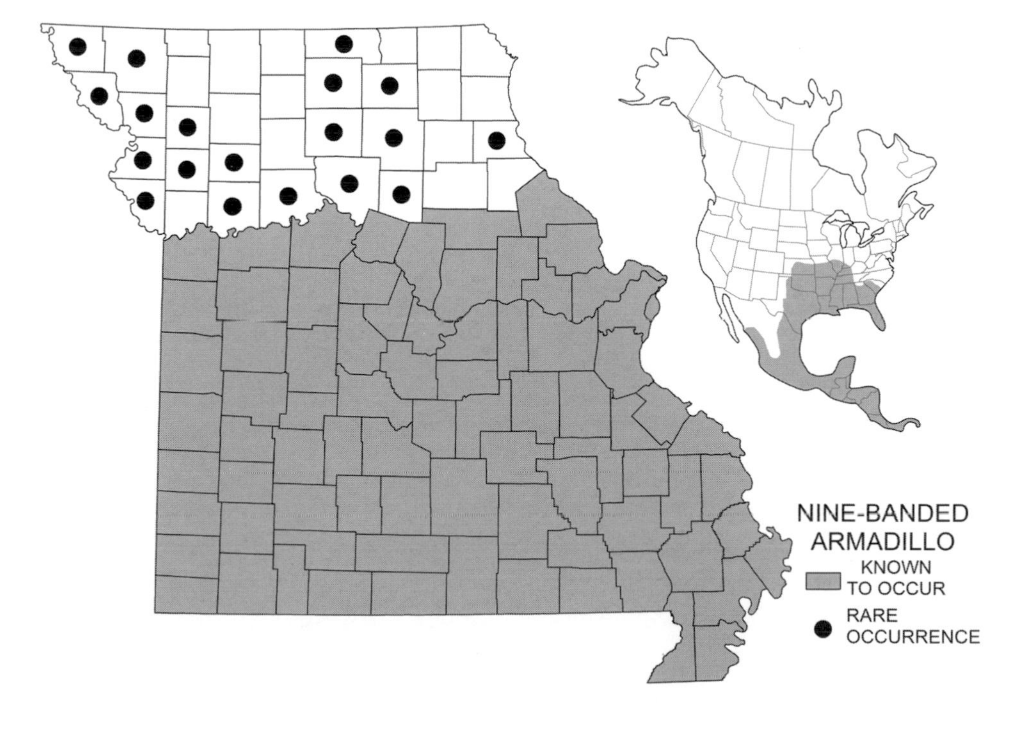

been reported as far north as the Missouri–Iowa state line. They are now common in south Missouri, with gradually fewer numbers farther northward in the state.

Further range expansion in the United States is not likely, as armadillos are limited to the north by extreme cold and to the west by lack of rainfall. They are not adapted for hibernation during cold winters, nor to the accompanying hardship of snow and ice preventing access to food. In addition, precipitation levels need to be sufficient enough to maintain a litter layer important for their food.

Habitat and Home

Armadillos live in a variety of habitats. They are terrestrial and seemingly prefer forests of oak and hickory or mature shortleaf pine. Because they dig burrows in the ground, they select wooded bottomlands, brushy areas, and fields with ground cover and loose soil. Rocky shelter and crevices and exposed tree roots are used for den sites.

Burrows are often the first sign that armadillos are present. Armadillos have strong feet and claws that are used in digging for food and shelter. During summer they are mostly nocturnal, but in winter they may be abroad more often during the day.

Once a suitable home site is selected, the armadillo digs one or more tunnels that lead to an enlargement that becomes the nest. When the armadillo wants to carry leaves inside the tunnel, it gathers a pile, pushes it under its body, and then clamps the shell over it. The armadillo then shuffles backward into the nest with the material in tow. This nest is where it rests daily and where the young are born. One animal may have several dens. Other animals, especially rabbits, squirrels, opossums, woodrats, and woodchucks, may utilize these dens when the armadillo is not present.

Habits

It is interesting to note how an armadillo crosses streams and rivers. When crossing small streams, it holds its breath and walks across the bottom, but when crossing larger waters, it holds air in its stomach and intestines, which makes it buoyant, and it floats across.

Armadillos commonly jump straight upward about 1 m (3 or 4 ft.) when frightened. This habit may help scare off predators but also explains why so many armadillos are hit by automobiles. Armadillos run fast and often elude predators by seeking shelter in a burrow. Contrary to popular belief, the nine-banded armadillo cannot roll itself into a ball to escape predators. Only the three-banded armadillo of South America can roll itself up.

Male armadillos travel over as much as 6 ha (15 ac.), but females have smaller home ranges. In general, armadillos are sociable, and often several are seen together. They are most active in winter because food is harder to get at that time.

Foods

Almost all the food of armadillos is of animal origin—mostly insects and other invertebrates. Adult and larval beetles, fly larvae, ants, earthworms, and other insects are their main food items. They also consume an occasional reptile, bird, and bird eggs. Some fungi and fruits are also eaten. Food items are located by the nose, which is held close to the ground; the sharp claws then dig to expose the food, which is then flicked into the mouth with the long sticky tongue.

Reproduction

Breeding occurs in summer followed by a delay of 2 to 3 months during which the embryo divides into four cells before each one becomes implanted in the uterus. This results in four young that are genetically identical, including sex. After 4 months' implantation, the young are born. Their shells are soft and flexible at birth, their eyes are open, and they can move about. They are weaned when 3 months old and become mature at 12–15 months of age.

Some Adverse Factors

Several factors limit armadillo populations, primarily cold weather with its accompanying scarcity of available food and adequate nesting sites. Their few predators include humans, domestic dogs, coyotes, black bears, bobcats, foxes, and raccoons.

Importance

Armadillos destroy harmful insects, are food for some people, and are important den makers for other animals. They are known to acquire the bacterium that causes leprosy, and some armadillos are used for biomedical research for growing the leprosy bacterium.

Conservation and Management

Armadillos are often considered undesirable where their burrowing and feeding habits undermine foundations of buildings or damage lawns, flower beds, and gardens. If you are experiencing problems with armadillos, contact a wildlife professional for advice, assistance, regulations, or special conditions for handling these animals.

In the United States, leprosy is rare but a few cases have been linked to contact with infected animals. These cases are concentrated in Louisiana and Texas, where some people hunt, skin, and eat armadillos.

SELECTED REFERENCES

See also discussion of this species in general references, page 23.

Humphrey, S. R. 1974. Zoogeography of the nine banded armadillo (*Dasypus novemcinctus*) in the United States. *BioScience* 24:457–462.

Lippert, K. J. 1994. Food habits, distribution, and impact of the nine-banded armadillo in Missouri. M.S. thesis, Southwest Missouri State University, Springfield. 88 pp.

McBee, K., and R. J. Baker. 1982. *Dasypus novemcinctus. Mammalian Species* 162. 9 pp.

Robbins, L. W., K. J. Lippert, and P. T. Schell. 1994. The ecology and impact of the armadillo (*Dasypus novemcinctus*) in Missouri. Unpublished Report, Department of Biology, Southwest Missouri State University, Springfield.

Schell, P. T. 1994. Home range, activity period, burrow use, and body temperatures of the nine-banded armadillo (*Dasypus novemcinctus*) on the northern edge of its range. M.S. thesis, Southwest Missouri State University, Springfield. 35 pp.

Sikes, R. S., G. A. Heidt, and D. A. Elrod. 1990. Seasonal diets of the nine-banded armadillo (*Dasypus novemcinctus*)in a northern part of its range. *American Midland Naturalist* 123:383–389.

Staller, E. L., W. E. Palmer, J. P. Carroll, R. P. Thornton, and D. C. Sisson. 2005. Identifying predators at northern bobwhite nests. *Journal of Wildlife Management* 69:124–132.

Talmage, R. V., and G. D. Buchanan. 1954. *The armadillo (*Dasypus novemcinctus*): A review of its natural history, ecology, anatomy, and reproductive physiology.* Rice Institute pamphlet 42. 135 pp.

Taulman, J. F., and L. W. Robbins. 1996. Recent range expansion and distributional limits of the nine-banded armadillo (*Dasypus novemcinctus*) in the United States. *Journal of Biogeography* 23:635–648.

———. 2014. Range expansion and distributional limits of the nine-banded armadillo in the United States: An update of Taulman and Robbins (1996). *Journal of Biogeography* 41:1626–1630.

Truman, R. W., P. Singh, R. Sharma, P. Brusso, J. Rougemont, A. Paniz-Mondolfi, A. Kapopoulou, S. Brisse, D. M. Scollard, T. P. Gillis, and S. T. Cole. 2011. Probable zoonotic leprosy in the southern United States. *New England Journal of Medicine* 364:1626–1633.

6

Hare-shaped Mammals
Order Lagomorpha

This small order includes the pikas, hares, and rabbits. The members typically possess two pairs of upper incisors, one directly behind the other. These are separated from the cheek teeth by a wide space, the *diastema*. They also have a bony network of openings (*fenestration*) on the rostrum, in front of the eye socket.

The name, Lagomorpha, comes from two Greek words meaning "hare shape" or "hare form" and describes the physical appearance of most members of this group, the pikas being a notable exception. This order is native to all continents except Antarctica and Australia. Members of the order have been successfully introduced to parts of Australia, where they have few predators and compete with native species for resources.

Key to the Species
by Whole Adult Animals

1a. Body lanky, with ears, hind legs, and feet very large in proportion to the body (see plate 22); hind foot 114 mm (4½ in.) or more in length; ear 101 mm (4 in.) or more in length; no rusty-colored area on back of neck. **Black-tailed Jackrabbit** (*Lepus californicus*) p. 144

1b. Body not lanky, with ears, hind legs, and feet moderately large in proportion to the body (see plates 20–21); hind foot less than 114 mm (4½ in.) in length; ear less than 101 mm (4 in.) in length; rusty-colored area on back of neck. **Go to 2**

2a. (From 1b) When fluffed, fur of rump tawny; tops of hind feet reddish brown; back of neck slightly

rust colored. **Swamp Rabbit** (*Sylvilagus aquaticus*) p. 139

2b. (From 1b) When fluffed, fur of rump grayish; tops of hind feet tan to whitish; back of neck bright rust colored. **Eastern Cottontail** (*Sylvilagus floridanus*) p. 128

Key to the Species
by Skulls of Adults

1a. Length of skull 92 mm (3⅝ in.) or more; interparietal bone without obvious borders (see plate 22); base of skull with many perforations. **Black-tailed Jackrabbit** (*Lepus californicus*) p. 144

1b. Length of skull usually less than 92 mm (3⅝ in.); interparietal bone usually with obvious borders (see plates 20–21); base of skull without many perforations. **Go to 2**

2a. (From 1b) Length of skull less than 82 mm (3¼ in.); postorbital processes usually not fused to skull and a distinct slitlike opening remains (see plate 20). **Eastern Cottontail** (*Sylvilagus floridanus*) p. 128

2b. (From 1b) Length of skull usually 82 mm (3¼ in.) or more; postorbital processes usually fused to skull and only a small hole remains (see plate 21). **Swamp Rabbit** (*Sylvilagus aquaticus*) p. 139

Hares and Rabbits (Family Leporidae)

The family name, Leporidae, is based on the Latin word meaning "hare." This family now occurs throughout most of the world, but only three species are found wild in Missouri.

Hares and rabbits have long ears; large hind legs and feet, the soles of which are well furred; a very short tail; and large eyes. The young of hares (jackrabbits) are born well furred, with their eyes open, and are able to run about shortly after birth (they are *precocial*), while the young of rabbits (eastern cottontail and swamp) are born nearly naked, blind, and helpless (they are *altricial*).

GENERAL REFERENCE

Hall, E. R. 1951. A synopsis of the North American Lagomorpha. *University of Kansas Publications, Museum of Natural History* 5:119–202.

Eastern Cottontail
(*Sylvilagus floridanus*)

Name

The first part of the scientific name, *Sylvilagus*, is of Latin and Greek origin and means a "wood hare" (*sylva*, "wood," and *lagos*, "hare"). The second part, *floridanus*, is a Latinized word meaning "of Florida," for the place from which the first specimen was collected and described.

The common name originated as follows: "eastern" refers to the range of the species and "cottontail" describes the characteristic appearance of the tail.

Description (Plate 20)

The eastern cottontail is a medium-sized mammal with long ears, large hind legs and feet, smaller front legs and feet, a short, fluffy tail, and soft fur. The ears, hind legs, and feet of this rabbit are smaller in proportion to the body than those of a jackrabbit. There are 5 toes on the front feet and 4 on the hind feet. The soles of the feet are densely furred. As in all hares and rabbits, the upper lip is divided.

Color. The upperparts vary from reddish to grayish brown sprinkled with black. This variegated color arises from the black and buff banding on the individual hairs. When fluffed, the fur of the rump is grayish compared to the tawny color of the swamp rabbit. The ears of the cottontail are dark grayish tan bordered with black. The back of the neck is rusty. There is a light ring around the eye and usually a white blaze on the forehead. The iris of the eye is deep brown; at night, in the beam of a bright light, the eyes reflect a brilliant red color. The underparts are grayish white except for a brownish chest. On the upper surface, the

Plate 20

Eastern Cottontail (*Sylvilagus floridanus*)

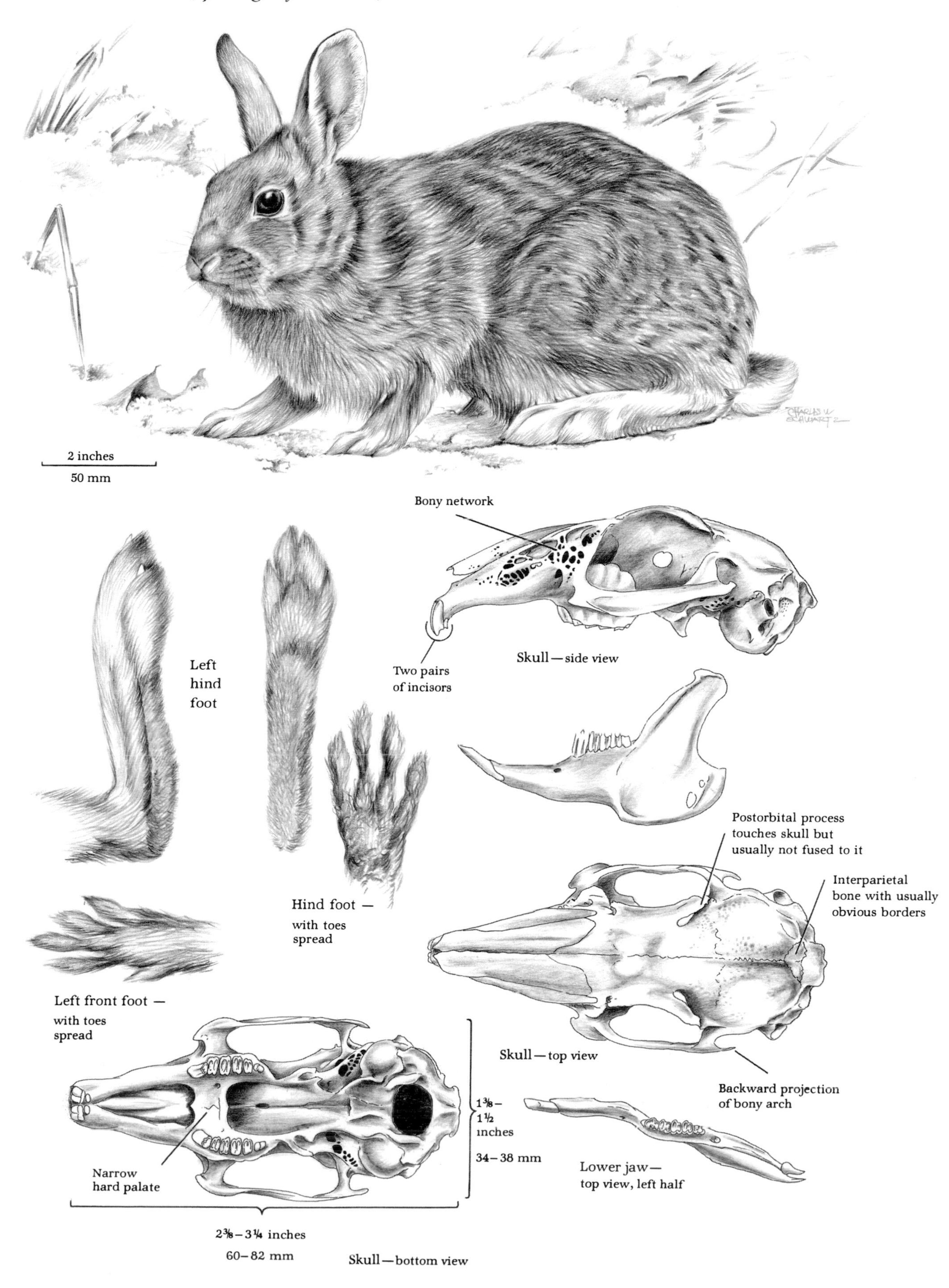

tail is brownish like the back, but on the undersurface it is white like the belly; because the tail is usually turned up when the rabbit runs, the white part is the most conspicuous. The tops of the hind feet are tan to whitish. Rarely, some individuals lack the normal black pigment and are white or reddish brown; a few may be unusually dark because of considerable black pigment.

The sexes of the cottontail are colored alike and do not greatly change in color during the year. In the spring and summer molt, the hair is replaced in "spots," while in the fall molt it is replaced in larger areas or "sheets."

Measurements

Total length	355–482 mm	14–19 in.
Tail	38–76 mm	1½–3 in.
Hind foot	82–107 mm	3¼–4¼ in.
Ear	50–76 mm	2–3 in.
Skull length	60–82 mm	2⅜–3¼ in.
Skull width	34–38 mm	1⅜–1½ in.
Weight		
Male	0.9–1.2 kg	2–2¾ lb.
Female	1.1–1.4 kg	2½–3¼ lb.

Teeth and skull. The dental formula of the eastern cottontail is:

$$I\ \frac{2}{1}\ C\ \frac{0}{0}\ P\ \frac{3}{2}\ M\ \frac{3}{3} = 28$$

Characters of the teeth and skull are described in detail in the account of the black-tailed jackrabbit (see page 144).

The skulls of the adult eastern cottontail and swamp rabbit are distinguished from that of the black-tailed jackrabbit in Missouri by their smaller size, by the borders of the interparietal bone located on the upper surface of the skull toward the rear, which are usually obvious, and by the absence of many perforations on the base of the skull. The skulls of adult cottontails and swamp rabbits are very similar and, because of individual variation, are sometimes hard to tell apart. The cottontail's skull is smaller. The postorbital process, or the backward extension of the prominent projection above the eye socket, touches the skull toward the rear in both species, but in the cottontail it usually is not fused to the skull and a distinct slitlike opening remains, while in the swamp rabbit it usually is fused and only a small hole remains.

Sex criteria and sex ratio. The determination of sex in rabbits is discussed under the black-tailed jackrabbit. There are four pairs of teats in female cottontails. The sex ratio is nearly even in young and adult cottontails.

Age criteria, age ratio, and longevity. The most commonly used method of age determination is based on the weight of the dried lens of the eye. This is especially valuable in estimating the age of animals older than 9 months (as determined by the absence of cartilage at the end of the long bones).

Age can be determined up to 9 months by the presence of a certain type of cartilage that progressively turns to bone (see accompanying illustration). This cartilage is best observed on the ends of the long bones of the front leg, specifically the humerus, radius, and ulna. X-rays can be used to demonstrate this cartilage in live or dead animals, while direct observations are used on the bones of dead animals. The simplest means of detecting this cartilage in dead animals is to scrape the front surface of the shoulder end of the

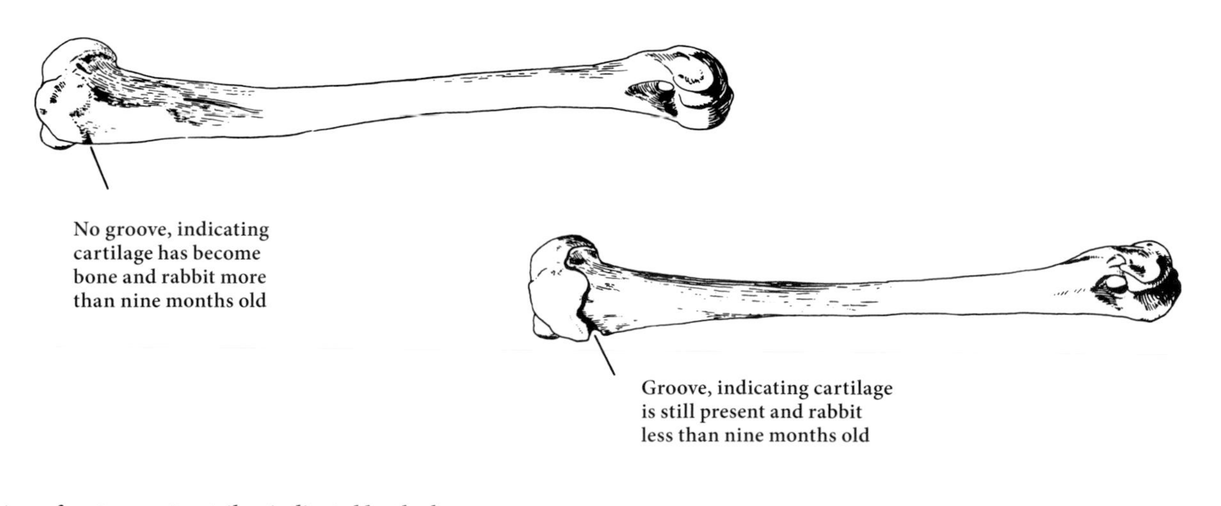

Age of eastern cottontail as indicated by the humerus (upper bone of the front leg)

humerus (upper bone of the front leg) with a pocket knife. If the surface is smooth, all cartilage has become bone and the animal is more than 9 months old; if the knife blade catches in a groove that is the edge of the cartilage plate, cartilage is still present and the animal is less than 9 months old.

In the first weeks of life, young cottontails are somewhat grayer than adults. They can be distinguished from adults by weight until 4½ to 5 months of age, when the young reach the minimum adult weight of 0.9 kg (2 lbs.) for males and 1.1 kg (2½ lbs.) for females.

During the breeding season, young of the year that have become fully grown can be distinguished from adults by their pelage: the young have a lighter buff underfur and a salt-and-pepper appearance because of the sparseness of dark, outer, long hairs; the adults are darker brown and more glossy because of the greater density of long hairs.

Young females have small, undeveloped teats that are difficult to locate. They are less than 1.6 mm (1/16 in.) in diameter and in height. Young females also have a complete membrane over the vagina and no placental scars (places where embryos were attached during a pregnancy) in the uterus. Adult females have teats about 5 mm (3/16 in.) in diameter and between 5 and 6 mm (3/16 and ¼ in.) high; these are whitish, blunt-tipped, and often flabby. Adult females have no membrane over the vagina and usually possess placental scars in the uterus. After December, the placental scars are reabsorbed and this criterion of age becomes unreliable.

Females that have produced young can be identified by having longitudinal striations on the uterus, as discussed under the black-tailed jackrabbit.

Age ratios in fall populations show from 3 to 8 young per adult depending upon local conditions. In spite of the large number of young produced each year, only a very few survive to breed. There are reports of marked wild individuals reaching 5 years of age and of a captive eastern cottontail still living at 9 years of age. The potential life is at least 10 years.

Glands. Scent glands on either side of the anus in both sexes give off a musky odor that is left on the ground where the rabbit sits. This odor serves as an identification to other rabbits.

Voice and sounds. Eastern cottontails are generally silent animals, although they may give a shrill cry when frightened, captured, or fighting. The female has a call for assembling her young. On occasion the large hind feet may be thumped as a signal.

Distribution and Abundance

The eastern cottontail has the widest distribution of any *Sylvilagus* species. In North America, it is found from southern Manitoba and Quebec through the eastern United States to the Great Plains in the west, and into Mexico and Central America. It has been introduced into Oregon, Washington, southwestern Canada, and other locations.

In the first four decades of the 1900s, the cottontail population in Missouri was extremely high and rabbits were shipped by the thousands to the live and dead rabbit markets in the eastern United States. Because the population showed a marked decline in

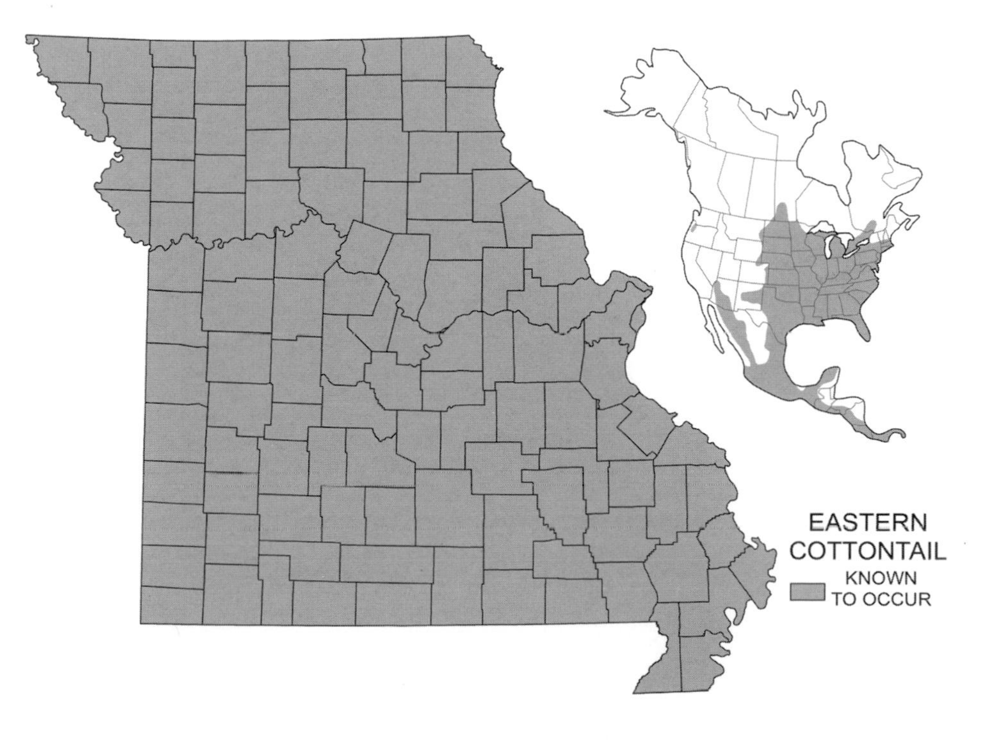

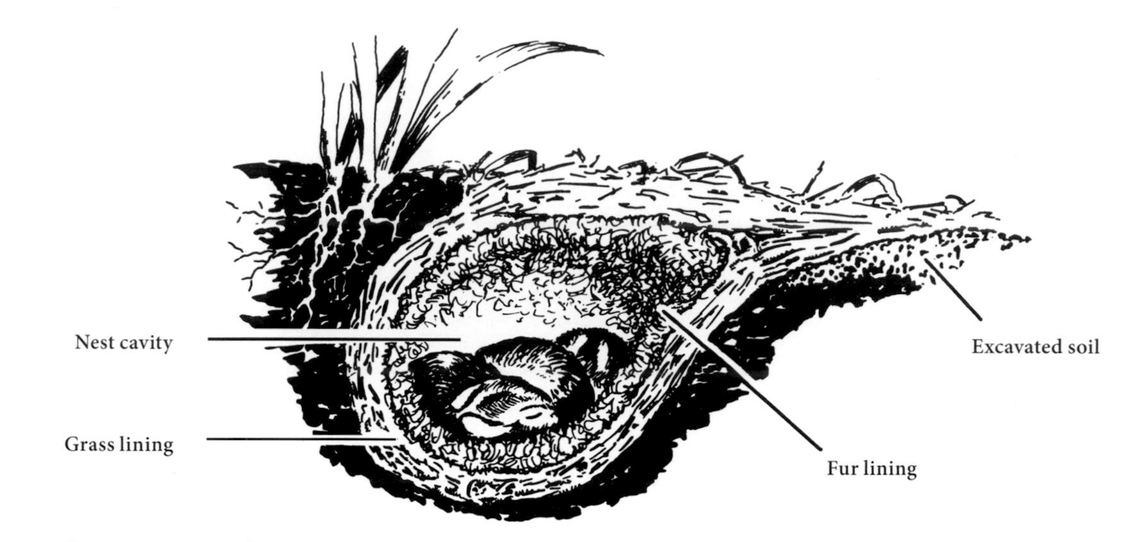

Diagram of eastern cottontail nest

the latter part of this period, it was realized that the resource could no longer withstand this heavy toll, and this commercialization was stopped in 1955. The harvest in 1957–1958 was 6 million. Since that time, the intensiveness and extensiveness of land use have accelerated, resulting in a great loss of cottontail habitat. Today, they are well distributed throughout Missouri and are considered common in most years. Their numbers fluctuate, sometimes greatly, from year to year and place to place.

Habitat and Home

Cottontails are highly adaptable and are found in practically all types of cover. They prefer, however, an open brushy or forest-border habitat. While they may venture into the open, they usually do not go far from the sanctuary of some good cover, such as brushy and herbaceous fencerows, hedge fences, thickets, dense high grass, weedy growth, or brush piles.

During most of the year the cottontail's home is a "form" or resting place concealed in a clump of grass, under a brush pile, or in a thicket. A form consists of a well-worn or slightly scratched-out depression, usually with a single opening toward which the rabbit faces. Most forms in an area are apparently used by all the cottontail residents, but there may be some ownership of specific ones. During periods of heavy snow, cottontails frequently use underground dens of other animals, particularly those of woodchucks, for temporary homes.

For her young, the female digs a nest cavity in the ground about 15.2 to 17.8 cm (6 to 7 in.) long, 12.7 cm

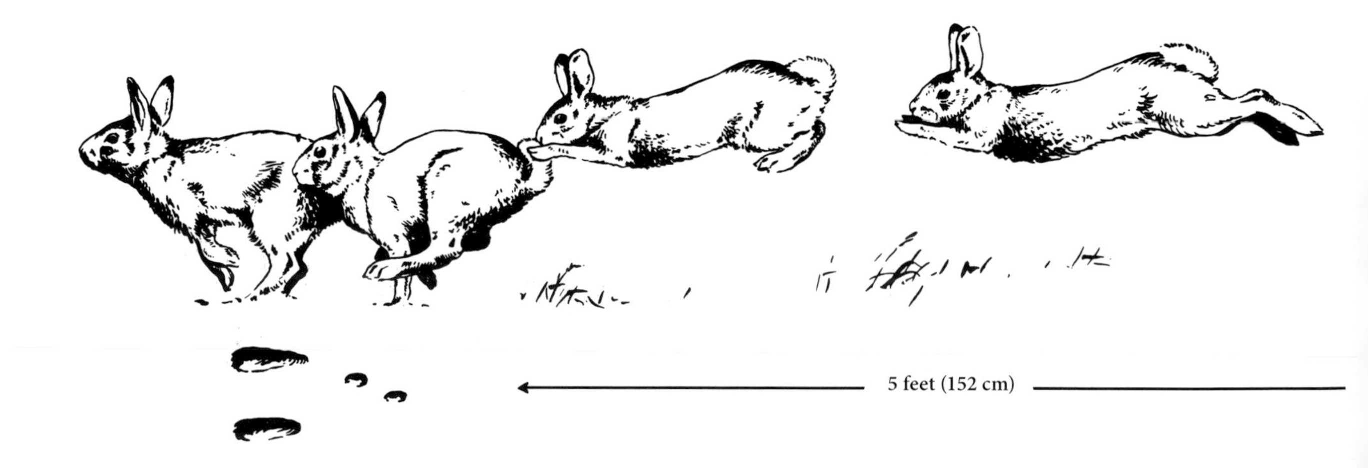

(5 in.) wide, and 7.6 to 10.1 cm (3 to 4 in.) deep. This is usually dug from one end, so the cavity is somewhat undercut on the side opposite the digging. A small pile of excavated soil is sometimes left at the site of the digging. The nest is lined and covered with grass and with fur the female plucks mostly from her belly. It is located usually in a well-drained spot in a variety of cover types such as grass in open pastures, broomsedge, legume fields, brushy fencerows, and city lawns and gardens. Sometimes the female digs a preliminary nest before she constructs the actual one used for her young.

Habits

In Missouri the home range varies from 0.4 to 2 ha (1 to 5 ac.) in good habitat but may include as much as 6 ha (15 ac.) where food and cover are of poorer quality. The home ranges of different eastern cottontails may overlap. Cottontails do not migrate, but during severe cold and deep snow those living in more open cover may move into nearby woodlands, heavily vegetated draws, or other retreats with heavier cover. They may remain inactive in their forms or temporary homes for a day following a heavy snowfall or sudden drop in temperature but do not "hole up" for any extended period. There is no indication that either sex is less active than the other. However, in northern parts of their range in the United States, female eastern cottontails especially and sometimes males are reported to hole up for a considerable time when the temperature is below −11°C (12°F).

There is an interesting report that one female eastern cottontail was trapped and taken 4 km (2½ mi.) away, where she was released. Nearly a year later she was retrapped in her original location.

Cottontails are active mostly at night and early in the morning. They spend the remainder of the day resting in their forms, occasionally taking a sunbath, and in summer stretching out in some cool spot in the shade. There are two main feeding periods, one

"Frozen" in its tracks

When a cottontail was surprised away from cover, it contracted without any perceptible movement over a period of five minutes from the above strained position to the hunched-up, more comfortable, and less conspicuous one below

The leaping gait of the cottontail

around sunrise and another in late afternoon and early evening. In feeding or traveling in their home range, cottontails often move along regular paths, wearing trails or runways in the vegetation, snow, or underbrush piles.

Cottontails usually hop along slowly, moving a few to many centimeters (a few inches to several feet) at a time. They frequently sit up on their hind legs to obtain a better view of their surroundings. When frightened or pursued by an enemy, they may cover between 3 and 4.5 m (10 and 15 ft.) for the first several leaps, and then hop shorter distances, often in a zigzag fashion. To avoid detection, it is not uncommon for a cottontail to remain motionless for 15 minutes or longer, or to take refuge in a form or perhaps a woodchuck's hole. When cornered, cottontails strike with their hind feet and attempt to rush past their opponent. They can swim but seemingly prefer not to.

In an eastern cottontail population, there is a regular behavior pattern of dominance. A dominant male is aggressive to other males, using threats or a charge to force them to crouch in submission or give way. A similar dominance is displayed by females to females. This behavior is a means of reducing physical conflict. Dominant males carry out most of the mating and occupy the most desirable habitat.

Their courtship behavior is very ritualized. The male approaches the female from the rear, and she quickly turns to face him. He then rushes at her, and she jumps over him. They may continue this pattern or reverse roles.

Cottontails groom themselves regularly, cleaning their faces with their front feet and licking other parts of their bodies. They are high-strung and often die of shock when caged. Adult cottontails are extremely difficult to tame.

Foods

Cottontails feed almost entirely on plants, preferring succulent green vegetation. Their choice depends upon the seasonal availability as well as palatability of the plant. The three most preferred foods during all seasons are bluegrass, wheat, and white clover. Other choice foods that are heavily used when available are red clover, Korean lespedeza, small and common crabgrass, timothy, and common chess. Some sedges, forbs, and cultivated plants are relished, such as ladino clover, alfalfa, soybeans, rye, fruits, and vegetables.

In winter when the ground is free of snow, dried forbs and grasses, some fruits and berries, waste corn, winter wheat, and rye constitute the main foods, but when heavy snow or ice covers their usual foods, they consume buds, twigs, bark, and sprouts of shrubs, vines, and trees. The most important woody foods are dwarf sumac, staghorn sumac, white oak, flowering dogwood, sassafras, black oak, buckbrush, and New Jersey tea. Rabbits can subsist on a diet of these woody plants but do not maintain a healthy condition. At times of extreme food scarcity, pupae of moths, snails, or even carcasses of their own kind are eaten. Moisture is obtained from the succulent diet, dew, snow, and surface water.

Cottontails, like jackrabbits, eat their own soft droppings and apparently obtain the same nutritional benefits of this practice (*coprophagy*). This type of feeding occurs in the daytime resting period.

Reproduction

Breeding of cottontails in Missouri usually begins in mid-February, but if there are long periods of snow and accompanying cold weather, it may be delayed until early March. The population of a given area starts to breed uniformly and produces young at regular successive periods throughout the season. Young are born up through September.

Pregnancy requires from 26 to 28 days, larger litters being born earlier than smaller ones. During the peak of the breeding season, females are both pregnant and nursing.

Stages of development of the embryo inside the female's body are shown on page 135.

With an early start to the breeding season due to mild winter weather, a female could produce as many as 8 litters per year. The number of young per litter varies from 1 to 9, with 4 to 6 being the most common. Largest litters are produced in May and June. It has been estimated that a female can bear about 35 young during one breeding season.

At birth the young are from 10.1 to 12.7 cm (4 to 5 in.) long and weigh about 28 g (1 oz.) each. They may be naked except for a sparse covering of light-colored guard hairs, but some, particularly those of small litters, may be further developed and possess very fine, gray fur. Their eyes and ears are closed, and the incisors are barely through the gums. The sex can be determined at birth.

The female nurses her young at dawn or dusk, covering them after each visit with a mat of grass and fur. In nursing she either squats over the opening of the nest with her feet spread apart or sits upright with hind legs spread on the pile of excavated soil at the edge of the nest. She usually spends the day in a form nearby and may defend the nest from an intruder. If it is necessary to move the young to a new nest, she carries them one at a time in her mouth.

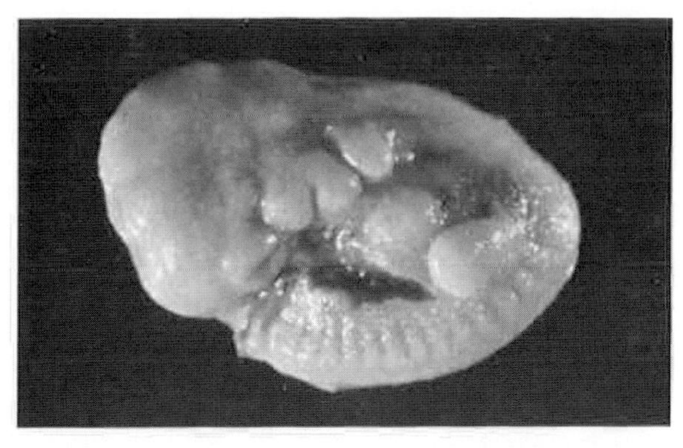

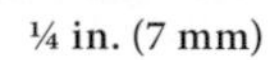

¼ in. (7 mm) about 11 days old

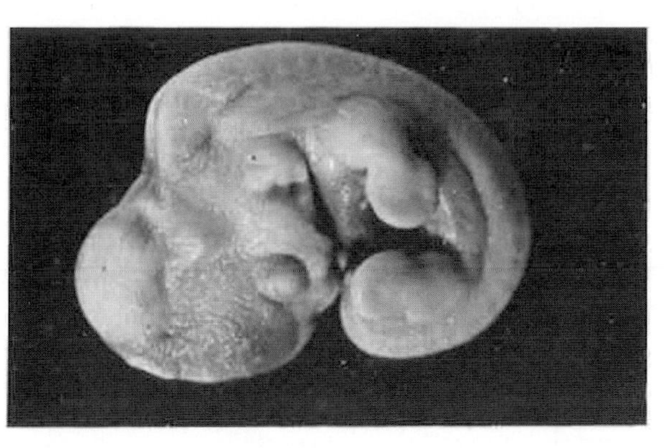

½ in. (13 mm) about 13 days old

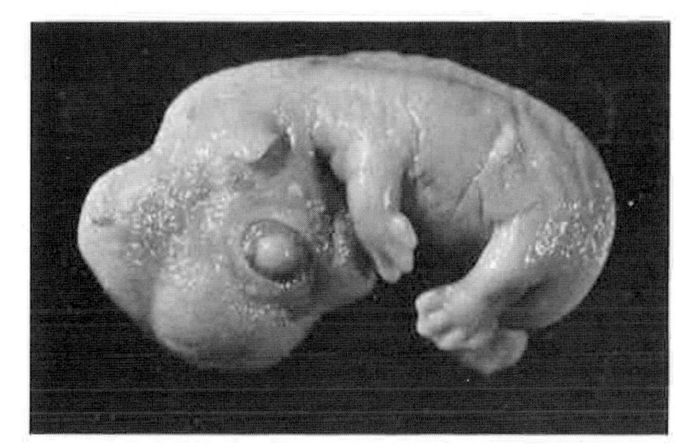

¾ in. (19 mm) about 15 days old

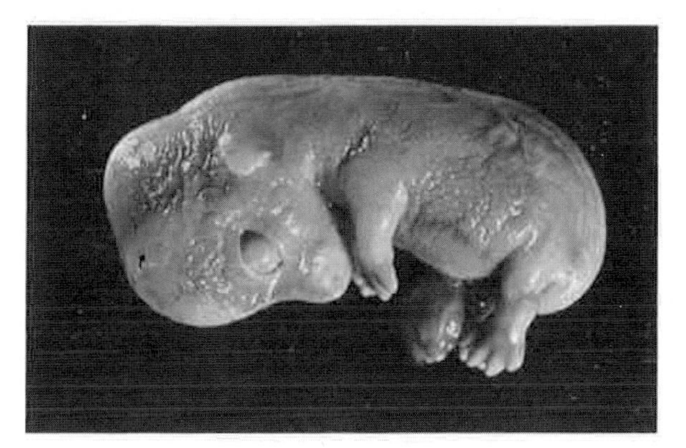

⅞ in. (22 mm) about 16½ days old

1½ in. (38 mm) about 20 days old

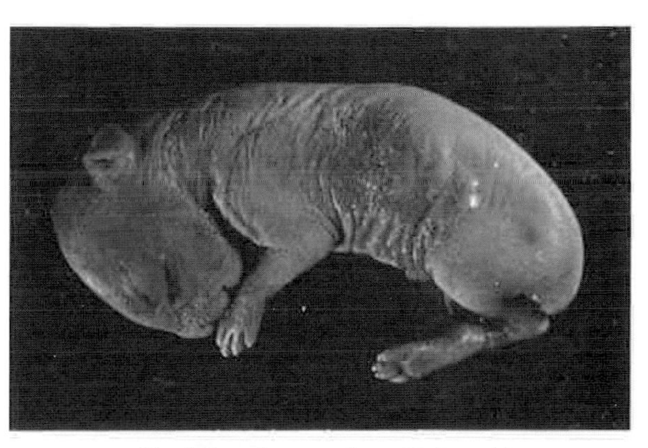

2 in. (50 mm) about 22 days old

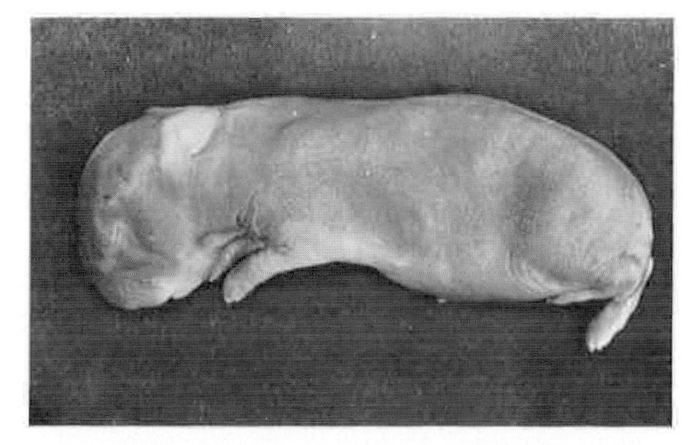

3⅜ in. (85 mm) about 24 days old

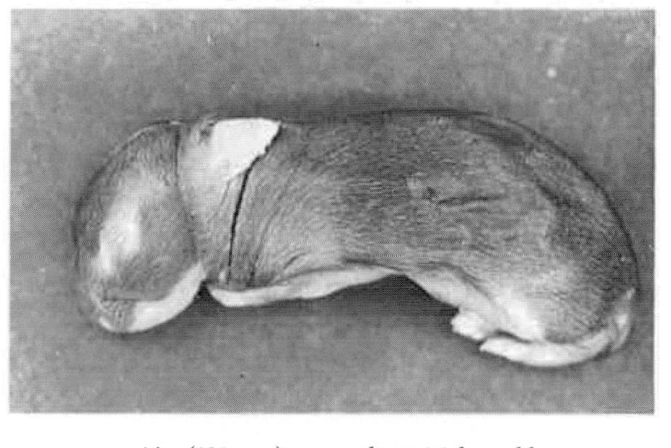

4 in. (101 mm) about 26 days old

Cottontail one day after birth

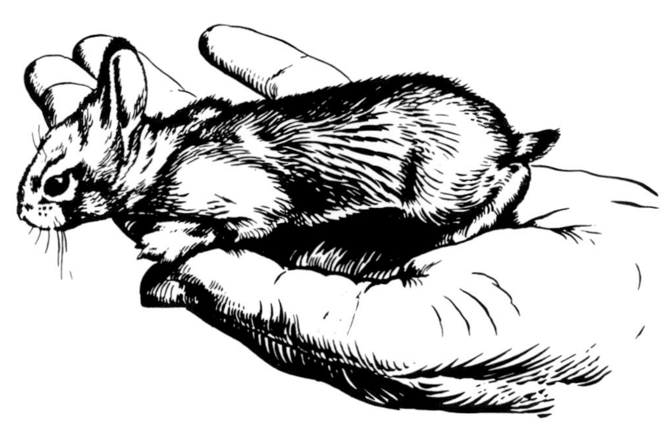

Cottontail about nine days old

Cottontail about two weeks old

During the first week the young become completely furred. Their eyes and ears open between 6 and 8 days of age, and the young are able to hop and squeal loudly by this time. They leave the nest between 13 and 16 days after birth but may return to it occasionally during the next few days or remain nearby in individual forms. The young continue to nurse but after a few days become independent of the female. They have a full set of teeth by 1 month of age. Most young breed for the first time in the spring following their birth, but a sizable part of the juvenile population breeds in the year of its birth. This contribution to the total population is significant.

Some Adverse Factors

Eastern cottontails face many adverse factors in their environment. Drought affects the growth of green vegetation—their main source of food. This and other effects of weather may influence periodic fluctuations of the population. But probably the most significant, long-term factor is change in land use. Economic conditions greatly impact agricultural practices, which can affect cottontail habitat. Increased cattle production puts a heavier toll on pastures, reducing even further winter food and cover for cottontails. The shift to fescue as the principal cattle forage caused a major decline in bluegrass, clover, and many herbs that cottontails prefer to feed upon. The cleaning or removal of fencerows, more intensive cultivation of former idle land, and the increase in size of cultivated fields further destroy their necessary cover. One new practice that is detrimental in several ways is the increased use of herbicides to kill weeds in row crops, such as soybeans. The chemicals kill many plants cottontails feed on and may have injurious effects on the rabbits themselves.

Cottontails are eaten by many animals including hawks, owls, crows, foxes, coyotes, bobcats, minks, weasels, domestic dogs, domestic cats, and snakes. Humans take them for food and destroy many young in their nests through mowing or plowing operations, or by burning waste areas during nesting periods. Heavy rains may wash out nests in poorly drained sites or chill the young and cause their death. Automobiles take a heavy toll. It has been estimated that 10 cottontails are killed annually per 1.6 km (1 mi.) of road in Missouri.

It has been estimated that 44 percent of newborn cottontails die during the first month of their life. In addition, only 20 to 25 percent of young live to 1 full year, and about 85 percent of the population dies each year.

The following parasites are known to occur on or in eastern cottontails: protozoa, roundworms, flukes, tapeworms, mites, ticks, fleas, and flies (see detailed discussion of some of these parasites on pages 11–12). The presence of moderate numbers of parasites does not affect the well-being of the individual. Excessive numbers of parasites, however, and some diseases, like tularemia, are fatal to cottontails or weaken them so that they die from another cause. A certain virus infection causes swellings, or "papillomas horns," on the nose, eyelids, lips, ears, back, and feet.

Importance

Around 300,000 rabbits are harvested annually in Missouri, eastern cottontails constituting most of this with swamp rabbits composing a smaller part. At 0.6 kg (1½ lb.) of meat per rabbit, this is a sizable amount.

Cottontails form much of the diet of many carnivorous animals and, because of their general availability, reduce predation on less common game species and livestock. By converting plant food into animal matter, they constitute an important link in the food chain. They serve as an intermediate host for many internal parasites of carnivorous species that feed upon them and can also transmit the bacterial disease tularemia to humans.

Cottontail fur is not very durable, and prices for pelts are very low. The fur is mostly used in the making of coats, hats, and clothing accessories like scarves, and for trimming and lining coats and gloves.

The edges of crop fields, especially those near brushy cover, are sometimes damaged by cottontails; additionally, in winter, orchard and ornamental trees, and shrubs may be barked or girdled when other foods are scarce. This latter kind of damage is often the work of woodland voles, which tunnel and gnaw beneath the snow. Cottontail cuttings are easily identified because they are made at a 45° angle from the vertical axis.

Conservation and Management

Cottontail populations have declined in some parts of Missouri, but fortunately these rabbits are so prolific and adaptable that, if given an opportunity, they will populate all suitable habitat and maintain good numbers. To that end one of the best means of improving the habitat for cottontails is to provide plenty of cover—for nesting, escape, and shelter from the weather—interspersed with desirable feeding areas. The following types of cover contribute to good cottontail habitat: brushy and herbaceous fencerows; idle land and its weedy growth; brush piles near other cover; field borders sown to lespedeza or heavy mixed stands of timothy, redtop, orchard grass, and improved pasture systems; and blackberry thickets. Woodchucks, whose burrows are used by cottontails for escape and shelter, should be encouraged in areas where cottontails are desired.

The best measure for controlling damage is to place rabbit-proof wire netting around individual trees, gardens, or orchards. Heavy shooting and persistent trapping in limited areas may alleviate the damage. If you are experiencing problems with cottontails, contact a wildlife professional for advice, assistance, regulations, or special conditions for handling these animals.

The bacterial disease tularemia affects rabbits, ticks, humans, and certain other animals. This disease is transmitted primarily by tick bites, but humans may

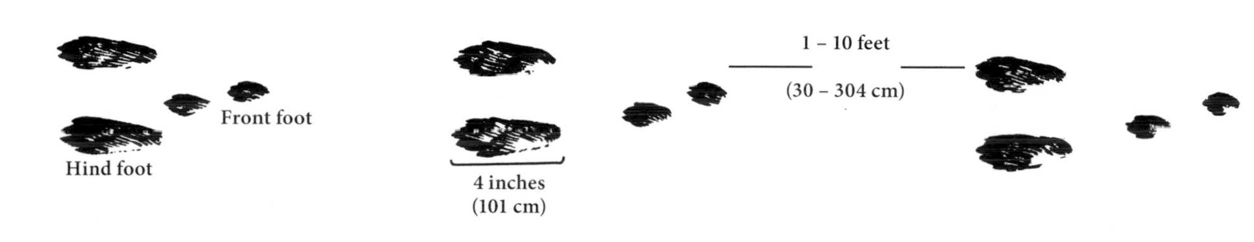

Eastern cottontail tracks—hopping

also become infected by contact with diseased animals or contaminated drinking water. To avoid contamination from rabbits, hunting should be done in cold weather because by this time sick animals will have perished and the healthy ones will have survived. The meat should be thoroughly cooked to eliminate chances of infection. Since new drugs have been found to be a ready cure for tularemia, this disease in humans is not as serious as it was in the past.

SELECTED REFERENCES

See also discussion of this species in general references, page 23.

Allen, D. L. 1939. Michigan cottontails in winter. *Journal of Wildlife Management* 3:307–322.

Beule, J. D. 1940. Cottontail nesting-study in Pennsylvania. *Transactions of the Fifth North American Wildlife Conference* 320–327.

Beule, J. D., and A. T. Studholme. 1942. Cottontail rabbit nests and nestlings. *Journal of Wildlife Management* 6:133–140.

Casteel, D. A. 1966. Nest building, parturition, and copulation in the cottontail rabbit. *American Midland Naturalist* 75:160–167.

Chapman, J. A., J. G. Hockman, and M. M. Ojeda C. 1980. *Sylvilagus floridanus. Mammalian Species* 136. 8 pp.

Conaway, C. H., and H. M. Wight. 1962. Onset of reproductive season and first pregnancy of the season in cottontails. *Journal of Wildlife Management* 26:278–290.

Conaway, C. H., H. M. Wight, and K. C. Sadler. 1941. Food habits of the eastern and New England cottontails. *Journal of Wildlife Management* 5:216–228.

———. 1963. Annual production by a cottontail population. *Journal of Wildlife Management* 27:171–174.

Conaway, C. H., K. C. Sadler, and D. H. Hazelwood. 1974. Geographic variation in litter size and onset of breeding in cottontails. *Journal of Wildlife Management* 38:473–481.

Dusi, J. L. 1952. Food habits of several populations of cottontail rabbits in Ohio. *Journal of Wildlife Management* 16:180–186.

Eisen, L. 2007. A call for renewed research on tick-borne *Francisella tularensis* in the Arkansas–Missouri primary national focus of tularemia in humans. *Journal of Medical Entomology* 44:389–397.

Evans, R. D., K. C. Sadler, C. H. Conaway, and T. S. Baskett. 1965. Regional comparisons of cottontail reproduction in Missouri. *American Midland Naturalist* 74:176–184.

Hamilton, W. J., Jr. 1940. Breeding habits of the cottontail rabbit in New York state. *Journal of Mammalogy* 21:8–11.

Haugen, A. O. 1942a. Life history studies of the cottontail rabbit in southwestern Michigan. *American Midland Naturalist* 28:204–244.

———. 1942b. Home range of the cottontail rabbit. *Ecology* 23:354–367.

Heisinger, J. F. 1962. Periodicity of reingestion in the cottontail. *American Midland Naturalist* 67:441–448.

Hill, E. P., III. 1967. Homing by a cottontail rabbit. *Journal of Mammalogy* 48:648.

Kibbe, D. P., and R. L. Kirkpatrick. 1971. Systematic evaluation of late summer breeding in juvenile cottontails, *Sylvilagus floridanus*. *Journal of Mammalogy* 52:465–467.

Korschgen, L. J. 1980. *Food and nutrition of cottontail rabbits in Missouri*. Terrestrial Series no. 6. Missouri Department of Conservation, Jefferson City. 16 pp.

Lord, R. D., Jr. 1959. The lens as an indicator of age in cottontail rabbits. *Journal of Wildlife Management* 23:358–360.

Marsden, H. M., and C. H. Conaway. 1963. Behavior and the reproductive cycle in the cottontail. *Journal of Wildlife Management* 27:161–171.

Marsden, H. M., and N. R. Holler. 1964. Social behavior in confined populations of the cottontail and the swamp rabbit. *Wildlife Monographs* 13:1–39.

McKinney, T. D. 1970. Behavioral interactions in grouped male cottontails. *Journal of Mammalogy* 51:402–403.

Negus, N. C. 1958. Pelage stages in the cottontail rabbit. *Journal of Mammalogy* 39:246–252.

Pelton, M. R. 1969. The relationship between epiphyseal groove closure and age of the cottontail rabbit (*Sylvilagus floridanus*). *Journal of Mammalogy* 50:624–625.

Rongstad, O. J. 1966. Biology of penned cottontail rabbits. *Journal of Wildlife Management* 30:312–319.

———. 1969. Gross prenatal development of cottontail rabbits. *Journal of Wildlife Management* 33:164–168.

Rowe, K. C. 1947. Population studies of cottontails in central Missouri, *Sylvilagus floridanus* (Allen), and effect of commercial shipping of live rabbits. M.S. thesis, University of Missouri, Columbia. 140 pp.

Sadler, K. 1980. Of rabbits and habitat, a long term look. *Missouri Conservationist* 41:4–8.

Sadler, K. C., and C. H. Conaway. 1971. Cold temperatures, snow and ice as reproductive inhibitors in cottontail rabbits. *Proceedings snow and ice in relation to wildlife and recreation symposium*, 197–202.

Schwartz, C. W. 1941. Home range of the cottontail in central Missouri. *Journal of Mammalogy* 22:386–392.

———. 1942. Breeding season of the cottontail in central Missouri. *Journal of Mammalogy* 23:1–16.

Wight, H. 1959. Eleven years of rabbit-population data in Missouri. *Journal of Wildlife Management* 23:34–39.

Wight, H. M., and C. H. Conaway. 1962. A comparison of methods for determining age of cottontails. *Journal of Wildlife Management* 26:160–163.

Swamp Rabbit (*Sylvilagus aquaticus*)

Name

The derivation of the first part of the scientific name, *Sylvilagus*, is the same as that of its close relative, the eastern cottontail. The last part, *aquaticus*, is the Latin word meaning “found in water.” Thus, part of the scientific name and part of the common name, “swamp,” refer to this species’ use of water. “Rabbit” is from the Middle English *rabet*. Swamp rabbits are also known as swampers, cane cutters, and cane jakes.

Description (Plate 21)

The swamp rabbit is similar to the eastern cottontail but is distinguished by its generally larger size; proportionately shorter and rounder ears; somewhat coarser body fur with a yellowish cast, particularly to the rump, and more black mottling; and the color of the tops of the hind feet and heavier toenails.

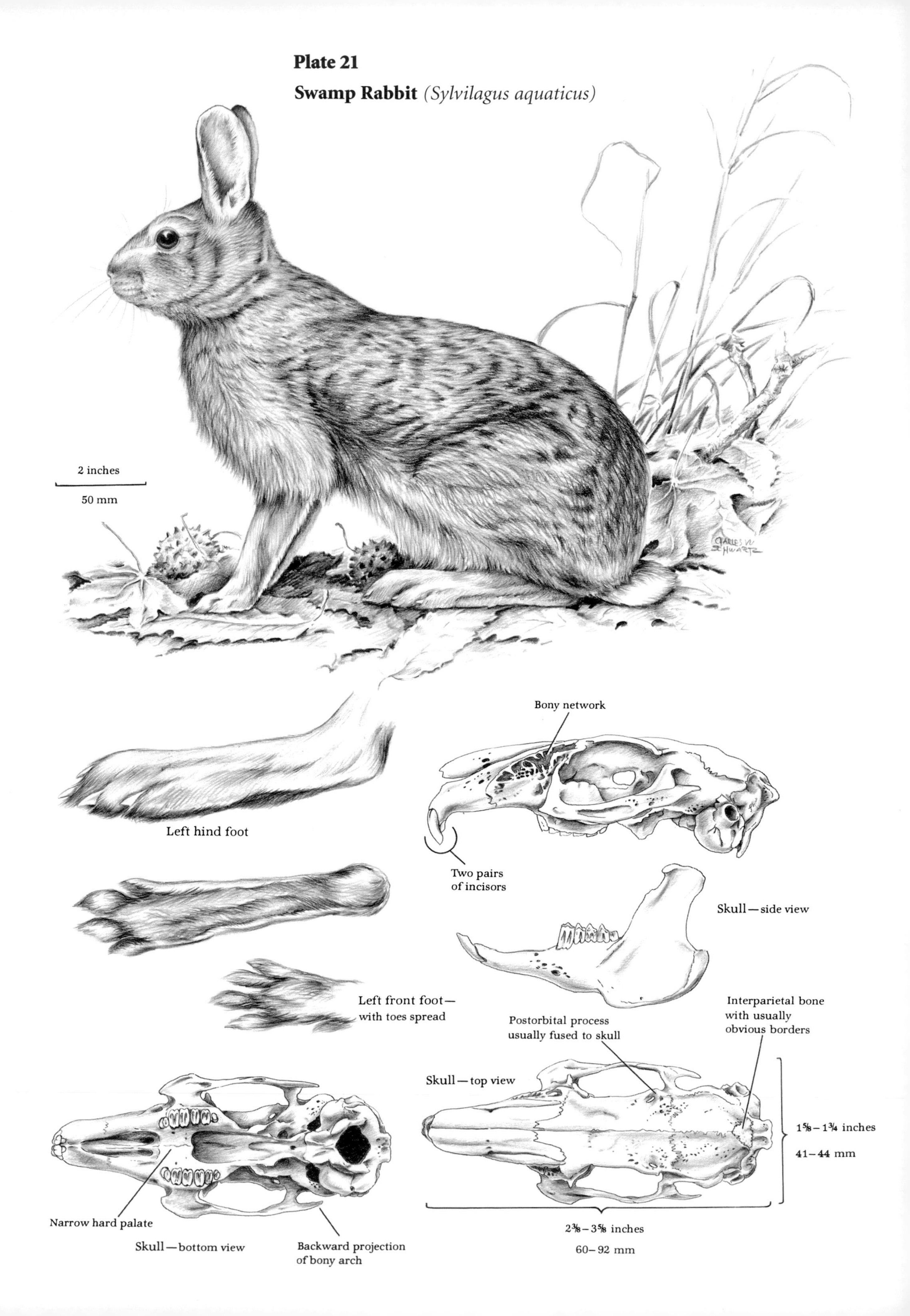

Plate 21

Swamp Rabbit *(Sylvilagus aquaticus)*

Swamp rabbit tracks

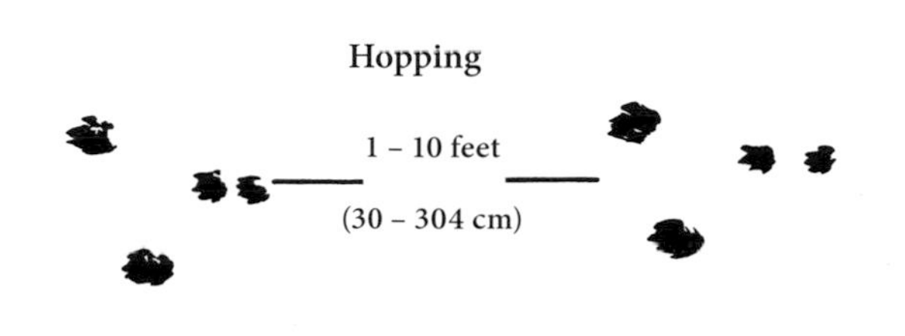

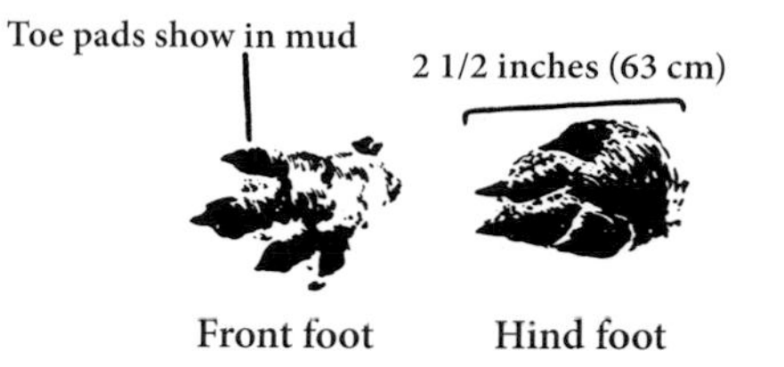

Color. The swamp rabbit is dark grayish or yellowish brown above with coarse black peppering or mottling. The underparts are white except for the chest, which is buffy gray. The back of the neck is slightly rusty. The tops of the hind feet are reddish brown.

Measurements

Total length	406–558 mm	16–22 in.
Tail	41–76 mm	1⅝–3 in.
Hind foot	98–114 mm	3⅞–4½ in.
Ear	63–95 mm	2½–3¾ in.
Skull length	60–92 mm	2⅜–3⅝ in.
Skull width	41–44 mm	1⅝–1¾ in.
Weight	1.1–2.7 kg	2½–6 lb.

Teeth and skull. The dental formula of the swamp rabbit is the same as that of all the members of this family in Missouri:

$$I\,\frac{2}{1}\;C\,\frac{0}{0}\;P\,\frac{3}{2}\;M\,\frac{3}{3} = 28$$

The skull of the swamp rabbit has been discussed with the eastern cottontail's skull.

Sex ratio. The sexes occur in approximately equal numbers.

Age criteria, age ratio, and longevity. As in cottontails, the age of swamp rabbits can be determined by the dry weight of the eye lens. Swamp rabbits can also be aged by X-raying the ends of the bones of the front leg for the presence of cartilage. This method identifies young swamp rabbits until 10 to 13 months of age.

Age ratios in January and February vary between 56 and 63 percent juveniles.

Distribution and Abundance

Swamp rabbits are found in most of the south-central United States and the Gulf Coast. They range from South Carolina through Georgia and into central Texas, and north to southern Missouri, Illinois, and Indiana.

Swamp rabbits were widely distributed throughout 21 Missouri counties in the 1930s but at that time were reported to be decreasing in number, mostly due to the drainage of the Mississippi River Alluvial Basin of southeast Missouri and conversion to agriculture.

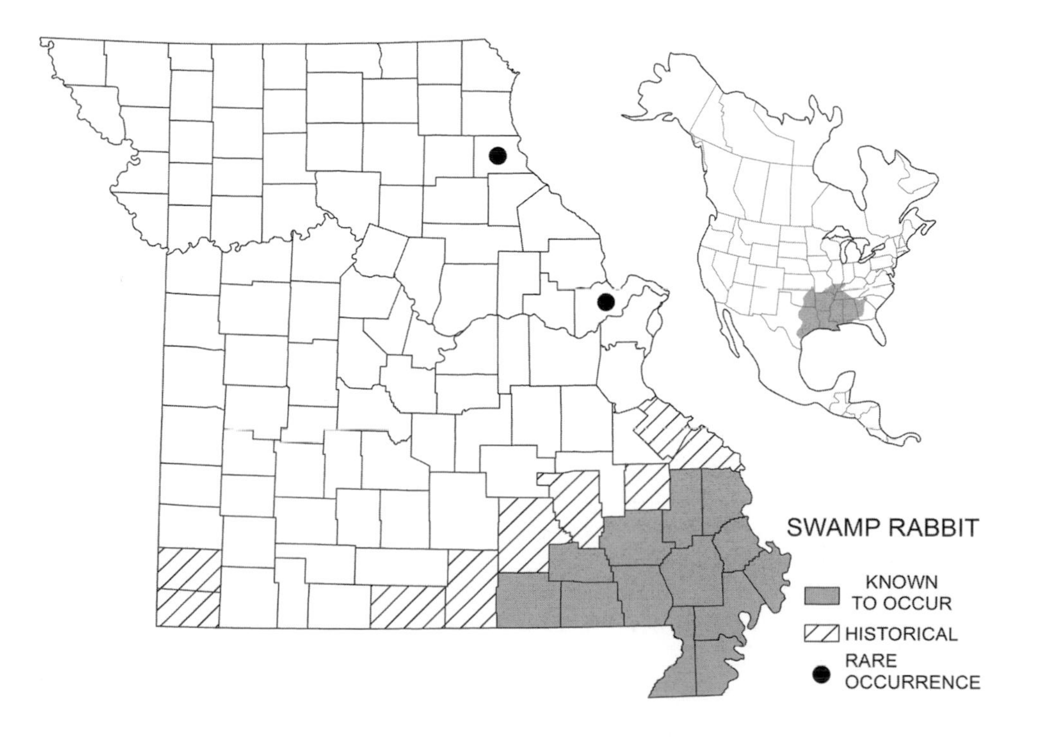

By the 1970s, only 10 counties had swamp rabbits, and in 5 of those counties, only a few small populations remained. In 1974, the swamp rabbit was listed as rare in Missouri based on continued habitat loss and fragmentation.

Extensive monitoring began in the 1990s and continues today. Swamp rabbits now occur in 13 southeast Missouri counties, but unfortunately their habitats continue to be lost, fragmented, isolated, and limited in extent. They are listed as an imperiled species of conservation concern in Missouri but are secure globally. Recent photographic and latrine evidence suggests that a few small pockets of swamp rabbits occur outside of their historical range in St. Charles County within the floodplain of the Missouri and Mississippi Rivers, but further surveys are needed to confirm their presence.

A reliable record for St. Charles County was reported in 1941 and for Marion County in 1967, but the origin of those animals is unknown and no populations were known to exist.

Habitat and Home

Swamp rabbits live in wet lowlands and along the banks of streams and drainage ditches. The nest is built in a slight depression in the ground. Grass is the primary nesting material, but dead leaves and twigs may also be used. The lining is of fur. Most nests are located in tall grass or piles of fallen branches under oak trees; some, however, are built in hollow logs.

The female builds the nest as many as three days before the young are born. She spends about 30 minutes digging the depression, then makes from 10 to 21 trips for nesting material that she collects within 4.5 to 6 m (15 to 20 ft.) of the nest. She carries the material to the nest in her mouth. After each trip she arranges the nest. The fur lining is added before the young are born. Sometimes dummy nests are built before the final one is made for the young.

Nursing times are usually dawn and dusk, but the female is not regular about nursing at each of these periods. When young are in the nest, the female sits in front of the nest and extends her body over the entrance. After the young have left the nest, she crouches on all four feet while they snuggle under her and nurse.

Habits

When pursued, swamp rabbits usually enter the water, swimming and diving well, and often come up under roots or other overhanging protection. They have resting forms on logs, stones, or other elevations projecting out of the water that are covered with vegetation, or in cane, open grassy patches, brush piles, cavities at the base of live trees, and downed hollow logs. Their droppings are usually left on logs and stumps. Fortunately, these conspicuous latrines are easy to find, especially in winter, and searches for these latrine sites are used across their range to determine their presence and abundance.

In Missouri, average home ranges for females vary up to 2.4 ha (6 ac.) and for males up to 1.8 ha (4½ ac.). Maximum home ranges were estimated to be 2.2 ha (5.5 ac.) for females and 3.2 ha (7.9 ac.) for males. Swamp rabbits develop a hierarchy in which certain males become dominant. All individuals have behavior patterns similar to those described for the eastern cottontail.

Foods

During spring, summer, and fall, the foods of swamp rabbits include many grasses, sedges, and herbs growing in moist localities, succulent aquatic plants, and the stems of cane. The common hop sedge is a preferred food by swamp rabbits in southeastern Missouri from the time it becomes green in spring until it turns brown in midwinter. Twigs and the bark of woody plants such as blackberry, hazelnut, deciduous holly, and spicebush are fed upon to a considerable extent in winter. Swamp rabbits eat their own droppings during the daytime.

Reproduction

The reproductive season extends generally from February through mid-July, but rarely litters are born in August and September. The gestation period varies from 35 to 40 days, and up to 5 litters are produced annually. From 1 to 6, but usually 3 to 4, young compose a litter. There is a synchrony in breeding as in the eastern cottontail.

At birth the young weigh about 57 g (2 oz.) each and are slightly over 12.7 cm (5 in.) long. They are well furred but blind. In 2 or 3 days their eyes open and they are able to walk feebly. The gestation period is longer than that of the eastern cottontail, and the young are correspondingly further developed at birth.

The young leave the nest in their second week but continue to nurse a while longer. There is no evidence of breeding by young of the year.

Some Adverse Factors; Importance; and Conservation and Management

These have been discussed for rabbits in general under the eastern cottontail.

Swamp rabbits are present throughout southeast Missouri, but their habitat continues to shrink. Most of the largest remaining habitats are publicly owned and are likely to remain in permanent forest cover. However, privately owned forest blocks continue to be lost, and the consequences to swamp rabbit populations are unknown. In addition, the effects of flooding, ice events, wind, and drought on swamp rabbits are also a concern due to changes in habitat.

The Missouri swamp rabbit population is operating as a *metapopulation*, where a few large forested blocks of habitat provide the "source" of swamp rabbits for smaller, more isolated forested areas. Without a larger forest block nearby, and adequate travel corridors, an isolated forest block is unlikely to be recolonized if the current swamp rabbit population is depleted or disappears. However, a small block close to a larger "source" block is likely to be recolonized more easily because of proximity. The Mississippi River Alluvial Basin in Missouri comprises a series of north-south and east-west ditches, and swamp rabbits are known to use at least the larger ditches for travel and to inhabit some of the small forested areas found along some ditches.

The more limited distribution and fewer numbers make this rabbit less important than the eastern cottontail as a game species in Missouri.

SELECTED REFERENCES

See also discussion of this species in general references, page 23.

Chapman, J. A., and G. A. Feldhamer. 1981. *Sylvilagus aquaticus. Mammalian Species* 151. 4 pp.

Dailey, T. V., T. M. Vangilder, and L. W. Burger Jr. 1993. Swamp rabbit distribution in Missouri. *Proceedings of the Annual Conference of the Southeast Association of Fish and Wildlife Agencies* 47:251–256.

Dumyahn, J. B., and P. A. Zollner. 2010. Home range characteristics of swamp rabbits (*Sylvilagus aquaticus*) in southwestern Indiana and northwestern Kentucky. *Proceedings of the Indiana Academy of Science* 119:80–86.

Fantz, D. K., B. Gillespie, and I. Vining. 2013. *Distribution and status of the swamp rabbit (*Sylvilagus aquaticus*) in Missouri, 2010–2012*. Final Report Missouri Department of Conservation, Jefferson City. 34 pp.

Holler, N. R., T. S. Baskett, and J. P. Ropers. 1963. Reproduction in confined swamp rabbits. *Journal of Wildlife Management* 27:179–183.

Hunt, T. P. 1959. Breeding habits of the swamp rabbit with notes on its life history. *Journal of Mammalogy* 40:82–91.

Korte, P. A., and L. H. Frederickson. 1977. Swamp rabbit distribution in Missouri. *Transactions of the Missouri Academy of Science* 10 & 11:72–77.

Lowe, C. E. 1958. Ecology of the swamp rabbit in Georgia. *Journal of Mammalogy* 39:116–127.

Marsden, H. M., and N. R. Holler. 1964. Social behavior in confined populations of the cottontail and the swamp rabbit. *Wildlife Monographs* 13:1–39.

Martinson, R. K., J. W. Holten, and G. K. Brakhage. 1961. Age criteria and population dynamics of the swamp rabbit in Missouri. *Journal of Wildlife Management* 25:271–281.

Porath, J. W. 1997. Swamp rabbit (*Sylvilagus aquaticus*) status, distribution, and habitat characteristics in southern Illinois. M.S. thesis, Southern Illinois University, Carbondale. 61 pp.

Sadler, K. C. 1969. Distribution of jackrabbits and swamp rabbits in Missouri. Unpublished memorandum, Missouri Department of Conservation, Jefferson City. 2 pp.

Scharine, P. D., C. K. Nielsen, E. M. Schauber, and L. Rubert. 2009. Swamp rabbits in floodplain ecosystems: Influence of landscape- and stand-level habitat on relative abundance. *Wetlands* 29:615–623.

Scheibe, J. S., and R. Henson. 2003. The distribution of swamp rabbits in southeast Missouri. *Southeastern Naturalist* 2:327–334.

Smith, C. C. 1940. Notes on the food and parasites of the rabbits of a lowland area in Oklahoma. *Journal of Wildlife Management* 4:429–431.

Sorensen, M. F., J. P. Ropers, and T. S. Baskett. 1968. Reproduction and development in confined swamp rabbits. *Journal of Wildlife Management* 32:520–531.

———. 1972. Parental behavior in swamp rabbits. *Journal of Mammalogy* 53:840–849.

Toll, J. E., T. S. Baskett, and C. H. Conaway. 1960. Home range, reproduction, and food of the swamp rabbit in Missouri. *American Midland Naturalist* 63:398–412.

Warwick, J. A. 2003. Distribution and abundance of swamp rabbits and bats in fragmented wetland forests of southeast Missouri. M.S. thesis, University of Missouri, Columbia. 127 pp.

Zollner, P. A., W. P. Smith, and L. A. Brennan. 1996. Characteristics and adaptive significance of latrines of swamp rabbits (*Sylvilagus aquaticus*). *Journal of Mammalogy* 77:1049–1058.

Black-tailed Jackrabbit
(Lepus californicus)

Name

The scientific name, *Lepus*, is the Latin word for "hare," and *californicus* is a Latinized word meaning "of California," for the place from which the first specimen was collected and described.

The common name originated from three sources: "black-tailed" is from the color of the upper surface of the tail; "jack" probably comes from the resemblance of the jackrabbit's ears to those of a jackass; and "rabbit" is from the Middle English name *rabet* for the common rabbit of Europe. The black-tailed jackrabbit is also known as the American desert hare.

Description (Plate 22)

Jackrabbits are true hares with extremely long ears, very large hind legs and feet in proportion to the body, smaller front legs and feet, a lanky body, a short fluffy tail, and soft fur. They have 5 toes on the front feet and 4 on the hind. The soles of the feet are densely furred. The upper lip is divided.

Jackrabbits, like other members of this family, possess acute hearing and a very keen sense of smell. The large ears and nose move almost constantly, analyzing the air currents for every sound and scent. The eyes are located on the sides of the head, and the angle of vision of one eye overlaps that of the other, both in front and behind. While one of these animals can thus see all around itself at once, objects are observed more by motion than by sharp visual perception or focus. The large ears and wide field of vision assist them in detecting the approach of a predator in time to employ their best means of defense, that of escape by running. The ears, with their extensive blood supply near the surface of the skin, serve to dissipate heat from the body.

Color. The back of the black-tailed jackrabbit is dark buffy gray with wavy, black markings. These wavy markings are caused by the variable color of the hairs, some being black tipped while others are buff tipped with a black band below the tip. There is a dense whitish undercoat. A black area lies along the middle of the rump and extends onto the upper surface of the tail; the border of the top of the tail and the undersurface are white. The insides of the ears are buff and the backs whitish, usually with about an inch of black at the tips. There is a light ring around the eye, and sometimes a small white blaze on the forehead. The iris of the eye is yellow ocher and the pupil black. The chin and underparts are generally white with a buff band across the chest and buffy white on the tops of the hind feet. The sexes are colored alike. Black-tailed jackrabbits molt in spring and fall but have no seasonal change in color.

Measurements

Total length	463–660 mm	18¼–26 in.
Tail	50–114 mm	2–4½ in.
Hind foot	114–46 mm	4½–5¾ in.
Ear	101–77 mm	4–7 in.
Skull length	92–101 mm	3⅝–4 in.
Skull width	44–50 mm	1¾–2 in.
Weight	1.8–3.6 kg	4–8 lb.

Females are larger than males.

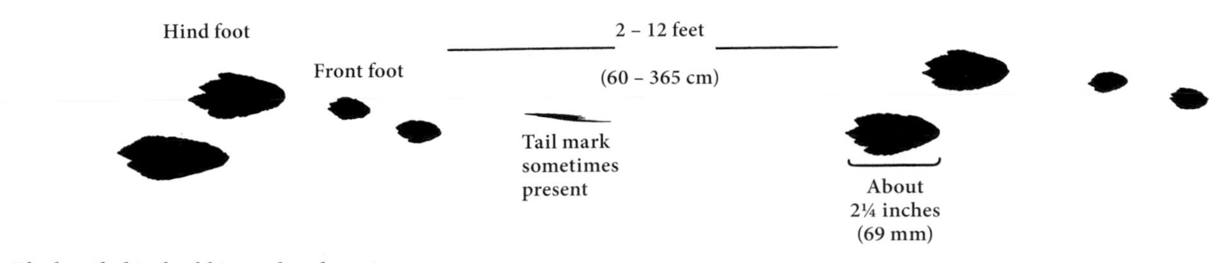

Black-tailed jackrabbit tracks—hopping

Plate 22

Black-tailed Jackrabbit (*Lepus californicus*)

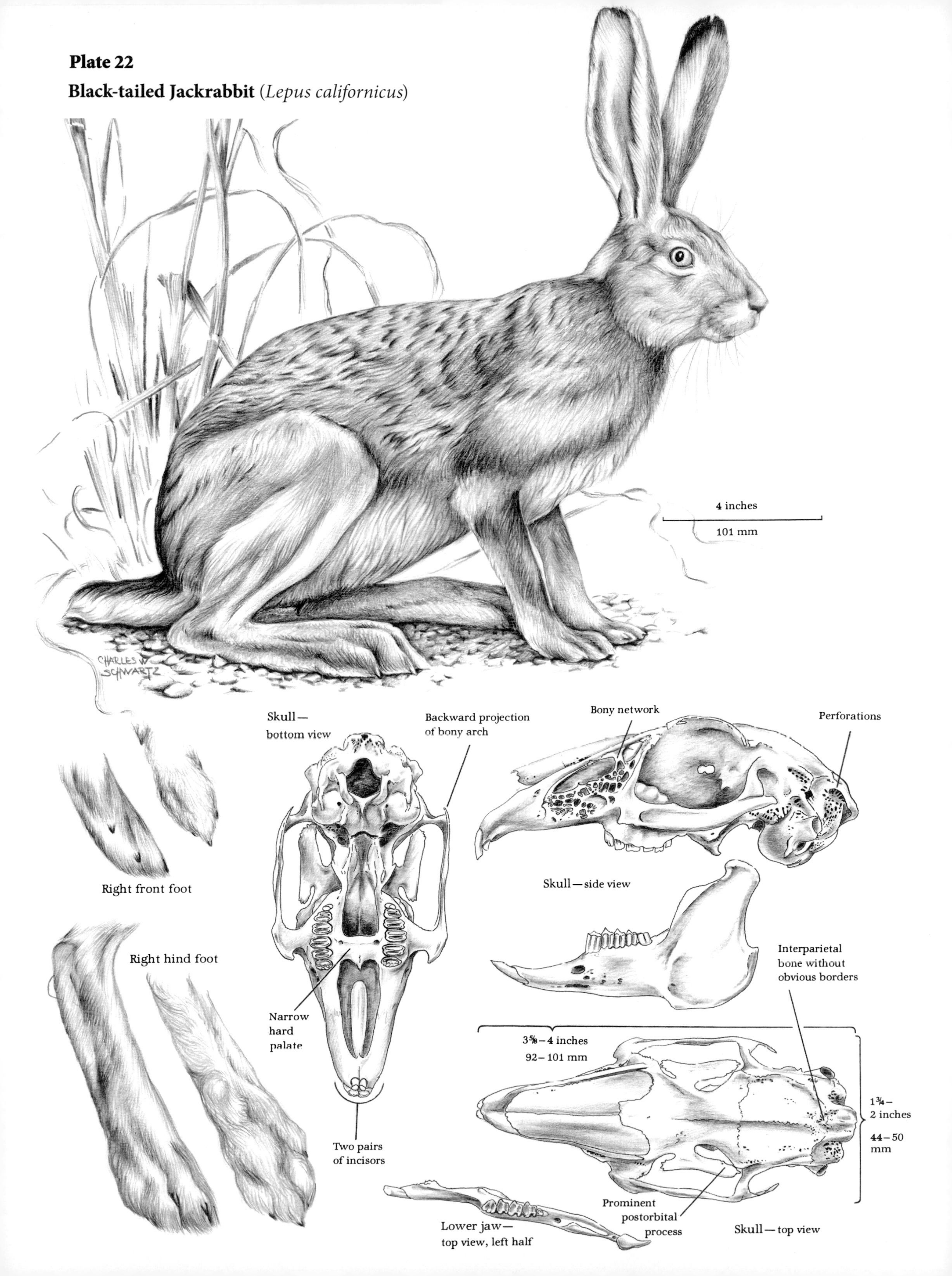

Teeth and skull. All the hares and rabbits in Missouri have the same dental formula:

$$I\ \frac{2}{1}\ C\ \frac{0}{0}\ P\ \frac{3}{2}\ M\ \frac{3}{3} = 28$$

They have two pairs of incisors in the front of the upper jaw, one directly behind the other. The front or outer pair is larger, has a broad groove on the front surface, and possesses a straight cutting edge; the rear or inner pair is smaller, is almost circular in outline, and has no cutting edge. When the jaws are closed, the cutting edge of the lower incisors fits in between the paired incisors of the upper jaw. These teeth grow throughout life, and the growth must be worn away by gnawing or by grinding the teeth against each other. A wide gap separates the incisors from the cheek or chewing teeth, which are high crowned and have cross crests. Because the upper and lower jaws oppose each other only on one side at a time, the motion of chewing is from side to side.

The skull of the adult black-tailed jackrabbit also has the following features in common with other hares and rabbits: a bony network on the side of the face in front of the eye socket; a backward projection of the zygomatic arch (the bony arch at the side of the eye socket); the hard palate forming a narrow bridge between the two sides of the upper jaw; and a prominent projection (the backward extension of which is the postorbital process) above the eye socket.

Black-tailed jackrabbit skulls are distinguished from the skulls of the two other Missouri members of this family by their larger size, the lack of obvious borders of the interparietal bone located on the upper surface of the skull toward the rear, and the presence of many perforations on the base of the skull.

Sex criteria and sex ratio. The sexes of hares and rabbits are sometimes difficult to distinguish, especially during the nonbreeding season and in the young. The penis of the male and the similar-appearing clitoris of the female can be exposed by applying slight pressure to the sides of the reproductive opening, which lies between the hind legs and in front of the anus. The penis is cylindrical and has an opening in the tip, while the clitoris is flattened and has no opening at the tip. Males can be identified easily by the large testes when they are in the scrotum during the breeding season. The testes move into the body cavity during the nonbreeding season, and sometimes as a result of shock during the breeding season. The scrotum lies in front of the penis as it does in the pouched mammals, instead of behind the penis as it does in most mammals. Certain conditions in the structure of the ovary and early embryonic development also show a close association between the hare-shaped mammals and pouched mammals. Female black-tails have three pairs of teats on the outer edge of the belly.

The sex ratio is approximately equal.

Age criteria and longevity. The best method for determining age of jackrabbits is by the dry weight of the eye lens.

Young jackrabbits can be identified up to 9 months of age by the presence of cartilage on the long bones of the front leg as described for the eastern cottontail.

A white blaze on the forehead is most conspicuous in 2- to 3-week-old jackrabbits; this begins to disappear at about 10 weeks of age and by 18 weeks is represented by only a white line. The young have a yellower to browner coat than the adults.

Females that have produced young can be identified by the presence of longitudinal striations on the uterus; these are present even if the uterus has shrunken to its nonbreeding size. Females that have not produced young lack these striations.

Black-tailed jackrabbits in the wild live up to 5 years. However, a few have been known to live for 8 years in the wild.

Glands. Paired glands about 1.2 cm (½ in.) deep on either side of the anus in both sexes secrete a strong musky odor that is left on the ground where the black-tailed jackrabbit sits. This scent probably serves as an identification sign to other jackrabbits.

Voice and sounds. Black-tailed jackrabbits are generally silent but give a loud, coarse squeal when caught. They may also utter a loud *qua-qua* when fighting. The female has a call for assembling her young.

Distribution and Abundance

The black-tailed jackrabbit is found throughout most of the western and midwestern United States, and northern Mexico. There are a few small introduced populations in the eastern United States, including Florida, Virginia, and New Jersey.

Black-tailed jackrabbits probably did not become a permanent member of Missouri's fauna until around 1900. Numbers increased as prairies and forests were converted to agricultural land more favorable to jackrabbits, and overgrazing of pasture provided the shorter vegetation they prefer. The population peaked during the 1920s and 1930s when there was considerable habitat disturbance and drought favorable to jackrabbits. They eventually inhabited 40 western prairie counties mostly south of the Missouri River with very few scattered reports just north of the Missouri River.

Beginning in the 1950s, their range and abundance in Missouri began to rapidly decrease mostly due to changes in farming practices. Concern over this rapid decline and the possible extirpation of the white-tailed jackrabbit (*Lepus townsendii*) led to both species being declared as state endangered and the closing of the hunting season on both jackrabbits in 1971. Surveys in the 1970s through 1990 reported scattered, isolated populations of black-tailed jackrabbits mostly around small dairy farms, orchards, and tree nurseries. Two of the last known Missouri populations were on small, isolated, dairy farms that were less than 32 ha (80 ac.) with no suitable black-tailed jackrabbit habitat in the surrounding areas; both disappeared when the farms were converted from dairy to beef operations or to houses.

Since 2000, very few jackrabbits have been reported (shown as "rare occurrence" counties on the Missouri range map), and they may now be extirpated from Missouri. They have also declined in states adjacent to Missouri. They are classified as a Missouri critically imperiled species of conservation concern and state endangered (since 1990), but they are considered secure across their entire range.

Habitat and Home

The black-tailed jackrabbit is a resident of the open plains and prefers short grass. During peak populations in Missouri, they were mostly found in pastures, hay lands, and cultivated areas, especially before the crops developed very high growth. Pasture with moderate grazing with mixed grasses and legumes, and associated pure legume hay fields are preferred. Cover more than 0.6 m (2 ft.) high is generally avoided.

A jackrabbit's home is a simple, unlined hollow or "form" that it scratches in the ground. Such a form is from 25 to 46 cm (10 to 18 in.) long, 10 to 15 cm (4 to 6 in.) wide, and up to 9 cm (3½ in.) deep. Some are made in the open but most are found at the base of a shrub, in a clump of tall grass or other vegetation, or next to wooden fence posts and telephone poles. There is usually no lining, but often the back end is lower than the front end, providing more comfort for the hindquarters. A form may be used only once or many times over a period of days.

For her nest, the female constructs a similar but deeper form, up to 20 cm (8 in.) in depth. It is well concealed in grassy cover. She sometimes plucks fur from her chest to line and cover it.

Habits

The home range of a black-tailed jackrabbit varies from 1 to 5 sq km (0.5 to 2 sq. mi.) and does not differ between sexes or among seasons. Much of this area may be covered each day in traveling to and from feeding and resting sites. Because individuals habitually use the same routes, trails are often worn in the vegetation.

Black-tailed jackrabbits have an easy, bounding gait and cover between 1.5 and 3 m (5 and 10 ft.) at a jump. This distance may be extended to 4.5 or 6 m (15 or 20 ft.) when speeding. Leaps are ordinarily about 0.6 m

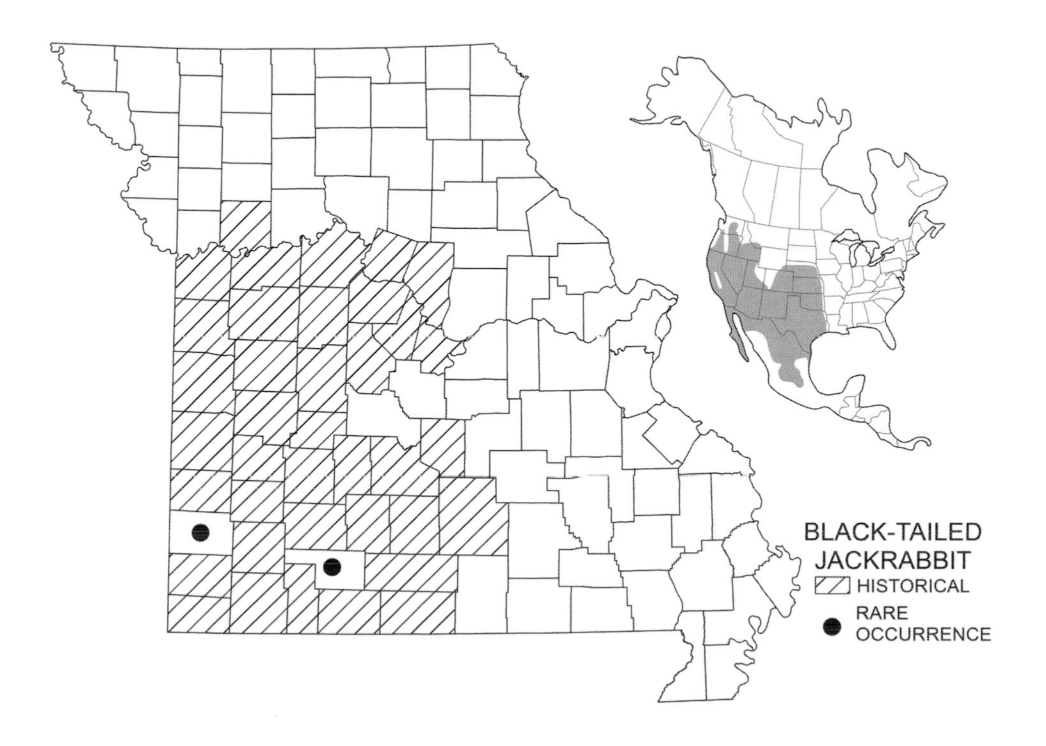

(2 ft.) high but may be as much as 1.7 m (5½ ft.) off the ground. At fairly regular intervals of 4 or 5 jumps, but modified by the height of the cover, blacktails perform a higher and longer leap than usual to obtain a better view of their surroundings. When pressed, jackrabbits can run between 48 to 56 km per hour (30 to 35 mi. per hour) for short distances. They can also swim but tend to avoid water.

Where food is abundant, black-tailed jackrabbits may feed in small groups but otherwise lead solitary lives. However, in some parts of their range in the western United States, they may live in extremely dense concentrations.

Black-tailed jackrabbits regularly feed during dusk or early evening and occasionally have a short feeding period during daybreak. During the day, they usually sit in their forms with the body hunched up and the ears either raised or lowered over the back. To relax they stretch out on their sides and dust-bathe occasionally. They keep their coats neat by licking themselves with the tongue or washing other parts with the front feet.

Black-tailed jackrabbits react to an alarm by "freezing," or moving away slowly or very rapidly. Where there are several jackrabbits feeding in the same area, the alarm spreads to the group.

During the nonbreeding season, black-tailed jackrabbits are generally tolerant of others. They apparently recognize certain individuals and allow them to feed nearby while others are resented and chased away. During the breeding season, more antagonism is evident. Pregnant females tend to chase others away, which may be a type of territorial behavior aimed at providing space for the young.

When fighting, jackrabbits rear up on their hind feet and "box" with their front feet. They may jump over each other and kick with their strong hind feet. They often bite their opponents, especially about the ears.

Foods

Black-tailed jackrabbits eat only vegetation. In spring and summer they prefer herbaceous plants and grasses but eat various cultivated crops such as cabbage, alfalfa, clover, and soybeans. In fall and winter they eat assorted dried grasses and other plants, buds, twigs, bark, fruits, and roots. They also browse on the short green growth of winter wheat.

The amount of vegetation consumed by jackrabbits is very great; it has been estimated that 30 blacktails take as much forage as 1 sheep.

Because moisture is obtained from the vegetation they eat, jackrabbits seldom drink surface water. Soil is sometimes licked for its minerals.

Hares and rabbits obtain some of their nourishment by regularly passing most of their food through their bodies twice. These animals normally have two kinds of droppings: soft, moist, mucous-coated ones first and then dry, fibrous ones. Following the periods of feeding on vegetation, the soft droppings are formed and, after several hours, are passed from the body. The animal customarily eats these as they come from the anus (a behavior called *coprophagy*). This food then goes through the body again and the remains are passed onto the ground as dry droppings when the hare or rabbit is actively moving about and feeding on vegetation. The soft droppings are high in protein and contain large quantities of certain B vitamins that are believed to be formed by bacteria in the intestine. These substances are utilized when the food is passed through the body for a second time. Young animals start to eat their own soft droppings about the time they are weaned. It has been suggested that they may take soft droppings of their mother and in this way inoculate their own digestive systems with these essential intestinal bacteria as well as with internal parasites.

Reproduction

Although black-tailed jackrabbits can breed throughout the year, most breeding takes place from late winter to midsummer (end of January through August in Kansas). Pregnancy lasts from 41 to 47 days. One to 4 litters are produced annually, containing from 1 to 5 young each (with extremes of 1 to 8 and an average of 3). Females may be both pregnant and nursing at the height of the breeding season.

Black-tailed jackrabbit two to three weeks old

Depending upon the vitality of the female and number in the litter, the young weigh from 57 to 170 g (2 to 6 oz.) at birth and are between 14 and 20.3 cm (5½ and 8 in.) in total length. They are well furred, have their eyes open, have their incisor teeth cut, and are able to take a few steps. The ears are very small, less than one-half the length of the head. For 3 to 4 days after birth, the young are hidden in one nest or placed individually or in pairs in different nests. The female visits each one nightly for nursing. By the fifteenth day, the ears are equal in length to the head; at 4 weeks, they are about 10.1 cm (4 in.) long. The young become independent of the female when 3 to 4 weeks old but do not reach adult size until approximately 2 months of age. A few may breed in the year of their birth, but most do not.

Some Adverse Factors

The major predators on black-tailed jackrabbits are coyotes, hawks, owls, foxes, humans, and snakes. Of these the coyote is the most important.

The following parasites are known to occur on or in black-tailed jackrabbits: ticks, lice, fleas, flies, roundworms, and tapeworms. Epidemics in jackrabbits of the bacterial disease tularemia have been reported in some portions of the range. Coccidiosis is often fatal, and many deaths result from a condition of shock.

Importance

Formerly blacktails were killed for the market by the thousands, but they can no longer be legally hunted in Missouri. The meat is wholesome but unpopular because of internal parasites and the threat of tularemia. However, since the recent discovery of new drugs as a remedy for tularemia, this disease in humans is now less serious. The fur is not used much commercially except in limited amounts for hatter's felt, gloves, or trimming and linings of inexpensive cloth coats. Most of the rabbit fur of today's trade is imported from Australia and England.

Where black-tailed jackrabbits are numerous, they may damage cultivated crops, girdle young orchard plantations, and compete with cattle for forage.

Conservation and Management

The black-tailed jackrabbit is one of the most endangered Missouri mammals and may be extirpated from the state. Managing for this species means protecting large expanses of shortgrass habitat with open vistas, and corridors to facilitate dispersal between populations or from an existing population to unoccupied habitat. Missouri prairies have historically been tallgrass prairies and are managed as such, not as shortgrass. In addition, changing agricultural practices, a growing interest in rural living, and development will continue to decrease habitat and increase habitat fragmentation. It is unlikely that the changed western Missouri landscape will ever support a sustainable population of black-tailed jackrabbits. If you see a black-tailed jackrabbit, please report it immediately to your local Missouri Department of Conservation office.

SELECTED REFERENCES

See also discussion of this species in general references, page 23.

Best, T. L. 1996. *Lepus californicus. Mammalian Species* 530. 10 pp.

Bronson, F. H., and O. W. Tiemeier. 1958. Reproduction and age distribution of black-tailed jack rabbits in Kansas. *Journal of Wildlife Management* 22:409–414.

Brown, H. L. 1947. Coaction of jack rabbit, cottontail, and vegetation in a mixed prairie. *Transactions of the Kansas Academy of Science* 50:28–44.

Dumke, R. T. 1973. *The white-tailed jack rabbit in Wisconsin.* Pittman-Robertson Study 106. Project W-141-R-8. Wisconsin Department Natural Resources, Madison. 24 pp.

Fantz, D. K. 2013. History and status of the black-tailed jackrabbit (*Lepus californicus melanotis*) in Missouri. Report to the Missouri Department of Conservation, Jefferson City. 11 pp.

Goodwin, D. L., and P. O. Currie. 1965. Growth and development of black-tailed jack rabbits. *Journal of Mammalogy* 46:96–98.

Hill, R. W., D. P. Christian, and J. H. Veghte. 1980. Pinna temperature in exercising jackrabbits, *Lepus californicus. Journal of Mammalogy* 61:30–38.

Lechleitner, R. R. 1957. Reingestion in the black-tailed jack rabbit. *Journal of Mammalogy* 38:481–485.

———. 1958. Certain aspects of behavior of the black-tailed jack rabbit. *American Midland Naturalist* 60:145–155.

———. 1959. Sex ratio, age classes, and reproduction of the black-tailed jack rabbit. *Journal of Mammalogy* 40:63–81.

Norton, J. L. 1987. The status of the black-tailed jackrabbit (*Lepus californicus*) in Missouri. M.S. thesis, Southwest Missouri State University, Springfield. 42 pp.

Palmer, T. S. 1897. *The jackrabbits of the United States.* Bulletin 8. U.S. Department of Agriculture, Division of Biological Survey, Washington, DC. 88 pp.

Plettner, R. G. 1984. Vital characteristics of the black-tailed jack rabbit in east-central Nebraska. M.S. thesis, University of Nebraska, Lincoln. 90 pp.

Robbins, L. W., and A. H. Hodge. 1990. Ecological requirements of the black-tailed jackrabbit *Lepus californicus* in Missouri. Unpublished report, Missouri Department of Conservation, Jefferson City. 45 pp.

Schmidt-Nielsen, K., T. J. Dawson, H. T. Hammel, D. Hinds, and D. C. Jackson. 1965. The jack rabbit—a study in its desert survival. *Hvalradets Skrifter* 48:125–142.

Skinner, R. M., and B. J. Roedner Skinner. 1979. The distribution and abundance of jackrabbits in Missouri. Unpublished report, Missouri Department of Conservation, Jefferson City. 35 pp.

Smith, G. W. 1990. Home range and activity patterns of black-tailed jackrabbits. *Great Basin Naturalist* 50:249–256.

Tiemeier, O. W. 1965. *The black-tailed jack rabbit in Kansas*. Contribution 336, Department of Zoology, and 418, Department of Bacteriology, Kansas Agricultural Experiment Station, Manhattan. 75 pp.

Tiemeier, O. W., and M. L. Plenert. 1964. A comparison of three methods for determining the age of black-tailed jack rabbits. *Journal of Mammalogy* 45:409–416.

7

Gnawing Mammals
Order Rodentia

Rodents are easily identified by their characteristic teeth. In the front of both upper and lower jaws there is a single pair of long, chisel-like incisors, or gnawing teeth, covered with hard, usually orange enamel. The incisors are separated from the cheek, or grinding, teeth by a wide space, the *diastema*. The name, Rodentia, comes from the Latin word meaning "gnawing" and refers to the use of the prominent incisor teeth.

This is the largest order of mammals both in number of species and number of individuals. They occur on all continents except Antarctica. In addition to the seven families found in Missouri, this order includes many other families that contain such members as the mountain beavers, porcupines, guinea pigs, pacas, capybaras, and agoutis.

Key to the Species
by Whole Adult Animals

1a. Tail large, horizontally flattened, naked, and scaly (see plate 32); hind feet entirely webbed. **American Beaver** (*Castor canadensis*) p. 211

1b. Tail not horizontally flattened, or if horizontally flattened, well furred; hind feet not webbed or at most with only short web at base of toes. **Go to 2**

2a. (From 1b) External fur-lined cheek pouches opening by slits on either side of the mouth (not to be confused with cheek pouches opening inside of mouth) (see plates 30–31). **Go to 3**

2b. (From 1b) No external cheek pouches. **Go to 4**

3a. (From 2a) Large (190–349 mm; 7½–13¾ in.); each upper incisor with 2 lengthwise grooves on front surface; front feet with extremely large claws. **Plains Pocket Gopher** (*Geomys bursarius*) p. 201

3b. (From 2a) Small (114–130 mm; 4½–5⅛ in.); each upper incisor with 1 lengthwise groove on front surface; front feet with small claws. **Plains Pocket Mouse** (*Perognathus flavescens*) p. 207

4a. (From 2b) Gliding membrane present (a loose fold of skin extending from body to wrist and

ankle); well-furred tail flattened horizontally (see plate 29). **Southern Flying Squirrel** (*Glaucomys volans*) p. 194

4b. (From 2b) No gliding membrane. **Go to 5**

5a. (From 4b) Pronounced lengthwise stripes on back and sides (see plates 23, 25). **Go to 6**

5b. (From 4b) No pronounced lengthwise stripes on back and sides. **Go to 7**

6a. (From 5a) Seven yellowish to white stripes alternating with 6 blackish to reddish brown stripes dotted with light spots (see plate 25). **Thirteen-lined Ground Squirrel** (*Ictidomys tridecemlineatus*) p. 168

6b. (From 5a) Five dark brown to blackish stripes with buff to whitish band between the 2 dark stripes on each side (see plate 23). **Eastern Chipmunk** (*Tamias striatus*) p. 155

7a. (From 5b) Tail bushy (see plates 24, 26–28). **Go to 8**

7b. (From 5b) Tail not bushy. **Go to 11**

8a. (From 7a) Tail about one-fourth total length (see plate 24); body stout. **Woodchuck** (*Marmota monax*) p. 162

8b. (From 7a) Tail more than one-fourth total length; body slender. **Go to 9**

9a. (From 8b) Obvious cheek pouches opening inside mouth (see plate 26); body brownish gray speckled with black. **Franklin's Ground Squirrel** (*Poliocitellus franklinii*) p. 174

9b. (From 8b) No obvious cheek pouches opening inside mouth. **Go to 10**

10a. (From 9b) Fringe of tail and belly reddish yellow; back and sides reddish yellow mixed with gray; body rarely all black; total length up to 736 mm (29 in.) (see plate 28). **Eastern Fox Squirrel** (*Sciurus niger*) p. 187

10b. (From 9b) Fringe of tail and belly white; back and sides gray; body rarely reddish or all black; total length up to 533 mm (21 in.) (see plate 27). **Eastern Gray Squirrel** (*Sciurus carolinensis*) p. 179

11a. (From 7b) Tail flattened vertically (see plate 44); hind feet with short web at base of toes. **Common Muskrat** (*Ondatra zibethicus*) p. 272

11b. (From 7b) Tail round; hind feet not webbed. **Go to 12**

12a. (From 11b) Tail about 1½ times length of head and body (see plate 49); hind feet greatly elongated in proportion to the body, adapted for jumping; hairs on belly white or pale yellow to their bases; back yellowish brown with blackish band from head to tail; sides yellowish orange; deep, lengthwise groove on front of each upper incisor; mouselike. **Meadow Jumping Mouse** (*Zapus hudsonius*) p. 298

12b. (From 11b) Tail less than 1½ times length of head and body; hairs on belly all gray, or gray at base. **Go to 13**

13a. (From 12b) Tail up to slightly more than twice the length of hind foot; ears not projecting much beyond fur. **Go to 14**

13b. (From 12b) Tail considerably more than twice the length of hind foot; ears projecting beyond fur. **Go to 17**

14a. (From 13a) Tail nearly same length as hind foot. **Go to 15**

14b. (From 13a) Tail nearly twice length of hind foot. **Go to 16**

15a. (From 14a) Upper incisors with shallow, lengthwise groove on outer edges (see plate 45). **Southern Bog Lemming** (*Synaptomys cooperi*) p. 280

15b. (From 14a) Upper incisors not grooved (see plate 43). **Woodland Vole** (*Microtus pinetorum*) p. 267

16a. (From 14b) Tail 2 or slightly more times length of hind foot; upperparts uniformly dark brown (see plate 41). **Meadow Vole** (*Microtus pennsylvanicus*) p. 258

16b. (From 14b) Tail slightly less than twice length of hind foot; upperparts grizzled grayish to blackish brown (see plate 42). **Prairie Vole** (*Microtus ochrogaster*) p. 261

17a. (From 13b) Upper incisors with deep lengthwise groove in middle (see plates 34–35). **Go to 18**

17b. (From 13b) Upper incisors not grooved. **Go to 19**

18a. (From 17a) Tail shorter than head and body (see plate 34); back brown, sides grayish tan; total length up to 155 mm (6⅛ in.). **Western Harvest Mouse** (*Reithrodontomys megalotis*) p. 224; **Plains Harvest Mouse** (*Reithrodontomys montanus*) p. 223

18b. (From 17a) Tail longer than head and body (see plate 35); back brown, sides reddish yellow; total length up to 203 mm (8 in.). **Fulvous Harvest Mouse** (*Reithrodontomys fulvescens*) p. 229

19a. (From 17b) Upperparts grizzled tan, brown, and black (see plate 39); ears extend only slightly beyond fur. **Hispid Cotton Rat** (*Sigmodon hispidus*) p. 248
19b. (From 17b) Upperparts not grizzled; ears extend well beyond fur. **Go to 20**

20a. (From 19b) Total length less than 228 mm (9 in.). **Go to 21**
20b. (From 19b) Total length 228 mm (9 in.) or more. **Go to 23**

21a. (From 20a) Color of upperparts sharply marked off from underparts (see plates 36–37). *Peromyscus* spp. **Deer Mouse** (*Peromyscus maniculatus*) p. 231; **White-footed Mouse** (*Peromyscus leucopus*) p. 239; **Cotton Mouse** (*Peromyscus gossypinus*) p. 242; **Texas Mouse** (*Peromyscus attwateri*) p. 243
21b. (From 20a) Color of upperparts not sharply marked off from underparts. **Go to 22**

22a. (From 21b) Back yellow to orange brown (see plate 38). **Golden Mouse** (*Ochrotomys nuttalli*) p. 245
22b. (From 21b) Back grayish brown (see plate 48). **House Mouse** (*Mus musculus*) p. 293

23a. (From 20b) Tail densely covered with short hairs and not obviously scaled (see plate 40). **Eastern Woodrat** (*Neotoma floridana*) p. 253
23b. (From 20b) Tail sparsely haired and obviously scaled (see plates 33, 46–47). **Go to 24**

24a. (From 23b) Tail shorter than head and body (see plate 47). **Brown Rat** (*Rattus norvegicus*) p. 287
24b. (From 23b) Tail as long or longer than head and body. **Go to 25**

25a. (From 24b) Toes very long and slender; soles of feet with obvious small tubercles in addition to large tubercles, or pads (see plate 33); tail distinctly lighter on undersurface. **Marsh Rice Rat** (*Oryzomys palustris*) p. 219
25b. (From 24b) Toes stubby; soles of feet smooth except for large tubercles, or pads (see plate 46); tail not distinctly lighter on undersurface. **Black Rat** (*Rattus rattus*) p. 285

Key to the Species
by Skulls of Adults

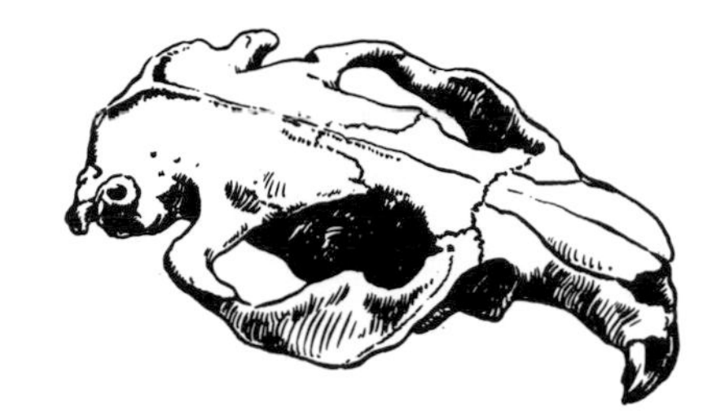

1a. Total teeth 22. **Go to 2**
1b. Total teeth less than 22. **Go to 6**

2a. (From 1a) Upper surface flat when seen in profile (see plate 24); postorbital processes project at right angles to length of skull; depressed area between postorbital processes. **Woodchuck** (*Marmota monax*) p. 162
2b. (From 1a) Upper surface not flat; postorbital processes extend backward and downward along side of braincase (see plates 25–27, 29); raised area between postorbital processes. **Go to 3**

3a. (From 2b) Braincase highly arched (see plate 29); front end of skull (nasal bones) slightly upturned or arched; deep notch in upper rim of eye socket. **Southern Flying Squirrel** (*Glaucomys volans*) p. 194
3b. (From 2b) Braincase not highly arched; front end of skull (nasal bones) not upturned or arched; no deep notch in upper rim of eye socket. **Go to 4**

4a. (From 3b) Length of skull less than 44 mm (1¾ in.)(see plate 25). **Thirteen-lined Ground Squirrel** (*Ictidomys tridecemlineatus*) p. 168
4b. (From 3b) Length of skull 44 mm (1¾ in.) or more. **Go to 5**

5a. (From 4b) Region between eye sockets on upper surface of skull relatively narrow; postorbital processes not prominent (see plate 26). **Franklin's Ground Squirrel** (*Poliocitellus franklinii*) p. 174
5b. (From 4b) Region between eye sockets relatively wide; postorbital processes prominent (see plate 27). **Eastern Gray Squirrel** (*Sciurus carolinensis*) p. 179

6a. (From 1b) Total teeth 20. **Go to 7**
6b. (From 1b) Total teeth less than 20. **Go to 11**

7a. (From 6a) Each upper incisor with 1 or 2 lengthwise grooves on front surface (see plates 30–31). **Go to 8**

7b. (From 6a) Upper incisors not grooved (see plates 23, 28, 32). **Go to 9**

8a. (From 7a) Each upper incisor with 2 lengthwise grooves on front surface; large (34–63 mm or 1⅜–2½ in.). **Plains Pocket Gopher** (*Geomys bursarius*) p. 201

8b. (From 7a) Each upper incisor with 1 lengthwise groove on front surface; small (20 mm or 13⁄16 in.). **Plains Pocket Mouse** (*Perognathus flavescens*) p. 207

9a. (From 7b) Length of skull 101 mm (4 in.) or more (see plate 32). **American Beaver** (*Castor canadensis*) p. 211

9b. (From 7b) Length of skull less than 101 mm (4 in.). **Go to 10**

10a. (From 9b) Length of skull 63 mm (2½ in.) ormore; cheek tooth rows do not converge slightly toward rear (see plate 28). **Eastern Fox Squirrel** (*Sciurus niger*) p. 187

10b. (From 9b) Length of skull less than 63 mm (2½ in.); cheek tooth rows converge slightly toward rear (see plate 23). **Eastern Chipmunk** (*Tamias striatus*) p. 155

11a. (From 6b) Total teeth 18 (see plate 49). **Meadow Jumping Mouse** (*Zapus hudsonius*) p. 298

11b. (From 6b) Total teeth 16. **Go to 12**

12a. (From 11b) Upper cheek teeth with cusps arranged in 3 lengthwise rows (obscure in very old individuals with worn teeth) (see plates 46–48). **Go to 13**

12b. (From 11b) Upper cheek teeth with cusps arranged in 2 lengthwise rows (obscure in very old individuals with worn cusps) or folds of enamel surrounding dentine (see plates 33, 36–45). **Go to 15**

13a. (From 12a) No paired ridges on upper surface of skull (see plate 48); notch in cutting surface of upper incisors (when viewed from side); first upper molar larger than second and third combined. **House Mouse** (*Mus musculus*) p. 293

13b. (From 12a) Paired ridges on upper surface of skull (see plates 46–47); no notch in cutting surface of upper incisors (when viewed from side); first upper molar not larger than second and third combined. **Go to 14**

14a. (From 13b) Ridges on upper surface of skull more or less parallel (see plate 47); length of parietal bone, measured along the temporal ridge, about equal to the greatest distance between temporal ridges (see illustration, p. 289); space between upper incisors and cheek teeth nearly twice the length of the cheek tooth row. **Brown Rat** (*Rattus norvegicus*) p. 287

14b. (From 13b) Ridges on upper surface of skull bowed decidedly outward on each side (see plate 46); length of parietal bone, measured along the temporal ridge, decidedly less than the greatest distance between temporal ridges (see illustration, p. 289); space between upper incisors and cheek teeth considerably less than twice the length of the cheek tooth row. **Black Rat** (*Rattus rattus*) p. 285

15a. (From 12b) Upper cheek teeth with cusps arranged in 2 lengthwise rows (see plates 33, 36–38). **Go to 16**

15b. (From 12b) Upper cheek teeth with folds of enamel surrounding dentine (see plates 39–45). **Go to 20**

16a. (From 15a) Deep, lengthwise groove in middle of front of each upper incisor (see plates 34–35). **Go to 17**

16b. (From 15a) Upper incisors not grooved (see plates 33, 36–38). **Go to 18**

17a. (From 16a) Dentine of third lower molar shows C pattern and dentine of third upper molar continuous and not separated into islands (see plate 34). **Western Harvest Mouse** (*Reithrodontomys megalotis*) p. 224; **Plains Harvest Mouse** (*Reithrodontomys montanus*) p. 223

17b. (From 16a) Dentine of third lower molar shows S pattern and dentine of third upper molar separated into islands (see plate 35). **Fulvous Harvest Mouse** (*Reithrodontomys fulvescens*) p. 229

18a. (From 16b) Cusps on upper molars tend to be opposite (see plate 33); distinct ridge on each side of skull above eye socket; hard palate extends beyond last upper molars. **Marsh Rice Rat** (*Oryzomys palustris*) p. 219

18b. (From 16b) Cusps on upper molars tend to be alternate (see plates 36–38); no ridge above eye socket; hard palate ends opposite last upper molars. **Go to 19**

19a. (From 18b) Paired openings (posterior palatine foramina) about halfway between back of hard palate and larger openings (anterior palatine foramina) in front of hard palate (see plates 36–37); front border of infraorbital plate bowed forward. *Peromyscus* spp. **Deer Mouse** (*Peromyscus maniculatus*) p. 231; **White-footed Mouse** (*Peromyscus leucopus*) p. 239; **Cotton Mouse** (*Peromyscus gossypinus*) p. 242; **Texas Mouse** (*Peromyscus attwateri*) p. 243

19b. (From 18b) Paired openings (posterior palatine foramina) nearer back of hard palate than to larger openings (anterior palatine foramina) in the front of hard palate (see plate 38); front border of infraorbital plate straight. **Golden Mouse** (*Ochrotomys nuttalli*) p. 245

20a. (From 15b) Second and third upper molars have S or modified S pattern of enamel folds surrounding dentine while second and third lower molars have typical S pattern, not conspicuous in young individuals with unworn teeth (see plate 39). **Hispid Cotton Rat** (*Sigmodon hispidus*) p. 248

20b. (From 15b) Cheek teeth with sharp-angled enamel folds surrounding dentine (see plates 40–45). **Go to 21**

21a. (From 20b) Length of skull 38 mm (1½ in.) or less. **Go to 23**

21b. (From 20b) Length of skull more than 38 mm (1½ in.). **Go to 22**

22a. (From 21b) Dentine in middle loop of first and second upper molars extends completely across tooth (see plate 40); region between eye sockets depressed; postorbital processes not apparent. **Eastern Woodrat** (*Neotoma floridana*) p. 253

22b. (From 21b) Four or more islands of dentine in each upper cheek tooth (see plate 44); region between eye sockets raised to form a slight ridge; postorbital processes prominent. **Common Muskrat** (*Ondatra zibethicus*) p. 272

23a. (From 21a) Each upper incisor with shallow, lengthwise groove on outer edge (see plate 45); third upper molar with 4 islands of dentine extending across tooth. **Southern Bog Lemming** (*Synaptomys cooperi*) p. 280

23b. (From 21a) Upper incisors not grooved. **Go to 24**

24a. (From 23b) Second upper molar with 5 islands of dentine (rarely 4) (see plate 41). **Meadow Vole** (*Microtus pennsylvanicus*) p. 258

24b. (From 23b) Second upper molar with 4 islands of dentine. **Go to 25**

25a. (From 24b) Second upper molar with front border of second island on tongue side rounded (see plate 42). **Prairie Vole** (*Microtus ochrogaster*) p. 261

25b. (From 24b) Second upper molar with front border of second island on tongue side squared (see plate 43). **Woodland Vole** (*Microtus pinetorum*) p. 267

Squirrels, Chipmunks, Marmots, and Relatives (Family Sciuridae)

The family name, Sciuridae, is based on the Latin word meaning "squirrel." Representatives of this family include tree squirrels, ground squirrels, chipmunks, marmots, flying squirrels, and prairie dogs. They are native to North and South America, Europe, Asia, and Africa. The eastern gray squirrel and northern palm squirrel were introduced into Australia in the late 1800s, but the gray squirrel quickly disappeared and only a few palm squirrels remain. Seven species are found in Missouri.

Typically, members of this family have thickly haired tails. Tails of tree-dwelling species, such as the eastern gray squirrel and the eastern fox squirrel, are usually long and bushy, but those of ground-dwelling species, such as the eastern chipmunk, thirteen-lined and Franklin's ground squirrels, and woodchuck, are usually shorter and less bushy. There are 4 or 5 cheek teeth above and 4 cheek teeth below on each side of the jaw.

Eastern Chipmunk (*Tamias striatus*)

Name

The first part of the scientific name, *Tamias*, is the Greek word for "a storer" and aptly describes the food-storing habit of this animal. The second part, *striatus*, is the Latin word for "striped" and refers to the conspicuous body stripes. The common name, "eastern," denotes the general range in eastern North America, while "chipmunk" comes from *chitmunk*, an Algonquian name that is associated with its chipping call.

Description (Plate 23)

The eastern chipmunk is a small, ground-dwelling squirrel with conspicuous lengthwise stripes on the back, sides, and cheeks. The ears are short, rounded, and erect. Paired, large, internal cheek pouches, or pockets, open into the mouth; these are used mainly for transporting food and sometimes for carrying excavated soil. The tail is well haired but not bushy, is somewhat flattened, and is shorter than the body. The legs are moderately short. On each front foot there are 4 toes with claws and a small thumb covered by a soft, roundish nail; each hind foot has 5 clawed toes. The fur is moderately long and soft.

Color. The background color of the upperparts is a grizzled reddish brown, grading into rusty on the rump and flanks. There are 5 dark brown to blackish lengthwise stripes from shoulders to rump (1 in the middle and 2 on each side) with a buff to whitish band between the 2 stripes on each side. The belly and sides are buff to white and the feet are tan. The tail is blackish above and rusty below with a narrow white or yellowish fringe. There is a tan or whitish stripe above and below the eye and a black stripe passing across the eye. The cheek has a slight reddish brown stripe. The winter coat is paler than the summer one. Albino or very dark individuals are extremely rare. The sexes are colored alike. The time of molting varies with the sex and age of individuals. Adult males molt in April and May and again in August and September. Adult females molt only in June. Spring-born young molt in June and July, while summer-born young molt in the fall.

Measurements		
Total length	203–304 mm	8–12 in.
Tail	63–114 mm	2½–4½ in.
Hind foot	31–38 mm	1¼–1½ in.
Ear	12–19 mm	½–¾ in.
Skull length	38–50 mm	1½–2 in.
Skull width	20–23 mm	13/16–15/16 in.
Weight	56–141 g	2–5 oz.

Teeth and skull. The dental formula of the eastern chipmunk is:

$$I\ \frac{1}{1}\ C\ \frac{0}{0}\ P\ \frac{1}{1}\ M\ \frac{3}{3} = 20$$

The chipmunk and the fox squirrel are the only Missouri members of the squirrel family with 20 teeth; their skulls can be distinguished by the respective sizes. The cheek tooth rows of the chipmunk tend to converge slightly toward the rear.

Sex criteria and sex ratio. The sexes are identified by the usual external sex organs. The testes are in a temporary scrotum. A penis bone is present. There are 4 pairs of teats on the belly: 1 pair is near the front legs, 2 pairs are behind this, and 1 pair is in the groin region. In litters, the sexes are about equal.

Age criteria and longevity. The permanent dentition is complete in animals 3 months of age. On the upper surface of the skull, the temporal ridges, which are absent during the first 2 months of life, gradually develop during the third and fourth months, and by the tenth month they are well developed. This is a structure common to practically all adult chipmunks. The development of these ridges and the replacement and wear of the teeth are used to place chipmunks into age classes.

Changes in the ends, or epiphyseal region, of certain bones (the humerus, femur, and tibia) also indicate the age of an eastern chipmunk. This method of age determination is somewhat similar to that given for the fox squirrel and the eastern cottontail.

Eastern chipmunks can live up to 5 years in the wild, and some have reached 8 years of age in captivity.

Glands. Musk glands are present on either side of the anus.

Voice and sounds. Chipmunks have several calls, but a loud *chip* is the most common. This is often sung continuously for many minutes at the rate of 130 *chips* per minute. A soft *cuck-cuck* is given frequently. When surprised, a chipmunk may give a trilling *chipp-r-r-r-r.* They are generally silent when they fight, but they may squeal when in pain. All the calls can be given with the mouth shut and even with the cheek pouches full. Sometimes the tail is twitched or jerked while calling. Chipmunks often sing at favorite sites, and several may sing together in spring and fall. Birds are sometimes attracted to a singing chipmunk.

Distribution and Abundance

Eastern chipmunks range from southeastern Canada throughout most of the eastern United States. They do not occur in the coastal plains region south of North Carolina or in the Florida peninsula. They were introduced into Newfoundland. The abundance of chipmunks throughout this range varies with the attractiveness of the habitat and fluctuates from year to year, probably in response to changes in the annual

Plate 23

Eastern Chipmunk *(Tamias striatus)*

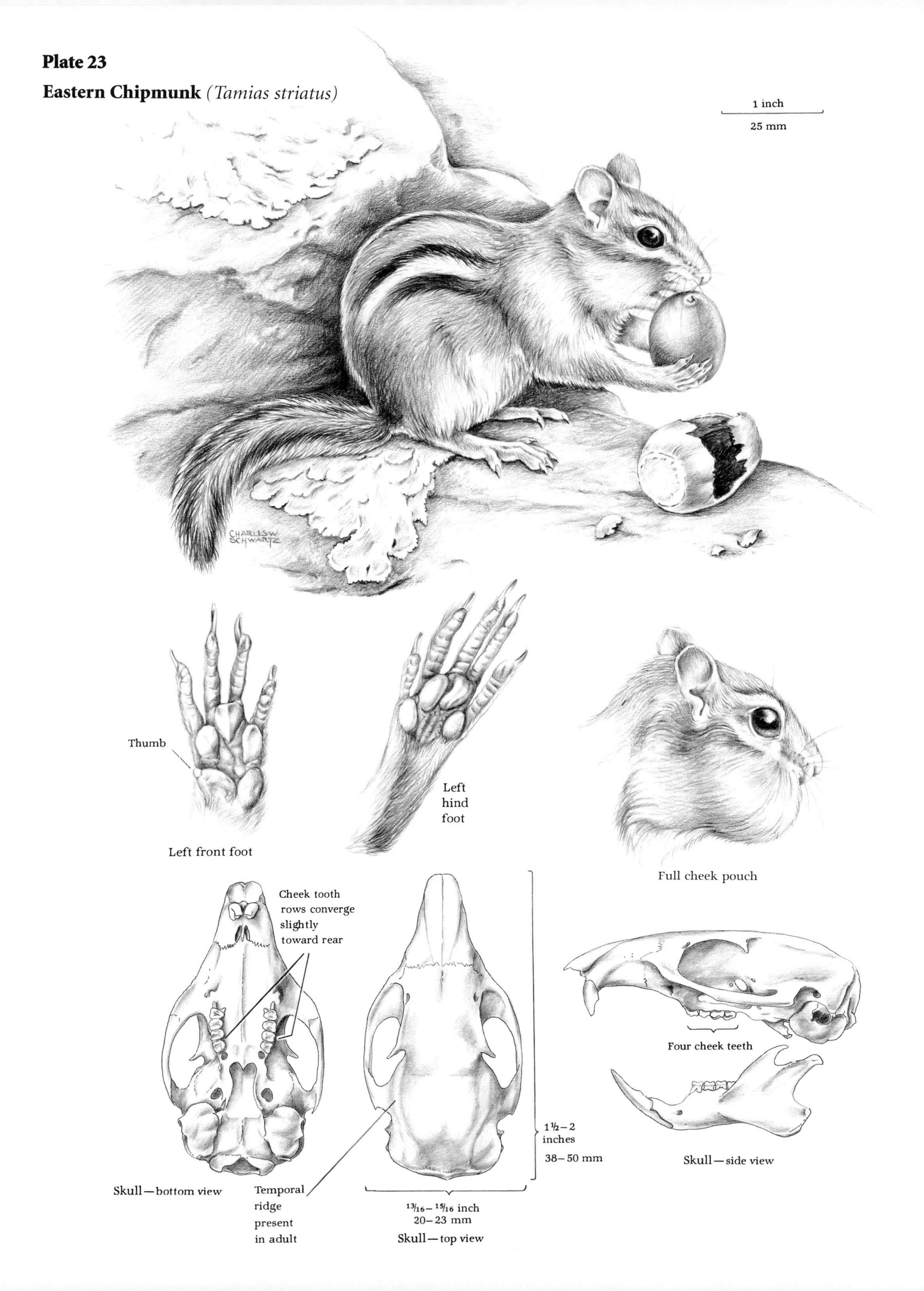

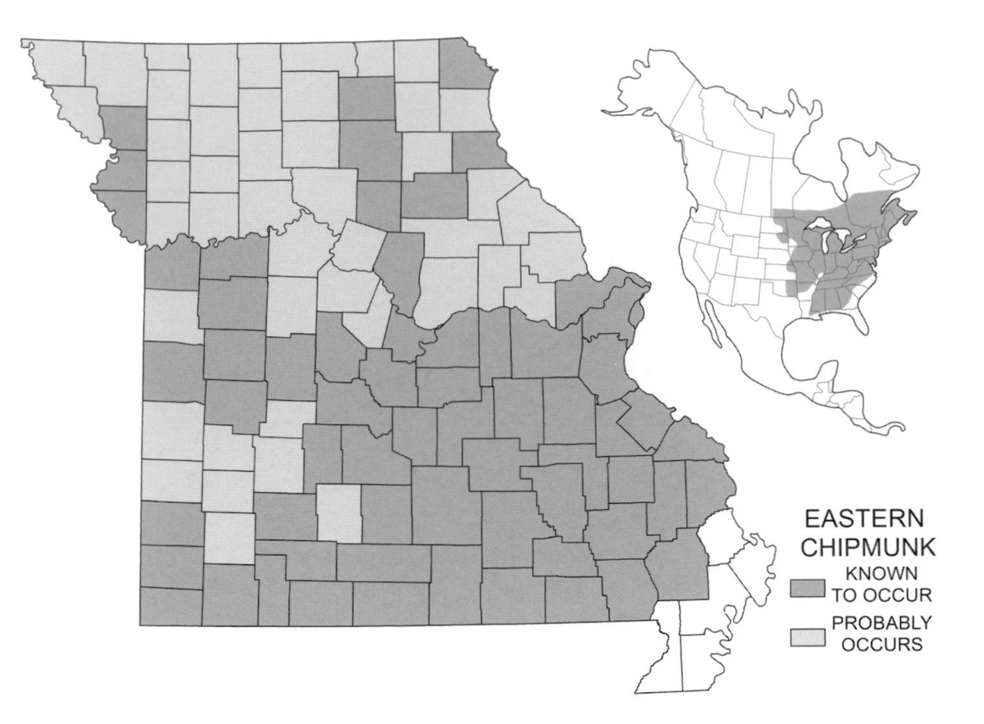

food supply and to population pressure. In some places it appears to fluctuate on a 3- to 4-year cycle. In good habitat densities vary from 4 to 15 per 0.4 ha (1 ac.).

Eastern chipmunks probably range across most of Missouri. They are likely absent from the southeastern corner of the state and are most common in the Ozark Highlands.

Habitat and Home

Eastern chipmunks inhabit primarily deciduous wooded areas and prefer timber borderland rather than deep forests. Here they select areas with logs, stumps, rocky outcrops, and crevices for refuge; elevated observation and vocalization posts; and wooded banks, log heaps, stone piles, broken rocky ridges, or rubbish heaps as sites for their tunnels and nest chambers. They are also found around city parks and near urban, suburban, and rural homes and other buildings, where they live in shrubbery, stone walls, and old outbuildings.

In general, the burrow system is greater in areas where cover is least. The simplest one consists of a short tunnel with a pocket in which to hide and someplace to store food. Usually the system is more extensive. The main entrance to such a tunnel system is well hidden under a stump, log, or rock; measures 5 cm (2 in.) across; and does not contain excavated soil. Side entrances may be conspicuous because of piles of excavated soil nearby; the 2 or more side entrances are often plugged with soil and debris and are seldom used. The main entrance leads to a slanting or nearly vertical passageway that goes down for about 25 cm (10 in.) or more, depending upon the nature of the

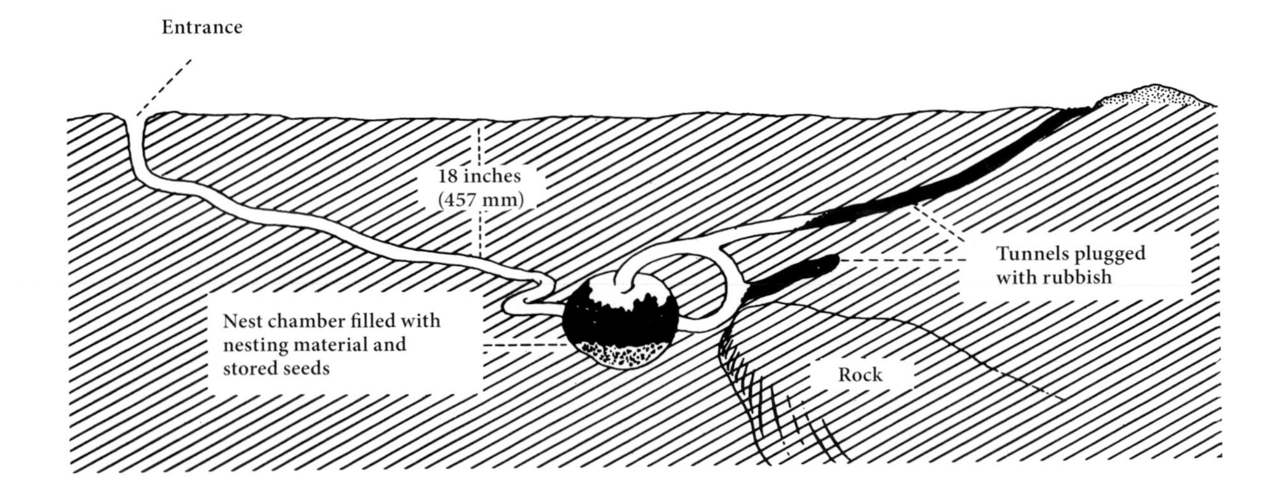

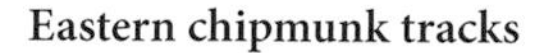

Eastern chipmunk tracks

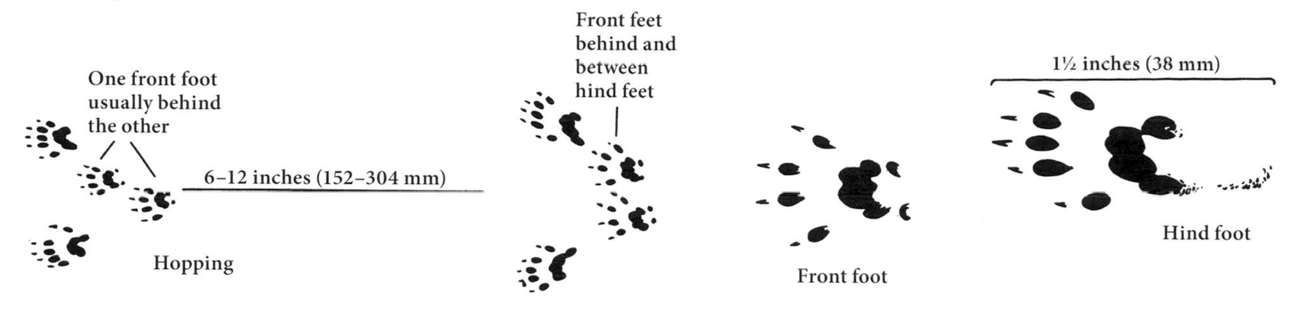

ground, then turns and meanders for 6 to 9 m (20 to 30 ft.) or so.

Somewhere in the tunnel system an enlarged chamber, roughly 30 cm (1 ft.) wide and 15 to 20 cm (6 to 8 in.) high, is built about 75 cm (2½ ft.) below the surface. This chamber is used as a storehouse for food and as a nest for the adult or young. The bottom of the chamber is filled with stored nuts and seeds that may amount to as much as 17 l (half a bushel) in volume. Over this pile of food is nesting material of dried leaves or grass. In the fall this nesting material is near the top of the chamber; but during the winter, as the stored food is eaten and not replaced, the nest level gradually drops to the bottom of the chamber. In addition to the main chamber, several small ones occasionally serve for food storage. One corner of the nest is frequently used as a latrine, or a special room may be built for this purpose.

There is an unusual report of a chipmunk rearing a litter in a tree nest.

Habits

Sometime in late fall or early winter, chipmunks begin to retire underground for the winter. They accumulate little fat, and different individuals exhibit varying degrees of hibernation. Some become completely torpid or dormant and have a lowered body temperature along with reduced body activity. Others are active during mild periods but become dormant in severe cold weather, while still others remain active all winter. This varying amount of activity in different individuals is probably related to their respective endocrine glands, which are unstable during the winter months. The hibernation period usually ends about the last of February or first of March, males coming out a little earlier than females. Inclement weather causes the animals to return temporarily to their nests.

A chipmunk generally occupies the same home for several years, but adult males and young animals tend to move more than adult females. In good habitat the home range is very small, averaging less than 0.2 ha (0.4 ac.), but in poorer areas it may cover 0.8 to 1.2 ha (2 to 3 ac.).

The permanency of use of a burrow system is dependent upon availability of food and social pressures. Adults defend a radius of 15 to 25 m (50 to 80 ft.) from the burrow system. Beyond this there is an undefended area that overlaps the home range of another chipmunk. In fall when the main activity is supplying food caches for the winter, the home range is expanded to reach more distant sources of food; when populations are densest, home ranges are restricted.

An eastern chipmunk released 0.4 km (¼ mi.) from home found its way back the same day. Two returned from 0.5 km (⅓ mi.).

Chipmunks are solitary and territorial animals. Each has its individual tunnel system; however, as many as four have been found hibernating together, probably representing a family group.

These mammals are active during the day, coming out mostly in early morning and late afternoon. They may change this pattern during the hottest part of summer by retiring to their cool nests and sleeping. A main daily activity is gathering and storing food. This

is especially pronounced in late summer and early fall when stores are being accumulated for the coming winter. Seeds and nuts are placed in the cheek pouches by the front feet, the two pouches being stuffed alternately. Chipmunks carefully bite off any sharp points before placing a nut or seed in the cheek pouches. These foods are carried to a storage spot where they are squeezed out of the pouches with a back-to-front motion of the front feet. Storage places may be located under leaves, in shallow holes in the ground, or in an underground chamber.

Another major activity is the digging of tunnels. New holes are opened to the surface and through them soil is either pushed outside by the nose and front feet or carried outside in the cheek pouches. After the soil has been deposited near the opening, the hole is usually plugged.

Chipmunks are primarily ground-dwelling animals, but they climb trees to obtain nuts and buds and sometimes to escape predators. However, they do not jump from limb to limb as tree squirrels do. Chipmunks swim on occasion, holding the tail high. On land the tail is normally held horizontally while running, but it is held straight up when the animal is frightened or very excited. Chipmunks have been timed to run at a speed of 3.3 m (11 ft.) per second.

These little mammals are alert, curious, and nervous. Although normally quite shy, they become accustomed to people and with a little encouragement learn to accept food from a person's hand. But even while displaying this tameness, they are wary, whisking away or darting out of sight upon the slightest provocation. It is best to *not* feed any wild mammals.

Foods

The food of chipmunks consists mainly of nuts, seeds, and fruits but occasionally includes small animals. Favorite nuts are hickory nuts, acorns, beechnuts, hazelnuts, and walnuts, while favorite seeds are corn and wheat. Although many perishable foods, such as mushrooms and many kinds of berries, are eaten, they are not usually stored. The animal foods include insects and their larvae, millipedes, earthworms, slugs, snails, young mice, small frogs, salamanders, small snakes, young birds, and bird eggs.

Chipmunks hunch up on their hind feet and hold food in their front feet. They have favorite sites such as a stump or log where they like to feed. These places are marked by accumulations of shelled seeds, fruit pits, or nut fragments. Chipmunks drink water from streams and ponds.

Reproduction

Breeding begins in the spring, which is usually upon resumption of activity following hibernation, and continues throughout the summer. Females may have 1 or 2 litters a year. Eastern chipmunks are unique among hibernating squirrels because they can have two breeding periods a year. However, the realization of 2 litters varies with the year. In a local population, in some years females produce in the spring only; in other years, some or all of the females have second litters. The cause is not understood but is thought to be related to differences in climate between years for a given population, and between latitudes for different populations.

Most young females breed first when 1 year old, but some may breed as early as 3 to 7 months old. All young males mature sometime in the spring following their birth. The gestation period is 31 to 32 days. There are two periods when most of the young are born—April and May, and July and August. Litters contain from 1 to 8 young, averaging 4 or 5.

At birth the young are blind and hairless. They weigh 3 g (1/10 oz.) and measure 6.4 cm (2½ in.) in length. The striped coat shows at 8 days of age, but until 3 weeks old the young are still scantily furred and have disproportionately large, long muzzles; by the fourth week they look more like adults. The eyes open at 30 to 31 days of age; at this time the young weigh a little over 28 g (an ounce) and measure 14 cm (5½ in.) in length. They are weaned around 40 days after birth and appear aboveground for the first time when 5 or 6 weeks old. At this time, the female will leave the young in the natal burrow until they disperse.

Some Adverse Factors

Domestic cats, bobcats, weasels, foxes, coyotes, rats, hawks, owls, and snakes are predators on chipmunks. The following parasitize chipmunks: mites, fleas, botfly larvae, and roundworms. Chipmunks may become entangled in certain vines, such as ground ivy, and ultimately die.

Importance

Chipmunks are beautiful and exquisite little mammals and have a high aesthetic value to us because they can be easily observed in daylight. Their life activities are very interesting and contribute much to the successful functioning of a woodland habitat. Their tunneling helps aerate the soil and check rain and snow runoff; their food habits influence the growth of certain plant species and act as a partial check on insect populations; their bodies furnish food for many carnivorous animals and thus serve as an important link in the food chain of the wildlife community; and their body wastes and decomposed remains contribute to the fertility of the soil. It is seldom that chipmunks constitute an economic problem, except in localities where they are very abundant and may cause some damage to sprouting corn, garden bulbs, or stored grain.

Conservation and Management

Maintaining good refuge sites—such as downed logs or rock piles—is favorable to chipmunks. There is evidence that suggests chipmunk populations may be less healthy in fragmented forest patches.

The digging habits of chipmunks can cause problems. If you are experiencing problems with chipmunks, contact a wildlife professional for advice, assistance, regulations, or special conditions for handling these animals.

SELECTED REFERENCES

See also discussion of this species in general references, page 23.

Allen, E. G. 1938. The habits and life history of the eastern chipmunk, *Tamias striatus lysteri*. *New York State Museum, Bulletin* 314:7–119.

Blair, W. F. 1942. Size of the home range and notes on the life history of the woodland deer mouse and eastern chipmunk in northern Michigan. *Journal of Mammalogy* 23:27–36.

Burt, W. H. 1940. Territorial behavior and populations of some small mammals of southern Michigan. *University of Michigan, Museum of Zoology, Miscellaneous Publication* 45:7–58.

Ellis, L. S. 1979. Systematics and life-history of the eastern chipmunk, *Tamias striatus*. Ph.D. dissertation, University of Illinois, Urbana. 320 pp.

Mahan, C. G., and T. J. O'Connell. 2005. Small mammal use of suburban and urban parks in central Pennsylvania. *Northeastern Naturalist* 12:307–314.

Mahan, C. G., and R. H. Yahner. 1999. Effects of forest fragmentation on behavior patterns in the eastern chipmunk (*Tamias striatus*). *Canadian Journal of Zoology* 77:1991–1997.

Munro, D., D. W. Thomas, and M. M. Humphries. 2008. Extreme suppression of aboveground activity by a food-storing hibernator, the eastern chipmunk (*Tamias striatus*). *Canadian Journal of Zoology* 86:364–370.

Nupp, T. E., and R. K. Swihart. 1998. Effects of forest fragmentation on population attributes of white-footed mice and eastern chipmunks. *Journal of Mammalogy* 79:1234–1243.

Pidduck, E. R., and J. B. Falls. 1973. Reproduction and emergence of juveniles in *Tamias striatus* (Rodentia: Sciuridae) at two localities in Ontario, Canada. *Journal of Mammalogy* 54:693–707.

Schooley, J. P. 1934. A summer breeding season in the eastern chipmunk, *Tamias striatus*. *Journal of Mammalogy* 15:194–196.

Snyder, D. P. 1982. *Tamias striatus*. *Mammalian Species* 168. 8 pp.

Thomas, K. R. 1974. Burrow systems of the eastern chipmunk (*Tamias striatus pipilans* Lowery) in Louisiana. *Journal of Mammalogy* 55:454–459.

Yahner, R. H. 1978. Burrow system and home range use by eastern chipmunks, *Tamias striatus*: Ecological and behavioral considerations. *Journal of Mammalogy* 59:324–329.

Yahner, R. H., and G. E. Svendsen. 1978. Effects of climate on the circannual rhythm of the eastern chipmunk, *Tamias striatus*. *Journal of Mammalogy* 59:109–117.

Yerger, R. W. 1953. Home range, territoriality, and populations of the chipmunk in central New York. *Journal of Mammalogy* 34:448–458.

Zollner, P. A., and K. J. Crane. 2003. Influence of canopy closure and shrub coverage on travel along coarse woody debris by eastern chipmunks (*Tamias striatus*). *American Midland Naturalist* 150:151–157.

Woodchuck (*Marmota monax*)

Name

The first part of the scientific name, *Marmota*, is the Latin word for "marmot." It was probably derived through corruption from two Latin words meaning "mouse of the mountain" and is the name given to the European marmot and the North American marmot, which are close relatives of the woodchuck. The last part, *monax*, is a Native American name for this rodent and means "the digger"; it alludes to the woodchuck's habit of excavating burrows.

The derivation of the common name, "woodchuck," is not clear but is believed to be a corruption of *wuchak*, a Native American name for this species. However, the origins of its other common names, "groundhog" and "whistle-pig," are obvious from the animal's squat appearance, waddling gait, and habit of living in the ground, and the high-pitched whistle they emit when alarmed.

Description (Plate 24)

The woodchuck is a medium-sized, stout mammal with short, powerful legs and a medium-long, bushy, and somewhat flattened tail. The broad head has large and conspicuous white or pale yellowish brown incisor teeth, a blunt nose, moderately sized eyes, and small, rounded ears that can be closed at will over the ear opening to exclude soil. There are 4 clawed toes and a small thumb with a flat nail on each front foot, and 5 clawed toes on each hind foot. Small internal cheek pouches are present. The body fur is long and coarse; it is much sparser on the underparts than on the back.

Color. The general color above is a grizzled grayish brown with a yellowish or reddish cast, the overall coloration varying in different individuals from light to dark. The head is a darker brown on top with buffy white on the sides of the face, nose, lips, and chin. The legs are black to dark brown while the feet are black. The hairs of the tail are black to dark brown with buffy white tips. On the underparts, the general color is light buff to white, although the hairs are blackish brown at the base. Occasionally a pure white or nearly black individual occurs. Both sexes are colored alike.

There is only one molt annually, which begins in May and continues until September. The molt starts at both the tail and face and works forward and backward, respectively. In general, adults start to molt first, then one-year-olds, and lastly young of the year. Sometimes the molt is not completed before hibernation.

Measurements

Total length	406–685 mm	16–27 in.
Tail	95–187 mm	3¾–7⅜ in.
Hind foot	60–114 mm	2⅜–4½ in.
Ear	25–34 mm	1–1⅜ in.
Skull length	73–95 mm	2⅞–3¾ in.
Skull width	47–60 mm	1⅞–2⅜ in.
Weight	1.8–6.3 kg	4–14 lb.

Woodchucks weigh least in spring when they are just out of hibernation and most in fall prior to hibernation. During hibernation, they lose between one-third and one-half of their autumn weight. Males are generally larger than females.

Teeth and skull. The dental formula of the woodchuck is:

$$I\ \frac{1}{1}\ C\ \frac{0}{0}\ P\ \frac{2}{1}\ M\ \frac{3}{3} = 22$$

The incisors of rodents grow continuously throughout life and must be used constantly to wear down the growth. Rather frequently woodchuck skulls are found with very long and curled incisor teeth. These have resulted from some deformity that prevented

Plate 24

Woodchuck *(Marmota monax)*

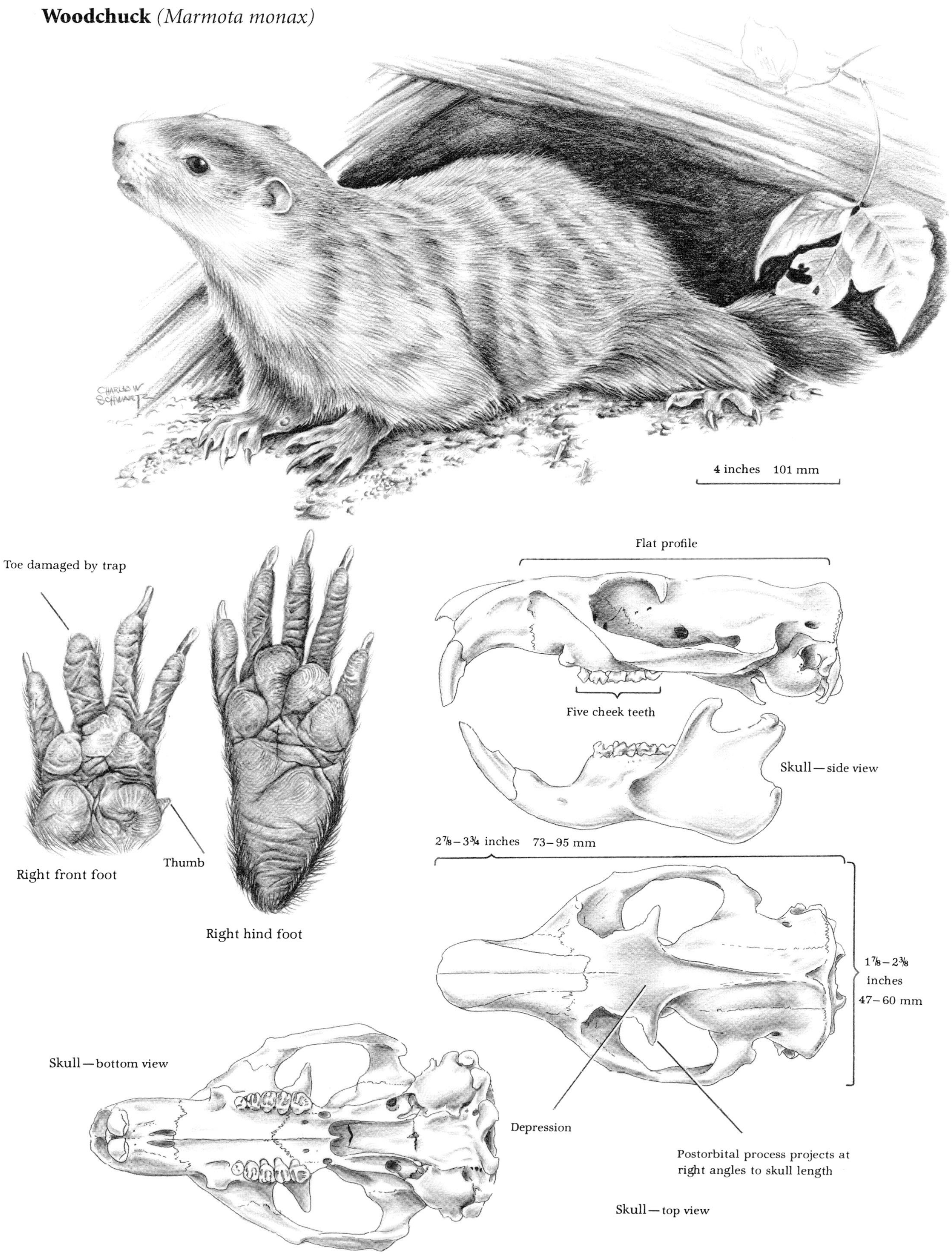

them from being ground off against each other. In chewing, the lower jaw of rodents moves forward and backward as opposed to that of the hare-shaped mammals, which moves from side to side. Woodchucks do not have orange-colored incisors characteristic of most rodents.

The woodchuck's skull is easily distinguished from skulls of other Missouri members of the squirrel family by the following characteristics: large size, a flat upper surface when seen in profile, postorbital processes projecting at right angles to the length of the skull, and a depressed area between the postorbital processes.

Sex criteria and sex ratio. The sexes are identified by the usual external organs. There are 4 pairs of teats on the belly in both sexes, of which 2 pairs are near the front legs, 1 pair behind this, and 1 pair in the groin region; occasionally an extra pair occurs and is functional. Males have a baculum. The sex ratio in 24 woodchucks collected in Missouri showed 46 percent males to 54 percent females.

Age criteria and longevity. The age of woodchucks can be estimated from a number of characteristics. In general, these hold until the young reach one year of age (then known as yearlings), but none is certain beyond this.

Size and weight distinguish young from adults until late summer.

Young are not as richly colored as adults. Because of different times of molt, young of the year can be told from older animals in the fall by their shorter and finer fur.

The external appearance of the incisors also helps identify age. In young of the year, the incisors are narrow, long, and pointed, and they lack a brown stain. As the animal ages, the incisors become broader, the point wears down, and they acquire a brown stain.

The head of a young woodchuck is small, and the muzzle is narrow and pointed.

The color of the testes also varies with age. Yearling males have white testes until May or June, while older animals have pigmented testes (graduating from light to dark brown).

The dried weight of the eye lens shows a general relationship to age and appears to be the best means for aging adults.

Annual rings in the cementum layer on the teeth are not reliable age criteria.

Wild individuals are known to have lived to 6 years of age, and captives have lived 10 years.

Glands. Three anal glands open just within the vent and are responsible for a characteristic, musky odor. This odor is used for purposes of communication.

Voice and sounds. When alarmed or disturbed suddenly, especially at its den, the woodchuck gives a loud, shrill whistle. When angered, the teeth are chattered or ground together. A muffled bark is emitted when the animal is handled. In fighting, woodchucks commonly squeal or growl.

Distribution and Abundance

When North America was first settled, woodchucks were scarce, but as timbered areas were opened and

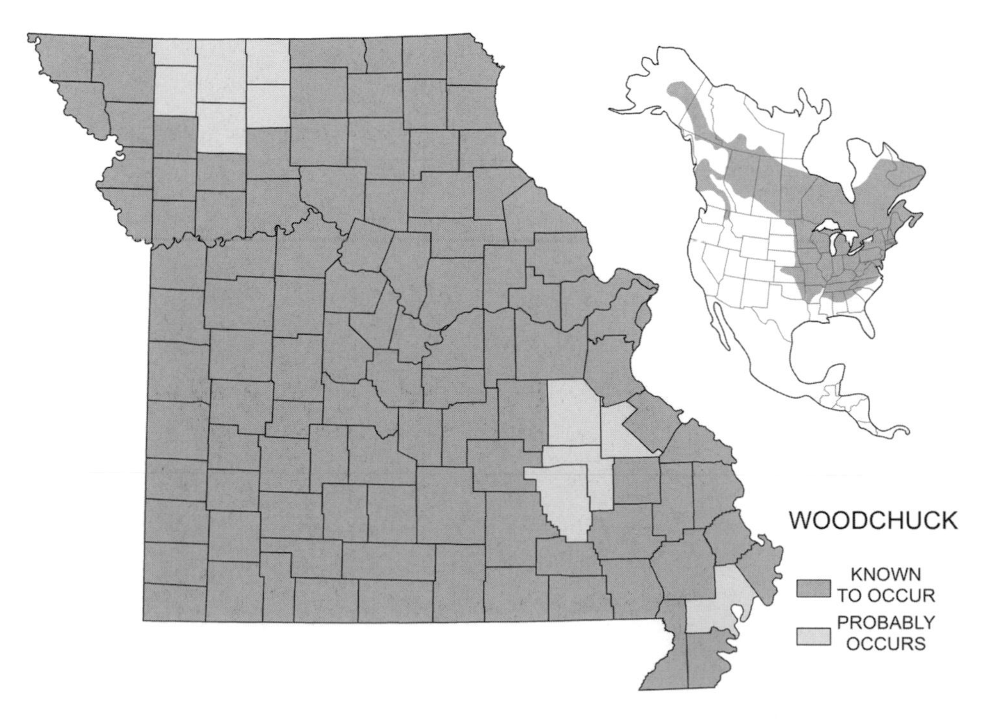

woodland edge, fencerows, and meadows increased (the present preferred habitat of woodchucks), their range expanded. Today they occur from central Alaska, across most of Canada, and into the central and eastern United States. There are years of population abundance and scarcity, but no regular pattern is evident. In times of abundance, there may be as many as 1 woodchuck to 0.8 to 1.2 ha (2 to 3 ac.) of suitable habitat.

The woodchuck is common everywhere in Missouri except in the Mississippi River Alluvial Basin section, where it is rare. Its infrequency there appears to be due to lack of suitable places to dig dens because of the high water table. But where levees and dikes have been created in this region, the woodchuck has been able to extend its range because these man-made structures afford good denning sites.

Habitat and Home

Woodchucks prefer to live in the parts of timber bordered by open land or along fencerows and heavily vegetated gullies or stream banks. Here they usually build their own burrows, showing a preference for rocky or sandy sloping land. They only rarely utilize natural caves or fallen hollow logs. Sometimes they build dens in man-made earth fills.

Different sites are usually selected for dens used at different seasons. Hibernating dens are located primarily in wooded or brushy areas with a southern exposure, while summer dens are often in open crop fields or grassland.

Woodchucks are a solitary species. One individual (but sometimes one family, or a small group of woodchucks) has a burrow system composed of several entrances, tunnels, and chambers. The main entrance to the burrow system is often located beneath a tree stump or rock and is usually conspicuous because of a pile of freshly excavated earth and stones in front. This mound may be up to 137 cm (4½ ft.) in diameter and is used as an observation platform, sunning site, and latrine. The tunnel opening, about 30.5 cm (1 ft.) in diameter, gradually narrows to about half this size; after approximately 3 to 14 m (10 to 45 ft.) it leads either directly to an enlarged chamber containing the nest or to passageways that lead to 1 or more nests, to blind tunnels, or to as many as 5 side entrances. All side entrances are generally small, sometimes well concealed in vegetation, and commonly lack an outside pile of soil.

The nest chamber is about 40 cm (16 in.) wide and up to 35 cm (14 in.) high. It is lined with dry leaves and grass, and it lies between 1 and 2 m (3 and 6 ft.) underground depending upon the slope and nature of the land. When building a nest, the woodchuck may make between 12 and 21 trips in one hour, carrying the material in its mouth. Just before hibernation, and also if

Diagram of woodchuck home

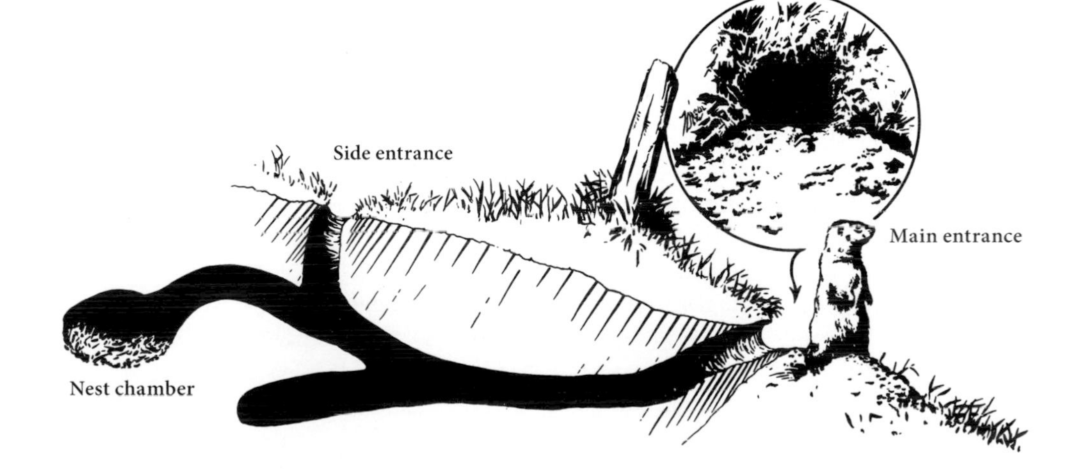

closely pursued by some predator, the woodchuck may wall off the nest with soil. The nest is used for sleep, hibernation, a home for the young, and escape from enemies.

Habits

Woodchucks are true hibernators. As the autumn days shorten and become cooler, woodchucks remain out of their dens for progressively briefer periods. The fattest and oldest adults gradually become less active and start to hibernate first. Lean adults and young animals are abroad latest in the fall.

By the end of October or mid-November, most woodchucks are curled up asleep in their underground nests. In this torpid state, breathing is very slow and the body temperature is between 6 and 14°C (43 and 57°F). So profound is this sleep that even if an animal is warmed up, it requires several hours to awaken. Only rarely is a woodchuck out of his burrow during the winter, and this mostly occurs following an unusually warm spell. Sometimes captive animals, even if kept outdoors, may not hibernate.

Woodchucks start to emerge from hibernation in Missouri as early as the first week of February, but severe cold weather may delay them. At first they come out only for short periods because little food is available; but as the daily temperatures rise and green growth increases, they spend a correspondingly longer time aboveground. Adult males tend to emerge from hibernation before females and subadults. In very early spring, male woodchucks may be abroad on moonlight nights in their search for mates. In April through mid-June, they are aboveground for about half of their daily activity, mostly during the middle of the day. From late June through September, they are aboveground for about two-thirds of their daily activity, which occurs in early morning and late afternoon. They stay underground during the heat of the day because they are susceptible to high temperatures and direct solar radiation. Their sensitivity is related to the thick skin and fur plus the layers of fat that are gradually deposited under the skin during the summer. In October and November their daily activity periods are similar to those of spring.

Their activities outside the burrow are mostly concerned with feeding, although a favorite pastime is basking in the sun on the earthen platform at the opening of the burrow or on a tree stump or convenient rock. Woodchucks are very alert and, when feeding, spend much of their time in watchfulness, alternately feeding and rising up on their hind legs to sniff the air or peer about. At the least alarm, the animal freezes momentarily, then dashes for a burrow.

A woodchuck is familiar with all the locations of burrows over a large area, up to 0.8 km (½ mi.) in diameter. In its daily foraging it moves from one burrow to the next nearest burrow. When moving into an area where burrows are scarce, the woodchuck travels slowly to about a halfway point then hurries to the closest burrow. The mean home range for woodchucks is 1.5 to 2 ha (3½ to 5 ac.). Female home ranges are smaller than those of males and are sometimes half the size.

The adult woodchuck has been considered a solitary animal, but because burrow systems may be grouped in clusters and there is movement between adjacent systems and clusters, there could be more of a social structure to the population than formerly thought.

Digging is done primarily with the front feet and claws, but teeth are sometimes used to move stones or cut roots. In removing earth from the tunnel, hind feet are used for propulsion and soil is pushed forward with the face or chest. The subsoil removed in the course of digging one burrow amounted to 325 kg (716 lb.). Digging is done so rapidly that a small burrow can be finished in one day. Burrows, however, are continuously being enlarged, and the fresh soil is piled outside the main entrance. Thus, fresh soil is usual evidence of occupancy. Also, conspicuous trails in the vegetation around a burrow indicate that it is being used.

The normal gait is a slow walk, but, when alarmed, woodchucks may lope or gallop. A woodchuck can run as fast as 16 km per hour (10 mi. per hour), but only for short distances. They swim voluntarily and on occasion climb trees, especially pawpaws, for fruit. When cornered, woodchucks are strong and capable fighters.

Defecation usually occurs in the pile of soil at the entrance of the burrow. There a hole is dug for the droppings, which are then buried with the front feet. The droppings are long and rounded or slightly coiled, measuring between 3.8 to 7.6 cm (1½ and 3 in.) in length. If the woodchuck does not leave the burrow, defecation occurs in a blind tunnel and the droppings are similarly covered with soil. The young defecate in the nest, and the female keeps the nest clean by changing the leaf and grass lining.

Foods

The woodchuck is almost a complete vegetarian, eating less than 1 percent animal foods. The plant foods consist of the leaves, flowers, and soft stems of various grasses, of field crops such as clover and alfalfa, and of wild herbs of many kinds. Sometimes woodchucks climb small elms and feed on the leaves. Certain garden crops such as peas, beans, and corn are favorite items, and fruits such as apples and pawpaws are likewise relished. The bark of some trees is taken occasionally, especially in spring. The animal matter consists of grasshoppers, May beetles (June bugs), snails, and rarely an egg or fledgling of a ground-nesting bird or of domestic poultry. Woodchucks eat sand and gravel from the berms of asphalt roads, apparently seeking and obtaining salt spread there for winter snow removal. This activity is most common in spring. Salt is an essential element in their diet.

The amount of food consumed varies with the season. The least is eaten in March just after coming out of hibernation, and the most is consumed in May, followed by a gradual decline to fall. However, in some individuals there is a slight increase in intake just prior to hibernation.

The amount of food eaten daily is often considerable and may be as much as 0.7 kg (1½ lb.). A great deal of moisture is obtained from succulent plants, dew, and rain. The fat that is stored in their bodies sustains them during hibernation and also in spring after emergence before succulent and nutritious foods are available.

Reproduction

The breeding season follows emergence from hibernation and in central Missouri begins in mid-February. Fighting between males is common at this time. A male may stay with one female for a while, but each male probably mates with several females. The gestation period is 31 to 33 days, and the young are born toward the end of March. From 2 to 9 young compose a litter, but 4 or 5 are average. Smaller litters are born to one-year-old females, while larger litters are produced by older females.

At birth the young are naked, pink, wrinkled, blind, and helpless. They measure about 10 cm (4 in.) in length and weigh between 28 and 45 g (1 and 1½ oz.). The eyes open at about 4 weeks of age; although the young come to the opening of the burrow at this time, they seldom venture outside until 6 or 7 weeks old. The female brings green food into the burrow and weans them about the time they first go outside.

The young are playful and often wrestle. They are usually entirely under the female's care, although some males have been known to help the female raise the family after the young begin to come outside the burrow. In nursing, the female either stands on all four feet or sits up. When transporting the young, she carries them in her mouth by the loose skin over the neck.

By midsummer the young are 51 cm (20 in.) long and weigh around 1.8 kg (4 lb.). They leave home about this time and, after digging temporary burrows near the nursery, often move some distance away and establish their own homes.

Only about one-fourth of the young are capable of breeding in the first spring following their birth; these probably mate later in the spring than do older animals. All are mature at two years of age.

Some Adverse Factors

Predators such as foxes, coyotes, domestic dogs, bobcats, weasels, minks, raccoons, hawks, owls, and large snakes prey on woodchucks, taking the young primarily. They can climb trees to escape predators. Parasites include ticks, fleas, adult flies, warbles (larvae of flies), and roundworms. The disease tularemia has been associated with woodchucks. Floods may destroy the young in creek-bottom burrows. While brushfires are a potential danger, woodchucks probably survive in their burrows. Highway traffic can take a heavy toll.

Importance

Woodchucks were formerly trapped for their fur, which was sometimes used for cheap fur coats. In the past, especially in rural areas, the hide was used for

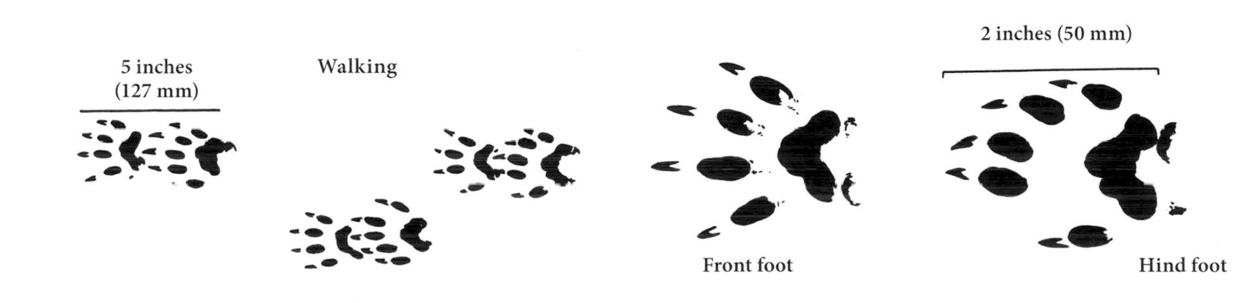

Woodchuck tracks

patching leatherwork, and for straps, laces, ball covers, and catchers' mitts. Some people eat woodchuck, especially the young and lean animals. They are sometimes used in biomedical research.

The role of the woodchuck as a builder of homes for other animals is significant; because of this, the woodchuck occupies an important niche in the wildlife community. Skunks, foxes, weasels, opossums, raccoons, and rabbits use woodchuck burrows and may den with hibernating woodchucks. Many other small to medium-sized mammals, snakes, and birds use abandoned or vacated woodchuck burrows. Also, because tremendous quantities of subsoil are moved in the course of burrow construction, the countless generations of woodchucks have contributed much to the aeration and mixing of the soil. Woodchucks are active during daylight hours and many people get enjoyment from seeing them.

Conservation and Management

Habitat can be provided by leaving hedgerows and windbreaks. The woodchuck's taste for crops often places it in an unfavorable position with farmers and gardeners. Burrows and mounds can be hazardous to farm equipment, livestock, and horses and riders. Burrows can undermine building foundations, driveways, porches, and other structures and can create problems when they occur in levees and pond dams. If you are experiencing problems with woodchucks, contact a wildlife professional for advice, assistance, regulations, or special conditions for handling these animals.

Woodchucks are a game species in Missouri and specific harvest regulations apply for their take. Their harvest is around 10,000 animals annually.

SELECTED REFERENCES

See also discussion of this species in general references, page 23.

Davis, D. E. 1964. Evaluation of characters for determining age of woodchucks. *Journal of Wildlife Management* 28:9–15.

———. 1967. The role of environmental factors in hibernation of woodchucks (*Marmota monax*). *Ecology* 48:683–689.

Fall, M. W. 1971. Seasonal variations in the food consumption of woodchucks (*Marmota monax*). *Journal of Mammalogy* 52:370–375.

Grizzell, R. A., Jr. 1955. A study of the southern woodchuck, *Marmota monax monax*. *American Midland Naturalist* 53:257–293.

Hayes, S. R. 1976. Daily activity and body temperature of the southern woodchuck, *Marmota monax monax*, in northwestern Arkansas. *Journal of Mammalogy* 57:291–299.

———. 1977. Home range of *Marmota monax* (Sciuridae) in Arkansas. *Southwestern Naturalist* 22:547–550.

Kwiecinski, G. G. 1998. *Marmota monax*. *Mammalian Species* 59. 8 pp.

Merriam, H. G. 1966. Temporal distribution of woodchuck interburrow movements. *Journal of Mammalogy* 47:103–110.

———. 1971. Woodchuck burrow distribution and related movement patterns. *Journal of Mammalogy* 52:732–746.

Murie, J. O., and G. R. Michener, eds. 1984. *The biology of ground-dwelling squirrels*. University of Nebraska Press, Lincoln. 459 pp.

Snyder, R. L., and J. J. Christian. 1960. Reproductive cycle and litter size of the woodchuck. *Ecology* 41:647–656.

Snyder, R. L., D. E. Davis, and J. J. Christian. 1961. Seasonal changes in the weights of woodchucks. *Journal of Mammalogy* 42:297–312.

Tumlison, R., and H. W. Robison. 2010. New records and notes on the natural history of selected vertebrates from southern Arkansas. *Journal of the Arkansas Academy of Science* 64:145–150.

Twichell, A. R. 1939. Notes on the southern woodchuck in Missouri. *Journal of Mammalogy* 20:71–74.

Thirteen-lined Ground Squirrel
(*Ictidomys tridecemlineatus*)

Name

The meaning of the scientific name, *Ictidomys*, is derived from the Greek word for "weasel" and *myos* for "mouse," referring to their slender body form. The last part, *tridecemlineatus*, is from two Latin words and means "thirteen-lined" (*tridecem*, "thirteen," and *lineatus*, "lined").

The common name, thirteen-lined ground squirrel, as well as the scientific name, aptly describe this animal since it is marked with about thirteen prominent lines or stripes and lives on and in the ground.

The name "squirrel" comes from the Old French *esquireul* or *escuriuel*. Although it is sometimes called the striped gopher, the name "gopher" is preferably used for a very different Missouri rodent, the plains pocket gopher. This species was formerly known as *Citellus tridecemlineatus*, then *Spermophilus tridecemlineatus*.

Description (Plate 25)

This ground squirrel is a small, slender mammal with large eyes, small ears set low on the head, well-developed cheek pouches opening into the mouth, and a slightly bushy tail about half as long as the head and body. The hind foot has 5 clawed toes, while the front foot has 4 clawed toes and a small thumb. The body fur is short but is overlain by long guard hairs.

Some of the obvious characteristics fitting this animal for its underground existence are the long-clawed feet, which do the digging; the small, close-lying ears that collect little soil and offer little resistance in the tunnels; and the slender body shape that permits easy travel in small passageways. Aboveground, the striped coat that blends with shadow-flecked grasslands is an important concealing factor.

The eyes of ground squirrels have a strong yellow to nearly orange lens. This permits sharp vision in the bright light of open grasslands.

Color. The general color pattern of both sexes is one of alternating light and dark stripes extending lengthwise from head to rump on the back and sides. The light stripes are yellowish to white, and the dark ones are blackish to reddish brown punctuated with a series of light spots. As the common name implies, there are usually about 13 stripes (7 light and 6 dark), but some individuals have a few more or a few less. The face and underparts vary from buff to white. The yellowish brown tail is fringed with coarse black hairs having yellowish white tips. Both albino and black individuals occur, but rarely. Molting occurs twice annually, once in late summer prior to hibernation and again in spring following hibernation.

Measurements

Total length	177–317 mm	7–12½ in.
Tail	60–127 mm	2⅜–5 in.
Hind foot	25–41 mm	1–1⅝ in.
Ear	6–12 mm	¼–½ in.
Skull length	31–41 mm	1¼–1⅝ in.
Skull width	19–23 mm	¾–15/16 in.
Weight		
Spring	113–141 g	4–5 oz.
Fall	226–255 g	8–9 oz.

Teeth and skull. The dental formula of the thirteen-lined ground squirrel is:

$$I\ \frac{1}{1}\ C\ \frac{0}{0}\ P\ \frac{2}{1}\ M\ \frac{3}{3} = 22$$

The skull is very similar to, but smaller than, that of the Franklin's ground squirrel. The thirteen-lined ground squirrel's skull is likely to be confused with the skulls of two other similar-sized Missouri members of the squirrel family, the eastern chipmunk and the southern flying squirrel. From the chipmunk, it is distinguished by having 5 cheek teeth on each side of the upper jaw; from the flying squirrel, it is distinguished by lacking a highly arched braincase, a slightly upturned or arched front end (nasal bones), and a deep notch in the upper rim of the eye socket.

Sex criteria and sex ratio. The sexes are identified by the usual external sex organs. Males possess a baculum. Six pairs of teats occur in this species. The sex ratio in adults is nearly 1 male to 2 females, while among the young it is 1 male to 1 female.

Age criteria and longevity. Up to 90 percent of young die from predation before their first hibernation. They only live to be a few years old in the wild but are known to live up to 8 years in captivity.

Glands. There are three anal musk glands.

Voice and sounds. The thirteen-lined ground squirrel gives a long, high, trilling whistle that in alarm changes to a short, sharp whistle. When angry or captured, a snarl is often given. The female has a special alarm call for her young. When the young are first out of the nest they do not respond to this call, but after three or four days they learn to dive for their burrows at this signal. The young have a distress call they give if they wander far from the nest burrow. The female responds to this and retrieves her young.

Distribution and Abundance

The thirteen-lined ground squirrel ranges from central Canada into the central United States. Historically, it was confined to prairie areas, but over the past two centuries it expanded its range, particularly in the north and east, by moving into lands that were once timbered but subsequently cleared for agriculture. In parts of the range, these ground squirrels are abundant; in areas of concentration, there may be from 5 to 20 per 0.4 ha (1 ac.). The range and abundance seem to be influenced to a large extent by the type of soil, because the soil is related to the kind of vegetation and the ease with which burrowing can be accomplished.

Plate 25

Thirteen-lined Ground Squirrel

(Ictidomys tridecemlineatus)

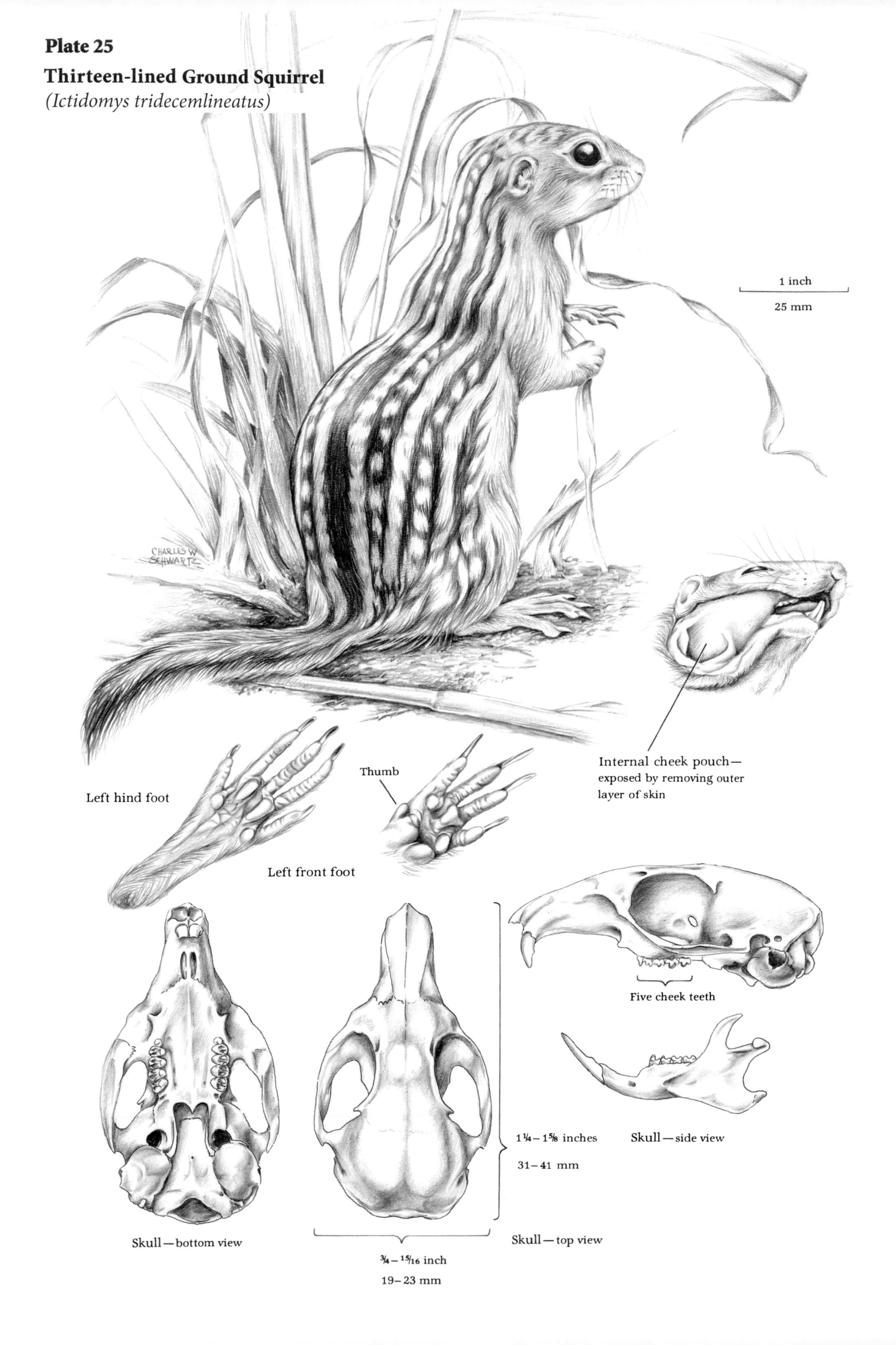

The thirteen-lined ground squirrel once occurred, in localized distributions, in several counties in the Central Dissected Till Plains in northern Missouri and in the Osage Plains in western Missouri. It also occurred in some small tongues of prairie that project into the western border of the Ozark Highlands. Today, it has been extirpated from some counties but still occurs in localized populations mostly in northwest Missouri. They are classified as a vulnerable species of conservation concern in Missouri due to their declining range and population numbers, but they are secure across their North American range.

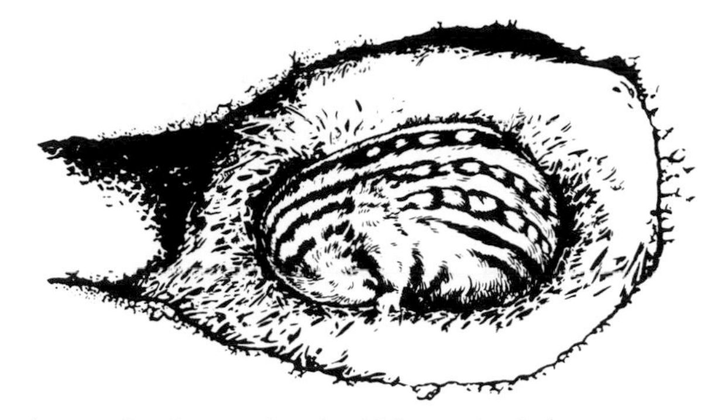

Thirteen-lined ground squirrel hibernating in burrow

Habitat and Home

The thirteen-lined ground squirrel is primarily a resident of flat, open grasslands or other dry, open fields, wherever cover is short. Only occasionally does it occupy brushy fencerows, woody borders, or sandy river bottomlands. It has adapted well to the neatly cropped lawns around our homes, golf courses, parks, cemeteries, schools, and businesses, and to well-grazed pastures and the mowed borders of roads. In one north Missouri study, thirteen-lined ground squirrels in a managed grassland also used areas with vegetation taller than 30 cm (12 in.), an indication of their ability to adapt to some tallgrass habitats.

The home is a burrow in the ground with several outside entrances. The main entrance is open during the day but plugged with sod or grass each night, while the side entrances are usually plugged at all times or are well concealed with vegetation. There is rarely a pile of earth around the openings because the soil is carried in the cheek pouches and scattered away from the site of excavation. The main entrance, about 5 cm (2 in.) in diameter, leads to a vertical tunnel that goes down from 15 to 122 cm (½ to 4 ft.), depending upon the nature of the soil, the age of the occupant, and use of the burrow. Older animals make deeper burrows than younger ones do, and burrows used for hibernation are deeper than those used for a nest or escape. The tunnel turns at right angles and takes an irregular course for varying distances, up to 6 m (20 ft.) or so. There are many blind branches, several passageways leading to the surface, and, somewhere below the frost line, an oval chamber up to 23 cm (9 in.) in diameter. This chamber is filled with fine, dry grass and is used for sleep, hibernation, escape, and as a nest for the young. There is usually a tunnel that serves as a drain leading downward from the chamber. One or more storehouses for food are also built in the tunnel system.

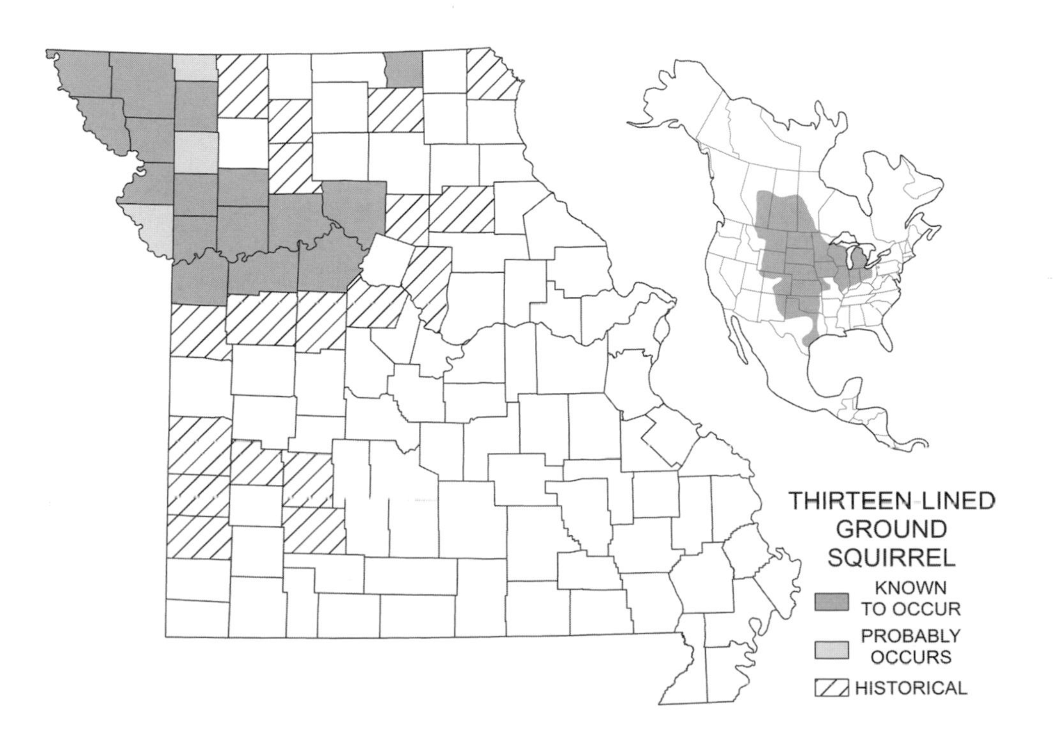

In addition to the main burrow, there are other tunnels just beneath the sod coverage. These are often near the feeding grounds and are used for escape. Inconspicuous trails usually connect these tunnels with each other and the main burrow.

Habits

The thirteen-lined ground squirrel is one of the few truly hibernating Missouri mammals. Toward the end of summer these squirrels become very fat. Some females may enter hibernation as early as late summer or early fall. A few adult males and juveniles of both sexes are active into October.

Each thirteen-lined ground squirrel retires to its burrow deep in the ground, plugs all openings to the outside, and then enters a profound sleep. The animal sits on its rump and curls up in a ball with its nose touching the abdomen and the tail over the head or to one side. The physical processes of the body are greatly reduced, as is indicated by a drop in body temperature from a normal of 30–41 to 3°C (86–106 to 37°F), a drop in heartbeat from a normal of 200–350 to as low as 5 times a minute, and a reduction in breathing rate from a normal of 50 times a minute to 4. The stored body fat is the principal source of energy during this period of inactivity, since no food is eaten; as this body fuel is used up, there is a loss of ⅓ to ½ of the animal's autumn weight. Although ground squirrels typically sleep through the entire winter, unseasonably warm temperatures may arouse a few, and freezing temperatures that penetrate deep into the burrows usually awaken all. This automatic thermal control in severe cold weather prevents the animals from freezing to death.

The end of hibernation depends upon the soil temperature. When the soil has become sufficiently thawed for the ground squirrels to dig their way out easily, which is usually in late March or early April, the males begin to emerge. The females appear two or three weeks later. The main activity in early spring is cleaning out the burrow and eating the stored food. New green growth is also sought.

The thirteen-lined ground squirrel regulates its daytime activities according to light intensity and comes out of its burrow only during bright daylight. In early spring they are aboveground mostly from noon until midafternoon. This time is expanded in May and June until they are out from about 9 until 5. After this, they tend to have peaks of activity in midmorning and late afternoon. On cloudy days they may not even appear aboveground. About 70 percent of the time outside the burrow is devoted to obtaining food, and a great deal of effort is expended in filling the underground storehouses. Much more food is stored than is ever eaten. Roughly 6 percent of their time is spent in digging, for each squirrel may dig three to four burrows a year. The remaining time is divided between alert behavior, sunning or resting in the shade, grooming, or investigating. While primarily terrestrial, ground squirrels can climb low, bushy trees.

Although each ground squirrel lives a solitary life and is generally antagonistic toward others of its own kind, many individuals usually build their burrows close together forming a colony. Males stay close to their home burrows after emergence in early spring, then expand their home range to about 4 ha (11 ac.) during the breeding season. After this they again restrict their activity range. The largest home range of a female is about 1.5 ha (3½ ac.), which occurs when she is pregnant and when the young are inside the burrow. Her home range is restricted when the young emerge. Females tend to use the same home range area year after year. In times of food shortage, the home ranges of both males and females are greater than during times of food abundance.

Ground squirrels are very curious. They often rear up on their hind feet, and, with front legs pressed against the body and the tail used as a brace, sit motionless for long periods. This position is the basis for another common name, picket pin, derived from the resemblance of these animals to stakes used by cattlemen on the prairie to tether their horses. When startled, ground squirrels scamper into their burrows but invariably turn around immediately and pop their heads out to see what is going on.

Foods

The thirteen-lined ground squirrel eats about equal amounts of animal and plant foods. The animal matter consists mostly of insects, particularly grasshoppers and the larvae of beetles, moths, and butterflies. Earthworms, lizards, eggs and young of ground-nesting birds, domestic chicks, mice, baby rabbits, and some carrion are also eaten. Seeds, fruits, nuts, roots, and foliage of cultivated grains, wild grasses, legumes, and other herbs compose the plant matter. Waste grain scattered around storage bins, cattle feedlots, and along roads is frequently eaten. Moisture is obtained from the succulent foods.

The front feet are used to handle the food while eating. Perishable foods are usually eaten on the spot, but seeds may be carried in the cheek pouches and stored in underground caches. The seeds are placed in the mouth with the front feet, then forced into the cheek pouches by the lowering of the head and contraction of the neck muscles. If seeds are not for storage but are

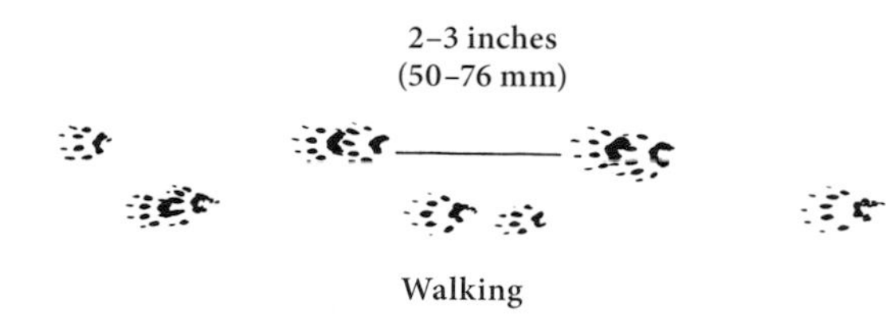

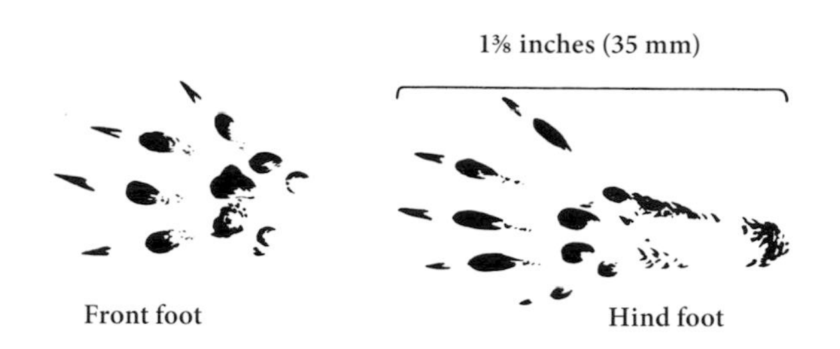

Thirteen-lined ground squirrel tracks

to be eaten leisurely, a few at a time are allowed to slip from the cheek pouches into the mouth where they are chewed and swallowed.

Reproduction

Hibernation is necessary for reproduction of female thirteen-lined ground squirrels but not for males. Shortly after coming out of hibernation, ground squirrels mate. The older animals emerge from hibernation in a more advanced reproductive state than younger ones do and hence mate first. The gestation period is 27 to 28 days, and the young are born about the middle of May. From 4 to 14 may compose the single annual litter, but the usual number is between 8 and 10. In Texas, some females have produced 2 litters a year.

At birth the young are blind, toothless, and helpless. The skin is pink and no fur is present except for the whiskers. Each young is about 5.7 cm (2¼ in.) long and weighs 3 g (1⁄10 oz.) or less. The sexes are readily identifiable. At 12 days of age, the young are covered with downy hair and the color pattern is obvious. Between 15 and 20 days, the incisor teeth appear and the young start to walk. From 23 to 26 days, the eyes open and the weight has increased to about 21 g (¾ oz.). The first solid foods are taken at this time, and weaning occurs shortly thereafter. The young come out of the nest between 5 and 6 weeks of age and after a week or two lead independent lives in the vicinity of their home burrow. Juvenile males disperse up to 267 m (876 ft.) from the home burrow, while juvenile females move slightly shorter distances. Both sexes mature the year following their birth.

Some Adverse Factors

Coyotes, foxes, skunks, weasels, minks, badgers, domestic cats, hawks, owls, snakes, and humans are predators of thirteen-lined ground squirrels. Botfly larvae (warbles) are sometimes found in the skin of the groin region. Mites, ticks, parasitic beetles, and fleas are common external parasites.

In parts of the range where they are common, highway traffic takes a toll.

Importance

Thirteen-lined ground squirrels do about as much economic harm as good. When abundant, they may damage cultivated crops by digging up sprouting corn or eating the choicest heads of ripe oats, but their feeding on insects and mice generally offsets this detrimental habit. They may also be undesirable on golf courses. If you are experiencing problems with these squirrels, contact a wildlife professional for advice, assistance, regulations, or special conditions for handling them.

They have many other values, which are usually overlooked. Their digging aerates the soil, conditions it for plant growth, and may attract earthworms, insects, and other soil-building organisms. Their plant-feeding and seed-storing activities influence the growth and distribution of certain plant species, which in turn may profoundly affect soil conditions and other plant and animal life. Their body wastes contribute to the organic structure of the soil, and their bodies provide food for predators. However, it is not the individual contribution but the sum total of all activities of many individuals of the species that makes them important members of a living community. In addition, the activities of ground squirrels can be enjoyed by interested observers because these animals are abroad in daylight and can be easily approached.

Conservation and Management

As with the Franklin's ground squirrel, if a population seems too large, it is usually best to let it run its course. Large numbers of ground squirrels will attract a diversity of wildlife including predatory raptors.

SELECTED REFERENCES

See also discussion of this species in general references, page 23.

Baldwin, F. M., and K. L. Johnson. 1941. Effects of hibernation on the rate of oxygen consumption in the thirteen-lined ground squirrel. *Journal of Mammalogy* 22:180–182.

Beer, J. R. 1962. Emergence of thirteen-lined ground squirrels from hibernation. *Journal of Mammalogy* 43:109.

Blair, W. F. 1942. Rate of development in young spotted ground squirrels. *Journal of Mammalogy* 23:342–343.

Evans, F. C. 1951. Notes on a population of the striped ground squirrel (*Citellus tridecemlineatus*) in an abandoned field in southeastern Michigan. *Journal of Mammalogy* 32:437–449.

Helgen, K. M., F. R. Cole, L. E. Helgen, and D. E. Wilson. 2009. Generic revision in the Holarctic ground squirrel genus *Spermophilus*. *Journal of Mammalogy*, 90:270–305.

McCarley, H. 1966. Annual cycle, population dynamics, and adaptive behavior of *Citellus tridecemlineatus*. *Journal of Mammalogy* 47:294–316.

Murie, J. O., and G. R. Michener, eds. 1984. *The biology of ground-dwelling squirrels*. University of Nebraska Press, Lincoln. 459 pp.

Niva, L. M. 2010. Habitat use and coexistence of ground squirrels in northern Missouri. M.S. thesis, University of Central Missouri, Warrensburg. 62 pp.

Rongstad, O. J. 1965. A life history study of the thirteen-lined ground squirrels in southern Wisconsin. *Journal of Mammalogy* 46:76–86.

Streubel, D. P., and J. P. Fitzgerald. 1978. *Spermophilus tridecemlineatus*. *Mammalian Species* 103. 5 pp.

Wade, O. 1950. Soil temperatures, weather conditions, and emergence of ground squirrels from hibernation. *Journal of Mammalogy* 31:158–161.

Whitaker, J. O., Jr. 1972. Food and external parasites of *Spermophilus tridecemlineatus* in Vigo County, Indiana. *Journal of Mammalogy* 53:644–648.

Franklin's Ground Squirrel
(*Poliocitellus franklinii*)

Name

The origin of the scientific name, *Poliocitellus*, is derived from the Greek *polios*, meaning "hoary" or "gray," and *citellus*, the Latinized word for "ground squirrel." The species name *franklinii* is a Latinized name meaning "of Franklin" and was given in honor of Sir John Franklin, commander of the Overland Expedition into Canada in 1818–1822, on which this species was collected.

The common name, Franklin's ground squirrel, refers to the person for whom this species was named and to its ground-dwelling habit. Another common name, gray ground squirrel, indicates the predominant color of the animal. Although it is sometimes called the gray gopher, the name "gopher" is preferably used for a very different Missouri rodent, the plains pocket gopher. This species was formerly known as *Citellus franklinii*, then *Spermophilus franklinii*.

Description (Plate 26)

The Franklin's ground squirrel is a medium-sized mammal somewhat similar in build to, but larger than and not quite as slender as, its close relative, the thirteen-lined ground squirrel. Franklin's ground squirrel superficially resembles the tree-dwelling eastern gray squirrel but is readily distinguished from it by having a shorter and less bushy tail, shorter and rounder ears, cheek pouches, longer and straighter claws, and a yellowish cast to the rump.

Color. The color of both sexes is rather uniformly brownish gray speckled with black, which gives a spotted or barred effect, particularly to the rump. The head is slightly darker than the rest of the upperparts and the sides are lighter. There is a yellowish tone to the rump. The underparts vary from yellowish white to gray or buff and the feet are dark gray. The tail is black and gray mixed, with black predominating toward the tip.

Measurements

Total length	349–419 mm	13¾–16½ in.
Tail	114–158 mm	4½–6¼ in.
Hind foot	34–66 mm	1⅜–2⅝ in.
Ear	15–19 mm	⅝–¾ in.
Skull length	50–57 mm	2–2¼ in.
Skull width	28–31 mm	1⅛–1¼ in.
Weight		
Spring	340–453 g	¾–1 lb.
Fall	453–680 g	1–1½ lb.

Males are generally heavier than females.

Teeth and skull. The dental formula of the Franklin's ground squirrel is:

$$I\,\frac{1}{1}\;C\,\frac{0}{0}\;P\,\frac{2}{1}\;M\,\frac{3}{3} = 22$$

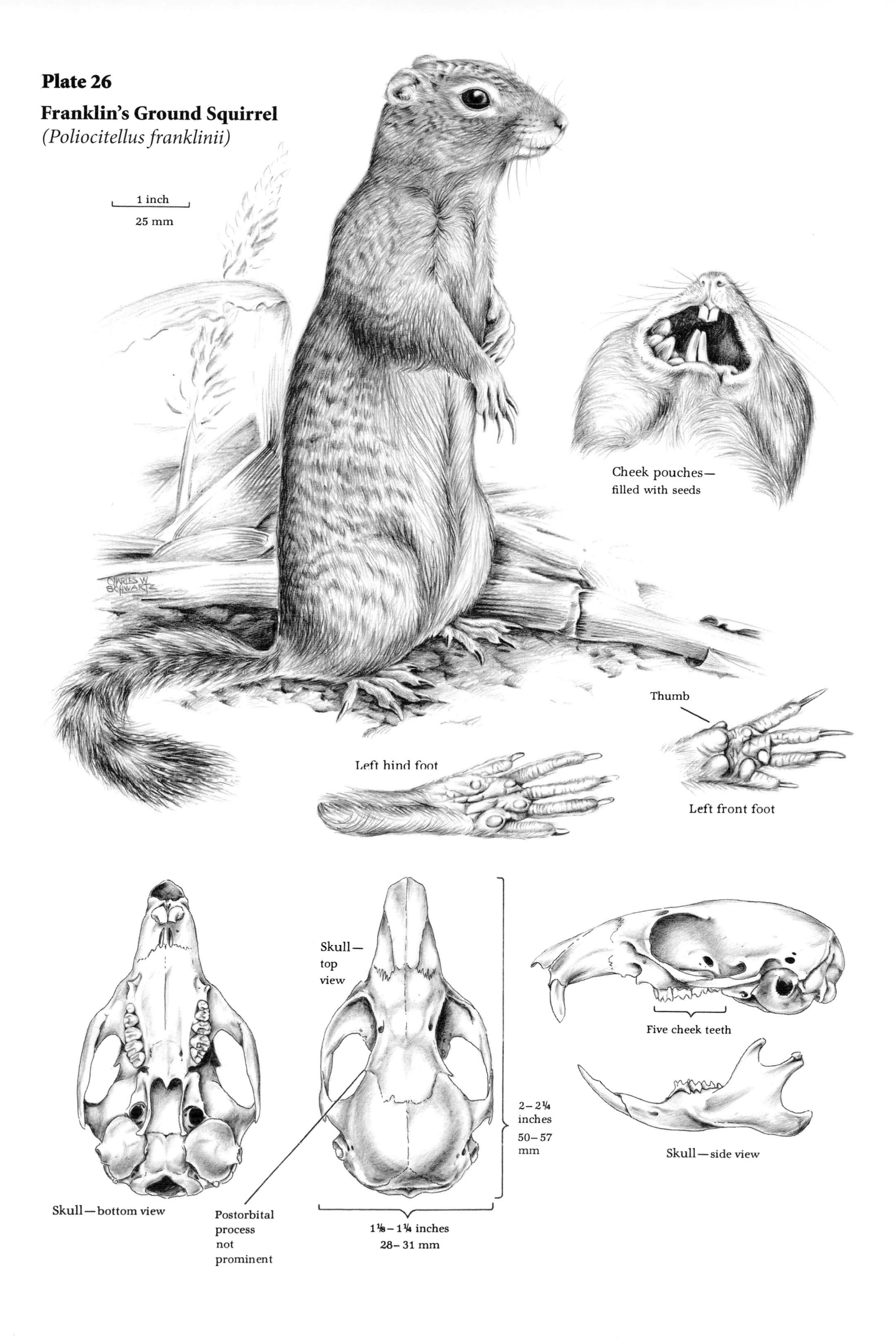

Plate 26

Franklin's Ground Squirrel
(Poliocitellus franklinii)

The teeth are similar in number and arrangement to those of the thirteen-lined ground squirrel. The skulls of these two ground squirrels are likewise very similar, but that of the Franklin's is larger. The skull of the Franklin's ground squirrel can be distinguished from that of the similar-sized eastern gray squirrel by being narrower between the eye sockets and not having prominent postorbital processes.

Sex criteria and sex ratio. The sexes are identified by the usual external sex organs. Sex ratios are almost always 1 to 1.

Age criteria and longevity. Life expectancy for females is 4–5 years compared with 1–2 years for males.

Glands. There are 3 anal glands, about 9 mm (3/8 in.) in diameter, embedded in the flesh around the rectum. When excited, the ends of the ducts show as 3 white spots, 1 above, and 1 on each side of the anus. These emit a powerful, musky scent.

Voice and sounds. A remarkably clear, musical whistle is given by both sexes, but females whistle less often than males. This call has given rise to another common name, whistling ground squirrel.

Distribution and Abundance

The Franklin's ground squirrel has a small range extending from central Canada into the north central United States. The population density fluctuates, showing a period of abundance every 4 to 6 years. They are becoming increasingly uncommon, especially in the eastern portion of their range, due to habitat loss and fragmentation and to pesticide use, but they may be abundant locally. A dense population consists of 5 to 30 per 0.4 ha (1 ac.). Individuals are rarely seen; therefore their presence and abundance is often indicated when they are seen scampering across roads or dead along highways.

The Franklin's ground squirrel once occurred in most counties in the Central Dissected Till Plains in northern Missouri. It has been extirpated from a few counties mostly along the Missouri River. It is generally rare but may be common locally. It is ranked as a vulnerable species of conservation concern in

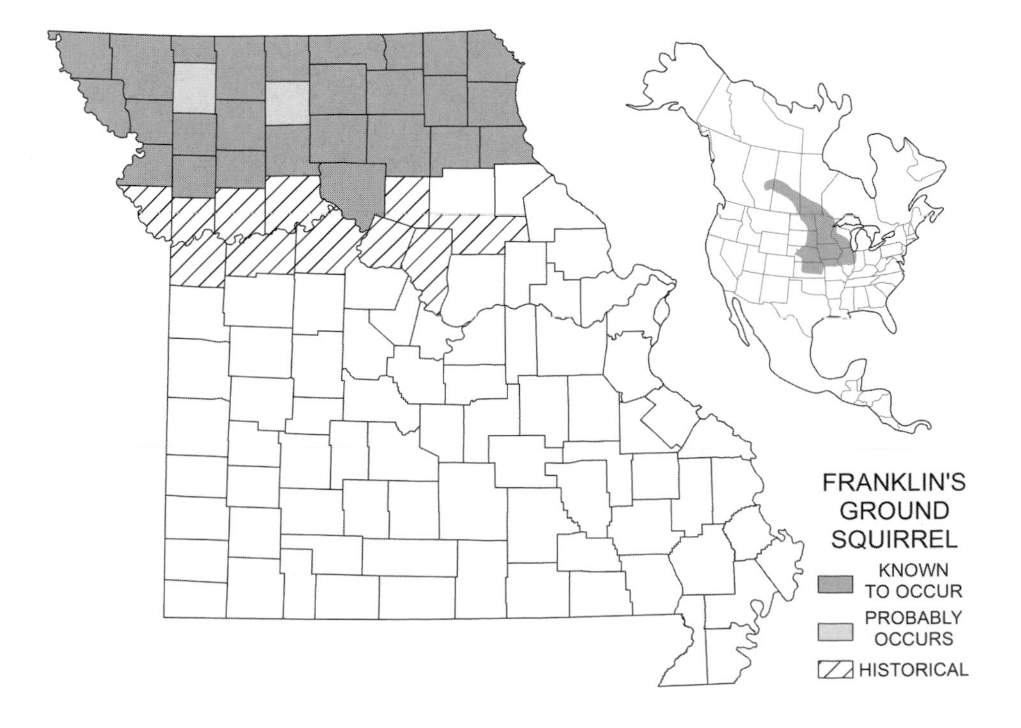

Missouri but it is secure across its North American range.

Habitat and Home

The Franklin's ground squirrel is primarily a resident of the borderland between woods and prairies and seldom uses either forests or strictly open prairies. It prefers tallgrass to midgrass prairies, with fencerows, wooded banks, gullies, and little-grazed sod. They also live along railroad rights-of-way and roadsides, which function as corridors linking islands of suitable habitat. The Franklin's ground squirrel frequently selects a brushy bank or draw for its burrows, but in times of abundance it may establish its home around farmyards, where it utilizes the shelter of shrubbery, trash piles, and buildings. The burrows lead to a den in the ground; they are similar to but larger, deeper, and better concealed than those made by the thirteen-lined ground squirrel.

Habits

During the summer, Franklin's ground squirrels accumulate a heavy layer of body fat; between late August and early October they start their long hibernation. Adults are the first to retire for the winter, the young becoming dormant a little later. Several animals may hibernate together. Emergence in spring is relatively late; the males come out first in early April and are followed by the females about a week later. The time of emergence can be noted by the resumption of the ground squirrels' typical whistling.

Franklin's ground squirrels are normally abroad only during bright daylight and often stay in their dens on windy, rainy, cloudy, or unseasonably cold days. They spend approximately 90 percent of their lives underground. The home range of an individual is about 91 m (100 yd.) in diameter. Franklin's ground squirrels live together in small colonies of 10 to 12 animals or more, which often stay in one locality for a year or part of a year and then move and set up a colony elsewhere.

Franklin's ground squirrels are good tree climbers and can swim. They are notoriously curious and, after being alarmed and taking refuge in their burrows, frequently turn around immediately and look out.

The droppings are deposited in small tunnels leading off the main passageway.

Foods

Vegetable matter forms about three-fourths of the diet, while animal matter makes up the remainder. Plant foods are the seeds, fruits, roots, and green vegetation of grasses, herbaceous plants, garden vegetables,

and cultivated grain. Some of the cultivated grain is taken from seed heads in the field, but a great deal is from waste sources including grain spilled from trucks along roads. Animal foods consist of adult and larval insects, toads, frogs, mice, crayfish, fish, earthworms, young chickens and turkeys, small birds, bird eggs, mice, young rabbits, other ground squirrels, and some carrion. Particularly in the northern states and Canada, the eggs, young, and more rarely adults of nesting ducks are preyed upon. Water is taken on occasion, but succulent plants provide considerable moisture.

Reproduction

Following emergence from hibernation, the males fight and chase each other, but when the females appear in a week or so, the males direct their attention to chasing them. During this period both sexes emit a powerful, musky scent from the anal glands. After a 28-day gestation period, the single annual litter is born in May or June and contains between 4 and 11 (usually 7) young.

The young ground squirrels are approximately 2.5 cm (1 in.) long at birth and are blind, naked, and pink. On the ninth day, a few short hairs are visible, and by the sixteenth day the young are fully furred. The

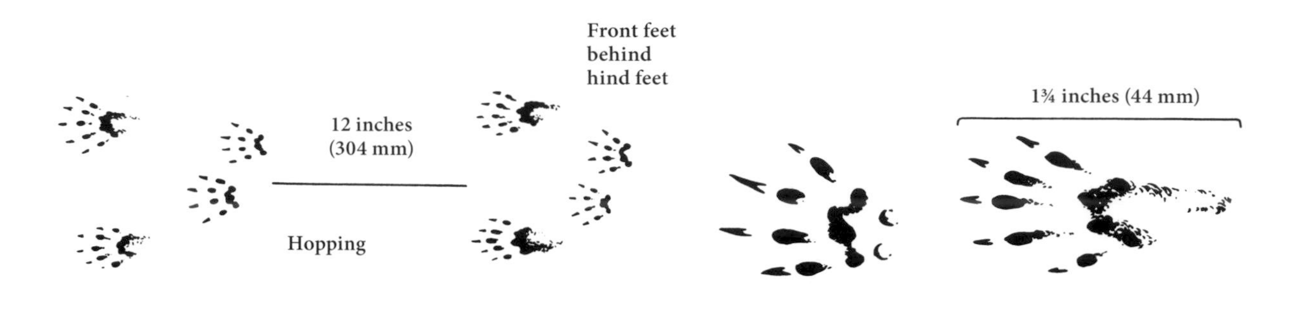

Franklin's ground squirrel tracks

eyes open between the eighteenth and twentieth day, and the first shrill whistle is given at this age. Short trips are made outside the burrow when the young are about 30 days old, and weaning soon follows. The young stay with the female for a few weeks after they come out of the burrow but soon disperse. By fall, they are nearly as large as adults. Both sexes mature in their first spring.

Some Adverse Factors

Foxes, coyotes, weasels, badgers, and snakes are probably the most important predators because they can dig out ground squirrels or seek them in the ground; hawks and owls take a small toll aboveground. In parts of the range where they are common, highway traffic takes a toll. Known parasites include fleas, roundworms, and tapeworms.

Importance

Although Franklin's ground squirrels cause some destruction to crops and some predation on wild ducks and domestic poultry, these losses are only important in localities or years where ground squirrels are abundant. If you are experiencing problems with these squirrels, contact a wildlife professional for advice, assistance, regulations, or special conditions for handling them. Other aspects of their ecological importance are similar to those discussed under the thirteen-lined ground squirrel.

Conservation and Management

The preference for tallgrass prairie habitat indicates that to maintain healthy Franklin's ground squirrel populations, these habitats must be preserved, restored, and managed. In addition, railroad rights-of-way can be managed in such a way to benefit Franklin's ground squirrels.

The disappearance of local populations may not always be attributed to local extirpation but could result from this species' tendency to move colonies from one area to another. If Missouri colonies do shift locations over time, this could account for some of the observed absence of the species from areas with past occupancy.

Because control is fairly difficult, the best suggestion is to let the population run its course, for sooner or later there will be a decline in abundance.

SELECTED REFERENCES

See also discussion of this species in general references, page 23.

Chromanski-Norris, J. F., and E. K. Fritzell. 1983. Status and distribution of ten Missouri mammals. A report to the Missouri Department of Conservation, Jefferson City. 38 pp.

DeSanty-Combes, J. 2002. Reports on the status of Franklin's ground squirrel in Missouri 2001 and 2002. A report to the Missouri Department of Conservation, Jefferson City. 43 pp.

Duggan, J. M. 2011. Occupancy dynamics, personality, and behavior of Franklin's ground squirrel in agricultural landscapes. Ph.D. dissertation, University of Illinois, Urbana. 123 pp.

Ellis, L. S. 1982. Life-history studies of Franklin's ground squirrel, *Spermophilus franklinii*, in Missouri. M.S. thesis, Northeast Missouri State University, Kirksville. 50 pp.

Erlien, D. A., and J. R. Tester. 1984. Population ecology of sciurids in northwestern Minnesota. *Canadian Field-Naturalist* 98:1–6.

Helgen, K. M., F. R. Cole, L. E. Helgen, and D. E. Wilson. 2009. Generic revision in the Holarctic ground squirrel genus *Spermophilus*. *Journal of Mammalogy*, 90:270–305.

Huebschman, J. J. 2003. A conservation assessment of Franklin's ground squirrel (*Spermophilus franklinii* Sabine 1822): Input from natural history, morphology, and genetics. Ph.D. dissertation, University of Nebraska, Lincoln. 230 pp.

Johnson, S. A., and J. Choromanski-Norris. 1992. Reduction in the eastern limit of the range of the Franklin's ground squirrel (*Spermophilus franklinii*). *American Midland Naturalist* 128:325–331.

Lyon, M. W. 1932. The Franklin ground squirrel and its distribution in Indiana. *American Midland Naturalist* 13:16–20.

Martin, J. M., E. J. Heske, and J. E. Hofmann. 2003. Franklin's ground squirrel (*Spermophilus franklinii*) in Illinois: A declining prairie mammal? *American Midland Naturalist* 150:130–138.

Murie, J. O., and G. R. Michener, eds. 1984. *The biology of ground-dwelling squirrels*. University of Nebraska Press, Lincoln. 459 pp.

Niva, L. M. 2010. Habitat use and coexistence of ground squirrels in northern Missouri. M.S. thesis, University of Central Missouri, Warrensburg. 62 pp.

Ostroff, A. C., and E. J. Finck. 2003. *Spermophilus franklinii*. *Mammalian Species* 724. 5 pp.

Sowls, L. K. 1948. The Franklin ground squirrel, *Citellus franklinii* (Sabine), and its relationship to nesting ducks. *Journal of Mammalogy* 29:133–137.

Eastern Gray Squirrel
(*Sciurus carolinensis*)

Name

The first part of the scientific name, *Sciurus*, is the Latin word for "squirrel." It is derived from the Greek words *skia*, "shadow," and *oura*, "tail." A liberal translation means "a creature that sits in the shadow of its tail." The second part, *carolinensis*, is a Latinized word meaning "of Carolina," for the place from which this species was first collected and described. The common name originated as follows: "eastern" denotes the general range in the United States; "gray" refers to the predominant color; and "squirrel" comes from the Old French *esquireul* or *escuriuel*.

Description (Plate 27)

This medium-sized, slender tree squirrel is distinguished from a close relative, the eastern fox squirrel, by its grayer back, white underparts, white edging of the tail, smaller size, more slender facial profile, and two additional upper cheek teeth.

The lens in the eye of the eastern gray squirrel and certain other diurnal squirrels develops a yellow color that serves as a filter. This gives the squirrel better sight in bright light and increases contrast in vision. The eyes are so located on the head that the squirrel has 40° binocular vision and good lateral sight. This helps the squirrel judge distance and depth when jumping and traveling through trees at high speed.

Color. The general color above is gray, washed with yellowish brown on the head, midback, sides, and upper surfaces of the feet. There is a white to buff ring around the eye and white on the back of the ears. The outsides of the legs and feet are predominantly gray; the chin and underparts are whitish; and the hairs of the tail are banded blackish and tan with broad white tips. The body fur is short, but the hairs of the tail are very long and slightly crinkled.

The sexes are colored alike, but the young tend to be grayer than adults. In winter the general appearance is a more silvery gray and, in common with the slightly longer body fur, the ears have a noticeable projecting fringe of white fur. Sometimes black animals occur in the same litter with gray ones; these may be entirely glossy black or show various gradations between black and gray. Albino squirrels occur occasionally and, in some instances where this characteristic is common in the heredity of a local population, small colonies of albinos may be formed. Reddish individuals occur very rarely.

Eastern gray squirrels have two molts annually. The fall molt in both young and adults begins in late September and is completed in October. Shedding starts at the rear of the body and works toward the head. The spring molt starts in late April and is completed in mid-June. Shedding begins at the head and proceeds toward the tail; the hairs of the tail are shed in the late stages of this molt.

Measurements

Total length	355–533 mm	14–21 in.
Tail	177–254 mm	7–10 in.
Hind foot	57–76 mm	2¼–3 in.
Ear	25–34 mm	1–1⅜ in.
Skull length	60–63 mm	2⅜–2½ in.
Skull width	31–34 mm	1¼–1⅜ in.
Weight	340–680 g	¾–1½ lb.

Teeth and skull. The dental formula of most eastern gray squirrels is:

$$I\,\frac{1}{1}\;C\,\frac{0}{0}\;P\,\frac{2}{1}\;M\,\frac{3}{3} = 22$$

In about 1 percent of the population, the first upper cheek tooth may be absent. The skull of this squirrel is distinguished from that of the eastern fox squirrel by its smaller size and an additional upper cheek tooth on

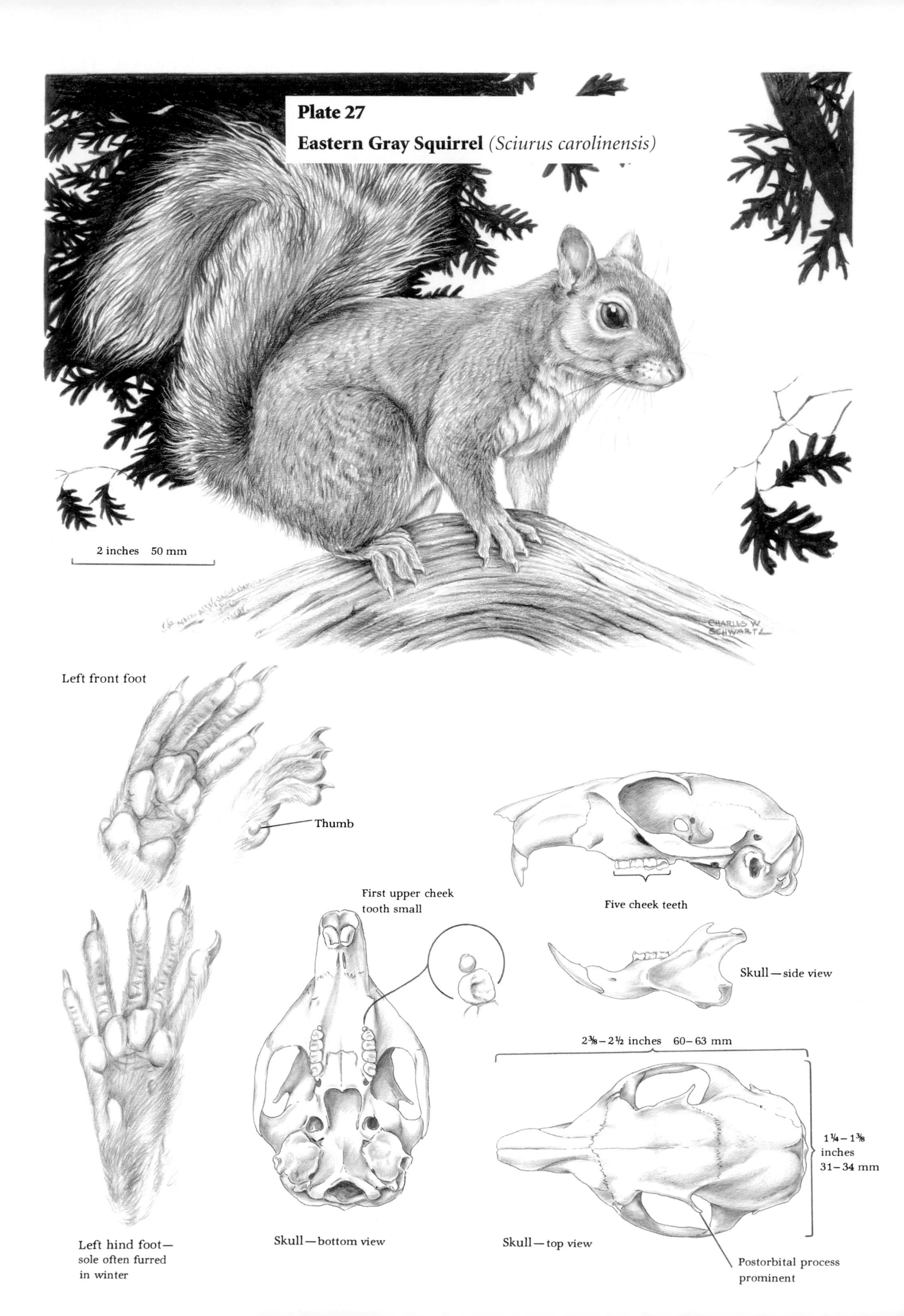
Plate 27
Eastern Gray Squirrel (Sciurus carolinensis)
2 inches 50 mm
CHARLES W. SCHWARTZ
Left front foot
Thumb
Five cheek teeth
First upper cheek tooth small
Skull—side view
2⅜–2½ inches 60–63 mm
1¼–1⅜ inches 31–34 mm
Left hind foot—sole often furred in winter
Skull—bottom view
Skull—top view
Postorbital process prominent

each side. It is easily distinguished from the skull of the similar-sized Franklin's ground squirrel by being wider between the eye sockets and having prominent postorbital processes.

Sex criteria and sex ratio. The sexes are identified as in the eastern fox squirrel. Missouri sex ratios from harvest records show that males compose 54 percent of the population. As in the fox squirrel, these figures are probably related to the method of collection because males are more active and more easily hunted than females.

Age criteria and longevity. Gray squirrels, in late fall, can be placed into three age classes on the basis of fur characteristics. By an examination of the underside of the tail, young less than 6 months old are identified by having the outline of the tail bone visible and several dark bars on the fur that form lengthwise lines around the tail; older young from 6 to 16 months old have the tail bone obscured for the third nearest the body but still have several dark, lengthwise bars visible around the tail; adults (over 16 months of age) have the tail bone outline obscured by hairs, and the lesser dark bars are no longer obvious (see accompanying illustration).

By parting the fur in the rump region, one can identify juveniles by the presence of some 5-banded, black-tipped guard hairs; older squirrels have 4-banded, white-tipped guard hairs. Among the older squirrels, adults have a distinctly yellow basal band, while subadults have this area indistinctly colored.

Other age criteria are discussed under the eastern fox squirrel.

In the wild, eastern gray squirrels rarely live longer than 6 years, but there is a record of one reaching 12½ years of age. It takes approximately 7 years for the population of an area to be replaced by new individuals. In captivity, some individuals have lived for 15 and 20 years.

Glands. No anal musk glands are present. There are sweat glands between the pads on the feet. When a gray squirrel is excited or hot, its tracks become wet from sweat.

Voice and sounds. Compared with most mammals, gray squirrels are noisy and have a large vocabulary. They call more often from the treetops than from the ground. One of the common calls is a *cherk-cherk-cherk* given repeatedly and rapidly. This expresses excitement and is a warning to other squirrels. Nasal throaty grunts, purrs, or chattering of the teeth are given when one squirrel approaches another.

Age in the eastern gray squirrel as indicated by the appearance of undersurface of the tail

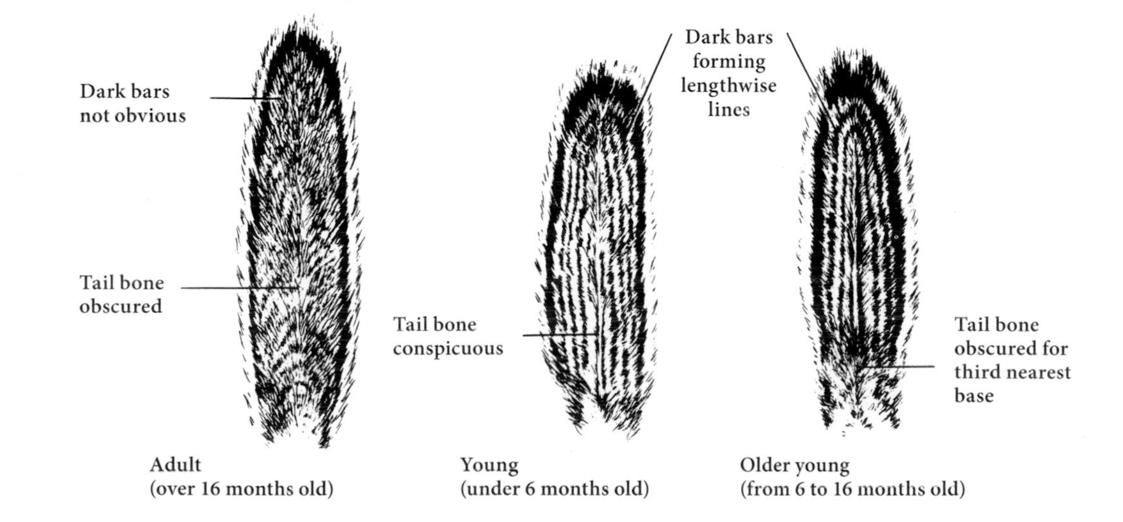

These sounds express mild stress. A *quack-quack-quackquaaaaaa* is given by both sexes and denotes contentment. A scream indicates distress.

Distribution and Abundance

The eastern gray squirrel ranges from south-central and southeastern Canada throughout the eastern half of the United States. They were introduced into California, Oregon, Washington, and Montana, and some provinces of Canada.

They are found throughout Missouri and compose about half of the tree squirrel population. This species is generally less abundant than the fox squirrel in the Central Dissected Till Plains and Osage Plains, and more common than the fox squirrel elsewhere.

There are periodic fluctuations in the population: a high squirrel population usually follows a year of good nut production, and a low one occurs after a poor nut crop.

Habitat and Home

In Missouri, gray squirrels occupy dense hardwood forests but show a preference for those with a bushy understory along river bluffs or in river bottoms and possessing natural or man-made windfalls of treetops. They live to a lesser extent in other timbered areas of the state. They also occur in cities, particularly in parks, where there are large nut and shade trees affording food and denning sites.

The gray squirrel's home is similar to that of the fox squirrel and is discussed with that species.

Habits

Gray squirrels usually live most of their lives in and around one nest tree and seldom travel farther than 183 m (200 yd.) from home in any one season. The minimum home range is around 0.6 ha (1½ ac.), males having larger home ranges than females. However, as different sources of food become available during the growing season, they may shift their home one or more times and range over as much as an 8 km (5 mi.) area. Sometimes because of a failure of the nut crop or pressure from a high population, or for reasons unknown, eastern gray squirrels may show a mass movement from one region to another. During such movements they have been observed to swim across rivers and large lakes. Many perish in their effort to reach the other shore. Such movements have been quite extensive in former times but are less so now. Eastern gray squirrels exhibit a certain homing instinct; some trapped individuals made their way home in less than four weeks after being released 4.4 km (2¾ mi.) away.

Gray squirrels are active all year and do not hibernate. They are busiest during the fall when the nut crop is ripening and are least likely to be abroad during unusually cold weather, heavy rains, or high winds.

Gray squirrels are early risers and leave their nests with the first light. Their periods of greatest daily activity occur around sunrise and shortly thereafter and again in late afternoon. They spend most of the day in the nest or lying on a limb or other platform where they sleep and sun themselves. They are abroad until dusk when they retire to their nests; some may come

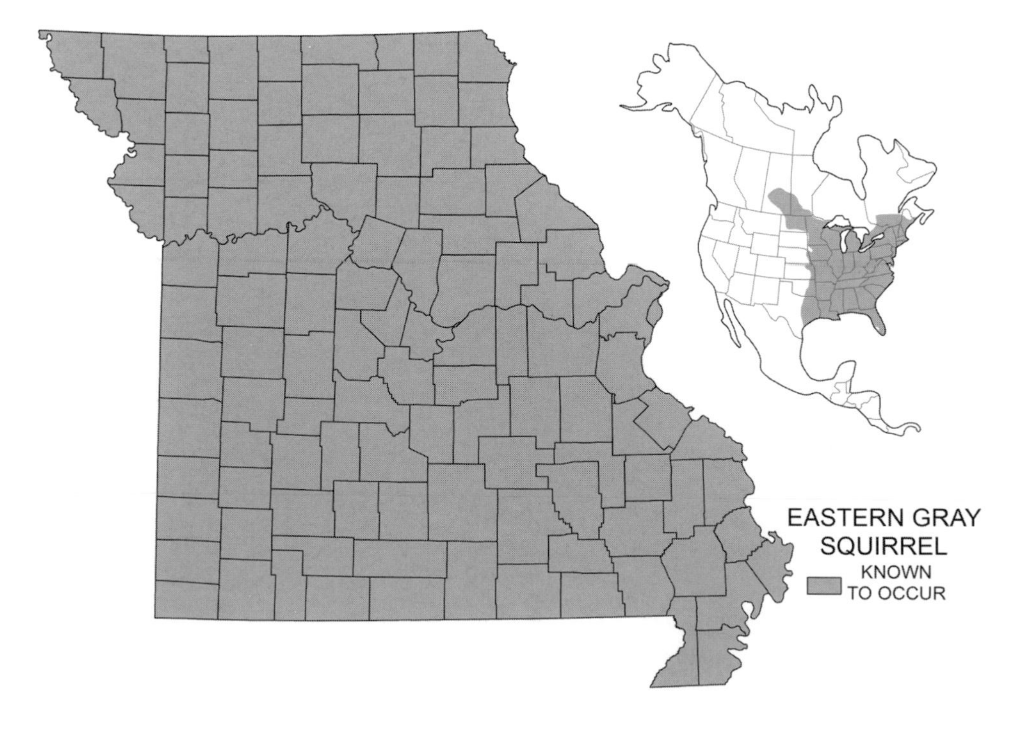

out on moonlight nights. Compared to fox squirrels, gray squirrels are active earlier and later in the day and are much less active in the middle of the day.

In addition to their tree life, gray squirrels spend a great deal of time on the ground. Mostly they hop along, but when foraging they usually walk. When traveling fast they make long leaps up to 1.5 m (5 ft.) in length and travel for short distances at rates up to 24 km (15 mi.) per hour. In trees they may jump as much as 2 m (6 ft.) between branches on the same level and leap for 4.5 m (15 ft.) or more downward. Gray squirrels are more agile jumpers than fox squirrels, but in spite of this they occasionally miscalculate a jump and fall. In escaping they often run on the ground until they reach their home tree, which they quickly ascend. Sometimes they flee through the treetops or climb to the top of a tree where they flatten out and "freeze."

The tail is used in many ways. It serves as a balance in jumping, climbing, and running along branches, and as a parachute in breaking an unexpected fall. On sunny days it is a sunshade, and in cold weather it is wrapped around the body for warmth. The tail is fluffed out and is very sensitive to touch and air currents. When flicked slowly either in a circular motion or from side to side or front to back, mild aggression is indicated. But if the tail is flicked rapidly, it communicates a greater threat. In swimming it acts as a rudder, but in swimming long distances it may become so wet as to weigh the squirrel down and hinder it.

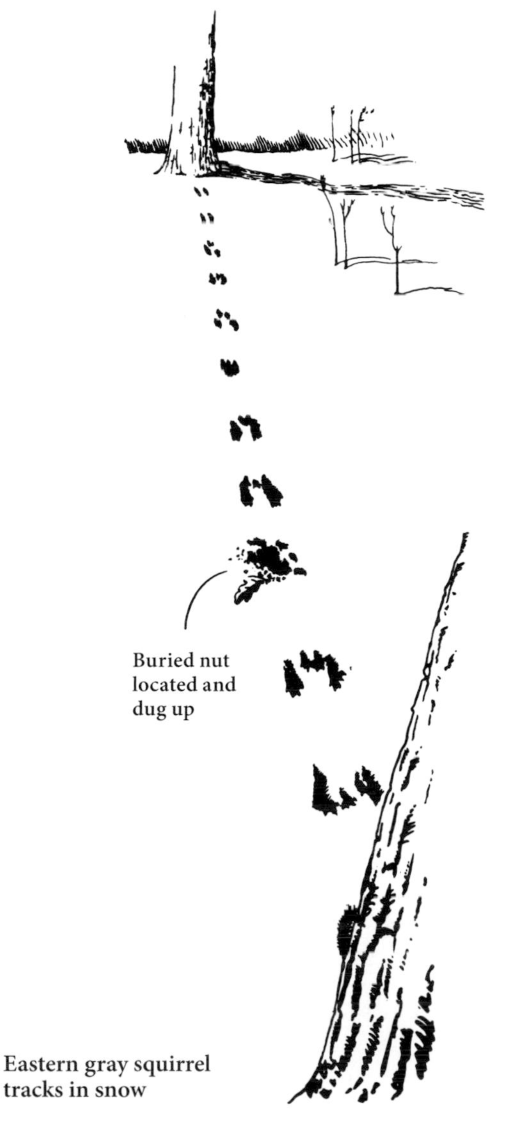

Eastern gray squirrel tracks in snow

Gray squirrels keep their fur well groomed. They have no regular latrine and continually soil their nest, which must either be relined or vacated. In summer, dust baths are taken often, to help rid the body fur of obnoxious insects.

Droppings are rough in texture, dark brown, and 5 mm (3⁄16 in.) long.

Squirrels live in loose colonies. All the members of one colony recognize each other, and they are familiar with all the nest sites and food sources in the area. Males are dominant over females, and older males are dominant over younger ones. Replacement occurs from young born within the group, but mating is usually by males from outside the group. Various behavioral signals and calls prevent antagonism.

While eastern gray squirrels usually den alone, during extremely cold weather several may huddle together in one nest for warmth. The calls of one squirrel serve as a warning to other squirrels, and several squirrels have been observed to cooperate in driving a predator away. Gray squirrels are nervous and easily frightened. They are more wary than fox squirrels and thus are regarded by most hunters as more challenging game.

Foods

The foods of squirrels are extremely varied, but of some 100 different species of plants eaten, about 7 compose the important staple food sources. These are hickory, pecan, oak, walnut, elm, and mulberry trees, and field corn. Locally, however, other foods may be important depending upon their abundance and the availability of preferred foods. Some woodlands may have one or two staple food items but lack enough variety to maintain sizable squirrel populations. Most forests of Missouri, including farm woodlots and their adjacent fields of corn, possess enough variety of food species for squirrels, but large river-bottom stands of elm, maple, or willows, even if bordered by cornfields, are deficient in suitable foods.

While the feeding habits of practically all Missouri mammals are influenced by the seasonal occurrence of available foods, the squirrels, perhaps as much as

Feeding site of squirrel is often marked by gnawed shell of nuts

any, exhibit a good example of their adaptability to the seasonal sequences in which wild fruits, nuts, berries, and the domestic corn crop ripen. It is largely because of knowledge of the squirrel's seasonal feeding habits that hunters select the time and place to locate them.

In late summer when the nut crop is maturing, squirrels start to cut hickory nuts of all varieties. By mid-August, the acorn crop is ripening, and they turn their attention to this food, taking white oak acorns first and leaving black oak acorns until later. Walnuts are readily eaten at this time when they are still green. Other foods, like wild grapes, pokeberries, wild cherries, hackberries, black gum, and Osage orange, are maturing at that time and are similarly utilized. Squirrels also make inroads upon ripening corn, especially in rows bordered by timber.

With the advance of fall, hickories, oaks, and walnuts become the dominant source of food and remain so throughout winter. Of these, walnuts are least preferred and are seldom taken in large quantities until the others are scarce. Because squirrels select the kernels of acorns to eat and discard the shells, they fail to obtain the salt that is most available in the shells.

Late winter is the most critical time for squirrels because much of the nut crop may be depleted and new spring growth has made little progress. Bark, immature buds, and twigs are then taken as emergency foods. Corn in open storage bins or waste corn can be very important in tiding squirrels over this difficult period, although as a food it lacks the mineral nutrition and energy source supplied by mast.

In early spring, the buds, seeds, tender twigs, and flowers of elms, oaks, maples, and other trees are fed upon lavishly, and sometimes the sap of these trees is licked. Even roots of herbs are dug up and eaten. From mid-May through summer, fruits are important in the diet. Wherever mulberry trees occur, their fruits are heavily utilized. Various other species of fruit, nonpoisonous mushrooms, and corn in the milk stage are likewise eaten during this season.

Insects and insect larvae are fed upon in small amounts, mostly in late spring and summer, and eggs and young birds may be eaten occasionally. Bones such as the shed antlers of deer and turtle shells may be gnawed, and cattle salt blocks and soil are licked on occasion. Squirrels may also obtain additional salts by licking the berms of asphalt highways that were treated by salt for snow removal. These animal foods and salts are especially sought by breeding females and presumably are essential sources of calcium, phosphorus, and other mineral salts.

In feeding, the squirrel commonly sits on its hind legs and holds the food item with its front feet. Squirrels frequently feed at the site where they pick up a particular food, but sometimes they have favorite stumps or perches where they go to eat; these are marked by an accumulation of shells, hulls, or other debris. The ground beneath a tree where a squirrel feeds is frequently littered with clippings of twigs, dislodged fruit, or the gnawed shells or hulls of nuts.

When selecting nuts to eat or store, the squirrel quickly detects and discards any unsound ones. In preparing nuts for storage, the cup is cut from acorns and the hulls are cleaned from walnuts. Frequently the squirrel carries the nut in its mouth 15 to 30 m (50 to 100 ft.) away from the food tree. Then with its front

feet it digs a hole in the ground about 4 cm (1½ in.) deep and puts the nut in it. Soil is replaced and leaves and twigs are brushed over the site. This entire operation requires from 1 to 3 minutes. On occasion, eastern gray squirrels have been observed to cache nuts in the uppermost whorl of branches in pine saplings.

Stored nuts have no particular ownership, and the members of a squirrel community share each other's efforts. The general position of stored food is probably located to a limited degree by a sense of memory, but the actual position of individual nuts is located by a keen sense of smell. Nuts are more easily detected in moist soil than in dry soil. Unburied nuts are preferred as long as they are available. Many buried nuts are not recovered, particularly in years of nut abundance, and a large percentage of them sprout and eventually become trees.

The nut crop varies from year to year because of many factors. Not all tree species bear each year, and even those that do seldom produce heavily every season. Unfavorable weather in the spring may prevent fertilization or development of flowers and thus influence the production of nuts. Sometimes adverse weather may affect only a local area, but at other times the condition is more widespread. Other factors, such as soil characteristics and the amount of shade, rainfall, and drainage likewise contribute to variability of nut production. Consequently, in some years the crop is abundant while in others it is inadequate for squirrels and other wildlife.

An average eastern gray squirrel consumes about 0.9 kg (2 lb.) of food a week, and an adequate range must provide 45 kg (100 lb.) of food a year to support one squirrel. In general, gray squirrels consume relatively more food than fox squirrels. This may be related to their smaller size and greater metabolic demands. Gray squirrels drink water daily, and available water makes a more desirable habitat.

Sometimes gray squirrels consume their own droppings—an adaptation that permits complete utilization of all nutrients in the food. The first time food passes through the body, bacteria digest the fibrous part of the food, and the second time, the nutrients can be used by the squirrel.

Reproduction

The onset of the breeding season is indicated by increased antagonism between males and by the active

Eastern gray squirrel carrying young

mating chase in which dominant males pursue females. It has been found that those males actively pursuing females are more successful than the younger subordinate males. While males are capable of breeding throughout the year, females generally mate only during two periods a year, each lasting 10 to 14 days. The first mating period begins in late December or early January and varies with the latitude; squirrels in the southern part of Missouri start to breed from 2 to 4 weeks before those in northern Missouri. The second general mating period occurs from late May to early July. However, a limited amount of breeding may occur in the population between these peaks.

Old females, and young females born in the preceding spring, breed at the first period. At the second, exceptionally vigorous females both old and young may produce second litters, and young females born in the preceding summer have their first litters. Rarely do young born in the spring have a litter in their first summer or do summer-born young have a litter in their first spring.

Pregnancy requires 44 to 45 days, and most litters are born in February or March, and July or August. From 1 to 8 young compose a litter, but 2 and 3 are the most usual. The female is solely responsible for care of the young and often moves them if the nest is disturbed. In transporting her young, she grasps it by the loose belly skin with her teeth, and the young clasps her around the neck with its legs and tail.

At birth the young are hairless, have their eyes and ears closed, and possess well-developed claws. They weigh about 14 g (½ oz.). The sexes can be identified by the external sex organs. By 3 weeks of age, the body is covered with short hair, the lower incisors are appearing, and the ears begin to open. The eyes open between 4 and 5 weeks of age. The young come out of the nest for the first time when 6 to 7 weeks old. At 8 weeks of age, they are half-grown, fully furred, and have a bushy tail. Weaning occurs about this time, and the young become self-sustaining in the following weeks. The litter born in the spring often remains with the female until summer, and the young born in late summer occasionally stay with the female during the winter. Spring and summer-born males attain sexual maturity about 10 or 11 months of age.

Some Adverse Factors

Predators of squirrels are coyotes, foxes, domestic cats and dogs, bobcats, raccoons, owls, hawks, some snakes, particularly the tree-climbing ones, and humans. Mites, ticks, lice, fleas, flies, roundworms, tapeworms, and protozoa parasitize gray squirrels. The chigger is important because it causes severe irritation of the skin. Scabies, or mange, caused by the *Sarcoptes* mite and promoted by malnutrition and insanitary nest conditions, begins to be serious in December and reaches a peak in spring. Mange results in the loss of fur over the body and weakens the host. Some mortality occurs from shock disease caused by a deficiency of sugar in the blood. Death from shock occurs more frequently among low-ranking members of the squirrel colony. The larvae of warble flies cause swellings in the shoulder or behind the forelegs and, while probably discomforting to the squirrel, do not affect the meat of the animal. Both eastern gray and eastern fox squirrels are known to be carriers of the West Nile virus.

Automobiles kill many squirrels annually.

Conservation and Management

The conservation and management of eastern gray squirrels are discussed in the account of the eastern fox squirrel.

It should be noted that acorn crops have been found to influence the population dynamics of these squirrels, and any population studies of gray squirrels should take this into account.

SELECTED REFERENCES

See also discussion of this species in general references, page 23.

Barkalow, F. S., Jr., and M. Shorten. 1973. *World of the gray squirrel*. Lippincott, Philadelphia, PA. 160 pp.

Barrier, M. J., and F. S. Barkalow Jr. 1967. A rapid technique for aging gray squirrels in winter pelage. *Journal of Wildlife Management* 31:715–719.

Doebel, J. H., and B. S. McGinnes. 1974. Home range and activity of a gray squirrel population. *Journal of Wildlife Management* 38:860–867.

Fitzwater, W. D., Jr., and W. J. Frank. 1944. Leaf nests of gray squirrel in Connecticut. *Journal of Mammalogy* 25:160–170.

Goodrum, P. D. 1940. *A population study of the gray squirrel in eastern Texas*. Texas Agricultural Experiment Station, Bulletin 591. 34 pp.

Kirkpatrick, C. M., and R. A. Hoffman. 1960. Ages and reproductive cycles in a male gray squirrel population. *Journal of Wildlife Management* 24:218–221.

Koprowski, J. L. 1993. Alternative reproductive tactics in male eastern gray squirrels: "Making the best of a bad job." *Behavioral Ecology* 4:165–171.

———. 1994. *Sciurius carolinensis. Mammalian Species* 480. 9 pp.

Korschgen, L. J. 1979. *Food habits of fox and gray squirrels in Missouri.* Federal Aid Project no. W-13-R-33. Missouri Department of Conservation, Jefferson City. 27 pp.

Lima, S. L., and T. J. Valone. 1986. Influence of predation risk on diet selection: A simple example in the grey squirrel. *Animal Behavior* 34:536–544.

McShae, W. J. 2000. The influence of acorn crops on annual variation in rodent and bird populations. *Ecology* 81:228–238.

Nixon, C. M., and M. W. McClain. 1969. Squirrel population decline following a late spring frost. *Journal of Wildlife Management* 33:353–357.

Nixon, C. M., M. W. McClain, and R. W. Donohoe. 1975. Effects of hunting and mast crops on a squirrel population. *Journal of Wildlife Management* 39:1–25.

Pack, J. C., H. S. Mosby, and P. B. Siegel. 1967. Influence of social hierarchy on gray squirrel behavior. *Journal of Wildlife Management* 31:720–728.

Padgett, Kerry A., et al. 2007. West Nile virus infection in tree squirrels (Rodentia: Sciuridae) in California, 2004–2005. *American Journal of Tropical Medicine and Hygiene* 76:810–813.

Sharp, W. M. 1958. Aging gray squirrels by use of tail-pelage characteristics. *Journal of Wildlife Management* 22:29–34.

Smith, N. B., and F. S. Barkalow Jr. 1967. Precocious breeding in the gray squirrel. *Journal of Mammalogy* 48:328–330.

Uhlig, H. G. 1956. *The gray squirrel in West Virginia.* Conservation Commission of West Virginia, Charleston. 83 pp.

Eastern Fox Squirrel (*Sciurus niger*)

Name

The derivation of the first part of the scientific name, *Sciurus*, is the same as that of its close relative, the eastern gray squirrel. The last part, *niger*, is the Latin word for "black" and describes the color of many members of this species, particularly in the southeastern United States. However, the fox squirrel in Missouri is classified in the subspecies *Sciurus niger rufiventer.* The subspecies name is from two Latin words meaning "reddish belly" (*rufus*, "reddish," and *venter*, "belly") and refers to the predominant reddish yellow color of this subspecies.

The common name originated as follows: "eastern" denotes the general range in the United States; "fox" refers to the reddish yellow color of this squirrel, which is similar to that of the red fox; and "squirrel" comes from the Old French *esquireul* or *escuriuel.*

Description (Plate 28)

The eastern fox squirrel is a medium to large, heavy-bodied tree squirrel; its long, very bushy tail distinguishes it from all other members of the squirrel family in Missouri except its close relative, the eastern gray squirrel. The fox squirrel is distinguished from the gray squirrel by its redder color, larger size, squarer facial profile, and 2 fewer upper cheek teeth. The bones of these two species can usually be distinguished by their color, those of the fox squirrel being pink while those of the gray squirrel are white. There are 4 clawed toes and a knoblike thumb on each front foot and 5 clawed toes on each hind foot. The soles of the feet are naked in summer but often furred in winter. Unlike the ground squirrels, no obvious cheek pouches occur.

Color. Fox squirrels vary greatly in color, even within one locality, but the prevailing color in Missouri is reddish yellow. This is mixed with gray on the back and sides but is more uniform on the belly, feet, and cheeks, around the ears, and on the fringe of the tail. In Missouri, black or albino individuals occur rarely. The sexes are colored alike.

There are two molts annually. Adult males and adult females, except those that are still nursing their spring litter, begin to molt in April while the young and those females that did not molt earlier start their molt in May or June. On the body, replacement of hairs begins at the head and progresses toward the tail, but on the tail molting begins at the tip and proceeds toward the body. Molting requires 3 to 4 weeks, and in most individuals there is a definite line between the old and new hair. In the fall molt, new hair growth begins on the rump and progresses toward the head and tip of the tail. All animals except females having summer litters begin their fall molt in September; these females do not molt until their nursing is completed.

Plate 28

Eastern Fox Squirrel *(Sciurus niger)*

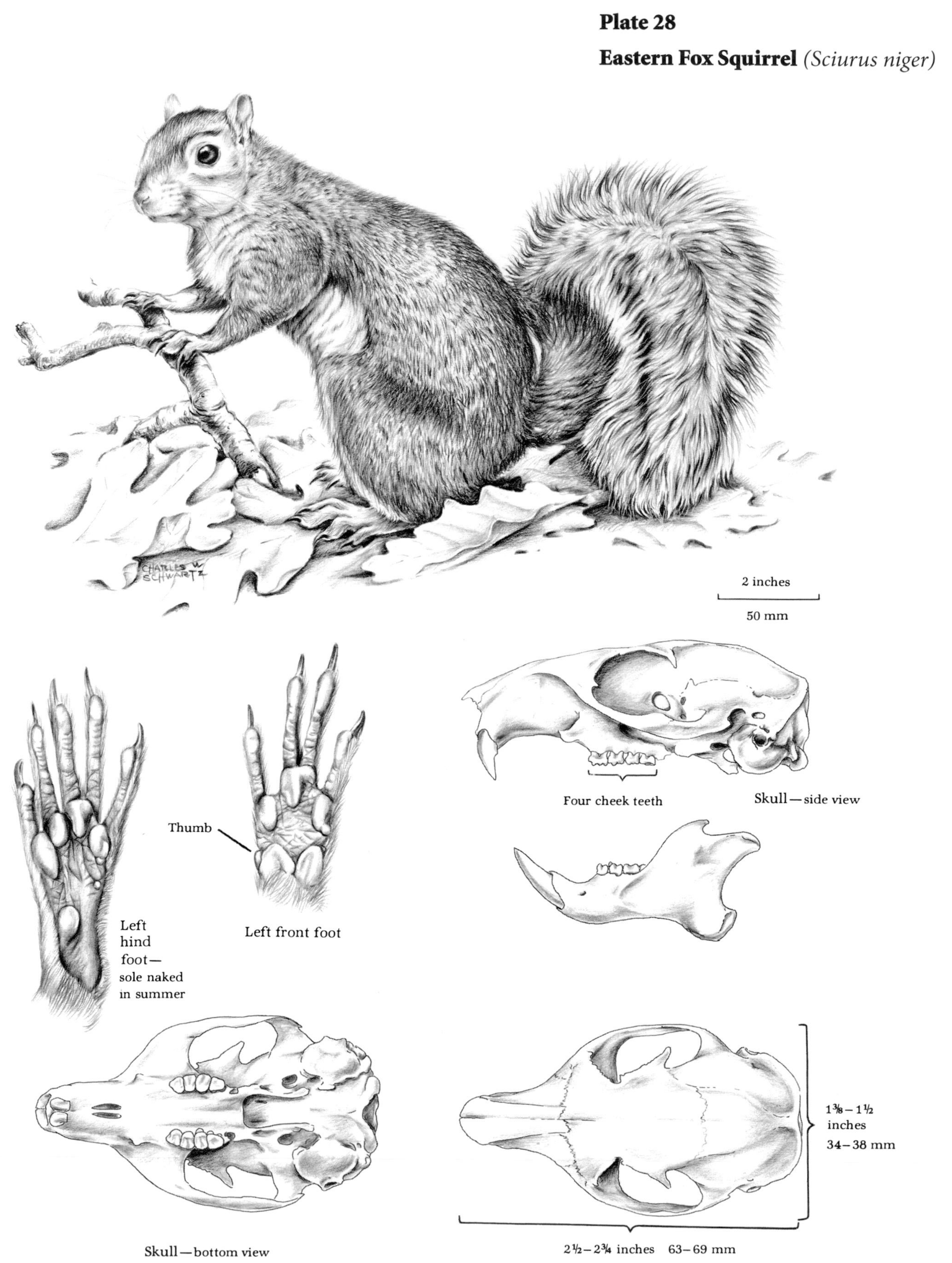

Measurements

Total length	482–736 mm	19–29 in.
Tail	177–355 mm	7–14 in.
Hind foot	63–88 mm	2½–3½ in.
Ear	19–31 mm	¾–1¼ in.
Skull length	63–69 mm	2½–2¾ in.
Skull width	34–38 mm	1⅜–1½ in.
Weight	453–1,360 g	1–3 lb.

Teeth and skull. The dental formula of the eastern fox squirrel is:

$$I\,\frac{1}{1}\;C\,\frac{0}{0}\;P\,\frac{1}{1}\;M\,\frac{3}{3} = 20$$

The eastern fox squirrel's skull is distinguished from that of the eastern gray squirrel by its larger size and by having 1 fewer upper cheek tooth on each side. In about 1 percent of the gray squirrel population, the first upper cheek tooth may be absent, making the total number of teeth 20. The larger size of the fox squirrel's skull distinguishes it from the eastern chipmunk's skull, the only other Missouri member of the squirrel family that also has 20 teeth.

Sex criteria and sex ratio. The sexes are identified by the penis in males and vagina in females. As in all rodents, the testes descend into a temporary scrotum during the breeding season. Males possess a baculum, or penis bone. Females have four pairs of teats.

Missouri sex ratios from harvest records show 56 percent males in the population. However, this figure may be influenced by the method of harvest because males are more active and more easily hunted than females.

Age criteria, age ratio, and longevity. Young fox squirrels can be identified in general by the deeper reddish yellow color of their belly fur. However, during the latter part of summer and fall when the young are not easily distinguishable by color, age can be determined by an examination of the wrist joint. Live animals must be X-rayed, but dead specimens can be examined by scraping away flesh from the bones of the wrist. In the young of both sexes, the ends, or epiphyses, of the two long bones, the radius and the ulna, of the front leg are either still distinctly separated by cartilage from the main portions, or shafts, of these bones, or a line of fusion is present between the ends and shafts. In adults, the ends and the shafts are completely united. The presence of cartilage on the shoulder end of the upper bone of the front leg also indicates a young animal. This method is reliable between 1 July and 15 January for fox squirrels and between 1 February and 1 December for gray squirrels.

In very young male fox or gray squirrels before their first breeding season, the testes are in the abdomen and the scrotum is small and furred. In adult males during their first breeding season, the testes are in the scrotum and the rear of the scrotum is brown to black and naked. In older adult males during the breeding season, both the undersurface and rear of the scrotum are blackened and generally free of hair. Following the breeding season, the pigmented skin is sloughed off, new hair grows in, and the testes may be retracted into the abdomen. From November to July, older adult males have large Cowper's glands (12 mm; ½ in. or more in diameter) at the base of the penis, while all younger males have small, undeveloped ones. Penis bones from older squirrels are slightly longer, more flaring at the base, and weigh more than those from younger squirrels.

Young female fox or gray squirrels that have not bred have small, light-colored teats that are concealed by hair. In the fall, adult females having had only spring litters show a small dot of black pigment at the tips of the teats; adult females that suckled summer litters have large, dark-tipped teats that usually protrude through the belly hair.

The dried weight of the eye lens has been used to establish age in fox and gray squirrels. By this method it is possible to identify spring-born young, summer-born young, and adults, but the specific age of adults is only reliable through 2½ years.

A count of the cementum layer on the roots of molar teeth of fox and gray squirrels is the most accurate means of identifying young and year classes of adults.

In fall populations of fox and gray squirrels in Missouri, the proportion of young varies from 50 to 55 percent of the total harvest.

Marked fox squirrels are known to have lived from 4 to 7½ years in the wild. Eighteen years is the record life span in captivity.

Glands. No anal musk glands are present.

Voice and sounds. The calls of fox squirrels are similar to those of gray squirrels.

Distribution and Abundance

The eastern fox squirrel originally ranged over approximately the eastern half of the United States. Today, it occurs throughout the central and southeastern United States, where it has spread a little north and west and into adjacent south central Canada and northeastern Mexico. It is nearly gone from the northeastern part of its range and is declining in the southeastern states. This species has been introduced

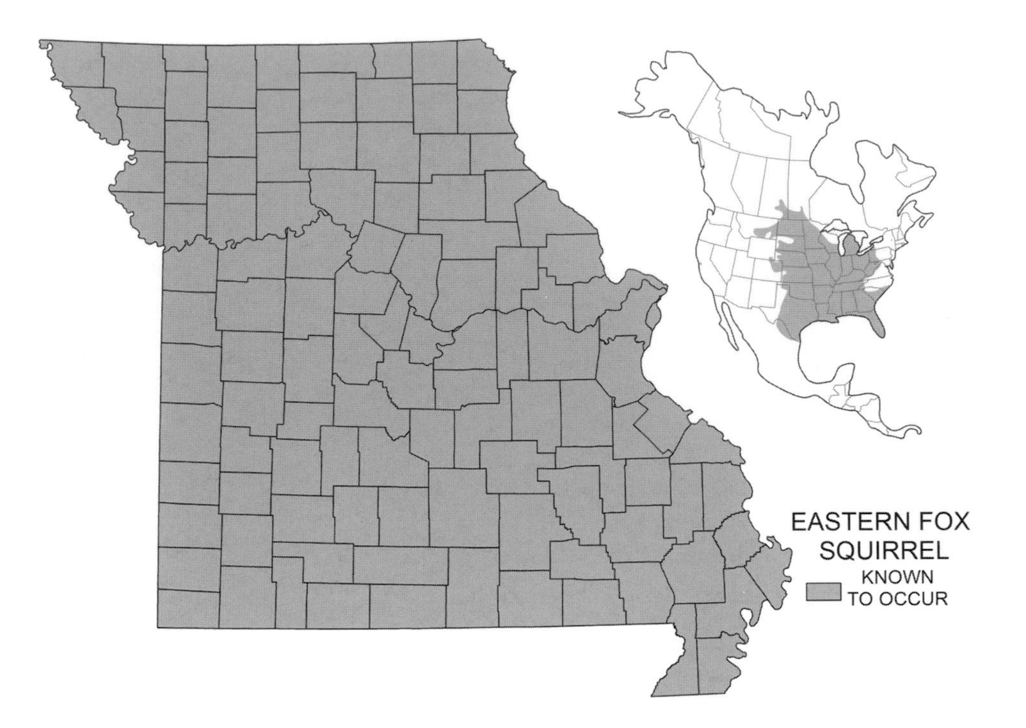

into California, Colorado, Idaho, New Mexico, North Dakota, Oregon, Texas, Washington, and Ontario. In small, ideal habitats the highest density is from 1 to 3 fox squirrels per 0.4 ha (1 ac.), but over large areas, the general population density is considerably less. Minor population fluctuations occur from year to year.

The eastern fox squirrel is found throughout Missouri and generally outnumbers the gray squirrel in the Central Dissected Till Plains and Osage Plains regions of the state. Its numbers have somewhat decreased in farming areas where Osage orange hedge fences and wooded thickets have been removed for cultivation; at the same time, the gray squirrel has shown some increase in numbers due to improved conditions in forested areas.

Habitat and Home

Throughout their ranges in Missouri, fox and gray squirrels occupy approximately the same habitat. This is primarily a mixed hardwood association with oak and hickory trees predominating. Generally speaking, however, the "grays" predominate in the bottomlands and the "foxes" along the higher ridges. In the grassland regions of the state, fox squirrels are found along Osage orange hedge fences, in farm woodlots, along timbered fencerows, and in timbered draws. They also live in urban areas, especially where large oak and hickory trees commonly occur.

The eastern fox squirrel's home is a leafy nest located in a tree fork or cavity. Cavities suitable for squirrel nests occur most often in older trees, particularly white oaks, elms, sycamores, and soft maples. These trees may have been shaded by forest cover until their lower limbs died and broke off or may have been damaged by wind or ice storms. The scars from such broken limbs usually heal over, but some fail to do so and rot because of weathering and the activities of decay-producing organisms and certain insects. Woodpeckers often enlarge these places by pecking out small cavities, and squirrels gnaw the openings larger and increase the size of the cavity. When the opening is about 76 mm (3 in.) in diameter, the squirrels maintain it at this size by repeatedly gnawing away the new growth. Squirrels also often enlarge and utilize the homes of woodpeckers. Cavities occupied by squirrels are about 15 cm (6 in.) wide, 35 to 41 cm (14 to 16 in.) deep, and have an opening about 7.6 cm (3 in.) in diameter. Large cavities are commonly filled with leaves in the fall and again in the spring. The leaves serve as good insulation from the winter's cold and as a warm lining for the young.

While leaf nests are usually built in the tops of large trees, they occasionally are located in other places like a grapevine tangle or the top of an Osage orange hedge fence. On the average, they are about 12 m (40 ft.) above the ground but may be as low as 3 m (10 ft.). The leaf nest consists of a rough twig framework, from 30 to 51 cm (12 to 20 in.) across, and a bulky pile of leaves heaped layer upon layer. The squirrel hollows out a nest cavity in the center of the leaves. Summer-built nests are usually of green twigs with green leaves

Leaf nest of fox squirrel

attached and are often flimsily constructed. Winter-built nests are made of bare twigs with separate leaves interlaced in the framework; they are usually very substantial. The inside cavity, about 15 to 20 cm (6 to 8 in.) in diameter, is reached through a hole in the side of the nest. The nest lining consists mostly of frayed leaves from the inside of the nest. The nest material is generally from the tree in which the nest is located, but sometimes leaves and twigs of other trees, grass, roots, moss, corn husks, or other items are added.

Leaf nests can be constructed in less than 12 hours; if well made and repaired, they usually last for 6 to 10 months. However, some large nests that are added to from time to time may last 2 to 3 years. Fragile or unused nests are easily destroyed by the wind. Leaf nests are advantageous as supplementary homes because they can be readily built near a source of food where hollow limbs may not be available. Also, they can be occupied when the home nest becomes badly infested with insect parasites or dirtied with droppings. A fox squirrel may utilize two or more tree cavities, or a tree cavity and a leaf nest concurrently. One female fox squirrel has been recorded as using a particular den tree for 3 years where she reared 2 and possibly 3 litters; she also was known to use a leaf nest nearby. In some places like scrub oak or heavily cut-over woodlands lacking tree cavities for dens, squirrels may live in leaf nests all year. However, tree cavities are preferred homesites, especially for winter and for nurseries, because they provide better protection from weather and predators.

Habits

In general, fox squirrels spend the entire year in the vicinity of a specific nest tree. They may use about 4 ha (10 ac.) during any one season, but over a year sometimes cover approximately 16 ha (40 ac.) by building nests near a new source of food as it becomes seasonally available. Daily activities are considerably more restricted than annual activities. In cold or stormy weather squirrels may move only 9 to 14 m (10 to 15 yd.) from home, providing a supply of food is within this radius, but in open weather they may cover up to 0.8 ha (2 ac.). During the breeding season, males tend to wander farther than females; the greatest movement occurs in fall when there may be some shifting because of high population pressure. Autumn records of tagged squirrels indicate some have moved from 1.6 to 22 km (1 to 14 mi.) from their previous home, with 64 km (40 mi.) a known maximum. At present there is no indication of a regular annual migration, although in years of drought or crop failure fox squirrels often abandon their homes in search of better fare.

While fox squirrels are active all year, they are exceptionally busy during the fall when the annual nut crop is ready to be harvested. During unusually cold weather, deep snow, heavy rains, or high winds, they usually stay in their nests. They are more active on clear days than cloudy ones, and on days when the

Squirrel carrying leafy nesting material

Fox squirrel tracks

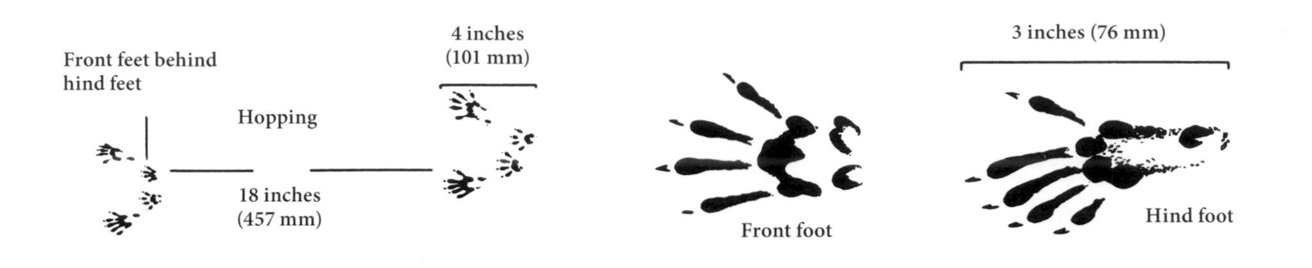

daily temperature ranges between 4 and 9°C (40 and 49°F).

There are three main periods of daily activity. Fox squirrels come out just before sunrise and are active for several hours. They then either return to their nests or idle on a limb or other spot. During the middle of the day, they move about some and then relax for several more hours. In late afternoon they emerge and forage again and in early dusk retire to their nests. Compared to gray squirrels they come out later and retire earlier and are frequently more active in the middle of the day.

Fox squirrels are essentially tree dwellers, but they also spend considerable time on the ground foraging. They travel principally by hopping but, when in a hurry, cover the ground in long leaps—sometimes at a rate of 16 to 19 km (10 to 12 mi.) per hour. In trees they run along the branches and leap from limb to limb and from tree to tree. In making these aerial leaps, they occasionally slip or miscalculate the distance and fall to the ground, even from heights as much as 12 m (40 ft.). They go both up and down a tree headfirst. Fox squirrels swim well but normally do not take to water.

Healthy squirrels groom themselves regularly and keep their fur in good condition. They exhibit little sanitation in the nest, which must either be relined frequently or vacated. The droppings are rough and about 6 mm (¼ in.) long. When fresh they are light brown but upon drying become dark brown or black.

Fox squirrels usually live alone, but pairs may stay together for a while at the beginning of the mating periods. A litter may stay with the female during the winter.

There is little or no social competition between individual fox and gray squirrels where they live in the same range. Where the environment favors both, however, it is possible that the two species may compete because their den requirements and food tastes are similar.

Foods

The foods and feeding habits of fox squirrels are essentially the same as those of gray squirrels. However, the fruit of Osage orange is probably more important to fox squirrels than to gray squirrels because fox squirrels are more common residents of the upland grassland country where Osage orange fencerows have been planted. Corn is likewise more common in the fare because of its greater prevalence in fox squirrel habitat. Both the spatial distribution of resources and predation risk influence the selection of the diet. Fox squirrels drink surface water on occasion, but succulent plants probably provide most of their moisture.

Reproduction

In breeding and rearing of young, fox squirrels are similar to gray squirrels, except that fox squirrels tend to begin breeding 10 days to 2 weeks earlier.

Some Adverse Factors

The predators, parasites, and diseases listed for the eastern gray squirrel also affect the eastern fox squirrel. Highway traffic takes a considerable toll.

Importance

Eastern fox and eastern gray squirrels are valuable game animals because of their sporting and food qualities. In Missouri from 600,000 to 700,000 squirrels are taken annually. At the rate of 0.5 Kg (1 lb.) of dressed meat per animal, squirrels supply a large amount of meat each year for Missouri hunters.

One of the many interrelationships of wild animals and their environment is evident in the nut-storing habit of squirrels. This practice of storing nuts in the ground results in eventual germination of many unrecovered nuts and their growth into trees that furnish not only food and shelter for subsequent squirrel populations, but also timber of economic value to humans.

Fox squirrels four days old

Squirrels also provide considerable pleasure for city dwellers who observe them around their homes and in parks. The hair of the tail is sometimes used commercially for "camel's hair" brushes and fishing lures.

Economically, squirrels do some harm by their inroads on cornfields. They may girdle ornamental trees, particularly in early spring when they gnaw the inner bark. Occasionally, they damage the insulation on outdoor electrical wiring and frequently become a nuisance when they gain access to the attics of homes.

Conservation and Management

Where it is desirable to increase squirrels, attempts should be made to improve their habitat. Because middle-aged or mature trees are the most valuable source of food and cover for squirrels, these trees should be saved in any timber stand. In forest management, proper thinning will permit the remaining trees to develop better crowns and produce a better nut crop; it also will create more openings where a variety of food plants can grow. Moderate grazing will allow some underbrush to occur and yet prevent too heavy herbaceous cover from growing. Other measures that promote better squirrel habitat are prevention of forest fires; leaving some trees with hollows for nesting sites; providing artificial nest boxes where necessary; planting windbreaks and encouraging shade trees; reforestation, especially of nut trees; maintaining Osage orange hedge fencerows (which are rapidly being bulldozed out of the landscape to provide another strip for cultivated crops); and encouraging wild fruit-bearing shrubs and vines along fencerows. Squirrels rank next to songbirds in value to nature watchers and photographers.

If you are experiencing problems with gray or fox squirrels, contact a wildlife professional for advice, assistance, regulations, or special conditions for handling these animals.

SELECTED REFERENCES

See also discussion of this species in general references, page 23.

Allen, D. L. 1942. Populations and habits of the fox squirrel in Allegan County, Michigan. *American Midland Naturalist* 27:338–379.

———. 1952. *Gray and fox squirrel management in Indiana.* Indiana Department of Conservation, Pittman-Robertson Bulletin 1. 112 pp.

Baumgartner, L. L. 1939. Fox squirrel dens. *Journal of Mammalogy* 20:456–465.

Beale, D. M. 1962. Growth of the eye lens in relation to age in fox squirrels. *Journal of Wildlife Management* 26:208–211.

Brown, J. S., and R. A. Morgan. 1995. Effects of foraging behavior and spatial scale on diet selectivity: A test with fox squirrels. *Oikos* 74:122–136.

Brown, L. G., and L. E. Yeager. 1945. Fox squirrels and gray squirrels in Illinois. *Illinois Natural History Survey, Bulletin* 23:449–536.

Carson, J. D. 1961. Epiphyseal cartilage as an age indicator in fox and gray squirrels. *Journal of Wildlife Management* 25:90–93.

Fisher, E. W., and A. E. Perry. 1970. Estimating ages of gray squirrels by lens-weights. *Journal of Wildlife Management* 34:825–828.

Fogl, J. G. 1978. Aging gray squirrels by cementum annuli in razor-sectioned teeth. *Journal of Wildlife Management* 42:444–448.

Geluso, K. 2004. Westward expansion of the eastern fox squirrel (*Sciurus niger*) in northeastern New Mexico and southeastern Colorado. *Southwestern Naturalist* 49:111–116.

Hicks, E. A. 1949. Ecological factors affecting the activity of the western fox squirrel, *Sciurus niger rufiventer* (Geoffroy). *Ecological Monographs* 19:287–302.

Kiupel, M., H. A. Simmons, S. D. Fitzgerald, A. Wise, J. G. Sikarskie, T. M. Cooley, S. R. Hollamby, and R. Maes. 2003. West Nile virus infection in eastern fox squirrels (*Sciurus niger*). *Veterinary Pathology* 40:703–707.

Koprowski, J. L. 1993. Behavior tactics, dominance, and copulatory success among male fox squirrels. *Ethology, Ecology, and Evolution* 5:169–176.

———. 1994. *Sciurus niger. Mammalian Species* 479. 9 pp.

Koprowski, J. L., J. L. Roseburg, and W. D. Klimstra. 1988. Longevity records for the fox squirrel. *Journal of Mammalogy* 69:383–384.

Korschgen, L. J. 1981. Foods of fox and gray squirrels in Missouri. *Journal of Wildlife Management* 45:260–266.

Packard, R. L. 1956. The tree squirrels of Kansas: Ecology and economic importance. University of Kansas, Museum of Natural History and State Biological Survey of Kansas, Miscellaneous Publication 11. 67 pp.

Shaw, W. W., and W. R. Magun. 1984. Nonconsumptive use of wildlife in the U.S. U.S. Fish and Wildlife Service Resource Publication 154. 20 pp.

Zelley, R. A. 1971. The sounds of the fox squirrel, *Sciurus niger rufiventer. Journal of Mammalogy* 52:597–604.

Southern Flying Squirrel
(Glaucomys volans)

Name

The first part of the scientific name, *Glaucomys*, is from two Greek words and means "gray mouse" (*glaukos*, "gray," and *mys*, "mouse"). The second part, *volans*, is the Latin word for "flying." The common name, "southern," indicates its general geographical distribution in North America compared to that of a close relative, the northern flying squirrel, which is not found in Missouri; "flying" refers to this rodent's habit of gliding or sailing from tree to tree.

Description (Plate 29)

The southern flying squirrel is readily distinguished from other squirrels by possessing a loose fold of skin, the *patagium*, which is continuous on each side with the skin of the body, extends from the outside of the wrist on the front leg to the ankle on the hind leg, and is used for gliding through the air. A cartilaginous process, or spur, supports the fold of skin at the wrist and permits it to extend beyond the outstretched leg. When both front and hind legs are spread, the patagium produces a winglike, or gliding, surface. The tail is smoothly furred, broad, and horizontally flattened; it functions as a rudder and stabilizer during the glide.

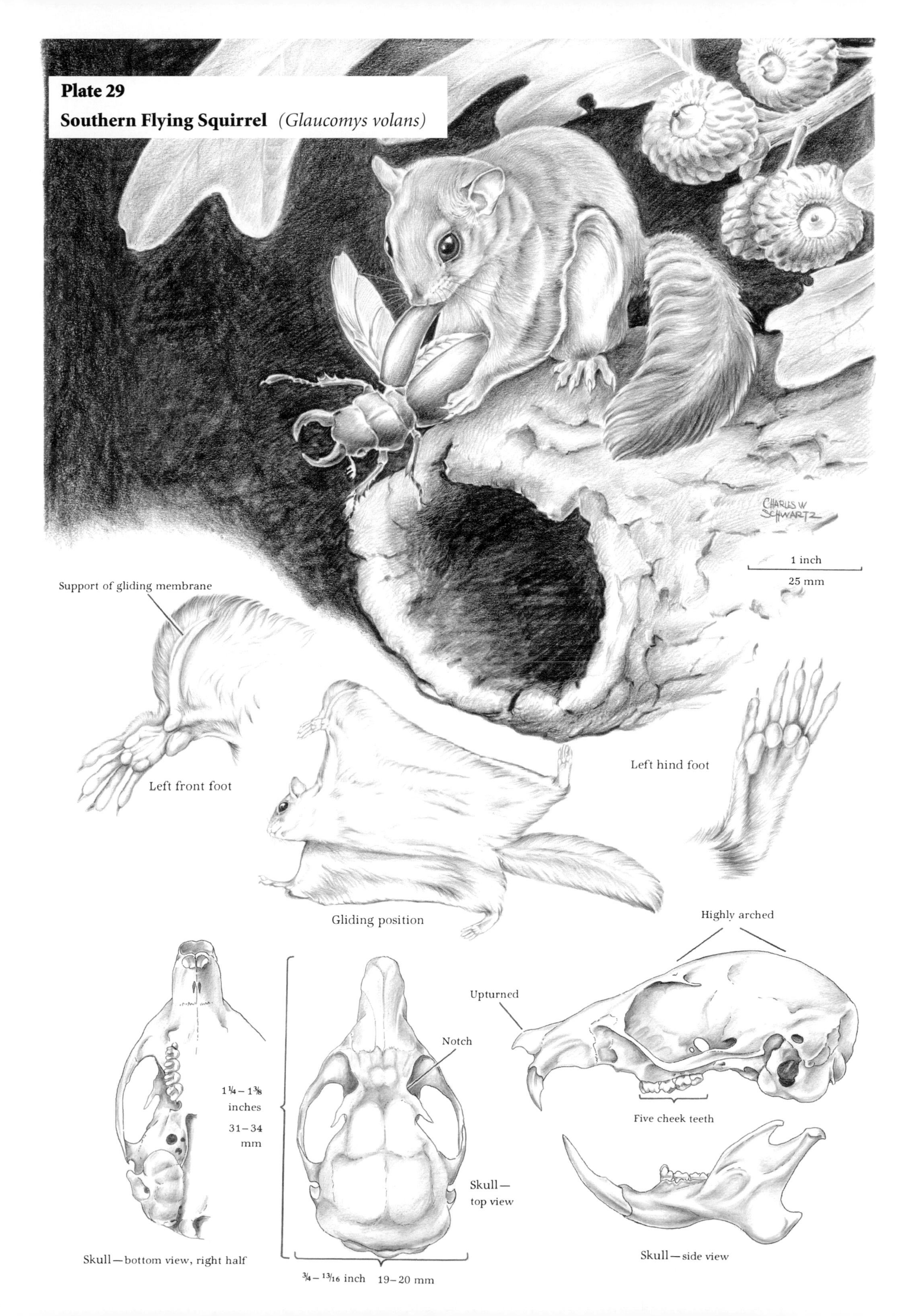

Plate 29
Southern Flying Squirrel (Glaucomys volans)
CHARLES W SCHWARTZ
1 inch
25 mm
Support of gliding membrane
Left front foot
Gliding position
Left hind foot
Highly arched
Upturned
Notch
Five cheek teeth
1¼ – 1⅜ inches
31 – 34 mm
Skull—top view
Skull—bottom view, right half
¾ – 13/16 inch 19 – 20 mm
Skull—side view

The eyes are conspicuously large and black but in the glare of a flashlight at night shine ruby red. The ears are prominent; the nose is short and slightly upturned. Long tactile whiskers project from the face. The body fur is short, thick, very soft, and silky. There are 4 clawed toes on the front foot and 5 on the hind.

Color. On the back, the flying squirrel is grayish to brownish with a blackish undercolor showing wherever the hairs are parted. On the sides, the darker tones grade into almost a pure black border along the edge of the gliding membrane. This contrasts sharply with the entirely white hairs of the underparts. The upperpart of the face is colored like the back and the lower part like the belly; there is a narrow black ring around the eye. The tail is brownish gray above, paler and grayer below. The feet are gray on top and whitish underneath. The sexes are colored alike. The general coloration in summer is darker and browner than in winter. There is one molt annually, beginning in September and taking approximately two months for completion.

Measurements

Total length	203–285 mm	8–11¼ in.
Tail	79–130 mm	3⅛–5⅛ in.
Hind foot	25–31 mm	1–1¼ in.
Ear	15–25 mm	⅝–1 in.
Skull length	31–34 mm	1¼–1⅜ in.
Skull width	19–20 mm	¾–13⁄16 in.
Weight	42–141 g	1½–5 oz.

Flying squirrel launching

Teeth and skull. The dental formula of the southern flying squirrel is:

$$I\ \frac{1}{1}\ C\ \frac{0}{0}\ P\ \frac{2}{1}\ M\ \frac{3}{3} = 22$$

The southern flying squirrel's skull is likely to be confused with the skulls of two other similar-sized Missouri members of the squirrel family: the thirteen-lined ground squirrel and the eastern chipmunk. From the latter, it is distinguished by having 5 cheek teeth on each side of the upper jaw; it can be told from the thirteen-lined ground squirrel by having a highly arched braincase, a slightly upturned or arched front end (nasal bones), and a deep notch in the upper rim of the eye socket.

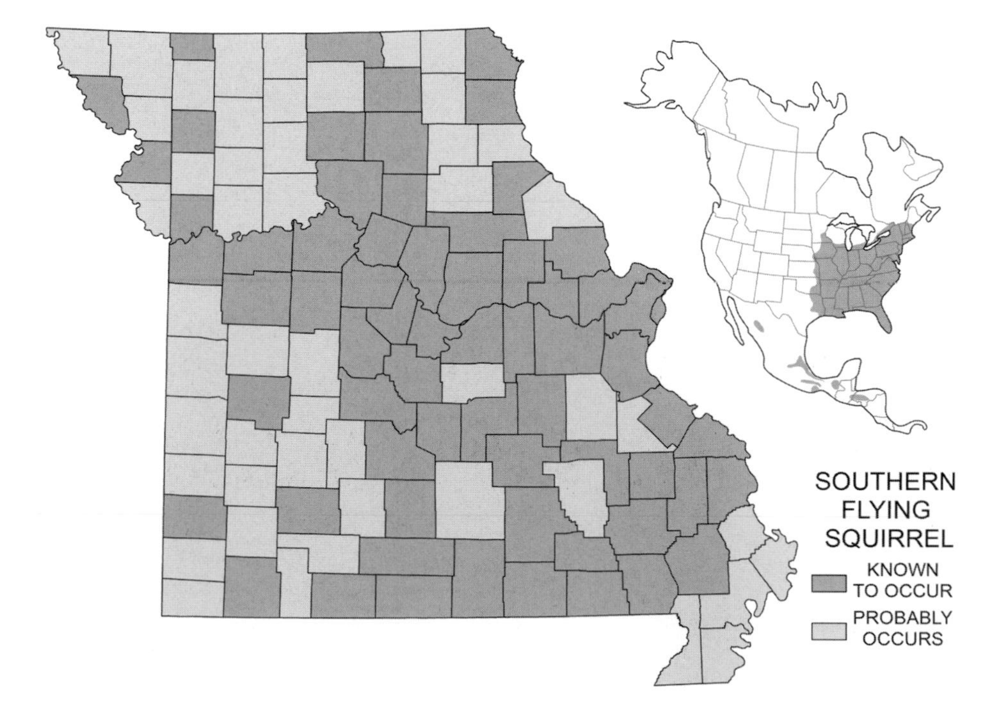

Glaucomys sabrinus. *Journal of Experimental Biology* 210:1413–1423.

Reynolds, M. G., et al. 2003. Flying squirrel–associated typhus, United States. *Emerging Infectious Diseases* 9:1341.

Scheibe, J. S., D. Figgs, and J. D. Heiland. 1990. Morphological attributes of gliding rodents: A preliminary analysis. *Transactions of the Missouri Academy of Science* 24:49–56.

Sollberger, D. E. 1940. Notes on the life history of the small eastern flying squirrel. *Journal of Mammalogy* 21:282–293.

———. 1943. Notes on the breeding habits of the eastern flying squirrel (*Glaucomys volans volans*). *Journal of Mammalogy* 24:163–173.

Stone, K. D., G. A. Heidt, P. T. Caster, and M. L. Kennedy. 1997. Using geographic information systems to determine home range of the southern flying squirrel (*Glaucomys volans*). *American Midland Naturalist* 137:106–111.

Taulman, J. F. 1999. Selection of nest trees by southern flying squirrels (Sciuridae: *Glaucomys volans*) in Arkansas. *Journal of Zoology* 248:369–377.

Pocket Gophers (Family Geomyidae)

One distinctive characteristic of this family is the pair of fur-lined cheek pouches that open by long slits on the sides of the face near the mouth. The front feet are large and have heavy claws, which aid in digging subterranean burrows. The eyes and ears are very small and the body is stocky. There are 4 cheek teeth both above and below on each side of the jaw.

The family name, Geomyidae, is based on two Greek words meaning "earth mouse." Around 35 species of pocket gophers live in North America, but only one species is found in Missouri.

Plains Pocket Gopher
(*Geomys bursarius*)

Name

The first part of the scientific name, *Geomys*, is from two Greek words and means "earth mouse" (*ge*, "earth," and *mys*, "mouse"). It alludes to the subterranean life of this rodent. The second part, *bursarius*, is of Latin origin and means "pertaining to a pouch of skin" (*bursa*, "pouch," and *-arias*, "pertaining to"); this refers to the gopher's large fur-lined external cheek pouches.

The common name originated as follows: "plains" denotes the range in the plains or prairie area of the central United States; "pocket" describes the prominent paired cheek pouches; and "gopher" is from the French *guafre* meaning "honeycomb" and refers to the animal's habit of digging an intricate network of tunnels in the ground.

Description (Plate 30)

The plains pocket gopher is a stocky rodent with a body well adapted for both digging and a subterranean life. The head is broad, the neck stout, and the legs and tail are short. A very distinctive characteristic is the pair of spacious, fur-lined pockets, or cheek pouches, opening by long slits on the sides of the face near the mouth and extending backward as far as the shoulders. The pockets are used for carrying food and nesting material.

The strong front limbs have 5 toes with long claws, the middle 3 toes having the longest and fastest-growing claws. A fringe of rather stiff and probably sensory hairs, mostly on the inner edges of the second and third toes, aids in excavating earth by acting as a brush. The hind limbs are not as well developed for excavating as the front ones and possess only short, blunt claws on the 5 toes.

The large, yellowish incisors are always exposed because the furry lips and furry front part of the mouth close behind them. This specialized tooth and mouth structure permits the gopher to dig hard soil and clip roots with the incisor teeth yet keeps soil out of the inner mouth cavity. The small, but quite apparent, beady eyes have a limited range of vision. The tiny, rounded, and nearly naked ears have valves that close to exclude soil during tunneling. Sensitive facial whiskers help guide the animal forward in its underground passages, while nerve endings on the short, scantily haired tail perform this service when the animal is moving backward. Along with the keen sense of touch, the sense of smell is well developed. The loose skin is covered with fairly short fur that is dense, glossy, and soft. They are average swimmers.

Color. The coloration of plains pocket gophers varies in individuals, ranging from light to dark brown or blackish above, with the top of the head and midback slightly darker than the rest of the upperparts. The fur in the cheek pouches and on the underparts is pale brown. Occasionally there is some whitish fur on the throat, midbelly, or even the back. The tail is brownish near the body but whitish to pale buff for the outer half. The feet are pale brown to white. Melanistic (blackish) individuals are rare, but albinos are more common. The sexes are colored alike.

Plate 30

Plains Pocket Gopher *(Geomys bursarius)*

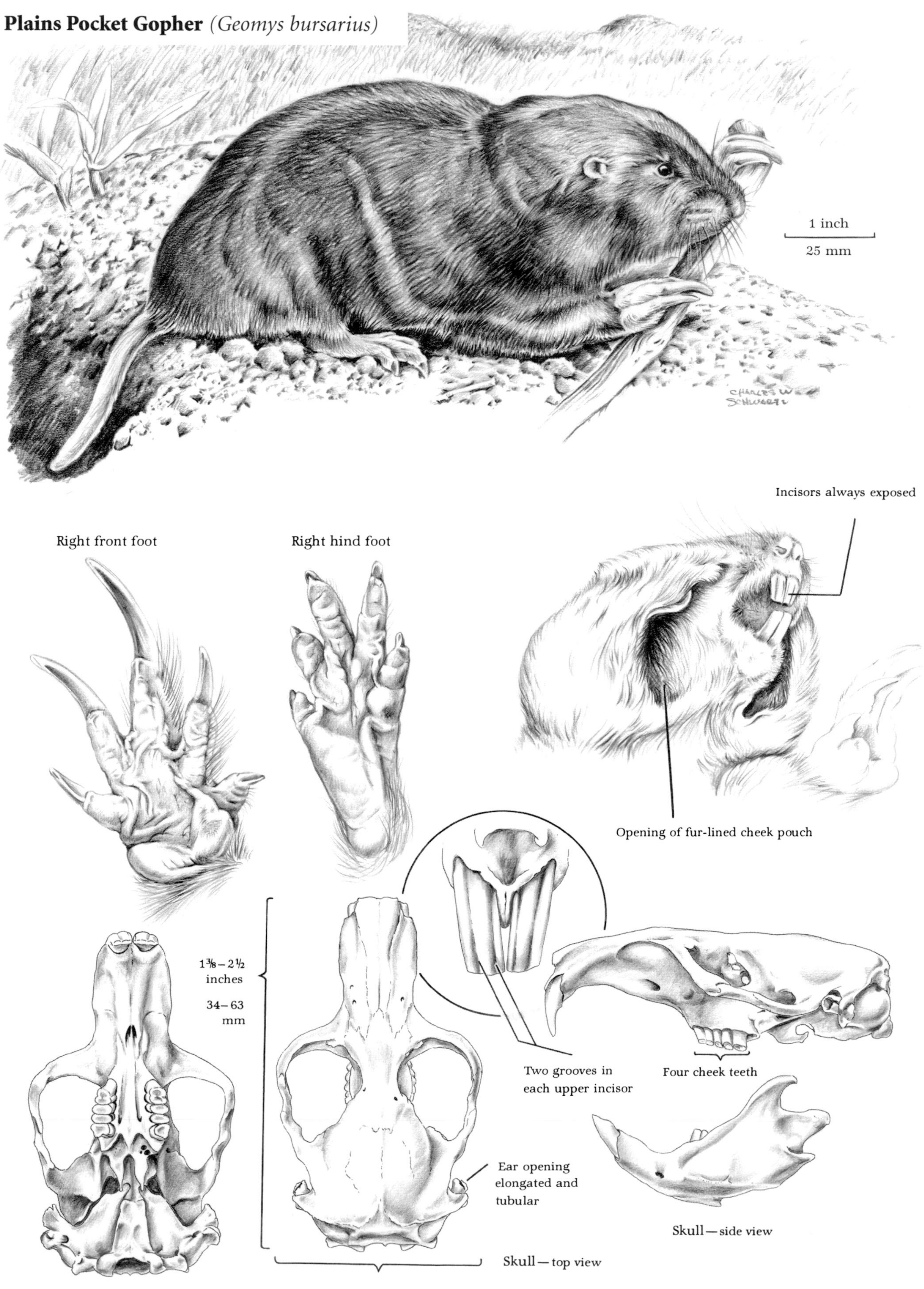

There is only one molt annually, which takes place from spring until fall. The new fur appears first near the rump and gradually replaces the worn fur in a forward direction. There is usually a very noticeable line across the back separating old and new fur.

Measurements

Total length	190–349 mm	7½–13¾ in.
Tail	50–114 mm	2–4½ in.
Hind foot	28–38 mm	1⅛–1½ in.
Ear	3–4 mm	⅛–3⁄16 in.
Skull length	34–63 mm	1⅜–2½ in.
Skull width	25–38 mm	1–1½ in.
Weight	141–510 g	5–18 oz.

Males are larger than females.

Teeth and skull. The dental formula for the plains pocket gopher is:

$$I\ \frac{1}{1}\ C\ \frac{0}{0}\ P\ \frac{1}{1}\ M\ \frac{3}{3} = 20$$

Each upper incisor possesses 2 lengthwise grooves on the front surface. The rate of growth of the pocket gopher's incisors is greater than that recorded for other rodents. The upper incisors grow as much as 23 cm (9 in.) a year and the lower ones 35 cm (14 in.). The heavily built skull is flat and broad with wide-spreading zygomatic arches (the bony arch on the outside of the eye socket). The ear opening is elongated and tubular. This is the only rodent in Missouri with 20 teeth and grooved upper incisors.

Sex criteria. In males the penis and anus open separately. There is a small penis bone about 9 mm (⅜ in.) long. In the groin region of females there are three openings from front to rear: the urethra, vagina, and anus. Females have 3 pairs of teats on the belly: 1 pair is near the front legs, 1 pair behind this, and 1 pair in the groin region.

There tend to be more females than males in a population. One Missouri study reported 38 percent males and 62 percent females captured.

Age criteria. Young animals tend to be grayer than adults and their hair is longer, softer, and more crinkled.

Voice and sounds. Pocket gophers are generally quiet animals. In captivity they grind their teeth, squeal in anger, and squeak when handled.

Distribution and Abundance

The plains pocket gopher ranges across the Great Plains from extreme southern Manitoba to northern Texas, and as far east as extreme western Indiana. It was recently added to the known fauna of Arkansas, occurring in two north-central counties.

Its Missouri range is not known with certainty. This gopher is fairly common in northwestern, northeastern, and east-central Missouri but is uncommon in other portions of the state. It is probably absent from several southern Missouri counties.

In some Missouri localities, gophers may be abundant for six to eight years, then become rare.

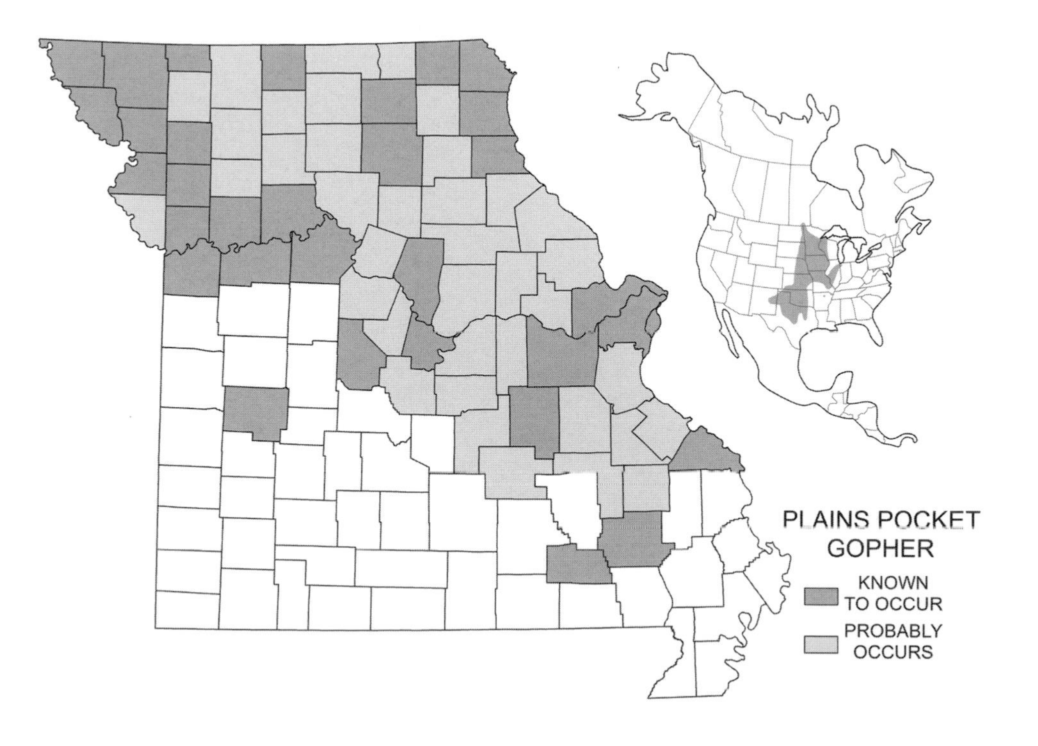

Pocket gopher pushing loosened dirt through tunnel

Frequently there are areas near active colonies that are unoccupied, although formerly pocket gophers inhabited them. These areas, when not used, will produce new food plants, permit old droppings and stored food to disintegrate, and allow the soil to reconsolidate. In Arkansas, numbers of plains pocket gophers averaged 20 per ha (2.5 ac.) with a range of 4 to 60 per ha depending on size of the area and quality of the habitat.

Habitat and Home

Plains pocket gophers live mostly in open lands such as prairie grasslands, pastures, cultivated areas of alfalfa and clover, and meadowlands. Within these types, they range from hilltops to river bottoms. Gophers show a preference for deep, moist soils such as wind-blown, or loess, soil; sandy soils of river floodplains; and fills for roads that are easy for them to work. They have also been found at airports and golf courses and in gardens and lawns.

The pocket gopher's home is an extensive system of underground tunnels marked on the surface of the ground by numerous mounds of excavated earth. Mounds are commonly from 46 to 61 cm (18 to 24 in.) in diameter and about 15 cm (6 in.) high but have been recorded up to 3 m (10 ft.) in diameter and more than 30 cm (1 ft.) high. One animal's tunnel system may cover 0.4 ha (1 ac.) or more. The main tunnel is located from 15 to 23 cm (6 to 9 in.) below the surface and is about 10 cm (4 in.) in diameter. It may be 152 m (500 ft.) or more long. At intervals along this main runway, side tunnels from 20 to 46 cm (8 to 18 in.) in length lead to the surface of the ground, to food storage chambers, to latrine locations, and to feeding sites. One branch leads to an enlarged nest chamber 0.6 to 0.9 m (2 to 3 ft.) and rarely 2 m (6 ft.) beneath the surface. The nest is from 18 to 25 cm (7 to 10 in.) in diameter and lined with cut pieces of fresh grass, fine dry grass, shredded stubble, or down feathers.

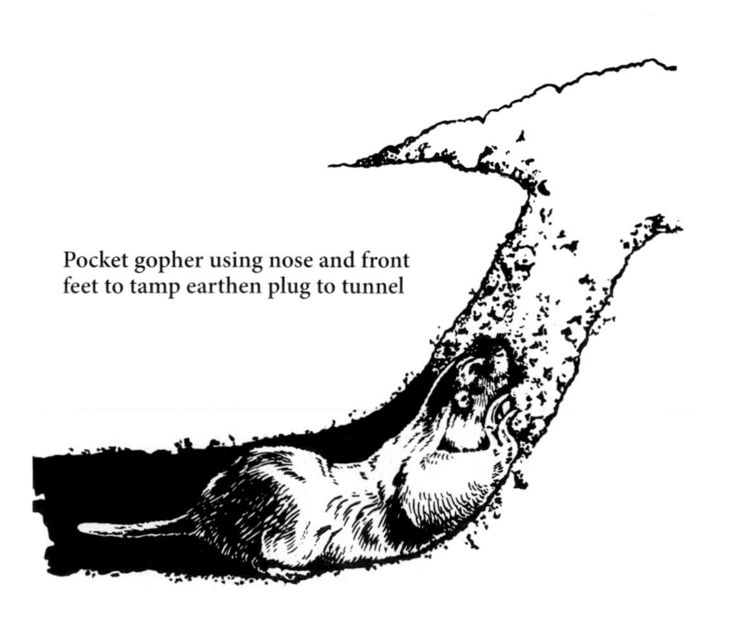

Pocket gopher using nose and front feet to tamp earthen plug to tunnel

Habits

In the course of excavating the tunnel system, the gopher pushes the soil to the surface through a side tunnel. Individual loads are pushed out to the front, right, and left of the opening, forming a flat, fan-shaped mound. As the heap grows, some soil is also pushed to the rear.

In digging, the strong, sharp claws of the front feet scratch and scrape the soil until it is finely crumbled, then pass it under the belly. Unusually hard soil lumps are cut with the incisor teeth. Periodically, the hind feet are used together to kick the loosened soil behind the animal.

When a pile of soil has accumulated, the gopher turns around by lowering its head between its hind legs and twisting its body. The load of soil is kept ahead of the gopher by the front feet, which are held against the sides of the face. The hind legs push the animal and the soil forward a few centimeters (inches) at a time.

When the gopher finishes using a side tunnel, it plugs up the opening. To do this, the gopher pushes successive loads of soil upward to fill the opening and holds each one briefly until the particles start to stick together. The gopher may also tamp the soil rapidly with its nose, chest, or feet. Only this shallow plug may be left to seal the tunnel, or the entire side tunnel may be filled with soil. The location of the plug is frequently visible in the mound of excavated earth.

Digging occurs at various speeds depending upon the nature and temperature of the soil. Most mounds are made in spring and fall. In summer the soil is often dry and digging is difficult. In Missouri, considerable digging often goes on below the frost line during winter, but the soil is stored in tunnels at higher levels and does not appear aboveground. Under optimum digging conditions, a gopher may build from one to two mounds a day.

The soil workings of the pocket gopher and another subterranean mammal, the mole, are often misidentified. The mound of the pocket gopher is roughly fan

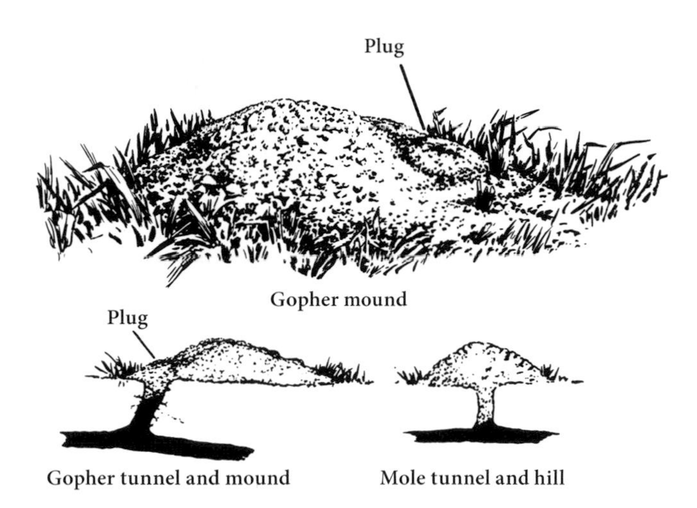

Comparison of gopher mound and molehill

shaped because the soil is pushed from the slanting tunnel to the front, right, and left. The mound of the mole is an irregular mass since the soil is pushed out through a vertical tunnel and allowed to fall freely in any direction. Moles frequently push up ridges as they progress just beneath the surface, while gophers make all of their tunnels deeper. In the western United States, pocket gophers of other species pack soil from under the ground into tunnels of snow on top of the ground. These cores of earth remain on the surface of the ground for some time after the snow has melted.

Gophers are active day and night but dig most energetically at night, early in the morning, and late in the evening. In midday they occasionally open a hole to let sunshine and fresh air into the tunnel system. Because of their subterranean habits they are seldom seen aboveground. However, they come out of their tunnels more at night and on cloudy days than at other times. They seldom go farther than 61 to 91 cm (2 to 3 ft.) from the opening; during this time they are very nervous and alert. In the breeding season males may wander overland to locate females; in the fall some animals travel aboveground looking for new homes. Because gophers do not swim far, rivers are a barrier to their dispersal.

In placing material in the pockets, the gopher uses first one front foot, then the other, and fills the pockets alternately. Pieces too long to fit are pushed inside until they coil, cut into convenient lengths before being placed in the pockets, or dragged to the tunnel system with the mouth. In emptying the pockets, the front feet are placed at the back of the pockets and, with a forward motion, press the contents onto the ground in front of the animal. Sometimes several strokes are required to remove all the material. The pockets can be turned inside out for cleaning and are pulled back in place by a special muscle.

Food is stored in chambers off the main tunnel or near the nest. Such stores may amount to as much as 17 l (half a bushel). Much of the food spoils; when this happens, the chamber is sealed off. A definite community of insects and other arthropods occurs in pocket gopher burrows; some of these animals live entirely on the stored food, while others consume the waste products of the gopher. Still others are parasitic upon the gophers themselves.

Many gophers may have their individual burrows in the same field, so "colonies" are common. However, they are solitary animals, and tunnels leading to those of another animal are usually sealed off. Several gophers are occasionally taken from a single trap set; in these instances, successive gophers probably take over the tunnel system of an absent neighbor. When two gophers get together, they fight ferociously, often until death.

Gophers groom themselves very carefully. They frequently shake the soil particles from their fur and clean soil from their claws with their teeth.

Foods

Pocket gophers are almost entirely vegetarians, only rarely taking a mouse or eating another gopher that has died. Their foods consist of fleshy roots, underground stems, succulent stems of grasses and legumes, leaves of livestock forage plants, small fruits, and grains. The gopher obtains almost all of its food underground by digging tunnels to the sources of food. They are also known to store acorns and black walnuts. A gopher

eats approximately half its weight in food daily. Sufficient moisture is provided by the succulent foods.

Pocket gophers, like some other rodents and rabbits, reingest their own droppings.

Reproduction

Mating usually takes place from spring through summer, although January breeding has been reported. Pregnancy requires 4 weeks, and the annual litter of 3 or 4 (with extremes of 1 and 5) is usually born in March, April, or May.

At birth the little gophers are about 5 cm (2 in.) long, have pink, loose, and wrinkled skins, and weigh about 5 g (1/5 oz.). Their eyes, ears, and cheek pouches are closed, and no teeth are visible. A few days after birth, the incisors come through the gums, but the lips do not close behind them. Their eyes open around 3 weeks of age and they are weaned at around 4 to 5 weeks. At around 5 weeks of age the cheek pouches and ears are open and the lips meet behind the incisors. The young disperse shortly afterward, digging a new tunnel system near their old home. They become sexually mature at 1 year of age.

An interesting adaptation to subterranean life is shown in the hip, or pelvic, girdle of female pocket gophers. The narrow girdle, which permits easy turning in the close confinement of a tunnel, is not large enough to allow the young to pass during birth. At the approach of the first breeding season, the underpart of the girdle (the pubic bones) begins to be resorbed. This process continues during pregnancy; by the time of giving birth, the pubic bones of the female have almost disappeared, permitting easy passage for the young. This loss is permanent and is due to a hormone secreted by the ovary. If females do not become pregnant at their first breeding season, resorption of the pubic bones is only partial. There is no comparable change in males.

Some Adverse Factors

Badgers, weasels, foxes, skunks, domestic cats, hawks, and owls take pocket gophers on occasion. Bullsnakes, hog-nosed snakes, and kingsnakes have been known to dig into side tunnels by bringing repeated loads of soil to the surface in a coil of the body. These snakes kill a gopher by coiling around it or by pressing it against the side of the tunnel. Mites, lice, fleas, roundworms, and flukes all parasitize pocket gophers.

Importance

Pocket gophers are extremely valuable mammals because of their contribution to the formation and conditioning of our soil. Countless generations of gophers have moved vast quantities of subsoil to the surface, exposing it to the action of sun, wind, and rain. They have mixed vegetation and their excrement with the soil, increasing its fertility. Their tunnels have served as passageways, carrying water and air to deeper soil layers. Where pocket gophers are numerous, it has been estimated that they completely turn over the surface of the ground at least once in two years. While the earthworm's soil-building activities are better known than those of the gopher, there are large parts of our country where earthworms are absent and pocket gophers are responsible for a deep, fertile topsoil. However, the activities of gophers in forming soil are perhaps less important at present than in the past due to the use of mechanized agricultural machinery to work the soil over large areas. But there still remain considerable areas not suited to agriculture where gophers continue to work.

Many other species utilize pocket gopher burrows and mounds both when occupied and unoccupied by gophers. These include toads, turtles, snakes, salamanders, spotted skunks, weasels, small mammals, and numerous insects.

The economic importance of this species is related to its population density. In those parts of Missouri where pocket gophers are common, no damage is reported. If very numerous, however, they may become pests by eating garden crops, alfalfa, clover, and the roots of fruit trees. Their digging activities may also be undesirable in alfalfa, hay, and grain fields.

Conservation and Management

Except in peak populations or localized areas, pocket gophers need no control in Missouri and should be tolerated and encouraged due to their beneficial soil enhancement activities. Where it is necessary to reduce the population in small areas, contact a wildlife professional for advice, assistance, regulations, or special conditions for handling these animals.

SELECTED REFERENCES

See also discussion of this species in general references, page 23.

Barrington, B. A., Jr. 1942. Description of birth and young of the pocket gopher, *Geomys floridanus*. *Journal of Mammalogy* 23:428–430.

Connier, M. B. 2011. *Geomys bursarius*. *Mammalian Species* 43(879). 14 pp.

Each upper incisor has a lengthwise groove on the front surface. The grinding surfaces of the cheek teeth have rounded cusps.

The only other rodent in Missouri with 20 teeth and grooved upper incisors is the plains pocket gopher. The difference in size easily distinguishes the respective species.

Longevity. The life span in the wild is very short, similar to that of other small rodents. In captivity, these mice may live more than 6 years.

Distribution and Abundance

Plains pocket mice range throughout the Great Plains region of the United States, portions of the southwest, and into northern Mexico. They have only been reported from two locations in northern Missouri: one in Atchison County and one in Clark County. Both locations are just south of the Iowa border where there are prevailing deposits of loess soil that provide limited habitat. Recent surveys have failed to capture any pocket mice at those two locations. This species is a critically imperiled species of conservation concern in Missouri but is secure in its overall range.

Habitat and Home

The preference of these mice for arid to semiarid conditions limits their distribution in Missouri. Throughout their range, they are generally confined to areas of sandy or sandy-loam soil. Some mice captured in Atchison County were at burrow entrances in loess mounds, and in Clark County they were captured in sandy, open habitat at the edge of a soybean field.

The entrance to the burrow opens into a vertical tunnel going into the ground for over 1.4 m (4½ ft.). The doorway is usually plugged with soil from the inside and may be marked on the outside by a mound of soil. Usually there are several openings into the tunnel system, which contains chambers for stored food and a nest.

Habits

Pocket mice are burrowing animals and they come aboveground at night. They are most active when temperatures are warm, becoming lethargic or torpid when cold or wet weather sets in. They are not deep hibernators because they put on little body fat and wake periodically in winter to feed. They are generally solitary. Dusting places are used during the warm part of the year. These mice are quite docile in captivity.

Foods

Pocket mice feed mostly on the seeds of foxtail, other grasses and sedges, and the composite family. Only small bits of green vegetation and insects are taken. Food is stored in special chambers in the burrow. Each cache contains from about 5 to 240 ml (1 teaspoon to 1 cup) of seeds. These are carried to the burrow in the fur-lined pockets, which hold from 0.6 to 2.5 ml (⅛

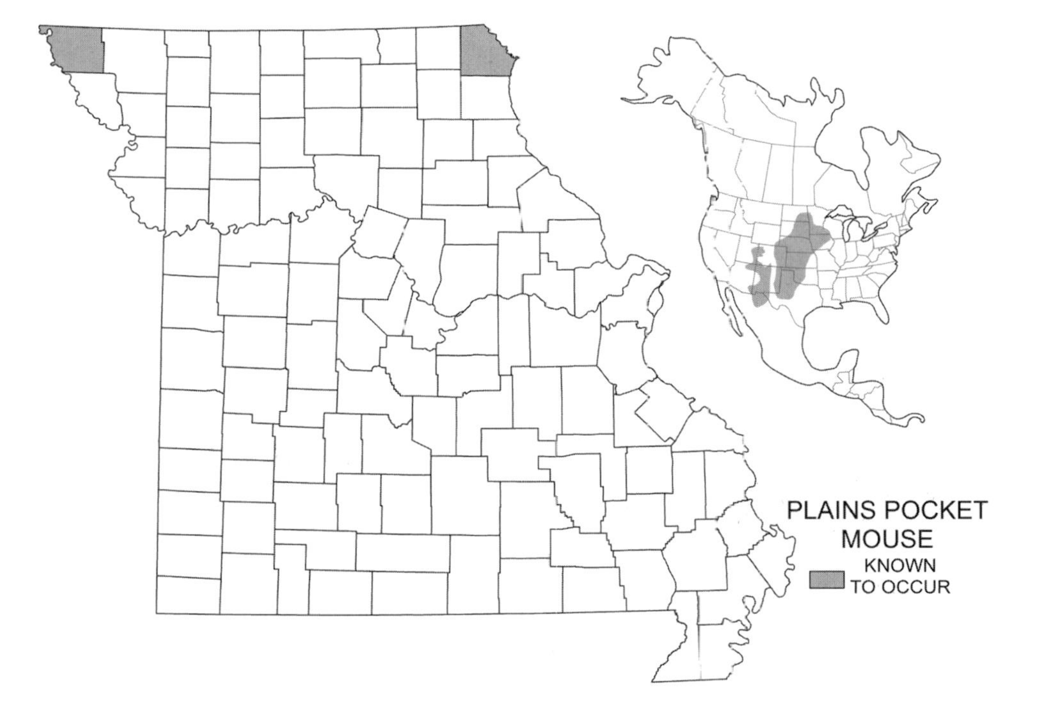

to ½ teaspoon) each. The mouse can turn the pockets inside out to clean them and then retract them by the use of a special muscle. Food is put into the pockets with the front feet.

These mice seldom drink water; their water requirements are satisfied through their food in several ways. Some water is obtained directly from their food source, while other water is manufactured in the mouse's stomach from starch in the seeds it eats. In some species of pocket mice, the excretory system is modified to extract the maximum liquid from food, and their bodies are physiologically adapted to conserve water in other ways, as well.

Reproduction

The breeding season extends from April to September, with June and July showing somewhat reduced activity. From 2 to 7 young are produced per litter, the average being 4. There may be from 1 to 2 litters a year. The gestation period is 3 to 4 weeks.

The young have soft, dull gray fur. This is molted in late summer.

Some Adverse Factors

Many predators feed on pocket mice, including coyotes, foxes, weasels, badgers, skunks, snakes, owls, and hawks. Even some *Peromyscus* spp. mice are known to eat pocket mice.

Importance

When pocket mice are plentiful, they may consume considerable grain, selecting early planted corn or peas and, later, the heads or shattered heads of wheat, barley, and oats. They are an interesting addition to our fauna and should be appreciated. Undoubtedly they contribute to the ecological balance of the arid habitat they live in.

SELECTED REFERENCES

See also discussion of this species in general references, page 23.

Beer, J. R. 1961. Hibernation in *Perognathus flavescens*. *Journal of Mammalogy* 42:103.

Chromanski-Norris, J. F., and E. K. Fritzell. 1983. Status and distribution of ten Missouri mammals. A report to the Missouri Department of Conservation, Jefferson City. 38 pp.

Coyner, B. S., T. E. Lee Jr., D. S. Rogers, and R. A. Van Den Bussche. 2010. Taxonomic status and species limits of *Perognathus* (Rodentia: Heteromyidae) in the southern great plains. *Southwestern Naturalist* 55:1–10.

Easterla, D. A. 1967. First specimens of plains pocket mouse from Missouri. *Journal of Mammalogy* 48:479–480.

Ellis, L. S. 1984. *Small mammal studies in northeast Missouri: Use of railroad rights-of-way and survey of ten sites*. Final report, Northeast Missouri State University, Kirksville. 40 pp.

Geluso, K., and G. D. Wright. 2012. Current status of the plains pocket mouse (*Perognathus flavescens*) in eastern Nebraska. *Western North American Naturalist* 72:554–562.

Monk, R. R., and J. K. Jones Jr. 1996. *Perognathus flavescens*. *Mammalian Species* 525. 4 pp.

Morrison, P., and F. A. Ryser. 1962. Hypothermic behavior in the hispid pocket mouse. *Journal of Mammalogy* 43:529–533.

Otto, H. W., and J. A. White. 2011. Late seasonal captures of the plains pocket mouse (*Perognathus flavescens*) in Iowa. *Prairie Naturalist* 43:124–126.

White, J. A. 2011. Distribution of the plains pocket mouse (*Perognathus flavescens perniger*) in the loess hills. Final report, University of Nebraska, Omaha. 23 pp.

Wilson, G. W., J. B. Bowles, and J. W. Van Zee. 1996. Current status of the plains pocket mouse, *Perognathus flavescens*, in Iowa. *Journal of the Iowa Academy of Science* 103:52–55.

Beavers (Family Castoridae)

The family name, Castoridae, is based on the Greek word, *castor*, meaning "beaver." There are only two living members in this family: the American beaver (*Castor canadensis*) native to North America, and the Eurasian beaver (*Castor fiber*) in Europe and Asia. The American beaver has been introduced into southern South America and some European countries.

Members of this family have a large, horizontally flattened, and scaly tail, and webbed hind feet. They are semiaquatic in habit. There are 4 cheek teeth both above and below on each side of the jaw.

American Beaver (*Castor canadensis*)

Name

The first part of the scientific name, *Castor*, is the Greek word for "beaver." The last part, *canadensis*, is the Latinized form "of Canada," for the country from which the first specimen was taken and named. The common name is from the Anglo-Saxon word *beofor*.

Description (Plate 32)

The American beaver is the largest North American rodent, an adult reaching 1.4 m (4½ ft.) in length. It is easily distinguished by its large size, webbed hind feet, and large, horizontally flattened tail. The beaver has a blunt head, small eyes, small ears nearly concealed in the fur, a short neck, and a stout body. The short legs have 5 clawed toes. The broad tail is furred at the base, but the rest is covered with leathery scales and a few coarse hairs. The pelage consists of very dense, soft underfur overlain by long, coarse outer hairs.

Many specializations adapt the beaver to its unique way of life. The eyes are protected by a nictitating membrane, or "third eyelid," so that vision is only fair above water but good in water. Both ears and nose have valves that close when the beaver submerges. As an aid in gnawing underwater, the lips meet behind the large, prominent incisors and the raised back part of the tongue closes the passageway to the lungs so water entering the mouth does not interfere with breathing. The tail is used as a rudder and propeller when swimming, as a support for the body when cutting a tree, and as a balance when walking. It is often vigorously slapped on the water as a warning to other beavers and as a means of startling an enemy. Physiologically, it plays an important role in the storage of fat and in the regulation of body temperature.

The front feet, in addition to walking, are used for digging, combing the fur, and handling food and construction material. The large hind feet are specialized for swimming, but the webs help support the body when walking on soft, muddy ground. The 3 outer toes on the hind feet have typical claws, but the 2 inner toes possess specialized claws that are used to comb the fur, remove parasites, and distribute oil. The innermost toe has a long, double-edged claw that clamps down over a long, soft lobe, forming a "coarse comb"; the second toe has a similar claw but possesses a horny growth with a sharp, finely cut upper edge between the claw and the soft lobe below, forming a "fine comb." Internally, large lungs hold a supply of air and an enlarged liver stores enough oxygenated blood to permit the beaver to stay underwater for 15 minutes at a time.

Color. The beaver is a uniform dark brown above with somewhat lighter underparts and a blackish tail. The face is paler than the top of the head, but the ears are blackish brown. The sexes are colored alike and show no seasonal variation. Black or white beavers are found rarely. There is only one molt annually, and the pelt is prime in late winter and spring.

Measurements

Total length	863–1,371 mm	34–54 in.
Tail	228–450 mm	9–17¾ in.
Hind foot	152–203 mm	6–8 in.
Ear	31–38 mm	1¼–1½ in.
Skull length	114–139 mm	4½–5½ in.
Skull width	79–101 mm	3⅛–4 in.
Weight	11.8–40.8 kg	26–90 lb.

A record skull from Missouri measured 16 cm (6¼ in.) and the world's record American beaver weighed 52.2 kg (115 lb.). The state furbearer record program was started in 2011.

Teeth and skull. The dental formula of the beaver is:

$$I\,\frac{1}{1}\;C\,\frac{0}{0}\;P\,\frac{1}{1}\;M\,\frac{3}{3} = 20$$

The front surface of the incisors is orange. The skull can be distinguished from that of all other Missouri mammals by its large size and number of teeth.

Sex criteria and sex ratio. The sexes are difficult to identify externally because there is only one common opening, the cloacal opening, for digestive, reproductive, and excretory systems. Specialists can tell the sex of beavers by identifying the openings into the cloaca and by feeling for the penis and testes. The two pairs of teats are visible in females only during late pregnancy and nursing. The lower pair gives more milk than the upper pair.

It is possible to identify the sex of a beaver by microscopically examining the white blood cells.

The color and viscosity of secretions from the anal glands can be used to distinguish the sexes in the field. Male secretions are brown and viscous, while female secretions are whitish to light yellow and runny.

The sex ratio varies. In some populations it is approximately equal, while in other populations males

Plate 32

American Beaver *(Castor canadensis)*

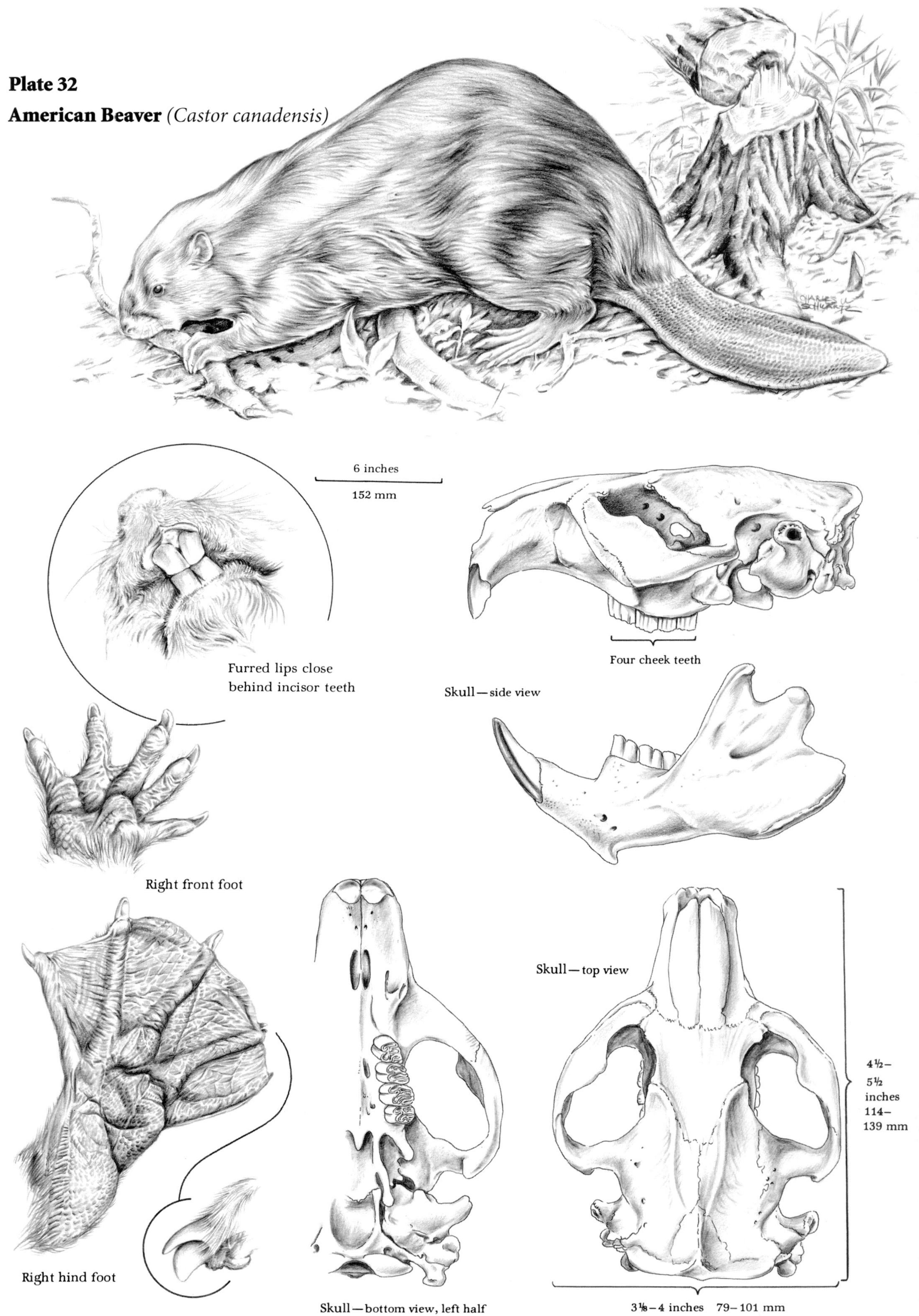

to resting within the den or lodge, while most of the night is spent foraging.

Beavers swim about 3 km (2 mi.) an hour and have been observed to swim continuously underwater for 0.8 km (½ mi.). In winter, they swim under the ice, obtaining oxygen from air holes or from air trapped in pockets between the ice and water. On land, they walk with a slow, shuffling gait or, if hurried, gallop at a moderate pace. The tail is dragged behind or held slightly off the ground and is permitted to sway from side to side. Beavers are very playful both on land and in water. They seldom fight except at mating time.

Foods

The feeding activities of beavers vary with the season. In early spring and fall about half of their food consists of woody vegetation and the other half of nonwoody vegetation. In summer, the amount of woody vegetation decreases to about 10 percent while in winter it increases to nearly 100 percent.

While beavers are primarily bark eaters, they consume mostly the bark of tender twigs and the cambium layer of trees (the new growth between the outer bark and the wood of branches and trunk). Along the Missouri River, beavers feed on cottonwood and willow almost entirely. In the Ozark Highlands they take blue beech, white oak, red and black oaks, alder, hickory, flowering dogwood, elm, sycamore, black gum, red maple, shrub willow, Ozark witch hazel, birch trees, and wild grapevines.

Their nonwoody foods consist of corn, from the milk stage on, whenever it is available; various water plants such as yellow pond lily, water willow, watercress, bur reed, and arrowhead; and small land plants and shrubs.

When foraging, a beaver wanders from place to place and leaves an accumulation of litter behind. In the fall, foods are stored for winter use and may be cached in a deep pool near the underwater entrance to a bank den or lodge, or hidden under the roots of trees growing along the shore. A beaver consumes between 0.7 and 0.9 kg (1½ and 2 lb.) of food daily; in addition, considerable food is wasted.

When feeding, a beaver grasps a branch with its front legs, pulls it to the mouth, and cuts it with the teeth.

Within the lower part of the digestive tract of the beaver, microorganisms aid in the breakdown of woody fibers (cellulose) permitting the beaver to utilize more of the nutrients in its food. In addition, the beaver periodically reingests its own fecal material, thus increasing the absorption of nutrients.

Reproduction

The breeding season starts in January or February. Although the adult male and female in a colony probably pair for life, they may take other mates. The gestation period is about 107 days. The single annual litter is usually born in April, May, or June and consists of 1 to 8, but mostly 3 to 4 young. Larger litters are born under more favorable environmental conditions and to older females.

At birth the young are completely furred and have their eyes open; the incisors are visible. Newborns are around 38 cm (15 in.) long, including a 9 cm (3½ in.) tail, and weigh about 0.5 kg (1 lb.). Although they are able to swim at once, they seldom come out of the den until 1 month old. Then they swim with their mother, who often carries them on her back in the water. The female takes entire care of the kits until this time, but the male soon returns to the family. The young are weaned when they are about 6 weeks old and weigh 2 kg (4 lb.). They live with their parents until approximately 2 years of age, when they apparently leave voluntarily. They become transients for a while, then establish new colonies at the fringe of occupied range or replace lost members of some colony. Although some young may breed when 1 or 2 years old, most beavers breed first in the spring of their third year.

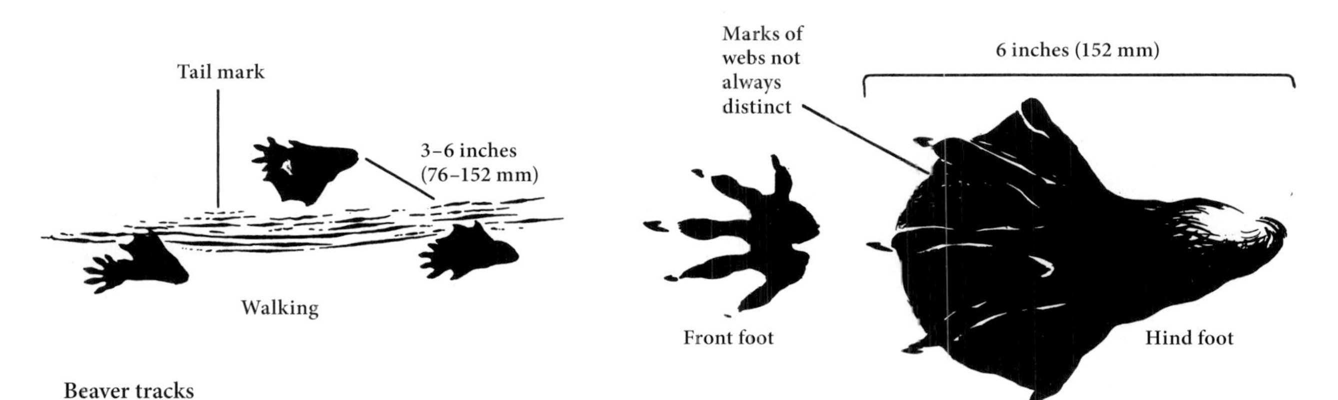

Beaver tracks

Some Adverse Factors

While coyotes, bobcats, river otters, and minks may prey on beavers, humans are their most important predators.

Lice, fleas, beetles, flies, roundworms, and flukes parasitize beavers. In other states, tularemia has caused large die-offs of beavers. Waters contaminated by the bodies of diseased animals are believed responsible for the spread of this disease. Rabies occurs very rarely.

Importance

The search for beaver pelts was one of the inducements for the exploration and settlement of this country. In 1763 Pierre de Laclède and Auguste Chouteau founded a fur-trading post below the convergence of the Missouri and Mississippi Rivers; by 1880 this settlement, St. Louis, was the raw fur center of the world.

In colonial times, the beaver pelt was the basis of value for commodities and other fur. Beaver pelts were manufactured into high hats—a standard of fashion at that time—and into jackets and coats. The hide of the tail was used as ornamental leather for pouches and purses. Currently, beaver pelts are used extensively in the manufacture of coats, jackets, and hats, and as trimming for other fur and cloth coats. Native Americans and early explorers used beaver for food, clothing, and medicine.

The first trapping season for beaver in recent years was in 1953 and during this and the succeeding six years the average number of pelts sold annually in Missouri was 2,753. This average annual harvest increased in the 1960s to 4,417, declined in the 1970s to 3,123, more than doubled in the 1980s to 6,532, declined again in the 1990s to 5,328, and increased to 7,100 in the first decade of the 2000s. During this six-decade period, the highest annual take was 11,715 in 1980 and the lowest was 1,893 in 1953. Between 1953 and 1959, the average annual price paid per pelt was $4.30. The average annual price was $5.94 in the 1960s, $7.28 in the 1970s, $10.36 in the 1980s, $10.06 in the 1990s and $11.53 in the first decade of the 2000s. The highest average annual price per pelt was $18.10 in 2006 and the lowest was $3.00 in 1956. It is hard to measure the relative effect of the several factors that influence harvest numbers and pelt prices. Obviously, population abundance is involved, but to what extent has not been determined. Changes in demand for furs and the fluctuating values of their pelts influence economic reasons for trapping beaver. In addition, the numbers of trappers working in Missouri influence the harvest.

Castoreum was formerly an ingredient in patent medicines but is used now as a fixative in high-quality perfumes and as a bait for traps because many animals are attracted to the scent.

Beavers are known as "ecosystem engineers" because their relationship to the land and wildlife is intricate and has profound influences. For centuries beaver dams have backed up silt-laden waters and subsequently formed many of the fertile valley floors in the wooded areas of northern North America. Their dams have also stabilized stream flow and slowed down runoff. Beaver dams create ponds, which change water temperatures and other conditions for fish and many kinds of aquatic life. Muskrats, raccoons, minks, waterfowl, reptiles, amphibians, and other mammals and birds are also affected by beaver activities and, in most cases, benefited by them.

In Missouri, the damage caused by beavers can be sufficient to arouse complaints. Beavers take some corn; dam some drainage ditches and small streams; burrow into some levees; damage boat docks; cause flooding due to their dam-building activities; and may destroy trees around homes, businesses, and other areas. If you are experiencing problems with beavers, contact a wildlife professional for advice, assistance, regulations, or special conditions for handling these animals.

Conservation and Management

Past management for beavers in Missouri has consisted of giving them complete protection supplemented by a program of live trapping and transplanting of surplus animals. This restoration program was so successful that beaver are now common across the state. Their successful restoration to rivers and streams has allowed for the return of a vital process of wetland patch creation as a result of their impoundments. Beaver ponds are important for facilitating diverse aquatic communities of plants, birds (including waterfowl), mammals, invertebrates, amphibians, reptiles, and fish. Today, the most important management measure is to regulate the harvest.

SELECTED REFERENCES

See also discussion of this species in general references, page 23.

Aleksuik, M. 1968. Scent-mound communication, territoriality, and population regulation in beaver (*Castor canadensis* Kuhl). *Journal of Mammalogy* 49:759–762.

Atwood, E. L., Jr. 1938. Some observations on adaptability of Michigan beaver released in Missouri. *Journal of Wildlife Management* 2:165–166.
Bloomquist, C. K., C. K. Nielsen, and J. J. Shrew. 2012. Spatial organization of unexploited beavers (*Castor canadensis*) in southern Illinois. *American Midland Naturalist* 167:188–197.
Bradt, G. W. 1938. A study of beaver colonies in Michigan. *Journal of Mammalogy* 19:139–162.
———. 1939. Breeding habits of beaver. *Journal of Mammalogy* 20 486–488.
———. 1947. *Michigan beaver management.* Michigan Department of Conservation, Lansing. 56 pp.
Brenner, F. J. 1964. Reproduction of the beaver in Crawford County, Pennsylvania. *Journal of Wildlife Management* 28:743–747.
Dalke, P. D. 1947. The beaver in Missouri. *Missouri Conservationist* 8:1–3.
Edwards, N. T., and D. L. Otis. 1999. Avian communities and habitat relationships in South Carolina Piedmont beaver ponds. *American Midland Naturalist* 141:158–171.
Hoover, W. H., and S. D. Clarke. 1972. Fiber digestion of the beaver. *Journal of Nutrition* 102:9–16.
Jenkins, S. H., and P. E. Busher. 1979. *Castor canadensis. Mammalian Species* 120. 8 pp.
Larson, J. S., and F. C. Van Nostrand. 1968. An evaluation of beaver aging techniques. *Journal of Wildlife Management* 32:99–103.
Leege, T. A. 1968. Natural movements of beavers in southeastern Idaho. *Journal of Wildlife Management* 32:973–976.
McNew, L. B., Jr., and A. Woolf. 2005. Dispersal and survival of juvenile beavers (*Castor canadensis*) in southern Illinois. *American Midland Naturalist* 154:217–228.
Payne, N. F. 1979. Relationship of pelt size, weight, and age for beaver. *Journal of Wildlife Management* 43:804–806.
Rosell, F., O. Bozsér, P. Collen, and H. Parker. 2005. Ecological impact of beavers *Castor fibre* and *Castor canadensis* and their ability to modify ecosystems. *Mammal Review* 35:248–276.
Russell, K. R., C. E. Moorman, J. K. Edwards, B. S. Metts, and D. C. Coynn Jr. 1999. Amphibian and reptile communities associated with beaver (*Castor canadensis*) ponds and unimpounded streams in the Piedmont of South Carolina. *Journal of Freshwater Ecology* 14:149–158.
Schulte, B. A., D. Müller-Schwarze, and L. Sun. 1995. Using anal gland secretion to determine sex in beaver. *Journal of Wildlife Management* 59:614–618.
Svendsen, G. E. 1978. Castor and anal glands of the beaver (*Castor canadensis*). *Journal of Mammalogy* 59:618–620.
———. 1980. Seasonal change in feeding patterns of beaver in southeastern Ohio. *Journal of Wildlife Management* 44:285–290.
Tevis, L., Jr. 1950. Summer behavior of a family of beavers in New York state. *Journal of Mammalogy* 31:40–65.
van Nostrand, F. C., and A. B. Stephenson. 1964. Age determination for beavers by tooth development. *Journal of Wildlife Management* 28:430–434.

The food pellets of bones and hair cast up by hawks and owls indicate the different kinds of small mammals, particularly rodents, living in an area

New World Rats and Mice, Voles, and Relatives (Family Cricetidae)

Cricetidae is a family of rodents that includes almost 700 species of New World rats and mice, true hamsters, voles, muskrats, and lemmings. It is the largest family of rodents and the second largest family of mammals, and members inhabit a wide range of habitats throughout North and South America, Europe, and most of Asia. In Missouri, this family contains 16 species of small- to medium-sized rodents.

Marsh Rice Rat (*Oryzomys palustris*)

Name

The first part of the scientific name, *Oryzomys*, is from two Greek words and means "rice mouse" (*oryza*, "rice," and *mys*, "mouse"). This name refers to the rice-feeding habits of this small rodent. The last part, *palustris*, is the Latin word for "marshy," indicating the preferred habitat of this species. The common name also refers to the habitat and food of this species.

Description (Plate 33)

The marsh rice rat is a small, slender rodent with a scaly, sparsely haired, and slender tail about as long as the head and body combined. The eyes are moderately

Plate 33
Marsh Rice Rat
(Oryzomys palustris)

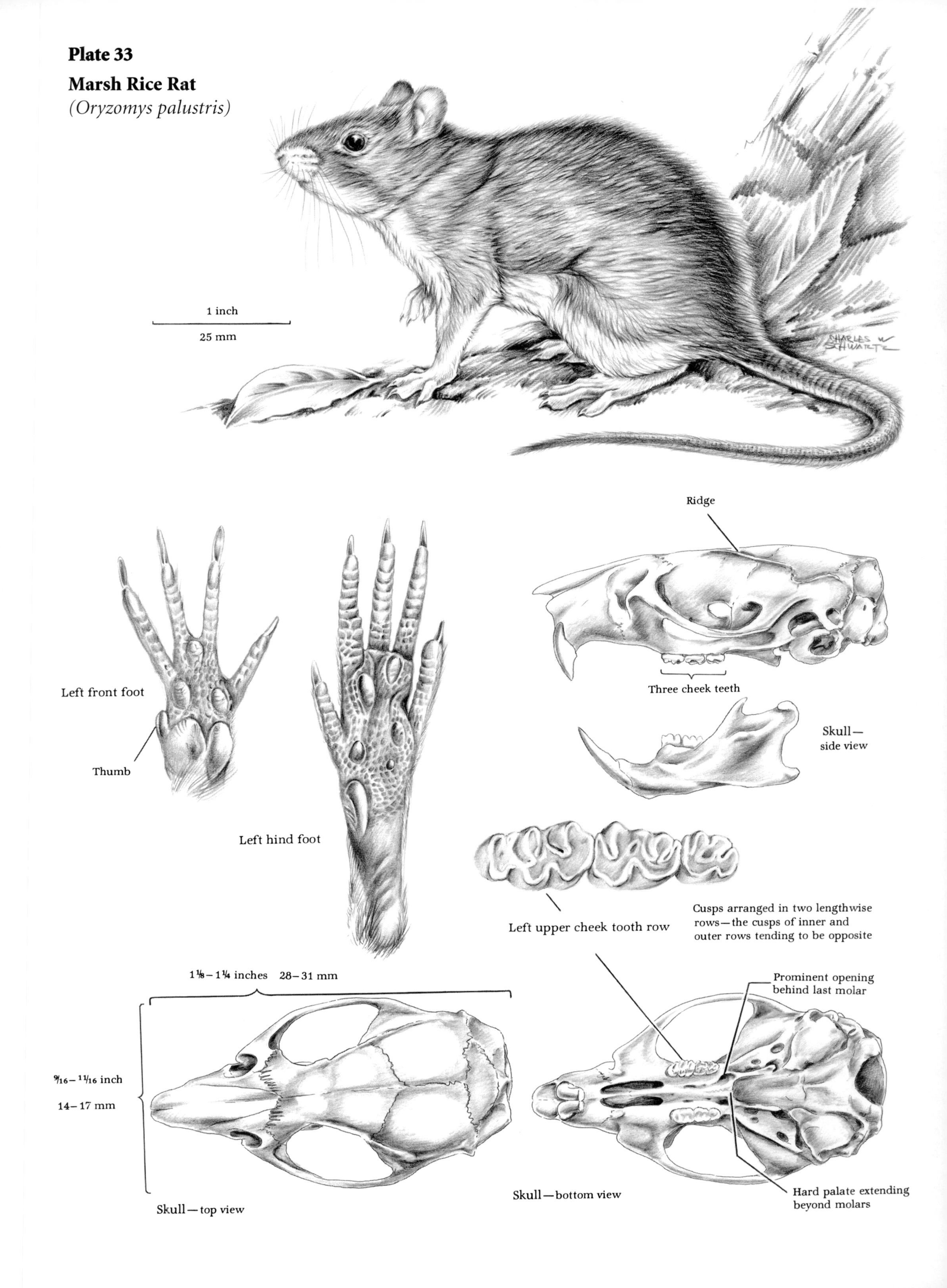

parasites as do other small rodents. The disease typhus has been found in this species in the southern states, and they are known to carry hantavirus (see discussion under the deer mouse).

Importance

In rice fields, pastures, or croplands, an abundant population of rice rats can be economically destructive. Because of their large numbers, they are an important source of food for many predators.

Conservation and Management

Maintenance and management of wetland areas with consideration of connectivity and adjacent grasslands should benefit marsh rice rats. If you are experiencing problems with rice rats, contact a wildlife professional for advice, assistance, regulations, or special conditions for handling these animals.

SELECTED REFERENCES

See also discussion of this species in general references, page 23.

Abuzeineh, A. A., R. D. Owen, N. E. McIntyre, C. W. Dick, R. E. Strauss, and T. Holsomback. 2007. Response of marsh rice rat (*Oryzomys palustris*) to inundation of habitat. *Southwestern Naturalist* 52:75–78.

Conaway, C. H. 1954. The reproductive cycle of rice rats (*Oryzomys palustris palustris*) in captivity. *Journal of Mammalogy* 35:263–266.

Esher, R. J., J. L. Wolfe, and J. N. Layne. 1978. Swimming behavior of rice rats (*Oryzomys palustris*) and cotton rats (*Sigmodon hispidus*). *Journal of Mammalogy* 59:551–558.

Eubanks, B. W., E. C. Hellgren, J. R. Nawrot, and R. D. Bluett. 2011. Habitat associations of the marsh rice rat (*Oryzomys palustris*) in freshwater wetlands of southern Illinois. *Journal of Mammalogy* 92:552–560.

Hamilton, W. J., Jr. 1946. Habits of the swamp rice rat, *Oryzomys palustris palustris* (Harlan). *American Midland Naturalist* 36:730–736.

Kruchek, B. L. 2004. Use of tidal marsh and upland habitats by the marsh rice rat (*Oryzomys palustris*). *Journal of Mammalogy* 85:569–575.

Negus, N. C., E. Gould, and R. K. Chapman. 1961. Ecology of the rice rat, *Oryzomys palustris* (Harlan) on Breton Island, Gulf of Mexico, with a critique on the social stress theory. *Tulane Studies of Zoology* 8:93–123.

Sharp, J. H. F. 1967. Food ecology of the rice rat, *Oryzomys palustris* (Harlan), in a Georgia salt marsh. *Journal of Mammalogy* 48:557–563.

Svihla, A. 1931. Life history of the Texas rice rat (*Oryzomys palustris texensis*). *Journal of Mammalogy* 12:238–242.

Wolfe, J. L. 1982. *Oryzomys palustris. Mammalian Species* 176. 5 pp.

Worth, C. B. 1950. Observations on the behavior and breeding of captive rice rats and woodrats. *Journal of Mammalogy* 31:421–426.

Plains Harvest Mouse
(*Reithrodontomys montanus*)

Name

The first part of the scientific name, *Reithrodontomys*, is the same as that of its close relative, the western harvest mouse. The last part, *montanus*, comes from the Latin word *mons* and means "belonging to the mountain"; this is, at best, an ambiguous term for this plains-dwelling mammal. The common name refers to the primary habitat of this species.

Description

The plains harvest mouse is so similar to the western harvest mouse that these two species cannot be distinguished easily. In general, the plains harvest mouse is smaller and has a shorter tail (slightly less than the length of head and body combined). Where these two species occur together, the plains harvest mouse tends to have a stripe down the middle of the back (lacking in the western harvest mouse), a narrower blackish stripe on the top of the tail (wider in the western harvest mouse), and a whiter undersurface (grayer in the western harvest mouse). Doubtful specimens should be submitted to an authority for comparison with museum specimens.

Measurements

Total length	101–139 mm	4–5½ in.
Tail	44–63 mm	1¾–2½ in.

Hind foot	12–19 mm	½–¾ in.
Ear	9–15 mm	⅜–⅝ in.
Skull length	19 mm	¾ in.
Skull width	9 mm	⅜ in.
Weight	5–14 g	⅕–½ oz.

Distribution and Abundance

The plains harvest mouse ranges in the dry upland areas of the central United States and northern Mexico. It has been found at only five locations in four southwestern Missouri counties. It is listed as an imperiled species of conservation concern in Missouri but is considered secure across its global range.

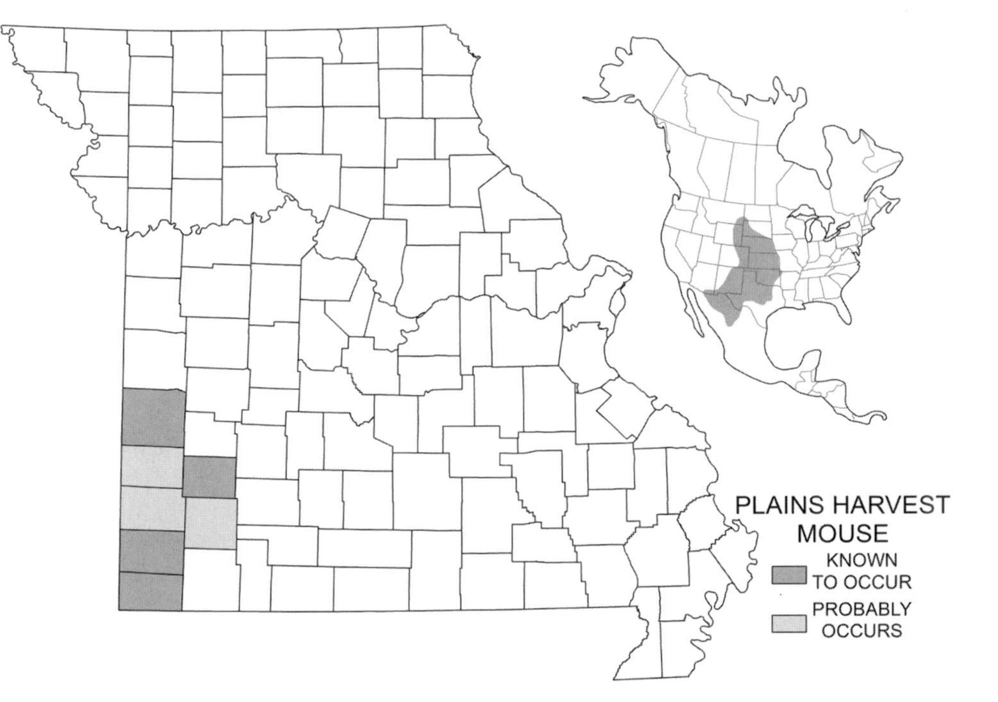

Life History

The plains harvest mouse's way of life is similar to that of the western harvest mouse.

SELECTED REFERENCES

See also discussion of this species in general references, page 23.

Chromanski-Norris, J. F., and E. K. Fritzell. 1983. Status and distribution of ten Missouri mammals. A report to the Missouri Department of Conservation, Jefferson City. 38 pp.

Long, C. A. 1961. *Reithrodontomys montanus griseus* in Missouri. *Journal of Mammalogy* 42:417–418.

Wilkins, K. T. 1986. *Reithrodontomys montanus. Mammalian Species* 257. 5 pp.

Western Harvest Mouse
(*Reithrodontomys megalotis*)

Name

The first part of the scientific name, *Reithrodontomys*, is from three Greek words and means "groove-toothed mouse" (*reithron*, "groove," *odous*, "tooth," and *mys*, "mouse"). This refers to the characteristic groove in each upper incisor. The last part, *megalotis*, is from two Greek words and means "large ear" (*megas*, "large," and *otis*, "ear"). This indicates a predominant feature of this mouse.

The common name originated as follows: "western" refers to the range in the western United States; "harvest" possibly comes from the observation of these mice during harvests, particularly of hay; and "mouse" describes the typical mouse shape.

Description (Plate 34)

The western harvest mouse is a very small rodent with large eyes, large ears, and a moderately furred and scaly tail about as long as the head and body combined. There are 4 clawed toes and a small thumb on each front foot and 5 clawed toes on each hind foot. The soles of the hind feet are slightly furred to the 6 pads, or tubercles. There is a deep, lengthwise groove in the middle of each upper incisor. The body fur is soft and fairly long.

The western harvest mouse can be distinguished from all other small mice in Missouri as follows: from the fulvous harvest mouse by the absence of reddish

Plate 34

Western Harvest Mouse

(Reithrodontomys megalotis)

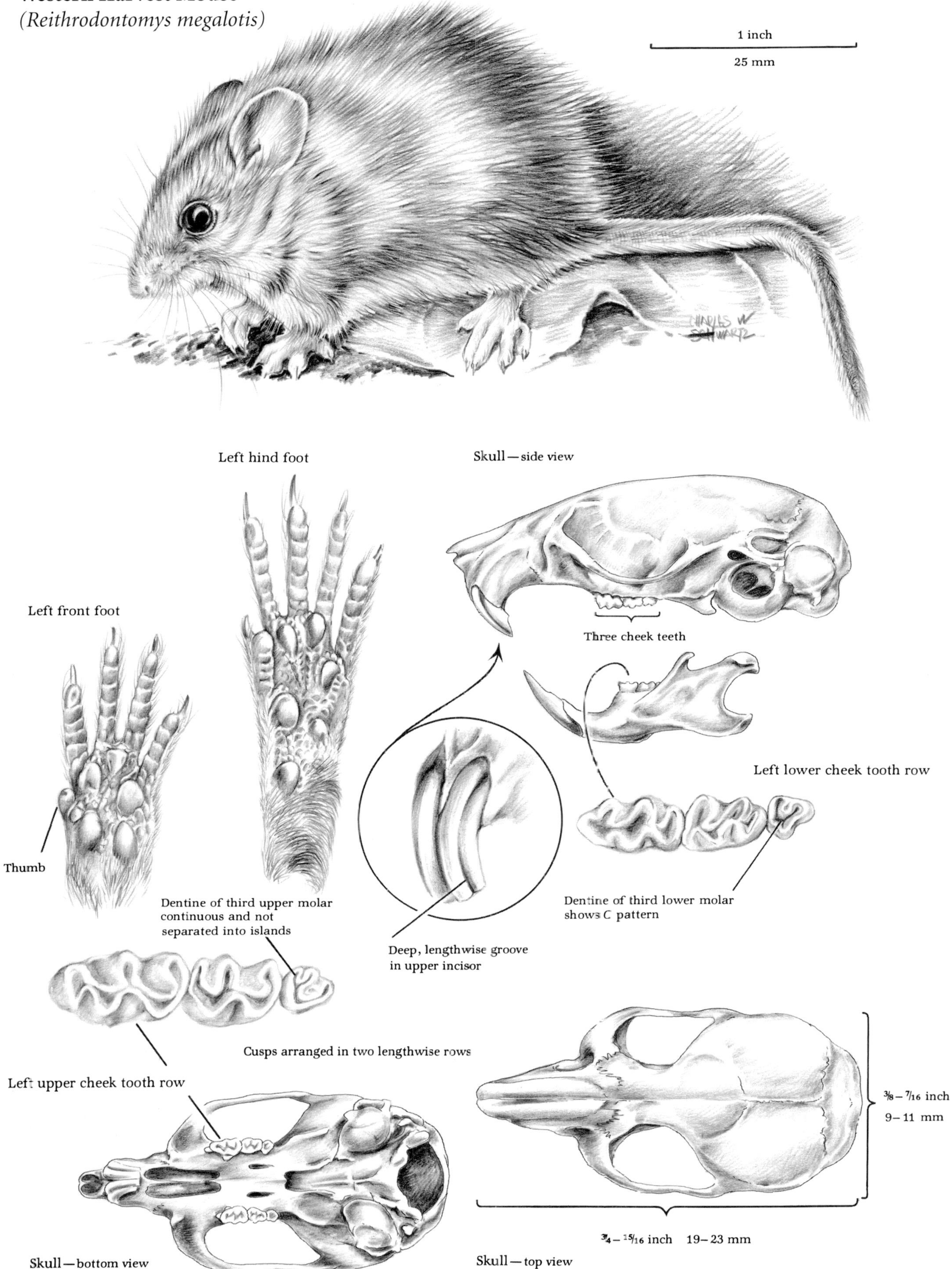

yellow (on the sides, tops of the hind feet, and upper surface of the tail), the proportionately shorter tail, and usually smaller size; from the house mouse, which it most closely resembles in appearance, and from the various *Peromyscus* mice and the golden mouse by the grooved upper incisors; from the meadow jumping mouse by the smaller hind feet, shorter tail, and 16 teeth; and from the plains pocket mouse by the absence of external, fur-lined cheek pouches, smaller ears, and differences in coloration.

Color. The upperparts are predominantly brown with numerous black-tipped hairs. The sides are grayish tan and the belly and feet are white. The tail is colored like the back above and like the belly below. The sexes are similar in color. The annual molt occurs in fall.

Measurements

Total length	101–155 mm	4–6⅛ in.
Tail	50–82 mm	2–3¼ in.
Hind foot	12–19 mm	½–¾ in.
Ear	9–15 mm	⅜–⅝ in.
Skull length	19–23 mm	¾–15/16 in.
Skull width	9–11 mm	⅜–7/16 in.
Weight	9–14 g	⅓–½ oz.

Teeth and skull. The dental formula of the western harvest mouse is:

$$I\,\frac{1}{1}\;C\,\frac{0}{0}\;P\,\frac{0}{0}\;M\,\frac{3}{3} = 16$$

Each upper incisor has a deep, lengthwise groove in the middle of the front surface. The grinding surfaces of the upper cheek teeth consist of small, rounded cusps capped with enamel that are arranged in 2 lengthwise rows. The arrangement of cusps and the presence of grooves in the upper incisors distinguish the skull of the western harvest mouse from all other Missouri rodents with 16 teeth except the fulvous harvest mouse. The skulls of these two species of harvest mice are distinguished by differences in the pattern of dentine on the grinding surfaces of the upper and lower third molar teeth. The dentine of the third upper molar is continuous and not separated into islands in the western harvest mouse, and the dentine of the third lower molar shows a C pattern.

Sex criteria. The sexes are identified as in the eastern woodrat. There are 3 pairs of teats on the belly: 1 pair near the front legs and 2 pairs in the groin region. Males tend to outnumber females except among the youngest and oldest age classes.

Age criteria and longevity. The young are identified by their duller fur and blackish to grayish backs. The amount of wear on the teeth has been used to place harvest mice in age classes. Harvest mice probably live little more than a year.

Voice and sounds. These mice have very high voices. The male "sings" at a pitch so high it is almost inaudible to human ears. The "song" has a ventriloquial quality.

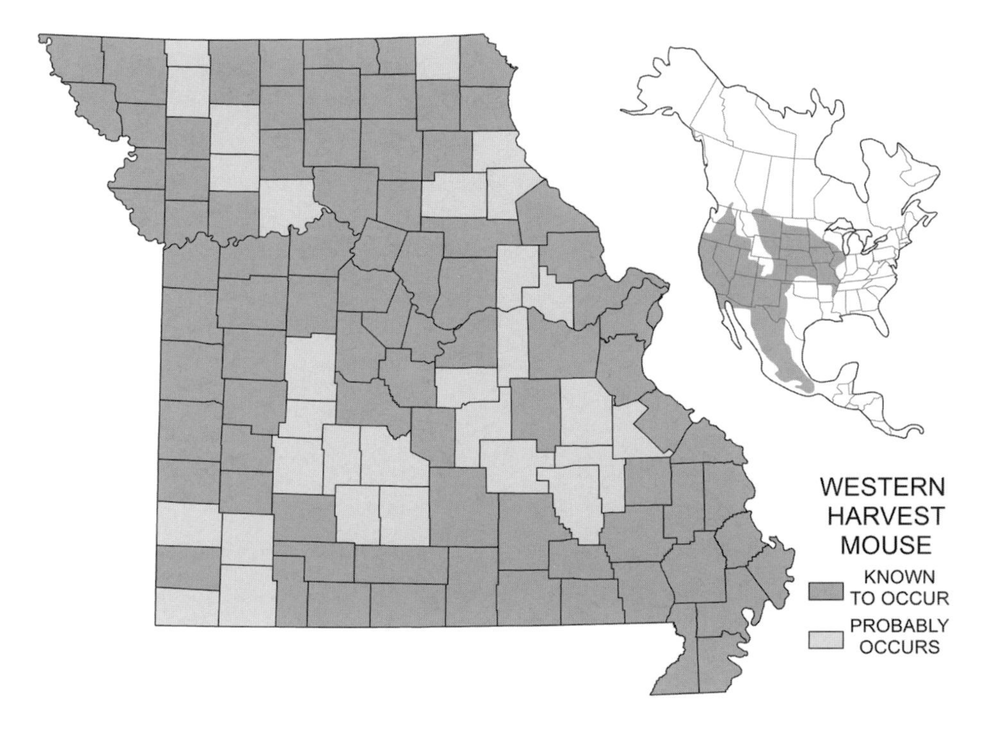

Distribution and Abundance

The western harvest mouse ranges from extreme southwestern Canada, throughout most of the western and midwestern United States, and southward into Mexico. This species probably occurs throughout Missouri. Although this mouse is not commonly taken by collectors, in places it is very abundant.

Habitat and Home

The western harvest mouse lives in abandoned fields, prairies, meadows, fencerows, weedy roadsides, and marsh borders with dense ground cover such as matted grasses, broomsedge, weeds, or briers. They were found to prefer open grasslands over areas with abundant woody cover in northeastern Kansas. It is most abundant in sites near water.

The nest, which is used all year, is usually located in matted grass or weeds or under bushes. Less often is it a few centimeters to meters (few inches to several feet) above the ground in a shrub, clump of weeds, or taller grass. The nest is woven of shredded grass and plant fibers, patted with the front feet and jerked with the teeth into a sphere about 8 cm (3 in.) in diameter. There is only one opening, about 1.3 cm (½ in.) in diameter, leading into a small cavity lined with cattail or thistledown, milkweed floss, or soft fine grass. Rarely is a deserted bird's nest or woodpecker's hole used for the home or is the nest built in a haystack or underground tunnel. Nests of all three species of harvest mice have been found inside aluminum drink cans discarded in roadside ditches.

Western harvest mouse nest

Habits

Harvest mice are active both day and night but show a preference for feeding from just after sunset to about midnight. They do not hibernate. In general, harvest mice probably live in a home range with a diameter of 76 m (250 ft.). The home range of two marked adult females varied from 0.2 to more than 0.6 ha (½ to more than 1½ ac.) in summer. Males probably have a larger range throughout life than females.

In a harvest mouse social unit, one male dominates other males and females with nests. Females are not antagonistic. The aggression between males is related to the breeding season and is lost during the winter months when mice of both sexes sleep together.

These mice climb grass stalks and shrubs in their search for food. They often use runways of other kinds of mice but make none of their own.

Harvest mice are fastidious and spend considerable time cleaning their faces and fur. A male and female usually live together in one nest.

Foods

The seeds of many grasses and legumes form the important foods of harvest mice. Occasionally new shoots and leaves, flowers, fruits, and insects, particularly larvae of moths and butterflies, are included in the diet. The seeds are usually eaten where they are found, but some may be carried to the nest where they are eaten or stored. Surface water is taken when available, but moisture is also obtained from the natural juices of the vegetation that they eat. In captivity one harvest mouse consumed one-third of its weight in food each day.

Reproduction

Breeding may occur all year but generally takes place from spring through fall. The gestation period is 23 to 24 days but may be up to 32. Up to 7 litters are born annually; they contain mostly 3 to 5 but sometimes 1 to 7 young. Two female wild-caught western harvest mice each produced 14 litters in one year in the laboratory, totaling 57 and 58 young respectively.

At birth the helpless young weigh about 1.1 g (1/25 oz.) and measure slightly more than 6.4 mm (¼ in.) in length. The eyes and ears are closed and the skin is naked and pink. By the end of the first week, the young are sparsely furred on both back and belly. Between 8 and 10 days after birth, the eyes and ears open and the incisors cut through the gums. The young begin to walk at this time and when 2 weeks old are able to run and jump. They are soon weaned; when 3 to 4 weeks old, they leave the nest to live alone. The young reach adult size and weight when 5 weeks old. Some breed as early as 2 months of age, but most probably mate first when 3 to 4 months old.

Some Adverse Factors

Hawks, owls, snakes, and carnivorous mammals are predators of harvest mice. Presumably the usual types of external and internal parasites of other mice occur on this species. Immature stages of dog ticks have been found on western harvest mice, but the risk of acquiring disease from these mice is deemed unlikely.

Importance

Harvest mice feed primarily on seeds of grasses and weeds and convert the nutrients from these plants into flesh. The bodies of harvest mice serve as food for carnivorous birds and mammals, supplying them with nutrients they could not obtain directly from plants. By changing plants into food for carnivorous animals, rodents play a very valuable role in the wildlife community.

Conservation and Management

Harvest mice live and feed largely in waste areas and are of little economic importance. There is no need for control unless they become unusually abundant, in which case a healthy predator population should be able keep the population in balance. There is typically a large variation in the population dynamics of this species both within and between years.

SELECTED REFERENCES

See also discussion of this species in general references, page 23.

Bancroft, W. L. 1967. Record fecundity for *Reithrodontomys megalotis. Journal of Mammalogy* 48:306–308.

Fisler, G. F. 1965. Adaptation and speciation in harvest mice in the marshes of San Francisco Bay. *University of California Publications in Zoology* 77:1–108.

———. 1966. Homing in the western harvest mouse, *Reithrodontomys megalotis. Journal of Mammalogy* 47:53–58.

———. 1971. Age structure and sex ratio in populations of *Reithrodontomys. Journal of Mammalogy* 52:653–662.

Hooper, E. T. 1952. A systematic review of the harvest mice (genus *Reithrodontomys*) of Latin America. *University of Michigan, Museum of Zoology, Miscellaneous Publication* 77:1–255.

Kaye, S. V. 1961. Movements of harvest mice tagged with gold-198. *Journal of Mammalogy* 42:323–337.

Matlack, R. S., D. W. Kaufman, and G. A. Kaufman. 2008. Influence of woody vegetation on small mammals in tallgrass prairie. *American Midland Naturalist* 160:7–19.

Meserve, P. L. 1977. Three-dimensional home ranges of cricetid rodents. *Journal of Mammalogy* 58:549–558.

Pearson, O. P. 1960. Habits of harvest mice revealed by automatic photographic recorders. *Journal of Mammalogy* 41:58–74.

Pitts, R. M., J. R. Choate, H. W. Garner, and R. D. Kagy Jr. 1994. Unusual nesting behavior of harvest mice. *Prairie Naturalist* 26:311.

Skupski, M. P. 1995. Population ecology of the western harvest mouse, *Reithrodontomys megalotis*: A long-term perspective. *Journal of Mammalogy* 76:358–367.

Smith, C. F. 1936. Notes on the habits of the long-tailed harvest mouse. *Journal of Mammalogy* 17:274–278.

Storm, J. J., and C. M. Ritzi. 2008. Ectoparasites of small mammals in western Iowa. *Northeastern Naturalist* 15:283–292.

Webster, W. D., and J. K. Jones Jr. 1982. *Reithrodontomys megalotis. Mammalian Species* 167. 5 pp.

Whitaker, J. O., Jr., and R. E. Mumford. 1972. Ecological studies on *Reithrodontomys megalotis* in Indiana. *Journal of Mammalogy* 53:850–860.

Fulvous Harvest Mouse
(*Reithrodontomys fulvescens*)

Name

The first part of the scientific name, *Reithrodontomys*, is the same as that of its close relative, the western harvest mouse. The last part, *fulvescens*, is from the Latin word *fulvus* and means "reddish yellow"; this refers to the color on the sides of the body. The common name also indicates this same color.

Description (Plate 35)

The fulvous harvest mouse closely resembles the western harvest mouse and the plains harvest mouse but is distinguished by its color, the proportionately longer tail, which is longer than head and body combined, and its usually larger size.

Color. The coloration is similar to that of the other harvest mice, but the face, the sides of the body, tops of the hind feet, and upper surface of the tail are more reddish yellow.

Measurements

Total length	133–203 mm	5¼–8 in.
Tail	73–117 mm	2⅞–4⅝ in.
Hind foot	15–22 mm	⅝–⅞ in.
Ear	12–19 mm	½–¾ in.
Skull length	22 mm	⅞ in.
Skull width	11 mm	7/16 in.
Weight	9–20 g	⅓–¾ oz.

Teeth and skull. The skull is very similar to that of the western harvest mouse and the plains harvest mouse, but a noticeable difference occurs in the pattern of dentine on the grinding surfaces of the upper and lower third molar teeth. In the fulvous harvest mouse the dentine of the third upper molar is separated into islands, and that of the third lower molar shows an S pattern.

Distribution and Abundance

The fulvous harvest mouse occurs in the south-central and extreme southwestern United States, and southward throughout Mexico and into Central America. In Missouri it has been found in the southern portion of the state, mostly in southwestern Missouri.

Habitat, Home, and Life History

The fulvous harvest mouse is very similar to the western harvest mouse in its way of life.

SELECTED REFERENCES

See also discussion of this species in general references, page 23.

Hooper, E. T. 1952. A systematic review of the harvest mice (genus *Reithrodontomys*) of Latin America. *University of Michigan, Museum of Zoology, Miscellaneous Publication* 77:1–255.

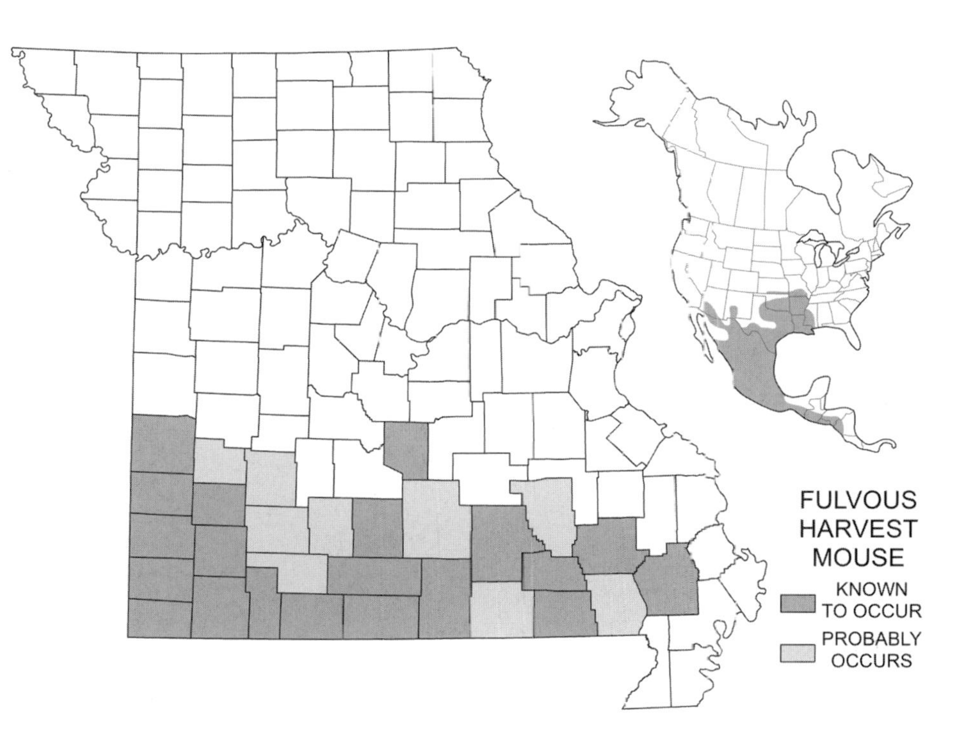

Plate 35

Fulvous Harvest Mouse

(Reithrodontomys fulvescens)

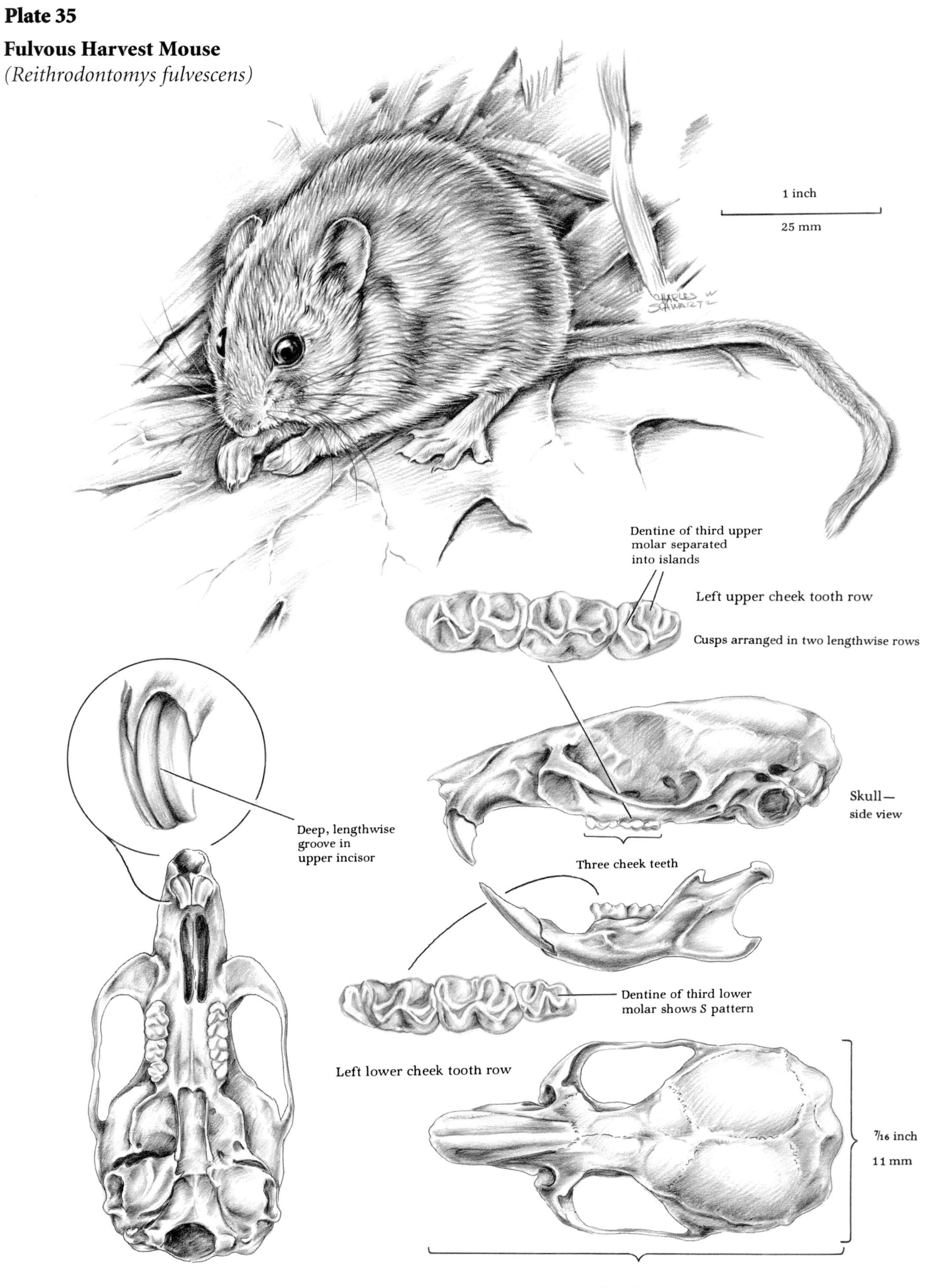

Long, C. A. 1965. Fulvous harvest mouse in Missouri. *Journal of Mammalogy* 46:506.
Spencer, S. R., and G. N. Cameron. 1982. *Reithrodontomys fulvescens. Mammalian Species* 174. 7 pp.

Peromyscus Mice

There are four species of closely related mice that are similar in appearance and habits. These are the deer mouse, *Peromyscus maniculatus*, white-footed mouse, *Peromyscus leucopus*, cotton mouse, *Peromyscus gossypinus*, and Texas mouse, *Peromyscus attwateri*. In areas where these species occur together they are known to be of intermediate size and very difficult to identify to species. For accurate species identification, specimens should be examined by a person familiar with these species.

Deer Mouse (*Peromyscus maniculatus*)

Name

The first part of the scientific name, *Peromyscus*, is from two Greek words and means "pouched little mouse" (*pera*, "pouch," and *myskos*, "little mouse"). This refers to the internal cheek pouches of this small rodent. The last part, *maniculatus*, is the Latin word for "small handed" and indicates the size of the front feet. The common name, "deer," refers to the similarity of color with the white-tailed deer (generally brownish back and sides with whitish underparts), and "mouse" describes the typical mouse shape. It is also known as the North American deermouse.

Because this species has such a wide distribution and variation in the structure, behavior, and habitat preferences of its members, the following discussion emphasizes the subspecies, *Peromyscus maniculatus bairdii*, that occurs in Missouri.

Description (Plate 36)

The deer mouse is a small rodent with large, protruding, black eyes; large, scantily furred ears; long, coarse whiskers; and a moderately to well-furred tail from one-third to less than one-half of the animal's total length with a slight tuft at the tip. There are 4 clawed toes and an inconspicuous nailed thumb on each front foot, and 5 clawed toes on each hind foot. The soles of the hind feet are thinly furred from the heel to the 6 pads, or tubercles. Small internal cheek pouches are present. The body fur is long and soft.

The species of *Peromyscus* in Missouri are difficult to distinguish but are best told by differences in their size and characters of the tail (see accounts of the other *Peromyscus*).

Missouri mice, other than members of the genus *Peromyscus*, that are similar in size and general appearance to the deer mouse are the golden mouse, the harvest mice, the house mouse, the meadow jumping mouse, and the plains pocket mouse. The adult deer mouse is readily told from the golden mouse by the color and by the smaller median pad on the outside of the sole on the hind foot; from the harvest mice by the absence of grooved upper incisors; from the adult house mouse by the sharp contrast in color between back and belly and between upper and lower surfaces of the moderately to well-furred tail, and by the large protruding eyes; from the meadow jumping mouse by the absence of grooved upper incisors, the shorter tail, and the smaller hind feet; and from the plains pocket mouse by the absence of external, fur-lined cheek pouches, smaller ears, differences in coloration, and the absence of grooved upper incisors.

Color. There is considerable color variation in individual deer mice, but in general the back and sides of the adults vary from grayish to reddish brown with or without a darker area in the middle of the back. This color is sharply marked off from the lower face and underparts, which are white or sometimes grayish. The base of the hairs on both back and belly is dark gray. The feet are white. The tail is dark like the back above and sharply contrasted to light like the belly below. The ears are dark brown for approximately the outer half with a very slight grayish to whitish margin but are whitish to pinkish for the inner half. The sexes are colored alike. Adults molt in late summer or early fall. Only a few hairs are replaced at a time, beginning at the head and working toward the tail.

Measurements

Total length	111–203 mm	4⅜–8 in.
Commonly	152 mm or less	6 in. or less
Tail	41–98 mm	1⅝–3⅞ in.
Commonly	63 mm or less	2½ in. or less
Hind foot	15–25 mm	⅝–1 in.
Commonly	20 mm or less	13⁄16 in. or less
Ear	12–22 mm	½–⅞ in.
Commonly	15 mm or less	⅝ in. or less
Skull length	22–25 mm	⅞–1 in.
Skull width	12 mm	½ in.
Weight	9–32 g	⅓–1 oz.

Teeth and skull. The dental formula of the deer mouse is:

$$I\ \frac{1}{1}\ C\ \frac{0}{0}\ P\ \frac{0}{0}\ M\ \frac{3}{3} = 16$$

The grinding surfaces of the upper cheek teeth consist of small, rounded cusps capped with enamel. The

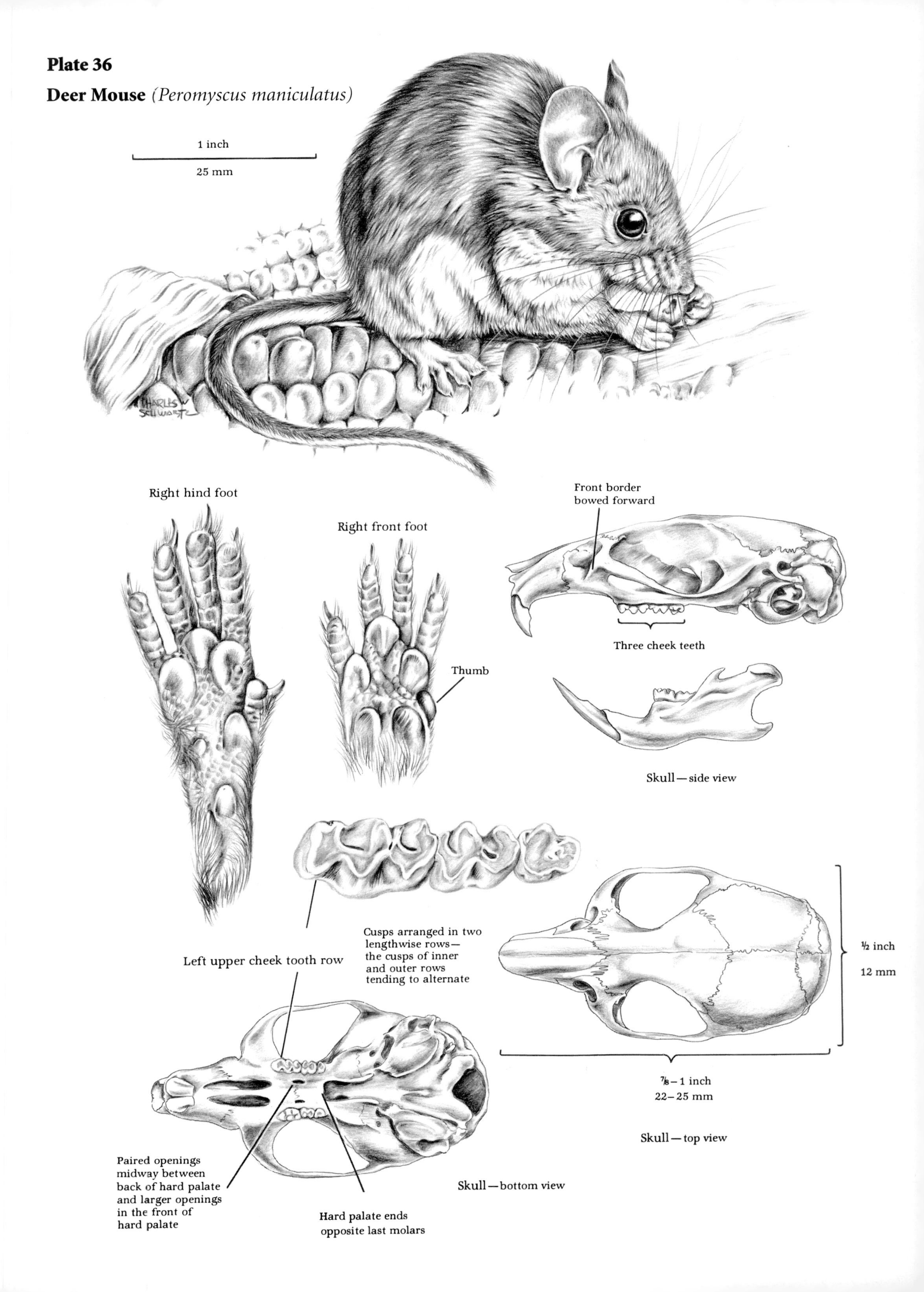
Plate 36
Deer Mouse (Peromyscus maniculatus)
1 inch
25 mm
Right hind foot
Right front foot
Thumb
Front border bowed forward
Three cheek teeth
Skull—side view
Left upper cheek tooth row
Cusps arranged in two lengthwise rows—the cusps of inner and outer rows tending to alternate
½ inch
12 mm
⅞–1 inch
22–25 mm
Skull—top view
Paired openings midway between back of hard palate and larger openings in the front of hard palate
Hard palate ends opposite last molars
Skull—bottom view

cusps are arranged in 2 lengthwise rows, those of the inner and outer rows tending to alternate.

The skull of the deer mouse is so similar to those of the other *Peromyscus* mice in Missouri that it is difficult to distinguish between them. However, the deer mouse skull can be told from that of the white-footed mouse because the deer mouse tends to have a narrower rostrum and to have the anterior palatine foramina (large paired openings in the front of the hard palate) parallel, while the white-footed mouse has a broader rostrum and the anterior palatine foramina bowed out at the middle.

Among Missouri mice with 16 teeth, other than *Peromyscus*, the skull of the deer mouse could be confused only with that of the golden mouse or the marsh rice rat. From the golden mouse, the deer mouse can be distinguished in two ways: in the deer mouse the posterior palatine foramina (tiny paired openings in the hard palate near the second molars) are about halfway between the back of the hard palate and the anterior palatine foramina (large paired openings in the front of the hard palate), and the front border of the infraorbital plate is bowed forward. From the marsh rice rat, it is distinguished by the following characteristics: size, the arrangement of cusps on the upper molar teeth, the absence of a ridge above the eye socket, the absence of paired openings in the hard palate behind the last molars, and the hard palate ending opposite the last upper molars.

Sex criteria and sex ratio. The sexes are identified as in the eastern woodrat. There are 3 pairs of teats on the belly: 1 pair near the front legs and 2 pairs in the groin region. Males outnumber females at birth and in older populations.

Age criteria, age ratio, and longevity. The young have gray to grayish black fur that changes to a duller and paler brown before adult coloration is acquired. Relative age is determined by the eruption and wear of the teeth.

In Missouri, 50 percent of the population of deer mice in winter and spring consists of young animals, resulting from the slow maturing of young during winter; in summer only 13 percent are young because the young mature rapidly at this time of year.

In the wild, deer mice may live as long as 1½ or 2 years. However, fewer than ⅕ of those born usually reach sexual maturity. The record of longevity in captivity is 8 years.

Glands. Certain salivary glands (the parotid and submaxillary glands) may contain a weak poison; extracts from these glands injected into laboratory mice induced hard breathing.

Voice and sounds. Deer mice utter high-pitched squeaks, trills, chatters, and a shrill buzz that lasts from 5 to 10 seconds and is audible for 15 m (50 ft.). In addition, some sounds are ultrasonic. The young have special calls for the female but no longer give them when they are old enough to care for themselves. When disturbed or excited, deer mice stamp their front feet up and down very rapidly.

Distribution and Abundance

The deer mouse is widespread across North America, ranging throughout central and southern Canada, the United States, and into southern Mexico. It is absent from the Atlantic and Gulf of Mexico coastal plains of the United States, but its range does extend to the coast in east Texas.

The deer mouse lives throughout Missouri and is one of the most abundant mammals on open lands. Highest populations in this state occur from March through June, lowest populations from July through October, and moderate ones from November through February. Local populations fluctuate greatly from year to year with peaks occurring about every 3 to 5 years.

Habitat and Home

In Missouri the deer mouse is usually found in an open habitat such as pastures, meadows, prairie, cultivated fields, and along field borders and fencerows. It may also live around and in human habitations but generally does not occur in heavy brush or wooded

Deer mouse tracks

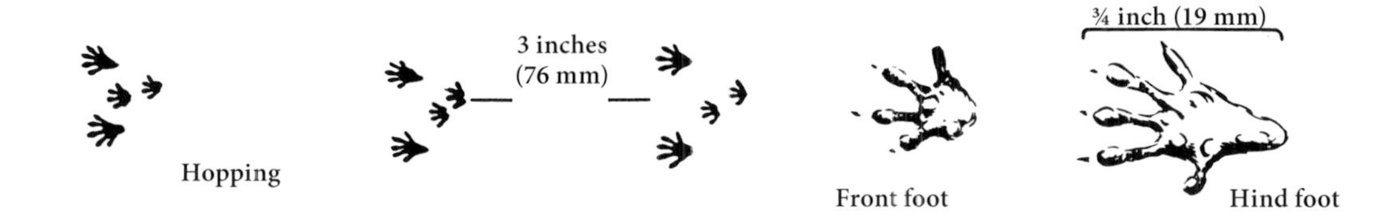

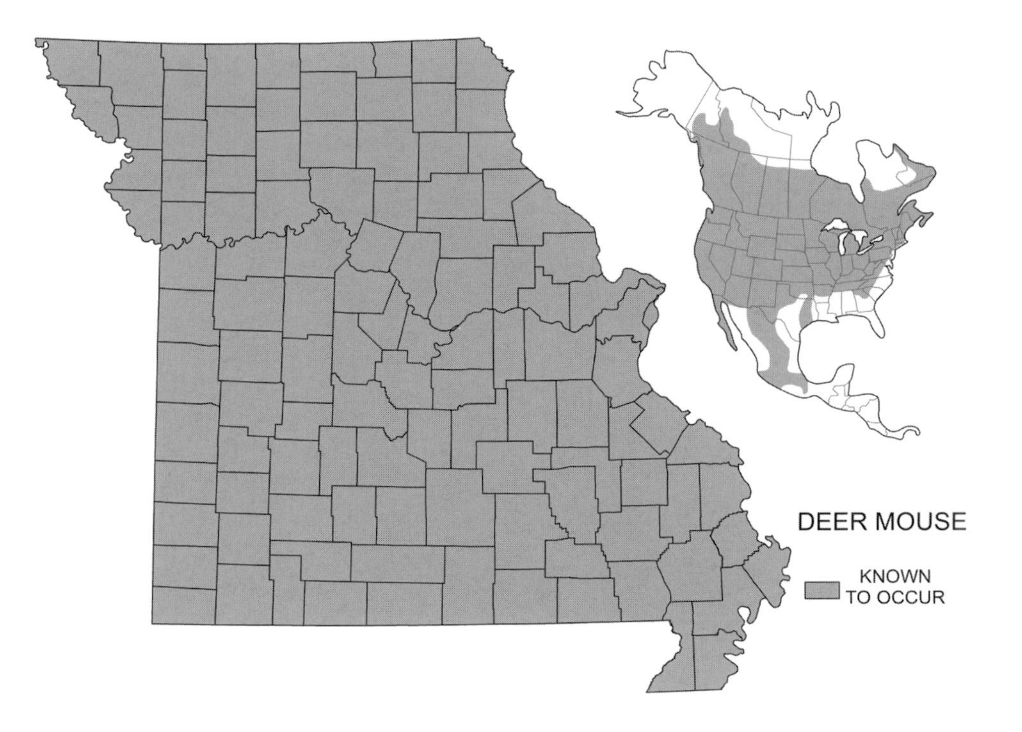

places. In Missouri it seldom occupies the same range as its close relative, the forest-dwelling white-footed mouse.

The several nests are generally located underground in cavities about the roots of trees or shrubs, beneath a log or board, or in a tunnel built either by another animal, but no longer used by it, or by the mouse itself. Less often are the nests aboveground in a hollow fence post, stump, log, or an old bird's or squirrel's nest.

The deer mouse's nest is usually spherical with a single side entrance that the mouse closes from the inside. Leaves, stems, and roots of grasses, sedges, and other plants and shredded bark are woven into the framework; thistledown, feathers, fur, and other soft material, such as cotton or rags, are fluffed inside for the lining. When a bird's nest is used for the nest base, a dome-shaped roof is added by the mouse.

Habits

Adult deer mice tend to spend their entire lives in one locality. The home range is usually from 0.2 to 0.6 ha (½ to 1½ ac.) in extent, but in rare cases may encompass from 2 to 4 ha (5 to 10 ac.). In general, males range over greater areas than females, and breeding females tend to stay close to their nests. Deer mice have a well-developed homing instinct, and some have returned to their homes when liberated up to 3.2 km (2 mi.) away. During winter, these mice live in a smaller area than during the rest of the year and may travel only about 9 m (30 ft.) from their nests, depending upon the food supply; in bad weather they may not leave their nests for several days at a time. The young often stay near the original nest site and rarely move farther than 1.7 km (550 ft.) before establishing their own homes.

Although deer mice are abroad mostly at night, they shun the hours of brightest moonlight. The customary feeding periods are during early evening and just before dawn.

During the breeding season, deer mice live singly or in pairs. A pregnant female may tolerate her mate in the nest with one litter of young, but she often chases both the male and young away before the birth of another litter; or she may leave the male with the weaned young in their nest and move away to establish a new home. In winter these mice often congregate in groups of up to 15, of mixed ages and sexes, and huddle together in a common nest for warmth. This huddling helps them survive in cold climates, especially where food is scarce in winter and they cannot obtain sufficient nourishment to maintain adequate body heat. Antagonism between individuals and sexes begins with the onset of the breeding season. Certain males display dominance over others. When white-footed mice occur in the same areas, they show more aggression and tend to dominate deer mice.

These mice do not make definite runways through grass or other vegetation but often use the runways and tunnels of other kinds of mice and of shrews. They have been observed to forage for aquatic organisms in water 1.3 cm (½ in.) deep and to swim across 3 m (10

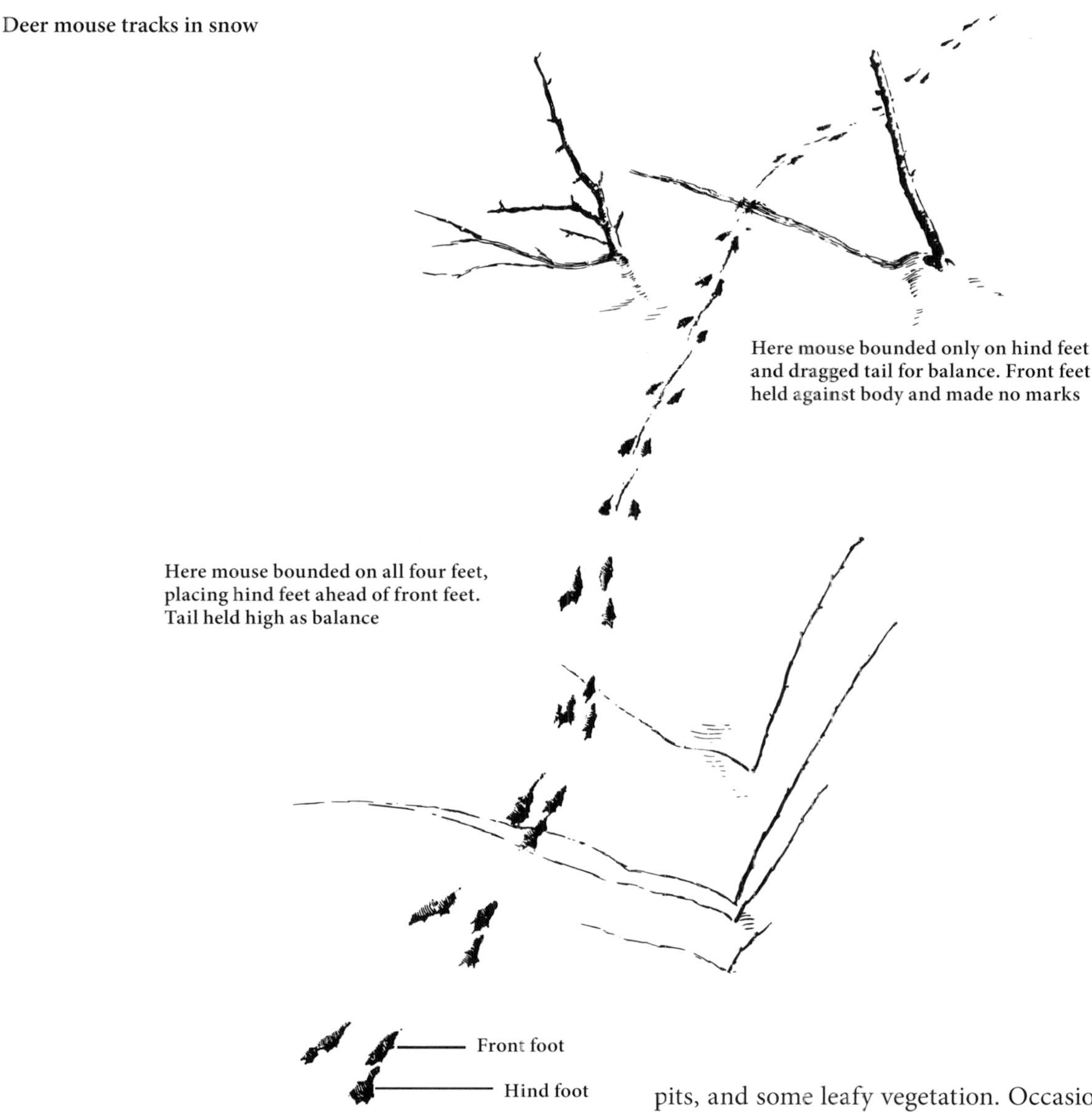

Deer mouse tracks in snow

ft.) of rough water. They climb trees readily, using their tails to help them keep balance. *Peromyscus* mice are known to survive periodic flooding in mature forests in the floodplain of the Missouri River. They usually leap rather than run and have been timed to travel at the rate of 2.4 m (8 ft.) per second.

Deer mice are very clean about their coats but unsanitary about their nests. Because of their habit of leaving scraps of food in the nest and using the nest for elimination purposes, they must move from one nest to another every few weeks.

Foods

The important foods of deer mice are insects (beetles and larvae of butterflies and moths), nuts, wild seeds, domestic grain (corn and soybeans), fruits and fruit pits, and some leafy vegetation. Occasionally they eat fungi, snails, slugs, worms, spiders, centipedes, millipedes, eggs and young of birds, and dead mice.

In the fall, seeds and nuts are stored in holes in the ground, in old birds' nests, or in trees. Such stores may contain as much as 0.5 l (1 pint) of food. Deer mice do much of their feeding at these storehouses. The food is carried in the small cheek pouches, which together hold about 5 ml (1 teaspoon) of seeds.

Reproduction

The principal mating periods of the deer mouse occur in spring and fall, but limited breeding takes place in summer and, under unusually favorable conditions of mild temperatures and abundant food, even in winter. Estrus, or heat, lasts for 26 hours and occurs every fifth day in unmated females. Estrus also occurs from 24 to 48 hours after birth of the young.

The gestation period is generally from 21 to 23 days but may be extended up to 37 days in nursing

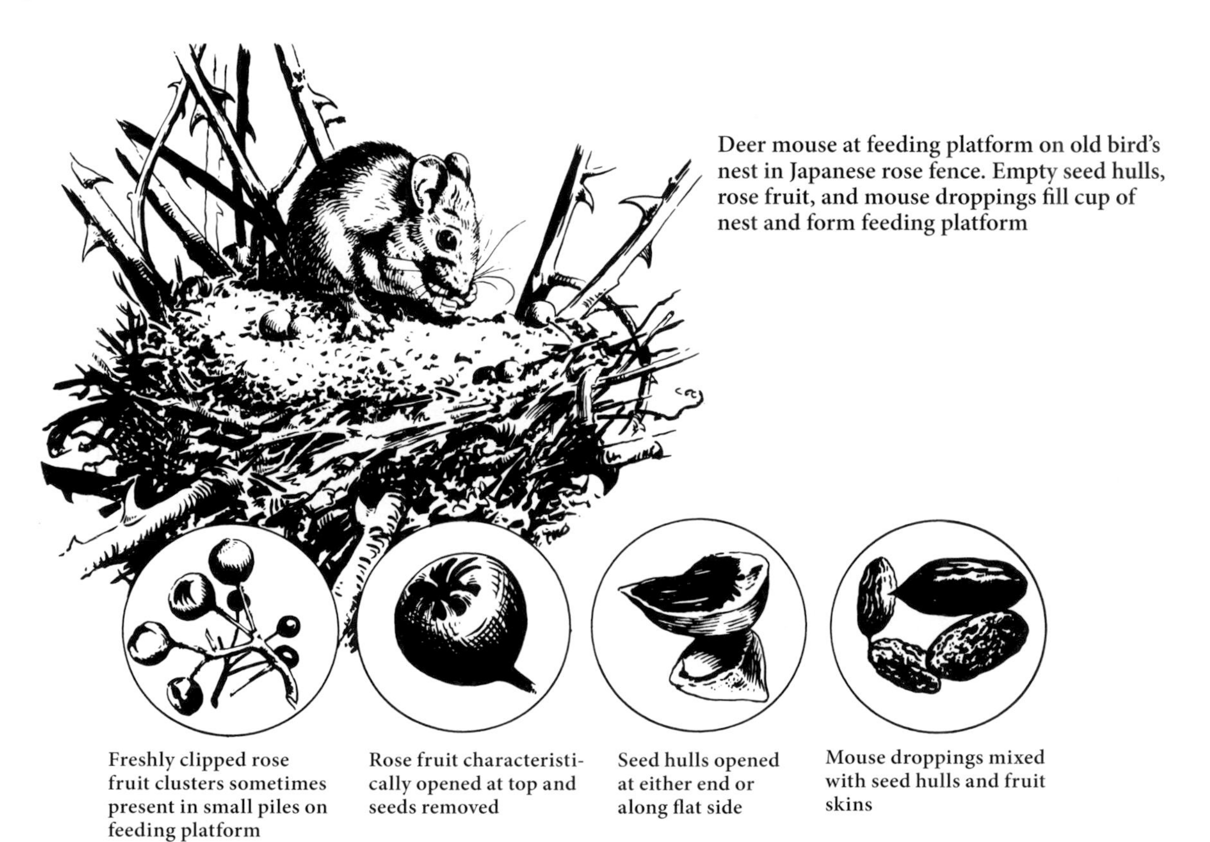

Deer mouse at feeding platform on old bird's nest in Japanese rose fence. Empty seed hulls, rose fruit, and mouse droppings fill cup of nest and form feeding platform

Freshly clipped rose fruit clusters sometimes present in small piles on feeding platform

Rose fruit characteristically opened at top and seeds removed

Seed hulls opened at either end or along flat side

Mouse droppings mixed with seed hulls and fruit skins

females. From 1 to 9 young are born per litter with 3 and 4 the most common. No female breeds continuously the year around, but a female may have 2 or more litters successively in the spring, followed by a rest period in summer, and 2 or more litters again in the fall. In captivity two females bore 10 and 11 litters with totals of 45 and 42 young, respectively, during one year.

Parturition takes from ½ to 1 hour and usually occurs in the morning. During birth, the female may assist the young by pulling them with her teeth or front feet and by prodding them with her nose. At birth the young are wrinkled, pink, naked, and weigh about 2 g (1/16 oz.) each. The eyes are closed and the ears folded. The toes and claws are formed, but the separation between the toes has not yet occurred.

The ears unfold at 3 to 4 days of age, and the incisors cut through the gums at 6 days of age. About 2 weeks after birth, the eyes open and the body is well furred. Weaning takes place when the young are 2 to 3 weeks old.

In brooding, the female straddles her young. When about one week old, they hold firmly to the female's nipples and are often dragged along if she moves suddenly and fails to disengage them. In transporting the young, the female rolls the young into a ball with her front feet and carries it in her mouth by the belly skin or back. She frequently permits the male to assist in caring for them.

In general, females become sexually mature between 46 to 51 days of age, while males become sexually mature when about 10 days older. The young born in early spring mature in 4½ to 9 weeks and may breed in the spring of their birth; young born late in spring

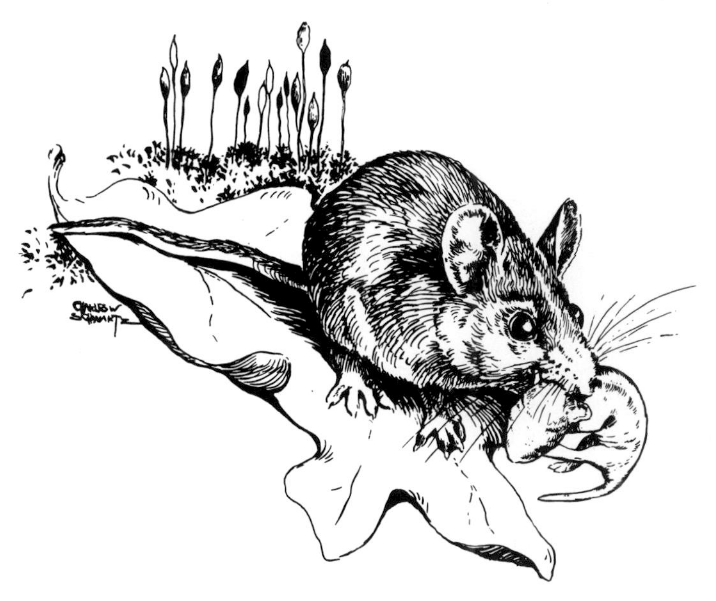

Young deer mice in nest inside rotten log

or in summer breed for the first time in the fall; and young born in the fall breed for the first time in the spring following their birth.

Mated mice usually stay together during the breeding season, if both survive; otherwise, new mates are acquired. If the young remain in the vicinity of the home nest, there may be considerable breeding among close relatives. In cases under laboratory conditions where two or more females have had their litters in the same nest, no antagonism was shown by the females, who nursed each other's young without apparent concern.

When deer mice are crossed experimentally with white-footed mice, no offspring are produced. However, in the wild it is remotely possible that interbreeding takes place because some individuals occur that appear to be hybrids.

Some Adverse Factors

Deer mice are preyed upon by opossums, short-tailed shrews, foxes, coyotes, weasels, skunks, minks, badgers, bobcats, domestic cats, hawks, owls, and snakes.

Parasites found on or in deer mice are mites, ticks, lice, fleas, botfly larvae, roundworms, and tapeworms. The scab mite produces swollen, scabby ears and tail and causes the hair to fall out, especially on the back. This condition is most common during the summer. Botfly larvae are found most often in late summer and early fall. They are usually located in the groin region of the mouse and may contribute to mortality by making the host awkward and easier prey. Cold weather with its accompanying food shortage is a principal cause of mortality.

Although not identified in Missouri, hantaviruses that cause hemorrhagic fever with renal syndrome and hantavirus pulmonary syndrome have been identified in the deer mouse, white-footed mouse, hispid cotton rat, and marsh rice rat within the United States. As of April 2014, there have been 639 cases of humans infected with hantavirus documented in 34 states, most notably in the desert Southwest. Because this is fatal in about 36 percent of the cases, humans should be careful about handling rodents. Since it is hard to tell if a mouse or a rat carries hantavirus, it is best to avoid all wild mice and rats and to safely clean up any rodent urine, droppings, or nesting materials with a disinfectant or a mixture of bleach and water. The deer mouse

and other mammals have also been associated with the spread of plague and Lyme disease.

Importance

Deer mice are very important as a prey species for flesh-eating animals. In fact, when they are abundant and form a ready supply of food, their predators likewise become abundant. These mice consume large quantities of weed seeds and insects. They are also important by returning their waste products to the soil as fertilizer, which, in the case of such a common animal, is considerable. Deer mice are reared in laboratories because they make good experimental animals for research on heredity, cancer, and many other subjects.

These mice damage some crops and stores of grain and in the western states dig up seeds planted for reforestation. In the fall, they commonly enter buildings, where they leave their droppings and become a nuisance.

Conservation and Management

Snaptraps can effectively eradicate a few deer mice. If you are experiencing problems with larger numbers of mice, contact a wildlife professional for advice, assistance, regulations, or special conditions for handling these animals. This is especially important in areas with large amounts of mouse droppings, where protective clothing and equipment are recommended.

SELECTED REFERENCES

See also discussion of this species in general references, page 23.

Blair, W. F. 1940. A study of prairie deer mouse populations in southern Michigan. *American Midland Naturalist* 24:273–305.

Brown, L. N. 1964. Ecology of three species of *Peromyscus* from southern Missouri. *Journal of Mammalogy* 45:189–202.

Centers for Disease Control (CDC). 2014. *Reported cases of HPS*. Centers of Disease Control and Prevention, Atlanta, GA.

Childs, J. E., et al. 1994. Serologic and genetic identification of *Peromyscus maniculatus* as the primary rodent reservoir for a new hantavirus in the southwestern United States. *Journal of Infectious Diseases* 169:1271–1280.

Clark, F. H. 1938. Age of sexual maturity in mice of the genus *Peromyscus*. *Journal of Mammalogy* 19:230–234.

Harris, V. T. 1941. The relation of small rodents to field borders on agricultural lands in central Missouri. M.S. thesis, University of Missouri, Columbia. 86 pp.

Howard, W. E. 1949. Dispersal, amount of inbreeding, and longevity in a local population of prairie deer mice on the George Reserve, southern Michigan. University of Michigan, Contribution of the Laboratory of Vertebrate Biology 43. 50 pp.

Kamler, J. F., D. S. Pennock, C. Welch, and R. J. Pierotti. 1998. Variation in morphological characteristics of the white-footed mouse (*Peromyscus leucopus*) and the deer mouse (*P. maniculatus*) under allotopic and syntopic conditions. *American Midland Naturalist* 140:170–197.

Moss, V. A. 1990. Morphological differences as a measure of convergence between *Peromyscus maniculatus* and *Peromyscus leucopus* in southeast Missouri. M.S. thesis, Southeast Missouri State University, Cape Girardeau. 217 pp.

Sheppe, W. 1963. Population structure of the deer mouse, *Peromyscus*, in the Pacific Northwest. *Journal of Mammalogy* 44:180–185.

Svihla, A. 1932. A comparative life history study of the mice of the genus *Peromyscus*. *University of Michigan, Museum of Zoology, Miscellaneous Publication* 24:6–39.

Whitaker, J. O., Jr. 1966. Foods of *Mus musculus*, *Peromyscus maniculatus bairdi*, and *Peromyscus leucopus* in Vigo County, Indiana. *Journal of Mammalogy* 47:473–486.

Williams, A. K., M. J. Ratnaswamy, and R. B. Renken. 2001. Impacts of a flood on small mammal populations of lower Missouri River floodplain forests. *American Midland Naturalist* 146:217–221.

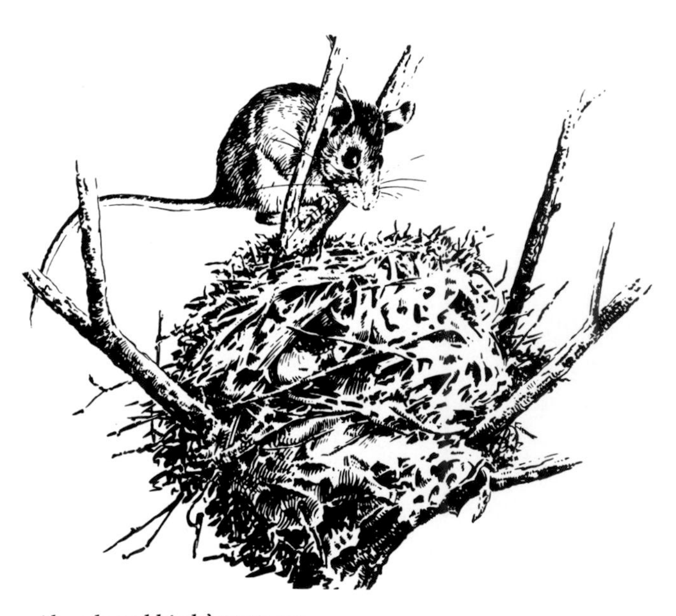

Abandoned birds' nests are frequently roofed and converted into white-footed mouse homes

White-footed Mouse
(Peromyscus leucopus)

Name

The first part of the scientific name, *Peromyscus,* is from two Greek words and means "pouched little mouse." The last part, *leucopus,* is from two Greek words and means "white-footed" (*leukon,* "white," and *pous,* "foot"). This and the common name describe the color of the feet. It is also known as the white-footed deermouse and wood mouse.

Description (Plate 37)

The white-footed mouse is very similar to the deer mouse. In general, the tail of the white-footed mouse tends to be slightly longer (about equal to or slightly less than one-half the total length), it lacks a sharp contrast in color between the upper and lower surfaces (but some tails show more contrast than others, especially in winter), and it is more sparsely furred and scaly and without a tuft of hairs at the tip. The white-footed mouse tends to be larger than the deer mouse and so has a larger hind foot and ear.

Compare this description with those of other *Peromyscus* species and see the account of the deer mouse for distinctions between *Peromyscus* and other mice in Missouri.

Color. The color is very similar to that of the deer mouse.

Measurements

Total length	139–212 mm	5½–8⅜ in.
Tail	63–101 mm	2½–4 in.
Hind foot	19–25 mm	¾–1 in.
Ear	15–19 mm	⅝–¾ in.
Skull length	25–28 mm	1–1⅛ in.
Skull width	12–14 mm	½–9/16 in.
Weight	11–28 g	⅖–1 oz.

Teeth and skull. The minor differences in teeth and skull between the white-footed mouse and the deer mouse are discussed with the deer mouse.

Distribution and Abundance

The white-footed mouse ranges from extreme southern Canada, throughout most of the eastern and central United States, and south into eastern and southern Mexico. It lives throughout Missouri and is one of the most abundant mammals in wooded regions. Population densities usually vary from 1 to 4 white-footed mice per 0.4 ha (1 ac.) but may reach 20 mice.

Habitat and Home

The white-footed mouse lives primarily in wooded areas, in brushy or weedy borders, and along fencerows. It occurs only in those grassy areas or fields with harvested grain that border woods or brush. The preferred habitat of this species rarely overlaps that of the deer mouse, and where ranges of these different species overlap, it is usually due to a change in the plant succession of the range.

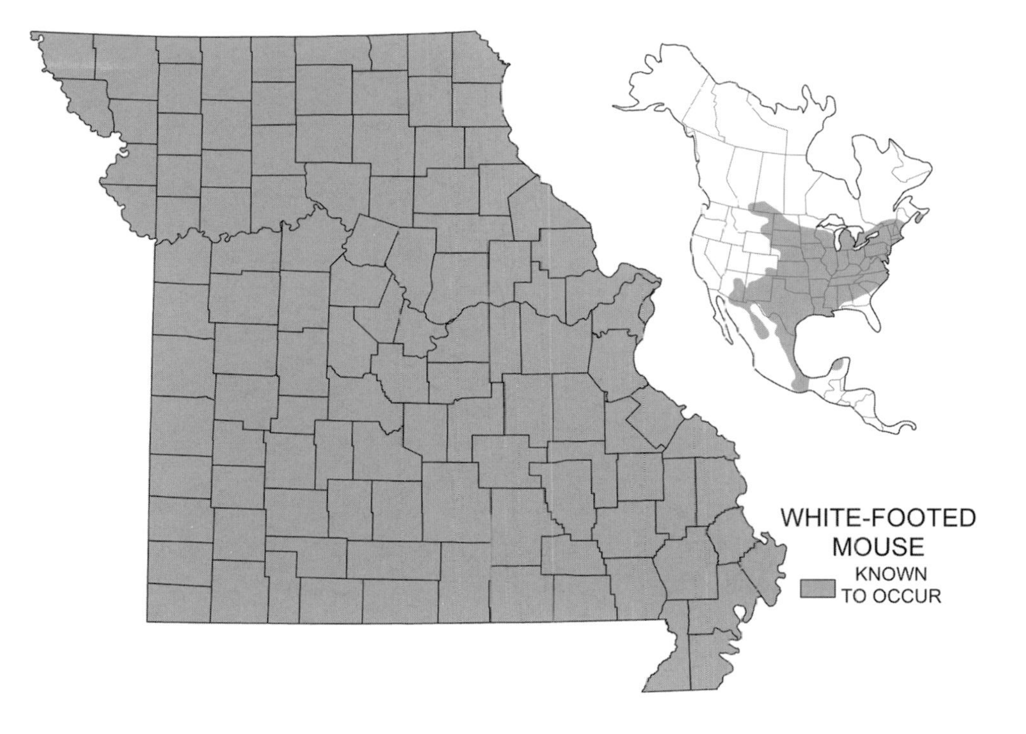

Plate 37

White-footed Mouse (*Peromyscus leucopus*)

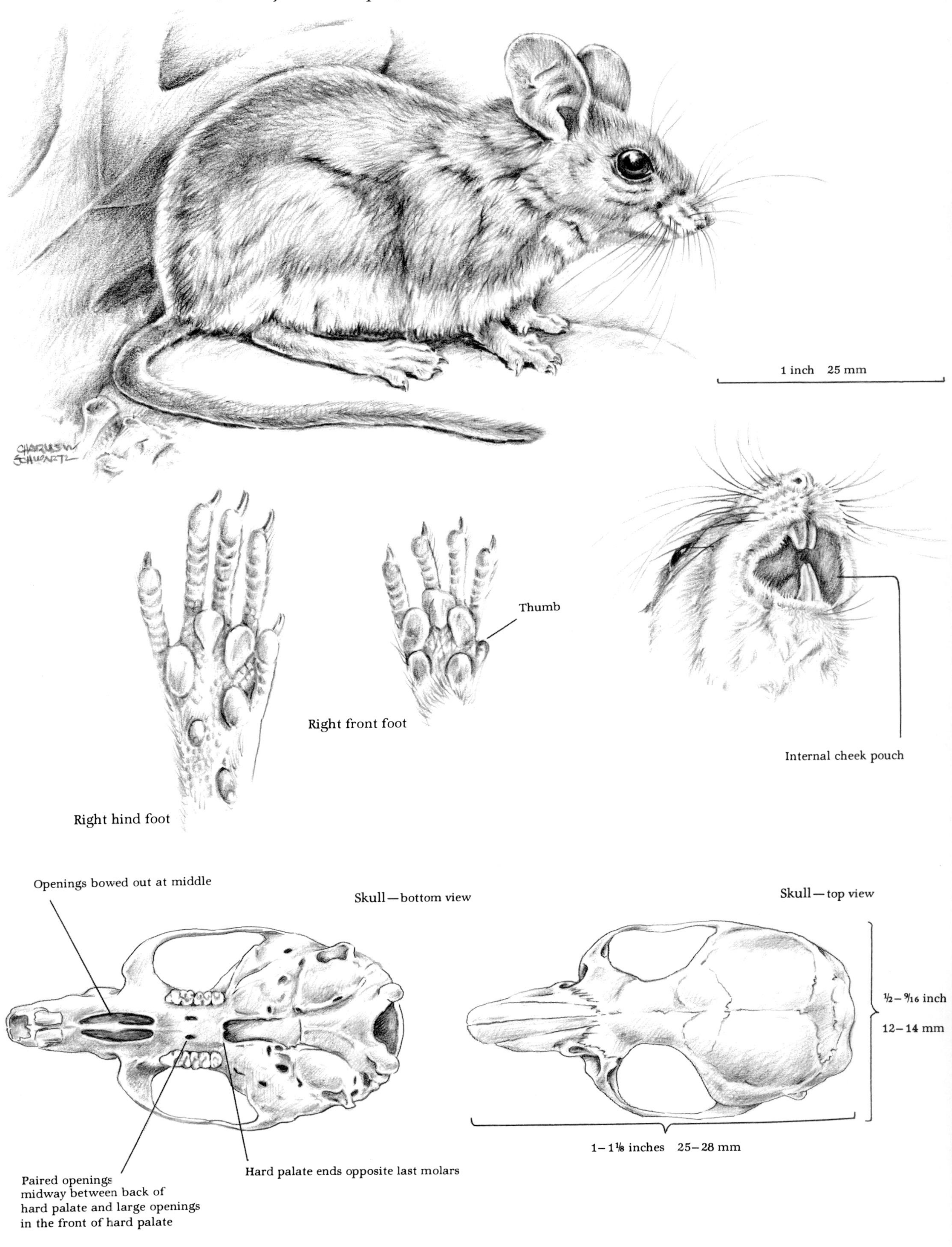

Nests are either in tree cavities, in old bird or old squirrel nests, or underground beneath some protective cover such as a log or the roots of a tree.

Habits

As this species is similar in appearance to the deer mouse, so is it similar in life history.

Many home ranges of a white-footed mouse tend to be circular; in other words, there is a center of activity with the mouse extending its travels in all directions from the center. Home ranges are generally small, 0.1 ha (1/5 acre) or more, but vary with the population density. They are large when the density is low and small when it is high. The different sexes tend to have home ranges that exclude the same sex but overlap the ranges of the opposite sex. Individuals are well acquainted with the landmarks in their home range and use the position of large trees, logs, and rocks to orient themselves.

There is no definite social hierarchy, which may be related to the high population turnover. In general, males are aggressive and some males tend to dominate others. Male white-footed mice actively fight with male deer mice and Texas mice where they occur in the same areas. White-footed mouse neighbors recognize each other and are aggressive toward strangers of the same species.

White-footed mice spend a great deal of their lives in trees. They swim well and voluntarily swim from one island to another, even when as far as 233 m (765 ft.) apart. Flooding of the home range has little influence on these semiarboreal mice. They tend to remain in trees and shrubs, then return to the ground as the water subsides.

In winter they become less aggressive and more communal. This is related to the cessation of breeding and to the need to concentrate in limited areas of good cover and available food.

Foods

Foods are generally the same as those of the deer mouse. The kinds and amounts of insects, seeds, fruits, and other food items reflect their availability.

Reproduction

Peaks of breeding occur in early spring and late summer, but restricted breeding may occur in midsummer. There is some evidence that certain females breed more than others and have up to eight litters a year. Most of the variation in litter size may be due to maternal age, younger females having smaller litters than older females.

Some Adverse Factors; Importance; and Conservation and Management

These are discussed under the deer mouse.

SELECTED REFERENCES

See also discussion of this species in general references, page 23.

Brown, L. N. 1964. Ecology of three species of *Peromyscus* from southern Missouri. *Journal of Mammalogy* 45:189–202.

Burt, W. H. 1940. Territorial behavior and populations of some small mammals in southern Michigan. *University of Michigan, Museum of Zoology, Miscellaneous Publication* 45:7–58.

Choate, J. R. 1973. Identification and recent distribution of white-footed mice (*Peromyscus*) in New England. *Journal of Mammalogy* 54:41–49.

Cornish, L. M., and W. N. Bradshaw. 1978. Patterns in twelve reproductive parameters for the white-footed mouse (*Peromyscus leucopus*). *Journal of Mammalogy* 59:731–739.

Fantz, D. K., and R. B. Renken. 2005. Short-term landscape-scale effects of forest management on *Peromyscus* spp. mice within Missouri Ozark forests. *Wildlife Society Bulletin* 33:293–301.

Hamilton, W. J., Jr. 1941. The food of small forest mammals in eastern United States. *Journal of Mammalogy* 22:250–263.

Harris, V. T. 1941. The relation of small rodents to field borders on agricultural lands in central Missouri. M.S. thesis, University of Missouri, Columbia. 86 pp.

Havelka, M. A., and J. S. Millar. 2004. Maternal age drives seasonal variation in litter size of *Peromyscus leucopus*. *Journal of Mammalogy* 85:940–947.

Lackey, J. A., D. G. Huckaby, and B. G. Ormiton. 1985. *Peromyscus leucopus*. *Mammalian Species* 247. 10 pp.

Metzgar, L. H., and R. Hill. 1971. The measurement of dispersion in small mammal populations. *Journal of Mammalogy* 52:12–20.

———. 1973. Home range shape and activity in *Peromyscus leucopus*. *Journal of Mammalogy* 54:383–390.

———. 1979. Dispersion patterns in a *Peromyscus* population. *Journal of Mammalogy* 60:129–145.

Nicholson, A. J. 1941. The homes and social habits of the wood mouse (*Peromyscus leucopus noveboracensis*) in southern Michigan. *American Midland Naturalist* 25:196–223.

Rintamaa, D. L., P. A. Mazur, and S. H. Vessey. 1976. Reproduction during two annual cycles in a population of *Peromyscus leucopus noveboracensis*. *Journal of Mammalogy* 57:593–595.

Ruffer, D. G. 1961. Effect of flooding on a population of mice. *Journal of Mammalogy* 42:494–502.
Sheppe, W. 1963. Population structure of the deer mouse, *Peromyscus*, in the Pacific Northwest. *Journal of Mammalogy* 44:180–185.
Stickel, L. F. 1960. *Peromyscus* ranges at high and low population densities. *Journal of Mammalogy* 41:433–441.
Vestal, B. M., and J. J. Hellack. 1978. Comparison of neighbor recognition in two species of deer mice (*Peromyscus*). *Journal of Mammalogy* 59:339–346.
Whitaker, J. O., Jr. 1963. Foods of 120 *Peromyscus leucopus* from Ithaca, New York. *Journal of Mammalogy* 44:418–419.
———. 1966. Foods of *Mus musculus*, *Peromyscus maniculatus bairdi*, and *Peromyscus leucopus* in Vigo County, Indiana. *Journal of Mammalogy* 47:473–486.

Cotton Mouse (*Peromyscus gossypinus*)

Name

The first part of the scientific name, *Peromyscus*, is from two Greek words and means "pouched little mouse." The last part, *gossypinus*, is derived from a Latin word meaning "cotton tree." This and the common name refer to the cottony material used in the nests of the first mice that were collected and named. It is also known as the cotton deermouse.

Description

The cotton mouse is very similar to the other *Peromyscus* mice in Missouri. It most closely resembles the white-footed mouse but is slightly larger. In some areas where these two species overlap, they are known to hybridize. The tail tends to lack a sharp contrast in color between the upper and lower surfaces, but some tails show more contrast than others.

See the descriptions of the other *Peromyscus* species and account of the deer mouse for distinctions between *Peromyscus* and other mice in Missouri.

Color. The upperparts of the body are grayish to yellowish brown, and the general coloration is very similar to that of the deer mouse.

Measurements

Total length	161–209 mm	6⅜–8¼ in.
Tail	69–101 mm	2¾–4 in.
Hind foot	20–25 mm	13/16–1 in.
Ear	17 mm	11/16 in.
Skull length	26 mm	1 1/16 in.
Skull width	12 mm	½ in.
Weight	19–25 g	7/10–9/10 oz.

Teeth and skull. The teeth and skull are similar to those of the deer mouse.

Distribution and Abundance

The cotton mouse occurs throughout the southeastern United States, with Missouri on the northern edge of its range. It is an imperiled species of conservation concern in Missouri, where it occurs only in the Mississippi River Alluvial Basin. The species is secure across its North American range. A high density is 4 cotton mice per 0.4 ha (1 ac.).

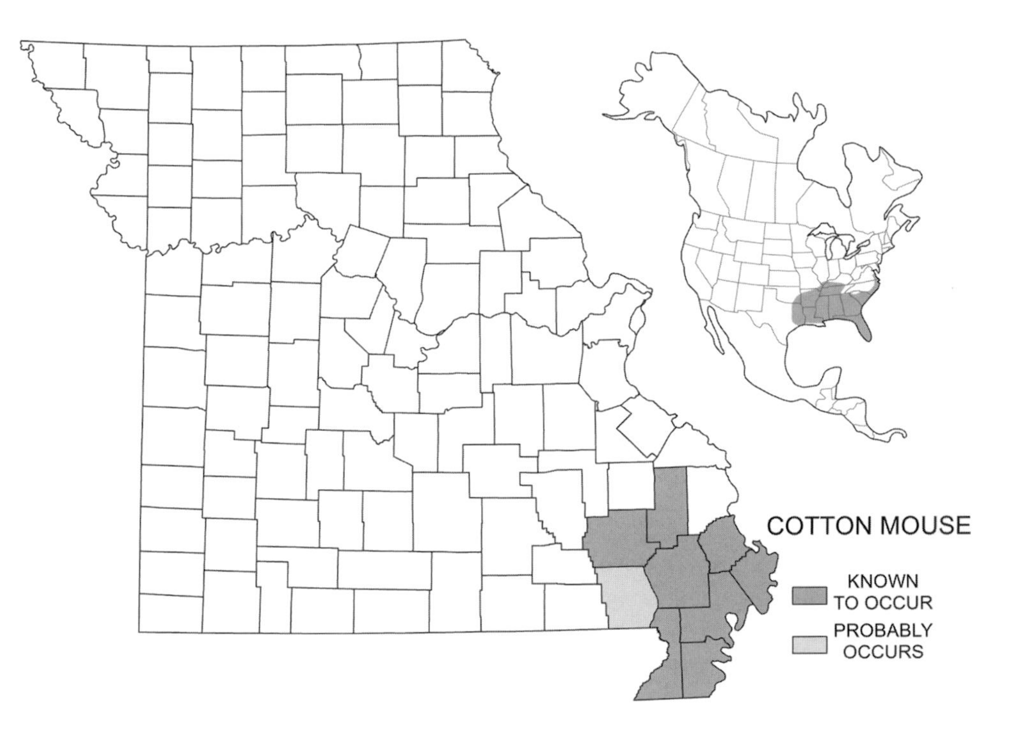

Habitat and Home

The cotton mouse is typically an inhabitant of moist, timbered areas, especially swamps and wet river bottoms where it lives in the dense underbrush. It is found in caves and in crevices around cliffs and rocky bluffs. It builds nests in logs, stumps, and trees, under brush piles, and in old buildings.

Habits

The cotton mouse may inhibit the white-footed mouse where they both live in the same area.

This mouse tends to run more than leap; it swims and dives well; and it shows great agility in climbing.

The home range varies from 0.2 to 0.8 ha (½ to 2 ac.).

Foods

The cotton mouse eats what is available. Animal matter may constitute as much as 68 percent of the diet.

Reproduction

In its southern range, the cotton mouse breeds mostly from fall through spring. Hot, dry summers reduce production. Whether it follows this pattern in Missouri is not known.

Gestation is usually 23 days but may be longer in nursing females. Some females have 3 to 4 litters a year with an average of 3 or 4 (with extremes of 1 to 7) young per litter.

The young and their pattern of growth are similar to those of the deer mouse.

In the laboratory, the cotton mouse crosses with the white-footed mouse, but interbreeding is probably rare in the wild.

Some Adverse Factors; Importance; and Conservation and Management

These are similar to those discussed for the deer mouse. However, large amounts of coarse woody debris greatly improve forest habitat quality for this species.

SELECTED REFERENCES

See also discussion of this species in general references, page 23.

Barko, V. A., and G. A. Feldhammer. 2002. Cotton mice (*Peromyscus gossypinus*) in southern Illinois: Evidence for hybridization with white-footed mice (*Peromyscus leucopus*). *American Midland Naturalist* 147:109–115.

Bekiares, N. 2000. Morphometric and allozyme variation in the cotton mouse (*Peromyscus gossypinus*) in southern Illinois, southwestern Kentucky, and southeastern Missouri. M.S. thesis, Southern Illinois University, Carbondale. 93 pp.

Bigler, W. J., and J. H. Jenkins. 1975. Population characteristics of *Peromyscus gossypinus* and *Sigmodon hispidus* in tropical hammocks of south Florida. *Journal of Mammalogy* 56:633–644.

Bradshaw, W. N. 1968. Progeny from experimental mating tests with mice of the *Peromyscus leucopus* group. *Journal of Mammalogy* 49:475–480.

Chromanski-Norris, J. F., and E. K. Fritzell. 1983. Status and distribution of ten Missouri mammals. A report to the Missouri Department of Conservation, Jefferson City. 38 pp.

Dice, L. R. 1940. Relationships between the wood-mouse and the cotton-mouse in eastern Virginia. *Journal of Mammalogy* 21:14–23.

Loeb, S. C. 1999. Responses of small mammals to coarse woody debris in a southeastern pine forest. *Journal of Mammalogy* 80:460–471.

Pearson, P. G. 1953. A field study of *Peromyscus* populations in Gulf Hammock, Florida. *Ecology* 34:199–207.

Pournelle, G. H. 1952. Reproduction and early postnatal development of the cotton mouse, *Peromyscus gossypinus gossypinus*. *Journal of Mammalogy* 33:1–20.

Wolfe, J. L., and A. V. Linzey. 1977. *Peromyscus gossypinus*. *Mammalian Species* 70. 5 pp.

Texas Mouse (*Peromyscus attwateri*)

Name

The first part of the scientific name, *Peromyscus*, is from two Greek words and means “pouched little mouse.” The last part, *attwateri*, is the Latinized name “of Attwater.” The common name, Texas, refers to the major portion of this mouse’s range.

This species is also known as the Texas deermouse and Attwater’s mouse. It was formerly known as *Peromyscus boylii*.

Description

The Texas mouse is very similar to the other *Peromyscus* mice in Missouri. It most closely resembles the deer mouse but is slightly larger and has a proportionately longer tail (about ½ or more than ½ of the total length). The tail is well haired and strongly bicolored. The Texas mouse also has a tuft of fur on the tip of the tail.

See the descriptions of other *Peromyscus* species and account of the deer mouse for distinctions between *Peromyscus* and other mice in Missouri.

Color. The flanks tend to be paler than those of the deer mouse; otherwise it is similar in color.

Measurements

Total length	111–228 mm	4⅜–9 in.
Tail	50–114 mm	2–4½ in.
Hind foot	17–25 mm	11⁄16–1 in.
Ear	15–20 mm	⅝–13⁄16 in.
Skull length	19–22 mm	¾–⅞ in.
Skull width	12–14 mm	½–9⁄16 in.
Weight	25–31 g	9⁄10–1 1⁄10 oz.

Teeth and skull. Teeth and skull are similar to those of the deer mouse.

Distribution and Abundance

The Texas mouse has a restricted range in the United States that includes parts of Missouri, Arkansas, Kansas, Oklahoma, and Texas. In Missouri, it occurs in the southwestern part of the Ozark Highlands, where it is confined to particular habitats in the White and Elk River drainages.

Habitat and Home

This species lives in the cedar and grass glade habitat with its accompanying rocky substratum. The nests are built under rocks and in crevices in rocky bluffs.

Life History

The Texas mouse is essentially similar in habits to the other Missouri *Peromyscus* mice. The individuals are generally docile and show little aggressive behavior. Where white-footed mice occur in the same or adjacent habitat, they tend to dominate the Texas mouse.

This mouse feeds on berries, acorns, seeds of cultivated plants like wheat, corn, and oats, and on beetles, grasshoppers, and other insects. Camel crickets, which occur commonly on rocky outcrops, compose a large percentage of its foods.

Breeding probably occurs throughout the year with peaks in spring and fall.

SELECTED REFERENCES

See also discussion of this species in general references, page 23.

Avise, J. C., M. H. Smith, and R. K. Selander. 1974. Biochemical polymorphism and systematics in the genus *Peromyscus*. VI. The *boylii* species group. *Journal of Mammalogy* 55:751–763.

Brown, L. N. 1964. Ecology of three species of *Peromyscus* from southern Missouri. *Journal of Mammalogy* 45:189–202.

Chromanski-Norris, J. F., and E. K. Fritzell. 1983. Status and distribution of ten Missouri mammals. A report to the Missouri Department of Conservation, Jefferson City. 38 pp.

Lee, M. R., D. J. Schmidly, and C. C. Huheey. 1972. Chromosomal variation in certain populations of

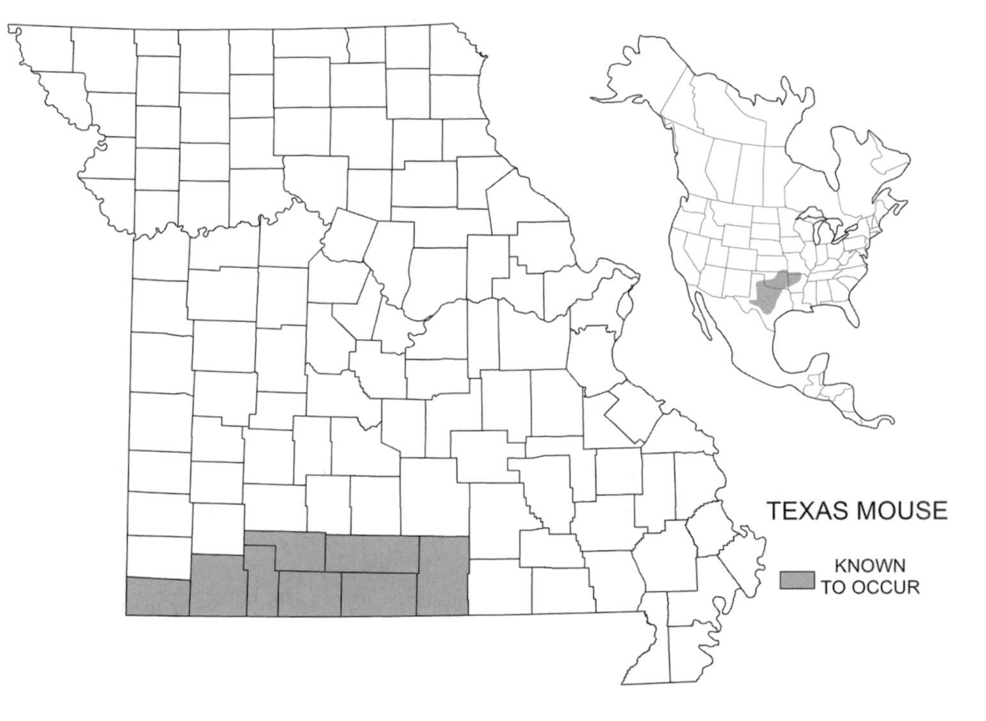

Peromyscus boylii and its systematic implications. *Journal of Mammalogy* 53:697–707.

Long, C. A. 1961. Natural history of the brush mouse (*Peromyscus boylii*) in Kansas with a description of a new subspecies. *University of Kansas, Publication of Museum of Natural History* 14:99–110.

Schmidly, D. J. 1973. Geographic variation and taxonomy of *Peromyscus boylii* from Mexico and the southern United States. *Journal of Mammalogy* 54:111–130.

———. 1974. *Peromyscus attwateri. Mammalian Species* 48. 3 pp.

Schnake-Greene, J. E., L. W. Robbins, and D. K. Tolliver. 1990. A comparison of genetic differentiation among populations of two species of mice (*Peromyscus*). *Southwestern Naturalist* 35:54–60.

Golden Mouse (*Ochrotomys nuttalli*)

Name

The first part of the scientific name, *Ochrotomys*, comes from two Greek words meaning "pale yellow mouse" (*ochra*, "pale yellow," and *mys*, "mouse"). The last part, *nuttalli*, is a Latinized name meaning "of Nuttall" and honors Thomas Nuttall, an early American naturalist. The common name describes the yellow to orange-brown color of the upperparts. This species was formerly known as *Peromyscus nuttalli*.

Description (Plate 38)

The golden mouse is a small, typically mouse-shaped rodent with big eyes; prominent, scantily furred ears; conspicuous, long whiskers; and a long, well-furred tail about one-half or slightly less than half the total length. The feet are small—an adaptation for climbing. Each front foot has 4 clawed toes and a small, barely nailed thumb, and each hind foot has 5 clawed toes. There are 6 pads, or tubercles, on the soles of the hind feet; the soles are furred from the heel to the pads. Small internal cheek pouches are present. The body fur is soft and thick.

This mouse is very similar to the *Peromyscus* mice in Missouri and formerly was included in the same genus with them. It can be distinguished by its coloration from all Missouri mice that are similar in size and general appearance.

Color. The golden mouse is yellow to orange brown on the head and ears, back, and sides, while the belly and feet are white and usually washed with the same color as that of the back. There is little to no sharp contrast in color between the back and belly or upper and lower surfaces of the tail. The winter coloration is darker than the summer.

The fur of the very young resembles that of the adult but is slightly darker and duller. The first molt can begin as early as 31 days of age and starts on the belly, spreads up the sides to the middle of the back, and then progresses forward to the head and backward to the tail. This molt takes between 25 and 29 days.

Measurements		
Total length	139–203 mm	5½–8 in.
Tail	60–101 mm	2⅜–4 in.
Hind foot	15–20 mm	⅝–13/16 in.
Ear	12–17 mm	½–11/16 in.
Skull length	23 mm	15/16 in.
Skull width	12 mm	½ in.
Weight	14–23 g	½–¾ oz.

Teeth and skull. The dental formula of the golden mouse is:

$$I\,\frac{1}{1}\;C\,\frac{0}{0}\;P\,\frac{0}{0}\;M\,\frac{3}{3} = 16$$

The teeth are similar to those of the deer mouse. The skull of the golden mouse can be distinguished from that of the deer mouse in two ways: in the golden mouse the posterior palatine foramina (tiny paired openings in the hard palate near the second molars) are nearer the back of the hard palate than they are to the anterior palatine foramina (large paired openings toward the front of the hard palate), and the front border of the infraorbital plate is straight. From the marsh rice rat, it is distinguished by the following characteristics: size, arrangement of cusps on the upper molar teeth, the absence of a ridge above the eye socket, the absence of paired openings in the hard palate behind the last molars, and the hard palate ending opposite the last upper molars.

Distribution and Abundance

The golden mouse occurs throughout the southeastern United States with Missouri on the northern edge of its range. In Missouri, it occurs in scattered populations in the Ozark Highlands and probably within the Mississippi River Alluvial Basin. Because of its semiarboreal nature and preferred habitats that are difficult to access and trap within, it is infrequently captured during small mammal surveys, so it may be more common than ordinarily presumed. It is listed as a vulnerable Missouri species of conservation concern but is secure across its North American range.

Habitat and Home

This species is considered a habitat specialist and lives in highly localized populations. Habitats include

Plate 38

Golden Mouse *(Ochrotomys nuttalli)*

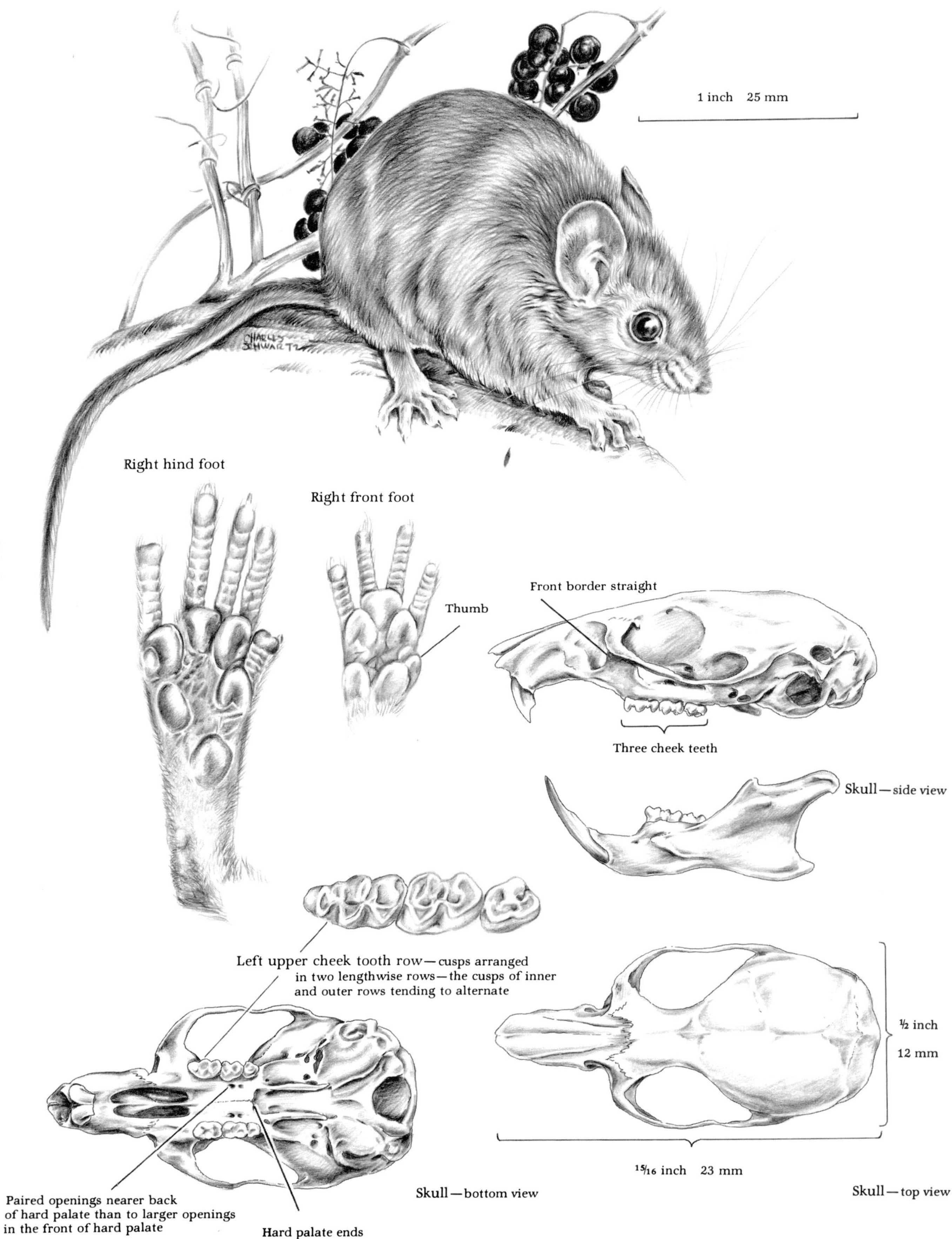

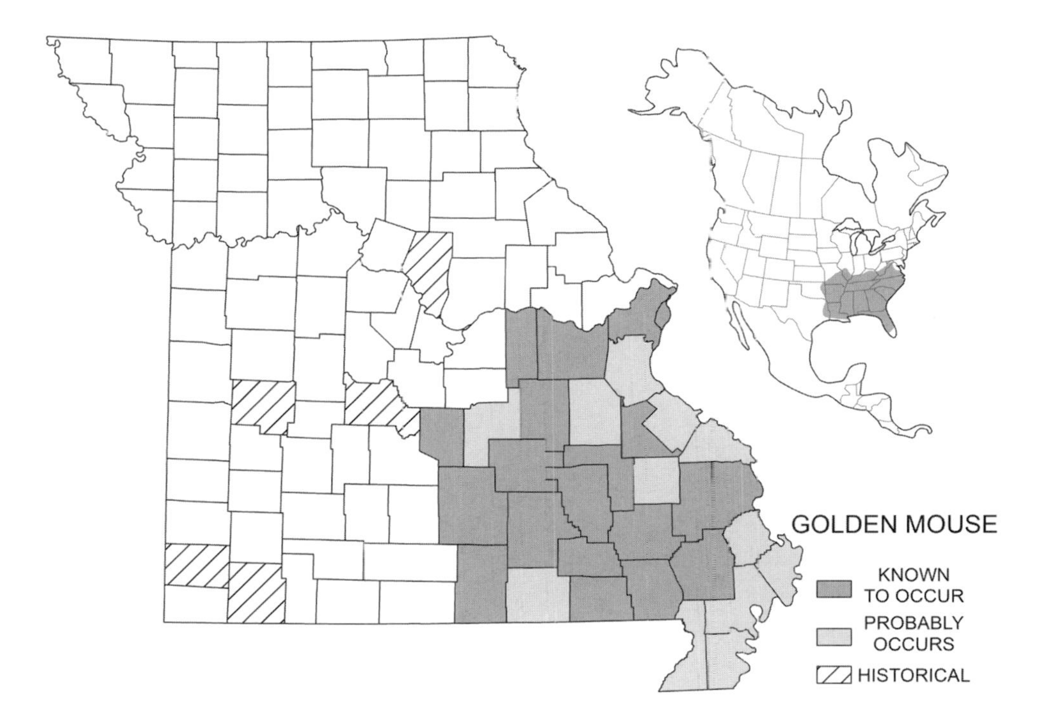

deciduous and coniferous forests, borders of old fields, swampy lowlands, and canebrakes. It is usually associated with an abundance of climbing vines, such as greenbrier, Japanese honeysuckle, grape, and poison ivy, and dense understory vegetation such as cane and blackberry. Only rarely does it live on dry hillsides.

Golden mice mostly use arboreal nests. These are usually from 15 to 20 cm (6 to 8 in.) in diameter and are built from near ground level to greater than 10 m (30 ft.) above the ground. Most are between 1.5 to 4.5 m (5 to 15 ft.) high in bushes, vines, or crotches of trees. Ground nests are also used and are built under a fallen log or leaf litter, or underground. An individual might use more than one arboreal nest in addition to using ground nests.

Each nest is a rather solid mass of leaves, inner bark, and grass with a lining of fur, feathers, shredded bark, or grass. It is slightly different than the nest of the white-footed mouse, which may be in similar localities. The nest of the white-footed mouse is more loosely constructed of shorter materials and appears rough and shaggy.

Habits

Golden mice are active during all seasons but are nocturnal and seldom seen. They are semiarboreal and often climb 9 m (30 ft.) or higher. They run easily along limbs, keeping their balance with the tail.

Golden mice are quite sociable. The nest is occupied by one or several mice. In winter, this number may increase to as many as eight as they huddle together for warmth.

In addition to nests, they may have up to six elevated feeding platforms scattered throughout their home range. These vary from a mere platform to a roofed-over shelter. Feeding platforms are recognized by the litter of seed hulls. Food is carried to these places in the cheek pouches. Home ranges vary from 0.2 to 2.8 ha (0.5 to 7 ac.).

Foods

Fruits, seeds, and some insects are main items in the diet. Seeds recovered from arboreal nests include wild cherry, dogwood, greenbrier, sumac, and oak.

Golden mouse nest

Reproduction

The breeding season extends from March until October. After a gestation period of 25 to 30 days, from 1 to 4 young are born. Litters tend to be larger in fall than in spring.

At birth the babies are blind, toothless, and naked. The ears unfold when they are 2 days old; some dark brown hair appears on the back and hindquarters on day 5; the lower incisors cut through the gums on day 6; the eyes open around day 13; and weaning is complete at about day 21.

Some Adverse Factors; Importance; and Conservation and Management

These subjects are similar to those discussed for the deer mouse.

SELECTED REFERENCES

See also discussion of this species in general references, page 23.

Barbour, R. W. 1942. Nests and habitat of the golden mouse in eastern Kentucky. *Journal of Mammalogy* 23:90–91.

Barrett, G. W., and G. A. Feldhamer, eds. 2008. *The golden mouse: Ecology and conservation.* Springer Science + Business Media, New York. 239 pp.

Christopher, C. C., and G. W. Barrett. 2006. Coexistence of white-footed mice (*Peromyscus leucopus*) and golden mice (*Ochrotomys nuttalli*) in a southeastern forest. *Journal of Mammalogy* 87:102–107.

Feldhamer, G. A., D. B. Lesmeister, J. C. Devine, and D. I. Stetson. 2012. Golden mice (*Ochrotomys nuttalli*) co-occurrence with *Peromyscus* and the abundant-center hypothesis. *Journal of Mammalogy* 93:1042–1050.

Gibbes, L. A., and G. W. Barrett. 2011. Diet resource partitioning between the golden mouse (*Ochrotomys nuttalli*) and the white-footed mouse (*Peromyscus leucopus*). *American Midland Naturalist* 166:139–146.

Goodpaster, W. W., and D. F. Hoffmeister. 1954. Life history of the golden mouse, *Peromyscus nuttalli*, in Kentucky. *Journal of Mammalogy* 35:16–27.

Layne, J. N. 1960. The growth and development of young golden mice, *Ochrotomys nuttalli. Quarterly Journal of Florida Academy of Science* 23:36–58.

Linzey, D. W. 1968. An ecological study of the golden mouse, *Ochrotomys nuttalli*, in the Great Smoky Mountains National Park. *American Midland Naturalist* 79:320–345.

Linzey, D. W., and A. V. Linzey. 1967a. Maturational and seasonal molts in the golden mouse, *Ochrotomys nuttalli. Journal of Mammalogy* 48:236–241.

———. 1967b. Growth and development of the golden mouse, *Ochrotomys nuttalli nuttalli. Journal of Mammalogy* 48:445–458.

Linzey, D. W., and R. L. Packard. 1977. *Ochrotomys nuttalli. Mammalian Species* 75. 6 pp.

Morzillo, A. T., G. A. Feldhamer, and M. C. Nicholson. 2003. Home range and nest use of the golden mouse (*Ochrotomys nuttalli*) in southern Illinois. *Journal of Mammalogy* 84:553–560.

Packard, R. L. 1969. Taxonomic review of the golden mouse, *Ochrotomys nuttalli. University of Kansas Museum of Natural History, Miscellaneous Publication* 51:373–406.

Patton, J. L., and T. C. Hsu. 1967. Chromosomes of the golden mouse, *Peromyscus (Ochrotomys) nuttalli* (Harlan). *Journal of Mammalogy* 48:637–639.

Wagner, D. M., G. A. Feldhamer, and J. A. Newman. 2000. Microhabitat selection by golden mice (*Ochrotomys nuttalli*) at arboreal nest sites. *American Midland Naturalist* 144:220–225.

Hispid Cotton Rat (*Sigmodon hispidus*)

Name

The first part of the scientific name, *Sigmodon*, is from two Greek words—*sigma*, which is the name of the Greek letter Σ, equivalent of the English letter *S*, and *odour*, meaning "tooth." These refer to the pattern of enamel on the grinding surfaces of certain molar teeth, which when worn resemble an *S* or Σ. The last part, *hispidus*, is the Latin word for "rough" and describes the texture of the fur.

The common name originated as follows: "hispid" refers to the rough or stiff hairs of the coat and sets this cotton rat apart from all other cotton rats; "cotton" refers to the cotton plantations in the southern United States where these rats are abundant and were first described; and "rat" comes from the Anglo-Saxon word *raet*.

Description (Plate 39)

The hispid cotton rat is a small, robust rodent with a scaly, sparsely haired tail that is shorter than the combined length of head and body. The eyes are moderately large, and the ears are large but nearly hidden in the fur. There are 4 toes and a small thumb on each front foot, and 5 toes on each hind foot. Six pads occur on the naked soles of the hind feet. Very small internal cheek pouches are present. The body fur is rough, fairly long, and coarse.

The hispid cotton rat is distinguished as follows: from the brown rat, by its smaller size, shorter tail, and longer, grizzled fur; from the marsh rice rat, by

Plate 39
Hispid Cotton Rat *(Sigmodon hispidus)*

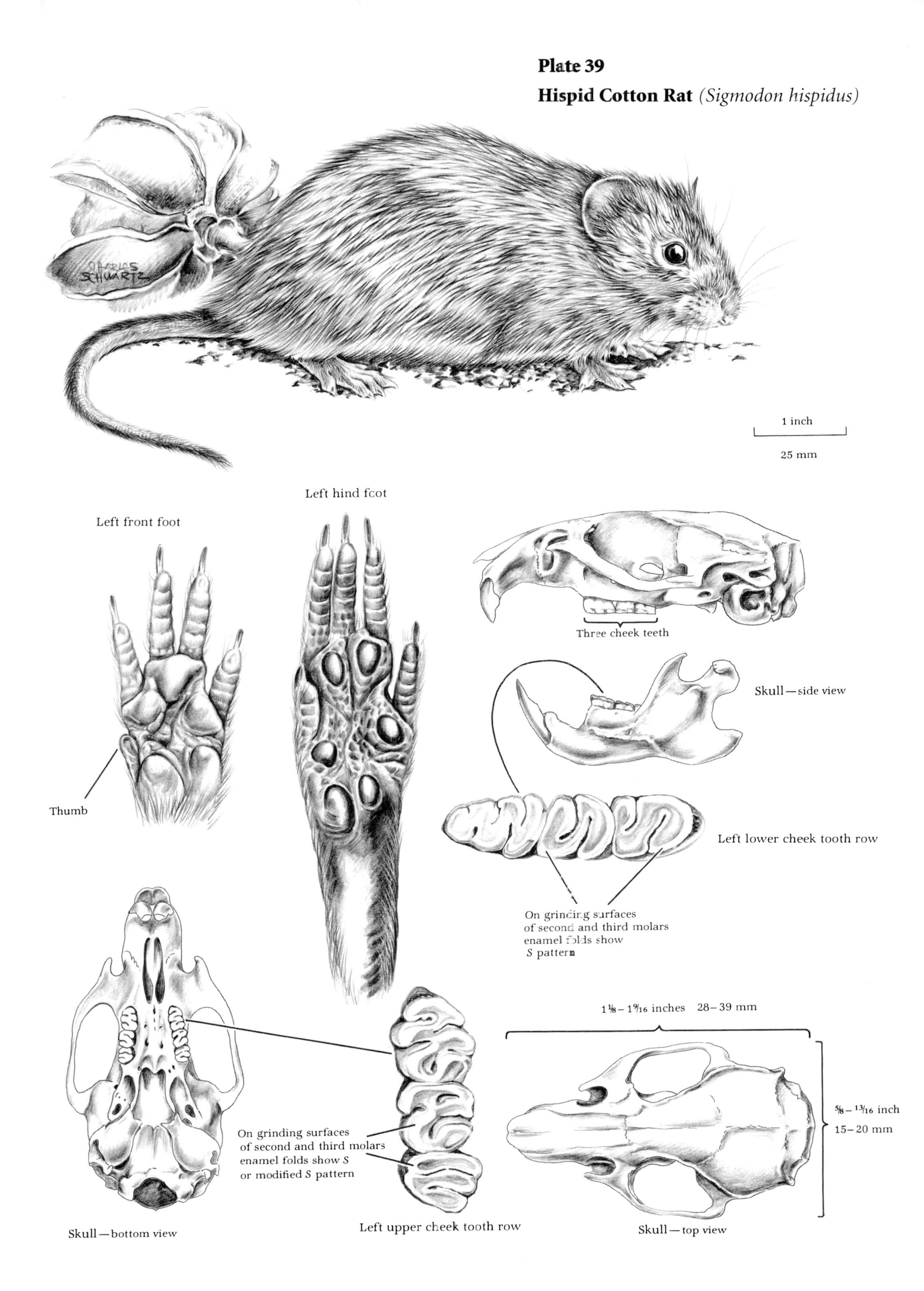

its more robust build, longer, grizzled fur, and shorter tail; and from the eastern woodrat, by its smaller size, coarser, grizzled fur, and shorter, scaly tail.

Color. The upperparts of both sexes are a mixture of tan, brown, and black with less black on the sides. The underparts are whitish to grayish or buff, and the feet are gray to dark brown. The tail is dark above, grading to light below. Albino or melanistic cotton rats are rare.

Measurements

Total length	203–371 mm	8–14⅝ in.
Tail	76–168 mm	3–6⅝ in.
Hind foot	25–38 mm	1–1½ in.
Ear	14–25 mm	9/16–1 in.
Skull length	28–39 mm	1⅛–1 9/16 in.
Skull width	15–20 mm	⅝–13/16 in.
Weight	56–240 g	2–8½ oz.
Commonly	85–113 g	3–4 oz.

Teeth and skull. The dental formula of the hispid cotton rat is:

$$I\ \frac{1}{1}\ C\ \frac{0}{0}\ P\ \frac{0}{0}\ M\ \frac{3}{3} = 16$$

The grinding surfaces of the upper and lower cheek teeth show a characteristic pattern of enamel folds surrounding dentine: the last 2 upper molars have an S or modified S pattern, while the last 2 lower molars have a typical S pattern. This enamel pattern is sufficient to distinguish the skull of the hispid cotton rat from all other rodents with the same number of teeth. This pattern is most evident in worn teeth.

Sex criteria and sex ratio. The sexes are distinguished as in the eastern woodrat. There are 4 or 5 pairs of teats. More females than males occur in high populations.

Age criteria, age ratio, and longevity. The young are identified by their color, which is darker and duller than that of adults.

The age of hispid cotton rats is best estimated by the growth of the lens of the eye. Wet weight of the lens, dry weight, and the measure of insoluble proteins in the lens can all be closely correlated with age.

Body weight is only a reliable indicator of age to 70 days. The presence of bone at the tip of the penis bone, or baculum, indicates an age of more than 7 months.

Individuals with a zygomatic arch breadth (greatest width of skull measured in either live or dead specimens) of 2 cm (¾ in.) or less are considered young; those larger than this are adults. However, size is not always an indication of sexual maturity because females born in the spring may breed when still small, while those born in the fall may not breed until the following spring when they are larger.

The age ratio varies with the population density. In years of dense populations, three-fourths of the population may be young; in years of low populations, young hispid cotton rats may compose only about one-half. Very few individuals survive longer than 6 months.

Voice and sounds. The voice is a high-pitched squeal.

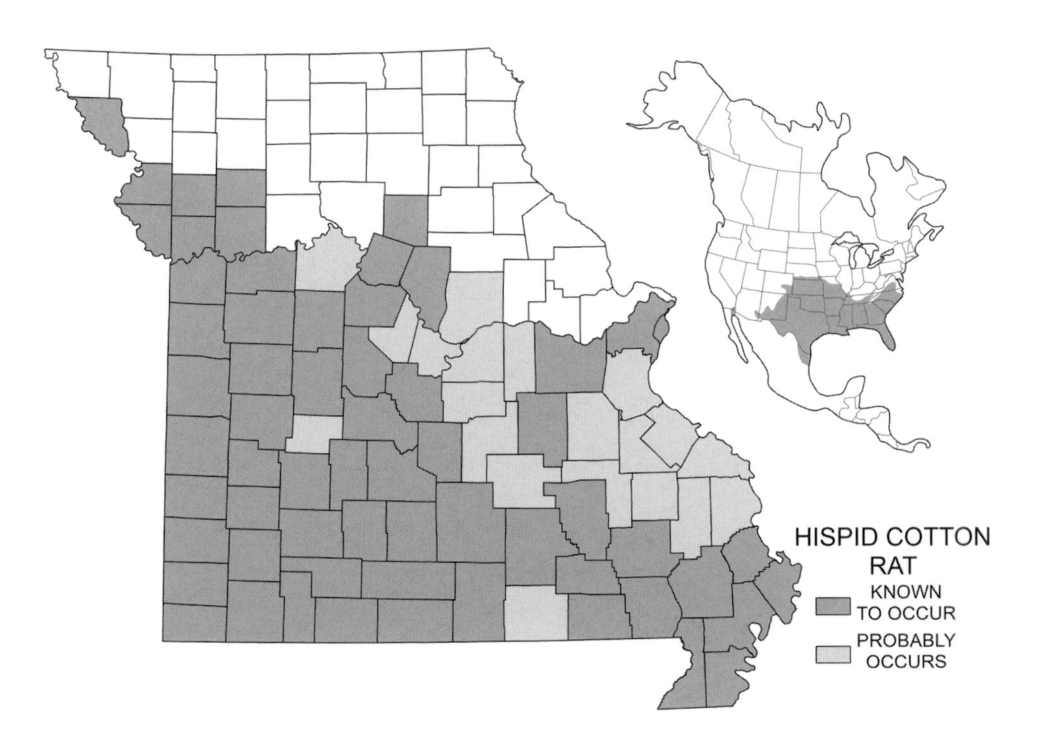

Distribution and Abundance

The hispid cotton rat is widely distributed in the southeastern and south-central United States and extreme northeastern Mexico, and an isolated population occurs in western Arizona and adjoining California. Until recently, it was thought that this species ranged into South America, but genetic analysis identified three different cotton rat species in Mexico. In the past 50 to 100 years, hispid cotton rats have been expanding their range northward and to higher elevations.

The first records of the hispid cotton rat in Missouri came from Ozark and Howell Counties in 1945. In 1949–1950 it was very abundant in southwestern Missouri but in the following years steadily decreased. Today is occurs in many Missouri counties, mostly south of the Missouri River.

The numbers of hispid cotton rats fluctuate from year to year, showing high densities every 2 to 5 years. Densities of 10 to 12 per 0.4 ha (1 ac.) are common. Sometimes, cotton rats occur in very dense populations, as indicated by the recovery of 513 cotton rats following the poisoning of 0.4 ha (a single acre). In open lands of the southern United States, the hispid cotton rat is the most abundant of all small mammals.

Habitat and Home

The habitat of the hispid cotton rat is dense, grassy fields and roadsides overgrown with broomsedge and weeds, and the waste borders of cultivated fields.

The small nest is built of dry grass, of fibers stripped from stems of larger plants, and of any other available material, such as cotton. Nests are usually built under logs or rocks, or in the ground at the end of a long, twisting, but shallow tunnel. Occasionally, abandoned dens of spotted skunks or ground squirrels serve as nest chambers. In winter, however, the nest may be built in a protected spot aboveground.

Habits

Hispid cotton rats make well-defined runways in the grass. The main runways are about 8 cm (3 in.) wide, while less-used ones are narrower. These are all kept shorn of new growth, and the cuttings are heaped in piles at irregular intervals. Tunnels, 2.5 to 5 cm (1 to 2 in.) below the surface of the ground, join the surface runway system and the underground nest.

The home range is small; females usually live within 0.1 to 0.3 ha (¼ to ¾ of an acre), while males occupy 0.4 to 0.5 ha (1 to 1¼ ac.). Territoriality results in nearly total separation of male home ranges. Although cotton rats are active during both day and night, they tend to be abroad more at night. They do not hibernate or cache stores of food. Cotton rats are able to swim. During the breeding season there is some dominance of older males over younger males, but during the nonbreeding season, this social organization is lost and various individuals huddle together for warmth. Cotton rats are better adapted to live in warm climates than in cold ones.

These rats groom their coats carefully. They are excitable, pugnacious, and aggressive toward other mice species living in the same fields. Cotton rats are reported to cause a decrease in voles by direct competition and by eating the young of voles.

Foods

Hispid cotton rats eat a wide variety of foods, including the stems, leaves, roots, and seeds of grasses and sedges, and cultivated plants like alfalfa, cotton, wheat, fruits, and vegetables. In addition, crayfish, insects, eggs and chicks of ground-nesting birds, and dead carcasses are all consumed. In feeding on tall plants, the rats sever the stems near the base and then cut them into sections. They drink water but do not need free water in their habitat.

Reproduction

Breeding occurs throughout the year but is less common during the winter months. Heat occurs every 7 to 9 days in unmated females and also following birth of a litter. The gestation period is 27 days and, unlike that of some other rodents, is not lengthened when the female

Hispid cotton rat tracks

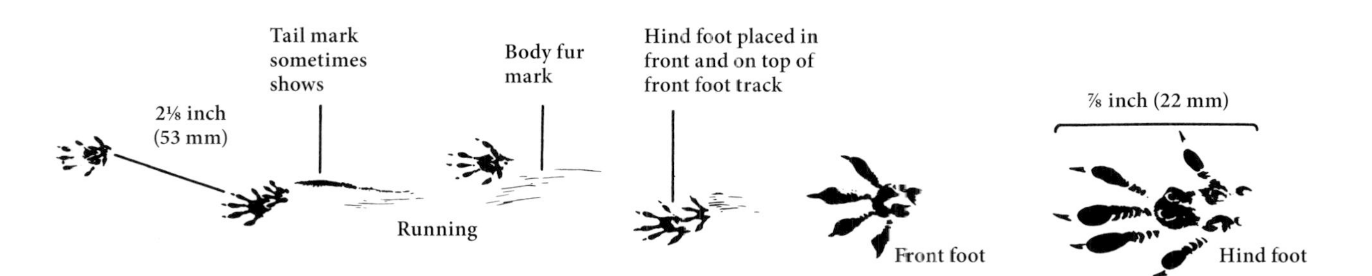

is suckling young. Several litters are produced annually and often in rapid succession. Litters may contain from 1 to 12 young but usually number from 5 to 7.

At birth the young weigh about 7 g (¼ oz.) and are about 7.6 cm (3 in.) long. They are precocial, being furred and able to run while still blind. Within 18 to 36 hours after birth, the eyes open and the ears unfold. Weaning normally occurs when the young are between 10 and 15 days old but may begin as early as 5 days of age.

The testes descend into the shallow scrotum in young males between 20 and 30 days of age, and the vaginas open in young females between 30 and 40 days old. Most young breed for the first time between 2 and 3 months of age, although some may breed when only 40 days old. Cotton rats are fully grown at 5 months.

Some Adverse Factors

Foxes, domestic dogs, coyotes, raccoons, weasels, minks, domestic cats, bobcats, owls, hawks, fire ants, and snakes prey on hispid cotton rats in Missouri. The external parasites of this rat are mites, ticks, lice, and fleas.

A fungus disease, aided by wet weather, often affects adults and even reaches epidemic proportions in high populations. Coccidiosis is common, and typhus has been reported in southern parts of the range. They are also known to carry hantavirus (see discussion under the deer mouse).

Importance

Hispid cotton rats are destructive to cultivated crops, causing as much as 75 percent loss in some fields of sugarcane, fruits, vegetables, cotton, grain, and alfalfa. However, cotton rats are important as food for many animals and, because of their abundance and availability, may act as a buffer between predators and game species. In the southern United States these rats are considered important in relation to bobwhite quail. They compete with quail for food and feed on quail eggs. However, occasional destruction of quail nests by these rats may have some beneficial aspects: it causes the hens to lay again and, because predation does not occur on all nests at one time, nest building is staggered throughout the breeding season. Thus, all nests are not vulnerable simultaneously to adverse weather or some other widespread agent of destruction. Hispid cotton rats are used in laboratories for research studies on poliomyelitis, diphtheria, tuberculosis, and typhus.

Conservation and Management

The best means for controlling these rodents is the systematic burning of heavy broomsedge and other grasses adjacent to cultivation. Fields with heavy cover can be plowed periodically or worked into some crop rotation system that reduces the heavy undergrowth. Continued protection of hawks and owls, important winged predators, will aid in controlling these rodents. If you are experiencing problems with cotton rats, contact a wildlife professional for advice, assistance, regulations, or special conditions for handling these animals.

SELECTED REFERENCES

See also discussion of this species in general references, page 23.

Birney, E. C., R. Jenness, and D. D. Baird. 1975. Eye lens protein as criteria for age in cotton rats. *Journal of Wildlife Management* 39:718–728.

Bowne, D. R., J. B. Peles, and G. W. Barrett. 1999. Effects of landscape spatial structure on movement patterns of the hispid cotton rat (*Sigmodon hispidus*). *Landscape Ecology* 14:58–65.

Bradley, R. D., D. D. Henson, and N. D. Durish. 2008. Re-evaluation of the geographic distribution and phylogeography of the *Sigmodon hispidus* complex based on mitochondrial DNA sequences. *Southwestern Naturalist* 53:301–310.

Cameron, G. N. 1981. *Sigmodon hispidus*. *Mammalian Species* 158. 9 pp.

———. 1995. Temporal use of home range by the hispid cotton rat. *Journal of Mammalogy* 76:819–827.

Chipman, R. K. 1965. Age determination of the cotton rat (*Sigmodon hispidus*). *Tulane University Studies in Zoology* 12:19–38.

Easterla, D. A. 1968. Hispid cotton rat north of the Missouri River. *Southwestern Association of Naturalists* 13:364–365.

Erickson, A. B. 1949. Summer populations and movements of the cotton rat and other rodents on the Savannah River refuge. *Journal of Mammalogy* 30:133–140.

Fleharty, E. D., and L. E. Olson. 1969. Summer food habits of *Microtus ochrogaster* and *Sigmodon hispidus*. *Journal of Mammalogy* 50:475–486.

Green, A., and D. L. Jameson. 1975. An evaluation of the zygomatic arch for separating juvenile from adult cotton rats (*Sigmodon hispidus*). *Journal of Mammalogy* 56:534–535.

Joule, J., and G. N. Cameron. 1974. Field estimation of demographic parameters: Influence of *Sigmodon hispidus* population structure. *Journal of Mammalogy* 55:309–318.

Martin, E. P. 1956. A population study of the prairie vole (*Microtus ochrogaster*) in northeastern Kansas. *University of Kansas Publication, Museum of Natural History* 8:361–416.

Meyer, B. J., and R. K. Meyer. 1944. Growth and reproduction of the cotton rat, *Sigmodon hispidus hispidus*, under laboratory conditions. *Journal of Mammalogy* 25:107–129.

Mohlhenrich, J. S. 1961. Distribution and ecology of the hispid and least cotton rats in New Mexico. *Journal of Mammalogy* 42:13–24.

Odum, E. P. 1955. An eleven year history of a *Sigmodon* population. *Journal of Mammalogy* 36:368–378.

Pournelle, G. H. 1950. Mammals of a north Florida swamp. *Journal of Mammalogy* 31:310–319.

Sealander, J. A., Jr., and B. Q. Walker. 1955. A study of the cotton rat in northwestern Arkansas. *Proceedings of the Arkansas Academy of Science* 8:153–162.

Stickel, L. F., and W. H. Stickel. 1949. A *Sigmodon* and *Baiomys* population in ungrazed and unburned Texas prairie. *Journal of Mammalogy* 30:141–150.

Terman, M. E. 1974. Behavioral interactions between *Microtus* and *Sigmodon*: A model for competitive exclusion. *Journal of Mammalogy* 55:705–719.

Worth, C. B. 1950. Observations on the behavior and breeding of captive rice rats and woodrats. *Journal of Mammalogy* 31:421–426.

Eastern Woodrat (*Neotoma floridana*)

Name

The first part of the scientific name, *Neotoma*, is from two Greek words and means "new," and "cut" (*neos*, "new," and *tomos*, "cut"). This name was given to indicate that this species was a new kind of animal with "cutting" teeth, or a rodent, that was distinct from another genus, *Mus*, as it was originally called. The last part, *floridana*, is a Latinized word meaning "of Florida," for the place from which the first specimen was collected and described.

The origin of the common name is as follows: "eastern" superficially separates this species from other species in western United States; "wood" indicates the timbered habitat preferred by this mammal; and "rat" comes from the Anglo-Saxon word *raet*. The origins of other common names, pack rat and trade rat, are discussed below.

Description (Plate 40)

The eastern woodrat is a medium-sized rodent with prominent, short-haired ears; bulging, black eyes; very long, conspicuous whiskers; and a moderately to well-haired tail less than half the total length of the animal. Small internal cheek pouches are present. Each front foot has 4 clawed toes and a small thumb, and each hind foot has 5 clawed toes. The soles of the hind feet are furred to the 6 tubercles. The body fur is rather long and soft.

Although the eastern woodrat superficially resembles the brown rat, these animals are easily distinguished by their appearance, habits, and habitats. The eyes of the brown rat are smaller and less protruding; the snout is more elongated; the fur is coarser; the tail is nearly naked, possesses obvious scaly rings, and its darker upper surface is not sharply marked off from the lighter undersurface; and there are six pairs of teats along the sides of the belly.

Color. The upperparts of the adult woodrat are brownish gray mixed with black; the sides are lighter brown with some dark hairs. The face and outsides of the legs are brownish gray; the throat, belly, and feet are white; and the tail is blackish brown above sharply contrasted to white below. The sexes are colored alike.

There is 1 annual molt in adults but the young have 3 molts during their first year.

Plate 40
Eastern Woodrat *(Neotoma floridana)*

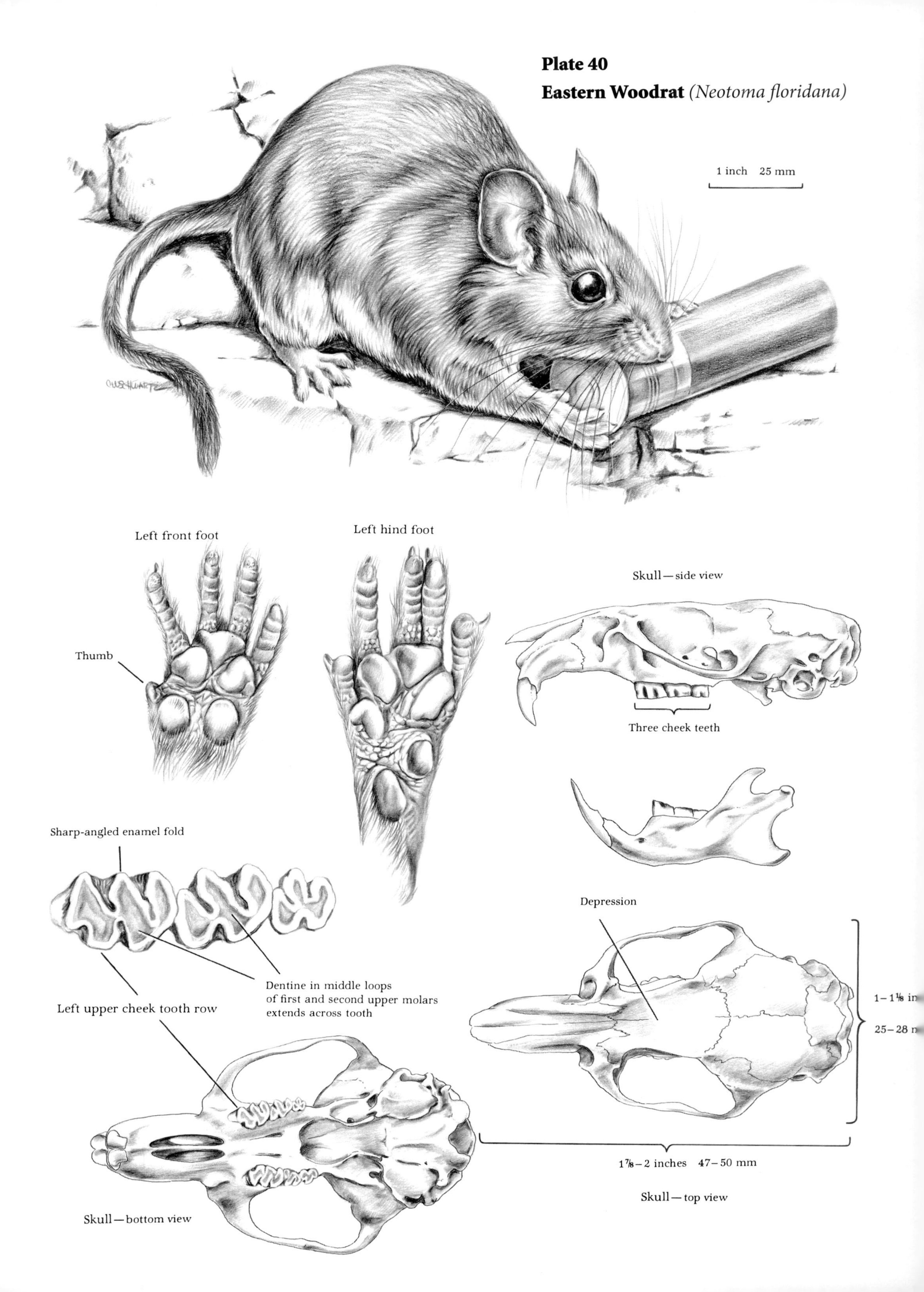

When this book was first published, the meadow vole was not known to occur in Missouri. The first records were reported in northwest Missouri in 1973. It is not known if this species had always occurred here or had recently extended its range southward from Iowa. However, because no specimens were taken here formerly and the area was trapped extensively, a recent expansion of range is considered the most likely explanation for its presence. Also, it is customary for this species to constantly invade new patches of habitat when they become available and where individuals, socially released from family ties, breed. It is probable that this vole occurs throughout much of northern Missouri in the low damp areas of stream valleys and floodplains.

The meadow vole is more cyclic in abundance than the prairie vole. There are peak populations about every four years, although these fluctuations may be local.

Habitat and Home

Meadow voles can be found in moist, low areas where there is a heavy growth of grasses, or in drier grasslands near streams, lakes, or swamps. In areas inhabited by both prairie voles and meadow voles, the meadow vole is generally in moister habitats. They construct runways in the rank cover and sometimes beneath the ground, where they build a nest of dry grasses and sedges for a resting site or for their young.

The meadow vole has extended its range continent-wide into otherwise highly cultivated areas by traveling and occupying continuous strips of dense grass along roadsides, drainage ditches, and railroads. In general, this vole occupies less dense cover than the prairie vole and is better adapted to moister situations than the prairie vole.

Life History

The life of the meadow vole is similar to that of the prairie vole. Recent studies have shown that multiple paternity exists in meadow voles, and some theories suggest that females allow this in an attempt to reduce infanticide through confusion over paternity.

SELECTED REFERENCES

See also discussion of this species in general references, page 23.

Ambrose, H. W., III. 1973. An experimental study of some factors affecting the spatial and temporal activity of *Microtus pennsylvanicus*. *Journal of Mammalogy* 54:79–110.

Berteaux, D., J. Bety, E. Rengifo, and J. Bergeron. 1999. Multiple paternity in meadow voles (*Microtus pennsylvanicus*): Investigating the role of the female. *Behavioral Ecology and Sociobiology* 45:283–291.

Christian, J. J. 1970. Social subordination, population density, and mammalian evolution. *Science* 168:84–90.

Easterla, D. A., and D. L. Damman. 1977. The masked shrew and meadow vole in Missouri. *Northwest Missouri State University Quarterly* 37. 26 pp.

Getz, L. L., F. R. Cole, and D. L. Gates. 1978. Interstate roadsides as dispersal routes for *Microtus pennsylvanicus*. *Journal of Mammalogy* 59:208–212.

Henterly, A. C., K. E. Mabry, N. G. Soloman, A. S. Chesh, and B. Keane. 2011. Comparison of morphological versus molecular characters for discriminating between sympatric meadow and prairie voles. *American Midland Naturalist* 165:412–420.

Reich, L. M. 1981. *Microtus pennsylvanicus*. *Mammalian Species* 159. 8 pp.

Prairie Vole (*Microtus ochrogaster*)

Name

The first part of the scientific name, *Microtus*, is from two Greek words and means "small ear" (*mikros*, "small," and *ous*, "ear"). This name is somewhat misleading because the ear is of medium size, although it is nearly hidden in the body fur. The last part, *ochrogaster*, is from two Greek words and means "yellow belly" (*ochro*, "yellow," and *gaster*, "belly"). It describes the yellowish tinge on the belly. The common name is self-explanatory.

Sometimes authorities refer to this species as *Pedomys ochrogaster*.

Description (Plate 42)

The prairie vole is a small, stocky rodent with a large head, short legs, and a short tail. The tail is moderately furred and is slightly less than twice the length of the hind foot. The black eyes are small and beady, and the well-furred ears project only slightly beyond the body fur. The lips close tightly behind the upper incisors, keeping soil out of the mouth cavity when the vole digs underground. The front foot has 4 clawed toes and a small thumb bearing a pointed nail. The hind foot has 5 clawed toes and usually 5 but sometimes 6 pads, or plantar tubercles, on the sole, which is furred from the heel to the pads. The long, loose body fur is glossy, coarse, and grizzled.

The prairie vole is distinguished from the closely related meadow vole by a shorter tail (tail slightly less than twice the length of the hind foot), 5 plantar tubercles on the hind foot, body fur coarser with more

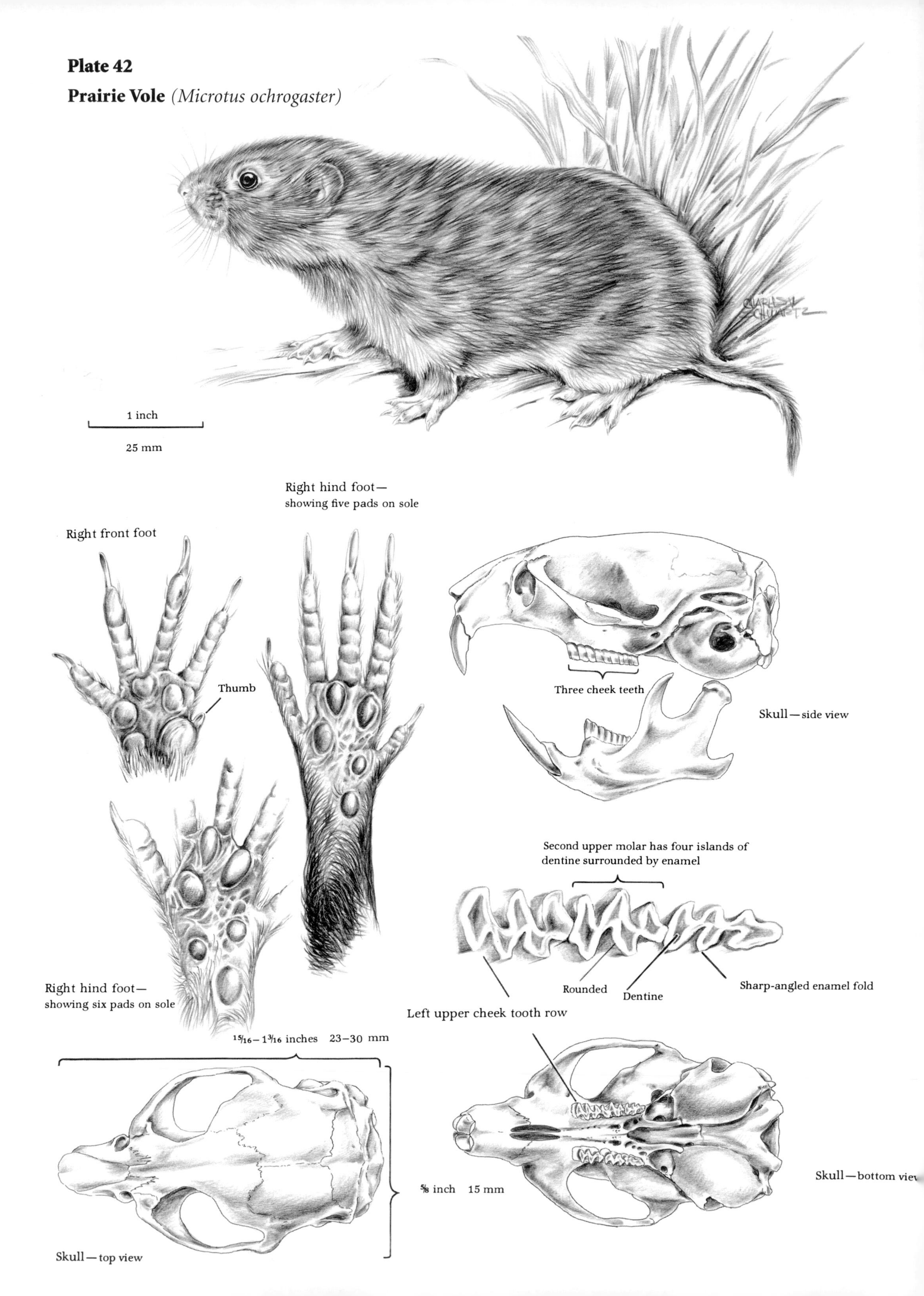

Plate 42
Prairie Vole (Microtus ochrogaster)
1 inch
25 mm
Right hind foot—
showing five pads on sole
Right front foot
Thumb
Three cheek teeth
Skull—side view
Second upper molar has four islands of
dentine surrounded by enamel
Rounded
Dentine
Sharp-angled enamel fold
Left upper cheek tooth row
Right hind foot—
showing six pads on sole
15/16–1 3/16 inches 23–30 mm
5/8 inch 15 mm
Skull—bottom view
Skull—top view

white, and belly fur usually yellowish or rusty. However, in places where these species occur together, the use of morphological characteristics alone may lead to misidentification due to intraspecific variability and overlapping characteristics. For differences in dentition, see *Teeth and Skull* below.

There are two other similar-appearing small mammals in Missouri, the woodland vole and the southern bog lemming. Both of these have a shorter tail (tail nearly same length as hind foot), and the southern bog lemming has grooved incisors.

Color. The upperparts of the prairie vole are grayish to blackish brown, mixed with a whitish, yellowish, or rusty hue, imparting a grizzled appearance; the color at the base of the hairs is dark gray. The sides are paler and the belly is tan or grayish, often washed with whitish, yellow, or a rusty color, especially around the base of the tail; the belly hairs are dark at the base. The feet are grayish tan. The tail is dark above and light below. Adults are rarely black, salmon colored, or albino. The sexes are colored alike. Molting occurs at any time of the year and requires three weeks for completion. In adults it begins in the chest region and works toward the head and tail.

Measurements

Total length	77–117 mm	4⅝–7 in.
Tail	22–47 mm	⅞–1⅞ in.
Hind foot	14–22 mm	9⁄16–⅞ in.
Ear	11–15 mm	7⁄16–⅝ in.
Skull length	23–30 mm	15⁄16–1 3⁄16 in.
Skull width	15 mm	⅝ in.
Weight	21–56 g	¾–2 oz.

Teeth and skull. The dental formula of the prairie vole is:

$$I\ \frac{1}{1}\ C\ \frac{0}{0}\ P\ \frac{0}{0}\ M\ \frac{3}{3} = 16$$

The grinding surfaces of the upper cheek teeth possess a pattern of sharp-angled enamel folds surrounding dentine. The second upper molar has 4 islands of dentine surrounded by enamel; the front border of the second island on the tongue side is rounded on most specimens. There is no lengthwise groove on the outer edge of the upper incisors.

These dental characteristics distinguish the skull of the prairie vole from the closely related meadow vole, the woodland vole, and the southern bog lemming.

Sex criteria. The sexes are distinguished as in the eastern woodrat. There are 3 pairs of teats in females: 2 pairs in the groin region and 1 pair in the chest region.

Age criteria, age ratio, and longevity. The young, until 8 or 9 weeks of age, are a dull gray to black with black feet and tails. They generally weigh less than 21 g (¾ oz.). Age can be determined, in general, by an examination of the skulls. Paired ridges on the upper surface between and behind the eye sockets are farther apart in younger animals but become closer with age. However, the best means for age determination in voles is by the eye-lens weight.

Young voles are more abundant than adults only when the population is increasing. In the wild, the life span is very short. Only a small proportion of the population exceeds 60 days of age. Few individuals become 16 months old. The heavy loss is largely

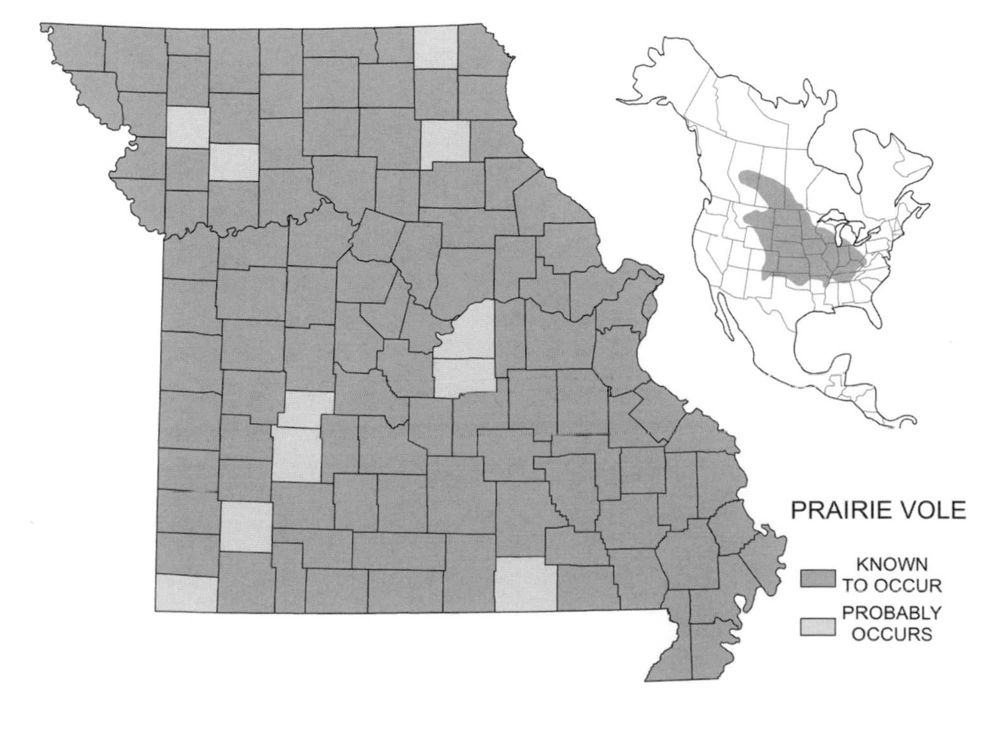

the result of predation by other animals. In captivity, some prairie voles lived for totals of 27 and 35 months.

Glands. Scent glands occur in the groin region of both sexes.

Voice and sounds. Squeaks, squeals, growls, and chattering of the teeth are typical sounds made by prairie voles. In addition, they stamp their front and hind feet.

Distribution and Abundance

The prairie vole ranges throughout the Central Plains of North America. In Missouri, it occurs statewide.

Voles show periods of abundance about every four years. An increasing population is accompanied by a longer breeding season, an increase in the number of young per litter, and proportionately more young per adult. Population decreases, which often occur suddenly in local areas of dense population, are caused by various factors such as drought in summer, severe winter weather, heavy parasitism and predation, epidemic diseases, changes in land use, or strife between individuals, which is intensified in higher concentrations.

Some indication of the size of the population can be ascertained from the number of runways in an area, since the number of runways is directly related to the population density. In years of low population, there may be only 15 to 40 prairie voles per 0.4 ha (1 ac.), but in times of abundance there may be from 60 to 429 per 0.4 ha (1 ac.).

Habitat and Home

Prairie voles live in upland herbaceous fields, grasslands, thickets, fallow fields, under shocks of corn and small grain, along fencerows, and in fields of alfalfa, bluegrass, clover, or lespedeza. They do not habitually dwell in timbered areas.

Nests are made in various locations—aboveground in clumps of vegetation or under debris, at the end of short tunnels off the main runway, or belowground from 5 to 46 cm (2 to 18 in.) deep. The nests are usually ellipsoid and are from 18 to 20 cm (7 to 8 in.) long, 10 to 15 cm (4 to 6 in.) wide, and about 10 cm (4 in.) high. They are woven of coarse grass and have a lining of finer grass or other soft material. There are 1 or 2 entrances to the nest.

Habits

Voles build a system of well-defined runways both on top of the ground and under the ground. Holes about 5 cm (2 in.) in diameter lead from the surface runways to the series of underground tunnels where the vole spends much of its life. Soil removed from the underground tunnels is piled near the hole or is carried and scattered along the runway, forming a pavement.

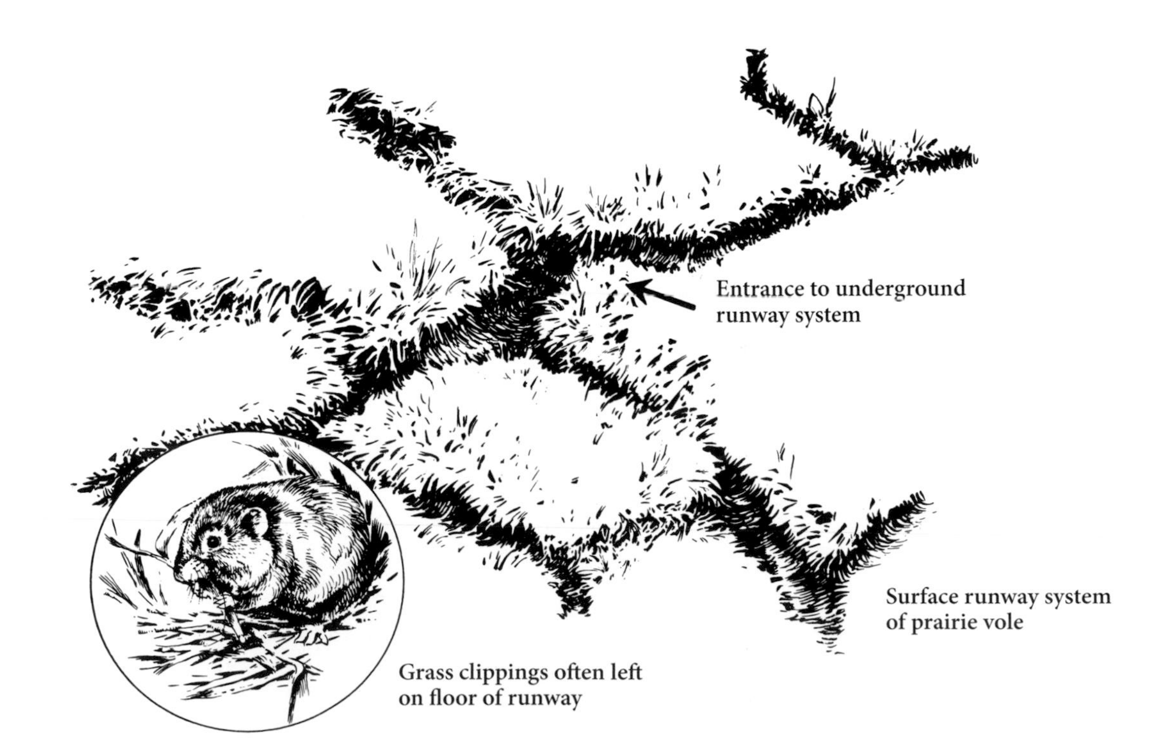

The surface runways are from 2.5 to 5 cm (1 to 2 in.) in diameter and are made by clipping the grass or other vegetation very close to the ground. The floor of the runways consists either of bare ground or a thin layer of trampled grass, stems, and an occasional leaf. The sides and roof are of living vegetation. These trails lead to feeding grounds where the vole may leave the trail and feed in the near vicinity.

Many voles live in the same general area, and the runway system belongs to the entire community. Yet individual voles tend to use specific portions and each has one or more nests of its own. However, sometimes they may share nests and stores of food. Other kinds of mice, shrews, and moles travel through the runway system on occasion. There is some evidence that southern bog lemmings may dominate voles when the ranges of these species overlap. However, when an area occupied by voles is taken over by lemmings, this probably is the result of successional changes in the vegetation that favor the lemmings rather than the result of physical conflict between the species.

In general, voles are sociable, but males show more aggression toward each other than toward females. On occasion, both sexes are strong fighters. Meadow voles tend to be more aggressive toward their own kind than are prairie voles.

Voles have a restricted home range and usually live within 0.1 to 0.2 ha (⅓ to ½ ac.). Individuals liberated up to 183 m (200 yd.) from their home have returned, but those taken farther failed to do so.

These voles are active at any time of the day or night but are most active in midday. They seemingly live on a four-hour schedule. This includes eating for the first part of the four-hour period and sleeping for the last part.

Voles are neat and clean and keep their fur carefully groomed. Their small brownish droppings are deposited along the runways and in blind alleys off the main runways.

Prairie voles swim voluntarily and have been observed to swim distances up to 27.4 m (90 ft.). When swimming on the surface of the water, most of the back is exposed. When swimming underwater, air bubbles become trapped in the fur and help keep the body from becoming wet.

Foods

The main foods of voles are the tender stems, leaves, roots, tubers, flowers, seeds, and fruits of grasses, sedges, and many other succulent plants. At times insects, snails, crayfish, and other mice are eaten and, when food is scarce, even the inner bark of trees, shrubs, and vines are consumed.

Food is stored in underground chambers near the nest and often aboveground in hollow stumps and similar localities. A cache may contain as much as 7.5 l (2 gal.) of tubers, roots, and small bulbs. A vole is a large eater, consuming its own weight in green food every 24 hours. In addition, it often cuts and wastes more growing vegetation than it eats. When feeding on tall plants, the vole clips the stem close to the ground. Piles of cut stems are left along the runways.

Water is probably not required in the wild because of the succulent diet, but captive animals fed on grain take large amounts of water, which they lap with their tongues.

Voles, like rabbits and some other rodent species, exhibit coprophagy, ingesting their own soft droppings when they rest following a period of feeding.

Reproduction

The breeding season may encompass the entire year, but peaks occur in spring and fall. The gestation period is 21 days, and a female can be both pregnant and nursing.

The prairie vole is one of the most prolific mammals known. Many litters are produced annually, the number being influenced by the food supply, temperature, amount of cover, including snow cover, abundance of mates, and other factors. Each litter contains from 1 to 7 young, but 3, 4, and 5 are the most common numbers. There are fewer young per litter at the beginning and end of the breeding season than at other times. Older and larger females average more young per litter than other females. In the closely related meadow vole, a captive female produced 17 litters in one year, and

Prairie vole tracks

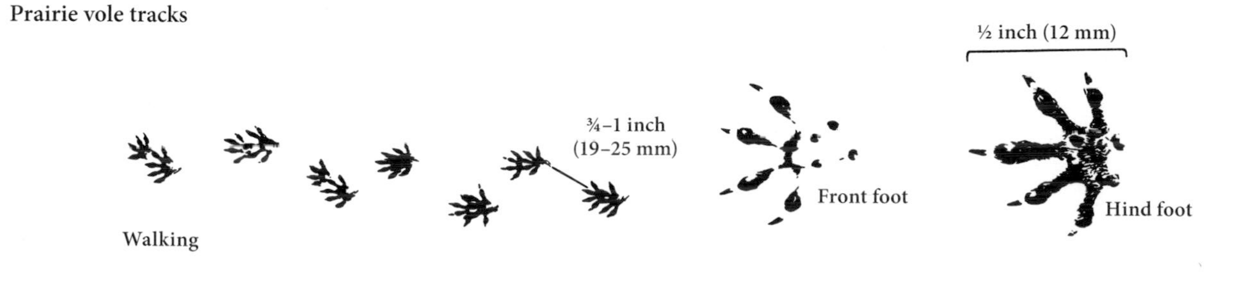

one of her daughters produced 13 families totaling 78 offspring before she was one year old.

A litter of 6 young was born over a period of 50 minutes, each delivery taking from 4 to 5 seconds. The female aided the young by pulling with her teeth and, following the births, ate the placentas.

At birth the young are pink and hairless, and weigh about 2.8 g (1/10 oz.). The eyes are closed and the ears folded against the head. When nursing, the young cling tightly to the female's teats, and are often carried in this way if she moves suddenly. The female protects her young and is very concerned about their welfare. If she needs to move them, she carries each one separately with her mouth.

About the fifth day after birth, the backs of the young are covered with velvety fur; in another day or two the incisors come through the gums. The eyes open and ears unfold at about 8 days of age. Weaning occurs when the young are from 2 to 3 weeks old.

At 8 or 9 weeks of age, the young are about 5 cm (2 in.) long and weigh between 28 to 42 g (1 and 1½ oz.). They soon begin to molt their juvenile fur and gradually acquire adult coloration. They reach adult size in 3 months.

Voles may breed at a very early age. Males are capable of breeding when 5 weeks old, but females can do so as early as 25 days of age, bearing young when they are only 45 days old. This difference in the time of reaching sexual maturity reduces the chances of mating by littermates.

Some Adverse Factors

Because the prairie vole is an important food of many animals, the list of predators is long. It includes opossums, shrews, raccoons, badgers, minks, skunks, domestic dogs, domestic cats, bobcats, foxes, coyotes, weasels, snakes, hawks, owls, crows, and even bullfrogs and snapping turtles.

These voles are parasitized by mites, ticks, lice, fleas, botfly larvae and other parasitic flies, flukes, and tapeworms. Some of these parasites carry organisms that cause several diseases in humans. These are sylvatic plague, spread by fleas; tularemia, spread by insects, especially the deerfly; Rocky Mountain spotted fever, spread by ticks; and rat-bite fever, spread by the bite of an infected mouse. In addition, prairie voles are the hosts for many parasites, such as ticks, whose later developmental stages parasitize many larger mammals, including game, furbearing, and domestic mammals, and even humans.

Occasionally voles become accidentally trapped in a tangle of weeds or briers where they may break a leg or die of starvation.

Importance

When voles are abundant, they often do considerable damage through their feeding habits. First, their incisors and molariform teeth are ever growing, permitting them to bite and chew tough plants (like grass); second, their digestive system, with the aid of microorganisms, can digest complex carbohydrates in the fibrous portion of grasses. Thus, they are direct competitors with humans for grasses (including cereals) and other flowering plants. They nibble on sprouting corn, eat the corn germ in winter shocks, damage vegetables in gardens, and consume large quantities of vegetation such as alfalfa. A theoretical population of 100 voles per 0.4 ha (1 ac.) in one year will eat 136 kg (300 lbs.) of alfalfa hay per 0.4 ha (1 ac.), or 87 metric tons (96 tons) per section (259 ha; 640 ac.), and waste at least twice this amount. Under conditions of prolonged snow they often girdle the base of young fruit trees or nursery stock.

Because of this damage, the contribution of voles to their environment is usually overlooked. They continually work the soil through their tunneling and other life activities, adding and mixing their stores of food and waste products with the soil. Their clipping of stems and leaves stimulates new tender growth of plants, and their working of the soil also favors better plant growth. They convert the vegetation they eat into their own flesh and pass the nutrients on to the many other animals that feed on their bodies. Because they are so common and form a ready food supply for so many predators, they actually reduce predation on other and more desirable animals, like game species. Also, when they are abundant, they feed large

numbers of predators and thus influence population trends in many animals, such as foxes.

Conservation and Management

Where it is desirable to keep voles away from cultivated fields or orchards, the elimination or close cutting of the ground cover around the crop will prove a partial barrier to their travel. Cinders or hardware cloth can be placed around individual trees to keep voles from girdling them. In small gardens, snaptraps will help eliminate these rodents. If you are experiencing problems with large numbers of voles, contact a wildlife professional for advice, assistance, regulations, or special conditions for handling these animals. Predators, especially hawks and owls, should be encouraged as a natural management option.

SELECTED REFERENCES

See also discussion of this species in general references, page 23.

Ambrose, H. W., III. 1973. An experimental study of some factors affecting the spatial and temporal activity of *Microtus pennsylvanicus*. *Journal of Mammalogy* 54:79–110.

Batzli, G. O., L. L. Getz, and S. S. Hurley. 1977. Suppression of growth and reproduction of microtine rodents by social factors. *Journal of Mammalogy* 58:583–591.

Beer, J. R., and C. F. MacLeod. 1961. Seasonal reproduction in the meadow vole. *Journal of Mammalogy* 42:483–489.

Blair, W. F. 1940. Home ranges and populations of the meadow vole in southern Michigan. *Journal of Wildlife Management* 4:149–161.

Carroll, D., and L. L. Getz. 1976. Runway use and population density in *Microtus ochrogaster*. *Journal of Mammalogy* 57:72–76.

Crawford, R. D. 1971. High population density of *Microtus ochrogaster*. *Journal of Mammalogy* 52:478.

Fisher, H. J. 1945. Notes on voles in central Missouri. *Journal of Mammalogy* 26:435–436.

Getz, L. L., B. McGuire, T. Pizzuto, J. E. Hofmann, and B. Frase. 1993. Social organization of the prairie vole (*Microtus ochrogaster*). *Journal of Mammalogy* 74:44–58.

Getz, L. L., M. K. Oli, J. E. Hofmann, and B. McGuire. 2006. Vole population fluctuations: Factors that initiate and determine intervals between them in *Microtus ochrogaster*. *Journal of Mammalogy* 87:387–393.

Getz, L. L., L. E. Simms, B. McGuire, and M. E. Snarski. 1997. Factors affecting life expectancy of the prairie vole, *Microtus ochrogastor*. *Oikos* 80:362–370.

Hall, E. R., and E. L. Cockrum. 1953. A synopsis of the North American microtine rodents. *University of Kansas Publications, Museum of Natural History* 5:373–498.

Hamilton, W. J., Jr. 1937. Activity and home range of the field mouse, *Microtus pennsylvanicus pennsylvanicus* (Ord). *Ecology* 18:255–263.

———. 1941. Reproduction of the field mouse, *Microtus pennsylvanicus* (Ord). *Cornell University Agricultural Experiment Station, Memoir* 237:1–23.

Harris, V. T. 1941. The relation of small rodents to field borders on agricultural lands in central Missouri. M.S. thesis, University of Missouri, Columbia. 86 pp.

Henterly, A. C., K. E. Mabry, N. G. Soloman, A. S. Chesh, and B. Keane. 2011. Comparison of morphological versus molecular characters for discriminating between sympatric meadow and prairie voles. *American Midland Naturalist* 165:412–420.

Jameson, E. W., Jr. 1947. Natural history of the prairie vole. *University of Kansas Publications, Museum of Natural History* 1:125–151.

Martin, E. P. 1956. A population study of the prairie vole (*Microtus ochrogaster*) in northeastern Kansas. *University of Kansas Publications, Museum of Natural History* 8:361–416.

Rose, R. K., and A. M. Spevak. 1978. Aggressive behavior in two sympatric microtine rodents. *Journal of Mammalogy* 59:213–216.

Stalling, D. T. 1990. *Microtus ochrogastor*. *Mammalian Species* 355. 9 pp.

Stump, W. Q., Jr., and R. G. Anthony. 1983. Use of eye lens protein for estimating age of *Microtus pennsylvanicus*. *Journal of Mammalogy* 64:697–700.

Terman, M. R. 1974. Behavioral interactions between *Microtus* and *Sigmodon*: A model for competitive exclusion. *Journal of Mammalogy* 55:705–719.

Thomas, R. E., and E. D. Bellis. 1980. An eye-lens weight curve for determining age in *Microtus pennsylvanicus*. *Journal of Mammalogy* 61:561–563.

Woodland Vole (*Microtus pinetorum*)

Name

The first part of the scientific name, *Microtus*, is from two Greek words and means “small ear” (*mikros*, “small,” and *ous*, “ear”). This refers to the nearly concealed ears. The last part, *pinetorum*, is of Latin origin and means “belonging to the pines” (*pinetum*, “a pine woods,” and *-orium*, “belonging to a place of”). This name refers to the Georgia pine forests where this species was first collected.

Other names used by different authorities are *Pitymys pinetorum* and *Pitymys nemoralis*. This species is also commonly referred to as the pine vole.

Plate 43

Woodland Vole *(Microtus pinetorum)*

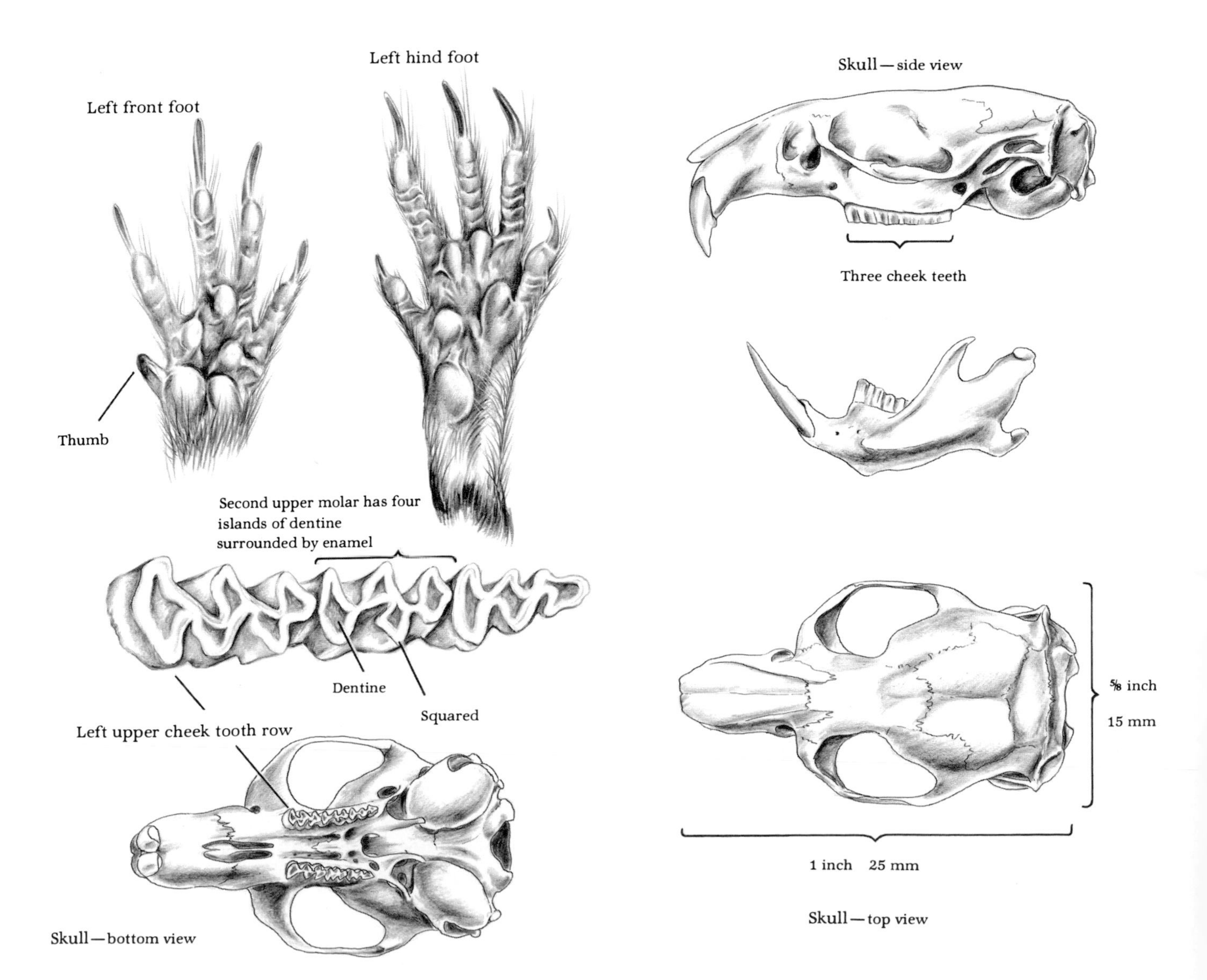

Description (Plate 43)

The woodland vole is a small, thickset rodent with a large head, short legs, and a short tail that is about the same length as the hind foot. The eyes are small, and the ears are nearly concealed in the fur. The lips close tightly behind the upper incisor teeth, helping to keep soil out of the mouth cavity when the vole digs underground. There are 4 toes and a small thumb on each front foot and 5 toes on each hind foot. The soles of the hind feet are furred from the heel to the 5 pads, or tubercles. The short, loose body fur is thick, soft, reddish, and glossy.

The woodland vole is distinguished from prairie and meadow voles by the shorter tail (tail nearly the same length as the hind foot) and from the southern bog lemming by the absence of grooved incisors.

Color. The woodland vole is predominately reddish brown above, but the fur, when parted, shows an undercolor of dark gray. The sides are paler than the back, and the belly is grayish washed with buff; the hairs of the belly are dark at the base. The feet are grayish tan; the tail is dark above and only slightly lighter below. Occasional individuals are buffy or have white spots. The sexes are colored alike. Adults molt in early spring, throughout the summer, and again in late fall and early winter.

Measurements

Total length	82–146 mm	3¼–5¾ in.
Tail	15–25 mm	⅝–1 in.
Hind foot	15–19 mm	⅝–¾ in.
Ear	7–11 mm	5/16–7/16 in.
Skull length	25 mm	1 in.
Skull width	15 mm	⅝ in.
Weight	21–56 g	¾–2 oz.

Teeth and skull. The dental formula of the woodland vole is:

$$I\,\frac{1}{1}\;C\,\frac{0}{0}\;P\,\frac{0}{0}\;M\,\frac{3}{3} = 16$$

The grinding surfaces of the upper cheek teeth possess a pattern of sharp-angled enamel folds surrounding dentine. The second upper molar has 4 islands of dentine surrounded by enamel; the front border of the second island on the tongue side is squared on most specimens. There is no lengthwise groove on the outer edge of the upper incisors.

These dental characteristics distinguish the skull of the woodland vole from the closely related meadow vole, the prairie vole, and the southern bog lemming.

Sex criteria. The sexes are identified as in the eastern woodrat. There are two pairs of teats in the abdominal region of females.

Age criteria and longevity. Compared to adults, the young have darker, fuzzier, and more lead-colored fur. Age is also told by the smaller size. Woodland voles seldom reach 10 to 12 months of age.

Glands. There are paired glands on the hips and in the groin region.

Voice and sounds. The woodland vole gives low, birdlike chirps when alarmed or harassed. When fighting, it emits harsh *chirrs*.

Distribution and Abundance

The woodland vole is fairly common in the eastern United States but is of special concern in Canada. The population shows periods of abundance every 3 to 4 years. Usually between 80 and 90 voles per 0.4 ha (1 ac.) is a high density, but in one locality during a population "high," densities reached 300 woodland voles per 0.4 ha (1 ac.).

The woodland vole occurs throughout Missouri. It is generally rare but is most common in the Ozark Highlands.

Habitat and Home

The woodland vole lives underground in oak-hickory forests and sometimes in mixed hardwood and pine forests where there is a heavy layer of dead

Woodland vole tracks

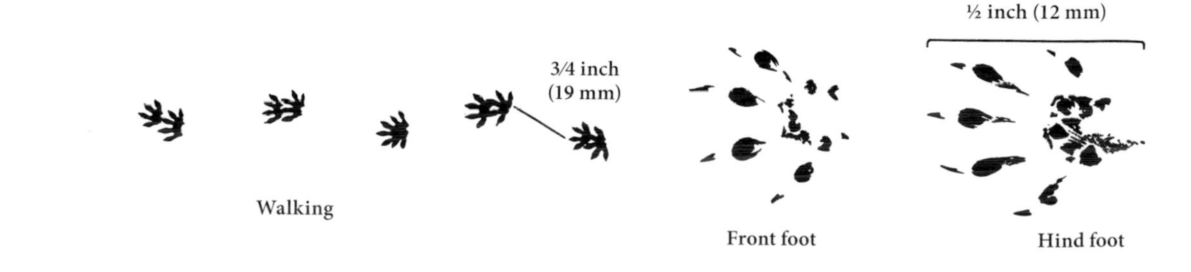

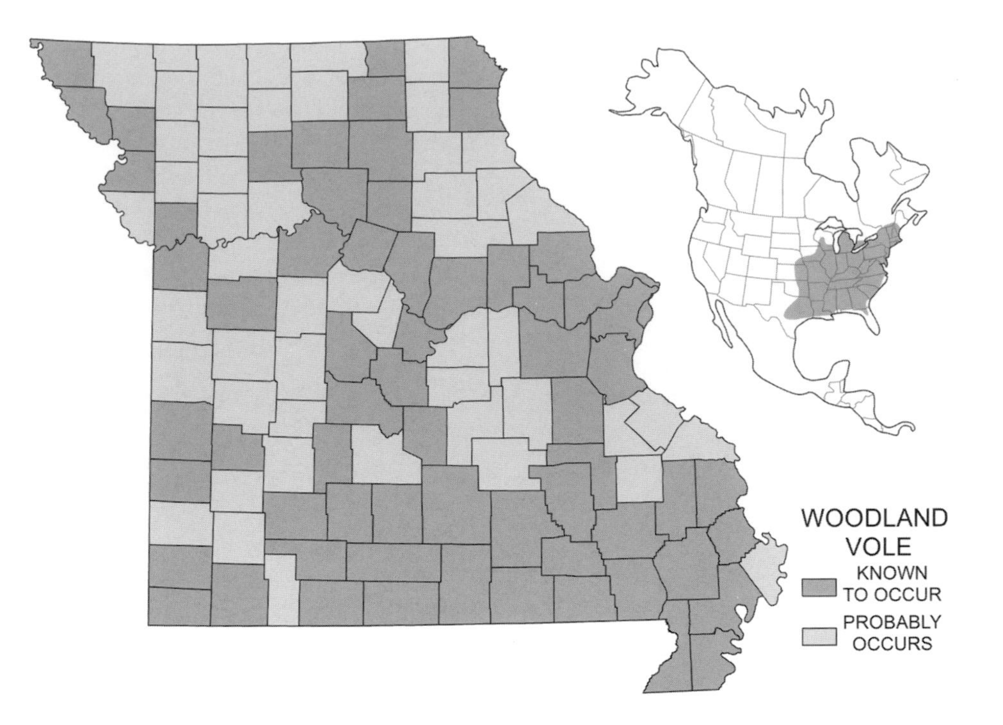

leaves or a dense mat of grass and other cover. It also lives in fields adjacent to timber, and in orchards, gardens, and scrub areas, providing they possess considerable ground litter. Loose, moist soils are preferred because they are easy to dig.

The spherical nest is built beneath a log, just below the surface litter, or several inches underground. It consists of shredded dead grass, leaves, and rootlets with a lining of short, fine pieces of grass. There may be 3 to 4 entrances.

Habits

The woodland vole makes tunnels from 2.5 to 5 cm (1 to 2 in.) in diameter. These are just under the carpet of leaves and grass and from 10 to 30 cm (4 to 12 in.) or more deep. Numerous holes open at intervals from these tunnels and lead to surface feeding grounds. Piles of soil, excavated from the tunnel system, may occur near these openings. Woodland voles seldom travel far on top of the ground; they feed mostly in the tunnels.

In digging, the woodland vole uses its teeth, head, and front feet to loosen the soil and the hind feet to push the soil behind. When a pile of soil has accumulated, the vole pushes it out of the tunnel with its head. Although voles dig their own tunnels, they may use nearby tunnels of moles, shrews, and other kinds of mice.

Woodland voles have an average home range of about 0.1 to 0.13 ha (¼ to ⅓ acre). They may remain in this area for their entire lives or gradually shift their home, occupying new areas. Voles released up to 46 m (150 ft.) from the point of capture have returned to the home area.

These voles tend to live in colonies. They are not very aggressive to members of their own species but are more so than the prairie vole. There is a report of one nest containing 3 litters, presumably belonging to 2 or 3 females.

Woodland voles do not hibernate and are slightly more active during the night than during the day. They are poor jumpers and climbers but are capable of

swimming. They normally walk slowly but have been timed to run at a rate 6 km per hour for nearly 8 m (3.8 mi. per hour for 25 ft.).

Foods

Because of its subterranean habits, much of this vole's food comes from below the surface of the ground. Succulent roots and tubers of many kinds of plants, sprouts, the tender bark of tree roots, stems, leaves, seeds, nuts, berries, apples, and an occasional insect or body of a dead woodland vole compose the diet. A fallen apple is consumed from the bottom by digging up underneath it.

Some food is stored in underground chambers that may contain as much as 3.8 l (1 gal.) of tubers. When filled, the chamber entrance is closed with soil. In captivity, woodland voles drink large amounts of water.

Reproduction

The breeding season encompasses most of the year, beginning in January and ending in November. The peak occurs in March and April. Several litters of 2 to 4 young, with extremes of 1 and 8, are born annually. The gestation period is about 21–24 days. From 1 to 6 litters may be born a year. The annual production is generally lower than that of meadow and prairie voles.

At birth the young are blind, naked, weigh about 2.8 g (1/10 oz.), and are between 38 and 42 cm (1¼ and 1⅜ in.) long. They hang onto the female's teats very tenaciously and are often dragged about when she moves suddenly. Young prefer to attach to the 2 rear teats, from which they are less likely to be dislodged. At 5 to 6 days of age, fur appears and the incisors cut through the gums. The ears unfold on the eighth day, and the eyes open between the ninth and twelfth days. The young leave the nest for short periods by 2 weeks of age and are weaned between 16 and 21 days of age. The young develop adult fur and coloration when between 7 and 10 weeks old; they are ready to breed at 2 months.

Some Adverse Factors

Owls, hawks, snakes, opossums, coyotes, foxes, domestic cats, raccoons, and minks are known predators on woodland voles, although shrews, weasels, and other carnivorous mammals may also prey on this species. Predation, however, is probably light because of the woodland vole's subterranean existence. The reproductive rate also indicates that mortality is

low in contrast to the prairie vole, which is more prolific and more heavily preyed upon.

The external parasites found on woodland voles are mites, ticks, lice, and fleas; the recorded internal parasites are tapeworm larvae, eggs of roundworms, and adult spiny-headed worms. A fatal skin disease frequently occurs in high populations.

Importance

Where woodland voles are abundant, they may damage orchard trees. They do this by their underground tunneling, severing of smaller roots, and girdling of larger roots. These voles may also reduce yields of some crops by digging along rows of potatoes and other root vegetables. Rabbits are often blamed for the work of woodland voles.

The tunneling by this species contributes to aeration of the soil and helps prevent the rapid runoff of rain. Other interrelations of voles in general with their environment are given under the prairie vole.

Conservation and Management

Because of their underground habits, the control of woodland voles is often difficult.

SELECTED REFERENCES

See also discussion of this species in general references, page 23.

Benton, A. H. 1955. Observations on the life history of the northern pine mouse. *Journal of Mammalogy* 36:52–62.

Burt, W. H. 1940. Territorial behavior and populations of some small mammals in southern Michigan. *University of Michigan, Museum of Zoology, Miscellaneous Publication* 45:7–58.

Hall, E. R., and E. L. Cockrum. 1953. A synopsis of the North American microtine rodents. *University of Kansas Publications, Museum of Natural History* 5:373–498.

Hamilton, W. J., Jr. 1938. Life history notes on the northern pine mouse. *Journal of Mammalogy* 19:163–170.

McGuire, B., and S. Sullivan. 2001. Suckling behavior of pine voles (*Microtus pinetorum*). *Journal of Mammalogy* 82:690–699.

Miller, D. H., and L. L. Getz. 1969. Life-history notes on *Microtus pinetorum* in central Connecticut. *Journal of Mammalogy* 50:777–784.

Novak, M. A., and L. L. Getz. 1969. Aggressive behavior of meadow voles and pine voles. *Journal of Mammalogy* 50:637–639.

Paul, J. R. 1970. *Observations on the ecology, populations, and reproductive biology of the pine vole in North Carolina.* Report of Investigations 20, Illinois State Museum, Springfield. 28 pp.

Powell, R. A., and J. J. Fried. 1992. Helping by juvenile pine voles (*Microtus pinetorum*), growth and survival of younger siblings, and the evolution of pine vole sociality. *Behavioral Ecology* 3:325–333.

Raynor, G. R. 1960. Three litters in a pine mouse nest. *Journal of Mammalogy* 41:275.

Schadler, M. H., and M. Butterstein. 1979. Reproduction in the pine vole, *Microtus pinetorum*. *Journal of Mammalogy* 60:841–844.

Smolen, M. J. 1981. *Microtus pinetorum*. *Mammalian Species* 147. 7 pp.

Common Muskrat (*Ondatra zibethicus*)

Name

The first part of the scientific name, *Ondatra*, is the Iroquois name for this animal. The last part, *zibethicus*, is the New Latin word for "musky-odored" and refers to this rodent's characteristic odor. Part of the common name, "musk," also refers to this scent, while the rest, "rat," comes from the Anglo-Saxon *raet*.

Description (Plate 44)

The muskrat is a medium-sized rodent with a broad head, stocky body, short legs, and a vertically flattened, sparsely haired, and scaly tail that is slightly shorter than the combined length of head and body. The eyes are small, and the ears barely project beyond the fur. The lips close behind the incisor teeth, permitting the muskrat to gnaw underwater. The small front feet have 4 clawed toes and a nailed thumb; the large hind feet have 5 clawed toes that are webbed at their bases. There is a fringe of stiff hairs on the edge of the web, the sides of the toes, and edge of the foot. Five tubercles are present on the soles of the hind feet. The pelage consists of a dense coat of waterproof underfur and long glossy overhairs.

Color. The back of the adult muskrat is dark blackish brown, while the sides are lighter brown with a reddish or sometimes yellowish tinge. The underparts are still lighter, shading to white on the throat, and have a silvery cast caused by the underfur showing through the overhairs. There is a small blackish spot on the chin and blackish fur at the wrists and ankles. The feet are dark brown and the tail is blackish brown. Some individuals are almost entirely black, tan, or white. The sexes are colored alike.

Adult muskrats molt continuously, but young ones molt twice during their first year. In fall, the fur of adults is more prime, or best for wearing quality and appearance, than the fur of young animals, but the fur of both ages is most prime in late winter or early spring.

Measurements

Total length	406–641 mm	16–25¼ in.
Tail	177–292 mm	7–11½ in.
Hind foot	63–92 mm	2½–3⅝ in.
Ear	19–25 mm	¾–1 in.
Skull length	60–69 mm	2⅜–2¾ in.
Skull width	38–44 mm	1½–1¾ in.
Weight	680–1,814 g	1½–4 lb.

Males are heavier than females. In 2013, a 1.6 kg (3.6 lb.) male became the official record-weight muskrat confirmed by the Missouri Department of Conservation. The state furbearer record program was started in 2011.

Teeth and skull. The dental formula of the muskrat is:

$$I\,\frac{1}{1}\;C\,\frac{0}{0}\;P\,\frac{0}{0}\;M\,\frac{3}{3} = 16$$

The grinding surfaces of the cheek teeth have a characteristic pattern of sharp-angled enamel folds

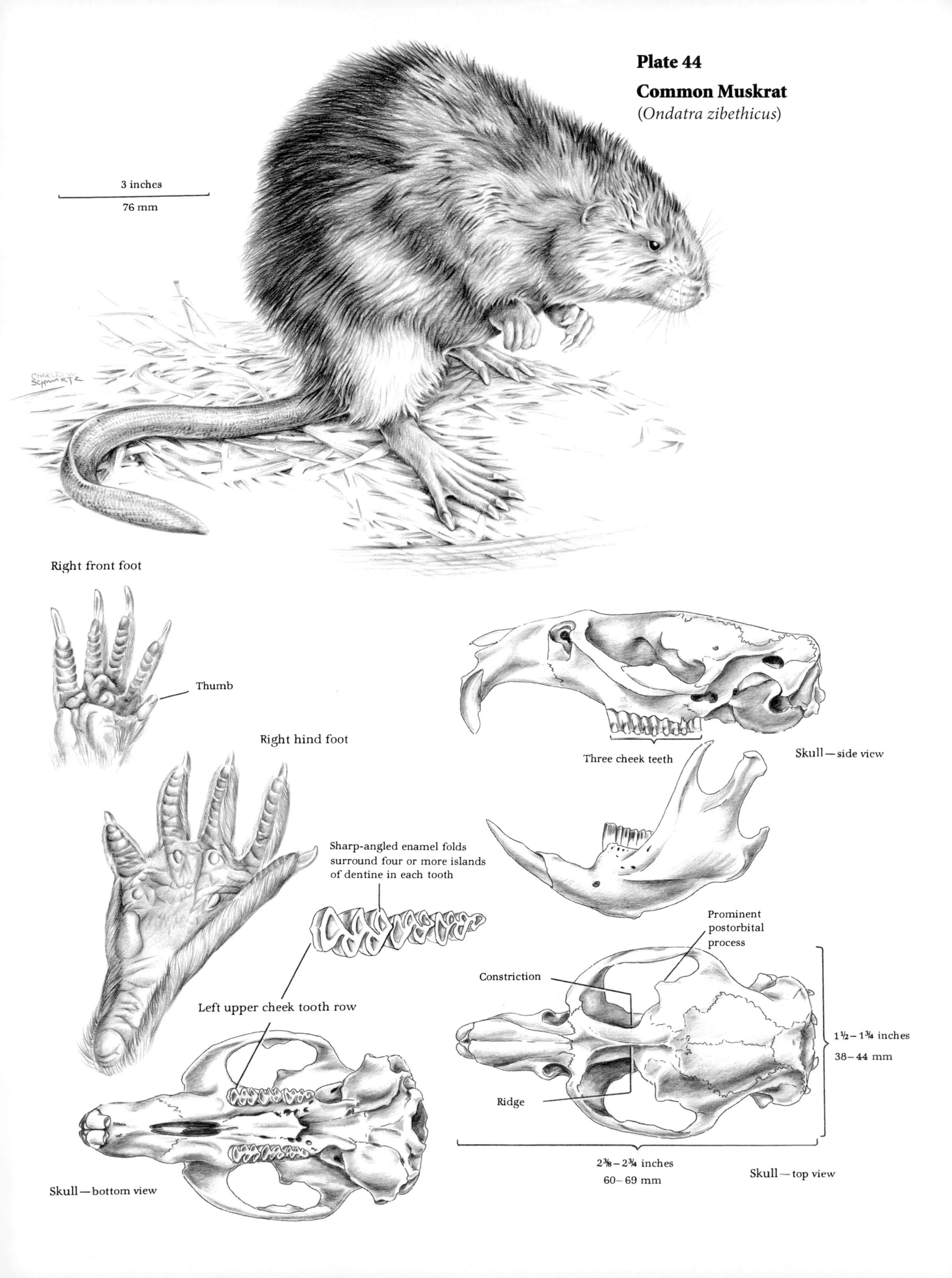
Plate 44
Common Muskrat
(Ondatra zibethicus)
3 inches
76 mm
Right front foot
Thumb
Right hind foot
Three cheek teeth
Skull—side view
Sharp-angled enamel folds surround four or more islands of dentine in each tooth
Prominent postorbital process
Constriction
Left upper cheek tooth row
1½–1¾ inches
38–44 mm
Ridge
2⅜–2¾ inches
60–69 mm
Skull—top view
Skull—bottom view

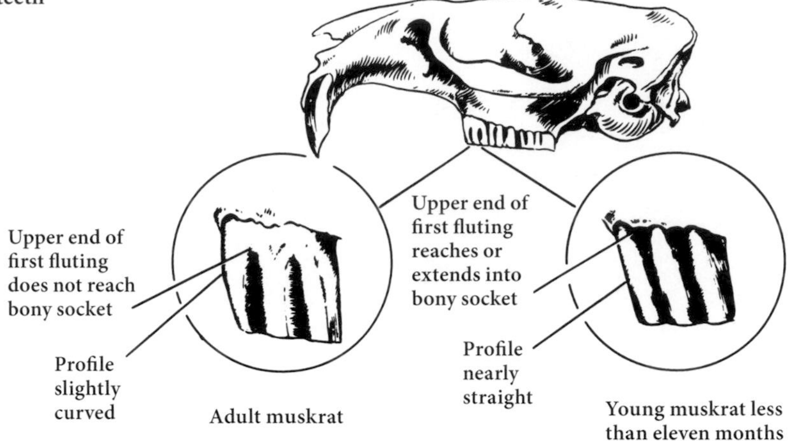

Age in the common muskrat as indicated by the first upper molar teeth

surrounding dentine. There are 4 or more islands of dentine in each tooth.

Among Missouri rodents with 16 teeth, the skull of the muskrat could be confused only with skulls of the brown rat, the black rat, or the eastern woodrat on the basis of size. One way to distinguish these skulls is by the pattern on the grinding surfaces of the upper cheek teeth. Other skull differences are as follows: a definite constriction occurs in the region between the eye sockets in the muskrat, but only a gradual narrowing occurs in this region in the woodrat, the brown rat, and the black rat; the bones between the eye sockets in muskrats are raised to form a slight central ridge but are depressed in the woodrat and rather flat in the brown rat and the black rat; and the postorbital processes are prominent in the muskrat but not prominent in the woodrat, the brown rat, or the black rat.

Sex criteria and sex ratio. The sexes are rather difficult to identify because both have a prominent urethral papilla. In males, the penis can usually be felt within the urethral papilla; in females, the urethral papilla lies just in front of the vagina, which may be open or covered by a membrane. The region between the urethral papilla and anus is greater and furred in males; less and naked near the urethral papilla in females. Males lack teats but females have from 3 to 5 pairs on the belly between the front and hind legs. In pelts, sex can be determined by the presence or absence of teats, which show up better in fall pelts than in spring ones.

Sex ratios tend to be nearly even among adults and in litters.

Age criteria, age ratio, and longevity. Young muskrats are duller in color than adults, being grayer on the back and paler on the sides.

The best means of distinguishing juvenile males from adult males during the winter trapping season is by the baculum, or penis bone. In young males up to 8 months of age, the distal (outer) end of the baculum is entirely cartilaginous, but in adults it has nearly completely turned to bone.

Internally, age is determined by an examination of the testes in males and uteri in females. In adult males, the testes are 1 cm (7/16 in.) or more long, flattened, wrinkled, and discolored; in males before their first breeding season, the testes are smaller, turgid, and cream colored. In adult females, the uteri are thickened and contain placental scars, or places of attachment of embryos; in females before their first breeding season the uteri are thin and transparent and lack placental scars. This method is reliable until early spring when the young muskrats start to breed.

The width of the skull may indicate age, but there is some question as to its reliability. If the greatest width of the skull is more than 4 cm (1 9/16 in.), the muskrat is usually an adult; if less than this, it is usually young.

Another method, but of doubtful value when applied to Missouri muskrats, is based on a tooth character, that of the first upper molar. In muskrats less than 11 months of age, the upper end of the first fluting extends deep into the bony socket and the front edge of the tooth is straight or nearly so in profile. In adult muskrats, the upper end of at least the first fluting does not reach the bony socket and the front edge of the tooth is slightly curved (see accompanying illustration).

An examination of the flesh side of pelts will reveal the age with reasonable accuracy. Those pelts with an unsymmetrical, mottled pattern of dark and light areas, indicating unprime and prime fur respectively, are from adults, while those with a bilaterally symmetrical pattern of dark and light areas are from young animals. In addition, pelts of adult females have

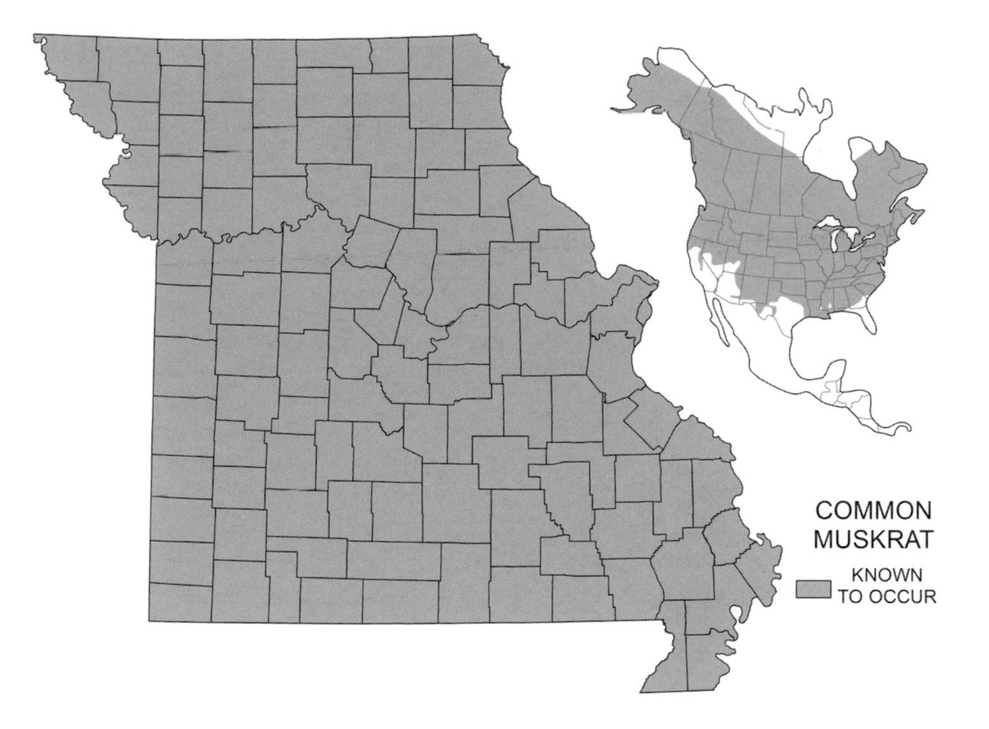

teats from 1.6 mm to 4.8 mm (1⁄16 to 3⁄16 in.) in diameter that are usually blackened, while pelts of young females have teats less than 1.6 mm (1⁄16 in.) in diameter that are usually unpigmented.

Age ratios in the harvests from 1947–1948 to 1977–1978 showed between 3.5 and 6.5 young per adult.

The life span is fairly short. Mortality is especially heavy among young muskrats, and only about a third of the young live into their first winter. One tagged muskrat is known to have lived 4 years in the wild.

Glands. Paired glands lie just under the skin on either side of the penis in males, their ducts opening within the foreskin of the penis. These glands enlarge during the breeding season and have a musky secretion. This becomes mixed with the urine and is deposited on feeding platforms and defecation posts and along trails. It is also expelled just before or at the time of mating. There are similar glands in females, but they do not enlarge as much as do those of males.

Voice and sounds. Adult muskrats give low squeaks, loud squeals, snarls, and moans and, when cornered or fighting, chatter their incisors. A high-pitched *n-n-n-n* is given during the breeding season by both sexes. The young have a characteristic squeaky cry.

Distribution and Abundance

The common muskrat is widely distributed throughout Alaska, Canada, and the United States. It has also been successfully introduced into Europe and Asia. In most of its North American range it is rather common.

Although the common muskrat is one of the most abundant furbearing mammals in Missouri, it has never been as numerous here as in some other parts of its range. The construction of thousands of farm ponds throughout the state has accounted for an increase in muskrat numbers and distribution.

Habitat and Home

Common muskrats prefer still or slowly running water with vegetation in the water and along the shore. In Missouri they live in marshes, sloughs, streams, rivers, ponds, and lakes. In coastal parts of the eastern and southern United States, they commonly inhabit saltwater marshes. In some parts of their range they are known to be tolerant of urbanization and invasive vegetation and may be urban adapters.

In Missouri the muskrat usually digs its home in a bank; if banks are not available, it builds a house out of vegetation in the water.

A bank den is about 15 cm (6 in.) wide and 20 cm (8 in.) high. It is reached by means of a tunnel 12 to 15 cm (5 to 6 in.) in diameter and 3 to 15 m (10 to 50 ft.) long, which usually opens underwater. During periods of low water, a canal may be built to connect the tunnel opening with deeper water. Bank dens may be from only a few centimeters to 1.5 or more meters (a few inches to 5 or more feet) below the ground surface. When dens are near the surface, some protection is offered by roots of trees or shrubs or by thickly matted sod. In heavy soil there may be an air shaft to the surface, but in most cases the air penetrates through the loose soil. When a female uses a bank den for her young, she sometimes builds 2 or 3 chambers with connecting tunnels and several openings to the water.

Common muskrat at bank den

In marshes or other shallow waters, the muskrat makes its house out of aquatic vegetation in the immediate vicinity. The foundation of such a house is placed on the bottom of the marsh or pond and consists of roots or some other support. Heaped upon this is a conical pile, up to 2.4 m (8 ft.) in diameter and 1.2 m (4 ft.) high, of grass, roots, or stems. There is usually only one nest chamber, but sometimes two or more are made. One or more tunnels connect them to the water. Occasionally, each member of the family works out an alcove for itself, and the original chamber may thus be divided into many parts. The walls of the house are about 0.3 meters (1 ft.) thick and are sometimes cemented with mud. The soggy walls of vegetation act as insulation by keeping the inside cool in summer and warm in winter.

Usually the bottom of the marsh or pond is deepened around the house and, if the water is too shallow, canals may be built to deeper water. Most of the house material comes from the bottom of the pond during the simultaneous construction of house and canal. The muskrat carries the excavated material in its mouth.

Habits

In addition to building homes, muskrats often undertake other construction. The canals leading from the home to deep water may connect with surface trails to feeding grounds or to other underground tunnels. When digging canals, muskrats loosen the material with the front feet and cast it up with the hind feet. All the debris is piled at one spot along the canal; where the pile emerges from the water, platforms are made for resting and feeding. Other shelters may be provided and used near the home, and sometimes floating rafts of bent-down vegetation are also utilized. Some of the feeding stations are large enough to be mistaken for a house but can be recognized by the accumulation of cut plants. Only rarely are jetties constructed along marshy shores for landing places or hauling docks.

Any rock, log, or hummock projecting from the water may be used as a defecating post. The muskrat usually crawls out of the water for this purpose, depositing from 3 to 12 black oval droppings at one time. Such defecating posts may measure 46 cm (18 in.) in diameter and have their entire surface covered with muskrat dung.

In the spring, both sexes may leave their winter homes and move about until a suitable new location is selected. Once established, they usually remain until the following spring. The home range of an individual depends to a certain extent upon the size and shape of the water area in which the animal lives. Common muskrats living in the center of a marsh usually occupy a circular area, while those along the shoreline live in a narrow area extending from bank burrows out several hundred meters into deeper water. Muskrats living in rivers extend their range along the bank up to 183 m (200 yds.) out into the river, even to the opposite bank in all except very large rivers like the Missouri or Mississippi.

If the water becomes too low or dries up, muskrats are forced to leave. They move first to the limits of the area with which they are familiar. Here they establish temporary homes before going farther. Muskrats may occupy woodchuck holes, cornfields, and many other localities during drought, but they return to the water area when it is again available.

In contrast to permanently established individuals, some drifting of both sexes may occur throughout the

Common muskrat house

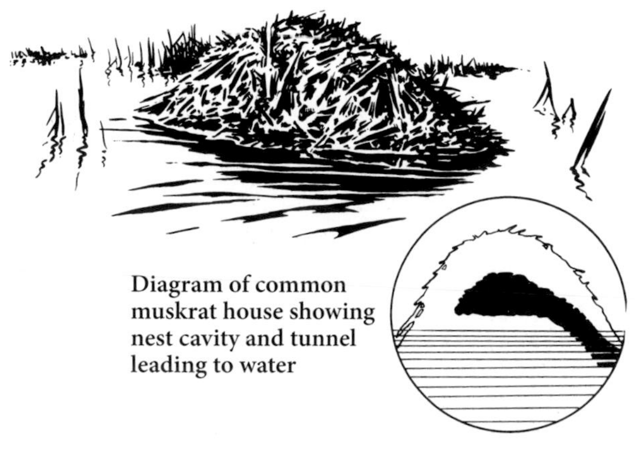

Diagram of common muskrat house showing nest cavity and tunnel leading to water

Waterfowl use common muskrat houses for idling

year. Once an animal moves out of the area with which it is familiar, it may go up to 34 km (21 mi.) away. This movement is most common in August and September when the pressure of young increases competition for a place to live. The maximum breeding concentration is approximately 2 pairs per 0.4 ha (1 ac.).

Muskrats are mostly nocturnal but during late spring and early summer may commonly come out in daytime. They are most active on days when it rains. Their gait on land is an amble or a rather slow hop. They usually enter the water silently but if alarmed may rush in with a splash. A muskrat swimming on the surface holds the front feet against the chin while the powerful hind feet alternately stroke the water in a vertical plane. On each return stroke, the hind feet are folded to reduce resistance to the water. A change of direction is made by altering the strokes of the hind feet. Although the tail is trailed in a wavy or straight line and not used in surface swimming, it may act as a rudder in turning. Muskrats swim at a speed of 3 to 5 km per hour (2 to 3 mi. per hour) and can swim backward as well as forward. They are very buoyant because of the air bubbles trapped in the body fur. When swimming underwater, the muskrat strokes the hind feet in a horizontal plane and uses the tail vigorously. A distance of 46 m (50 yd.) underwater is not unusual. One timed muskrat submerged for 17 minutes, came up for 3 seconds, and submerged again for 10 minutes.

Muskrats are very pugnacious and readily fight one another. However, where there is an abundance of food, there is less fighting than where food is scarce. Fighting among members of the species is one way the population of an area is kept within the limits of the food supply. Defeated individuals must leave, while the victorious remain. Competition for breeding sites can be intense, with females being more aggressive than males when defending their breeding territory.

During much of the year, muskrats live alone or with their mates, but several may sometimes spend the winter together in one bank den or house. Their homes are seldom built closer together than 7.6 m (25 ft.). Although not sociable animals, muskrats may warn each other of impending danger by slapping their tails on the water.

Many other kinds of animals are associated with muskrat habitation. Birds, particularly waterfowl, plus snakes, snapping turtles, frogs, toads, ants, spiders, and skunks, may bask on the top of muskrat houses or live inside, even when the houses are occupied by muskrats.

Common muskrat feeding place

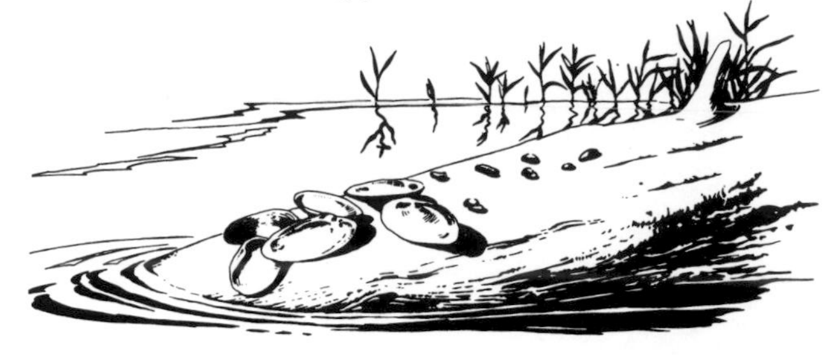

Foods

The foods of common muskrats depend to a large extent upon their availability. In the marshy areas of Missouri, muskrats eat the rootstocks and stems of cattail and three-square bulrush and the seeds of lotus. In other areas of the state, white clover, corn, and bluegrass are preferred foods, although wild celery, musk grass, cheatgrass, and even garden plants are eaten. The muskrat is chiefly a vegetarian but eats some animal foods. Muskrats dwelling in Ozark streams feed to a great extent upon freshwater clams. Snails, crayfish, fish, frogs, reptiles, young birds, and dead muskrats may also be eaten.

Summer food consists mostly of the parts of plants growing above the ground. Muskrats relish corn when the plants are approximately 1 to 1.2 m (3 to 4 ft.) high, and again in late August when the ears ripen. During winter, food is more scarce, and then muskrats may eat hibernating animals buried in the mud at the bottom of marshes or ponds, the roots of aquatic plants, dried grass and weeds, dried leaves and stalks of corn, corncobs, stems and down of cattails, tops of bulrushes, rotting vegetation, and even their houses and bedding. There is practically no storage of food for winter, although sometimes corn and a few bulbs are saved in the nest chamber.

In feeding, the front feet are used to obtain the food and convey it to the mouth. Cattails are reached by rearing on the hind legs and using the tail for support. The stems are cut into sections 12 to 25 cm (5 to 10 in.) long. The food is eaten mostly in shallow water or in some spot with protective cover often provided by the muskrat, and only rarely in the nest itself. Food is not washed before it is eaten. In opening clamshells, the muskrat bites the hinge between the two valves. Piles of clamshells are frequently left at feeding spots.

Where the main body of water dries up in summer, adult muskrats can survive without drinking water, providing succulent vegetation is abundant. Young animals, however, die of thirst on the same fare that suffices for adults.

Reproduction

The breeding season in Missouri begins in late winter and extends until the middle of September. However, three peaks of mating occur: the last of March, the last of April, and the last of May. The gestation period varies between 32 and 30 days with an average of 28 days.

From 1 to 5 litters may be produced annually by a female, but 2 or 3 are most common. Several litters can be born in rapid succession since mating may immediately follow birth. Most young are born in the spring. The litters contain from 1 to 11 young, with most having between 4 and 7.

At birth the young are blind, nearly helpless, and practically naked. They are about 10 cm (4 in.) long and weigh about 21 g (¾ oz.). The young cling very strongly to the female's nipples and, if she plunges unexpectedly, may be pulled into the water, where they often die. By 1 week of age the young are covered with coarse gray brown fur. The eyes normally open between 14 and 16 days of age. About this time the young can swim and dive and climb on low floating objects. When they are older, the female carries them by the skin of the belly. When swimming she holds them high out of the water, yet she can dive while carrying one. The male rarely shares the cares of raising the family.

Weaning occurs between 3 and 4 weeks of age. The female may build a new chamber in the same house or bank for her next litter and leave her older young in their old nest, or the young may move from 9 to 55 m (10 to 60 yd.) from their home and establish new living quarters. They find whatever shelter is available in neighboring muskrat houses or heaps of floating vegetation. Where populations are dense, competition is keen at this time, and older young may even eat newborn young unless prevented by the female. When about 6 months old, the young are the same size as adults. It is possible for young born in early spring to breed in late summer, but most breed for the first time in the spring following their birth.

Some Adverse Factors

Minks and humans are the most important predators on common muskrats in Missouri. Predators of lesser importance are the larger hawks and owls, coyotes, foxes, domestic dogs, raccoons, weasels, large snakes, snapping turtles, and certain predaceous fishes. The following parasites live on or in muskrats: mites, fleas, roundworms, flukes, and tapeworms.

Many diseases such as abscesses, septicemia, coccidiosis, leukemia, and gallstones are known in

American mink and common muskrat in combat

muskrats; some occur as epidemics. A fungal disease affects young to 2 months of age and is often fatal in individuals less than 2 weeks old. Humans may contract this fungal disease by handling infected animals. The muskrat is known to transmit tularemia.

Drowning is an important factor causing the loss of young. When the water level rises or the marsh burns, muskrats may be trapped and die in their dens or houses. Severe cold and prolonged frozen water contribute to the mortality of muskrats by preventing them from foraging or by drowning some under the ice. In summer, drought forces many to leave their homes and makes the population more vulnerable to predation and disease. In overpopulated areas, fighting between muskrats may be very severe and result in some mortality. Prolonged flooding may also reduce their food supply.

Importance

Common muskrat pelts are one of the most common furs on the market and are used to make coats, jackets, hats, and trim for other garments. For commercial use, the pelt is cut into several strips separating the belly, the sides, and the back. One of the former trade names for muskrat fur was "Hudson seal"; this referred to those pelts that had their guard hairs sheared and the remaining fur dyed black or brown to imitate seal. Muskrat fur is of good wearing quality. The skin makes strong leather and takes dye well.

The average annual number of pelts harvested in Missouri has fluctuated drastically between decades. In the 1940s, the average annual number of pelts harvested peaked at 140,721; the number declined to 80,267 in the 1950s but increased in the 1960s to 95,523. The 1970s showed a decrease to 76,374, followed by drastic decreases in subsequent decades to 57,136 in the 1980s, 13,867 in the 1990s, and 10,684 in the first decade of the 2000s. During this seven-decade period, the highest annual take was 217,847 in 1946, and the lowest was 5,225 in 2000.

The average annual price per pelt was $1.66 in the 1940s, $1.05 in the 1950s, $0.94 in the 1960s, $3.10 in the 1970s, $2.93 in the 1980s, $1.86 in the 1990s, and $3.39 in the first decade of the 2000s. The average annual price paid per pelt varied from a low of $0.60 in 1958 to a high of $6.91 in 2009.

It is hard to measure the relative effects of the several factors that influence pelt prices and average harvests. Obviously, population abundance is involved, but to what extent has not been determined. Changes in demand for furs and the fluctuating values of their pelts influence economic reasons for trapping muskrats. In addition, the numbers of trappers working in Missouri influence the harvest.

In some parts of the United States the carcasses of fall and winter muskrats are sold on the market as "marsh rabbit" or "marsh hare." The flesh has a gamey flavor. During the breeding season, the musk often impregnates the flesh, making it repellent as human food. Dried musk is used in the manufacture of perfumes and in preparing scent for trapping animals.

Common muskrat tracks

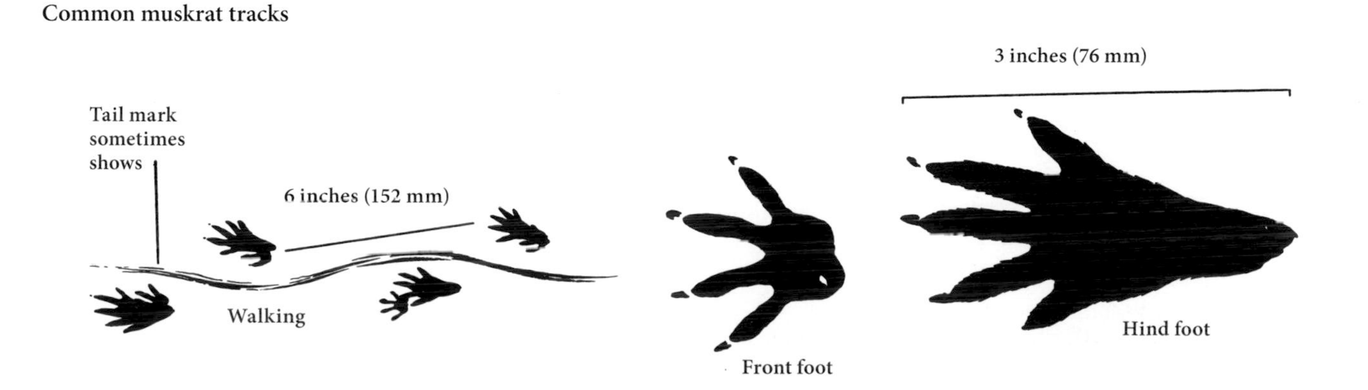

Tunneling by muskrats may damage dikes and pond dams. Muskrats may damage a corn crop by feeding on the grain.

Conservation and Management

In Missouri the most important management measure is to regulate the harvest. Where common muskrats are too numerous, trapping is the most satisfactory means of control. There is no practical way to keep muskrats out of ponds, but ponds can be constructed to minimize damage. If you are experiencing problems with muskrats, contact a wildlife professional for advice, assistance, regulations, or special conditions for handling these animals.

Numerous other species of wildlife use muskrat dens, and therefore muskrats should be tolerated to provide increased habitat for a diversity of wildlife.

SELECTED REFERENCES

See also discussion of this species in general references, page 23.

Alexander, M. M. 1951. The aging of muskrats on the Montezuma National Wildlife Refuge. *Journal of Wildlife Management* 15:175–186.

Applegate, V. C., and H. E. Predmore Jr. 1947. Age classes and patterns of primeness in a fall collection of muskrat pelts. *Journal of Wildlife Management* 11:324–330.

Baumgartner, L., and F. C. Bellrose Jr. 1943. Determination of sex and age in muskrats. *Journal of Wildlife Management* 7:77–81.

Buss, I. O. 1941. Sex ratios and weights of muskrats from Wisconsin. *Journal of Mammalogy* 22:403–406.

Cotner, L. A., and R. L. Schooley. 2011. Habitat occupancy by riparian muskrats reveals tolerance to urbanization and invasive vegetation. *Journal of Wildlife Management* 75:1637–1645.

Dozier, H. L. 1942. Identification of sex in live muskrats. *Journal of Wildlife Management* 6:292–293.

Elder, W. H., and C. E. Shanks. 1962. Age changes in tooth wear and morphology of the baculum in muskrats. *Journal of Mammalogy* 43:144–150.

Erickson, D. W. 1978. *Muskrat, raccoon, and mink productivity research.* Federal Aid Project no. W-13-R-32, Missouri Department of Conservation, Jefferson City. 18 pp.

———. 1979. Studies in the age and sex compositions of Missouri raccoon, muskrat, and mink harvests. Midwest Furbearer Workshop, Manhattan, Kansas. Mimeograph. 15 pp.

Errington, P. L. 1937. The breeding season of the muskrat in northwest Iowa. *Journal of Mammalogy* 18:333–337.

———. 1939a. Observations on young muskrats in Iowa. *Journal of Mammalogy* 20:465–478.

———. 1939b. Reactions of muskrat populations to drought. *Ecology* 20:168–186.

———. 1940. Natural restocking of muskrat-vacant habitats. *Journal of Wildlife Management* 4:173–185.

———. 1941. Versatility in feeding and population maintenance of the muskrat. *Journal of Wildlife Management* 5:68–89.

Forbes, T. R. 1942. The period of gonadal activity in the Maryland muskrat. *Science* 95:382–383.

Galbreath, E. C. 1954. Growth and development of teeth in the muskrat. *Transactions of the Kansas Academy of Science* 57:238–241.

Hall, E. R., and E. L. Cockrum. 1953. A synopsis of the North American microtine rodents. *University of Kansas Publications, Museum of Natural History* 5:373–498.

Olsen, P. F. 1959. Muskrat breeding biology at Delta, Manitoba. *Journal of Wildlife Management* 23:40–53.

Petrides, G. A. 1950. The determination of sex and age ratios in fur animals. *American Midland Naturalist* 43:355–382.

Sather, J. H. 1958. *Biology of the Great Plains muskrat in Nebraska.* Wildlife Monograph 2. 35 pp.

Schofield, R. D. 1955. Analysis of muskrat age determination methods and their application in Michigan. *Journal of Wildlife Management* 19:463–466.

Shanks, C. E. 1947. Populations, productivity, movements, and food habits of Missouri muskrats. M.S. thesis, University of Missouri, Columbia. 123 pp.

———. 1948. The pelt-primeness method of aging muskrats. *American Midland Naturalist* 39:179–187.

Shanks, C. E., and G. C. Arthur. 1952. Muskrat movements and population dynamics in Missouri farm ponds and streams. *Journal of Wildlife Management* 16:138–148.

Willner, G. R., G. A. Feldhammer, E. E. Zucker, and J. A. Chapman. 1980. *Ondatra zibethicus. Mammalian Species* 141. 8 pp.

Southern Bog Lemming
(*Synaptomys cooperi*)

Name

The first part of the scientific name, *Synaptomys*, is of Greek origin and means "unite" and "mouse" (*synapto*, "to unite," and *mys*, "mouse"). This implies that the bog lemming is a "link"—many of its physical characters being intermediate between the voles and another group of rodents, the true lemmings. The second part, *cooperi*, is a Latinized name meaning "of Cooper" and honors William Cooper, who collected the first recorded specimen.

The common name originated as follows: "southern" denotes the southern range when compared to the range of the northern bog lemming (*Synaptomys borealis*), which does not occur in Missouri; "bog" refers to the typical habitat where this animal lives; and "lemming," of Danish and Norwegian origin, is the name given to this type of rodent because of its slight resemblance to the lemmings of the arctic.

Description (Plate 45)

The southern bog lemming is small and thickset with a large head, short legs, and a short tail that is about the same length as the hind foot. The eyes are small and the ears are nearly concealed in the fur. There are 4 toes and a small, nailed thumb on each front foot and 5 toes on each hind foot. The soles of the hind feet are sparsely furred to the 6 pads, or tubercles. Each of the upper incisors possesses a shallow, lengthwise groove on the outer edge. The incisors are prominent because the lips close behind them, permitting the mouse to gnaw without getting particles in the mouth cavity. The long, loose body fur is glossy and coarse.

The bog lemming is distinguished from the three voles in Missouri by grooved incisors and from the prairie and meadow voles by the shorter tail (tail nearly same length as hind foot).

Color. The southern bog lemming's back is brownish, mixed with gray, black, and yellow; the undercolor is dark gray. The sides and underparts are silvery but have dark bases to the hairs. The feet are grayish tan. The tail is brownish above and slightly lighter below. The sexes are colored alike.

Measurements

Total length	95–152 mm	3¼–6 in.
Tail	12–31 mm	½–1¼ in.
Hind foot	15–25 mm	⅝–1 in.
Ear	7–14 mm	5/16–9/16 in.
Skull length	22–30 mm	⅞–1 3/16 in.
Skull width	14–19 mm	9/16–¾ in.
Weight	14–56 g	½–2 oz.

Teeth and skull. The dental formula of the southern bog lemming is:

$$I\ \frac{1}{1}\ C\ \frac{0}{0}\ P\ \frac{0}{0}\ M\ \frac{3}{3} = 16$$

The shallow, lengthwise groove on the outer edge of each upper incisor is sufficient to distinguish the skull of the southern bog lemming from all other Missouri rodents with 16 teeth. (In *Reithrodontomys* mice there is a deep lengthwise groove in the middle of each upper incisor.) The grinding surfaces of the upper cheek teeth possess a pattern of sharp-angled enamel folds surrounding dentine. The third upper molar has 4 islands of dentine extending across the tooth. The lower molars have moderately deep notches on the cheek side.

Sex criteria. The sexes are identified as in the eastern woodrat. There are three pairs of teats.

Age criteria and longevity. Young bog lemmings are recognized by their grayish fur. As with many small prolific rodents, the life span in the wild is probably under one year.

Glands. Small glands on the hips of males become well developed during the breeding season. Large glands occur in the groin region of both sexes.

Voice and sounds. Bog lemmings have a variety of calls. When quarreling or threatening, they give a harsh, grating, or rasping series of notes. Courting calls are mellow and rather high pitched and are also given in a series. Females call to their young in low, short notes; the young give high-pitched peeps.

Distribution and Abundance

The southern bog lemming extends from the Midwest through the northeastern United States and into southeastern Canada. It is common only in certain localities and may be absent from regions that seemingly have suitable habitat. This species varies in abundance from year to year.

The southern bog lemming is thought to occur in all parts of Missouri except the extreme southwest. Further collecting, however, may reveal it to exist in this part of the state, too.

Habitat and Home

The bog lemming prefers heavy stands of bluegrass growing in low, moist places, but it also lives in bogs, swamps, and damp woods with an accumulation of leaf mold. It occurs less often on high ground. One feature common to all these habitats is a thick mat of vegetation.

The globular nest is from 15 to 20 cm (6 to 8 in.) in diameter and has from 2 to 4 entrances. It is made of dry leaves and grass with a lining of some soft material such as fur. In winter, the nest may be from 10 to 15 cm (4 to 6 in.) below the surface of the ground, while in summer it is often aboveground.

Plate 45

Southern Bog Lemming *(Synaptomys cooperi)*

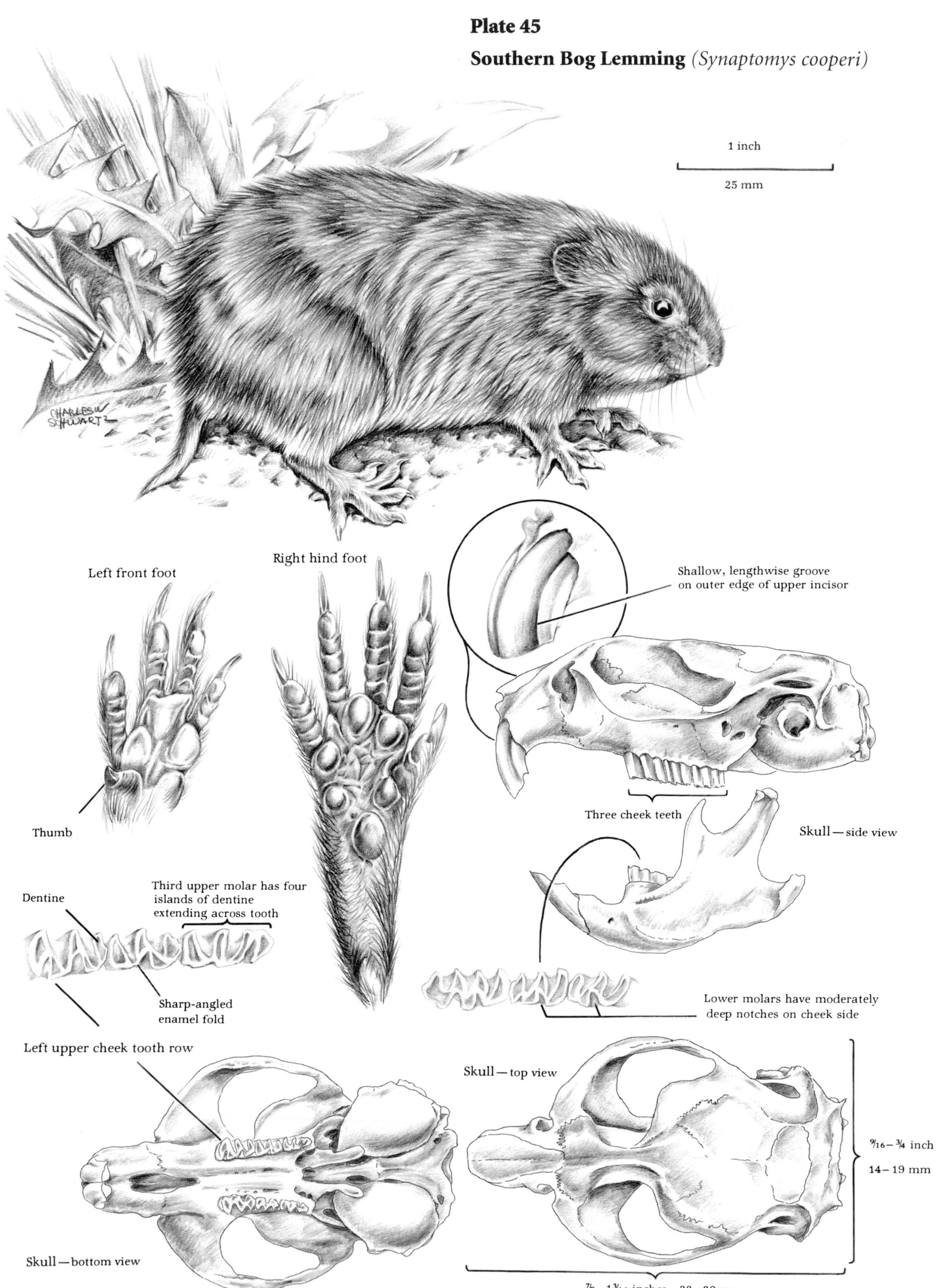

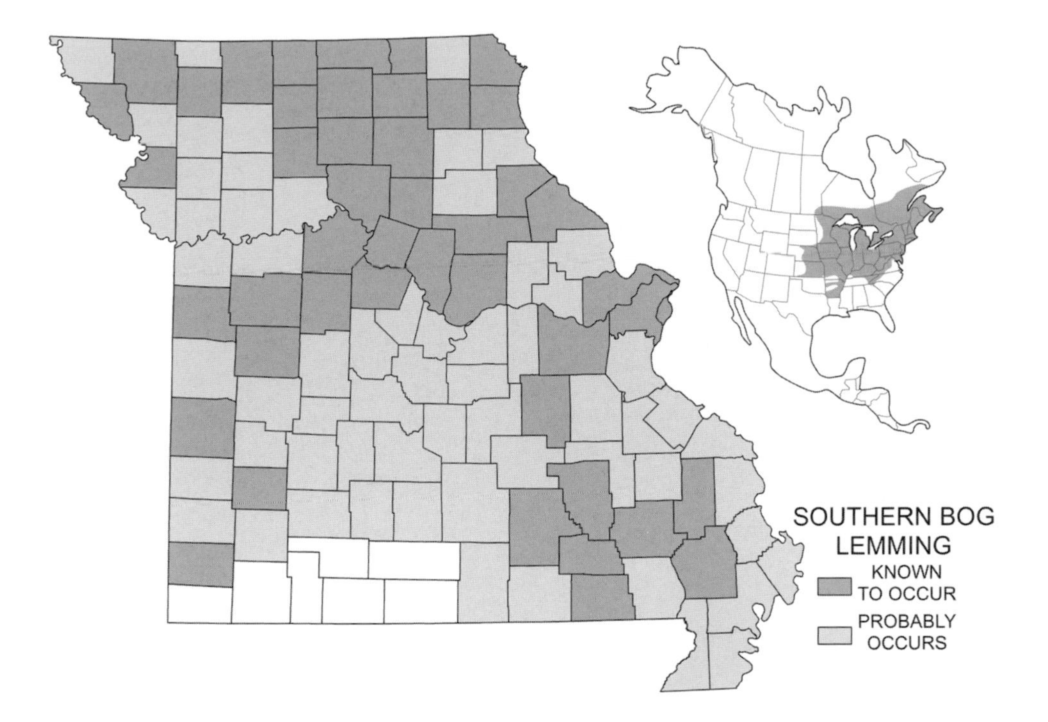

Habits

Southern bog lemmings make runways, from 2.5 to 5 cm (1 to 2 in.) in diameter, in the heavy vegetation where they live and excavate burrows under the ground to a depth of 15 or more cm (6 or more in.). The surface runways are kept closely cropped of new growth; the floor is generally covered with dead grass. Piles of clipped grass stems, about 8 cm (3 in.) long, and green droppings are often left along the trails. Side chambers, used for feeding, resting, food storage, or a nest, occur along the underground runways. Where bog lemmings live in wooded areas, runways may be made by pushing up the leaf mold.

Bog lemmings use the runways at all hours of the day and night but are most active in the afternoon and night. Although they are abroad during all seasons, there is restricted activity at temperatures below −6°C (20°F).

Bog lemmings are colonial and live in groups of a few to several dozen individuals. The home range is limited, ranging from 0.04 ha (1⁄10 ac.) to an area of 30 m (100 ft.) in diameter. One banded female lived within a radius of 37 m (120 ft.) from the last of August to the last of September but moved a distance of 105 m (345 ft.) during the early part of October. These bog lemmings often live in the same general area with

Paved runway of southern bog lemming

Droppings and grass cuttings on trail

voles, white-footed and deer mice, shrews, and moles. When they actually encounter each other, southern bog lemmings dominate vole species.

Foods

Stems, leaves, and seeds of bluegrass, white clover, and various other grasses are the most important foods of the southern bog lemming. However, fungi, moss, bark, and beetles (both adult and larval) are occasionally included in the diet. All the food is finely chewed and mixed; because of the large amount of green vegetation consumed, the intestinal contents and droppings are characteristically a uniform light green. Captive animals lap water with the tongue.

Reproduction

Breeding takes place all year, but peaks probably occur in spring and fall. There are several litters annually containing from 1 to 7 young with 3 to 5 being the most usual. A female can be both pregnant and nursing. The gestation period is from 21 to 23 days.

At birth the young are naked, blind, and weigh about 3 g (1/10 oz.). They show some hair by the fifth day, and their eyes open about the twelfth day after birth. The young breed before they reach maximum adult size.

Some Adverse Factors

The predators of bog lemmings are weasels, foxes, coyotes, bobcats, domestic cats, other carnivorous mammals, hawks, owls, and snakes. The common external parasites are mites, lice, and fleas.

Importance

Southern bog lemmings have seemingly little direct influence on human economic interests. However, in years and places of high populations, their tunneling activities are more extensive and may be objectionable, especially on lawns in wet areas.

These bog lemmings contribute in many ways to the ecosystem in which they live. Through their tunneling activities, they work and mix the soil, permit rain and air to penetrate into deeper layers, and mix vegetation and droppings with the soil, increasing its fertility. Through their feeding, they convert grass into the flesh of their bodies and pass the nutrients to flesh-eating animals that prey upon them.

Conservation and Management

If it is necessary to control a locally abundant population, the same measures can be employed as for the prairie vole.

SELECTED REFERENCES

See also discussion of this species in general references, page 23.

Burt, W. H. 1940. Territorial behavior and populations of some small mammals in southern Michigan. *University of Michigan, Museum of Zoology, Miscellaneous Publication* 45:7–58.

Connor, P. F. 1959. The bog lemming *Synaptomys cooperi* in southern New Jersey. *Michigan State University, Publications of the Museum, Biological Series* 1:161–248.

Fisher, H. J. 1945. Notes on voles in central Missouri. *Journal of Mammalogy* 26:435–436.

Getz, L. L. 1960. Home ranges of the bog lemming, *Synaptomys cooperi*. *Journal of Mammalogy* 41:404–405.

Hall, E. R., and E. L. Cockrum. 1953. A synopsis of the North American microtine rodents. *University of Kansas Publications, Museum of Natural History* 5:373–498.

Hamilton, W. J., Jr. 1941. The food of small forest mammals in eastern United States. *Journal of Mammalogy* 22:250–263.

Harris, V. T. 1941. The relation of small rodents to field borders on agricultural lands in central Missouri. M.S. thesis, University of Missouri, Columbia. 86 pp.

Linzey, A. V. 1983. *Synaptomys cooperi*. *Mammalian Species* 210. 5 pp.

Oehler, C. 1942. Notes on lemming mice at Cincinnati, Ohio. *Journal of Mammalogy* 23:341–342.

Rose, R. K., and A. M. Spevak. 1978. Aggressive behavior in two sympatric microtine rodents. *Journal of Mammalogy* 59:213–216.

Terman, M. R. 1974. Behavioral interactions between *Microtus* and *Sigmodon*: A model for competitive exclusion. *Journal of Mammalogy* 55:705–719.

Old World Rats, Mice, and Relatives
(Family Muridae)

The taxonomy related to the Muridae and Cricetidae families has undergone (and probably will continue to undergo) repeated changes that result in considerable confusion when reviewing the literature over the years. In the older tradition, the families Muridae (Old World rats, mice, and relatives) and Cricetidae (New World rats and mice, voles, and relatives) were separate. Somewhat more recently, the Cricetidae were lumped into the Muridae. Most recently, the Muridae and Cricetidae have again been recognized as separate families.

The murids today are the largest family of mammals, containing over 700 species. They are small mammals;

include true mice and rats, gerbils, and relatives; live in a variety of habitats; and are native to Europe, Asia, Africa, and Australia. A few species, including the black rat, brown rat, and house mouse, have been introduced nearly worldwide and are widespread, prolific urban pests. They are distinguished from the Cricetidae by typically lacking cheek pouches.

The family name Muridae comes from the Latin *mus* and *muris*, meaning "mouse."

Black Rat (*Rattus rattus*)

Name

The scientific name of the black rat is the medieval Latin word *rattus* for "rat." The common name, "black," describes the general color. It is also known as the roof rat and ship rat.

Description (Plate 46)

The black rat closely resembles the brown rat but is distinguished by its general coloration; more slender build; larger and broader ears; and more slender, finer-scaled tail, which is longer than the combined length of head and body. There are usually ten teats.

Color. The black rat is grayish black above, being nearly pure black in the middle of the back, which grades to a lighter gray, pale yellow, or white on the belly.

Measurements

Total length	327–463 mm	12⅞–18¼ in.
Tail	165–254 mm	6½–10 in.
Hind foot	31–44 mm	1¼–1¾ in.
Ear	23–26 mm	15/16–1 1/16 in.
Skull length	47 mm	1⅞ in.
Skull width	20 mm	15/16 in.
Weight	141–283 g	5–10 oz.

Teeth and skull. The teeth are similar to those of the brown rat. The skull can be distinguished from that of the brown rat by the supraorbital and temporal ridges on the upper surface, which are bowed decidedly outward on each side; by the length of the parietal bone, measured along the temporal ridge, which is decidedly less than the greatest distance between the ridges (see illustration of skull measurement, page 289); and by

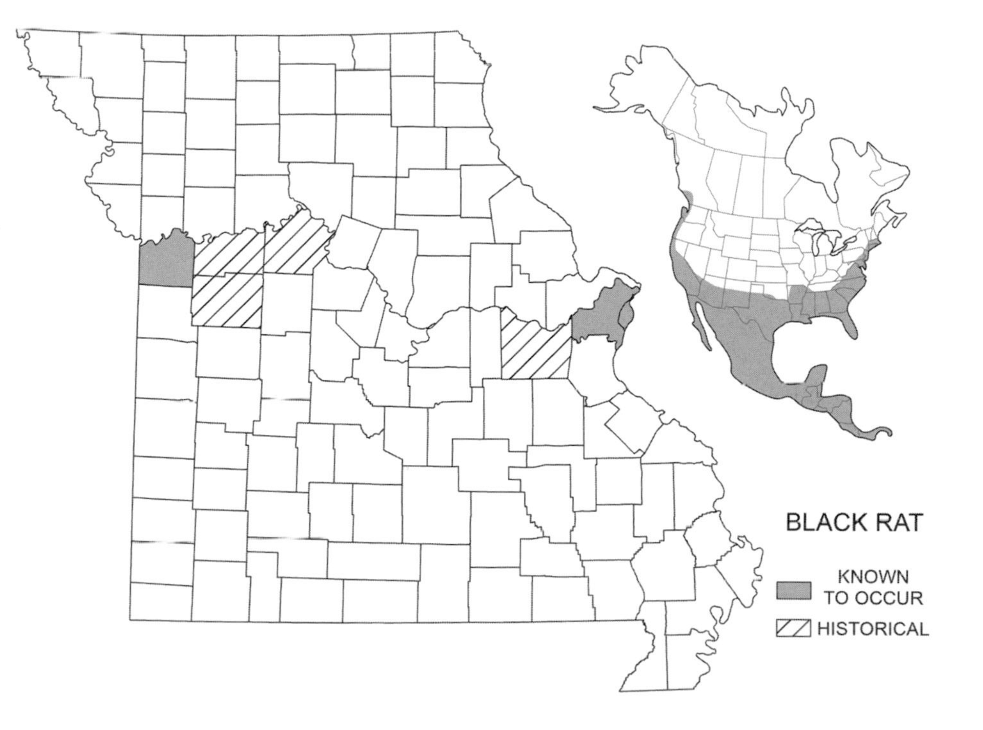

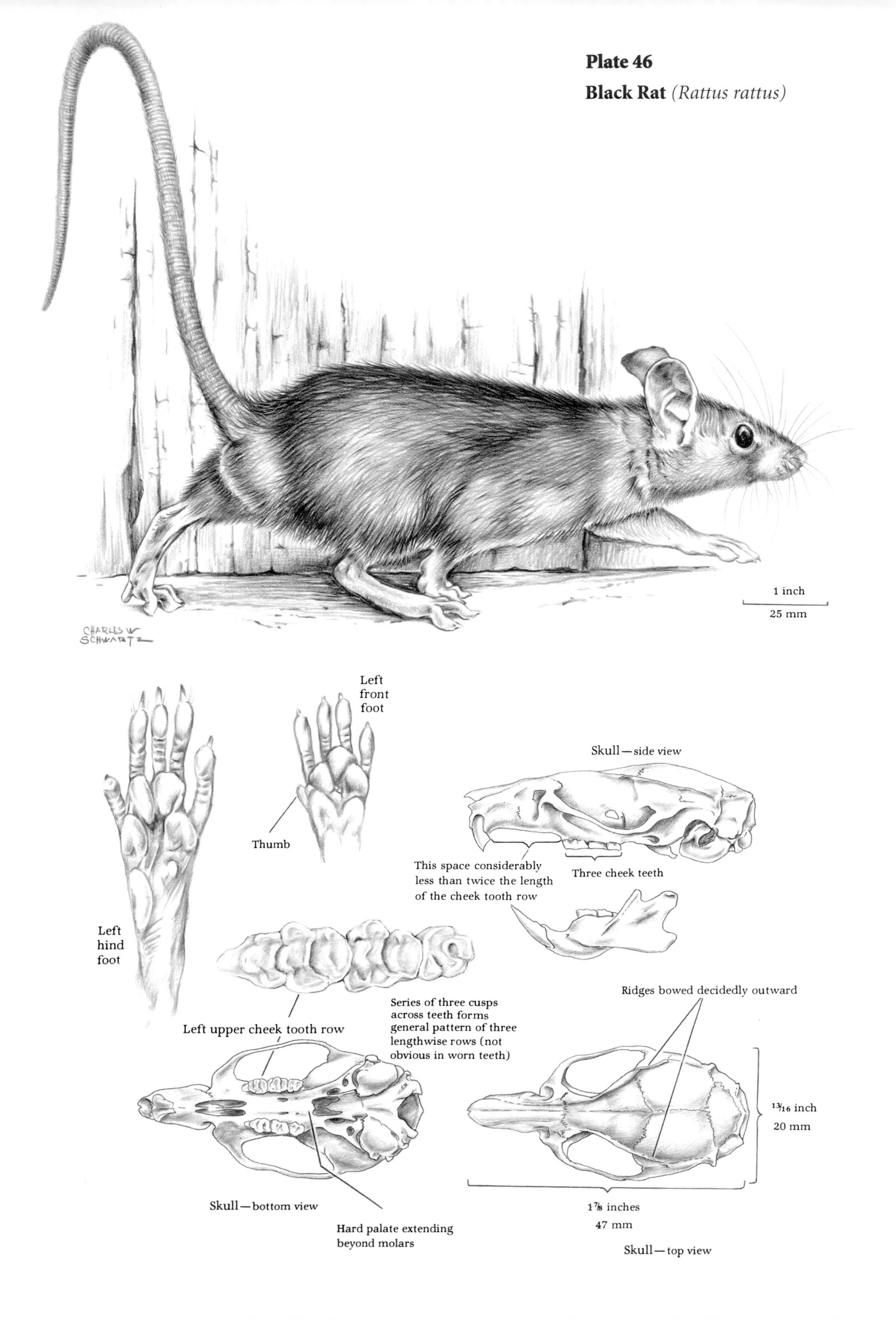

Plate 46

Black Rat *(Rattus rattus)*

the space between the incisors and cheek teeth, which is considerably less than twice the length of the cheek tooth row.

Distribution and Abundance

This native of Asia came to South America and Central America from Europe about 1554. In North America it is recorded as coming ashore with the early colonists at Jamestown in 1609. The more aggressive brown rat reached North America in 1775 and drove the black rat from many localities.

In Missouri the black rat occurs in St. Louis and Kansas City, presumably reaching these cities by way of river traffic from southern states. It occurs locally in such places as grain elevators, stockyards, and poultry houses. It is scarce in comparison to the brown rat.

Habitat and Home

The black rat, more than the brown rat, prefers to live in close proximity to humans. Black rats live and nest in the upper stories of buildings, in trees, and in ships; they are found less often on the ground or in open fields.

Life History

In its general way of life, the black rat resembles the brown rat. The black rat is a better climber than the brown rat. Breeding takes place throughout the year, but in general the black rat has fewer litters and fewer young per litter than the brown rat. It does not interbreed with the brown rat. Major food items are fruits, nuts, and seeds. The chief predator on the black rat is the brown rat.

Some Adverse Factors; Importance; and Management

These subjects are discussed under the brown rat.

SELECTED REFERENCES

See also discussion of this species in general references, page 23.

Davis, D. E. 1947. Notes on commensal rats in Lavaca County, Texas. *Journal of Mammalogy* 28:241–244.

———. 1953. The characteristics of rat populations. *Quarterly Review of Biology* 28:373–401.

Feng, A. Y. T, and C. G. Himsworth. 2014. The secret life of the city rat: A review of the ecology of urban Norway and black rats (*Rattus norvegicus* and *Rattus rattus*). *Urban Ecosystems* 17:149–162.

Solanke, A. K., V. H. Singh, and K. M. Kulkarni. 1996. Patterns of breeding habits, sex ratio, and sexual maturity in *Rattus rattus*. *Journal of Ecobiology* 8:67–69.

Spencer, H. J., and D. E. Davis. 1950. Movements and survival of rats in Hawaii. *Journal of Mammalogy* 31:154–157.

Worth, C. B. 1950. Field and laboratory observations on roof rats, *Rattus rattus* (Linnaeus), in Florida. *Journal of Mammalogy* 31:293–304.

Brown Rat (*Rattus norvegicus*)

Name

The first part of the scientific name, *Rattus*, is the medieval Latin word for "rat." The last part, *norvegicus*, is a Latinized word meaning "of Norway," for the country from which the first specimen was named. However, the native home of this species is Asia. The common name, "brown," refers to the predominant color of this rodent. This rat is also known as the Norway rat.

Description (Plate 47)

The general appearance of this medium-sized rodent is well-known. It has a somewhat elongated snout; small eyes; moderately sized, nearly naked ears; and a nearly naked tail with scaly rings. The tail is slightly shorter than the combined length of head and body. There are 4 clawed toes and a very small thumb on each front foot and 5 clawed toes on each hind foot. The soles of the hind feet are naked and have 6 prominent pads, or tubercles. The body fur is rather coarse and short.

The brown rat might be confused with several other Missouri rats but can be distinguished by the following more obvious characteristics: from the eastern woodrat by the grayer underparts of the body and tail, coarser fur, and nearly naked, scaly tail; from the black rat by a more robust build, grayish brown coloration, and a coarser-scaled tail slightly shorter than the head and body; from the marsh rice rat by the larger size, shorter and coarser fur, thicker tail, stubby toes, and soles of the feet that are smooth except for large tubercles; and from the hispid cotton rat by the larger size, longer tail, shorter fur, and lack of grizzling.

Color. The upperparts of the brown rat are grayish brown with scattered black hairs. This color grades to pale gray or yellowish white on the underparts. The tail is grayish above and lighter below, but the colors are not distinctly separated from each other. The feet

Plate 47
Brown Rat *(Rattus norvegicus)*

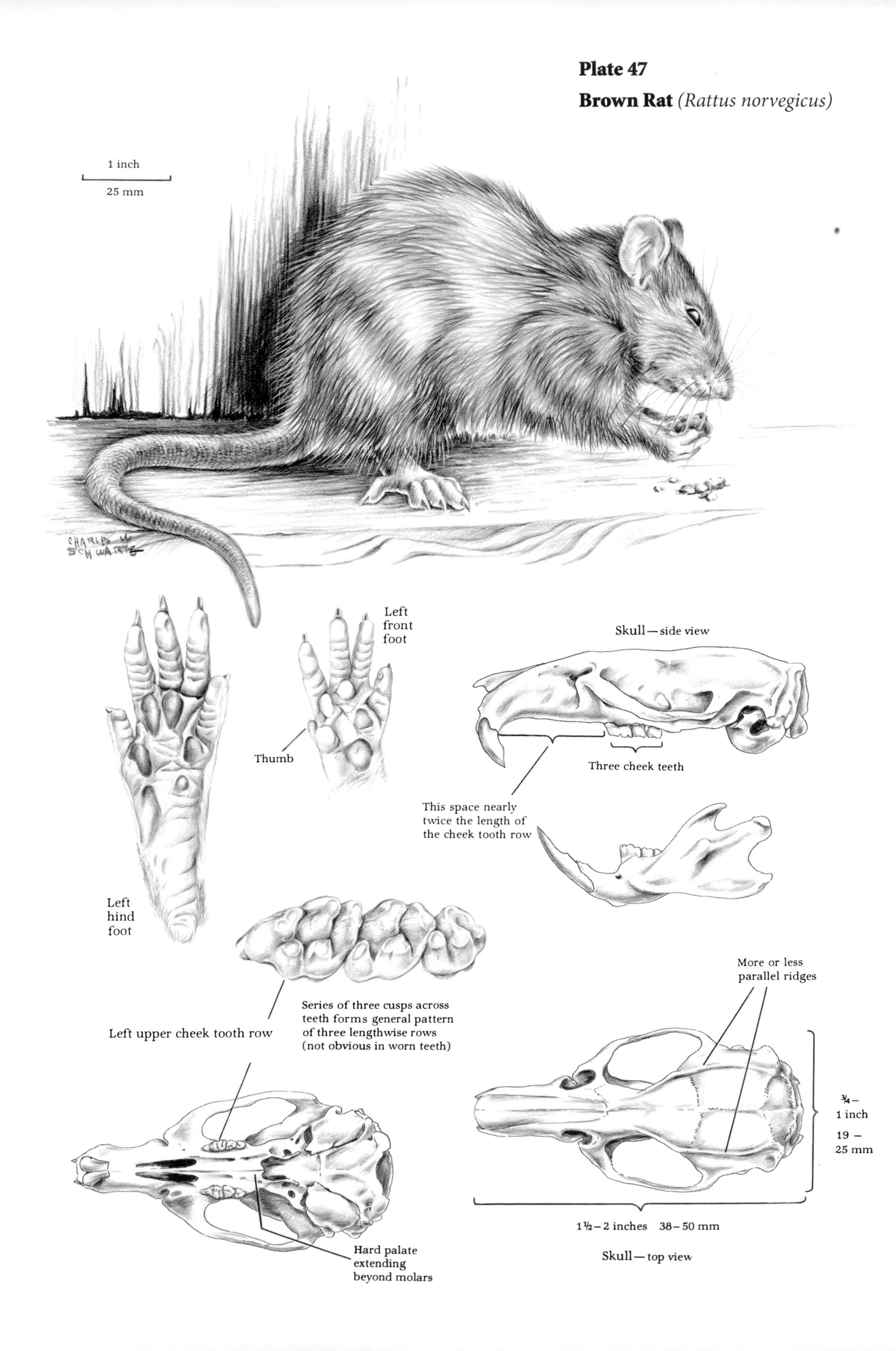

are grayish or whitish. The sexes are colored alike. Some wild brown rats are blackish or variously spotted. Strains of the brown rat commonly bred and used in laboratories, sold as pets, and sold as food for pet reptiles and other animals are either entirely white (albino) or show many variations in coat color. Wildlife rehabilitators rely on rats and mice to feed recuperating raptors, owls, and other animals.

Measurements

Total length	295–482 mm	11⅝–19 in.
Tail	120–215 mm	4¾–8½ in.
Hind foot	31–44 mm	1¼–1¾ in.
Ear	15–22 mm	⅝–⅞ in.
Skull length	38–50 mm	1½–2 in.
Skull width	19–25 mm	¾–1 in.
Weight	170–850 g	6–30 oz.
Commonly	283–340 g	10–12 oz.

Teeth and skull. The dental formula of the brown rat is:

$$I\ \frac{1}{1}\ C\ \frac{0}{0}\ P\ \frac{0}{0}\ M\ \frac{3}{3} = 16$$

The grinding surfaces of cheek teeth consist of small, rounded cusps capped with enamel. The arrangement of these cusps forms a general pattern of 3 lengthwise rows. However, this pattern is sometimes obscured in old teeth that have their cusps worn down. There is no notch in the cutting surface of the upper incisors, visible from the side, as occurs in the house mouse.

Among the Missouri rodents possessing 16 teeth, the skull of a brown rat might be confused with that of a young muskrat, the eastern woodrat, or the black rat on the basis of size. From the muskrat and woodrat, the skull of the brown rat can be distinguished easily by the cheek tooth pattern described above; by the presence of supraorbital and temporal ridges on the upper surface of the skull; and by the hard palate that ends well behind the last molars. From the black rat, the skull of the brown rat is identified by the more or less parallel ridges on the upper surface of the skull; by the length of the parietal bone, measured along the temporal ridge, which is about equal to the greatest distance between the ridges (see accompanying illustration); and by the space between the incisor and cheek teeth, which is nearly twice the length of the cheek tooth row.

A distinction between brown and black rat skulls

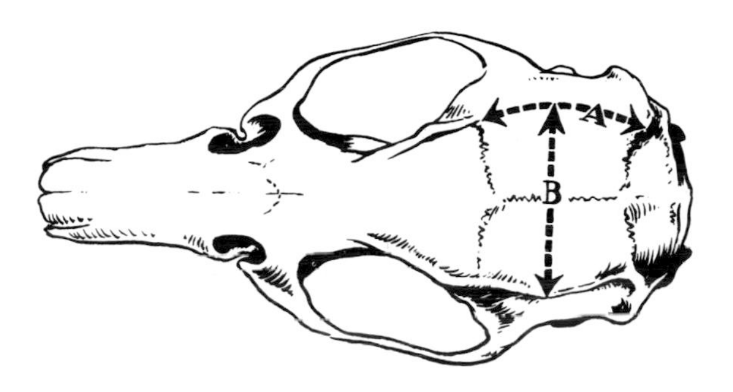

In the brown rat, A is about equal to B
In the black rat, A is decidedly less than B

Sex criteria and sex ratio. The sexes are identified as in the eastern woodrat. A penis bone is present in males. There are usually 6 pairs of teats along the sides of the belly, but 5 pairs sometimes occur. The sex ratio varies with age, locality, and method of collection. In general, sex ratios show more females than males among adults. Studies indicate that males die off more rapidly than females.

Age criteria and longevity. Young animals are grayer than adults and have duller fur.

Externally, males can be identified as sexually mature if the testes are in the shallow scrotum. Females are sexually mature when the vagina is open. The age and size at which sexual maturity is reached vary with rats living under different conditions.

Although wild brown rats may reach 2 or 3 years of age, probably less than 5 percent live for even 1 year. There is heavy mortality among the young and considerable fighting among all ages, which often leads to disease and death.

Glands. Paired scent glands lie just inside the anus. The odor is unpleasant but not as musky as that of the house mouse.

Voice and sounds. In fighting, brown rats give loud squeals, sharp screams, and high squeaks. Adults and young have sounds in the ultrasound range.

Distribution and Abundance

The brown rat first appeared in the eastern United States about 1775, probably arriving on ships from England, but did not occur on the Pacific Coast until 1851. Since the black rat was well established in North America by the time the brown rat arrived, there was considerable conflict between these two species. The brown rat succeeded in driving out the black rat from many localities and is now widespread and abundant nearly everywhere that humans live.

The population varies according to the habitat. In well-kept residential districts or clean farm units, the brown rat population may be low; in untidy urban or rural localities, however, the population may be high.

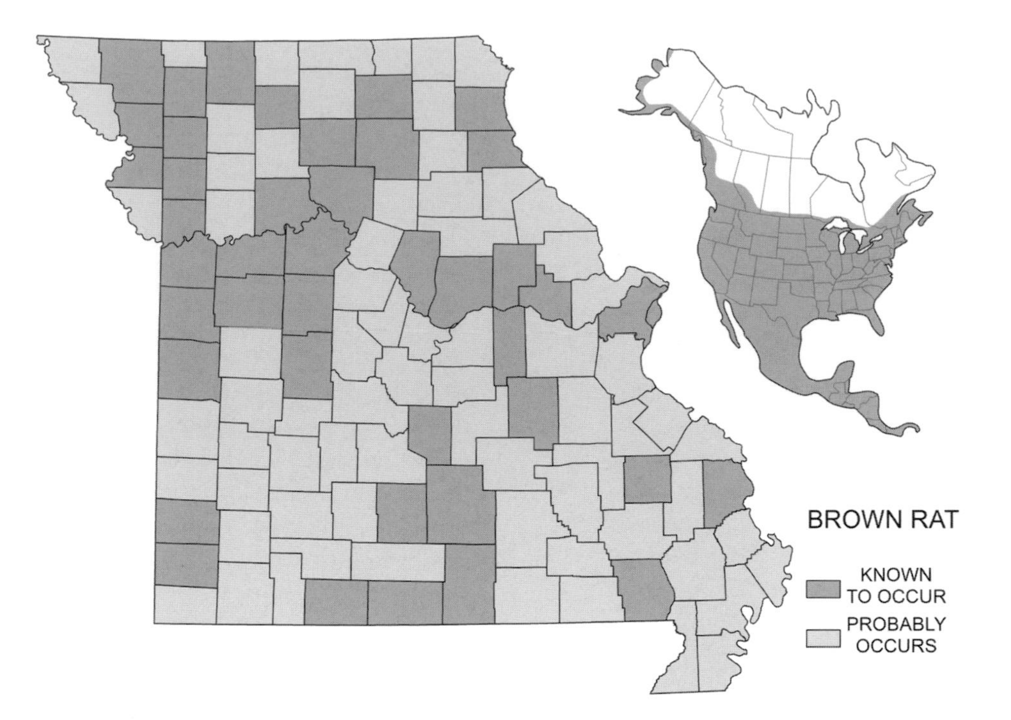

It has been estimated that the average farm harbors 50 rats. By contrast, 4,000 rats were taken from 1.2 ha (3 ac.) of a midwestern farm with good habitat for rats. When unusually dense populations of rats occur, a food shortage sometimes develops and results in a mass movement from the area. Such an emigration is reported to have occurred in parts of Missouri in 1877. Today they are known from several counties and probably occur statewide.

Habitat and Home

Brown rats live around human habitation wherever food and shelter are available. They live mostly on the ground floors of dwellings and buildings, in tunnels they make in the ground under buildings, in sewers, or around dumps. Although they generally stay near human habitation, they are also found in rural areas and may live in fields.

The nest is made of shredded grass, leaves, paper, cloth, or whatever material is available. It is usually well concealed in hollow walls, at the end of a tunnel in the ground, under rocks or rubbish, or in some other good hiding place.

Habits

Brown rats commonly dig tunnels in the ground to depths of 0.5 m (1½ ft.). They are from 5 to 7.6 cm (2 to 3 in.) in diameter and vary in length from 0.5 to 2 m (1½ to 6½ ft.), with 1 m (3 ft.) being average. One or more chambers, about 15 cm (6 in.) in diameter, are built in the tunnel system. These are used for a nest or place to eat. The network of tunnels has one or more main entrances, usually under a board or some kind of shelter, and several escape holes that are kept covered with only shallow soil or weeds.

In digging, a rat cuts roots with its teeth, loosens the soil and pushes it under the belly with its front feet, and passes the accumulation to the rear with its hind feet. In order to remove the soil from the tunnel, the rat must turn around and push it with its front feet and head.

In large cities, brown rats remain in the same locality all year. Once more or less established, individuals have a restricted range and probably stay within 30 to 60 m (100 to 200 ft.) of their nests. Around villages and farms, however, brown rats tend to leave areas of human habitation in spring to invade grain fields, pastures, and waste areas where they breed and feed. In the fall, most return to the farm units and other buildings, although some stay in shocks of corn and small grain.

Brown rats commonly live in colonies of 10 to 12 individuals. A male, usually the largest and oldest, becomes the dominant individual. He achieves this social status by aggressive behavior in driving others away from food or good nesting places. As members of a colony, the individuals are assured of sharing in the advantages or disadvantages of belonging to the group. They frequently cooperate to drive a strange rat away. When the rat population is high, there is considerable competition between individuals. The resulting strain and strife cause physical combat, poor health, less successful breeding, and exclusion.

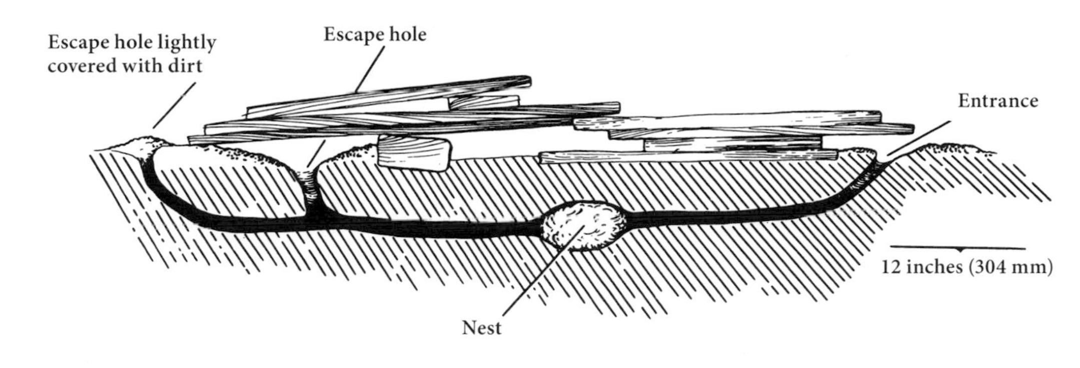

Diagram of brown rat tunnel system beneath pile of boards

While brown rats may be abroad at any hour, night is the period of their usual activity. Young animals generally come out earlier than adults, and adult males before adult (pregnant) females. There are two main periods of feeding: just after dark and just before daylight.

Brown rats establish definite pathways from nests to feeding places and use these habitually. When traveling on the surface of the ground they seldom leave some form of protective cover and stay along fences and buildings. While brown rats commonly cross alleys because these narrow thoroughfares are bordered with suitable habitat, they seldom cross streets, which are generally wider and lack nearby protective cover. Brown rats climb and jump readily and are good swimmers.

Secluded spots near runways or tunnels are often used for eating places, and piles of debris accumulate in these localities. Smears of grease, soot, and soil from their bellies and feet commonly mark surface runways. Brown rats are very shy and wary of anything new in their surroundings, even food.

Droppings are left anywhere, but rarely in the nest. The spindle-shaped droppings are blackish, soft, and shiny when fresh but hard and dull when dry. They measure about 2 cm (¾ in.) in length.

Foods

Brown rats are probably the most omnivorous of all mammals, feeding on both dead and living material. They eat vegetable matter such as grain, green

vegetation, fruits, roots, and garbage. Animal foods such as eggs, milk, fish, and other animals, and miscellaneous articles like soap, leather, and books, are included in their diet. Animals known to be killed by brown rats are chickens, young pigs, lambs, wild and domestic rabbits, rattlesnakes, and wild birds. Brown rats prey on black rats and even their own kind on occasion. Males often eat unattended young in their nests.

When eating, the rat sits on its haunches and holds food in its front feet. Rats eat as much as one-third of their weight in 24 hours. In addition, they often waste more than they consume. They frequently kill for lust and leave the remains to spoil. Rats need a great deal of water and have been known to gnaw through lead pipes to get moisture.

Reproduction

Brown rats breed all during the year, but the heaviest production of young occurs in spring and fall. In very favorable locations they reproduce more often than in poorer ones. Males are capable of breeding all year, but the female's more limited reproductive capacity determines how many litters will be produced annually. The gestation period varies between 21 and 26 days; the longer periods occur when the female is nursing another litter and are due to a delay in attachment of the embryos to the uterus in the early stages of development. The maximum number of litters known per year is 12, but the average is about 5. From 2 to 22 young may be born in one litter, but from 7 to 11 are the most common. One captive pair and their subsequent young produced more than 1,500 rats in a single year.

At birth the young are blind, naked, and helpless. They weigh about 6 g (⅕ oz.) and are nearly 50 mm (2 in.) long. The eyes open between 14 and 17 days of age, and weaning occurs at about 3 weeks. The young quickly learn to fend for themselves. While some brown rats become sexually mature as early as 28 days of age, most probably become mature when they are between 3 and 5 months old.

Female rats living in groups have been known to breed at the same time, resulting in population booms as large numbers of young enter the population almost simultaneously. Sometimes two or more females have their young in the same nest.

Some Adverse Factors

Hawks, domestic cats, owls, snakes, skunks, weasels, foxes, minks, and many other carnivorous animals prey on brown rats. Terrier dogs can be excellent ratters. Humans also destroy rats, either by killing them directly or by removing their habitat.

Mites, ticks, lice, fleas, tapeworms, and roundworms parasitize brown rats. Flies lay their eggs on the fur of newly weaned young and in wounds of adults; the developing maggots eat the flesh and may even cause death of the rat.

Brown rat tracks

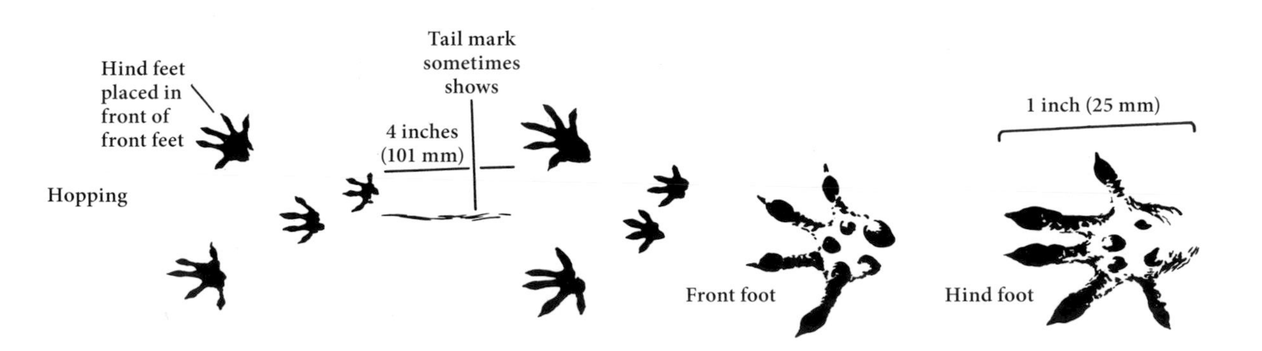

Importance

The brown rat is the most important of all mammals from the standpoint of destruction to human life and property. Each year the damage by rats to food and merchandise amounts to hundreds of millions of dollars. Both the black and the brown rat are particularly destructive agricultural pests. Since a pair of rats will eat the equivalent of a 45 kg (100 lb.) sack of grain in a year, it is obvious how much it costs a farmer to support an average population of 50 rats.

Rats not only consume vast quantities and varieties of food, but through their gnawing activities destroy considerable property. They may be a fire hazard by gnawing wires and insulation. They kill poultry and other domestic animals and destroy crops such as corn and sugarcane in the field.

Brown and black rats are claimed to have caused more human deaths than all the wars in history. They contaminate our food, water, and soil and carry communicable diseases to us and domestic animals. Some of the diseases transmitted directly by rats or by their parasites are bubonic plague, murine typhus, Seoul hantavirus, infectious jaundice, rat-bite fever, trichinosis, spotted fever, scarlet fever, typhoid fever, diphtheria, tularemia, rabies, and certain food poisoning.

On the credit side, mention must be made that rats contribute to human welfare by serving as experimental animals in scientific laboratories.

Conservation and Management

Both the black and brown rats are problematic invasive species that can have devastating effects on natural ecosystems and native species. The most important single way to control rats is to exclude them from the premises by eliminating protective cover and sources of food. An efficient cleanup campaign can reduce rat damage in cities by two-thirds and lower the population so much that it will take from 1½ to 4 years to return to the original level. Trapping is the safest method for use around homes. If you are experiencing problems with brown or black rats, contact a wildlife professional for advice, assistance, regulations, or special conditions for handling these animals.

SELECTED REFERENCES

See also discussion of this species in general references, page 23.

Calhoun, J. E. 1962. *The ecology and sociology of the Norway rat.* U.S. Department of Health, Education, and Welfare. Public Health Service. 288 pp.

Davis, D. E. 1953. The characteristics of rat populations. *Quarterly Review of Biology* 28:373–401.

Davis, D. E., J. T. Emlen Jr., and A. W. Stokes. 1948. Studies on home range in the brown rat. *Journal of Mammalogy* 29:207–225.

Emlen, J. T., Jr., and D. E. Davis. 1948. Determination of reproductive rates in rat populations by examination of carcasses. *Physiological Zoology* 21:59–65.

Feng, A. Y. T., and C. G. Himsworth. 2014. The secret life of the city rat: A review of the ecology of urban Norway and black rats (*Rattus norvegicus* and *Rattus rattus*). *Urban Ecosystems* 17:149–162.

McClintock, M. K., and N. T. Adler. 1978. The role of the female during copulation in wild and domestic Norway rats (*Rattus norvegicus*). *Behaviour* 67:67–96.

Perry, J. S. 1944. The reproduction of the wild brown rat (*Rattus norvegicus* Erxleben). *Proceedings Zoological Society* 115 (1 and 2):19–46.

Pisano, R. G., and T. I. Storer. 1948. Burrows and feeding of the Norway rat. *Journal of Mammalogy* 29:374–383.

House Mouse (*Mus musculus*)

Name

The first part of the scientific name, *Mus*, is the Latin word for "mouse," the name given long ago to this common rodent. The last part, *musculus*, is the Latin word for "small mouse," indicating the size. The common name originated as follows: "house" refers to the human habitations where this animal is commonly found; and "mouse" is from the Anglo-Saxon word *mus*, which descended through Latin from the ancient Sanskrit word *musha*, meaning "thief."

Description (Plate 48)

This small mammal is our most familiar mouse. It has a slightly elongated snout; small, black eyes that are somewhat protruding; large, scantily haired ears; and a nearly naked tail with conspicuous scaly rings that is about half the total length of the animal. The front feet have 4 clawed toes and a very small thumb with a short, flat, and pointed nail. The hind feet have 5 clawed toes and 6 pads, or tubercles, on the naked soles. The body fur is short.

The house mouse somewhat resembles harvest mice, *Peromyscus* spp. mice, the golden mouse, the meadow jumping mouse, and the plains pocket mouse, but there are many differences between them. The adult house mouse is distinguished from harvest mice, the meadow jumping mouse, and the plains pocket mouse by the absence of a groove in each upper incisor; from *Peromyscus* spp. mice by the absence of a sharp contrast in color between back and belly and between

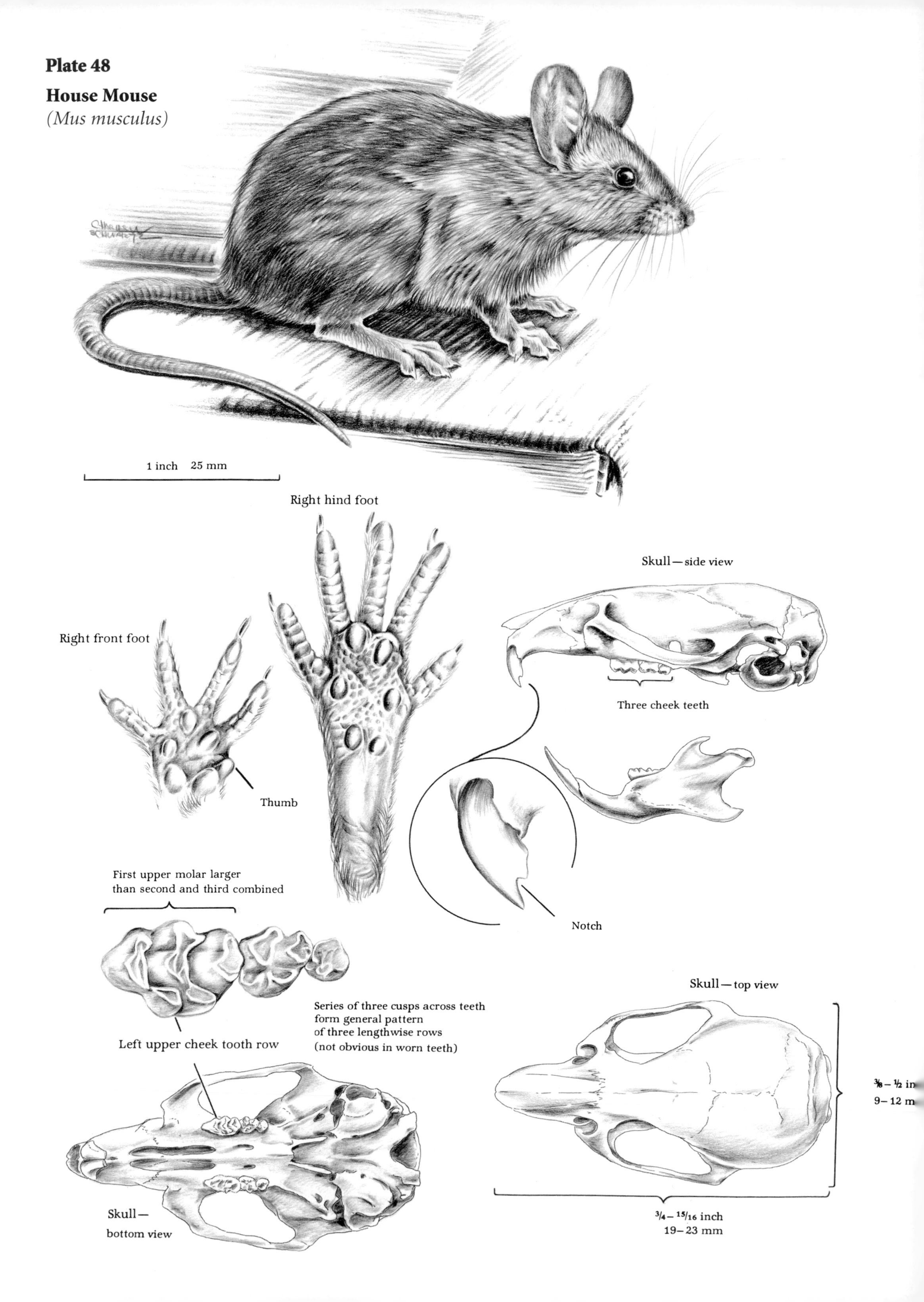

Plate 48
House Mouse
(Mus musculus)
1 inch 25 mm
Right hind foot
Skull—side view
Right front foot
Three cheek teeth
Thumb
First upper molar larger
than second and third combined
Notch
Skull—top view
Series of three cusps across teeth
form general pattern
of three lengthwise rows
(not obvious in worn teeth)
Left upper cheek tooth row
3/8 – 1/2 in
9–12 m
Skull—
bottom view
3/4 – 15/16 inch
19–23 mm

upper and lower surfaces of the scantily haired tail; and from the golden mouse by the coloration.

Color. The back of the house mouse is grayish brown with scattered black hairs. This general color grades to lighter brown, buff, gray, or whitish on the belly. The feet vary from brownish to whitish, and the tail grades from dark above to light below. The sexes are colored alike. There are many variations in this general coloration, including the albino mice reared in laboratories, sold as pets, and sold as food for pet reptiles and other animals. Wildlife rehabilitators rely on rats and mice to feed recuperating raptors, owls, and other animals.

Measurements

Total length	127–206 mm	5–8⅛ in.
Tail	60–101 mm	2⅜–4 in.
Hind foot	15–22 mm	⅝–⅞ in.
Ear	11–19 mm	7/16–¾ in.
Skull length	19–23 mm	¾–15/16 in.
Skull width	9–12 mm	⅜–½ in.
Weight	14–28 g	½–1 oz.

Teeth and skull. The dental formula of the house mouse is:

$$I\,\frac{1}{1}\;C\,\frac{0}{0}\;P\,\frac{0}{0}\;M\,\frac{3}{3} = 16$$

The grinding surfaces of the cheek teeth consist of small, rounded cusps capped with enamel. The arrangement of these cusps forms a general pattern of 3 lengthwise rows. However, this pattern is somewhat obscured in old teeth with worn cusps.

The pattern of cusps on the upper molars is sufficient to distinguish the skull of the house mouse from that of all other rodents in Missouri with 16 teeth except the brown and black rats. From the skulls of these rats, that of the house mouse is distinguished by its smaller size; the presence of a notch in the cutting surface of the upper incisors, which is visible from the side; the absence of paired ridges on the upper surface of the skull; and by the size of the first upper molar, which is larger than the second and third combined.

Sex criteria and sex ratio. The sexes are identified as in the eastern woodrat. There are five pairs of teats along the sides of the belly between the front and hind legs. The sex ratio is approximately equal.

Age criteria and longevity. Young animals are grayer and have softer fur than adults. They generally weigh less than 14 g (½ oz.). Young females have closed vaginas and young males show no obvious enlargement of the testes.

The amount of wear on the upper molar teeth is used as a means of aging house mice. This is a better method than weight or body length.

In the wild, few house mice live more than 1 year, although some may exist up to 15 or 18 months of age. In captivity the record life span is 6 years.

Glands. Scent glands lie just inside the anus. They secrete a sharp, musky odor. The presence of mice can frequently be detected by this odor. Sweat glands on the feet secrete another scent. Males have a general odor for their sex but, in addition, there is a particular variation that permits females to identify individual males.

Voice and sounds. House mice usually give various rapid squeaks, chirps, and chatters. In addition, both sexes have a high-pitched "song" that covers two octaves, is uttered at the rate of 2 to 6 notes per second, and can be heard from 3 to 7 m (10 to 25 ft.) away. They give other calls that are above the range of human reception.

Distribution and Abundance

The house mouse lives in most human-inhabited parts of the world. It is a native of central Asia and probably reached North America from Europe. Under unusually favorable conditions, house mice may reach tremendous densities of thousands per hectare (acre). Such populations eventually undergo drastic reductions from lack of food and accompanying epidemic diseases.

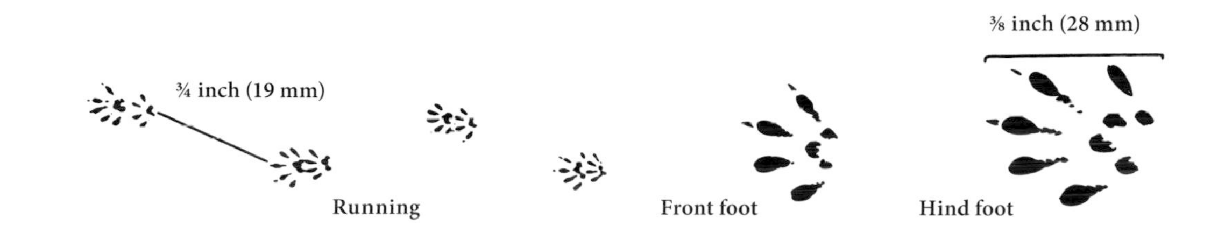

House mouse tracks

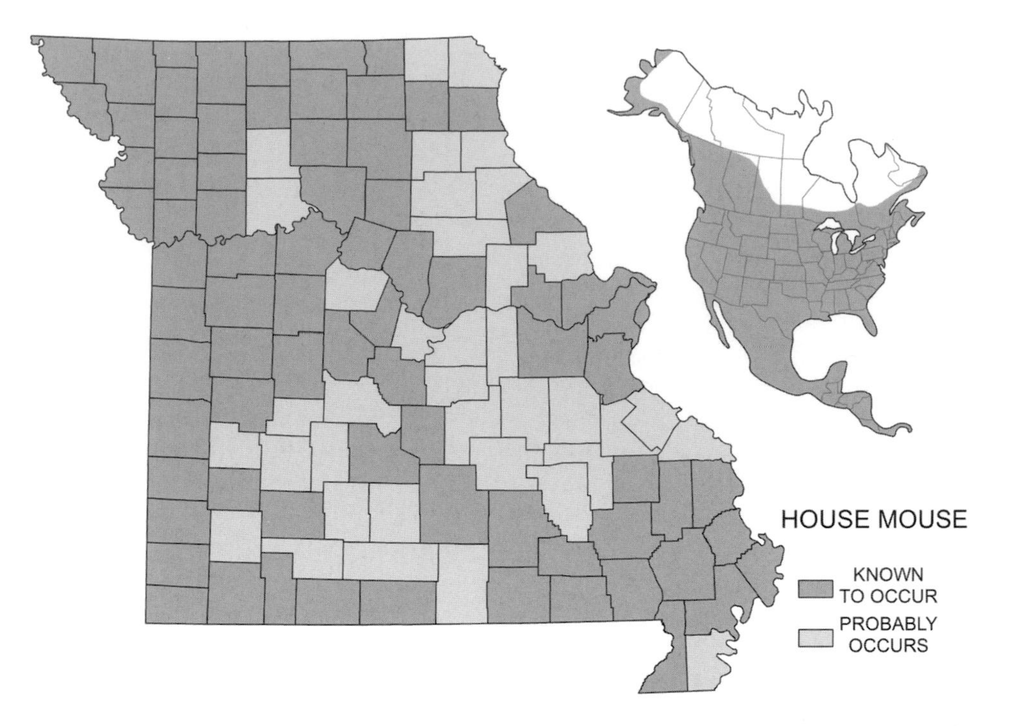

In Missouri, they most likely occur statewide.

Habitat and Home

House mice live in houses and buildings occupied or unoccupied by humans, in abandoned fields, fencerows, weedy roadsides, and cultivated grain fields, especially after the harvest.

The nests are hidden in any well-concealed place and are made from scraps of paper, shredded fabrics, grass, and feathers.

Habits

Although there is some shifting from indoors to outdoors in spring and the reverse in fall, house mice tend to be fairly sedentary and generally show a restricted home range. Indoors, where there are ample hiding places and food, they travel only an average of 3.3 m (11 ft.) from the nest; where there are few hiding places and little food, they go greater distances. Males range slightly farther than females.

Home ranges are first established when the mouse is young and are usually maintained as long as the animal remains in the population. House mice are very aggressive. They rapidly invade disturbed habitats in the field and compete actively with other species of mice living there.

These mice are very social animals and form colonies, both in buildings and in the wild. The family unit consists of a male and one or more females and their offspring. The male drives other males away from the family's nest. All members of the family are aggressive toward strange mice of either sex.

During winter, house mice spend considerable time huddling together in their nests. The warmth derived from their many bodies helps them to survive in cold weather; nonsocial mammals must eat increased amounts of food to supply their bodies with warmth.

House mice are mostly nocturnal and venture forth during daylight only in secluded places. They establish travel routes in buildings along walls or some other protective cover. They can go through openings as small as 1.3 cm (½ in.) in diameter. In fields they make trails through the vegetation and use the runways of other kinds of mice, shrews, and eastern moles. One house mouse was timed to travel at the rate of 4 m (12 ft.) per second. They climb and jump well and are capable swimmers.

The droppings are spindle shaped and measure about 6 mm long by 1.5 mm wide (¼ in. long by 1⁄16 in. wide). They are deposited wherever the mouse frequents. When living under restricted conditions, house mice tend to urinate at special spots.

Foods

House mice consume practically everything edible but prefer grain and various vegetable products. They eat a variety of insects and insect larvae, including those in stored grain; other animal matter; and foxtail and other grass and weed seeds; they also nibble paste and glue on articles such as books, boxes, and leather.

Some food is cached in or about the nests. An adult house mouse will eat 3 g (1⁄10 oz.) of corn a day but will waste more food than it eats. House mice do not require water but drink water when it is available.

Reproduction

Although female house mice come into heat every 4 to 6 days throughout the year, they breed mostly from early spring to late fall. The gestation period is between 19 and 21 days but may be 2 or 3 days longer if the female is nursing.

Thirteen or 14 litters are possible in one year, but from 5 to 10 are normally produced depending upon the amount of available food and other environmental factors. A litter may contain from 2 to 13 young, but from 5 to 7 are most common. Individual captive female house mice have produced 100 young in a year.

At birth the young are blind, naked, and helpless (altricial). They are furred when 10 days old and their eyes open when they are 14 days old. Weaning takes place at 3 weeks of age. The young become sexually mature as early as 6 weeks of age.

Some Adverse Factors

Domestic cats, rats, and many other carnivorous mammals, hawks, owls, and snakes are the common predators of this species. The house mouse is a host for ticks, including the American dog tick, which carries spotted fever, and for organisms causing typhus fever and certain infectious diseases. It also harbors roundworms and other internal parasites.

Importance

The house mouse is an extremely destructive mammal, second in importance only to the brown rat. It is difficult to estimate the economic loss caused by these pests, because they damage stored grain and seed stock, take food, spread filth and disease, cut up fabrics to secure nesting material, and are a general nuisance.

On the credit side, domestic strains of albino mice are used as experimental animals in research laboratories.

Management

The house mouse is a problematic invasive species that can have devastating effects on natural ecosystems and native species. The most efficient means of

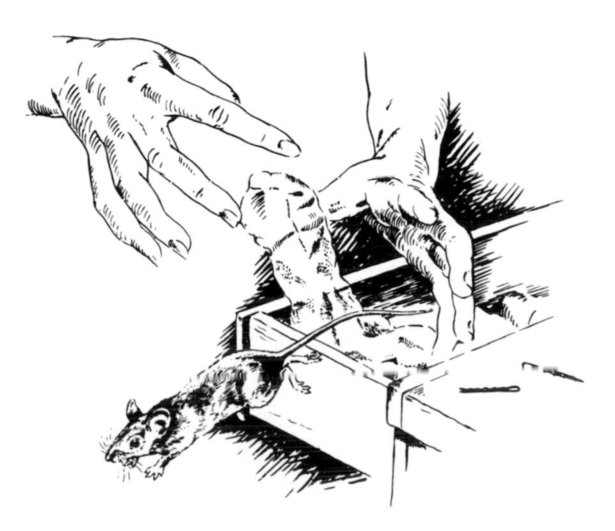

control is to use snap traps baited with peanut butter, cheese, rolled oats, or bacon. If you are experiencing problems with large numbers of mice, contact a wildlife professional for advice, assistance, regulations, or special conditions for handling these animals.

SELECTED REFERENCES

See also discussion of this species in general references, page 23.

Archer, J. 1968. The effect of strange male odor on aggressive behavior in male mice. *Journal of Mammalogy* 49:572–575.

Breakey, D. R. 1963. The breeding season and age structure of feral house mouse populations near San Francisco Bay, California. *Journal of Mammalogy* 44:153–168.

Crowcroft, P., and F. P. Rowe. 1963. Social organization and territorial behavior in the wild house mouse (*Mus musculatus* L.). *Proceedings of the Zoological Society of London* 140:517–531.

DeLong, K. T. 1967. Population ecology of feral house mice. *Ecology* 48:611–634.

Evans, F. C. 1949. A population study of house mice (*Mus musculus*) following a period of local abundance. *Journal of Mammalogy* 30:351–363.

Kaufman, D. W., and G. A. Kaufman. 1990. House mice (*Mus musculus*) in natural and disturbed habitats in Kansas. *Journal of Mammalogy* 71:428–432.

Laurie, E. M. O. 1946. The reproduction of the house mouse (*Mus musculus*) living in different environments. *Proceeding of the Royal Society of London. Series B, Biological Sciences* 133:248–281.

Schwarz, E., and H. K. Schwarz. 1943. The wild and commensal stocks of the house mouse, *Mus musculus* Linnaeus. *Journal of Mammalogy* 24:59–72.

Smith, W. W. 1954. Reproduction in the house mouse, *Mus musculus* L., in Mississippi. *Journal of Mammalogy* 35:509–515.

Southern, H. N., and E. M. O. Laurie. 1946. The house mouse (*Mus musculus*) in corn ricks. *Journal of Animal Ecology* 15:134–149.

Whitaker J. O., Jr. 1966. Food of *Mus musculus*, *Peromyscus maniculatus bairdi*, and *Peromyscus leucopus* in Vigo County, Indiana. *Journal of Mammalogy* 47:473–486.

Young, H., R. L. Strecker, and J. T. Emlen Jr. 1950. Localization of activity in two indoor populations of house mice, *Mus musculus*. *Journal of Mammalogy* 31:403–410.

Jumping Mice and Relatives (Family Dipodidae)

The family Dipodidae contains about 50 species, including jumping mice, jerboas, and birch mice, that occur across the northern hemisphere. Dipodidae is from Greek origin and means "two" and "footed" (*di*, "two," and *podos*, "footed"), which refers to these rodents' tendency to hop using their two enlarged hind feet and legs. Taxonomists reviewing this group are not in agreement as to whether they should be lumped into one family, the Dipodidae, or split into two families, the Dipodidae and Zapodidae. In previous editions of this book, the meadow jumping mouse was assigned to the family Zapodidae, but many taxonomists today agree that treating this group as the single family Dipodidae is a better reflection biologically than dividing them into two families.

The jumping mice are assigned to the subfamily Zapodinae, which refers to the enlarged hind feet that are typical of this group. Additional characteristics are the very long tail, which is more than 1½ times the length of the head and body, and the deep, lengthwise groove on the front of each upper incisor. Only one member occurs in Missouri.

Meadow Jumping Mouse (*Zapus hudsonius*)

Name

The first part of the scientific name, *Zapus*, is from Greek origin and means "very" and "foot" (*za*, "very," and *pous*, "foot"). Liberally interpreted, this name refers to the very large hind feet. The last part, *hudsonius*, is a Latinized word meaning "of Hudson," for Hudson Bay, the locality from which the first specimen was collected and named.

The common name originated as follows: "meadow" describes the habitat of this species; "jumping" refers to the predominant gait; and "mouse" denotes the general body shape.

Description (Plate 49)

The meadow jumping mouse is a small rodent with elongated hind feet and a sparsely haired, scaly tail about 1½ times the combined length of head and body, or approximately 60 percent of the total length. On each front foot there are 4 clawed toes and a small thumb with a blunt nail, and on each hind foot there are 5 clawed toes and 5 pads, or tubercles, on the naked

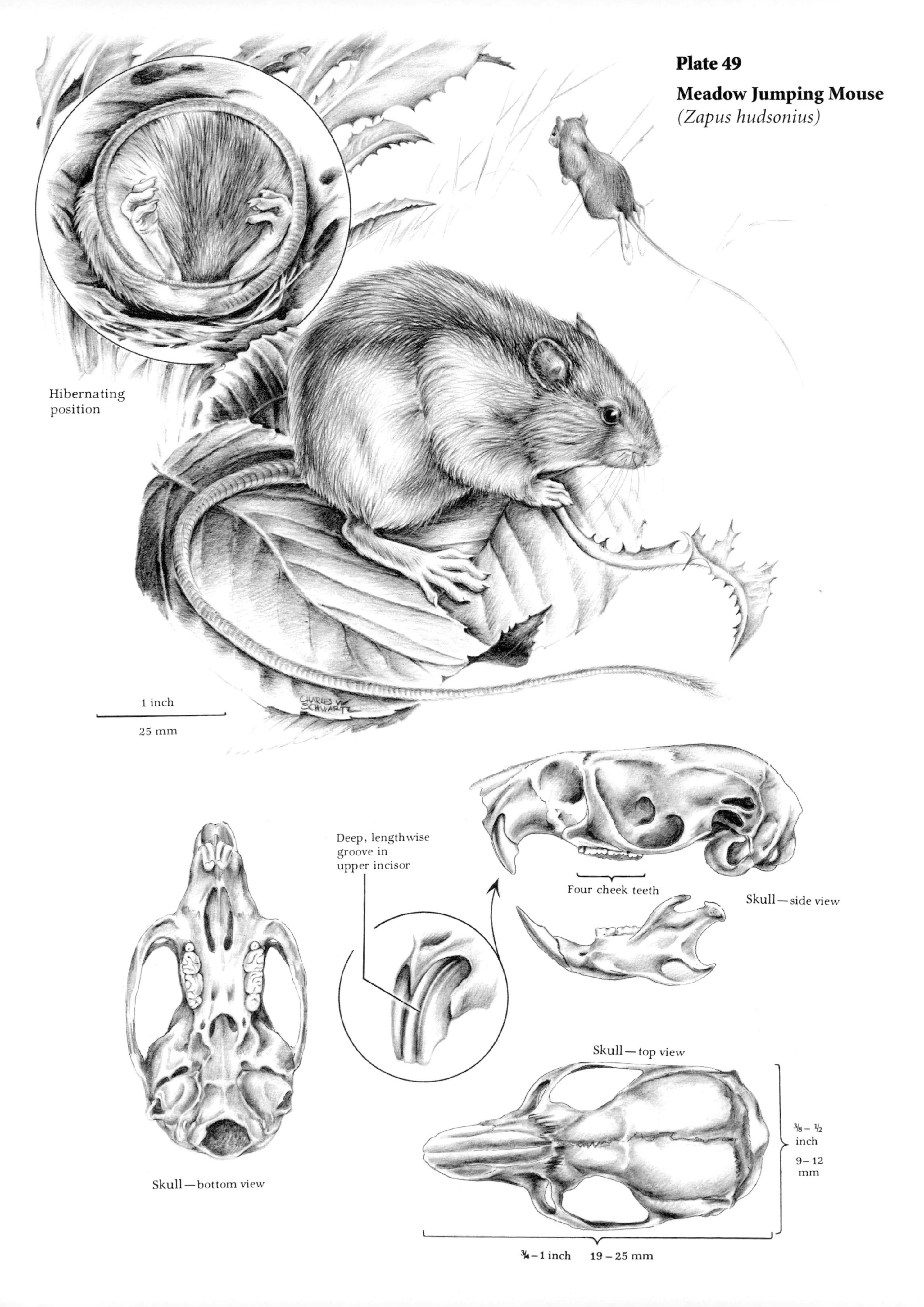
Plate 49
Meadow Jumping Mouse
(Zapus hudsonius)
Hibernating position
1 inch
25 mm
Deep, lengthwise groove in upper incisor
Four cheek teeth
Skull—side view
Skull—bottom view
Skull—top view
3/8–1/2 inch
9–12 mm
3/4–1 inch 19–25 mm

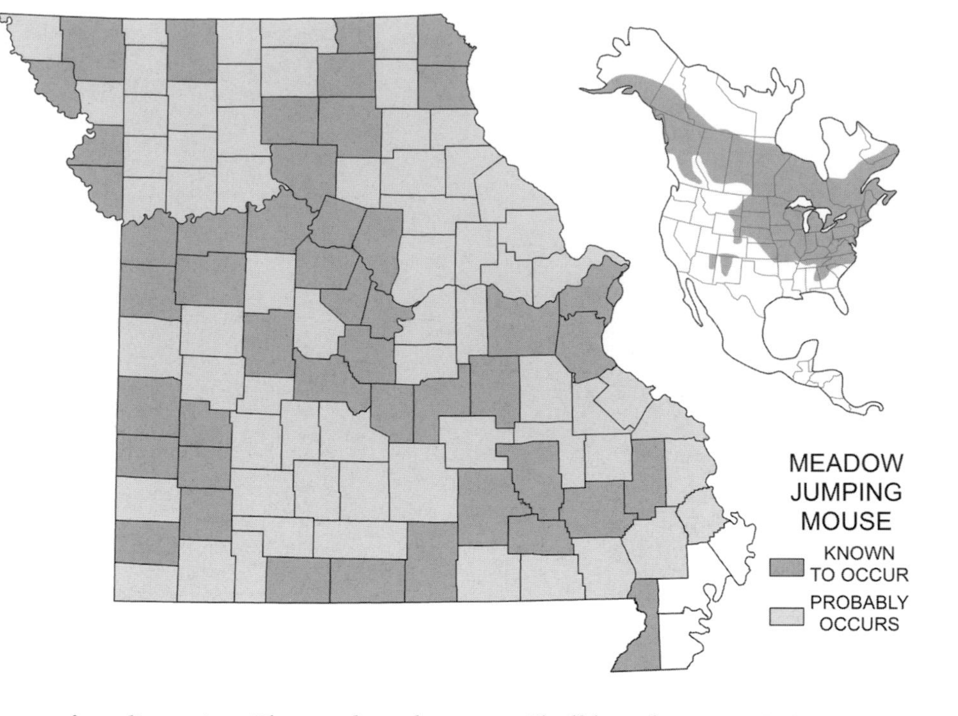

soles. The eyes are of medium size. The moderately large ears are somewhat concealed in the fur and can be closed over the ear opening. The front of each upper incisor has a deep, lengthwise groove in the middle. The body fur is long and rather coarse.

The meadow jumping mouse might be confused with several other small mice (the house mouse, the *Peromyscus* spp. mice, the golden mouse, harvest mice, and the plains pocket mouse) but can readily be distinguished by its longer tail, longer hind feet, 18 teeth, and absence of external, fur-lined cheek pouches.

Color. The upperparts are predominantly yellowish brown but many black-tipped hairs are interspersed, making a darkish band from head to tail. The sides are yellowish orange with some black hairs, and the underparts and feet are white to pale yellow. The hairs of the belly are light to their bases. The tail is grayish brown above contrasted to yellowish white below. A tiny tuft of black hair at the tip of the tail is characteristic of this species. The sexes are colored alike. Rare individuals are almost entirely black, have a white spot on the front part of the back, or lack all black coloring. Adults molt once annually, at any time when they are not in hibernation; the process takes about 3 weeks for completion.

Measurements

Total length	177–234 mm	7–9¼ in.
Tail	101–146 mm	4–5¾ in.
Hind foot	25–31 mm	1–1¼ in.
Ear	12–19 mm	½–¾ in.
Skull length	19–25 mm	¾–1 in.
Skull width	9–12 mm	⅜–½ in.
Weight	14–28 g	½–1 oz.

The greatest weight occurs just before hibernation.

Teeth and skull. The dental formula of the meadow jumping mouse is:

$$I\ \frac{1}{1}\ C\ \frac{0}{0}\ P\ \frac{1}{0}\ M\ \frac{3}{3} = 18$$

The number of teeth and the grooved yellowish orange upper incisors serve to distinguish the skull of the meadow jumping mouse from the skulls of all other mammals in Missouri.

Sex criteria and sex ratio. The sexes are identified as in the eastern woodrat. Females have four pairs of teats along the sides of the belly from front to hind legs. Males have a penis bone. Sex ratios obtained from jumping mice trapped in different parts of their range do not vary far from an even ratio.

Age criteria and longevity. Young animals are duller and yellower than adults and have softer fur. Jumping mice are fairly short-lived. The population that has lived over the winter is believed to be replaced by young during the summer months. However, two tagged jumping mice are known to have lived two years in the wild.

Voice and sounds. Adult jumping mice are usually silent but may *cluck* in a pleasant deep tone, chatter their teeth, and give birdlike chirps. They also squeak when fighting or disturbed. By vibrating their long tails against the ground or leaves, they make a drumming sound. The young give a high-pitched squeak.

Distribution and Abundance

The meadow jumping mouse is widely distributed from Alaska, across Canada, and into the Great Plains and the eastern United States; it has disjunct populations in Arizona and New Mexico. Within this general area it is unevenly distributed and fluctuates in number from year to year. Population densities in some parts of this range vary from 2 to 26 per 0.4 ha (1 ac.).

In the first edition of this book, only a few specimens of the meadow jumping mouse had been documented in 7 counties in 100 years prior to that book being published in 1959. Today, several have been collected in 46 counties across the state, and they probably occur statewide except for a few counties within the Mississippi River Alluvial Basin. It is not known how abundant this mouse is in Missouri, but from the paucity of records it can be inferred to be rare. Given that these mice hibernate for up to 7 months a year and populations probably occur in scattered pockets, they may be more common than is apparent.

Habitat and Home

Meadow jumping mice prefer an open grassy habitat but also live in grain and hay fields, shrubby or weedy fields, fencerows, and along the edge of woods. They frequent moist areas more than dry ones and are generally found near a stream or other water. Just before hibernation they show a tendency to shift to higher ground where hibernation sites will be drier.

During summer they use a spherical or oval nest of grass and leaves built either on the ground in a clump of shrubs or grass; above the ground in a hollow tree or suspended from vegetation; or under the ground from 7.6 to 15 cm (3 to 6 in.) deep. The nest is 10 to 15 cm (4 to 6 in.) in diameter with an inside cavity 5 to 7.6 cm (2 to 3 in.) across. There is one entrance on the side.

The winter, or hibernating, nest is similar to the summer nest but has no opening to the outside. It is always underground.

Habits

Meadow jumping mice are the only Missouri mice that undergo prolonged winter sleep. The other deep hibernating mammals in this state are the Franklin's ground squirrel, the thirteen-lined ground squirrel, the woodchuck, and some bats.

In September or October, jumping mice start to increase rapidly in weight. Two weeks before going into hibernation they acquire a layer of body fat and, as they do so, become slower in responses and actions. The fattest mice start to hibernate in mid-September, followed by the others as they accumulate their store of fat. All are in hibernation by the end of October. Although jumping mice usually hibernate singly, sometimes two may occupy the same nest.

Jumping mice prepare their winter hibernating quarters by digging a tunnel into a small mound, a well-drained slope, or an earthen bank. The soil is loosened with the front feet and thrown out by the hind feet. At the end of the tunnel they make a nest.

When a jumping mouse goes into the nest for its winter sleep, it closes the tunnel from within and rearranges the nesting material to cover the nest entrance.

Cross section of cinder pile showing leaf and dried grass nests of hibernating jumping mice

The mouse curls up in a ball with the head buried between the hind legs and the tail curled around the body. As it enters this deep sleep, the body temperature gradually becomes lower and the rate of heartbeat and circulation of blood through the body are greatly reduced. The mouse exists entirely on its body fat; utilization of this fat is greatest during the first few weeks of hibernation, as indicated by the rapid loss of weight, but decreases thereafter to a minimum level at which a rather steady body weight is maintained. In this inactive condition the only energy required is just enough to keep the animal's body temperature above freezing. However, about two-thirds of the population fails to survive hibernation. Apparently the young that have

Meadow jumping mouse tracks

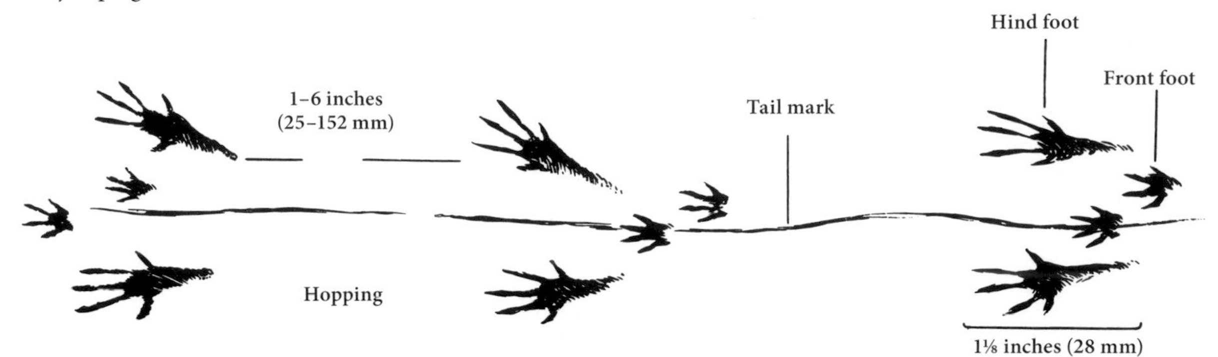

reached adult size in fall and have had enough time to accumulate sufficient fat are the most successful hibernators.

If handled during hibernation, the mouse feels very cold and appears dead. If taken from the nest near the beginning or end of the hibernating period and warmed up, the mouse becomes active in 15 minutes or so; if the warming up occurs near the middle of the hibernating period, several hours may be required for it to become active.

Meadow jumping mice are long hibernators, spending up to seven months in this torpid condition. Hibernation is usually continuous, but some mice may become active for short periods and even leave their winter nests temporarily. However, they usually do not come out of hibernation until spring is well advanced. The males first appear aboveground in late April or early May, while the females emerge in mid- or late May.

During their active life, jumping mice seldom stay long in one locality but move around from month to month. They often return to previously inhabited areas following absences of several weeks. Trapping records indicate that the average home range is about 0.4 ha (1 ac.), although it varies with relation to the terrain, cover, available moisture, and land use. Some jumping mice live for a while in an area as small as 0.04 ha (1⁄10 ac.), while others occupy 0.8 to 1.6 ha (2 to 4 ac.). The ranges of males and females overlap, males having larger home ranges than females. Neither sex shows a well-developed homing instinct.

Meadow jumping mice make no trails of their own, although they may use those of other kinds of mice. When traveling about they usually hop in a zigzag direction, covering distances of 2.5 to 15 cm (1 to 6 in.) per hop. Sometimes they make jumps of 0.6 to 1 m (2 or 3 ft.); when frightened they have been reported to leap as much as 3 to 3.6 m (10 or 12 ft.) at a time. One jumping mouse was timed to travel at the rate of 2.5 m (8 ft.) per second. In addition to the bounding gait, jumping mice often walk or run. They climb easily but usually not much above the lower branches of shrubs.

The tail serves as a balance in jumping, and any damage to the tail is extremely detrimental to the welfare of the mouse. More males have a portion of their tails missing than do females, probably as a result of fighting.

Meadow jumping mice swim well, both on the surface of the water and below the surface to a depth of 46 cm (18 in.). They have been observed to stay underwater for one minute without coming up for air. In swimming, the hind feet propel the mouse through the water, assisted on some occasions by the front feet. The tail is trailed behind.

Although they prefer to feed under cover of darkness, jumping mice may be active on cloudy, damp days. They are very wary and nervous and are seldom seen by people. One mouse, surprised by a human intruder, sat motionless, flattened against the ground, for 40 minutes to avoid detection.

In captivity, these mice spend considerable time dressing their coats and grooming themselves. The tail is cleansed by passing it through the mouth.

Foods

Grass seeds are the main foods of jumping mice, although fruits, roots, and the tender parts of many plants are included in the diet. Animal foods, consisting of insect larvae and adults, spiders, millipedes, snails, slugs, and dead mice, are also consumed. The choice of food varies with the season. After coming out of hibernation, a variety of foods is eaten of which animal matter constitutes about 50 percent and seeds 20 percent. During the summer, less animal matter is taken and more plant matter, particularly seeds, and a subterranean fungus, is eaten.

Jumping mice frequently bend down grass stems to reach the seed heads and usually cut the stems into lengths of 7.6 to 10 cm (3 to 4 in.). These cut pieces of stems are often dropped in a crisscross fashion,

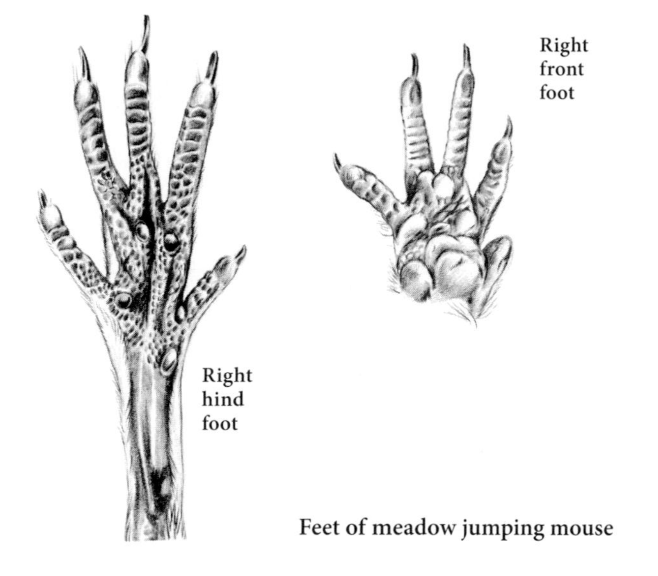

Feet of meadow jumping mouse

forming small piles where the mice feed. Jumping mice also climb up strong-stemmed plants until they reach the flower or seed head, which they cut. In eating, the jumping mouse sits on its haunches and handles the food with its front feet.

Jumping mice eat one-half of their weight daily. They can live without water if dew and succulent foods are available. However, in captivity they drink water regularly.

Reproduction

The breeding season begins immediately upon emergence from hibernation, and young are produced until the end of August. There are three main periods of breeding, with the young being born in the following three peaks: the latter part of June; mid- or late July; and mid-August. On the basis of recaptured wild females, it is suspected that some females have 3 litters in one breeding season. The gestation period is usually 18 days but may be slightly longer in nursing females. The litters contain 5 or 6 young, but extremes of 1 and 9 occur.

At birth the young are blind and naked. They weigh about 1 g (1⁄32 oz.) and are 32 mm (1¼ in.) long. The tail is relatively short (about half the combined length of head and body) and the hind feet are not elongated. At 1 week of age, the upperparts of the body and tail are dark; the ears are unfolded; and the claws are present. They are able to crawl but cannot support themselves. By 2 weeks of age, the backs are covered with fur; the tails are as long as the head and body; the incisors are cut through the gums; and the facial whiskers are prominent. They can now walk and make short hops.

In another week, the hind feet and tail are much longer and the eyes are open. By 4 weeks after birth the young have acquired adult coloration but are still slightly smaller than adults. They are weaned about this time and soon lead an independent existence. Mice born early in the breeding season may breed the same year, but those born late probably do not breed until the spring following their birth.

Some Adverse Factors

Carnivorous mammals, hawks, owls, snakes, and bullfrogs feed on jumping mice. The parasites include mites, ticks, lice, fleas, botfly larvae, roundworms, and flukes.

Importance

Wherever jumping mice live, they contribute in many ways to their environment through their life activities. They do not damage domestic crops.

Conservation and Management

This species is uncommon in Missouri and requires no control.

SELECTED REFERENCES

See also discussion of this species in general references, page 23.

Adler, G. H., L. M. Reich, and R. H. Tamarin. 1984. Demography of the meadow jumping mouse (*Zapus hudsonius*) in eastern Massachusetts. *American Midland Naturalist* 112:387–391.

Blair, W. F. 1940. Home ranges and populations of the jumping mouse. *American Midland Naturalist* 23:244–250.

Brown, L. N., and R. B. McMillan. 1964. Meadow jumping mouse in southern Missouri. *Journal of Mammalogy* 45:150–151.

Chromanski-Norris, J. F., and E. K. Fritzell. 1983. Status and distribution of ten Missouri mammals. A report to the Missouri Department of Conservation, Jefferson City. 38 pp.

Hoyle, J. A., and R. Boonstra. 1986. Life history traits of the meadow jumping mouse, *Zapus hudsonius*, in southern Ontario. *Canadian Field Naturalist* 100:537–544.

Krutzsch, P. H. 1954. North American jumping mice (genus *Zapus*). *University of Kansas Publications, Museum of Natural History* 7:349–472.

Quimby, D. C. 1951. The life history and ecology of the jumping mouse, *Zapus hudsonius. Ecological Monographs* 21:61–95.

Schwartz, C. W. 1951. A new record of *Zapus hudsonius* in Missouri and notes on its hibernation. *Journal of Mammalogy* 32:227–228.

Whitaker, J. O., Jr. 1963. A study of the meadow jumping mouse, *Zapus hudsonius* (Zimmerman), in central New York. *Ecological Monographs* 33:215–254.

———. 1972. *Zapus hudsonius. Mammalian Species* 11. 7 pp.

Red fox hunting mice

8

Flesh-eating Mammals

Order Carnivora

The name, Carnivora, is from two Latin words and means "flesh eating"; it refers to the general meat-eating food habits of this group. The teeth of carnivores are especially well adapted for securing and feeding on flesh. The canine teeth are large and pointed and serve to seize and hold the prey. As a further adaptation to this meat-feeding habit, specialized flesh-cutting teeth, the *carnassials*, are usually developed from the last upper premolar and first lower molar teeth. The brain is highly developed, and the group as a whole exhibits a fairly high level of intelligence.

This large order is distributed throughout the world. In addition to the six families whose members are now living wild in Missouri, the order Carnivora includes nine other families.

Key to the Species
by Whole Adult Animals

1a. Front and hind feet each with 5 toes. **Go to 2**
1b. Hind foot with 4 toes; front foot with 5 toes of which the "thumb" is high on the foot. **Go to 10**

2a. (From 1a) Face marked with black mask across eyes and cheeks (see plate 54); tail marked with alternate yellowish gray and brownish black rings. **Raccoon** (*Procyon lotor*) p. 333
2b. (From 1a) Face not marked with black mask across eyes; tail not marked with rings. **Go to 3**

3a. (From 2b) Body black with white stripes or spots on back. **Go to 4**
3b. (From 2b) Body not black with white spots or stripes on back. **Go to 5**

4a. (From 3a) Thin white stripe down center of face and broad white stripe on back of head, which may fork on shoulders and extend down back and onto tail (see plate 61). **Striped Skunk** (*Mephitis mephitis*) p. 375
4b. (From 3a) White spot on forehead; upperparts of body with two pairs of broken white stripes (see plate 60); flanks and sides with white stripes or spots. **Eastern Spotted Skunk** (*Spilogale putorius*) p. 369

5a. (From 3b) Total length usually 152 cm (60 in.) or more; weight over 90.8 kg (200 lb.); body black or reddish brown with brown face and white patch usually on chest; tail very short, not bushy, and nearly concealed in fur (see plate 53). **American Black Bear** (*Ursus americanus*) p. 326
5b. (From 3b) Total length less than 152 cm (60 in.); weight less than 22.7 kg (50 lb.). **Go to 6**

6a. (From 5b) Body brown on back and yellowish white on belly; body fur sometimes turns white in winter. **Go to 7**
6b. (From 5b) Body gray or all brown. **Go to 8**

7a. (From 6a) Tail brown with terminal one-third or one-fourth black; total length 29 cm (11½ in.) or more (see plate 56). **Long-tailed Weasel** (*Mustela frenata*) p. 344
7b. (From 6a) Tail brown without black tip or with only a few black hairs at tip; total length less than 29 cm (11½ in.) (see plate 55). **Least Weasel** (*Mustela nivalis*) p. 341

8a. (From 6b) Body gray; face brownish, marked with a single white stripe from nose to crown and onto neck and back; paired white cheek patches, and vertical black bars in front of ears (see plate 58); very strong, prominent claws on front feet; short, bushy tail. **American Badger** (*Taxidea taxus*) p. 356
8b. (From 6b) Body all brown. **Go to 9**

9a. (From 8b) Total length of both sexes up to 135 cm (53 in.); tail thick at base, flat on bottom, and obviously tapering from body toward tip (see plate 59); toes fully webbed. **North American River Otter** (*Lontra canadensis*) p. 362
9b. (From 8b) Total length of males up to 69 cm (27 in.); females up to 53 cm (21 in.); tail not obviously thick at base, not flat on bottom, and not obviously tapering from body toward tip (see plate 57); toes joined by short webs at their bases. **American Mink** (*Neovison vison*) p. 350

10a. (From 1b) Build catlike with short, broad face (see plates 62–63); retractile, strongly curved claws that can be concealed in the fur. **Go to 11**
10b. (From 1b) Build doglike, usually with elongated muzzle (see plates 50–52); nonretractile, moderately curved claws, which are well exposed. **Go to 12**

11a. (From 10a) Tail short, less than one-third total length (see plate 63); back and sides yellowish to reddish brown streaked and spotted with black; total length up to 127 cm (50 in.). **Bobcat** (*Lynx rufus*) p. 387
11b. (From 10a) Tail long, one-third or more of total length (see plate 62); back and sides uniform tan to grayish; total length 152 to 259 cm (60 to 102 in.). **Mountain Lion** (*Puma concolor*) p. 382

12a. (From 10b) Tail usually less than half as long as head and body; total length usually 107 cm (42 in.) or more; weight 8.2 kg (18 lb.) or more; hind foot usually 18 cm (7 in.) or more in length (see plate 50); if specimen is alive, pupil of eye round. **Coyote** (*Canis latrans*) p. 308

12b. (From 10b) Tail usually more than half as long as head and body; total length usually less than 107 cm (42 in.); weight less than 8.2 kg (18 lb.); hind foot usually less than 18 cm (7 in.) in length (see plates 51–52); if specimen is alive, pupil of eye vertically elliptical. **Go to 13**

13a. (From 12b) Body reddish yellow; tail reddish yellow mixed with black and tipped with white; legs and feet black (see plate 51). **Red Fox** (*Vulpes vulpes*) p. 315

13b. (From 12b) Body mixed gray and black with reddish brown on sides of neck and backs of ears; tail gray above with black mane for entire length of upper surface and with black tip; legs and feet reddish brown (see plate 52). **Gray Fox** (*Urocyon cinereoargenteus*) p. 322

Key to the Species
by Skulls of Adults[1]

1a. Length of skull 25 cm (10 in.) or more; total teeth usually 42, but some premolars often lost in old individuals (see plate 53). **American Black Bear** (*Ursus americanus*) p. 326

1b. Length of skull less than 25 cm (10 in.). **Go to 2**

2a. (From 1b) Total teeth 28 (see plate 63). **Bobcat** (*Lynx rufus*) p. 387

2b. (From 1b) Total teeth not 28. **Go to 3**

3a. (From 2b) Total teeth 30. **Go to 4**

3b. (From 2b) Total teeth not 30. **Go to 5**

1. Because skulls of the domestic dog and domestic cat are frequently found and might be confused with those of wild mammals, they are included here.

4a. (From 3a) Length of skull less than 10 cm (4 in.). **Domestic Cat** (*Felis catus*)

4b. (From 3a) Length of skull 10 cm (4 in.) or more (see plate 62). **Mountain Lion** (*Puma concolor*) p. 382

5a. (From 3b) Total teeth 40 (see plate 54). **Raccoon** (*Procyon lotor*) p. 333

5b. (From 3b) Total teeth not 40. **Go to 6**

6a. (From 5b) Total teeth 36 and rarely 38 (see plate 59). **North American River Otter** (*Lontra canadensis*) p. 362

6b. (From 5b) Total teeth not 36 or 38. **Go to 7**

7a. (From 6b) Total teeth 42. **Go to 8**

7b. (From 6b) Total teeth 34. **Go to 10**

8a. (From 7a) Upper surface of skull with prominent crest formed by paired ridges that converge in midline; upper surface of postorbital process convex (see plate 50 and illustration p. 310). *Canis*: **Coyote** (*Canis latrans*) p. 308; **Domestic Dog** (*Canis lupus familiaris*)

8b. (From 7a) Upper surface of skull smooth or possessing paired ridges that if they converge do so at rear of skull and do not form prominent crest (see plates 51–52); upper surface of postorbital process with shallow to deep depression. **Go to 9**

9a. (From 8b) Upper surface of skull smooth or possessing indistinct, paired ridges that start just behind the eye socket and either converge toward the rear to form a slight crest or come close to each other at the midline where they are separated by a space less than 9 mm (⅜ in.) wide (see plate 51); upper surface of postorbital process with shallow depression; lower jaw without obvious notch on bottom edge toward the rear. **Red Fox** (*Vulpes vulpes*) p. 315

9b. (From 8b) Upper surface of skull with prominent, paired ridges up to 3 mm (⅛ in.) high that start just behind the eye socket and enclose a U-shaped area but do not come closer to each other at the midline than 9 mm (⅜ in.) (see plate 52); upper surface of postorbital process with deep depression; lower jaw with prominent notch on bottom edge toward the rear. **Gray Fox** (*Urocyon cinereoargenteus*) p. 322

10a. (From 7b) Hard palate not extending beyond last upper molars (see plates 60–61). **Go to 11**

10b. (From 7b) Hard palate extending beyond last upper molars (see plates 55–58). **Go to 12**

11a. (From 10a) Region above eye sockets well rounded and possessing definite arch when seen in profile (see plate 61); mastoid region not inflated; auditory bulla not inflated; obvious notch on bottom edge of lower jaw toward the rear. **Striped Skunk** (*Mephitis mephitis*) p. 375

11b. (From 10a) Region above eye sockets nearly flat and lacking definite arch when seen in profile (see plate 60); mastoid region inflated; auditory bulla slightly inflated; no obvious notch on bottom edge of lower jaw toward the rear. **Eastern Spotted Skunk** (*Spilogale putorius*) p. 369

12a. (From 10b) Upper molars triangular (see plate 58); braincase triangular; auditory bulla moderately inflated and elongated. **American Badger** (*Taxidea taxus*) p. 356

12b. (From 10b) Upper molars dumbbell shaped (see plates 55–57); auditory bulla greatly inflated and noticeably longer than wide. **Go to 13**

13a. (From 12b) Length of skull more than 5 cm (2 in.); auditory bulla nearly as long as upper row of premolar and molar teeth (see plate 57). **American Mink** (*Neovison vison*) p. 350

13b. (From 12b) Length of skull 5 cm (2 in.) or less; auditory bulla longer than upper row of premolar and molar teeth. **Go to 14**

14a. (From 13b) Length of skull 3 cm (1¼ in.) or less (see plate 55). **Least Weasel** (*Mustela nivalis*) p. 341

14b. (From 13b) Length of skull 3 to 5 cm (1¼ to 2 in.) (see plate 56). **Long-tailed Weasel** (*Mustela frenata*) p. 344

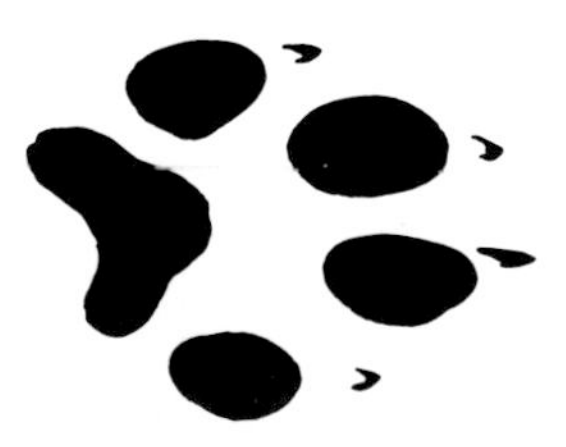

Dogs, Coyotes, Foxes, and Relatives (Family Canidae)

Members of this family are typically doglike and, except for the highly specialized breeds of domestic dogs, generally have an elongated muzzle, rather long legs, and a long, well-furred tail. There are usually 5 toes on each front foot and 4 on each hind foot. The carnassial teeth (the last upper premolar and first lower molar) are highly developed for flesh shearing.

The family name, Canidae, is based on the Latin word for "dog." Members of this family are found on all continents except Antarctica. Only one species occurs in Australia, the dingo, which was probably not native but was introduced by aboriginal humans. Within historic times, five wild species have lived in Missouri. The gray wolf and red wolf are no longer here. The three remaining members of our fauna are the coyote, red fox, and gray fox.

Coyote (*Canis latrans*)

Name

The coyote's scientific name, *Canis*, is the Latin word for "dog"; *latrans* is the Latin word for "barker" and refers to this animal's habitual barking. The common name comes originally from the Nahuatl (Aztec) word *coyotl*. Some people pronounce the common name "ki-ot" and others pronounce it "ki-o-tee" or "-tay"; the latter is closer to the Mexican Spanish pronunciation.

Description (Plate 50)

The coyote is extremely doglike; compared to the various domestic breeds, it most closely resembles a small German shepherd in general form. A callus is usually obvious on the front leg in the region of the elbow. The pelage is fairly long, coarse, and heavy.

The coyote is easily distinguished from the red and gray foxes by its larger size, coloration, shorter tail, and round pupil of the eye, which is apparent in living animals. Certain strains of the domestic dog, and particularly some hybrids between coyotes and dogs, are so similar to coyotes that their identification is difficult.

Color. Typically, the upperparts are light gray to dull yellow with the outer hairs broadly tipped with black. This black tipping creates wavy crossbands of variable width on the back and sides. The backs of the ears are reddish and the muzzle yellowish. Above, the tail is colored like the back; below, it is whitish near the base, then pale yellowish; the tip is black, sometimes with a few white hairs. The front legs are whitish; the outer sides of the hind legs are reddish, while the inner sides are whitish. The throat and belly are white to pale gray. The iris of the eye is tawny. Considerable variation in overall coloration occurs within the species, with extremes from nearly black to nearly white. The

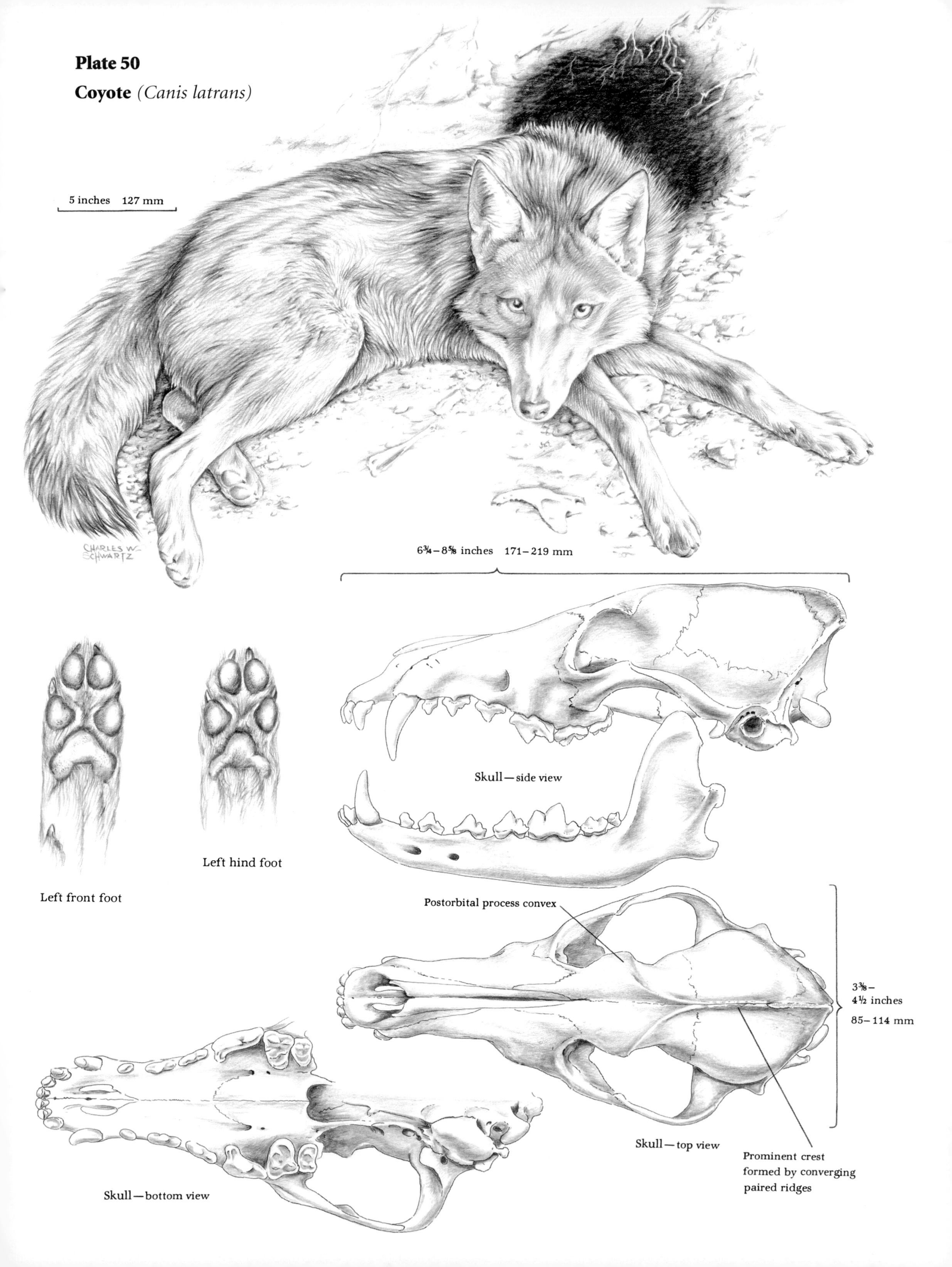
Plate 50
Coyote (Canis latrans)
5 inches 127 mm
CHARLES W. SCHWARTZ
6¾–8⅝ inches 171–219 mm
Skull—side view
Left hind foot
Left front foot
Postorbital process convex
3⅜–4½ inches
85–114 mm
Skull—top view
Prominent crest formed by converging paired ridges
Skull—bottom view

sexes are colored alike. Molting starts in late spring, and the lost fur is replaced gradually during the summer months. The pelage is prime from the end of November into February.

Measurements

Total length	100–137 cm	39½–54 in.
Tail	273–406 mm	10¾–16 in.
Hind foot	174–222 mm	6⅞–8¾ in.
Ear	101–120 mm	4–4¾ in.
Skull length	171–219 mm	6¾ – 8⅝ in.
Skull width	85–114 mm	3⅜–4½ in.
Weight	8.1–13.6 kg	18–30 lb.

Females are slightly smaller than males.

Two exceptionally heavy males in Missouri weighed 21.3 and 21.7 kg (47 and 48 lbs.), respectively.

In 2012, an 18.4 kg (40.6 lb.) male became the official record-weight coyote confirmed by the Missouri Department of Conservation. The state furbearer record program was started in 2011.

Teeth and skull. The coyote has 42 teeth, as do all members of the dog family, in the following dental formula:

$$I\,\frac{3}{3}\;C\,\frac{1}{1}\;P\,\frac{4}{4}\;M\,\frac{2}{3} = 42$$

The skull of an adult coyote is distinguished from a red or gray fox skull by the following combination of characteristics: size; a prominent crest on the rear of the upper surface that is formed by paired ridges starting just behind the eye socket and converging in the midline (particularly developed in males); and the convex surface of the postorbital process above the eye socket.

Because coyote, domestic dog, and red wolf skulls are often so similar and show such a wide range of individual variation, they are hard to separate and must be identified by some authority. However, it is usually possible to distinguish coyote from dog skulls by the coyote's relatively longer, narrower muzzle; but some dog skulls, particularly elongated-muzzle breeds like the collie and borzoi, may be indistinguishable by this means. In a coyote skull, the length of the upper molar tooth row (from the front of the socket of the first premolar to the rear of the socket of the last molar) is 3.1 or more times the palatal width (the distance between the inner borders of the sockets of the first upper premolars); in a dog skull, the length of this tooth row is less than 2.7 times the palatal width. Coyotes can also be identified by the relatively longer, narrower upper canine teeth, while dogs and red wolves have relatively shorter, broader upper canines. In a coyote skull, the tips of the upper canine teeth usually fall below a line drawn through the anterior mental foramina in the lower jaw (see accompanying illustration). In a dog or red wolf skull, the tips of the upper canines usually fall well above this line.

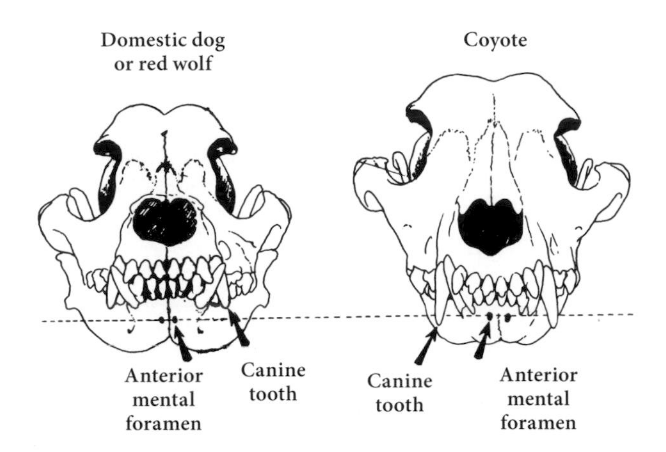

Coyote skulls can usually be distinguished from domestic dog or red wolf skulls by the length of the canine teeth in relation to a line drawn through the anterior mental foramina

There is no single, reliable way of distinguishing a coyote skull from a red wolf skull but, in general, the red wolf skull is longer and broader; has a heavier bone structure and a more pronounced crest; the constriction is relatively narrower behind the eyes; and the braincase is relatively smaller.

Sex criteria and sex ratio. Males are recognized by the penis and scrotum. They also possess a penis bone, or baculum. There are four pairs of teats. The sex ratio shows an equal number of males and females.

Age criteria and longevity. The young, until 6 months old, have duller and grayer coats than adults.

The amount of wear on the tips of the incisor and canine teeth of both upper and lower jaws serves as a means of age determination in coyotes. In coyotes 4 years of age, the tips of the central upper incisors

Coyote tracks

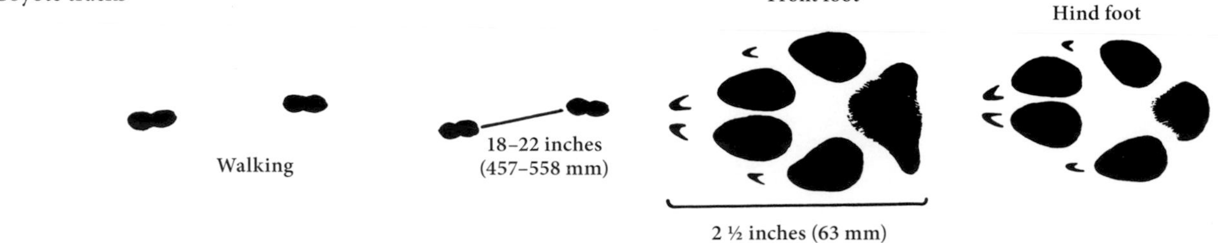

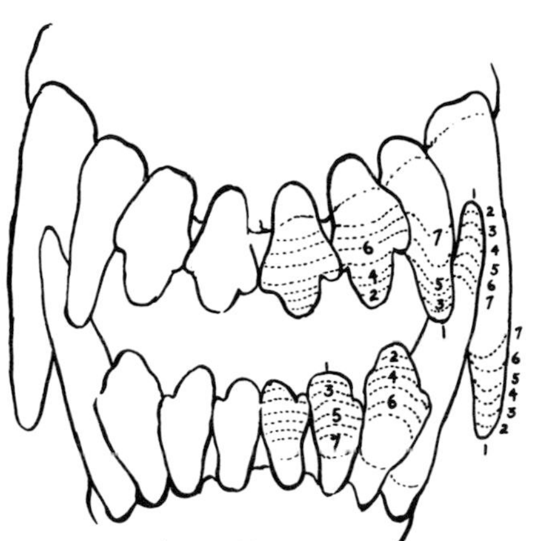

Age in the coyote as indicated by annual wear on incisor and canine teeth (dotted lines show limit of annual wear)

are worn even with the lobes (see accompanying illustration).

A count of the cementum layers on the teeth, particularly the canines, provides a fairly accurate estimate of the age of adult individuals.

Another aging technique is by the "tightness of the canine tooth socket." As an animal ages, the bone around the canine tooth grows down around the tooth itself. The relative amount of growth provides a method of separating juveniles from older animals and determining relative age classes.

Coyotes typically have been known to live up to 6–10 years in the wild and 18 years in captivity.

Glands. The scent gland on top of the tail, which is typical of many members of the dog family, is elliptical and measures about 38 mm (1½ in.) long by 9.5 mm (⅜ in.) wide. It begins about 5 cm (2 in.) from the base of the tail. When coyotes meet, the odor from this gland serves to identify the individual. The strongly scented urine is used to mark food caches and certain spots along the trails.

Voice and sounds. Coyotes have a rich vocabulary. They give various short sounds such as barks, yips, growls, and whimpers. The barks and yips commonly increase in power and pitch and end in a long, flat howl. Howling occurs at any time of the year, but more so during the mating period and less so when there are young. It is heard most often from sunset to sunrise, but occasionally in the daytime—for example, before a storm or as the result of some stimulus like a siren or whistle. Coyotes may bark alone or together; often, when one starts, others take up the call until it becomes a chorus. This song has a ventriloquial quality and carries for 2 or 3 miles. The young have higher pitched voices than adults. Coyotes doubtless howl for pleasure as well as to contact other coyotes.

In addition to sound, they also communicate with other coyotes by various means such as stance, erection of ears, gaze, position of tail, position of head, erection of fur, curling of lips, and exposure of teeth.

Distribution and Abundance

The coyote's range prior to European settlement was restricted to the western and plains regions of the United States and Canada, and northern and central Mexico. Their range expansion eastward began in the 1800s coinciding with land conversion that created more habitats, and the extirpation of wolves and mountain lions. Today they are widespread and common across most of North America and range into southern Central America. The overall population

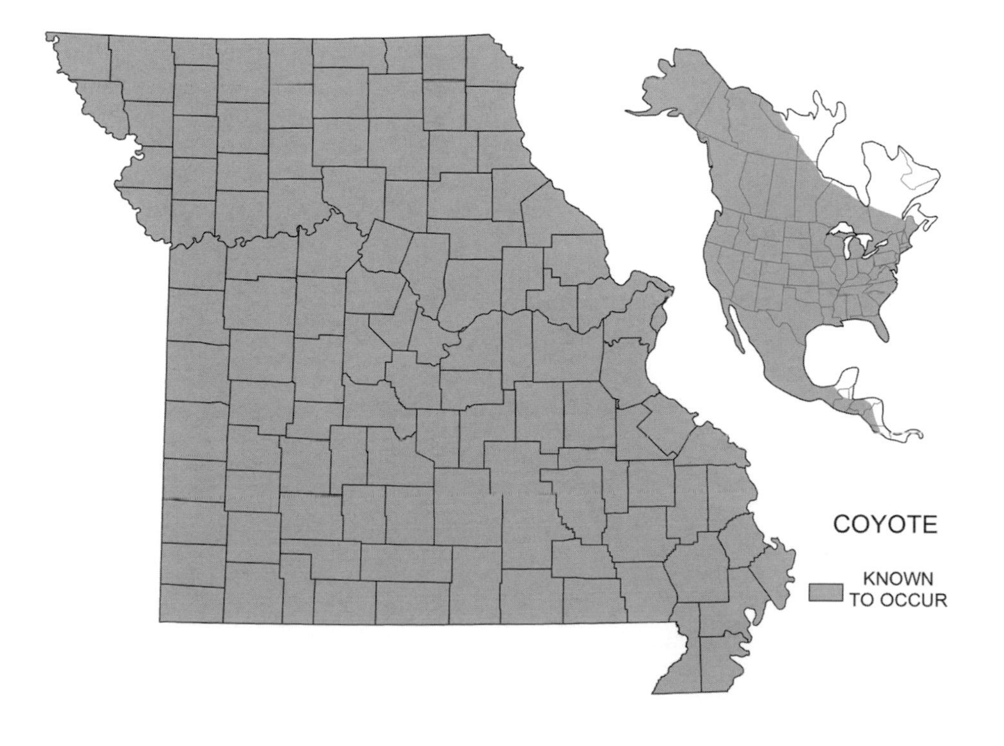

continues to increase and they are expected to expand their range into South America.

The coyote occurs throughout Missouri, being most abundant in the Central Dissected Till Plains and Osage Plains of northern and western Missouri. In Missouri, coyotes generally share the distinctive habitats of the gray and red foxes. Since 1970 the coyote population has been increasing, and it is now the most abundant wild member of the canid family in the state. This has been due partially to the adaptability of the coyote and its aggressive nature. As the red fox population decreased, coyotes filled the vacant range; they were also able to dominate the less aggressive gray fox.

Habitat and Home

Coyotes prefer to live in brushy country, along the edge of timber, and in open farmlands. During most of the year they merely sleep on the ground in some concealed, protected spot, but in the breeding season they have dens for the young. There may be several dens belonging to one family, one for the young and others for escape or for use in case of a heavy flea infestation in the home den.

Dens are usually located in unused fields and are often close to timber. They may be in a bank, under a hollow tree or log, in a rock cavity, or even under a deserted building. The den is frequently remodeled from one used by a fox, skunk, woodchuck, or American badger, but the coyote sometimes digs its own den in loose soil. The one or more openings, usually well concealed by brush, are approximately 25 cm (10 in.) wide by 50 cm (20 in.) high. The one or more tunnels are from 1.5 to 9 m (5 to 30 ft.) long and 30 to 60 cm (1 to 2 ft.) in diameter with an enlarged chamber at the end. This may contain some nesting material such as grass or fur. The pups occasionally dig pockets for themselves, considerably enlarging the den.

Habits

The home range of a coyote is variable across the species' geographic distribution. Home ranges may be as small as 800 ha (8 sq km or 1,976 ac.) when the young are being fed, or as great as 8,000 ha (80 sq km or 19,768 ac.) during the rest of the year. Authorities differ on whether adult males have larger home ranges than adult females and on whether their home ranges overlap. In Arkansas, home ranges of males were reported to be from 20.8 to 41.6 sq km (8 to 16 sq. mi.) and females from 8 to 9.6 sq km (3 to 3.7 sq. mi.). It is likely that a number of factors, primarily habitat quality and presence of other animals, contribute to the size of individual home ranges.

Coyotes like semiopen country and prefer to travel on ridges or old trails. Their travel lanes may cover 16 linear km (10 mi.). At particular places along their trails, they deposit urine, feces, and glandular scent, possibly to mark territories.

Social organization in coyotes varies from those living singly, in pairs of a male and female, or as a family. They are less social than wolves. A major factor influencing such social organization may be food availability.

Coyotes exhibit a complex series of facial expressions and body postures as indications of their social role. Adults and pups express their dominance by approaching another individual with a stiff-legged gait, ears forward and erect, fur erect on the back, and tail held at about a 45° angle from the vertical. They snarl frequently, grimace, and expose their teeth. Submission is displayed by running away, looking away, moving slowly away, or rolling over and urinating.

Coyotes are mostly nocturnal but occasionally are active in the daytime. They show a peak of activity at sunset and in early evening with a minor peak at daybreak. In summer they tend to come out most in daytime when the pups are most active. They can run as fast as 72 km (45 mi.) per hour for short distances. Water is no barrier, as they swim well. They seem to enjoy outwitting other animals and are even playful on occasion.

In securing small prey like rabbits or rodents, the coyote usually creeps stealthily for some distance, "freezes" momentarily, and then pounces with all four feet. In running down a large victim, two or more coyotes may chase the animal in relays and make the final kill by surprising it from behind protective cover. A large animal is killed by biting its throat and suffocating it.

Foods

Coyotes eat animal foods primarily, relying upon rabbits and mice for almost two-thirds of their diet. A study of the food habits of 770 coyotes in Missouri showed the following major food groups and their percentages by volume: rabbits, 53.7; mice and rats, 8.7; other wild mammals, 7.5; livestock, 8.9; poultry, 11.3; wild birds, 0.5; known carrion, 5.8; insects, 0.8; plants, 2.0; and miscellaneous, 0.8. Plant matter may be important seasonally, as for example persimmon fruit and seeds.

It is difficult to evaluate accurately the coyote's food habits because it is often impossible to tell whether the food item was secured as a kill or eaten as carrion. The take of domestic poultry is greatest during the period when pups are in the den or being trained to hunt.

Between 10 and 20 percent of the total foods of the coyote probably constitutes a financial loss to humans, while the rest is neutral or beneficial. When food is abundant, the surplus is cached in a hole the coyote digs in the ground with its front feet. The coyote uses its nose to cover the food with soil and tamp it down.

Reproduction

The height of the mating season is late February or early March. Unlike male dogs, male coyotes have a limited season of reproductive activity that coincides with the period when females are receptive. However, courtship behavior may begin a couple of months before the actual mating season. While the females in the population breed over a two-month span, individual females mate only within a period of 2 to 5 days once a year. The gestation period requires between 58 and 63 days; the young, ranging from 2 to 19 but usually 5 to 7, are born in late April or May. Largest litters are produced in years with good food supply from a high rodent population. Some pairs stay mated for a year, others for life. Only rarely do two females share a den with their respective litters. In such instances, the litters are nearly always of different ages; it is believed the female parents represent a mother and her daughter of the previous season. Sometimes an additional animal is associated with the den. This male or female acts as a nursemaid for the pups.

The young are blind and helpless at birth and are covered with brownish gray woolly fur. Their eyes open between 8 and 14 days of age. Coyote pups can be distinguished from fox pups by the shape of the eye pupil; the pupils are round in coyote pups and vertically elliptical in both red and gray fox pups. The young come out of the den for the first time when about 3 weeks old but do not remain outside for long periods until 5 or 6 weeks old.

Both parents care for the young. When the pups are tiny, the male assists by bringing food for the female.

Coyote pup has round eye pupil, distinguishing it from fox pup with spindle-shaped pupil

Later he brings food that the female tears into small bits for the young. About weaning time, at 6–8 weeks of age, both parents carry food in their stomachs and disgorge it outside the den in a partially digested state. The outside of a coyote's den is usually clean except at weaning time, when disgorged food may accumulate, putrify, and attract carrion-feeding birds and insects.

During their interactions as pups, the littermates establish a dominance relationship between themselves. Higher ranking pups tend to keep their distance from others and to be more independent about their individual activity. These, and the lowest ranking individuals, are probably the first to disperse from the family.

The pups are taught to hunt when between 8 and 12 weeks old. The family moves away from the den about this time and disperses in late summer or early fall. Some young are known to have gone as far as 190 km (120 mi.) from their home den. A few young may breed when one year old, but most young of both sexes mature in their second year. Coyotes occasionally hybridize with dogs or wolves and produce fertile offspring. The resulting young may resemble one or both parents and, in some instances, may look so much like the coyote parent as to mask its mixed inheritance.

Some Adverse Factors

Humans are the most important predators of coyotes. Domestic dogs and great horned owls may take some young, and a female deer has been known to kill an adult with her front feet.

The following parasites occur on or in coyotes: mites, ticks, lice, fleas, roundworms (particularly hookworms), flukes, heartworms, tapeworms, and protozoa. The most frequent diseases are rabies, distemper, and tularemia. Others, such as leptospirosis, encephalitis, and hepatitis, have been reported. Mange, caused by a mite, frequently is troublesome.

The relation between tapeworm infestation and coyote feeding habits is interesting and important. The tapeworm *Taenia pisiformes*, commonly found in the coyote, utilizes the cottontail as an intermediate host. In the rabbit, the eggs develop into larvae, or bladderworms, and form cysts in the liver, abdominal cavity, and lymphatics. When the coyote eats an infected rabbit, the cysts are ingested and the larvae are freed. They become attached to the intestinal wall of the coyote, develop into sexually mature tapeworms, and lay eggs that pass out of the coyote's body with the feces. A rabbit eats the eggs along with vegetation, and the cycle is completed.

The effects of such parasitism are felt in heavy infestations. The intestinal tract of the coyote can be blocked by the tapeworms, toxins given off by the parasites may affect the coyote adversely, and nutrients are drained from the host.

Importance

Although coyotes kill some livestock and poultry, they are often blamed unjustly for the large amount of damage done to domestic stock by free-running dogs. From the studies of stomach contents of coyotes in Missouri as well as elsewhere in their range, coyote depredations on livestock and poultry are believed to be confined only to certain individuals that compose about one-fourth of the population.

Because cottontails form a large part of the coyote's diet, many hunters consider the coyote detrimental to their interests. However, the take by coyotes has not been shown to limit seriously human rabbit harvests or to lower the rabbit population.

It is significant that coyotes show a great adaptability to live near human developments and to survive in less than pristine conditions. Aside from any conflict with people, the coyote is a valuable member of the wildlife community. It feeds on rodents and thus helps prevent the damage these abundant and often undesirable animals might otherwise cause. It kills and eats old, sick, or injured wild animals unfit to survive. As a scavenger on dead animals, both wild and domestic, it helps clean up the woods and fields.

Coyote fur is durable and attractive, but until recently pelts have had limited use for trimming coats and for scarves. Since 1970 there has been an increased demand for long-haired furs and, accompanying the increase in coyote populations, the harvest has accelerated. Today, coyote is used as a luxurious fur to make coats, jackets, hats, and other garments.

The sale of coyote pelts over the last seven decades has been variable: in the 1940s the average harvest was 472 pelts; it decreased in the 1950s to 36; it increased in the 1960s to 492 and in the 1970s to 13,060; and then it declined during the 1980s to 9,057, in the 1990s to 3,147 and in the first decade of the 2000s to 2,616. The highest annual take was 24,801 in 1976 and the lowest was 7 in 1958. The average annual price paid per pelt was $1.06 in the 1940s, $0.33 in the 1950s, $1.41 in the 1960s, $11.85 in the 1970s, $8.97 in the 1980s, $7.70 in the 1990s, and $12.13 in the first decade of the 2000s. The highest average annual price paid per pelt was $27.70 in 1978 and the lowest was $0.20 in 1957. It is hard to measure the relative effects of the several factors that influence harvest numbers and pelt prices. Obviously, population abundance is involved, but to what extent has not been determined. Changes in demand for furs and the fluctuating values of their pelts influence economic reasons for trapping coyote. In addition, the numbers of trappers working in Missouri influence the harvest.

Conservation and Management

Management of the coyote in Missouri is that of controlling the harvest. Because certain coyotes develop a habit of taking livestock, poultry, and pets, control, to be effective, should be directed toward the particular troublemakers. If you are experiencing problems with coyotes, contact a wildlife professional for advice, assistance, regulations, or special conditions for handling these animals.

In certain parts of the United States in the past, coyote populations have been checked by poisoning campaigns, but this method has its limitations because it destroys other animals such as pets, livestock, or furbearers that cause no harm. Innumerable small birds and mammals that feed upon the poisoned food are later fed upon by scavengers, and these in turn die from the poison. Indiscriminate poisoning campaigns are deplorable practices. Another control measure used in the past was bounties. Missouri established one of the first bounties on coyotes in 1825. However, the number of coyotes presented for bounty payments

was too small to affect the whole population or to curtail damage complaints, and the program proved too expensive to be justified economically. Antifertility chemicals placed into foods attractive to coyotes have been studied as an aid in controlling reproduction. However, nonlethal control measures require significant time and initial expense; furthermore, these treatments are not selective and may affect other animal species as well. Research has indicated that selective control measures, those that focus on nuisance individuals, have the best potential to work.

SELECTED REFERENCES

See also discussion of this species in general references, page 23.

Andelt, W. F., and P. S. Gipson. 1979. Home range, activity, and daily movements of coyotes. *Journal of Wildlife Management* 43:944–951.

Bekoff, M. 1977. *Canis latrans. Mammalian Species* 79. 9 pp.

Bekoff, M., ed. 1978. *Coyotes: Biology, behavior, and management.* Academic Press, New York. 384 pp.

Dobie, J. F. 1949. *The voice of the coyote.* Little, Brown, Boston. 38 pp.

Eads, R. B. 1948. Ectoparasites from a series of Texas coyotes. *Journal of Mammalogy* 29:268–271.

Gese, E. M., and S. Grothe. 1995. Analysis of coyote predation on deer and elk during winter in Yellowstone National Park, Wyoming. *American Midland Naturalist* 133:36–43.

Gier, H. T. 1957. *Coyotes in Kansas.* Agricultural Experiment Station, Kansas State College of Agriculture and Applied Science, Manhattan, Bulletin 393. 96 pp.

Gipson, P. S., and J. A. Sealander. 1972. Home range and activity of the coyote (*Canis latrans frustror*) in Arkansas. *Proceedings of the Annual Conference of the Southeastern Association of Game and Fish Commission.* 26:82–95.

Howard, W. E. 1949. A means of distinguishing skulls of coyotes and domestic dogs. *Journal of Mammalogy* 30:169–171.

Korschgen, L. J. 1957. Food habits of the coyote in Missouri. *Journal of Wildlife Management* 21:424–435.

Linhart, S. B., and F. F. Knowlton. 1967. Determining age of coyotes by tooth cementum layers. *Journal of Wildlife Management* 31:362–365.

McCarley, H. 1975. Long-distance vocalizations of coyotes (*Canis latrans*). *Journal of Mammalogy* 56:847–856.

Mitchell, B. R., M. M. Jaeger, and R. H. Barrett. 2004. Coyote depredation management: Current methods and research needs. *Wildlife Society Bulletin* 32:1209–1218.

Murie, A. 1940. Ecology of the coyote in the Yellowstone. *U.S.D.I., National Parks Fauna Series* 4. 206 pp.

Nellis, C. H., and L. B. Keith. 1976. Population dynamics of coyotes in central Alberta, 1964–1968. *Journal of Wildlife Management* 40:389–399.

Nellis, C. H., S. P. Wetmore, and L. B. Keith. 1978. Age-related characteristics of coyote canines. *Journal of Wildlife Management* 42:680–683.

Sampson, F. W., and W. O. Nagel. 1949. *Controlling coyote and fox damage on the farm.* Missouri Conservation Commission Bulletin 18. 22 pp.

Whiteman, E. E. 1940. Habits and pelage changes in captive coyotes. *Journal of Mammalogy* 21:435–438.

Wixsom, M. J., S. P. Green, R. M. Corwin, and E. K. Fritzell. 1991. *Dirofilaria immitis* in coyotes and foxes in Missouri. *Journal of Wildlife Diseases* 27:166–169.

Young, S. P., and H. Jackson. 1951. *The clever coyote.* Stackpole Books, Harrisburg, PA and Wildlife Management Institute, Washington, DC. 411 pp.

Red Fox (*Vulpes vulpes*)

Name

The scientific name, *Vulpes*, is the Latin word for "fox." The common name, "red," refers to the fur color, while "fox" is the Anglo-Saxon name for this animal and refers to its crafty behavior. This species was formerly known as *Vulpes fulva*.

Description (Plate 51)

The red fox is doglike in appearance with an elongated, pointed muzzle; large, pointed ears, which are usually held erect and forward; moderately long legs; a long, heavily furred, and bushy tail that is circular in cross section; and long, thick, soft body fur. The pupil of the eye is vertically elliptical in living animals. The front foot has 5 toes (the inside one is high), while the hind foot has 4. All the toes possess rather blunt claws that are not retractile. In winter, the pads of the feet are nearly concealed by fur, but in summer they are more exposed. The weight of the fox is supported on the toes.

Color. The upperparts are reddish yellow, becoming slightly darker on the back. The tail is also reddish

Plate 51

Red Fox *(Vulpes vulpes)*

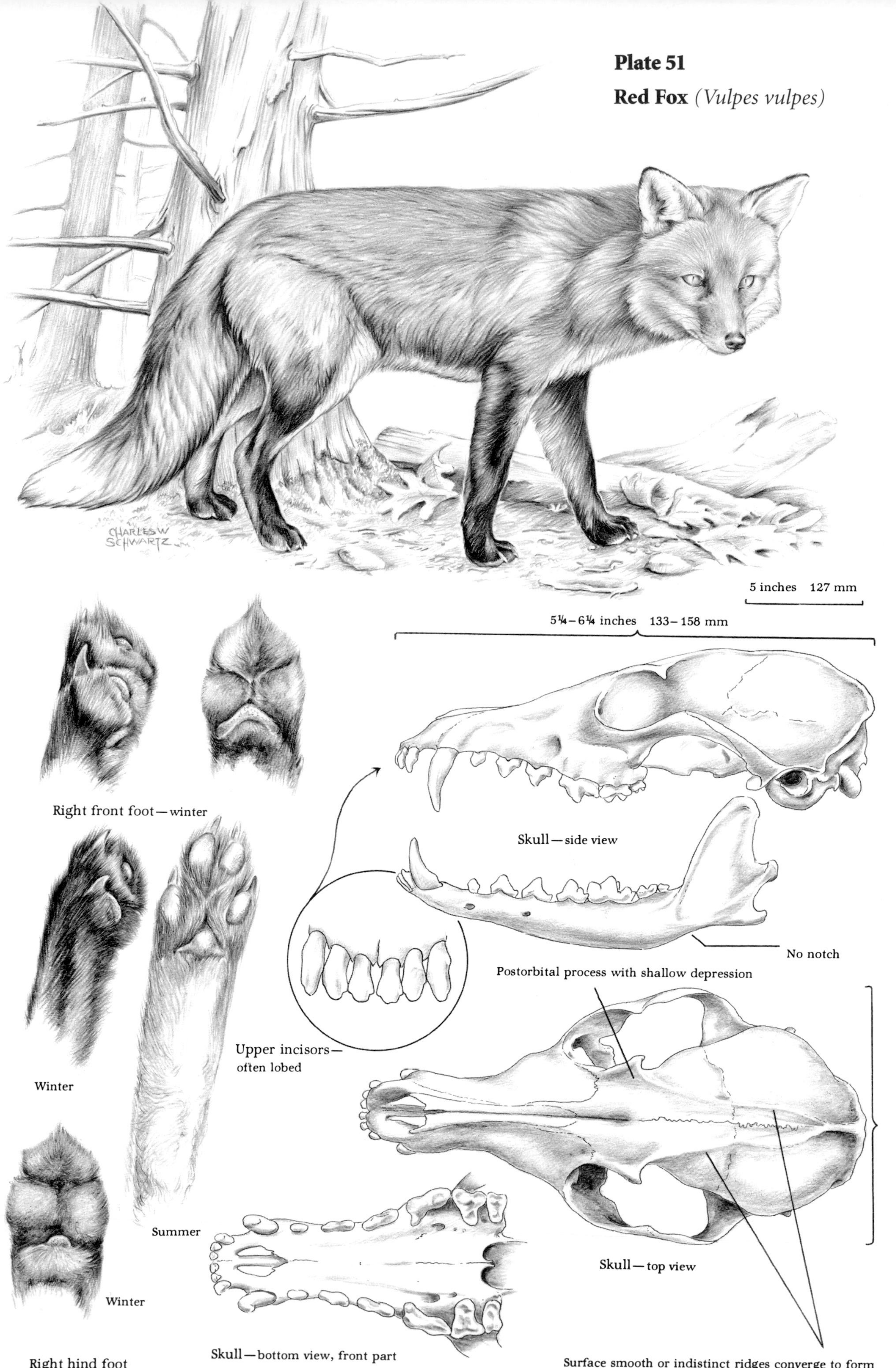

yellow but mixed with black and tipped with white. The nose pad is black and the backs of the ears blackish. The cheeks, throat, and belly are whitish, but the legs and feet are black. The iris of the eye is tawny. The sexes are colored alike, and little seasonal variation occurs.

In addition to the usual red, there are several other color phases, but all have a white tail tip. Black foxes are black, silver foxes are black frosted with white, cross foxes have a dark band down the back and across the shoulders, "bastard" foxes have a dark smoky coat, and "Samson" foxes have a woolly coat that lacks guard hairs. It is possible for any one or several color phases to occur in a single litter, but certain ones tend to predominate in different parts of the range. In Missouri nearly all individuals exhibit the typical red color; the only other phase represented is the very rare cross fox. Pure strains of some phases, like the silver fox, are used in the commercial fur-farm industry.

Measurements

Total length	32–117 cm	12½–46 in.
Tail	292–406 mm	11½–16 in.
Hind foot	100–180 mm	3⅞–7 in.
Ear	85–88 mm	3⅜–3 ½ in.
Skull length	133–158 mm	5¼–6¼ in.
Skull width	69–76 mm	2¾–3 in.
Weight	3.6–6.8 kg	7½–15 lb.

Males are slightly larger and heavier than females.

In 2013, a 5.4 kg (11.9 lb.) male became the official record-weight red fox confirmed by the Missouri Department of Conservation. The state furbearer record program was started in 2011.

Teeth and skull. The dental formula of the red fox is:

$$I\ \frac{3}{3}\ C\ \frac{1}{1}\ P\ \frac{4}{4}\ M\ \frac{2}{3} = 42$$

The skull of an adult red fox is identified by the following combination of characteristics: size; the upper surface smooth or possessing indistinct, paired ridges that start just behind the eye socket and either converge toward the rear to form a crest or come close to each other at the midline, where they are separated by a space less than 9 mm (⅜ in.) wide; a shallow depression on the upper surface of the postorbital process above the eye socket; the lower jaw without an obvious notch on the bottom edge toward the rear; and, in many skulls, the presence of lobes on the cutting edge of the upper incisor teeth.

Sex criteria and sex ratio. Males are recognized by the external sex organs, the penis and scrotum. Pelts of males are identified by the penis scar and the whorl of hair marking the scar. There are four pairs of teats on the sides of the belly extending from front to hind legs.

Of 1,009 red foxes taken in Missouri, 59 percent were males and 41 percent were females. Most trapping and kill records show a similar preponderance of males. However, this may not represent the true sex ratio because most collections are made in the fall and winter when males are ranging more widely than females.

The sex ratio of embryos is equal, but more male pups than female pups are counted at dens.

Age criteria and longevity. Young red foxes tend to be grayer and browner than adults. A pup about 4 weeks of age has a compact tail that tapers to a point. The presence of milk teeth distinguishes a young red fox until 4 to 5 months of age when adult dentition is acquired. About the middle of September, the young have reached full growth but are still identifiable by their glossy coats, compared to the duller coats of adults. Through October young animals have grayer belly fur than do adults.

Certain skull characters are useful in determining age in foxes. One of these is by the closure of sutures between certain bones. Foxes with an open or recently closed suture between two bones (the basisphenoid and presphenoid) on the undersurface of the skull are less than 12 months of age; those with an incompletely developed vomer (a bone also on the undersurface of the skull) are less than 24 months of age.

Dental characters, likewise, help age foxes. The count of cementum layers on the roots of teeth, particularly canines, is applicable as in the coyote. The height of the canines above the socket in the skull is significant, and the distance from the enamel line of the canine tooth to the socket is an additional aid. Wear on the teeth is progressive, and the relative amount of wear on the inner cusps of the first upper molar teeth has been correlated with age.

The dried weight of the lens of the eye as a means of age determination is fairly accurate, showing the closest agreement with tooth wear when used to separate juvenile and adult foxes.

By means of X-rays, both red and gray foxes less than 8 to 9 months of age can be identified by the presence of cartilage at the wrist end of the long bones (the radius and ulna) of the front leg. In adult foxes, the cartilage in this region has turned to bone.

The penis bones of adult and young males are not useful for age determination as they are in many other mammal species.

Pelts of adult red and gray fox females that have borne young have dark-colored, raised nipples more than 1.6 mm (1⁄16 in.) in diameter, while those of young females have light-colored and smaller nipples.

Female red and gray foxes with placental scars (places of attachment of embryos) in the branches of the uterus are readily identified as adults. Females without placental scars can be separated into two age groups on the basis of size and degree of transparency of the uterine branches: young of the year have uterine branches that are less than 3 mm (⅛ in.) in diameter and are thin, pink, and translucent; older females have uterine branches that are more than 3 mm in diameter and are darker.

In the wild, red foxes may live for 6 to 10 years; in captivity some have lived for 15 years.

Glands. Many members of the dog family have a scent gland at the base of the tail on the upper surface. The scent is used to communicate with other members of the same kind and to identify the individual. In red foxes, this gland is elliptical and measures 25 mm (1 in.) long by 6 mm (¼ in.) wide. Another means of communication is by "scent posts," which are certain small, isolated objects such as rocks, fence posts, ends of logs, or tree trunks; these are scented with urine or marked by droppings.

Voice and sounds. Red foxes have a large vocabulary of calls. They commonly give short yaps or barks followed by a single squall. They also emit long yells, yowls, screeches, and *chums*. During the mating season, females give a shrill squall, which males answer with two or three short barks.

Distribution and Abundance

The red fox, a native of North America, originally lived almost entirely north of latitude 40–45°; south of this region, in the original unbroken timber, it occurred only sparingly. With colonization of the eastern United States, the European red fox was introduced between 1650 and 1750 for fox hunting. At the same time the dense forest was opened in colonial development; this change permitted more annual plants to grow that are fed upon by rodents and rabbits, the principal prey of foxes, and it broke up the dense cover that red foxes generally avoid. It is not known whether the descendants of matings between the European red fox and the native red fox populated this area, as has commonly been claimed, or whether the European red fox died out and the native red fox was able to extend its range under the changed habitat conditions. At any rate, both of these are now considered the same species.

The North American range of the red fox today includes all of Alaska and Canada, and most of the United States. It is more common in the northern part of this range than in the southern. Although populations fluctuate in number, these are generally local occurrences and not widespread cycles. Such fluctuations are apparently influenced by the abundance or scarcity of the food supply. Disease and parasites could also have an effect.

The red fox lives throughout Missouri but is more common in the northern and western sections of the state. The statewide population has been in a long-term decline and remains at a low level. This long-term decline may be the result of interspecific competition with coyotes and bobcats.

Habitat and Home

Red foxes prefer the borders of forested areas and adjacent open lands for their range, avoiding dense

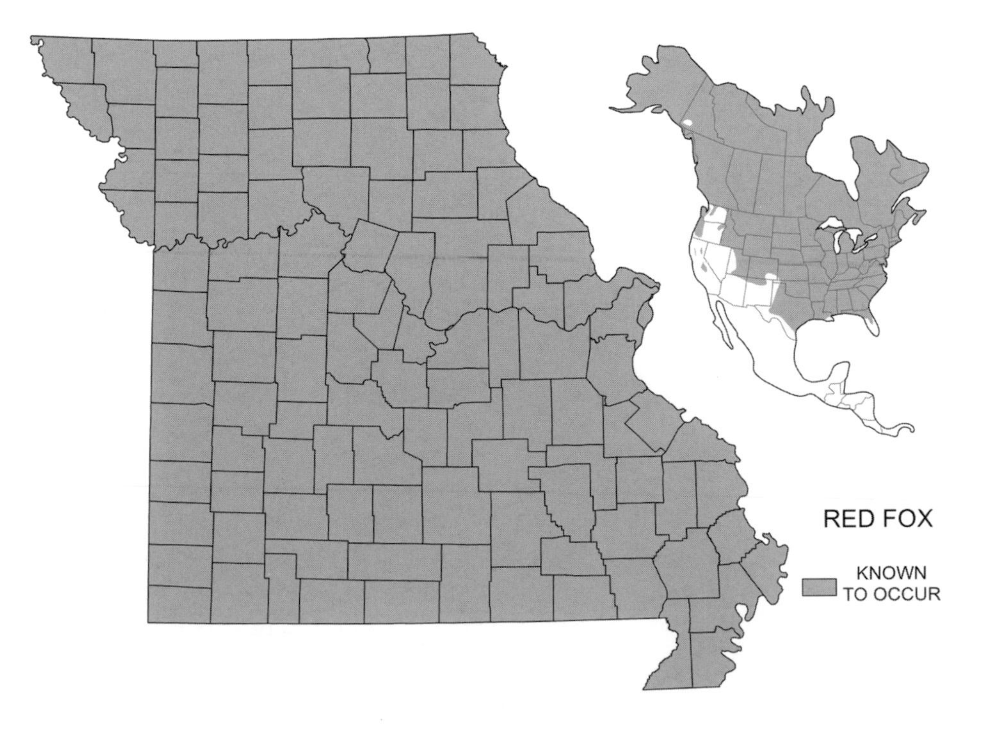

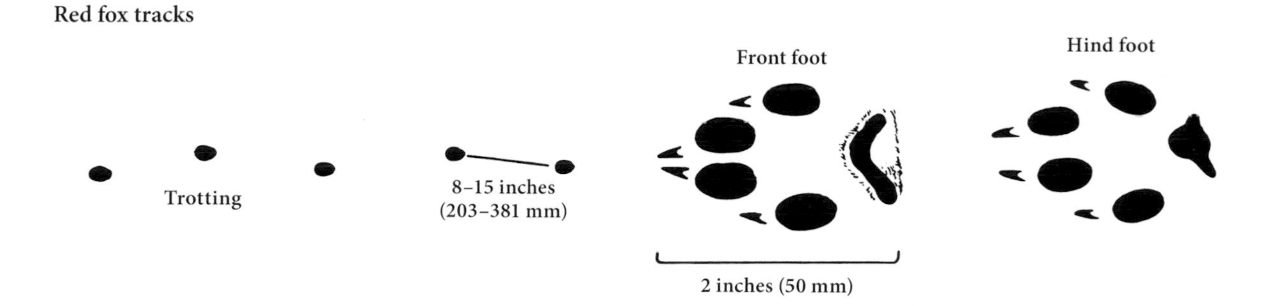

and extensive forests. Throughout most of the year they have no special home but sleep on the ground. However, during the breeding season a den is used for the young. This is often a modified woodchuck or former fox den but may be dug by the female fox. It is usually located in sandy soil on the sunny side of a hill or bank in an open field, along a fencerow, at the edge of timber, or in a natural rock cavity; occasionally it is in deep woods, in a hollow log, or under some abandoned building. Sometimes four or five trails approach the den from different directions.

A den may have many entrances, varying from 2 to 27, which are between 20 and 38 cm (8 and 15 in.) in diameter and lead to burrows up to 23 m (75 ft.) in length. About 1.2 m (4 ft.) underground there is a chamber containing a grass bed for the young. Sometimes there are additional rooms that are used for food storage. Several smaller, temporary dens may be located nearby where the pups can be moved in case of danger.

It is not unusual for a fox family to have two dens as much as 2 km (1¼ mi.) apart. Some of the young are moved to the second den when they are about 5 or 6 weeks old and thereafter the parents maintain both dens. The young learn to travel between the dens when 10 to 15 weeks of age.

Habits

The family unit is the male and female, their pups of the year, and maybe one or two adult females that serve as nursemaids for the pups.

The size of the home range of the family varies with the habitat, which can range from 60 to 600 ha (150 to 1,500 ac.). The home ranges of fox families do not overlap, and due to their small sizes, they are boldly defended. Foxes also appear to establish home ranges outside coyote territories, often in boundary areas between coyote groups. Within their home range, the foxes travel from 1.6 to 4.8 km (1 to 3 mi.) per night. In the daytime they return to a preferred resting area but do not bed in the same place each day.

Red foxes are chiefly nocturnal but also come out in daylight, especially in early morning and early evening when their prey are active. They often travel the same routes, which may become worn into trails. At places along their trails, they frequently lie down and comb their fur with the teeth, leaving the discarded fur on the spot. They also stop to roll on strong-smelling objects such as dead carcasses. In running and fast maneuvering, the fox uses its large, bushy tail as a balance; in sleep the tail covers and warms the feet and nose.

During the season when the young are being fed and food is plentiful, the adults usually travel less than 1.6 km (1 mi.) in any direction from the den; throughout the rest of the year, they go farther afield. By the fall, some adult males have moved as much as 64 km (40 mi.) from their former home range. Red foxes are not swift runners and travel only about 42 km (26 mi.) per hour at top speed. They seem to enjoy outwitting dogs and do so by devious tricks. Because they cannot get the best of large dogs, they avoid fighting if possible. Foxes are good swimmers.

When stalking prey, the fox either takes very high, deliberate steps or crouches low and wriggles along. It then rushes or pounces on the unwary victim, which is killed by a bite from the powerful jaws.

Foods

The bulk of the red fox's diet is animal matter, with rabbits and mice constituting the staple items. A study of the food habits of 1,006 red foxes in Missouri showed the following major food groups and their percentages by volume: rabbits, 36.8; mice and rats, 22.0; other wild mammals, 7.7; livestock, 5.1; poultry, 13.7; wild birds, 3.4; known carrion (but part of the livestock and poultry take undoubtedly belongs here), 7.7; insects, 0.7; plants, mostly persimmon fruits, 2.7; and undetermined animal matter, 0.1. Clay and gravel are occasionally licked for their minerals.

Red foxes eat about 0.5 kg (1 lb.) of meat at a feeding. Mice are consumed entirely after the bones are crushed, but birds are sheared of their feathers. When food is plentiful, a red fox kills more than it eats. This surplus is usually buried in the ground or covered with grass or leaves and sprinkled with urine. The fox

may return from time to time to its food cache, even if not hungry, to look at and play with some of the items. Shrews and moles are often caught and stored but rarely eaten. Some of the cached food is eaten by skunks, crows, owls, hawks, or other foxes.

Reproduction

The onset of the breeding season is indicated by an increased amount of nocturnal barking. While mating may occur from late December to March, January and February are the customary months in Missouri. Individual females have a very short period of heat, lasting only 1 to 6 days once a year. Adult females breed 1 to 3 weeks earlier than the young of the previous year do. Gestation requires between 49 and 56 days with an average of 53. The single annual litter is generally born in March or April and consists of 1 to 10 young with the usual number between 4 and 7. The number of young per litter increases with the female's age up to 5 to 7 years; litters born earlier in the season tend to be larger than those born later.

The female, or vixen, establishes the den site for the young, but both parents live together during the season when the young are being raised. The female stays in the den with the young for the first few days after their birth and the male, or dog fox, brings food to her. Later on, she hunts at night and nurses by day while he hunts more by day. Only rarely do two females have their respective litters in the same den or do males adopt stray young.

At birth the pups, or kits, are blind and helpless. They are dark grayish brown and weigh about 100 g (3½ oz.). Their eyes open around 10 days of age, but the pups stay within the den until 4 or 5 weeks old. At this time they begin eating solid food and begin fighting and establishing a hierarchy. As they grow older, the pups play in front of the den with such objects as bones, horse dung, and leftover food items and are fed there by the adults. Although the parents carry away the droppings and foods that spoil, the outside of a fox den has an untidy appearance and often an unpleasant odor. If the young are moved to another den, the parents frequently take the playthings along.

When the pups are 7 weeks old, they are pale yellowish brown. They soon lose their yellow and at 8 to 10 weeks are pale reddish brown. At this time they are weaned and start to accompany their parents on hunting trips. The milk teeth are lost between 16 and 20 weeks of age; shortly thereafter, the young, now reddish like their parents, are on their own.

They stay within 0.8 km (½ mi.) of the home den for the next month or so, meeting occasionally with the parents; in the fall, when fully grown, they disperse rather widely. Males disperse earlier and generally farther than females. It is not uncommon for a dispersing fox to move several miles the first night. Distances moved are usually around 10 to 23 km (6 to 14 mi.), but some extremes have been reported. One juvenile male moved 394 km (245 mi.) from the natal den while a male littermate stayed within 0.3 km (300 yd.) of the original site. A female was taken in a den with her pups two years and 203 km (126 mi.) away from her natal den.

Young foxes are sexually mature in the spring following their birth but continue to increase in weight during the next few years.

Some Adverse Factors

The most important predators on foxes in Missouri are humans, domestic dogs, and possibly the coyote.

Fox pup has spindle-shaped eye pupil, distinguishing it from coyote pup with round pupil

The following parasites are known to occur on or in red foxes: mites, ticks, lice, fleas, fly larvae, roundworms (heartworms), tapeworms, and flukes. The common diseases are coccidiosis, distemper, and rabies. Epidemics of rabies occur occasionally and affect both dense and sparse fox populations. Mange, caused by a mite, frequently is troublesome.

It is interesting that fleas specific to ground squirrels are common in many fox dens, yet ground squirrels are not a staple food item of foxes. It is thought that these fleas are accidentally acquired by rabbits (one of the fox's principal foods) or foxes that are traveling through areas where ground squirrels live and are thus transported to fox dens.

Importance

Red foxes are trapped for their fur, which is used for trimming, scarves, coats, jackets, and hats. The average annual harvest of red fox pelts has fluctuated over the decades. During the 1940s Missouri had the highest average annual harvest at 9,443. Then harvest rates decreased to 2,371 in the 1950s and 1,350 in the 1960s. A brief increase in harvest rates during the next two decades (2,643 in the 1970s and 2,909 in the 1980s), was followed by another decrease during the 1990s (1,138) and the first decade of the 2000s (1,022). The highest annual take was 14,674 in 1943, and the lowest was 409 in 1958. The average annual price for red fox pelts was $3.17 in the 1940s; $0.39 in the 1950s; $2.56 in the 1960s; $27.23 in the 1970s; $20.72 in the 1980s; $11.26 in the 1990s; and $16.05 in the first decade of the 2000s. The highest average annual price paid for pelts harvested in Missouri was $38.70 in 1976 and the lowest was $0.25 in 1953–1958. It is hard to measure the relative effects of the several factors that influence harvest numbers and pelt prices. Obviously, population abundance is involved, but to what extent has not been determined. Changes in demand for furs and the fluctuating values of their pelts influence economic reasons for trapping red foxes. In addition, the numbers of trappers working in Missouri influence the harvest. Most fox harvest today occurs when trappers are targeting bobcats or coyotes.

Silver foxes are reared on farms for their fur. Fox hunting with hounds is considered great sport, and the raising of foxhounds is an accessory business and pleasure. Still another type of hunting is by calling up foxes with a decoy call. Red foxes feed upon rodents and help check such abundant forms. Although about one-fifth of the red fox's diet consists of livestock and poultry, the economic loss is not as great as it appears because doubtless some of this is carrion.

Conservation and Management

Management of the red fox in Missouri is that of controlling the harvest. Where foxes are numerous, damage can be avoided by reducing the vegetation around poultry houses, providing an enclosed area for chickens, having an alert and aggressive dog, and trapping the offending individuals. If you are experiencing problems with red foxes, contact a wildlife professional for advice, assistance, regulations, or special conditions for handling these animals.

SELECTED REFERENCES

See also discussion of this species in general references, page 23.

Allen, S. H. 1974. Modified techniques for aging red fox using canine teeth. *Journal of Wildlife Management* 38:152–154.

Allen, S. H., and A. B. Sargeant. 1993. Dispersal patterns of red foxes relative to population density. *Journal of Wildlife Management* 57:526–533.

Churcher, C. W. 1959. The specific status of the New World red fox. *Journal of Mammalogy* 40:513–520.

———. 1960. Cranial variation of the North American red fox. *Journal of Mammalogy* 41:349–360.

Fisher, H. I. 1951. Notes on the red fox (*Vulpes fulva*) in Missouri. *Journal of Mammalogy* 32:296–299.

Gilmore, R. M. 1946. Mammals in archeological collections from southwestern Pennsylvania. *Journal of Mammalogy* 27:227–234.

Harrison, D. J., J. A. Bissonette, and J. A. Sherburne. 1989. Spatial relationships between coyotes and red foxes in eastern Maine. *Journal of Wildlife Management* 53:181–185.

Henry, J. D. 1986. *Red fox: The catlike canine.* Smithsonian Institution Press, Washington, DC. 174 pp.

Korschgen, L. J. 1959. Food habits of the red fox in Missouri. *Journal of Wildlife Management* 23:168–176.

Larivière, S., and M. Pasitschniak-Arts. 1996. *Vulpes vulpes. Mammalian Species* 537. 11 pp.

Linhart, S. B. 1968. Dentition and pelage in the juvenile red fox (*Vulpes vulpes*). *Journal of Mammalogy* 49:526–528.

Lord, R. D., Jr. 1961. The lens as an indicator of age in the gray fox. *Journal of Mammalogy* 42:109–111.

Phillips, R. L., R. D. Andrews, G. L. Storm, and R. A. Bishop. 1972. Dispersal and mortality of red foxes. *Journal of Wildlife Management* 36:237–248.

Richards, S. H., and R. L. Hine. 1953. *Wisconsin fox populations.* Wisconsin Conservation Department, Technical Wildlife Bulletin 6. 78 pp.

Sampson, F. W., and W. O. Nagel. 1949. *Controlling coyote and fox damage on the farm.* Missouri Conservation Commission Bulletin 18. 22 pp.

Sargeant, A. B. 1972. Red fox spatial characteristics in relation to waterfowl predation. *Journal of Wildlife Management* 36:225–236.
Scott, T. G. 1943. Some food coactions of the northern plains red fox. *Ecological Monographs* 13:427–479.
Seagears, C. B. 1944. *The fox in New York*. New York Conservation Department. 85 pp.
Sheldon, W. G. 1949. Reproductive behavior of foxes in New York state. *Journal of Mammalogy* 30:236–246.
———. 1950. Denning habits and home range of red foxes in New York state. *Journal of Wildlife Management* 14:33–42.
Sullivan, E. G., and A. O. Haugen. 1956. Age determination of foxes by X-ray of forefeet. *Journal of Wildlife Management* 20:210–212.
Storm, G. L. 1965. Movements and activities of foxes as determined by radio-tracking. *Journal of Wildlife Management* 29:1–13.
Storm, G. L., et al. 1976. Morphology, reproduction, dispersal, and mortality of midwestern red fox populations. *Wildlife Monograph* 49. 82 pp.
Wood, J. E. 1958. Age structure and productivity of a gray fox population. *Journal of Mammalogy* 39:74–86.

Gray Fox (*Urocyon cinereoargenteus*)

Name

The first part of the scientific name, *Urocyon*, is from two Greek words and means "tailed dog" (*oura*, "tail," and *kyon*, "dog"). The second part, *cinereoargenteus*, is from two Latin words and means "silvery gray" (*cinereus*, "ash colored" or "gray," and *argenteus*, "silvery"). Thus, parts of both the scientific name and the common name refer to the color of the fur. "Fox" is the Anglo-Saxon name for this animal and refers to its crafty behavior.

Description (Plate 52)

In general build, the gray fox is somewhat similar to the red fox. It is distinguished by the grayish coloration, slightly smaller size, black-tipped tail that is triangular in cross section, six teats, dark brown iris of the eye, and coarse body fur.

Color. The upperparts are mixed gray and black caused by alternate black and white bands on the guard hairs, which overlie a grayish brown undercoat. There is a considerable amount of reddish brown fur, which occurs mainly on the sides of the neck, the backs of the ears, the insides and backs of the legs, the feet, sides of the belly, the chest, and the undersurface of the tail. The tail is gray above, possesses a mane of coarse black hairs for the entire length, and terminates in a black tip. The nose pad is black and the muzzle blackish; light brownish patches occur above and below the eyes. The cheeks, throat, insides of the ears, and belly are whitish. The sexes are colored alike, and no seasonal change occurs in coloration.

Measurements

Total length	80–113 cm	31³⁄₁₆–44½ in.
Tail	220–433 mm	8½–16⅞ in.
Hind foot	101–152 mm	4–6 in.
Ear	63–76 mm	2½–3 in.
Skull length	120–130 mm	4¾–5⅛ in.
Skull width	66–73 mm	2⅝–2⅞ in.
Weight	2.2–7 kg	5–15½ lb.

There is a record weight of 8.6 kg (19 lbs.).

In 2011, a 4.4 kg (9.8 lb.) male became the official record-weight gray fox confirmed by the Missouri Department of Conservation. The state furbearer record program was started in 2011.

Teeth and skull. The dental formula of the gray fox is:

$$I\ \frac{3}{3}\ C\ \frac{1}{1}\ P\ \frac{4}{4}\ M\ \frac{2}{3} = 42$$

The skull of an adult gray fox is identified by a combination of characteristics, including the dental formula; size; prominent, paired ridges up to 3 mm (⅛ in.) high on the upper surface that start just behind the eye socket and enclose a U-shaped area but do not come closer to each other at the midline than 9.5 mm (⅜ in.); a deep depression on the upper surface of the postorbital process above the eye socket; a prominent notch on the bottom edge of the lower jaw toward the rear; and no obvious lobes on the cutting edge of the upper incisor teeth.

Sex criteria and sex ratio. Females are identified by the teats; males by the penis and scrotum. Pelts of

Plate 52

Gray Fox

(Urocyon cinereoargenteus)

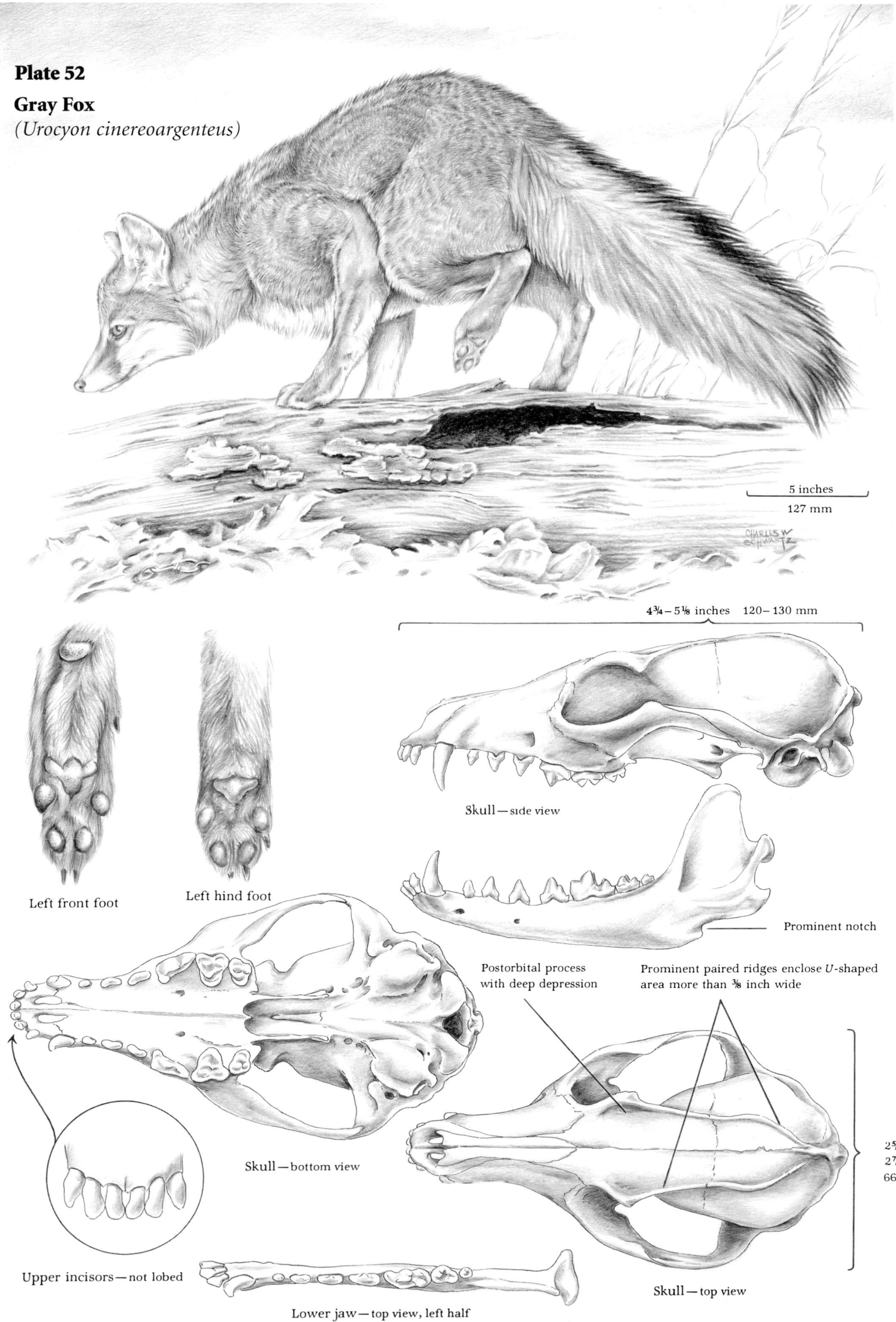

Gray fox tracks

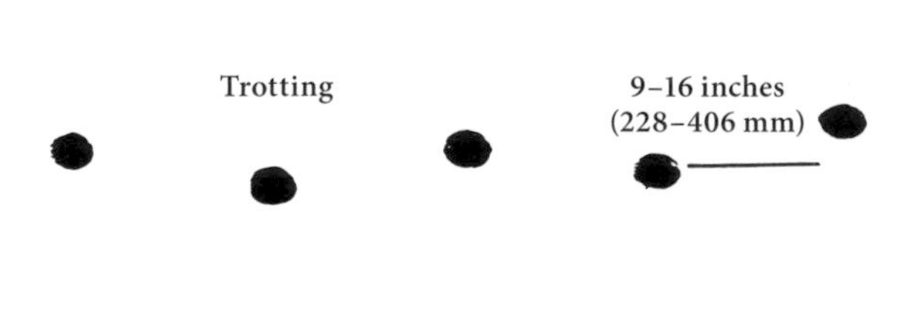

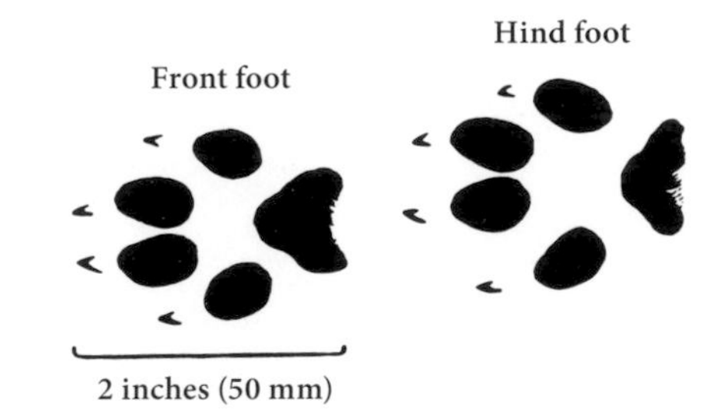

males are recognized by the penis scar and the whorl of hair marking the scar. The sex ratio in 374 gray foxes killed in Missouri was 52 percent males to 48 percent females. However, since males travel more widely than females these figures may not represent the true ratio of the sexes. The actual sex ratio is believed to be nearly equal.

Age criteria and longevity. After about a month of age, the young start to resemble adults in coloration. Body weight distinguishes young from adults only until the young are 5 to 6 months old. Likewise the presence of milk teeth identifies a young gray fox until it is 5 months old, when adult dentition is complete. Other means of age determination are discussed under the red fox.

Gray foxes may reach 6 to 10 years of age in the wild, but most die before the age of 2.

Glands. The scent gland on the upper surface of the tail, which is typical of many members of the dog family, begins between 4 and 5 cm (1⅝ and 2 in.) from the base of the tail and measures between 10.5 and 12 cm (4⅛ and 4¾ in.) in total length. It is from 6 to 9 mm (¼ to ⅜ in.) wide at the base and tapers to a point. When angry or provoked, gray foxes emit a strong, nauseating odor.

Voice and sounds. Gray foxes have a yapping bark that they give four or five times in succession. This is louder and harsher than the bark of the red fox. Gray foxes also growl, squeal, and chuckle.

Distribution and Abundance

The gray fox ranges throughout most of the southern half of North America. Its range is from extreme southern Canada throughout much of the United States, through Mexico, and into southern Central America. They are also known to occur in the extreme northern part of South America.

The gray fox is found throughout Missouri but is most common in the Ozark Highlands. The statewide population has been in a long-term decline and remains at a low level. This long-term decline may be the result of interspecific competition with coyotes and

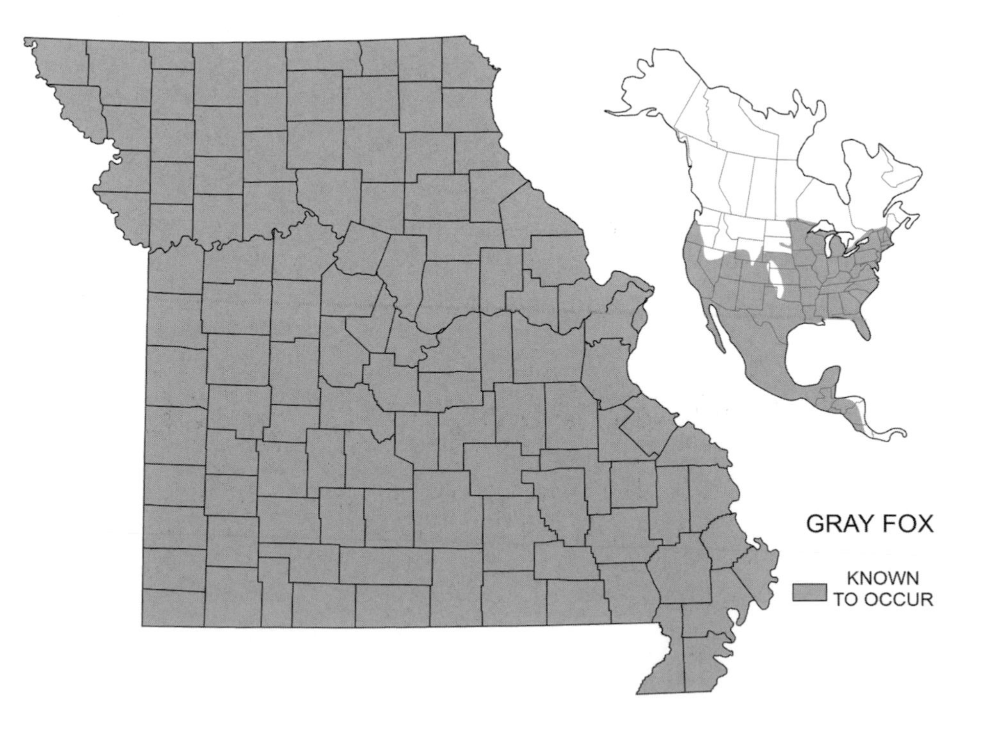

Gray fox four days old

bobcats, or the increase in raccoons and their associated distemper virus; the gray fox seems especially vulnerable to the distemper virus. There are fewer gray foxes than red foxes statewide.

Gray fox populations increase when the breeding season is wetter and warmer than usual and decrease when it is colder and drier.

Habitat and Home

The gray fox lives in wooded areas and fairly open brushland, preferring mature forest at night and young dense forest stands in the day. Gray foxes prefer a fragmented landscape providing both forest and grassland habitats and are known to avoid heavily agricultural areas. The gray fox is essentially an animal of warm climates; hence, in the northern part of its range, it uses dens for warmth more than does the red fox. The dens are located in hollow logs, hollow trees, under rock piles, or occasionally in the ground. They are filled with grass, leaves, or shredded bark.

Habits

The size and shape of gray fox home ranges reflect the distribution and abundance of timber and brush in a given area. Home ranges are generally maintained at greater than 200 ha (494 ac.) and up to 3.1 sq km (1.2 sq. mi.). However, only a small portion of this may be used in a given day or month.

The gray fox is primarily nocturnal but often is abroad in the daytime. In contrast to the red fox, the gray fox readily climbs trees, using the front feet to grasp the tree trunk and the hind feet to push upward. Gray foxes sun in trees, forage on fruits, and take refuge there from dogs. They are not as cunning as red foxes and are easier to trap. Gray foxes are very secretive and shy, but when necessary are fierce fighters. They can run about 42 km (26 mi.) per hour at top speed but slow down after the initial spurt.

A typical gray fox family consists of an adult male and female pair and their offspring; however, other unpaired adults may be present. Ranges of families do not overlap.

Foods

The gray fox feeds on much the same foods as the red fox, rabbits and mice forming the bulk of the diet. A study of the food habits of 305 gray foxes in Missouri showed the following major food groups and their percentages by volume: rabbits, 47.1; mice and rats, 20.7; other wild mammals, 3.6; livestock, 0.8; poultry, 9.7; wild birds, 6.6; known carrion (but part of the livestock and poultry take undoubtedly belongs here), 0.8; insects, 1.2; plants, 8.8; and miscellaneous, 0.7. Extra food is sometimes buried. Fruits are a relished and important item when available.

Reproduction

Although the breeding season may extend from January to May, the peak of mating occurs in February or the first week of March. The gestation period averages 53 days with extremes of 51 and 63 days. The single annual litter is usually born sometime from March to mid-May. There are from 1 to 7 young in a litter with 3 to 5 being the most common.

At birth the pups are blackish, blind, and scantily furred. They weigh about 85 g (3 oz.). Their eyes open between the ninth and twelfth day after birth. When the young are approximately 3 months old, they leave the den for the first time to accompany their parents on hunting trips. The family breaks up in late summer, when the pups are approximately 6–7 months of age. The young breed the first year following birth.

Some Adverse Factors

There are few predators on gray foxes except humans, domestic dogs, and possibly coyotes. The

Gray fox three weeks old

following parasites are known to occur on or in gray foxes: mites, ticks, lice, fleas, and roundworms. The disease rabies sometimes occurs, and they are especially vulnerable to the distemper virus.

Importance

The fur of gray foxes is coarse and thin and is used for collars and trimming on coats.

The average annual harvest of gray foxes for the past seven decades in Missouri was 6,947 in the 1940s, decreased to 1,326 during the 1950s and 1,309 in the 1960s, increased to 6,575 in the 1970s, but then decreased again to 5,043 in the 1980s, 1,234 in the 1990s, and 781 in the first decade of the 2000s. The highest annual harvest was 13,329 in 1944 and the lowest was 219 in 1958. The average annual price paid per pelt was $1.38 in the 1940s, $0.27 in the 1950s, $1.19 in the 1960s, $19.63 in the 1970s, $22.22 in the 1980s, $7.43 in the 1990s, and $17.73 in the first decade of the 2000s. The highest average annual price paid per pelt was $44.95 in 1979 and the lowest was $0.20 in 1953–1957. It is hard to measure the relative effects of the several factors that influence harvest numbers and pelt prices. Obviously, population abundance is involved, but to what extent has not been determined. Changes in demand for furs and the fluctuating values of their pelts influence economic reasons for trapping gray foxes. In addition, the numbers of trappers working in Missouri influence the harvest. Most fox harvest today occurs when trappers are targeting bobcats or coyotes.

The gray fox is important because it eats many rodents. It takes far less livestock and poultry than the red fox and thus causes little economic loss. The gray fox is not esteemed as highly by sportsmen as the object of a chase as is the red fox.

Conservation and Management

Management of the gray fox in Missouri is that of controlling the harvest. If you are experiencing problems with gray foxes, contact a wildlife professional for advice, assistance, regulations, or special conditions for handling these animals.

SELECTED REFERENCES

See also discussion of this species to general references, page 23.

Cooper, S. E., C. K. Nielsen, and P. T. McDonald. 2012. Landscape factors affecting relative abundance of gray foxes *Urocyon cinereoargenteus* at large scales in Illinois, USA. *Wildlife Biology* 18:366–373.

Fritzell, E. K., and K. J. Haroldson. 1982. *Urocyon cinereoargenteus. Mammalian Species* 189. 8 pp.

Haroldson, K. J., and E. K. Fritzell. 1984. Home ranges, activity, and habitat use by gray foxes in an oak-hickory forest. *Journal of Wildlife Management* 48:222–227.

Layne, J. N. 1958. Reproductive characteristics of the gray fox in southern Illinois. *Journal of Wildlife Management* 22:157–163.

Nicholson, W. S., E. P. Hill, and D. Briggs. 1985. Denning, pup-rearing, and dispersal in the gray fox in east-central Alabama. *Journal of Wildlife Management* 49:33–37.

Sampson, F. W., and W. O. Nagel. 1949. *Controlling coyote and fox damage on the farm*. Missouri Conservation Commission Bulletin 18. 22 pp.

Seagears, C. B. 1944. *The fox in New York*. New York Conservation Department. 85 pp.

Sheldon, W. G. 1949. Reproductive behavior of foxes in New York state. *Journal of Mammalogy* 30:236–246.

Sullivan, E. G. 1956. Gray fox reproduction, denning, range, and weights in Alabama. *Journal of Mammalogy* 37:346–351.

Bears (Family Ursidae)

The family name, Ursidae, is from the Latin word for "bear." This family contains eight species of the largest living members of the flesh-eating mammals and is represented in Europe, Asia, North America, and South America. The characteristics of this family are exemplified by the only species found in Missouri, the American black bear.

American Black Bear
(*Ursus americanus*)

Name

The first part of the scientific name, *Ursus*, is the same as for the family name. The last part, *americanus*, is the Latinized form "of America," for eastern North America, the locality cited in the first description. For the common name, "American" indicates its range in North America; "black," indicates the predominant color of most individuals; and "bear" comes from the Anglo-Saxon word *bera*. This species was formerly known as *Euarctos americanus*.

Description (Plate 53)

The American black bear is the largest and heaviest carnivore now living in Missouri. It has a long muzzle with a straight facial profile; medium-sized, rounded, erect ears; rather short, stout legs; and a very short tail practically concealed in the long, heavy fur. There are

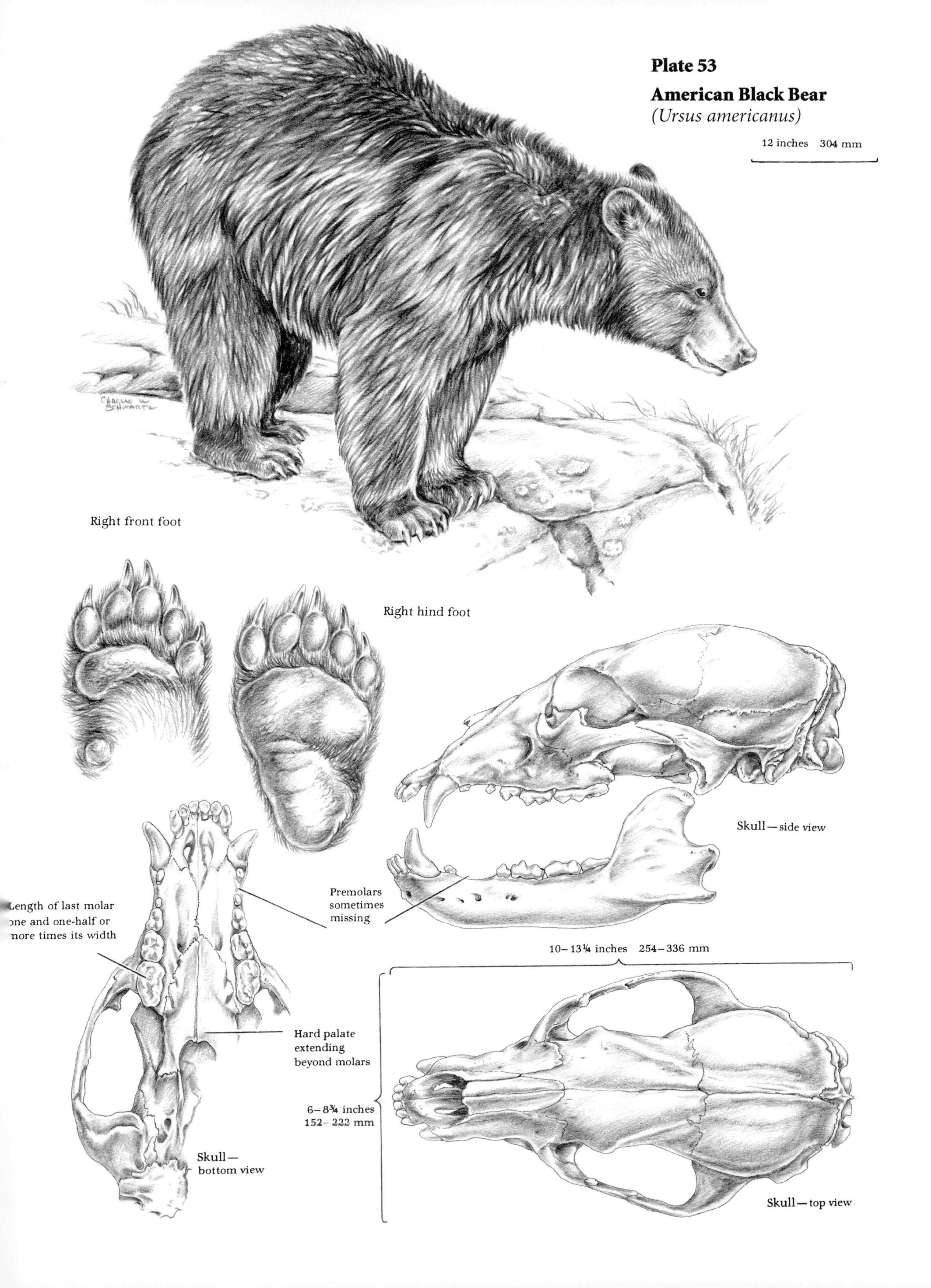
Plate 53
American Black Bear
(Ursus americanus)
12 inches 304 mm
Right front foot
Right hind foot
Skull—side view
Premolars sometimes missing
Length of last molar one and one-half or more times its width
10–13¼ inches 254–336 mm
Hard palate extending beyond molars
6–8¾ inches 152–222 mm
Skull—bottom view
Skull—top view

5 toes on both front and hind feet with short, curved claws that are about the same length on each foot. A bear has the small toe on the inside of its foot whereas a human has it on the outside. Black bears have very good eyesight, and the senses of hearing and smell are excellent.

Color. In eastern North America, black bears are predominantly glossy black with a brown muzzle and usually a small white patch on the chest. Primarily in the Rocky Mountain region and other western parts of the range, a reddish brown or cinnamon color phase is common, and a litter may contain both black and cinnamon young. In northwestern Canada and Alaska, other color phases occur—chocolate brown, whitish blue, and creamy white. Most bears in Missouri are black, but a brown color phase also occurs. About one-third of our bears are the brown color phase or a mixed black and brown. Eastern Missouri has more bears of the brown phase, and western Missouri has more of the black phase. There is no difference in color between the sexes. The fur is longest and glossiest in fall; in summer the coat is raked and dull due to wear and shedding.

Measurements

Total length	119–198 cm	46¾–78 in.
Tail	101–127 mm	4–5 in.
Hind foot	177–355 mm	7–14 in.
Ear	136 mm	5⅜ in.
Skull length	254–336 mm	10–13¼ in.
Skull width	152–222 mm	6–8¾ in.
Weight	39–408 kg	86–900 lb.

Adult males generally weigh 90 to 272 kg (200 to 600 lb.), and adult females weigh 45 to 136 kg (100 to 300 lb.). The largest recorded wild American black bear was a male from New Brunswick in 1972 that measured 2.41 m (7.9 ft.) long and weighed 409 kg (902 lb.) after it had been dressed; it weighed an estimated 500 kg (1,100 lb.) in life.

Teeth and skull. The dental formula of the black bear is:

$$I\ \frac{3}{3}\ C\ \frac{1}{1}\ P\ \frac{4}{4}\ M\ \frac{2}{3} = 42$$

The canine teeth are large, long, and pointed. The last upper molar is long (its length 1½ or more times its width) and is much larger than the molar in front of it. The molar teeth are the crushing type with broad flat crowns. There are no carnassial, or flesh-cutting, teeth. The hard palate extends beyond the upper molars.

Sex criteria and sex ratio. The male has a penis bone, or baculum. The sex of a bear skull, like that of a raccoon, can be determined by the size of the lower canine and molar teeth. The sex ratio is equal.

Age determination. The best means of age estimation is by counting the cementum layers on the roots of the canine or first premolar teeth. Age classes can be established by combining information from cementum layer counts, skull structure, skull and body measurements, tooth replacement and wear, epiphyseal closure in the long bones, and baculum growth.

Longevity. From 12 to 15 years is probably the normal life span in the wild, but captive individuals are known to have lived to 30 years of age.

Voice and sounds. Black bears are usually silent but on occasion utter a variety of sounds. They may grunt, mumble, squeak, roar, huff, bellow, hum, moan, or purr. Sometimes the young whimper. In addition, they use body language and scent to communicate.

Distribution and Abundance

Formerly the American black bear lived in most of the forested regions of North America including most of Alaska, Canada, the United States, and the central plateau of Mexico. By the early 1900s, however, unregulated and overhunting for meat, fat, and skins, plus habitat destruction through logging and clearing, caused the black bear to be extirpated from a large section of the north-central and central United States, and from parts of the eastern range. Today, it

American black bear tracks

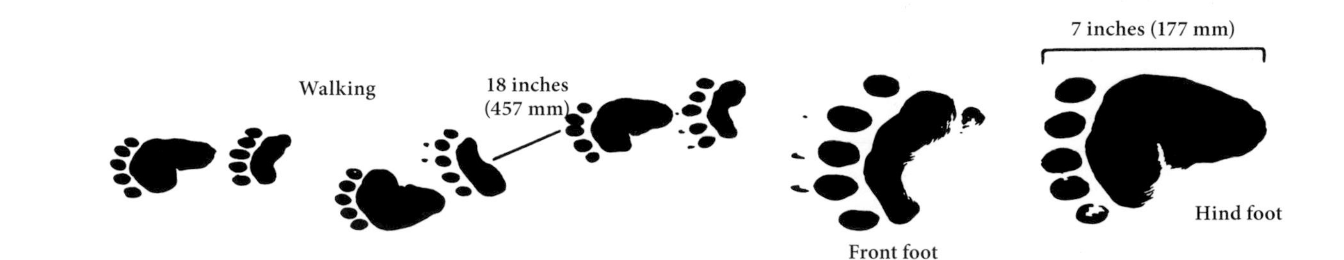

occurs from Alaska across much of Canada, into the northeast United States and then southward to Virginia and West Virginia, the northern Midwest, the Rocky Mountain region into central Mexico, and the northern west coast. Fragmented populations occur in some other states, including a combined estimate of 5,900 black bears from Arkansas (3,500), Missouri (400), and Oklahoma (2,000). The American black bear is the world's most common bear species.

Black bears were common residents throughout Missouri's forests and woodlands in the early 1800s. By 1850 they were becoming rare, but some still lived in northern Missouri until about 1880. A few survivors were reported in the swamps of the Mississippi River Alluvial Basin until 1931. By the 1950s, they were thought to be gone from Missouri, but rumors of their existence and a few reliable reports continued to occur in southern Missouri.

Many habitats capable of supporting black bears have since recovered. In 1958, Arkansas began reintroducing black bears into that state, and 254 bears procured from Minnesota and Manitoba, Canada, had been released by 1968. Sightings in Missouri increased beginning in the 1960s mostly due to this successful reintroduction program. In 1991, a bait station survey was started to better determine the distribution of black bears in Missouri and the status of their habitat. In 2010, a larger-scale black bear study in Missouri was initiated using live trapping, satellite tracking of bears, and collection of biological information including DNA to help determine the population size, structure, and lineage of Missouri's bears. DNA evidence to date suggests that the largest population, in south-central Missouri, located in Webster and Douglas Counties, may represent a small remnant of the historical population of that region combined with bears from Arkansas. Other Missouri bears are presumably descended from bears released into Arkansas that later dispersed into Missouri. Today, most Missouri bears live south of Interstate 44, but wandering individuals, mostly subadult males, have been seen as far north as the Iowa line. Counties adjoining the core bear range have more sightings that those north of Interstate 44.

Habitat and Home

American black bears live in heavily wooded areas. In winter they den in hollow trees, rock caves and crevices, sheltered spots under the roots of a tree or under a fallen tree, slash piles, or in slightly excavated hollows in the ground. Here they make a bed of grass, leaves, twigs, and bark. They have no permanent summer home and sleep either in a tree or on the ground.

Habits

American black bear populations are increasing in Missouri, and new information about the species' habits in the state is continually being learned. The following sections contain information on Missouri black bears and about black bears in general.

Black bears customarily eat heavily in the fall, a behavior called *hyperphagia*, and accumulate a layer of

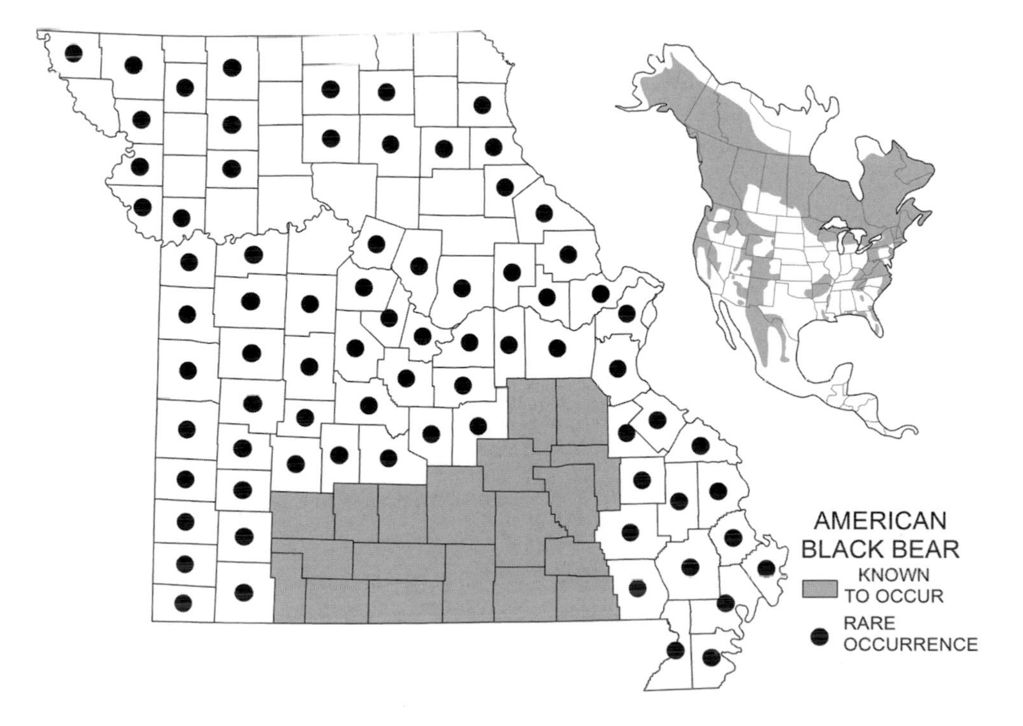

body fat resulting in a 30 percent weight gain. This fat provides nourishment and insulation during winter. In years of abundant food they become fat earlier than in other years and consequently become sluggish and may den sooner.

In Missouri, black bears usually retire to their winter dens between mid-November and mid-December and go into a deep torpor. Females with cubs tend to den earliest, females with yearlings next, and adult males followed by subadult males den last. Bodily functions are greatly slowed; they do not eat, drink or defecate. Their metabolic rate is reduced by as much as 50 percent, body temperature drops only 1 to 4°C (1 to 7°F) below normal, the pulse rate is lowered from 66 to 140 beats per minute in summer to 8 to 22 beats per minute, and breathing is slowed to the rate of 2 to 5 times a minute. There are alternate periods of deep and light sleep and, during warm spells, some bears may leave their dens for short periods. The females remain responsive to the needs of their cubs while in this state.

This period of winter inactivity usually extends until April. Bears generally deposit a "fecal plug" or enlarged dropping that contains the waste of their previous fall feeding after emerging from their winter dens. Bears lose weight during torpor and continue to do so for weeks after emerging from winter dens. Newly emerged bears generally eat green vegetation for nourishment and to activate their dormant digestive system.

The family unit is a mother and her cubs or yearlings. Sibling bears from one to two years old will also travel together prior to dispersal by males. Adult males travel extensively during the spring and summer breeding season, and often throughout the fall. Males typically have larger home ranges than females, sometimes twice or triple their size, which may overlap with 7–15 female ranges. In Missouri, female bears have annual home ranges of over 52 sq km (20 sq. mi.) and males over 259 sq km (100 sq. mi.).

Bears feed and move mostly during evening and early morning, but they can be active all day especially when feeding or during breeding periods.

Black bears may scar the trunks of trees by gouging out chunks with their teeth or making deep scratches with their claws. These marked trees serve in some way as signposts to make bears aware of one another's presence.

Bears usually walk with a lumbering gait but can run for short distances at a speed of 48 km (30 mi.) per hour. They often rear up on their hind legs to get a better view or scent of their surroundings. They are expert tree climbers. Bears descend a tree rear end first. They swim well and have been observed to swim as far as 8 km (5 mi.) at a time.

Bears are attracted to carrion and like to roll in it. They sometimes take dust and mud baths. In hot weather they show a preference for lying in damp places.

In general male bears are solitary and females travel in family units. While territorial, bears are tolerant of each other at feeding places like berry patches or perhaps salmon runs. Antagonistic bears frequently rear up on their hind legs and wrestle. At such times severe fights may develop.

Foods

Black bears eat a wide variety of food, including cool-season grasses and forbs such as stinging nettle, clover, and jewelweed; soft mast from sassafras, dogwood, persimmon, and witch-hazel; blackberries, blueberries, and other fruits; all kinds of seeds; and hard mast including acorns and pecans. The animal foods are ants, ground bees and wasps, beetles in hard and grub stages, crickets, grasshoppers, fish, frogs, small rodents, fawns, bird eggs, and carrion. In Missouri, the acorn crop in fall is a particularly important resource in preparation for the coming winter.

In searching for food, bears claw open rotten stumps and turn over logs and rocks. When they find a supply of extra food, they often cover it with debris and return to feed on it again. They can be attracted to human foods and may visit campsites, garbage disposal places, and bird feeders.

Reproduction

Mating occurs from May through July, when ovulation is induced by copulation. DNA analysis has demonstrated both single and multiple paternity within a single litter. After eggs are fertilized, their development is arrested for 5 months. About the time the female goes into hibernation, the eggs become implanted in the uterus and development resumes. By early December the embryos are less than 1.9 cm (¾ in.) long. The cubs are born in late January or early February while females are still in their winter dens.

Compared to most mammals at birth, the cubs are unusually small in proportion to the size of the mother, weighing 170 to 227 g (6 to 8 oz.) and being about 23 cm (9 in.) long. They are blind, toothless, and covered with fine hair. When they are about 6 weeks old their eyes open and their teeth begin to cut through the gums. At that time they are about 30 cm (12 in.) long, weigh around 0.9 kg (2 lb.), and are well furred.

A female was known to have moved her cubs to another den as much as 4 km (2½ mi.) from the first den. The mother made two trips, carrying one cub each time in her mouth.

At 2 months of age cubs leave the den with their mother for the first time. They stay with her, continue to nurse throughout the summer, and den with her during the following winter. Young males disperse at 1–3 years of age and may travel hundreds of kilometers (miles) before setting up an adult range. One young adult Missouri bear was observed to travel nearly 805 km (500 mi.). Young females may remain in or adjacent to their mother's territory through adulthood.

Males reach sexual maturity at 3–4 years but continue growing until they are 10–12 years of age. A female may have her first litter at 3 to 7 years of age. She has her cubs and dens with them during her second year, then following emergence from hibernation she is able to breed again. Thus an adult female can raise young every other year. Most litters contain 2 young, sometimes 3. Rare cases are reported of litters with 4 or 5.

The female takes entire care of the young and disciplines them very strictly. At any sign of danger she sends them up a tree and expects them to remain there until she returns.

Some Adverse Factors

The primary sources of mortality for American black bears across their North American range are human-related and include legal hunting, collisions with vehicles, and illegal killings. Cubs may be killed by coyotes, domestic dogs, or bobcats, but adults are not vulnerable to predation. Adult male bears sometimes attack and kill cubs, but the mother is a ferocious protector. Mortality is greatest among the young after they leave their mother. However, cubs deprived of their mother at 5½ months of age have survived on their own.

Bears may be afflicted with fleas, large mites, ticks, and mange mites, as well as numerous internal parasites such as nematodes and protozoans.

Importance

Black bears are hunted in some portions of their range as the meat, especially that of young animals, is very good.

Some bears may kill pigs, chickens, or sheep. They may also cause some damage to corn and beehives. Where campers leave food within a bear's reach, they can become troublesome and raid camps and cabins. Black bear attacks on humans are rare, with only 63 reported fatalities in the United States and Canada in a 109-year period (1900–2009).

Conservation and Management

As black bears continue to establish themselves in Missouri, steps must be taken to minimize human-bear conflicts so that they will be accepted and considered as a positive, successful return of a member of our native fauna. Black bears are normally shy and frightened of people, but they can become dependent

on non-natural foods supplied by humans, and people must be discouraged from allowing them access to such foods. Some necessary steps include feeding pets indoors; feeding wild birds only during the winter when bears are much less active; locking food inside vehicles, securing it in bear-proof containers, or hanging it from ropes between trees when camping or floating; keeping garbage secure; and fencing in gardens, orchards, and beehives where bears live. Under no circumstance should black bears be fed intentionally. Harassment and other scare tactics often convince a bear to stay away from people, but they sometimes do not get the message, keep going back for easy food, and may eventually have to be destroyed. If you are experiencing problems with black bears, contact your local Missouri Department of Conservation office.

Black bears use a wide variety of timbered habitat types. Habitat management for black bears consists simply of providing them large blocks of forested habitat with connecting travel corridors, a stable food source with abundant hard and soft mast–producing trees and other soft mast–producing plants, and adequate escape cover. Trees capable of serving as den sites should be protected from timber harvest and stand improvement activities. In addition, roadless areas should be preserved, as bears frequently in contact with roads are vulnerable to being struck by cars. Specially designed highway wildlife crossings may prevent many collisions between black bears and vehicles.

The black bear population in Missouri is slowly growing, and with good management and measures to reduce human-bear conflicts, a small sustainable bear population will continue to call this state home.

SELECTED REFERENCES

See also discussion of this species in general references, page 23.

Amstrup, S. C., and J. Beecham. 1976. Activity patterns of radio-collared black bears in Idaho. *Journal of Wildlife Management* 40:340–348.

Beringer, J., et al. 2008. Management plan for the black bear in Missouri. A report to the Missouri Department of Conservation, Jefferson City. 27 pp.

Beringer, J. 2012. Living with large carnivores. *Missouri Conservationist* 73:22–27.

Brody, A. J., and M. R. Pelton. 1989. Effects of roads on black bear movements in western North Carolina. *Wildlife Society Bulletin* 17:5–10.

Eiler, J. H., W. G. Wathen, and M. R. Pelton. 1989. Reproduction in black bears in the southern Appalachian mountains. *Journal of Wildlife Management* 53:353–360.

Faries, K. M., T. V. Kristensen, J. Beringer, J. D. Clark, D. W. White Jr., and L. S. Eggert. 2013. Origins and genetic structure of black bears in the interior highlands of North America. *Journal of Mammalogy* 94:369–377.

Garshelis, D. L., and M. R. Pelton. 1980. Activity of black bears in the Great Smoky Mountains National Park. *Journal of Mammalogy* 61:8–19.

Gordon, K. R., and G. V. Morejohn. 1975. Sexing black bear skulls using lower canine and lower molar measurements. *Journal of Wildlife Management* 39:40–44.

Herrero, S., A. Higgins, J. E. Cardoza, L. I. Hajduk, and T. S. Smith. 2011. Fatal attacks by American black bear on people: 1900–2009. *Journal of Wildlife Management* 75:596–603.

Horner, M. A., and R. A. Powell. 1990. Internal structure of home ranges of black bears and analyses of home range overlap. *Journal of Mammalogy* 71:402–410.

Larivière, S. 2001. *Ursus americanus. Mammalian Species* 647. 11 pp.

Marks, S. A., and A. W. Erickson. 1966. Age determination in the black bear. *Journal of Wildlife Management* 30:389–410.

Matson, J. R. 1954. Observations on the dormant phase of a female black bear. *Journal of Mammalogy* 35:28–35.

McKinley, D. 1962. The history of the black bear in Missouri. *Bluebird* 29:1–15.

Merkle, J. A., H. S. Robinson, P. R. Krausman, and P. Alaback. 2013. Food availability and foraging near human developments by black bears. *Journal of Mammalogy* 94:378–385.

Rogers, L. L. 1987. Effects of food supply and kinship on social behavior, movements, and population growth of black bears in northern Minnesota. *Wildlife Monograph* 97:1–72.

Rogers, M. J. 1973. Movements and reproductive success of black bears introduced into Arkansas. *Proceedings of the Southeastern Association of Fish and Wildlife Agencies* 27:307–308.

Schenk, A., and K. M. Kovacs. 1995. Multiple mating between black bears revealed by DNA fingerprinting. *Animal Behavior* 50:1483–1490.

Schooley, R. L., C. R. McLaughlin, G. J. Matula Jr., and W. B. Krohn. 1994. Denning chronology of female black bears: Effects of food, weather, and reproduction. *Journal of Mammalogy* 75:466–477.

Smith, K. G., and J. D. Clark. 1994. Black bears in Arkansas: Characteristics of a successful translocation. *Journal of Mammalogy* 75:309–320.

Titus, R., et al. 1993. Management plan for the black bear in Missouri. Missouri Department of Conservation, Jefferson City. 50 pp.

Willey, C. H. 1974. Aging black bears from first premolar tooth sections. *Journal of Wildlife Management* 38:97–100.

Wilton, C. M., J. L. Belant, and J. Beringer. 2014. Distribution of American black bear occurrences and human-bear incidents in Missouri. *Ursus* 25:53–60.

Raccoons and Relatives
(Family Procyonidae)

This family occurs in North and South America and includes the raccoons, coatis, kinkajous, olingos, olinguitos, ringtails, and cacomistles. The raccoon is the only representative in Missouri.

In this family there are 5 toes on both front and hind feet. The tail is long and well furred. It is often marked with a ringed color pattern and may be somewhat prehensile. Procyonids are generally omnivorous, the flesh-eating adaptations of the teeth are largely lost, the molars being low crowned with rounded cusps.

Raccoon (*Procyon lotor*)

Name

The first part of the scientific name, *Procyon*, is of Greek origin and means "before the dog" (*pro*, "before," and *kyon*, "dog"). Why this name was given to the raccoon is not known. The last part, *lotor*, is from the Latin word *lutor*, meaning "a washer." It refers to this animal's habit of sometimes washing items of food before eating.

The common name, "raccoon," is from the Algonquian name, variously spelled as *arocoun*, *arakun*, *arathkone*, and *aroughcun*.

Description (Plate 54)

The raccoon is a medium-sized, stocky mammal with a prominent black "mask" over the eyes and a heavily furred, ringed tail about half the length of head and body. The muzzle is pointed, but the head is broad across the jowls. The dark eyes are of medium size, and the short ears are prominent and pointed. The feet are rather long and slender, with naked soles. Both front and hind feet have 5 toes each with short, curved claws. The thick body fur is composed of long, coarse guard hairs and short, soft underfur.

Color. On the upperparts adults are grizzled brown and black, strongly washed with yellow. The sides are grayer than the back, and the underparts are dull brownish, grizzled with yellowish gray. The face is whitish with a prominent black band across the eyes and cheeks; the ears are grayish with black at the base of the back; and the feet are whitish or yellowish gray. The tail is distinctly marked with alternating rings of yellowish gray and brownish black: the dark rings vary in number from 4 to 7 and are more pronounced above and less defined below. Pure white or albino, very dark or melanistic, and "red" individuals are rare. The sexes are similar in color and show little seasonal variation.

Raccoons molt from the first of March until the end of May, either by gradually shedding hair from all parts of the body at the same time or by shedding first in the head region, then from the sides and belly, and last from the back. A lighter-weight coat is worn during the summer. In October and November, new underfur and guard hairs appear; this new growth begins on the belly and proceeds toward the back and tail. It takes about 6 weeks for completion of this process. The fur of most raccoons is prime by 1 December, but some females that have borne young do not become prime until late winter, and an occasional one may not become prime at all.

Measurements

Total length	55–97 cm	21½–38 in.
Tail	146–304 mm	5⅝–12 in.
Hind foot	85–177 mm	3 5/16–7 in.
Ear	47–63 mm	1⅞–2½ in.
Skull length	107–27 mm	4¼–5 in.
Skull width	69–73 mm	2¾–2⅞ in.
Weight		
Male	3.6–11.3 kg	8–25 lb.
Female	3.0–7.9 kg	6¾–17½ lb.

All animals weigh most in the fall and least in the spring. There is a record weight of 28.3 kg (62 lbs.).

A 12.3 kg (27.2 lb.) male is the official record-weight raccoon confirmed by the Missouri Department of

Plate 54

Raccoon *(Procyon lotor)*

4 inches 101 mm

Skull—bottom view, left half

Hard palate extending beyond molars

Lower jaw—top view, left half

$4\frac{1}{4}$–5 inches 107–127 mm

Skull—side view

Left front foot

Left hind foot

Skull—top view

$2\frac{3}{4}$–$2\frac{7}{8}$ inches 69–73 mm

Conservation. The state furbearer record program was started in 2011.

Teeth and skull. The dental formula of the raccoon is:

$$I\ \frac{3}{3}\ C\ \frac{1}{1}\ P\ \frac{4}{4}\ M\ \frac{2}{2} = 40$$

This number of teeth is sufficient to distinguish the skull of the raccoon from any other Missouri mammal. The skull is broad and massive and the hard palate extends noticeably beyond the last molars.

Sex criteria and sex ratio. Males are recognized by the testes (but in some young males the testes may be still within the body cavity) and by the penis. They also have a penis bone, or baculum. Females usually possess 3 pairs of teats on the belly (1 pair near the front legs, 1 pair behind this, and 1 pair in the groin region), but some have an additional teat or two.

In cased pelts, the sex can be determined by the location of the single opening of the urinary and reproductive systems. If the pelt is reversed with the flesh side out, males are identified by a lump near the middle of the belly marking the position of this urinary-reproductive opening. On female pelts, this opening is seldom found, as it is commonly trimmed off in skinning. When present, it is located considerably closer to the tail than it is in males.

The sex of a raccoon skull can be identified by the thickness of the root of the lower canine teeth. Males have roots thicker than 5 mm (3⁄16 in.); most females have thinner roots. This is applicable from 5 months to at least 4 years of age.

The sex ratio tends to vary with the population level. When the population is increasing, there are more females than males; when it is decreasing, the ratio of the sexes is reversed. During the increase in recent years, from 1951 on, the percentage of females in the harvest varied from 49.4 to 55.

Age criteria, age ratio, and longevity. Young raccoons before 4 months of age can be easily distinguished from adults by their smaller size and weight. As they get older, however, they become increasingly difficult to identify.

The permanent teeth are acquired from late August until October of the first year. Animals with wear on the permanent teeth, particularly on the tips of the upper canines, are more than 2 years old. Those with considerable wear are 3 or more years old.

Young can also be separated from older animals by the presence and amount of epiphyseal cartilage at the ends of the long bones of the front legs, similar to that described for the eastern cottontail.

Until February the age of males can be told easily by the shape of the penis bone, or baculum, and by the presence of cartilage or bone on the tip. The baculum of a male less than 1 year old has a porous, slender base, a flattened or truncated tip, and cartilage on the tip; also, it is 8.9 cm (3½ in.) or less in length and weighs 1.9 g (1⁄16 oz.) or less. The baculum of an adult has a nonporous, massive base, a rounded or pointed tip, and bone on the tip; also, it is more than 8.9 cm (3½ in.) long and weighs more than 1.9 g (1⁄16 oz.).

Either in whole animals or pelts, females that have not borne young are identified by their undeveloped, pinkish teats, about 3 mm (⅛ in.) long, while females that have had young possess enlarged, dark teats, 6 mm (¼ in.) or longer. Since only about 40 percent of the young females breed at 1 year of age, the first group includes all young less than 1 year old and about 60 percent of the young between 1 and 2 years old.

Internally, females that have borne young can be identified until February by the presence of placental scars (places of attachment of embryos) in the branches of the uterus, and by the uterine branches, which are about 6 mm (¼ in.) in diameter and are opaque; females that have not had young lack placental scars and the branches of the uterus are less than 3 mm (⅛ in.) in diameter and are translucent.

By the first of January, a male weighing 6.8 kg (15 lbs.) or more is usually adult; one weighing less than 6.8 kg is usually young. Weight alone is not sufficient to determine the age of females but in combination with characteristics of the teats and uterus aids in age determination. Females weighing less than 5.2 kg (11½ lbs.) and having small teats and no evidence of breeding in the uterus are young. Females weighing 5.2 kg or more and having small teats and no evidence of breeding in the uterus are adults that have not borne young; usually these are individuals between 1 and 2 years of age. Regardless of weight, females with enlarged teats and evidence of breeding in the uterus are

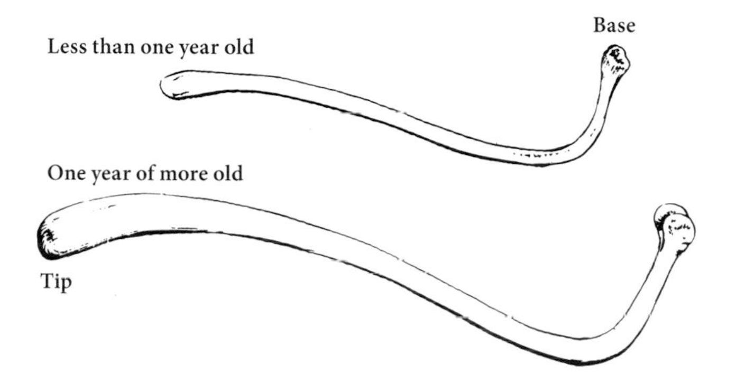

Age in the raccoon as indicated by shape of the baculum

Raccoon tracks

adults that have borne young; these are individuals 1 year of age and older.

X-rays of the feet assist in the identification of age. There is no fusion between the ends and shafts of the long bones in young raccoons during their first winter, but in older animals there is a fusion.

The much-used laboratory technique of dried weight of the eye lens can be used to identify the month of birth of a young raccoon until 1 year of age.

Probably the most accurate method of age estimation is by counting the cementum layers of the roots of incisor teeth. This is considered accurate until 4 years of age but probably underestimates ages over 4 years.

Other techniques used to age raccoons involve the time of closure of certain bones of the skull and the size and wear of certain teeth.

Most raccoons live less than 5 years in the wild, although some have reached 16 years of age. In captivity, 17 years is the oldest record for a raccoon. It has been estimated that in Missouri it takes 6.5 years to replace all members of the raccoon population.

Voice and sounds. When undisturbed, raccoons may utter a chuckling sound; when annoyed or fighting, they snarl, growl, and hiss. A low, soft snort is given in recognition of unsocial individuals. To call her young, the female uses a low, grumbling purr. The young, when separated from their mother, give a call very similar to that of a tree frog. In the autumn especially, raccoons may call to each other with a shrill whistle like that of a screech owl.

Distribution and Abundance

The range of the raccoon in North America was much smaller prior to European settlement. Beginning in the late 1800s raccoons began moving into new areas, and populations greatly increased after the 1930s. Today they are common mammals within their range from central Canada to Panama in Central America, and into northern South America. Their ability to adapt well to human environments, the conversion of forests to agricultural lands, introductions, and the extirpation of some of their natural

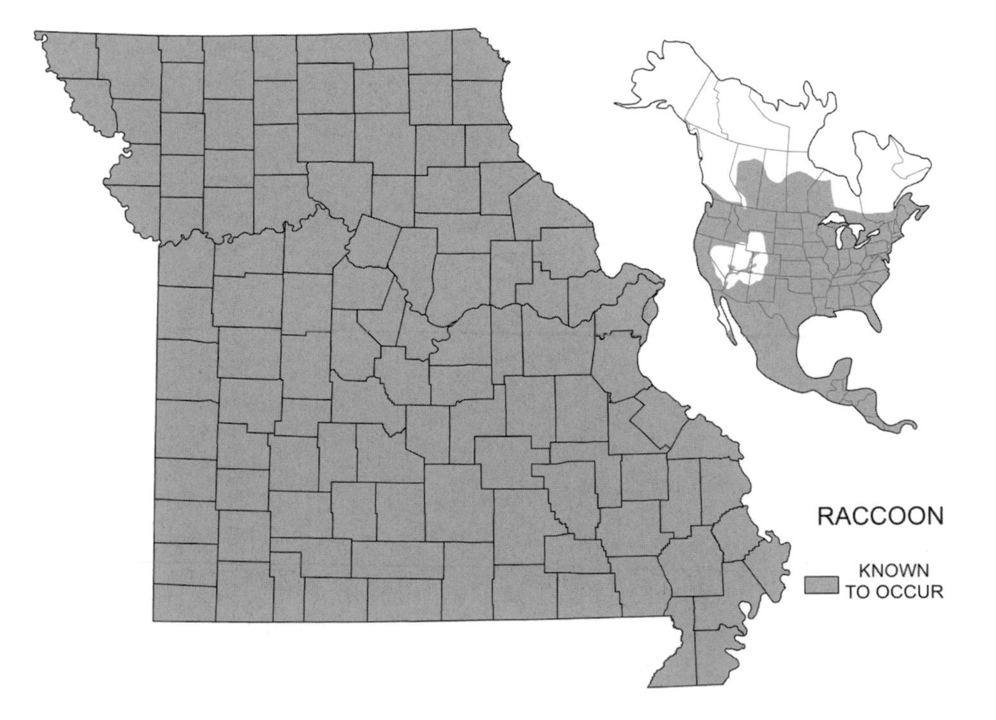

predators contributed to their increase in abundance and distribution. As a result of escapes and deliberate introductions in the mid-20th century, the raccoon is now widely distributed in several European and Asian countries, including Germany and its bordering countries, Japan, and the former Soviet Union.

Raccoons occur throughout Missouri but are most common in the Central Dissected Till Plains and Osage Plains of northern and western Missouri, and less abundant in the Ozark Highlands and the Mississippi River Alluvial Basin. Prior to 1940 the population in Missouri had steadily diminished; in the succeeding years, it started to increase and has continued to do so. How dense the population in any given area can become is unknown, but in 1949 a reliable estimate was made of 1 raccoon per 0.25 ha for 41 ha (0.63 ac. for 102 ac.) of excellent habitat and 1 raccoon per 0.8 ha for 809 ha (2 ac. for 2,000 ac.) of good habitat in north-central Missouri. A more recent study in Illinois found densities ranged from 36.6 to 72.6 raccoons per sq km (247 ac.) at urban sites, 41.1 to 93.0 raccoons per sq km at suburban sites, and 3.1 to 14.6 raccoons per sq km at rural sites.

Habitat and Home

In Missouri raccoons prefer habitat with hardwood timber, which may be either a dense forest or only a narrow stand of trees bordering a river or some watered area such as a lake, pond, swamp, or marsh. Sometimes raccoons show an adaptability to urban life and can even occupy range at the edges of large cities.

The home is usually a den in a hollow tree, caves, crevices in rocky ledges, abandoned woodchuck burrows, cavities under tree roots, stands of slough grass, corn shocks, haystacks, squirrel nests, muskrat houses, or abandoned farm buildings. Each raccoon has several dens in its range and does not necessarily use the same one continuously. Dens are generally located in maple, elm, oak, and sycamore trees, but nearly any other kind of tree is inhabited if it contains suitable hollows. Limbs, burls, or old squirrel nests are often used for sunning spots. Dens are used most in spring when the female has a litter, least in summer, and frequently in winter.

Habits

Raccoons are nocturnal, doing most of their foraging and prowling from dusk to dawn. On occasion, they come out in daytime. The distance an animal travels each night varies with weather conditions, food availability, and sex and age of the animal. During much of the year adult males occupy a home range about 1.6 km (1 mi.) in diameter, although during the breeding season they may travel farther in search of a mate. Adult females and their young, which compose the family unit, live in a smaller area, about 1.5 km (¾ mi.) in diameter. In parts of their range where the habitat is not as favorable, adult males may have an average home range of 26 sq km (10 sq. mi.) and adult females of 7.8 sq km (3 sq. mi.).

In fall, raccoons use certain areas more than others but generally select a different resting site each day. These spots may be as much as 1.6 km (1 mi.) apart.

The average distance traveled between dens is 436 m (1,430 ft.). Raccoons that have been marked and released in strange surroundings have moved as far as 120 km (75 mi.) from the site of their release. One long movement is reported by a yearling male who traveled 265 km (165 mi.) in 164 days.

It is interesting that the size of the home range and activity patterns were the same for a blind raccoon as for normal animals.

During periods of snow and ice storms, raccoons usually den up for a few days, either singly or in groups up to 9 or even 23 in number, including both sexes and all ages. During these periods of inactivity raccoons do not hibernate but remain responsive to touch and weather changes. Adults seemingly prefer to lead solitary lives but may use communal dens during unusually severe winter weather, in periods of high population, or in places with an abundant food supply.

Raccoons are expert climbers. In descending trees they come down either head or tail first and often jump. They frequent watercourses and swim well. On land they walk with a lumbering, flat-footed gait (ambulatory plantigrade locomotion). A human can outrun a raccoon on level terrain, but the animal usually climbs a tree or seeks some refuge to evade capture. Raccoons will fight if cornered but prefer to escape or conceal themselves.

On summer days raccoons spend much of their time on the ground; on sunny fall and winter days, they lie on limbs or other high sunning spots. They leave their droppings on branches, sunning spots, stumps, or on the ground, but not in the nest.

Raccoons are curious, clever, and cunning; in captivity, they want their own way and may become dangerous if not humored.

Foods

Raccoons eat plant and animal matter, depending upon what is available. The plant foods are mainly wild fruits such as persimmons, grapes, plums,

chokecherries, Osage oranges, greenbriers, and blackberries; grasses and sedges; corn in the milk stage or when ripe and hard; and acorns, pecans, and other nuts. Most of the animal fare is obtained on the ground by hunting in shallow pools, turning over rocks, digging into rotten logs, or coming upon some bird's nest or hapless animal. The animal foods consist of crayfish, clams, fish, various insects (including water-dwelling species, many kinds of beetles, grasshoppers, crickets, yellow jackets, and grubs), spiders, frogs, snakes, turtles and their eggs, snails, earthworms, eggs or young of ground-nesting birds, mice, squirrels, rabbits, and muskrats. They also eat ducks and geese killed or crippled during the hunting season.

In captivity, raccoons are fond of softening their food in water but do not always do so before eating. Rather than depend upon sight as a method of close examination, they feel the food with their sensitive front feet. The nose is likewise very sensitive to touch. In fall, raccoons eat great quantities of food and put on considerable body fat prior to cold weather. This supply of fat is the major source of energy during winter. By late winter they may lose as much as 50 percent of their fall weight.

Reproduction

Adult males come into breeding condition in December or January and remain sexually active until late June or July. They are promiscuous in their mating. Most breeding occurs in February, but some may take place later in the spring. There is usually only 1 litter annually with an average of 3 to 4 young and extremes of 1 to 8. The gestation period is 63 days. Most litters are born in April or early May, but some, the result of late matings, may arrive in June, July, or August. Late-breeding females are either those that did not become pregnant at an earlier mating or those that lost a litter early in the season.

The young, weighing about 71 g (2½ oz.) each, are furred at birth and either have the typical mask across their faces or develop it within the next 10 days. They are born blind, but their eyes open between 18 and 29 days following birth. The tail rings are fully furred at 3 weeks. Females may move the young to a different den, carrying them one at a time by the nape of the neck. The kits, or pups, or cubs, stay in the den until about 8 to 10 weeks of age, when they learn to eat solid foods and start foraging with their mother. The young are weaned in August and shed their milk teeth between then and October. Although some young may move away in the fall, most stay near the female until the following spring.

About 40–80 percent of young females breed the spring following their birth, while the rest do not breed until their second year. This annual variation could be a factor affecting population dynamics. Although young males are able to breed in the spring following their birth, they probably have little opportunity to mate. This is because they become sexually active late in the season and the adult males have already bred most of the available females.

Raccoon one week old

Some Adverse Factors

Humans and their dogs are the most important predators on raccoons, although great horned owls, foxes, bobcats, and coyotes may take some young. Ticks, lice, fleas, botfly larvae, roundworms, flukes, and tapeworms parasitize this species. One species of roundworm, *Baylisascaris procyonis*, is especially noteworthy as human infections may lead to severe central nervous system disease or even death; however, human infections are rare. Raccoons are known to have rabies, distemper, tuberculosis, tularemia, and a skin disease caused by a fungus. Occasional deaths are reported due to starvation, physiological stress resulting from high population densities, and to accidental falling trees. Automobiles take a high toll of raccoons crossing highways.

Importance

The raccoon is a valuable fur and game species. Many hunters enjoy the sport of pursuing and taking raccoons with their hounds and they are a favorite during trapping season. The durable fur is used for coats, jackets, collars, hats, muffs, and trimmings. In harvest, it outranks all other furbearing species in Missouri. Raccoon harvest records show the average annual take of pelts in the 1940s was 47,028, which increased to 110,072 in the 1950s, 151,195 in the 1960s, and 214,647 in the 1970s, then decreased to 161,866 in the 1980s, 115,377 in the 1990s, and 96,664 in the first decade of the 2000s. The highest annual harvest was 276,524 in 1975 and the lowest was 11,000 in 1940. The average annual price paid per pelt was $2.17 in the 1940s, $1.19 in the 1950s, $1.86 in the 1960s, $13.44 in the 1970s, $13.78 in the 1980s, $7.63 in the 1990s, and $9.84 in the first decade of the 2000s. The highest average annual price was $27.50 paid in 1978 and the lowest was $0.65 in 1958. It is hard to measure the relative effects of the several factors that influence harvest numbers and pelt prices. Obviously, population abundance is involved, but to what extent has not been determined. Changes in demand for furs and the fluctuating values of their pelts influence economic reasons for trapping raccoons. In addition, the numbers of trappers working in Missouri influence the harvest.

The meat of young animals is delicious when roasted, and many thousands of raccoons are eaten each year. Raccoons eat insects and mice. Extensive damage to corn, gardens, or chickens occurs only rarely.

In pioneer days, raccoon pelts were used for articles of clothing and for barter. The thin oil from the fat was used to keep leather in good condition and to lubricate machinery. The baculum has been used as a ripping tool by tailors.

A raccoon family living in the cavities of big timber is a valuable addition to the wildlife community. Their additional role as seed dispersers emphasizes their larger role in ecological systems.

Conservation and Management

The most important management measure for raccoons in Missouri is to preserve den trees. This involves a tolerance for large, dead, or decaying trees; selective lumbering; restriction of burning that might destroy den trees; curtailment of cutting of den trees by night hunters; and judicious placement of artificial den boxes where necessary. Food is generally ample, but discriminate burning in timbered areas will save the acorn crop, fruits of shrubs and trees, and insects dwelling in the leaf mold, all of which are fed upon by raccoons. Prevention of pollution in streams promotes a better aquatic food supply, and farm ponds built near timber provide a supplementary source of aquatic food. Although the population is increasing, the intensified demand created by high prices for the pelts makes a continual control of the harvest essential.

Raccoons occasionally cause problems for property owners, including in urban and suburban areas where they sometimes take refuge in attics and chimneys, and during their constant search for food. If you are experiencing problems with raccoons, contact a wildlife professional for advice, assistance, regulations, or

SELECTED REFERENCES

See also discussion of this species in general references, page 23.

Baker, R. H., C. C. Newman, and F. Wilke. 1945. Food habits of the raccoon in eastern Texas. *Journal of Wildlife Management* 9:45–48.

Berner, A., and L. W. Gysel. 1967. Raccoon use of large tree cavities and ground burrows. *Journal of Wildlife Management* 31:706–714.

Cypher, B. L., and E. A. Cypher. 1999. Germination rates of tree seeds ingested by coyotes and raccoons. *American Midland Naturalist* 142:71–76.

Dellinger, G. P. 1954. Breeding season, productivity, and population trends of raccoons in Missouri. M.S. thesis, University of Missouri, Columbia. 86 pp.

Erickson, D. W. 1978. *Muskrat, raccoon, and mink productivity research*. Federal Aid Project no. W-13-R-32. Missouri Department of Conservation. 18 pp.
———. 1979. Studies of the age and sex compositions of Missouri raccoon, muskrat, and mink harvests. Midwest Furbearer Workshop, Manhattan, Kansas. Mimeograph. 15 pp.
Fiero, B. C., and B. J. Verts. 1986. Comparison of techniques for estimating age in raccoons. *Journal of Mammalogy* 67:392–395.
Fritzell, E. K. 1978. Habitat use by prairie raccoons during the waterfowl breeding season. *Journal of Wildlife Management* 42:118–127.
Fritzell, E. K., G. F. Hubert, B. E. Meyen, and G. C. Senderson. 1985. Age-specific reproduction in Illinois and Missouri raccoons. *Journal of Wildlife Management* 49:901–905.
Gehrt, S. D., and E. K. Fritzell. 1997. Sexual differences in home ranges of raccoons. *Journal of Mammalogy* 78:921–931.
Giles, L. W. 1939. Fall food habits of the raccoon in central Iowa. *Journal of Mammalogy* 20:68–70.
———. 1940. Food habits of raccoon in eastern Iowa. *Journal of Wildlife Management* 4:375–382.
———. 1942. Utilization of rock exposures for dens and escape cover by raccoons. *American Midland Naturalist* 27:171–176.
———. 1943. Evidences of raccoon mobility obtained by tagging. *Journal of Wildlife Management* 7:235.
Goldman, E. A., and H. H. T. Jackson. 1950. Raccoons of North and Middle America. *U.S.D.I., North American Fauna* 60. 153 pp.
Grau, G. A., G. C. Sanderson, and J. P. Ropers. 1970. Age determination in raccoons. *Journal of Wildlife Management* 34:364–372.
Junge, R., and D. F. Hoffmeister. 1980. Age determination in raccoons from cranial suture obliteration. *Journal of Wildlife Management* 44:725–729.
Larivière, S. 2004. Range expansion of raccoons in the Canadian prairies: Review of hypotheses. *Wildlife Society Bulletin* 32:955–963.
Lotze, J-H., and S. Anderson. 1979. *Procyon lotor. Mammalian Species* 119. 8 pp.
Lynch, G. M. 1967. Long-range movement of a raccoon in Manitoba. *Journal of Mammalogy* 48:659–660.
Mech, L. D., J. R. Tester, and D. W. Warner. 1966. Fall daytime resting habits of raccoons as determined by telemetry. *Journal of Mammalogy* 47:450–466.
Montgomery, G. G. 1968. Pelage development of young raccoons. *Journal of Wildlife Management* 49:142–145.
Page, L. K., S. D. Gehrt, A. Cascione, and K. F. Kellner. 2009. The relationship between *Baylisascaris procyonis* prevalence and raccoon population structure. *Journal of Parasitology* 95:1314–1320.
Petrides, G. A. 1950. The determination of sex and age ratios in fur animals. *American Midland Naturalist* 43:355–382.
———. 1959. Age ratios in raccoons. *Journal of Mammalogy* 40:249.
Prange, S., S. D. Gehrt, and E. P. Wiggers. 2003. Demographic factors contributing to high raccoon densities in urban landscapes. *Journal of Wildlife Management* 67:324–333.
Sanderson, G. C. 1961a. Techniques for determining age of raccoons. *Biological Notes, Illinois Natural History Survey* 43:1–16.
———. 1961b. The lens as an indicator of age in the raccoon. *American Midland Naturalist* 65:481–485.
Schneider, D. G., L. D. Mech, and J. R. Jester. 1971. Movements of female raccoons and their young as determined by radiotracking. *Animal Behavior Monographs* 4:1–43.
Shirer, H. W., and H. S. Fitch. 1970. Comparison from radiotracking of movements and denning habits of the raccoon, striped skunk, and opossum in northeast Kansas. *Journal of Mammalogy* 51:491–503.
Stains, H. J. 1956. The raccoon in Kansas: Natural history, management, and economic importance. *University of Kansas, Museum of Natural History, and State Biological Survey of Kansas, Miscellaneous Publication* 10. 76 pp.
Stuewer, F. W. 1942. Studies of molting and priming of fur of the eastern raccoon. *Journal of Mammalogy* 23:399–404.
———. 1943a. Raccoons: Their habits and management in Michigan. *Ecological Monographs* 13:203–257.
———. 1943b. Reproduction of raccoons in Michigan. *Journal of Wildlife Management* 7:60–73.
Sunquist, M. E., G. G. Montgomery, and G. L. Storm. 1969. Movements of a blind raccoon. *Journal of Mammalogy* 50:145–147.
Twichell, A. R., and H. H. Dill. 1949. One hundred raccoons from one hundred and two acres. *Journal of Mammalogy* 30:130–133.

Weasels, Badgers, Otters, Minks, and Relatives (Family Mustelidae)

The family name, Mustelidae, is based on the Latin word for "weasel." This is the largest family within Carnivora and includes weasels, mink, otters, badgers, stoats, polecats, marten, fishers, wolverines, and others. They inhabit all continents except Australia and Antarctica. There are five species in Missouri.

Most members of this large and varied family have an elongated body, but some have a squat or broad body. Typically, the legs are short and possess 5 well-clawed toes on each foot. A pair of musk glands characteristically occurs in the anal region. There is only 1 molar on each side of the upper jaw, but 2 occur on each side of the lower jaw. Often a great difference exists in the size of males and females.

Least Weasel (*Mustela nivalis*)

Name

The first part of the scientific name, *Mustela*, is the Latin word for "weasel." The last part, *nivalis*, is from the Latin word *nix* meaning "snowy"; this undoubtedly refers to the white winter coat. The common name, "least," is very appropriate for this animal, the smallest weasel and, in fact, the smallest living carnivore. "Weasel" is from the Anglo-Saxon word *wesle*. Formerly, the least weasel was known as *Mustela rixosa*.

Description (Plate 55)

This mouse-sized animal has the typical weasel conformation—a long, slender body with short legs and a small, flattened head only slightly larger in diameter than the long neck. The ears are short and rounded, the whiskers prominent, and the eyes small and beady. The tail is very short, being less than one-fifth as long as head and body. The body fur consists of soft, close underhair and long guard hairs. The soles of the feet are furred in winter but only sparsely so in summer.

Color. The summer coat of adults is brown above and white below, including white on the chin and toes. Occasionally, there is some brown mottling between the forelegs. The tail is brown; it lacks the black tip of the long-tailed weasel but may have a few black hairs at the extreme end. In winter, there are variations in color from the typical summer coat to completely brown (brown on upperparts and most of underparts) to an entirely white coat with a few black hairs at the tip of the tail. In the southern range, including Missouri, the winter coat is generally a variation of brown.

Measurements

Total length	15–25 cm	5¾–9¾ in.
Tail	19–38 mm	¾–1½ in.
Hind foot	18–26 mm	¾–1 in.
Ear	9 mm	⅜ in.
Skull length	26–30 mm	1¹⁄₁₆–1³⁄₁₆ in.
Skull width	12–14 mm	½–⁹⁄₁₆ in.
Weight	35–56 g	1¼–2 oz.

Adult males are larger than adult females.

Teeth and skull. The dental formula of the least weasel is:

$$I\,\frac{3}{3}\;C\,\frac{1}{1}\;P\,\frac{3}{3}\;M\,\frac{1}{2} = 34$$

As in the long-tailed weasel, the upper molars are dumbbell-shaped; the hard palate extends beyond the upper molars; and the auditory bulla surrounding the inner ear is greatly inflated and noticeably longer than wide.

The skull is easily distinguished from those of the long-tailed weasel and mink by its smaller size, wider braincase, and curved upper surface of the braincase.

Sex criteria. Males are identified by the penis. Females have three pairs of teats.

Age criteria and longevity. The size and shape of the penis bone, or baculum, and the amount of cartilage replaced by bone can be used as indicators of age. Although most individuals die in the first year, a few individuals may live up to three years.

Glands. Anal glands are present as in the long-tailed weasel and, when excited, a pungent odor is emitted.

Voice and sounds. Least weasels have four typical vocalizations. They give a chirp when disturbed, a hiss when afraid or threatened, a trill when engaged in a friendly encounter, and a squeal in a stressful situation such as pain. This vocabulary is acquired gradually by the young as they grow up in the nest.

Distribution and Abundance

The least weasel has a wide range in North America, extending all across the arctic region south to northern Kansas and Missouri, and east to Tennessee and North Carolina. It is distributed sporadically throughout this range and is generally rare.

No least weasels had been recorded in Missouri when this book was first published, but since that time several animals have been reported from northern Missouri. These weasels are probably present in many of the northern counties of the state. However, their abundance may vary locally and seasonally and may be related to fluctuations in rodent populations, their main food supply. Least weasels are a vulnerable species of conservation concern in Missouri but are globally secure.

Habitat and Home

The least weasel lives in low, sparse ground cover such as pastures, stubble fields, and marshy areas, sometimes even in mouse-infested barns. Their selection of habitat is generally dictated by the distribution and abundance of prey. Their homes are usually appropriated mole runs or pocket gopher burrows. Within these burrows, they have a nest of grass, corn silk, and mouse fur (usually contributed by a former mouse resident). One weasel may have several shelters within the home range.

Plate 55

Least Weasel *(Mustela nivalis)*

White winter coat

Common annual coat

1 inch

25 mm

Right front foot—winter

Right hind foot—winter

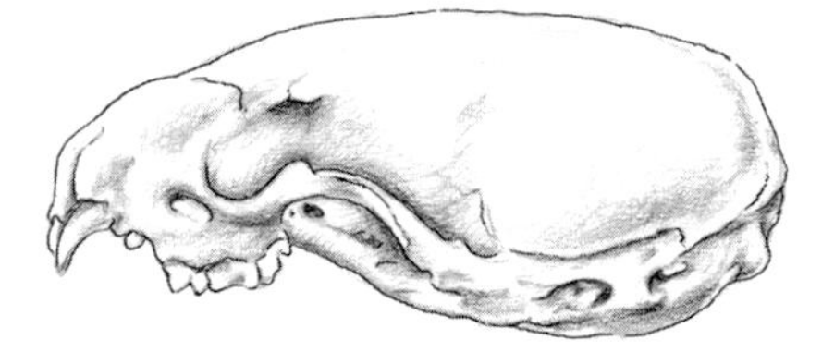

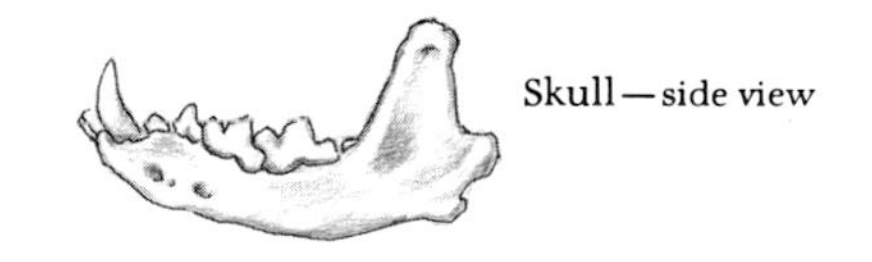

Skull—side view

Skull—bottom view

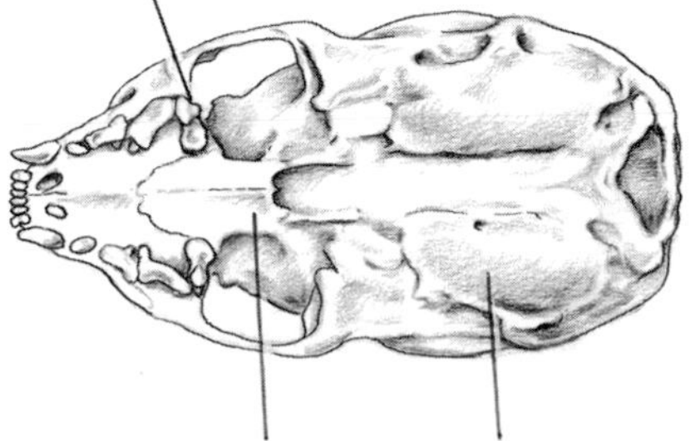

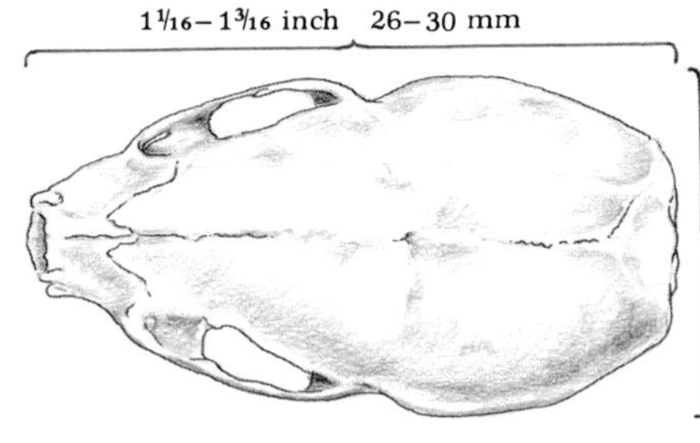

Skull—top view

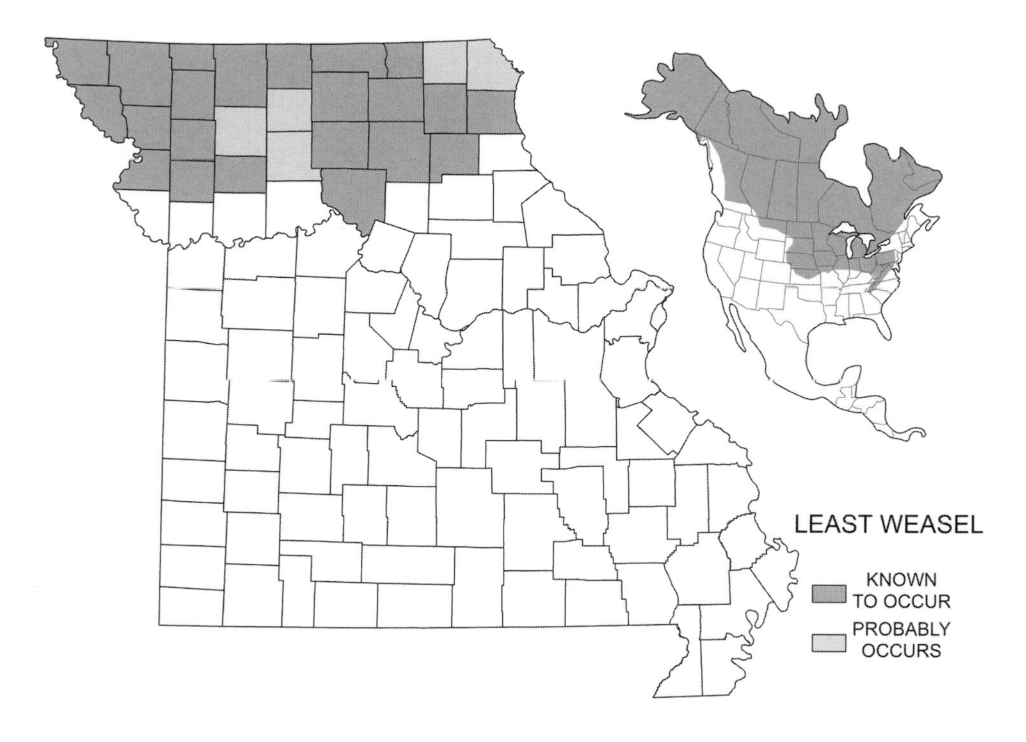

Habits

Because of their small size, least weasels can enter holes in the ground less than 2.5 cm (1 in.) in diameter, allowing them to pursue prey into their burrows. It is not known whether the weasel digs these holes or if the holes are excavated by other animals.

Little is known about their movements. During winter, four male weasels lived in one fencerow, had individual homes, and ranged over about 0.8 ha (2 ac.) in their foraging. Except during breeding, males and females remain spatially separate. Individuals of both sexes defend a home range, and home ranges of males (0.6 to 26.2 ha; 1.4 to 65 ac.) are slightly larger and overlap those of females (0.2 to 7.0 ha; 0.5 to 17.3 ac.). Males tend to dominate interactions with females except during late stages of pregnancy and lactation when females defend their home ranges against males and other females.

Though secretive and wary, least weasels are very aggressive and do not hesitate to attack an adversary, even one larger than themselves.

Foods

Least weasels are very energetic and eat more than half of their body weight each day. This is the equivalent of 1 to 1½ mice a day. Mice and other small rodents form their major food items, but insects and small birds are probably taken when available. Excess food is often killed and stored in a cache or in their burrows for future use.

In killing live, adult mice, which are frequently the same size as themselves, the least weasel grasps the prey by the back of the head then bites through the skull several times in succession. A tight hold is maintained by the teeth and front feet. In all, it only takes 30 seconds to kill a mouse. The head and brain are eaten first.

Reproduction

In the wild, breeding occurs throughout the year but is probably least in winter. Two or more litters are commonly produced annually with 1 to 6 young per litter. Least weasels have more litters and more young per year than any other American species of weasel. Due to the high energetic demand of pregnancy and lactation, a minimum of 10 to 15 voles per ha (2.5 ac.) is required to successfully rear offspring. When resources are higher, reproductive output can increase.

In captivity, one least weasel bore 3 litters in 1 year. The litters were born in March, June, and October. The gestation period in 9 litters born in captivity was 35 days (with extremes of 34 and 36), indicating that there is no delay in embryonic development as occurs in the long-tailed weasel.

At birth, the young weigh between 1.1 to 1.7 g (1⁄32 to 1⁄16 oz.), are pink, hairless, and toothless and have their eyes and ears closed. They make squeaking sounds and have fairly good control of the front legs but not of the back legs. At 4 days of age, fine white fur appears on the back. The teeth start to erupt at 11 days, and shortly thereafter the little weasels begin to eat some solid food brought to them by their mother. At 18 days, some brown hairs first show on the back, and the back soon becomes brown, but the tips of the ears and the underparts stay white. The ears open from 21 to 28 days of age, the eyes open from 26 to 30 days of age, and the young soon begin to utter hissing sounds followed by a sharp, threatening chirp. They make their first kill at about 6 weeks of age, and weaning is complete shortly after this.

At 4 weeks of age, males start to grow faster than females; they can breed first at 8 months of age. Females are sexually mature at 4 months and can have 2 litters during their first year.

Some Adverse Factors

Although least weasels are predators, they are also prey. A number of animals are known to kill least weasels, including hawks and owls, long-tailed weasels, foxes, coyotes, domestic cats, and snakes.

They also contain several parasitic internal roundworms as well as many ectoparasites including biting lice, mites, fleas, and ticks.

Secondary poisoning from exposure to rodenticides can be a factor impacting least weasels.

Importance and Conservation and Management

It is difficult to assess the economic value of this animal, although it does help control the rodent population. It should be afforded protection and appreciated because of its uniqueness—our smallest carnivore—and as a component in the overall ecological picture.

SELECTED REFERENCES

See also discussion of this species in general references, page 23.

Chromanski-Norris, J. F., and E. K. Fritzell. 1983. Status and distribution of ten Missouri mammals. A report to the Missouri Department of Conservation, Jefferson City. 38 pp.

Churchfield, S. 1990. *The Natural History of Shrews.* Comstock Publishing, Ithaca, NY. 178 pp.

Easterla, D. A. 1970. First records of the least weasel, *Mustela nivalis,* from Missouri and southwestern Iowa. *Journal of Mammalogy* 51:333–340.

Heidt, G. A. 1970. The least weasel, *Mustela nivalis* Linnaeus: Developmental biology in comparison with other North American *Mustela. Publications of the Museum, Michigan State University, Biological Series* 4:227–282.

Heidt, G. A., M. K. Petersen, and G. L. Kirkland Jr. 1968. Mating behavior and development of least weasels (*Mustela nivalis*) in captivity. *Journal of Mammalogy* 49:413–419.

Huff, J. N., and E. O. Price. 1968. Vocalizations of the least weasel. *Journal of Mammalogy* 49:548–550.

Llewellyn, L. M. 1942. Notes on the Alleghenian least weasel. *Journal of Mammalogy* 23:439–441.

Polderboer, E. B. 1942. Habits of the least weasel (*Mustela rixosa*) in northeastern Iowa. *Journal of Mammalogy* 23:145–147.

Sheffield, S. R., and C. M. King. 1994. *Mustela nivalis. Mammalian Species* 454. 10 pp.

Short, H. L. 1961. Food habits of captive least weasel. *Journal of Mammalogy* 42:273–274.

Long-tailed Weasel (*Mustela frenata*)

Name

The first part of the scientific name, *Mustela,* is the Latin word for "weasel." The last part, *frenata,* meaning "bridled," is from the Latin word *frenum* for "bridle"; this refers to particular facial markings in some forms, or subspecies, occurring in southern parts of the range.

"Long" aptly describes the tail of this animal. "Weasel" is from the Anglo-Saxon word *wesle.*

Description (Plate 56)

The long-tailed weasel is a slender, long-bodied mammal with short legs and a well-furred tail about half the length of the head and body. The head is small, flattened, and only slightly larger in diameter than the long neck. The ears are short and rounded, and the whiskers are prominent. The small eyes are beady. Both front and hind feet have 5 slightly webbed toes that support the body's weight. There are several bare pads on the soles of the feet and a pad on the undersurface of each toe. The pads are more exposed in summer when the undersurface of the feet is sparsely furred, but they are almost concealed in winter when the feet are well furred. The body fur is composed of soft, close underhair and long, glistening guard hairs.

Color. There are two color phases annually. In summer, adults are dark brown on the upperparts and yellowish white on the underparts, with a white chin. Occasionally some pale brown spots occur on the

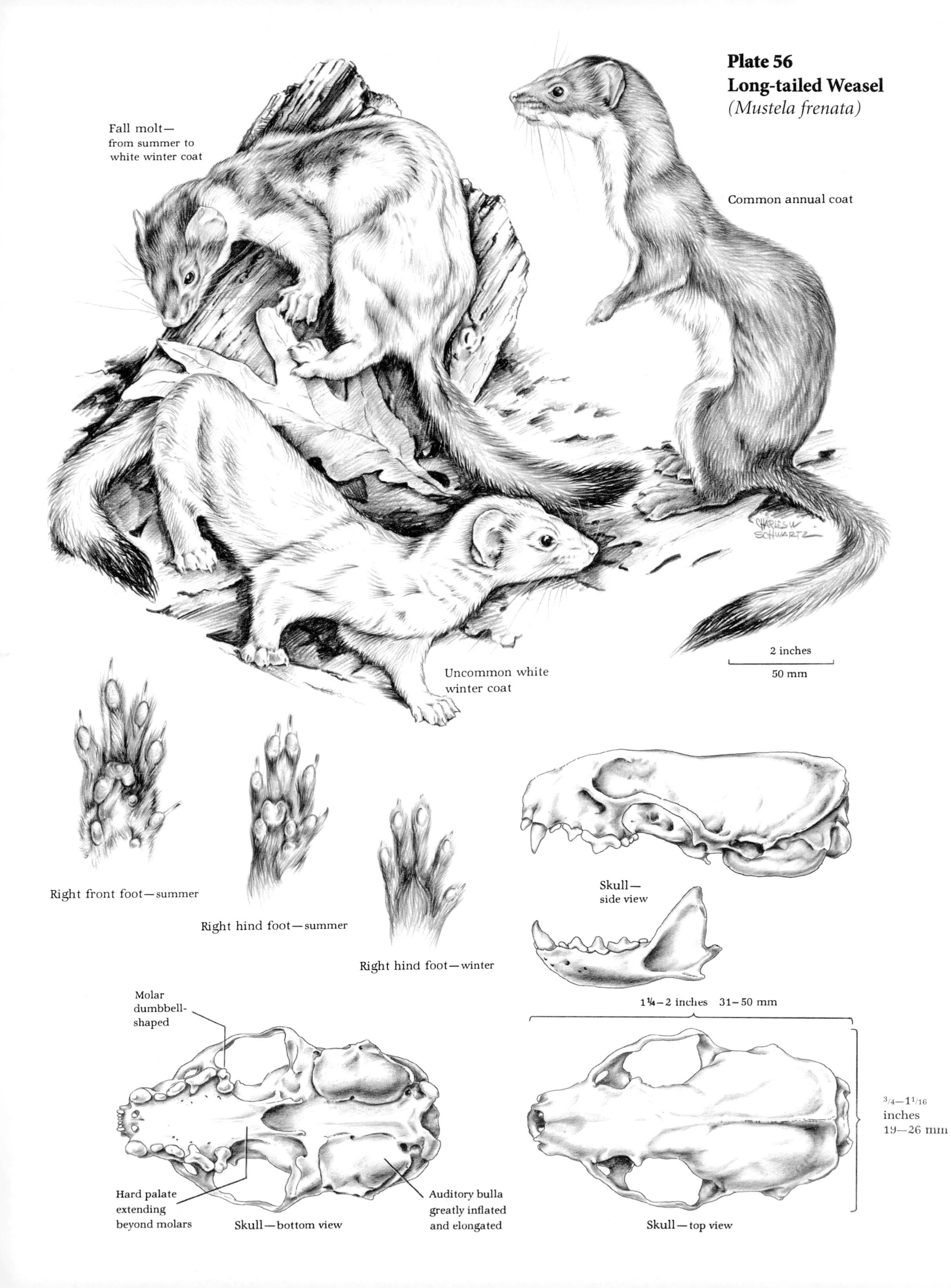

Plate 56
Long-tailed Weasel
(Mustela frenata)
Fall molt—
from summer to
white winter coat
Common annual coat
Uncommon white
winter coat
2 inches
50 mm
Right front foot—summer
Right hind foot—summer
Right hind foot—winter
Skull—
side view
1¼–2 inches 31–50 mm
Molar
dumbbell-
shaped
Hard palate
extending
beyond molars
Skull—bottom view
Auditory bulla
greatly inflated
and elongated
Skull—top view
3/4—1 1/16
inches
19—26 mm

underparts. The tail is brown except for the terminal third or fourth, which is black. The feet are usually brown but the light color of the underparts may extend onto them. In daylight the eyes are black, but at night in the glare of a spotlight they reflect a brilliant emerald green. The usual winter coat of long-tailed weasels in Missouri is merely paler than the summer one, but an occasional weasel in northern Missouri, like others in more northern latitudes or higher altitudes, has an all-white coat except for the tail tip, which remains black.

Each of the two annual molts takes 3 to 4 weeks for completion. The fall molt starts on the belly and works upward, while the spring molt begins on the back and grades downward. Those individuals that become white in winter have a mottled appearance during the fall molt but show a sharp line between the new brown fur and old white fur during the spring molt.

Measurements

Males		
Total length	34–44 cm	13½–17½ in.
Tail	114–158 mm	4½–6¼ in.
Hind foot	38–50 mm	1½–2 in.
Ear	19 mm	¾ in.
Skull length	44–50 mm	1¾–2 in.
Skull width	22–26 mm	⅞–1¹⁄₁₆ in.
Weight	170–269 g	6–9½ oz.
Females		
Total length	29–39 cm	11½–15½ in.
Tail	79–127 mm	3⅛–5 in.
Hind foot	25–38 mm	1–1½ in.
Ear	15–19 mm	⅝–¾ in.
Skull length	31–44 mm	1¼–1¾ in.
Skull width	19–23 mm	¾–⁵⁄₁₆ in.
Weight	70–127 g	2½–4½ oz.

Teeth and skull. The dental formula of the long-tailed weasel is:

$$I\ \frac{3}{3}\ C\ \frac{1}{1}\ P\ \frac{3}{3}\ M\ \frac{1}{2} = 34$$

The skull can be identified by the following combination of characteristics: size; number of teeth and the typical dumbbell-shaped upper molars; hard palate extending beyond the upper molars; and auditory bulla surrounding the inner ear greatly inflated and noticeably longer than wide. The long-tailed weasel's skull is distinguished from the very similar mink's skull by its smaller size and proportionately larger auditory bulla, which is longer than the upper row of premolar and molar teeth. From the least weasel, it is told by its larger size, narrower braincase, and flatter upper surface of the braincase.

Sex criteria and sex ratio. In addition to the larger body size, adult males are identified by the penis and scrotum. The testes descend into the scrotum in young males and remain there throughout life. Pelts of males are identified by the penis scar. Females have 4 or 5 pairs of teats on the belly toward the rear.

In trapped adults, there are usually more males than females, yet the sex ratio is believed to be equal. Males are more likely to be trapped than females because their greater weight springs the large traps set for other furbearers and because they range more widely.

Age criteria and longevity. Young long-tailed weasels are much like adults in coloration. The small testes distinguish young born in the spring from adult males until September when the larger testes of adults become inactive and diminish in size. Although adult males can be identified by the greatly expanded base of the penis bone, or baculum (that of a young male has no such base), the weight of this bone is the best means of age determination. The baculum of a young male, until the animal becomes 11 months old, weighs less than 28 mg (¹⁄₁₀₀₀ oz.), while that of an adult weighs over 56 mg (²⁄₁₀₀₀ oz.).

Young females reach adult weight and sexual maturity in 3 or 4 months and can be distinguished from adult females during summer and fall by the lack of well-developed teats. After the winter coat grows in, the teats of adults are well hidden, and then it is not possible to distinguish the two age groups by this method.

In the wild, the life span is probably very short; in captivity, it may be 5 years or longer.

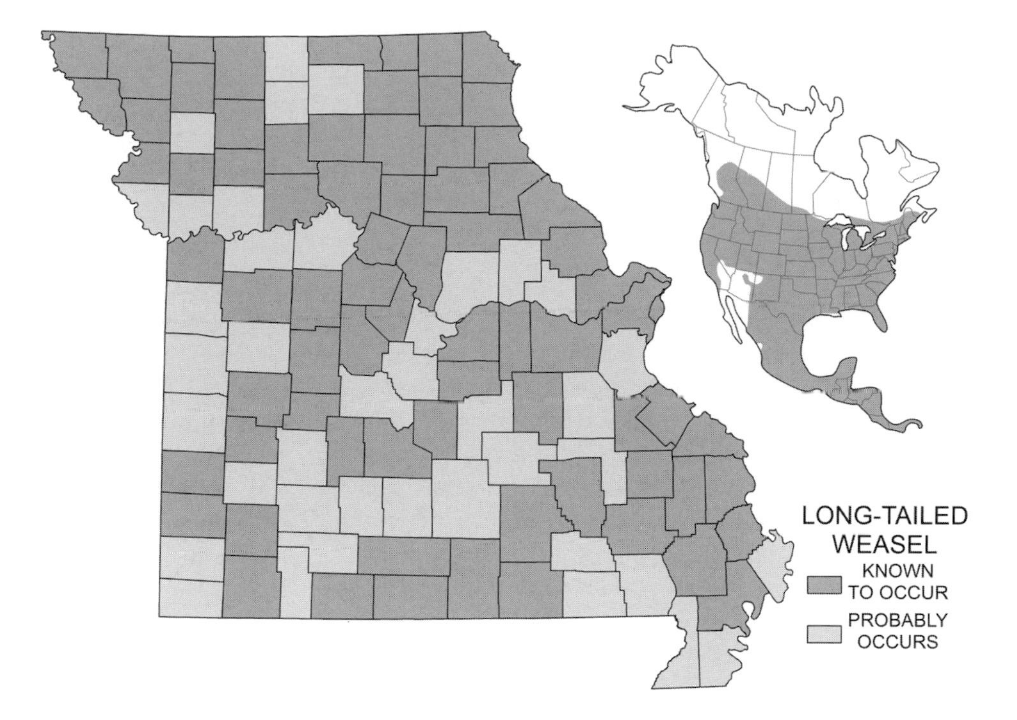

Glands. A pair of musk glands in the anal region secretes an extremely nauseating odor. These glands first begin to function when weasels are about 6 weeks old.

Voice and sounds. Long-tailed weasels purr and call a characteristic *took-took-took* rapidly and often. They hiss when hunting or disturbed and give a raucous screech when alarmed or attacking an aggressor. A female approached or pursued by a male gives a high-pitched reedy note. A squeal indicates distress. When annoyed, weasels may stamp their hind feet.

Distribution and Abundance

The range of the long-tail weasel includes much of North America, extending from western Canada throughout most of the United States, and south throughout Mexico and Central America. Long-tailed weasels also occur in northern South America. In general, they are numerous throughout the range when mice, their major food, are plentiful; they are less abundant when mice are scarce.

Long-tailed weasels range throughout Missouri but are generally rare. They are a vulnerable species of conservation concern in Missouri but are secure across their entire range.

Habitat and Home

Long-tailed weasels live in a variety of habitats but prefer woodlands, brushy fencerows, and thickets along watercourses. Their home is a shallow burrow that was usually the former abode of a mole, ground squirrel, or mouse. Weasels may also live in rock piles, under the roots of trees, in dense brushy vegetation, and, on occasion, in an old barn if mice are plentiful there. Inside the burrow they construct a nest of mouse and rabbit fur, grass, and sometimes feathers. A special place near the entrance may be used for a latrine. Usually a cache of food or a pile of discarded bones is located within the burrow. Depending upon conditions of their habitat, weasels have one or several homes.

Habits

These mammals are very suspicious and inquisitive and are continually investigating their surroundings. They hunt during both day and night but are abroad more at night. Home ranges can vary in size due to the density of prey species. Males have larger home

ranges than females; during the breeding season they increase in the size of their ranges in order to come into contact with more females. Home ranges have been reported to vary from 10 to 160 ha (25 to 395 ac.). So persistent are long-tailed weasels in their searchings that in a single night they may actually travel linear distances up to 5.6 km (3½ mi.) and still not go far from the den. Males tend to range farther than females. In some parts of their range, they may have a large hunting circuit, about 3.2 km (2 mi.) in length and 1.6 km (1 mi.) in width. They may spend 7 to 12 days covering this route.

In spite of their small size, long-tailed weasels are extremely aggressive and fearless and may attack animals larger than themselves. In stalking, a weasel waves its head from side to side in an effort to detect a scent since it seems to rely more upon its sense of smell than upon sight. When within pouncing distance of a small animal, such as a mouse, the weasel rushes and kills its victim with one swift bite, usually in the head region. If the mouse struggles, the weasel may embrace it with all four feet. When taking large prey, such as a rabbit, the weasel attacks so quickly that the victim is taken off guard. The weasel grasps the nearest part, then literally climbs onto the rabbit, maintaining a tight hold with both front and hind feet during the ensuing struggle, even though the contestants may roll over and over. While still clinging to its prey, the weasel maneuvers until it is in a position to inflict a fatal bite. So rapidly does the weasel bite the back of the neck or jugular vein that the action can hardly be followed. But in spite of the weasel's tenacity, the larger animal frequently dislodges its attacker and escapes.

Long-tailed weasels normally bound or lope along with the back arched and the tail extended straight out or slightly raised. Less often do they run or walk. Weasels swim well and climb trees easily. Their agility and speed enable them to follow their prey over all sorts of terrain regardless of obstacles, and their small girth, about 3.8 cm (1½ in.) in diameter, permits them to pass through tiny knotholes, crevices, and tunnels in hunting or in pursuing victims.

Long-tailed weasels are active all year and show no more tendency to "hole up" during winter than at any other time. On occasion they may stay in their dens for 24 to 48 hours regardless of weather conditions. Weasels are mostly solitary, but sometimes two individuals may frolic together.

Foods

Long-tailed weasels eat animal food entirely, preferring their prey alive and quivering. The only carrion consumed consists of victims they have stored in their burrows. As long as rodents are available, they are eaten almost exclusively. The major food items are mice, rats, squirrels, chipmunks, shrews, moles, and rabbits. Males are more likely to take larger prey, such as rabbits, than females. Occasionally small birds, bird eggs, reptiles, amphibians, earthworms, and some insects are eaten. They have even been observed to take bats from a nursery colony in an old barn.

After an animal is killed, the blood may be licked at the wound, but it is not sucked as is commonly believed. All the fur, feathers, and bones of small prey are consumed along with the flesh, but usually only some of the flesh is taken from large prey. When a nest is pilfered, one egg is removed at a time and carried in the mouth to some nearby cover where the long-tailed weasel bites off the top and licks out the contents.

Long-tailed weasels are voracious killers and often kill more than they can eat. This excessive take usually occurs in the spring when the young are being fed, and again in the fall. Part of the surplus kill in fall may be stored in the den or in the ground, but even a great deal of this spoils without being eaten. In captivity, young animals eat from one-fourth to one-half of their body weight in 24 hours while adults eat only about one-fifth to one-third of their weight in the same period. In general, more food is taken in summer than winter. Drinking water is essential both in the wild and in captivity.

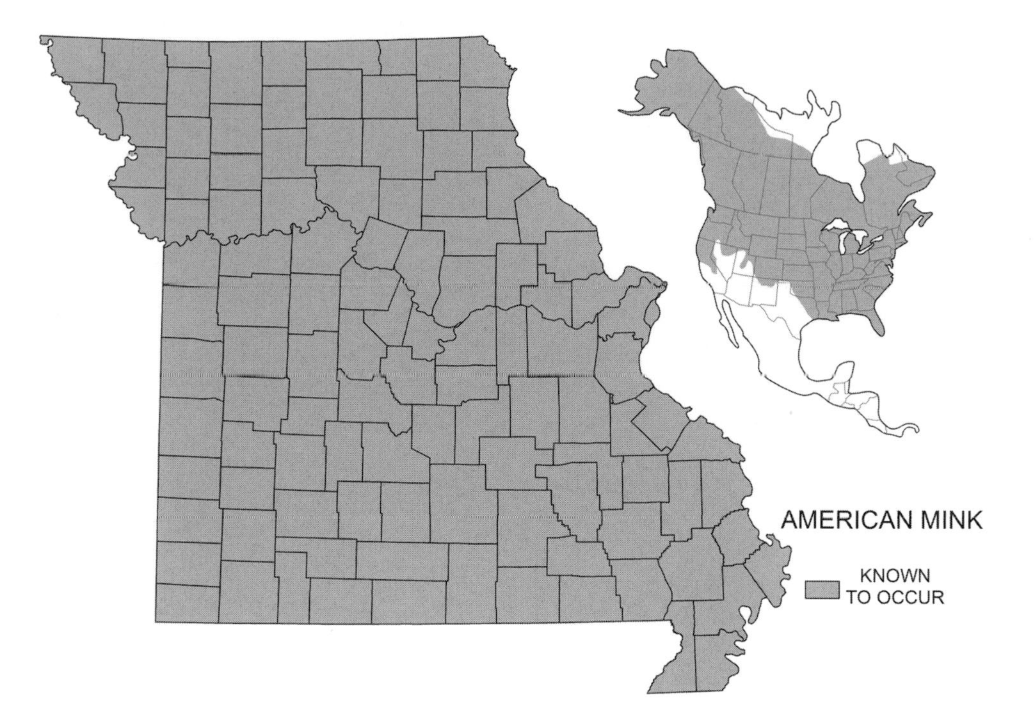

Glands. A pair of musk glands in the anal region, typical of the weasel family, secretes a strong odor, considered by many to be more obnoxious than that of either weasel or skunk. This odor is given off particularly during the breeding season but also at any period of intense excitement.

Voice and sounds. Minks are not noisy but do have a varied vocabulary of chuckles, growls, barks, hisses, squeals, and screeches.

Distribution and Abundance

The mink ranges from Alaska, through almost all of Canada, and most of the United States. It does not occur in the most arid areas of the southwestern United States. As a result of escapes and deliberate introductions in the mid-20th century, the mink has expanded its range to many parts of Europe and South America.

Minks are found throughout Missouri but are most common in the Mississippi River Alluvial Basin where there is a systematic network of drainage canals and ditches. In some places and years, populations vary from 3 to 8 minks per sq km (8 to 22 per sq. mi.).

Habitat and Home

The one basic requirement for mink habitat is permanent water. The presence of standing timber adjacent to water is attractive but not necessary for their environment. Minks dwell along the banks of streams and rivers or shorelines of lakes and marshes. The increase in number of farm ponds in Missouri has provided additional habitat.

Minks make their homes under the roots of trees, in cavities in banks, under logs or stumps, in hollow trees, or in muskrat burrows and lodges, usually within 183 m (600 ft.) of open water. A tunnel, 10 to 15 cm (4 to 6 in.) in diameter and about 30 cm (1 ft.) deep, leads to the nest chamber 25 to 30 cm (10–12 in.) in diameter. This chamber, which may have several entrances, contains grass or leaves and usually fur or feathers.

Habits

Male minks travel widely. They have a large home range, with an average linear range over 72 km (45 mi.), and take approximately 2 weeks to cover the entire area. Within this range there is a series of temporary homes used variously for a day or two at a time. The territories of different males sometimes overlap, and several males may use the different dens in succession. Adult female minks have a small home range, between 8 and 20 ha (20 and 50 ac.), and usually occupy only 1 or 2 homes throughout the year. However, a young female followed by radio tracking used 20 different dens in a month's time. The average straight-line distance between consecutive dens was 355 m (1,164 ft.). Although tending to stay along streams or lakes, minks may cut across country from one body of water to another. One wide-ranging mink is recorded to have traveled 24 km (15 mi.) in a night.

Minks are chiefly nocturnal but often come out at dawn or dusk and less frequently during the day. They are active all year, although during periods of low temperatures or following snow they may stay in their dens and sleep for a day or so. When snow is on the ground, minks have been observed to slide down slopes on their bellies in a fashion similar to the well-known antics of their relative, the river otter. Minks are not social animals and live alone except during the season when young are being raised. Their droppings and anal scent are left on flat stones or logs along watercourses.

Minks are well adapted to life both on land and in water. They usually walk or take low bounds of 25 to 60 cm (10 to 24 in.) in length, arching their backs at each bound. They occasionally run at a speed between 11 and 13 km (7 and 8 mi.) per hour. In order to obtain a better view of their surroundings, they often rear up on their hind legs. When disturbed, minks dodge into the nearest shelter and sometimes escape by climbing trees. They have been observed to swim underwater for 15 m (50 ft.) and to swim on the surface at the rate of 1.6 to 2.4 km (1 to 1½ mi.) per hour. On occasion they float downstream curled up in a ball, apparently asleep.

Minks are very aggressive and often attack animals larger than themselves. They are good fighters and put up a good defense. Their eyesight is not acute, and they rely heavily on their sense of smell to locate prey. Most of their victims are killed by a bite in the neck region. Their agility underwater permits them to pursue and easily capture fish.

Foods

Just as the mink is semiaquatic in its habits, dwelling on land like the weasel and in water like the river otter, so is it intermediate between these two other members of the mustelid family in its food habits. The mink preys upon mice, rabbits, and other terrestrial animals, as does the weasel, and feeds on fish, crayfish, and other aquatic forms, as does the river otter. The principal foods of 372 minks taken by trappers in Missouri are given here by percentage of the total volume: frogs, 24.9; mice and rats, 23.9; fish, 19.9; rabbits, 10.2; crayfish, 9.3; birds, 5.6; fox squirrels, 2.2; and muskrats, 1.3. Miscellaneous items (2.7 percent) include insects, spiders, snails, domestic cats, shrews, moles, bats, turtles and their eggs, snakes, bird eggs, blood, grass, and leaves. The mink is also an important predator of waterfowl and their eggs during the breeding season.

Minks do not kill wantonly. Most food is preferably carried to a den where it is eaten. The surplus is cached in the den but frequently spoils and is not used.

Reproduction

The breeding season begins in late February, and matings occur until early April. Pregnancy varies from 40 to 75 days with an average of 51. This pattern

Mink tracks

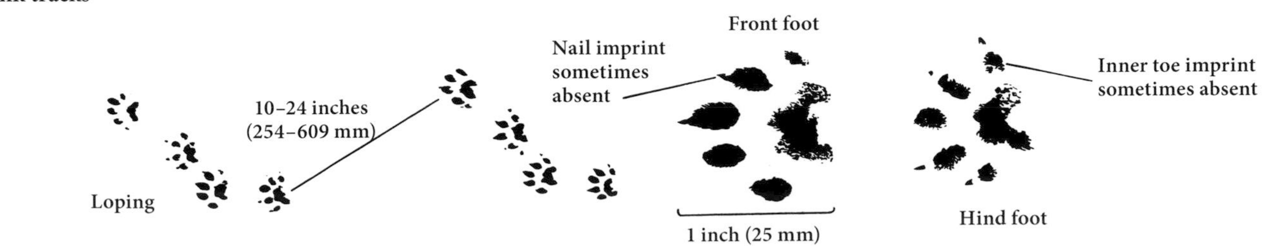

Minks at birth

of reproduction in the mink is somewhat similar to that in the long-tailed weasel because both of these species carry their young in a state of suspended development for some time. After fertilization there is a short period of development, followed by a variable period of dormancy when the embryos are free in the uterus. From 28 to 30 days before birth, the embryos become implanted in the uterus and complete their development. The duration of pregnancy is influenced by several factors. In general, the earlier in the breeding season the female mates, the longer is her pregnancy; the later she conceives, the shorter is her pregnancy. Furthermore, larger litters have a tendency to be born sooner than smaller litters, and yearling females have slightly longer pregnancies than older ones.

The single annual litter is born around the first of May. From 2 to 8 young may compose this litter, but the average is 4 or 5. At birth the kits are about 9 cm (3½ in.) long and weigh 6 g (⅕ oz.). They are blind, toothless, and may either be naked or possess a fine covering of silvery white hair. Males are only slightly larger than females at birth but become noticeably larger as they grow older. At 2 weeks of age the fur becomes pale reddish gray. At 3 weeks of age, the teeth begin to cut through the gums. The young resemble the adults now but soon become darker than their parents. The eyes open when the kits are about 5 weeks old, and weaning begins at this time. At 7 weeks, the permanent teeth are coming in.

Although a male may mate with many females, he usually stays with the last one and assists in caring for the young. When the young are between 6 and 8 weeks of age, the adults take them on foraging trips for the first time. On land, the parents occasionally carry the kits by the scruff of the neck; in water they frequently let the kits ride on their backs. The family stays together until the end of August when all go their own way. Both sexes reach maturity at approximately 10 months of age.

Some Adverse Factors

The principal predators on minks are humans, domestic dogs, owls, foxes, coyotes, and bobcats. The following parasites occur in or on minks: roundworms, flukes, tapeworms, protozoa, mites, lice, fleas, and flies. Minks are known to have convulsions and other nervous disorders. Anthrax, distemper, and encephalitis are diseases occurring in fur-farm minks.

Additionally, minks can be quite sensitive to pollution of aquatic environments.

Importance

Mink fur is durable and of excellent texture, but most mink pelts used in the garment industry today are from ranched minks. Pelts are made into coats, jackets, hats, gloves, scarves, or trimming. The harvest record from trapping shows a fairly consistent decline in the average annual number of pelts from 13,528 during the 1940s, to 13,079 in the 1950s, 7,747 in the 1960s, and 5,365 in the 1970s, but then harvest increased briefly to 5,716 in the 1980s, and then declined again to 2,661 in the 1990s and 1,196 in the first decade of the 2000s. The highest annual take was 22,658 in 1946 and the lowest was 702 in 2008. The average annual price paid per pelt was $13.30 in the 1940s, $15.25 in the 1950s, $7.14 in the 1960s, $11.33 in the 1970s, $20.00 in the 1980s, $12.83 in the 1990s, and $10.00 in the first decade of the 2000s. The highest annual average price paid per pelt was $29.69 in 1988 and the lowest was $3.95 in 1970. It is hard to measure the relative effects of the several factors that influence harvest numbers and pelt prices. Obviously, population abundance is involved, but to what extent has not been determined. Changes in demand for furs and the fluctuating values of their pelts influence economic reasons for trapping mink. In addition, the numbers of trappers working in Missouri influence the harvest. The rearing of minks in captivity for fur production is practiced on a limited scale in this state.

In general the mink's food habits are neither beneficial nor harmful to human interests. At times minks prey heavily on muskrats, but this occurs primarily when the muskrat population is overcrowded, suffering from drought, or enduring some other adverse factor. Minks occasionally take chickens, but as a rule they are not destructive. If you are experiencing problems with mink, contact a wildlife professional for advice, assistance, regulations, or special conditions for handling these animals.

Conservation and Management

Regulation of the harvest must be continued in accordance with the mink population density. Maintaining logjams and brush piles along streams will attract minks.

SELECTED REFERENCES

See also discussion of this species in general references, page 23.

Abramov, A. V. 2000. A taxonomic review of the genus *Mustela* (Mammalia, Carnivora). *Zoosystematica Rossica* 8:357–364.

Allen, J. A. 1940. *The principles of mink ranching*. Canada Department of Mines and Natural Resources, Game and Fisheries Branch. 148 pp.

Arnold, T. W., and E. K. Fritzell. 1990. Habitat use by male mink in relation to wetland characteristics and avian prey abundances. *Canadian Journal of Zoology* 68:2205–2208.

Bassett, C. F., and F. M. Llewellyn. 1949. The molting and fur growth pattern in the adult mink. *American Midland Naturalist* 42:751–756.

Buskirk, S. W., and S. L. Lindstedt. 1989. Sex biases in trapped samples of Mustelidae. *Journal of Mammalogy* 70:88–97.

Enders, R. K. 1952. Reproduction in the mink (*Mustela vison*). *Proceedings of the American Philosophical Society* 96:691– 755.

Erickson, D. W. 1978. *Muskrat, raccoon, and mink productivity research*. Federal Aid Project no. W-13-R-32. Missouri Department of Conservation, Jefferson City. 18 pp.

———. 1979. Studies in the age and sex compositions of Missouri raccoon, muskrat, and mink harvests. Midwest Furbearer Workshop, Manhattan, Kansas. Mimeograph. 15 pp.

Errington, P. L. 1943. An analysis of mink predation upon muskrats in north-central United States. *Iowa State College of Agriculture and Mechanical Arts, Research Bulletin* 320:798–924.

Greer, K. R. 1957. Some osteological characters of known-age ranch minks. *Journal of Mammalogy* 38:319–330.

Hamilton, W. J., Jr. 1940. The summer foods of minks and raccoons on the Montezuma Marsh, New York. *Journal of Wildlife Management* 4:80–84.

Hansson, A. 1947. The physiology of reproduction in mink (*Mustela vison* Schreb.) with special reference to delayed implantation. *Acta Zoologica* 28:1–136.

Korschgen, L. J. 1958. December food habits of mink in Missouri. *Journal of Mammalogy* 39:521–527.

Kurose, N., A. V. Abramov, and R. Masuda. 2008. Molecular phylogeny and taxonomy of the genus *Mustela* (Mustelidae, Carnivora), inferred from mitochondrial DNA sequences: New perspectives on phylogenetic status of the black-striped weasel and American mink. *Mammal Study* 33:25–33.

Larivière, S. 1999. *Mustela vison. Mammalian Species* 608. 9 pp.

Marshall, W. M. 1936. A study of the winter activities of the mink. *Journal of Mammalogy* 17:382–392.

Mitchell, J. L. 1961. Mink movements and populations on a Montana river. *Journal of Wildlife Management* 25:48–54.

Osowski, S. L., L. W. Brewer, O. E. Baker, G. P. Cobb. 1995. The decline of mink in Georgia, North Carolina, and South Carolina: The role of contaminants. *Archives of Environmental Contamination and Toxicology* 29:418–423.

Petrides, G. A. 1950. The determination of sex and age ratios in fur animals. *American Midland Naturalist* 43:355–382.

Schladweiler, J. L., and G. L. Storm. 1969. Den-use by mink. *Journal of Wildlife Management* 33:1025–1026.

Sealander, J. A. 1943. Winter food habits of mink in southern Michigan. *Journal of Wildlife Management* 7:411–417.

Snyder, D. H. 1962. Aging wild Missouri mink by examination of the baculum. M.S. thesis, University of Missouri, Columbia. 73 pp.

Stevens, R. T., T. L. Ashwood, and J. M. Sleeman. 1997. Fall–early winter home ranges, movements, and den use of male mink, *Mustela vison,* in eastern Tennessee. *Canadian Field-Naturalist* 111:312–314.

Svihla, A. 1931. Habits of the Louisiana mink (*Mustela vison vulgivagus*). *Journal of Mammalogy* 12:366–368.

American Badger (*Taxidea taxus*)

Name

The first part of the scientific name, *Taxidea*, is from Latin and Greek origins and means "badger-like" (*taxus*, "badger" and *eidos*, "like"). The last part, *taxus*, is the New Latin word for "badger"; it was originally given to the European badger, an animal somewhat resembling the American badger. The first part of its common name refers to its occurrence in North America, and "badger" is thought to be derived from the French word *bageard*, referring to the animal's badgelike white blaze.

Description (Plate 58)

The American badger is a heavy-bodied, medium-sized mammal with a broad head; a short, thick neck the same width as the head; short legs; and a short, bushy tail. The ears are low and rounded. There are 5 toes with prominent claws on each foot, the claws on the front feet being very long and exceeding 2.5 cm (1 in.) in length. The fur is rather short on the back but moderately long on the sides, which contributes to the badger's characteristic squat appearance. The skin is somewhat loose and tough. A third eyelid, or nictitating membrane, protects each eye and is likely an adaptation to life underground.

Color. The general coloration is gray with a slight yellowish cast. Each hair on the back is yellowish white at the base, then banded with black and tipped with white. The brownish face is marked with a white stripe reaching from near the nose to the crown of the head and sometimes onto the neck and back. Paired white areas extend from around the mouth onto the cheeks and insides of the ears, and a prominent, vertical black bar, or "badge," occurs in front of each ear. The backs of the ears and the feet are black; the tail is yellowish brown; and the underparts are predominantly yellowish white. The sexes are colored alike and show no marked change in coloration with the seasons. The annual molt in adults occurs in summer or fall. It begins on the head and progresses toward the tail along the middle of the back.

Measurements

Total length	61–89 cm	23¾–35 in.
Tail	98–174 mm	3⅞–6⅞ in.
Hind foot	88–155 mm	3½–6⅛ in.
Ear	50–53 mm	2–2⅛ in.
Skull length	107–130 mm	4¼–5⅛ in.
Skull width	77–88 mm	3¹⁄₁₆–3½ in.
Weight	5.9–13.6 kg	13–30 lb.

Males are heavier than females. In 2011, a 12.2 kg (27 lb.) male became the official record-weight badger confirmed by the Missouri Department of Conservation. The state furbearer record program was started in 2011.

Teeth and skull. The dental formula of the badger is:

$$I\ \frac{3}{3}\ C\ \frac{1}{1}\ P\ \frac{3}{3}\ M\ \frac{1}{2} = 34$$

This is the same as for all other members of the weasel family in Missouri except the river otter.

The skull can be identified by the following combination of characteristics: size; number of teeth and the typical triangular upper molars; triangular braincase; hard palate extending beyond the upper molars; and auditory bulla surrounding the inner ear moderately inflated and elongated.

Sex criteria and sex ratio. Males are identified by the penis and scrotum. Females have four pairs of teats. The sexes are approximately equal in number.

Age criteria, age ratio, and longevity. The young are similar to adults but are generally less grizzled. The age of males can be determined by the penis bone, or baculum. The short (6 to 8.2 cm; 2⅜ to 3¼ in.), lightweight baculum of a male less than 1 year old has only a shallow groove and slightly enlarged base, while the long (9.5 to 10.1 cm; 3¾ to 4 in.), heavy baculum of an adult has a prominent groove and much enlarged base that is often sharply ridged.

As in other carnivores, the dried weight of the eye lens can be used to estimate age, along with wear on the canine teeth and the closure of certain bones of the skull. All of these, however, place animals in age classes only. The most promising technique for estimating the age of individual animals is the count of rings in the cementum layers of the upper or lower canine teeth. This correlates well with body weight.

Pelts of adult, bred females have large teats (3 mm; ⅛ in. in diameter, and 3 mm to 9.5 mm; ⅛ to ⅜ in. long) that are very obvious in the sparse belly hair, while those of young, unbred females have small teats that are difficult to locate in the dense belly hair. The number of adults to young of the year is approximately equal.

In captivity, American badgers have lived up to 15 years of age. In the wild, the maximum life span has been reported from 4 to 10 years. It takes approximately 7 years for the population of an area to be replaced by new individuals.

Glands. American badgers have two pairs of scent glands, one pair on the belly and the other pair near the anus. The latter, or anal scent glands, secrete a strong but inoffensive odor. To release this scent the badger must raise its tail, but it cannot direct the scent stream accurately as do the skunks. These glands are probably used mostly during the breeding season for sexual attraction, and at other times for defense.

Voice and sounds. American badgers hiss, grunt, growl, and snarl when fighting or cornered by an aggressor. They have been heard to purr in captivity.

Distribution and Abundance

The American badger occurs throughout most of south-central Canada and into the central and western United States, and central Mexico. The present range

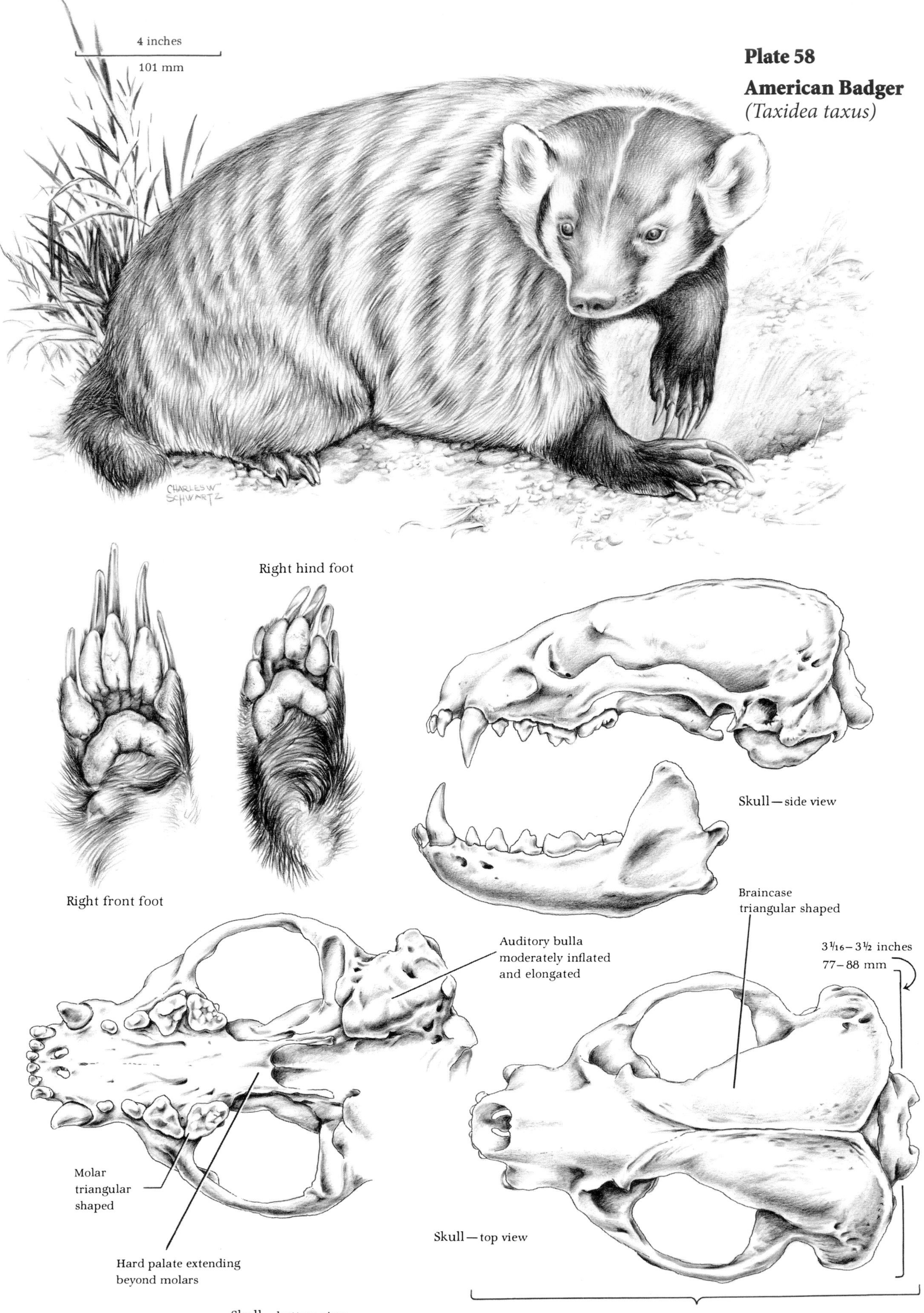

Plate 58

American Badger

(Taxidea taxus)

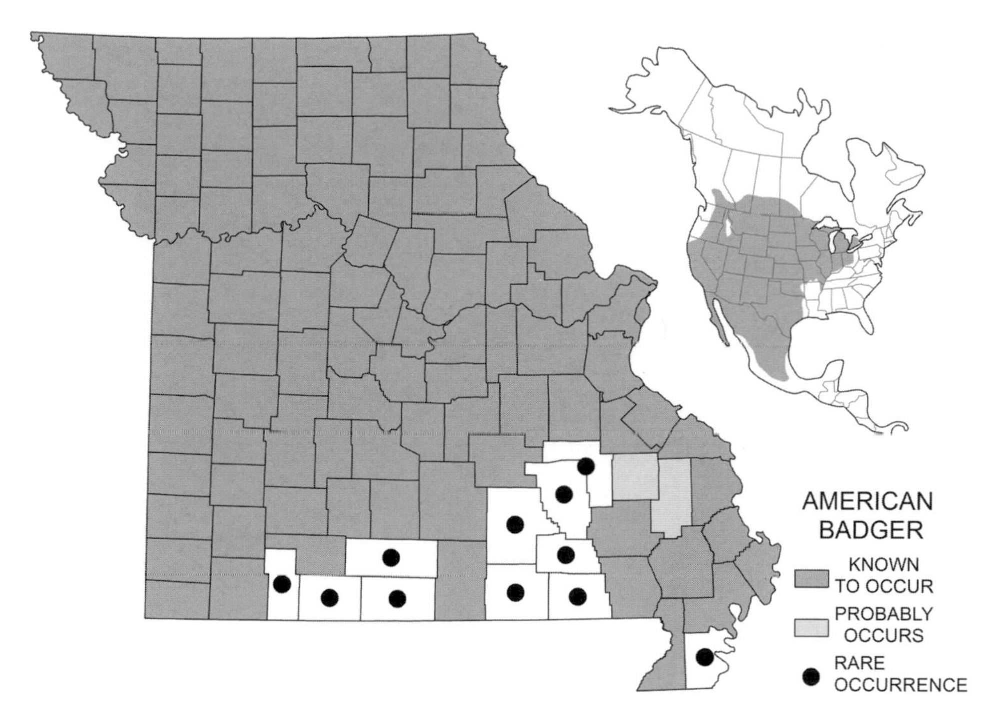

and numbers in the western United States are somewhat reduced from those of former times. The reasons for this are the plowing of prairies, which has restricted the habitat; the extensive poisoning campaigns for burrowing rodents, which eliminated the food supply; and the heavy trapping and poisoning of badgers themselves. In the East, however, badgers are extending their range and numbers; this may be related to lumbering operations that make more habitat available. A recent expansion into northeastern Arkansas is believed to have originated from Missouri badgers inhabiting the Mississippi River Alluvial Basin.

The American badger was probably never common in Missouri in early times and by 1900 had practically disappeared from the state. It was noted in western Missouri in 1926 and in the northern half of the state in 1938. The badger is uncommon throughout most of the state but is most common in the Central Dissected Till Plains of northern and Osage Plains of western Missouri. The badger is listed as a species of conservation concern in Missouri but it is secure across its North American range.

Habitat and Home

Throughout most of its North American range the American badger prefers open country, living in the prairies and plains where ground squirrels, prairie dogs, and other burrowing animals, their principal foods, abound. They find shelter along roadways, fencerows, ditches, banks, and field edges.

In northern and western Missouri, badgers occur primarily in prairies, other open grasslands, and croplands. They also inhabit the sand prairies of the Mississippi River Alluvial Basin in southeastern Missouri. Other open areas with grasslands are also used, including fields, pastures, yards around homes, parklands, and farms, wherever there is a good supply of rodent prey.

During most of the year the badger's home is a shallow burrow about 30 cm (1 ft.) in diameter, dug by the badger in its daily quest for food. During the breeding season, a deeper burrow from 1.5 to 9 m (5 to 30 ft.) in length is built, leading to an enlarged chamber 0.6 to 0.9 m (2 to 3 ft.) below the surface. This chamber is wider than high, reflecting the squat shape of the

animal, and contains a nest of grass. The dens usually have a single entrance marked by a mound of soil in front. The entrance is often partially plugged with loose soil.

Habits

American badgers commonly spend their entire lives within a small area, but, during the breeding season or if food is scarce, they may cover a much wider territory. Adult males have an average home range of about 6.5 sq km (2½ sq. mi.), although they are known to have taken trips from 0.8 to 2 km (½ to 1¼ mi.) out of their home range on occasion and returned. Males defend their home ranges from other males, but their ranges overlap with those of females. Adult females have a home range of approximately 2.5 sq km (1 sq. mi.); the ranges overlap in some locations. One female captured in a Missouri homeowner's backyard and relocated traveled 5 km (3 mi.) back to her den.

Badgers normally do not go far from some hole into which they can make a hasty retreat. They have a series of dens in their home range, which they use at different times, but they seldom return to the same one on consecutive days. Where home ranges overlap, dens may be used by different badgers on different days.

Badgers are active mostly at night but sometimes forage in early morning or late evening and sunbathe near the entrance of their burrows.

In the northern parts of their range, badgers acquire a store of body fat during late summer that serves as insulation and a reserve of energy during winter when they spend considerable time (up to 21 continuous days) sleeping in their burrows. Energy is conserved during these inactive periods, when the body temperature falls slightly and the heartbeat slows to half its normal rate. This state of torpor results in conditions not found in truly hibernating mammals. Badgers frequently come out of their burrows in winter, seek and dig up some hibernating animal upon which they feed, and then return to their burrows for another period of sleep. In the southern parts of their range, they are active all winter.

An adult female was monitored by radio from July to January in Minnesota. During late summer, she occupied 761 ha (1,880 ac.) and used 46 different dens of which only 2 were reused during this period. The average distance between these dens was 1.1 km (7⁄10 mi.). During fall, her nightly activity lessened and she used only 53 ha (130 ac.) where there was a dense population of pocket gophers. Many dens were reused, and the average distance between them was 0.2 km (1⁄10 mi.). Winter showed even greater restrictions. Only 1 den was used, nightly travels were infrequent, the area covered was 2 ha (5 ac.), and the most extensive hunting trip was 274 m (300 yd.).

Badgers are excellent diggers, their heavy body, powerful muscles, strong front feet, and long claws being well adapted for this activity. Since they secure a large part of their food by digging, it is often essential that they dig faster than their burrowing prey. Furthermore, the rapidity with which they dig helps them escape a pursuing enemy. Badgers dig at a faster rate than a person can dig with a shovel. In digging, badgers loosen the soil with their front feet, pass the soil under the belly, and kick it out of the hole with their hind feet. Sometimes the soil is removed so vigorously that it is thrown 1.2 to 1.5 m (4 to 5 ft.) into the air. The mouth may also aid in digging. Badgers often clean their claws and toes to remove the soil and may sharpen their claws on trees. Their diggings are the most conspicuous sign of their presence; where badgers are abundant, burrows are numerous.

Badgers apparently select a ground squirrel burrow with evidence of a lot of use as a place where good returns can be obtained from a minimum of digging effort. But they do not always dig up the burrow; instead, they have been reported to enter a ground squirrel tunnel, partially block up the entrance, and lie in wait for the occupant to return.

On the ground badgers walk or run, attaining a topmost speed of 16 to 24 km (10 to 15 mi.) per hour. They can walk backward for short distances as well as forward and can enter a hole either way. Once in a burrow they are difficult to dislodge unless flooded out with water. These animals are capable swimmers and have been found as far as a 0.8 km (½ mi.) from shore.

Badgers usually dig a shallow hole for their droppings and cover them with soil. In late summer, when digging is difficult in the dry, hard soil, the droppings are commonly left on the surface of the ground or merely covered by a tuft of sod.

Adults generally lead solitary lives. However, in the western part of the range, single badgers and coyotes have been reported as sometimes hunting together. The crafty coyote profits from this relationship since, in the course of a badger's digging operations, an elusive rodent may get away only to be pounced upon by the agile coyote. In reciprocation, the presence of the coyote at various burrow entrances increases the success of the badger by preventing the escape of weary ground squirrels. Hawks have also been observed to follow a badger, presumably on the alert for some escaping rodent.

Badgers are difficult to raise in captivity because of their tremendous appetites. They are strong fighters

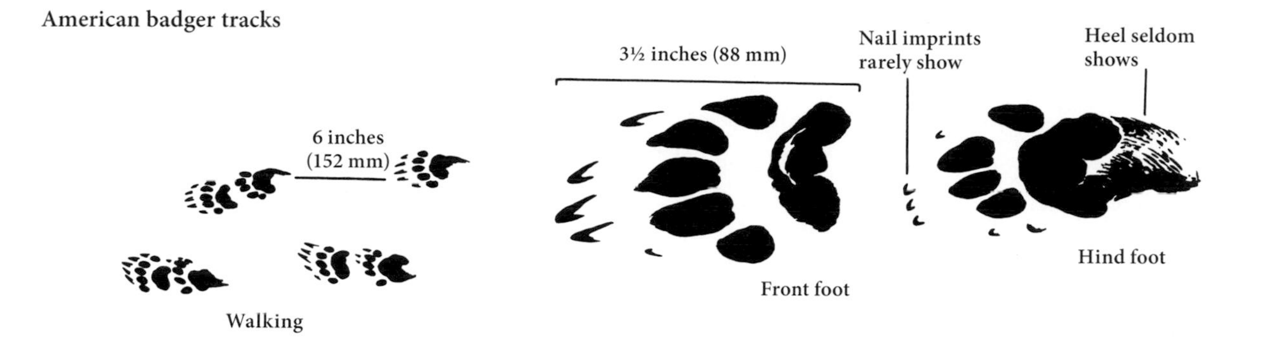

and can easily defeat most dogs. An adversary has difficulty in overcoming a badger because of the capable defense of the badger's strong claws and sharp teeth. The loose but tough skin and thick fur of the badger also prevent an enemy from securing a crippling or lethal grip.

Foods

Badgers are strictly animal eaters, their most important foods being rodents and rabbits. Ground squirrels, pocket gophers, and mice are their major prey items and vary in importance based on the season; mice form the bulk of the diet when ground squirrels are hibernating or are not plentiful. Some cottontails are eaten year around; those taken in spring and summer are mostly young. Insects, including beetles, grasshoppers, and bees, are taken predominantly in late summer when the ground is dry and digging is difficult. Other animal foods include moles; lizards; snakes (including hibernating rattlesnakes, which are eaten except for their heads); turtle eggs; scorpions; ground-nesting birds and their eggs; snails; and fish, when available.

The badger catches most animals alive but will also take them as carrion. Animals that appropriate badger burrows for their homes, such as ground squirrels and rabbits, are often captured when the badger returns to look over former diggings, as is its custom. A badger prefers to drag a large animal into a burrow where it can eat leisurely in shelter. If no hole is sufficiently near, one may be dug for the occasion. Only rarely is surplus food cached.

In the wild, badgers do not seem to require water to drink since they often live far from any surface source; in captivity, they drink water regularly.

Reproduction

Mating takes place in August or September, and there is a delay in the development of the young as occurs in some other members of the mustelid family. Following fertilization of the eggs, the resulting embryos develop slightly before they become dormant in the uterus for several months. Around February, the embryos become implanted in the uterus and complete their development in approximately 5 weeks. This makes a total of only about 6 weeks for the entire period of development, although the whole gestation period may last from 6 to 9 months. The single annual litter is born in March or early April and usually consists of 2 or 3 young (with extremes of 1 to 7).

At birth the young are furred and blind. Their eyes open between 4 and 6 weeks of age. When the young are a month or so old, the mother may move them to a new den. This permits her to hunt a new area and still be near her young. Weaning occurs when the young are about half-grown, but the female continues to bring food to them for some time. They are taught to hunt when about two-thirds grown. The young molt about this time and resemble adults. They stay in and around their home burrow until fall. Some females breed before they are 1 year old, but males are 1 year old before they become sexually active.

Some Adverse Factors

Most damage caused by badgers results from their digging activities. Open burrows create a hazard for livestock and horseback riders; agricultural machinery when burrows are located in crop fields; levees, dams, dikes, and irrigation canals; and roadsides and road surfaces when they dig along the shoulders of roads. They also dig in yards around homes, golf courses, and cemeteries.

Badgers occasionally prey on livestock or poultry.

The badger is an aggressive animal with humans being their most important predator. They are occasionally killed by vehicles. Coyotes, dogs, and bobcats may take some young, and bears and mountain lions are known to kill badgers.

Badgers have the usual complement of parasites: fleas, ticks, lice, roundworms, tapeworms, and flukes. They are susceptible to rabies and tularemia.

Importance

Badger fur is used to make coats and hats, and trim cloth coats. Formerly the fur was used in making shaving brushes and the tough hide was made into rugs.

Because badgers were not plentiful in Missouri, they were protected until 1960 when the season was opened for trapping. The average annual harvest was 27 in the 1960s; 101 in the 1970s; 81 in the 1980s; 27 in the 1990s; and 46 in the first decade of the 2000s. During this five-decade period, the highest annual take was 323 in 1979 and the lowest was 3 in 1961. The average annual price per pelt was $1.14 in the 1960s; $8.83 in the 1970s; $4.70 in the 1980s; $4.15 in the 1990s; and $12.53 in the first decade of the 2000s. The highest average annual price per pelt was $30.45 in 1978 and the lowest was $0.75 in 1967. It is hard to measure the relative effects of the several factors that influence harvest numbers and pelt prices. Obviously, population abundance is involved, but to what extent has not been determined. Changes in demand for furs and the fluctuating values of their pelts influence economic reasons for trapping badgers. In addition, the numbers of trappers working in Missouri influence the harvest.

Badgers are valuable because they feed on abundant burrowing rodents; by their digging activities they also aerate and mix the soil. Their excavations are sometimes objectionable if they damage dikes in irrigation systems. Usually, however, the diggings merely represent a search for some other animal that has already burrowed into the man-made structures. In parts of their range, badgers are considered a liability because of their burrows; horses sometimes step into the holes and throw their riders.

Conservation and Management

Because of their uncommon status in Missouri and a regulated harvest, badgers need no general management in Missouri. If you are experiencing problems with badgers, contact a wildlife professional for advice, assistance, regulations, or special conditions for handling these animals.

SELECTED REFERENCES

See also discussion of this species in general references, page 23.

Crowe, D. M., and M. D. Strickland. 1975. Dental annulation in the American badger. *Journal of Mammalogy* 56:269–272.

Goodrich, J. M., and S. W. Buskirk. 1998. Spacing and ecology of North American badgers (*Taxidea taxus*) in a prairie-dog (*Cynomys leucurus*) complex. *Journal of Mammalogy* 79:171–179.

Harlow, H. J. 1979. A photocell monitor to measure winter activity of confined badgers. *Journal of Wildlife Management* 43:997–1001.

Ingles, L. G. 1947. *Mammals of California*. Stanford University Press, Stanford, CA. Pp. 72–74.

Lindzey, F. G. 1978. Movement patterns of badgers in northwestern Utah. *Journal of Wildlife Management* 42:418–422.

Long, C. A. 1973. *Taxidea taxus*. *Mammalian Species* 26. 4 pp.

Messick, J. P., and M. G. Hornocker. 1981. Ecology of the badger in southwestern Idaho. *Wildlife Monograph* 76. 53 pp.

Minta, S. C. 1993. Sexual difference in spatio-temporal interaction among badgers. *Oecologia* 96:402–409.

Minta, S. C., K. A. Minta, and D. F. Lott. 1992. Hunting associations between badgers (*Taxidea taxus*) and coyotes (*Canis latrans*). *Journal of Mammalogy* 73:814–820.

Petrides, G. A. 1950. The determination of sex and age ratios in fur animals. *American Midland Naturalist* 43:355–382.

Sargeant, A. B., and D. W. Warner. 1972. Movements and denning habits of a badger. *Journal of Mammalogy* 53:207–210.

Snead, E., and G. O. Hendrickson. 1942. Food habits of the badger in Iowa. *Journal of Mammalogy* 23:380–391.

Tumlison, R., and R. Bastarache. 2007. New records of the badger (*Taxidea taxus*) in southeastern Oklahoma. *Proceedings of the Oklahoma Academy of Science* 87:107–109.

Tumlison, R., D. B. Sasse, M. E. Cartwright, S. C. Brandebura, and T. Klotz. 2012. The American badger (*Taxidea taxus*) in Arkansas, with emphasis on expansion of its range into northeastern Arkansas. *Southwestern Naturalist* 57:467–471.

Ver Steeg, B., and R. E. Warner. 1999. The distribution of badgers (*Taxidea taxus*) in Illinois. *Transactions of the Illinois State Academy of Science* 93:151–163.

Wright, P. L. 1966. Observations on the reproductive cycles of the American badger (*Taxidea taxus*). *Symposium Zoological Society of London* 15:27–45.

———. 1969. The reproductive cycle of the male American badger (*Taxidea taxus*). *Journal of Reproduction and Fertility*, supp. 6:435–445.

North American River Otter (*Lontra canadensis*)

Name

The scientific name, *Lontra*, is the Latin word for "otter," and *canadensis* is the Latinized word meaning "of Canada"; the latter is for the country from

which the first specimen was collected and technically named. In the common name, "North American" refers to the continent on which it occurs; "river" denotes the habitat of the animal; and "otter" comes from the Anglo-Saxon words *oter* or *otor.* This species was formerly known as *Lutra canadensis.*

Description (Plate 59)

The North American river otter is a large, elongated mammal having a broad, flattened head with a conspicuous nose pad, prominent whiskers, moderately sized eyes, and small ears. The body is almost cylindrical, with a stout neck that is nearly the same diameter as the head; and a long, heavy tail that is flat on the bottom, thick at the base, and tapers from the body toward the tip. The legs are short and possess 5 fully webbed toes on each foot. The body's weight is supported on the toes. Except for the pads on the toes and soles, the undersurfaces of the feet are furred. The dense, oily underfur is overlain by glossy guard hairs that are usually straight but in some individuals may be curly.

While the otter lives both on land and in water, it is particularly well fitted for an aquatic existence. Some of the obvious external characteristics adapting it for life in the water are the streamlined body, webbed feet, and long, tapering tail. The ears and nose close when the animal goes underwater, making them watertight. The eyes are near the top of the head, permitting the otter to see above water while it swims along nearly submerged, and they have a nictitating membrane, or third eyelid. The dense, oily fur and heavy layer of body fat under the skin serve as insulators in waters of all temperatures. The prominent facial whiskers, which are extremely sensitive to touch, and the acute sense of smell perhaps compensate for the less-developed senses of sight and hearing.

Color. The color of the upperparts is dark brown and sometimes nearly black when wet, while the underparts are pale brown or gray. The muzzle and throat are silvery. Albinos occur only rarely. The sexes are colored alike and show no seasonal variation.

Measurements		
Total length	89–135 cm	34¾–53 in.
Tail	298–507 mm	11¾–19⅞ in.
Hind foot	101–133 mm	4–5¼ in.
Ear	22–28 mm	⅞–1⅛ in.
Skull length	98–107 mm	3⅞–4¼ in.
Skull width	63–76 mm	2½–3 in.
Weight	4.5–14 kg	10–30⅞ lb.

Males are larger than females. There is a record weight of 22.7 kg (50 lb.). Recently, a 14.2 kg (31.2 lb.) male became the official record-weight North American river otter confirmed by the Missouri Department of Conservation. The state furbearer record program was started in 2011.

Teeth and skull. The river otter is the only member of the weasel family in Missouri normally possessing 36 teeth; an occasional specimen has 38. The dental formula is:

$$I\,\frac{3}{3}\;C\,\frac{1}{1}\;P\,\frac{4}{3}\;M\,\frac{1}{2} = 36$$

or

$$I\,\frac{3}{3}\;C\,\frac{1}{1}\;P\,\frac{4}{4}\;M\,\frac{1}{2} = 38$$

The characteristics further diagnostic of an otter's skull are: size; the rostrum, or that portion of the skull in front of the eye sockets, broad and short; a narrow constriction on the upper surface of the skull between the eye sockets; and the hard palate extending beyond the upper molars.

Sex criteria. Males are identified by the penis, which can be felt under the belly skin between the anus and the opening for the penis, which is well forward of the anus. Females are recognized by the vaginal opening, which is just forward of the anus. The three pairs of teats are conspicuous in females during late pregnancy or while nursing.

Age criteria and longevity. The age of males can be told by the shape of the penis bone, or baculum. The baculum of an adult possesses a "flare" or widening of the shaft toward the base; that of a young male has a small, poorly developed base.

Plate 59

North American River Otter *(Lontra canadensis)*

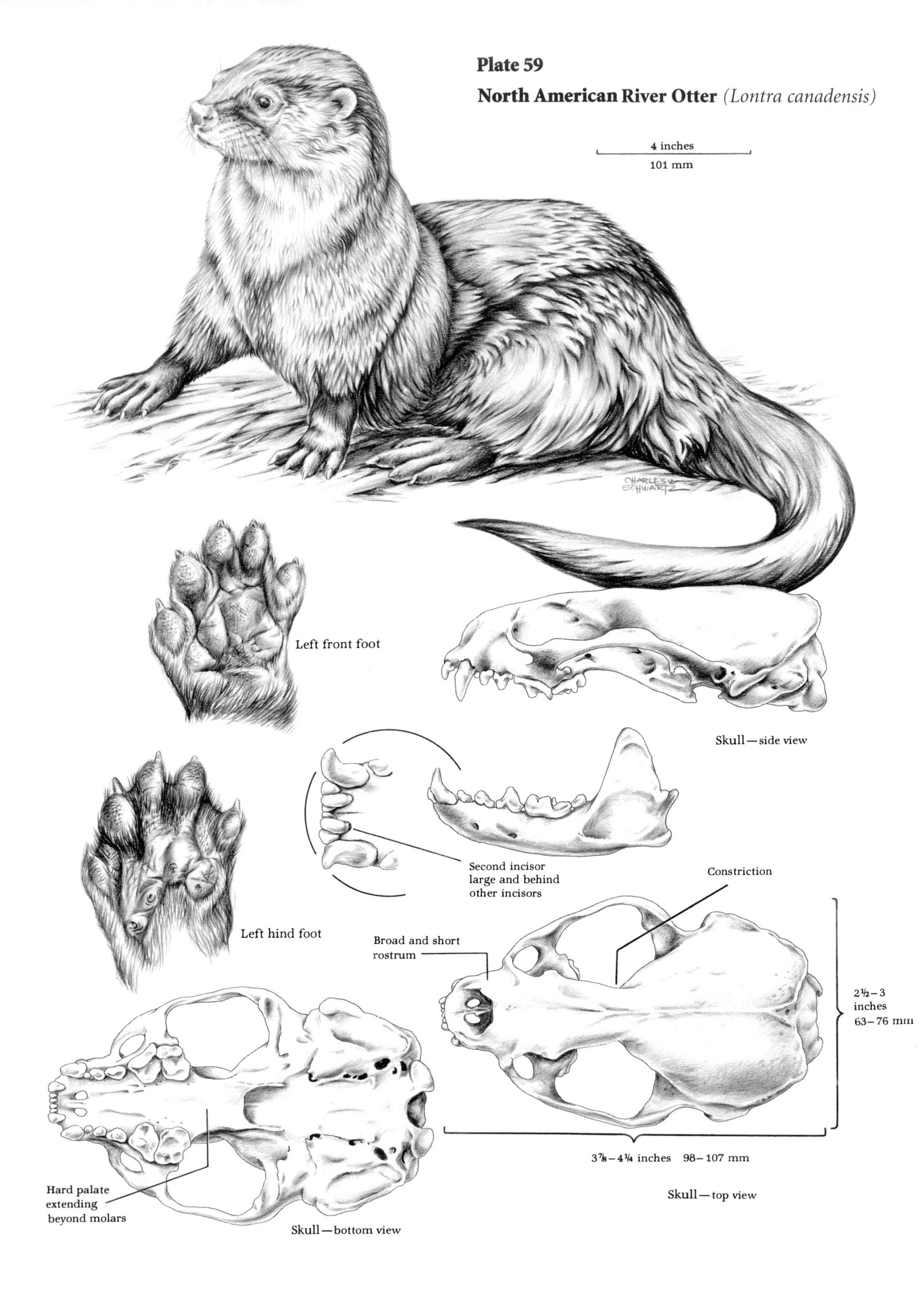

River otters are relatively long-lived. In captivity some individuals still bred at 17 years and lived to 19 years of age.

Glands. A pair of musk glands is present in the anal region. River otters twist tufts of grass along streams that they scent with the secretion from these glands and pile with their droppings. This scent is also given off in fright. The odor of otters is not as offensive as that emitted by other members of the weasel family.

Voice and sounds. Otters chirp, chuckle, grunt, growl, snarl, whistle, and scream. They also sniff loudly and blow through the nose. The female caterwauls while mating. Members of a family, when traveling together, keep in touch with one another by frequently giving birdlike chirps.

Distribution and Abundance

River otters were once common throughout North America with the exception of the southwestern United States, but numbers declined and many populations were extirpated, including most of Missouri's otters. Today they range from Alaska throughout most of Canada and the United States, except for areas of the southwestern United States, thanks to successful reintroductions and conservation efforts that have helped restore and stabilize populations.

The North American river otter historically occurred in all the major watersheds of Missouri. In 1935 the statewide otter population was estimated at only 70 individuals, all of which occurred in the Mississippi River Alluvial Basin except for a few in one locality on the Missouri River in the extreme northwest. By 1980, numbers had reportedly increased but only two major populations were known; one around the Mingo National Wildlife Area and Duck Creek Conservation Area in the southeast, and one in Henry and adjoining counties in west-central Missouri. Three other areas had recurring reports of otters; they were thought to be transient otters.

Translocations of otters into Missouri occurred during a restoration program from 1982 to 1992. Otters were acquired from Louisiana, Arkansas, and Ontario, Canada, and 845 were released at 45 locations. The reintroduction was a huge success, and the Missouri river otter's status was upgraded in 1991 from "rare" to a "watch list" species. It was removed from the "watch list" in 1997. Today an estimated 15,000 otters inhabit all major watersheds throughout the state.

Habitat and Home

River otter habitat consists of streams, rivers, and lakes, which are frequently but not always bordered with timber. The home is a burrow in a bank, under the roots of large trees, beneath rocky ledges, under fallen trees, or even in thickets of vegetation. These burrows are rarely built by the otter but represent former homes of muskrats, beavers, or woodchucks. Dens on the water's edge have an opening above water in summer, but in winter this is closed and the only

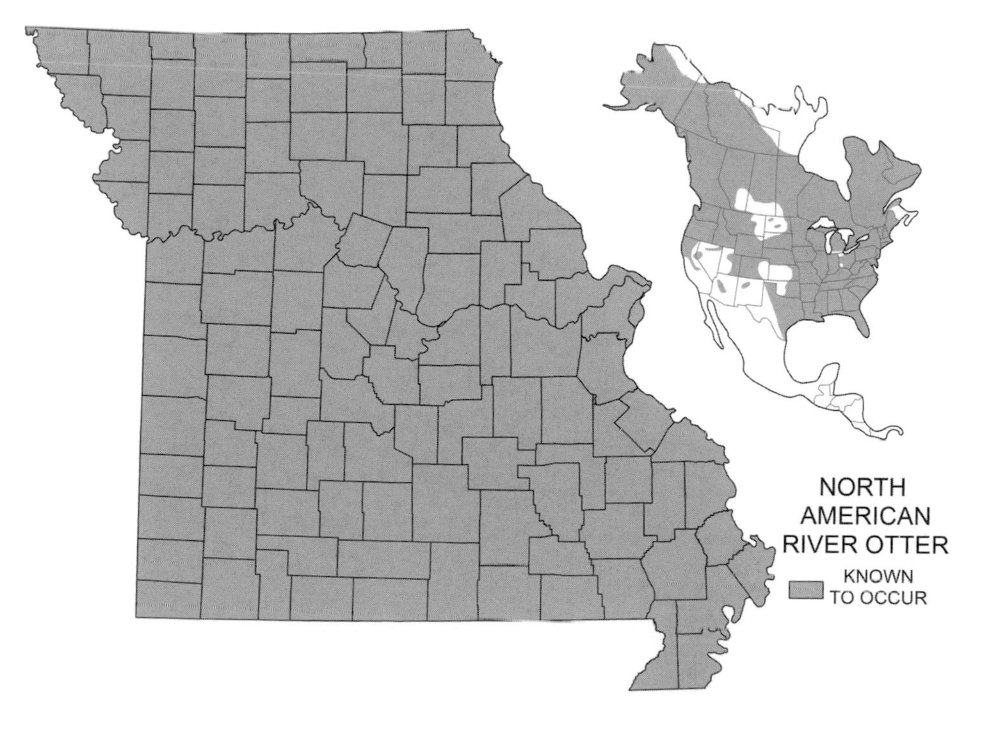

entrance is below water. The nest chamber may have a bare floor or a slight accumulation of leaves and grass.

Habits

River otters have a large yearly home range that may include between 80 and 160 km (50 and 100 mi.) of shoreline. However, during any one season a family may confine its activities to only a small section of stream amounting to between 5 and 16 linear km (3 and 10 linear mi.). While tending to follow watercourses, otters occasionally cut across land between two parts of a stream or from one body of water to another. Sometimes these crossings are used so often they become trails. Results of a study of southern Missouri rivers indicated an average otter density of 1 otter per 4 km (2.5 mi.).

Otters are mostly nocturnal but come out during the day on occasion. They are active all year and are not inhibited by changes in temperature or weather.

Otters are graceful and powerful swimmers. They may swim with just the head and shoulders showing above the surface, or completely underwater, or in an undulating pattern alternately going above and below the surface. They can swim 0.4 km (¼ mi.) under open water or ice and can remain submerged for 3 to 4 minutes. On the surface they can swim at least 9 km (6 mi.) per hour.

Along the shore, otters have regular "pulling out" places. Upon emerging from the water, they dry and dress their coats by shaking briskly and sometimes by rolling in the snow, grass, or leaves. They also take dust baths and wallow in the mud occasionally. In addition to the scent posts where they leave their droppings, latrines are made about 0.9 m (1 yd.) from the den entrance. Their large, coarse droppings contain fish scales, crayfish skeletal parts, and other indigestible foods.

When foraging for fish, the otter cruises slowly on the surface of the water, paddling only with its front feet. When fish are sighted, the otter arches its back and dives; then, with strong strokes of the feet, assisted by the sculling and rudder action of the tail, it outmaneuvers and captures even the most agile of quarry. In shallow water otters may root in the mud and debris for crayfish and other aquatic forms.

On land, otters commonly travel with a loping gait, but on snow or ice they alternate this with a series of slides. After a few steps forward, they slide on their bellies for 3 to 6 m (10 to 20 ft.) while holding all their feet backward. By running and sliding they can cover 4 to 5 km (15 to 18 mi.) in an hour.

One outstanding trait of river otters is their apparent zest for sliding down steep slopes. These slides may be on clay banks made slippery by their wet bodies or on slopes covered by snow or ice. The slide may terminate in a snowdrift or a deep pool of water. Otters slide on the chest and belly with all four feet folded out of the way. Sliding is probably indulged in as a social sport: otters seem to enjoy one another's company in this pastime.

In addition to the fun of sliding, they play with rocks and clamshells in the water, tossing and diving for them. In contrast to their relatives, the weasels and the mink, which are mostly solitary, otters generally live together all year in family groups and are very sociable animals.

Foods

The river otter feeds predominantly on animal foods. Fish and crayfish are favorite items, but on occasion frogs, salamanders, snails, shellfish, snakes, turtles, muskrats, birds, larvae of aquatic insects, and earthworms are added to the diet. Results from a study

conducted on Missouri Ozark streams indicated that crayfish are important year-round food, while fish are eaten more in the winter and less in the summer. Otters catch and kill most of their food but sometimes consume carrion. Captive animals do not fare well on an exclusive diet of fish.

Otters customarily eat the head of their prey first. A hungry otter may consume an entire fish if it weighs less than 0.5 kg (1 lb.), but they often leave behind parts of larger fish, especially the tail fin. When feeding on catfish, the otter will usually leave the head and major spines, eating only the softer body parts. After eating, an otter cleans its face and whiskers by rubbing them on grass or snow.

Digestion is very rapid. In captive animals, crayfish remains appear in the droppings one hour after the crayfish are eaten.

Reproduction

The testes descend into the scrotum in late November, indicating the approach of breeding condition in males. Females come into breeding condition between December and early April. Mating takes place in winter or early spring through May.

The length of pregnancy is apparently between 8 and 12½ months with a period of dormancy for the embryos, similar to the pattern of reproduction in weasels and the mink. One southern Missouri study reported implantation dates between early December and mid-January. In captive animals, females do not appear pregnant externally until about a month before the young are born. The single litter is generally born from January to May and usually contains from 2 to 4 young, with extremes of 1 and 6. The southern Missouri study reported from 1 to 4 young (with an average of 3) being born from early February to mid-March. Adults remate immediately following birth of the young. In captivity, males 5 to 7 years of age are the most successful breeders.

The young are born over a period of 3 to 8 hours, after which the female curls tightly around them in a "doughnut" shape and may put her head over the "hole" above them. They can thus nurse unmolested and be protected from the outside air. At birth the pups, or kits, are blind, toothless, and dark brown. One young reared by captive parents weighed 170 g (6 oz.) when a week or 10 days old, and weighed 454 g (1 lb.) 10 days later. The eyes open around 35 days of age, but the pups do not come out of the den much before they are 10 to 11 weeks old. Weaning occurs about 4 months after birth.

Although the female takes the major share of responsibility in caring for the young, the male may assist after the young leave the nest. Captive but free-roaming otters do not introduce their young to water until they are about 14 weeks old, and then the young have to be coaxed to swim. Before the young swim by themselves, the adults often carry them on their backs in the water. The pups stay with their parents during the first winter but leave in the spring. They are

Otter traveling on snow by alternately running and sliding

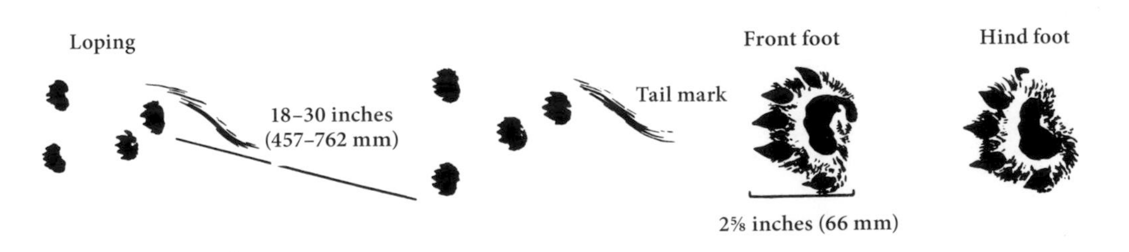

North American river otter tracks

capable of breeding at 1 year of age, but most do not breed before they are 2.

Some Adverse Factors

Humans are the most important predators on North American river otters, although bobcats, coyotes, and mountain lions will also take them. Tapeworms are known to parasitize them.

Importance

North American river otter fur is one of the most beautiful and durable of American furs, being used for coats, jackets, and hats. Otter pelts consequently bring a relatively good price. Thanks to a very successful reintroduction program, otter densities greatly increased across the state and a managed harvest was initiated again in Missouri in 1996. In 1977 it was listed in Appendix II of the Convention on International Trade in Endangered Species of Wild Fauna and Flora (CITES), and was included not because it is considered threatened with extinction at this time, but because it is similar in appearance to other otter species that are threatened or endangered, and hunting and trading must be closely monitored to avoid the possibility of confusion among North American river otter pelts and the pelts of endangered otters.

Annual harvest averaged 1,028 otters during the four-year period of the 1990s at an average of $34.10 per pelt. During the first decade of the 2000s, the average number of pelts sold annually was 2,066 at an average pelt price of $75.35. During this time, the highest annual take was 3,281 in 2005 and the lowest was 852 in 1998. The highest average annual price per pelt was $124.92 in 2005 and the lowest was $25.00 in 1998.

While the fishing habits of otters do not endear them to fishermen, it must be realized that otters eat rough as well as game fish and take many other kinds of food besides fish, especially crayfish. Research suggests that they have minimal impacts on fish populations in large streams, rivers, and lakes but may impact fish populations in small streams and ponds, and farmed fish.

Conservation and Management

Predation by humans and the declining productivity of streams brought about the reduction of river otters in Missouri. Full protection of otters was implemented in 1937. Improvement of stream conditions and subsequent restocking of otters have restored this species to all watersheds within the state.

A regulated trapping season was approved by the Conservation Commission beginning in 1996 and, since that time, an annual average of about 1,800 animals have been harvested for their fur and to help alleviate depredation by otters in certain areas where their densities are high. If you are experiencing problems with otters, contact a wildlife professional for advice, assistance, regulations, or special conditions for handling these animals.

SELECTED REFERENCES

See also discussion of this species in general references, page 23.

Bischof, R. 2006. Status of the northern river otter in Nebraska. *Nebraska Game and Parks Commission—Staff Research Publications.* Paper 35:117–120.

Chanin, P. 1985. *The natural history of otters.* Facts on File, New York. 179 pp.

Chromanski, J. F., and E. K. Fritzell. 1982. Status of the river otter (*Lutra canadensis*) in Missouri. *Transactions of the Missouri Academy of Science* 16:43–48.

Erickson, D. W., and C. R. McCullough. 1987. Fates of translocated river otters in Missouri. *Wildlife Society Bulletin* 15:511–517.

Friley, C. E., Jr. 1949. Age determination, by use of the baculum, in the river otter, *Lutra c. canadensis* Schreber. *Journal of Mammalogy* 30:102–110.

Gallagher, E. L. 1999. Monitoring trends in reintroduced river otter populations. M.S. thesis, University of Missouri, Columbia. 76 pp.

Hamilton, D. A. 1998. Missouri otter population assessment. A report to the Missouri Department of Conservation, Jefferson City. 23 pp.
Hamilton, W. J., Jr., and W. R. Eadie. 1964. Reproduction in the otter, *Lutra canadensis*. *Journal of Mammalogy* 45:242–252.
Lagler, K. F., and T. O. Ostenson. 1942. Early spring food of the otter in Michigan. *Journal of Wildlife Management* 6:244–254.
Larivière, S., and L. R. Walton. 1998. *Lontra canadensis*. *Mammalian Species* 587. 8 pp.
Liers, E. E. 1951a. Notes on the river otter (*Lutra canadensis*). *Journal of Mammalogy* 32:1–9.
———. 1951b. My friends—the land otters. *Natural History* 60:320–326.
Missouri Department of Conservation. 1993. River otter management plan: Its implementation and final report. Jefferson City, MO. 5 pp.
Mowry, R. A., M. E. Gompper, J. Beringer, and L. S. Eggert. 2011. River otter population size estimation using noninvasive latrine surveys. *Journal of Wildlife Management* 75:1625–1636.
Severinghaus, C. W., and J. E. Tanck. 1948. Speed and gait of an otter. *Journal of Mammalogy* 29:71.
Potechla, P. J., Jr. 1987. Status of the river otter (*Lutra canadensis*) population in Arkansas with special references to reproductive biology. Ph.D. dissertation, University of Arkansas, Monticello. 222 pp.
Raesly, E. J. 2001. Progress and status of river otter reintroduction projects in the United States. *Wildlife Society Bulletin* 29:856–862.
Roberts, N. M. 2003. River otter food habits in the Missouri Ozarks. M.S. thesis, University of Missouri, Columbia. 69 pp.
Roberts, N. M., S. M. Crimmins, D. A. Hamilton, and E. Gallagher. 2012. Implantation and parturition dates of North American river otters, *Lontra canadensis*, in southern Missouri. *Canadian Field-Naturalist* 126:28–30.
Roberts, N. M., C. F. Rabeni, J. S. Stanovick, and D. A. Hamilton. 2008. River otter, *Lontra canadensis*, food habits in the Missouri Ozarks. *Canadian Field-Naturalist* 122:303–311.

Skunks and Relatives
(Family Mephitidae)

Until the late 1990s skunks were considered members of the family Mustelidae. Although a taxonomic distinction for skunks at the family level had long been suspected, molecular systematic studies have provided the evidence necessary to group skunks within their own family, Mephitidae.

North American skunks are known for their distinctive black-and-white coloring and pungent anal scent glands. These two traits combined with behavioral displays serve as both warning and antipredator tactics.

The family name, Mephitidae, is based on the Latin word meaning "bad odor" and refers to the strong scent secreted when members of this family are provoked. Two skunk species are found in Missouri.

Eastern Spotted Skunk
(*Spilogale putorius*)

Name

The first part of the scientific name, *Spilogale*, is derived from two Greek words and means "spotted weasel" (*spilos*, "spot," and *gale*, "weasel"). This refers to the predominant spots on the body and to the close relationship between skunks and weasels. The last part, *putorius*, is from the Latin word *putor* for "a foul odor."

The common names have the following origins: "eastern" refers to the range in the eastern United States; "spotted" describes the pattern of the more or less connected spots on the body; and "skunk" is of Algonquian origin. The spotted skunk is also called the civet cat or polecat. These names are misleading and incorrect because this mammal is not closely related either to the true civets of the Old World or to cats.

Formerly the eastern and western forms of this spotted skunk were considered one species, *Spilogale interrupta*.

Description (Plate 60)

The eastern spotted skunk is a medium-sized, slender mammal with a small head, short legs, and a prominent, long-haired tail. The eyes are small and the ears short. Both front and hind feet have 5 slightly webbed toes and 4 pads on the soles at the base of the toes. The claws on the front feet are moderately long, and those on the hind feet are shorter. The weight of the body is

Plate 60

Eastern Spotted Skunk *(Spilogale putorius)*

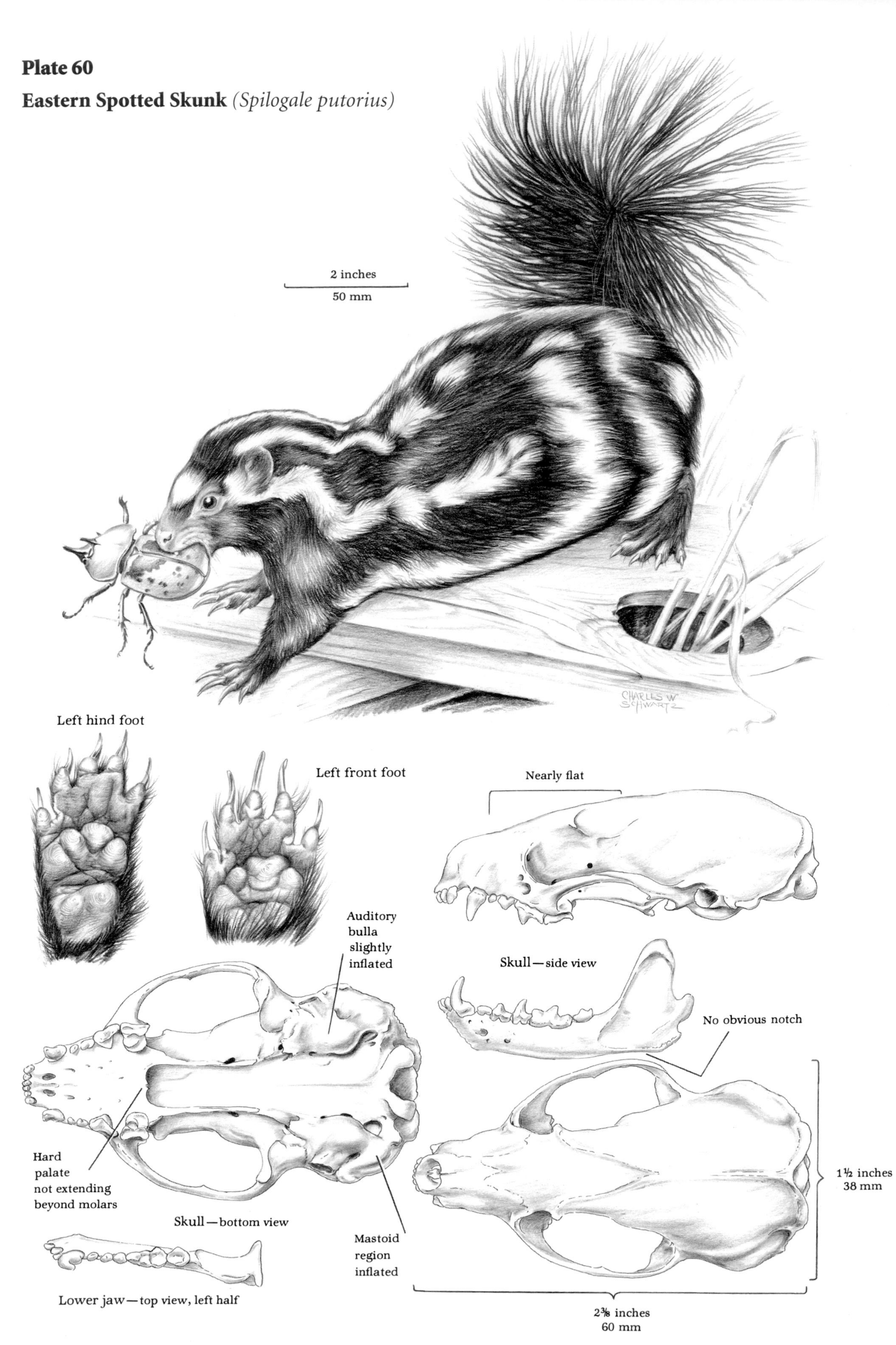

supported on most of the foot. The fur is rather long, soft, and glossy.

Color. The color pattern is very distinctive. The overall color is black with conspicuous white stripes and spots. A white spot occurs on the forehead and in front of each ear. Four white stripes along the neck, back, and sides extend from the head to about the middle of the body. Below and behind these are other white stripes with more or less connected spots. The tail of spotted skunks in Missouri is usually all black; rarely it may have a white tip. The sexes are colored alike.

Measurements

Total length	36–61 cm	14–23¾ in.
Tail	114–280 mm	4½–10⅞ in.
Hind foot	38–69 mm	1½–2¾ in.
Ear	25–28 mm	1–1⅛ in.
Skull length	60 mm	2⅜ in.
Skull width	38 mm	1½ in.
Weight	340–1,247 g	¾–2¾ lb.

Males are heavier than females.

Teeth and skull. The dental formula of the eastern spotted skunk is:

$$I\ \frac{3}{3}\ C\ \frac{1}{1}\ P\ \frac{3}{3}\ M\ \frac{1}{2} = 34$$

This is the same as for all the members of the weasel and skunk families in Missouri except the river otter. Skulls of the spotted skunk and its relative, the striped skunk, can be distinguished from those of other Missouri mammals by the following combination of characteristics: size; number of teeth; length and width of upper molars nearly equal; and hard palate that does not extend beyond the upper molars. The skull of the spotted skunk is distinguished from that of the striped skunk as follows: smaller size; region above eye sockets nearly flat and lacking a definite arch; mastoid region inflated; auditory bulla surrounding the inner ear slightly inflated; and no obvious notch on the bottom edge of the lower jaw toward the rear.

Sex criteria. Males are identified by the penis and females by their prominent 3 to 5 pairs of teats.

Age criteria and longevity. Of the various methods for age determination, the weight of the dried lens of the eye is the most useful. The weight of the penis bone, or baculum, is not reliable, but when the weight is considered along with changes in shape, it is a fairly reliable age indicator.

The time of closure of specific cranial sutures separates young of 10 or 11 months of age from older individuals. The ends of long bones lose their cartilage and become bone at 8 months and can be used as an age criterion until this time.

In the uterus, the presence of scars indicates that a female has produced young and is presumably older than 1 year.

It probably takes 5 years for all members of a wild population to be replaced.

Glands. While the scent glands are similar to those of the striped skunk, the scent of the spotted skunk is stronger and more disagreeable than that of the striped skunk. See the description in the account of the striped skunk.

Voice and sounds. Spotted skunks squeal, growl, hiss, snarl, and click their teeth upon occasion but usually are quiet animals. They also stamp their feet loudly.

Distribution and Abundance

The eastern spotted skunk was once common throughout the midwestern and southeastern United States. In the 1940s, populations declined suddenly and steeply, and today the spotted skunk is listed as a species of conservation concern across its range. Fewer numbers are found today throughout a much reduced, patchy range occurring throughout the Great Plains, the southeastern and east-central United States, and northeastern Mexico. Nowhere in its range has it ever been as abundant as the striped skunk.

The eastern spotted skunk once occurred across Missouri but was probably absent from extreme southeastern Missouri. Formerly it was most common in the western half of the state, but recently it has declined drastically in most of western and northern Missouri, and within those regions only six counties are known with relict populations. The Ozark Highlands are today the stronghold for spotted skunks, but populations are scattered where pockets of suitable habitat occur. Spotted skunks are now considered extremely rare in the state, and the subspecies occurring in Missouri, the plains spotted skunk (*Spilogale putorius interrupta*), is listed as critically imperiled and state endangered, but the species is apparently secure across its entire range.

The reasons for the decline of the spotted skunk remain unclear. The decrease may be related to changes in agriculture within the range that emphasize "clean" farming and leave little cover in odd places in which these skunks can live. As recently as the late 1940s, the practice of making huge haystacks on the prairies provided ideal denning sites for these carnivores. This

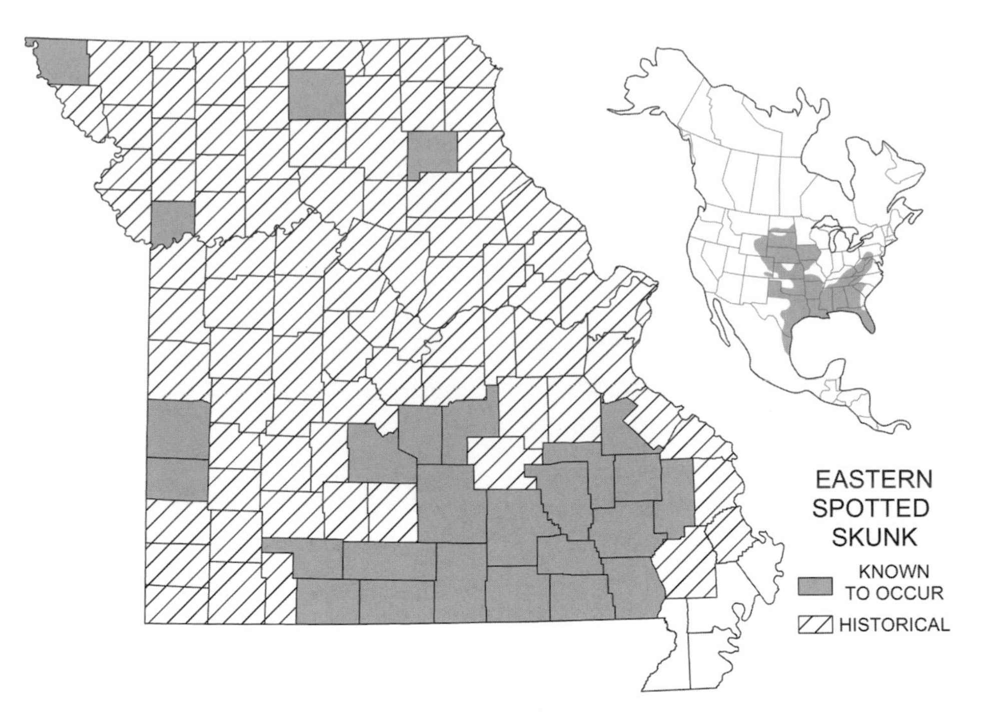

agricultural practice has declined, eliminating these valuable homesites. It is also possible that the prevailing use of pesticides in agricultural areas affected the food chain of these largely insectivorous mammals, with disastrous results. Or perhaps a disease-related decline due to a skunk parvovirus or other pathogen occurred.

Habitat and Home

Across their range, eastern spotted skunks prefer open prairies, brushy areas, and cultivated land. They seemingly require some form of cover such as a brushy field border, fencerow, or heavily vegetated gully between the den and foraging areas. However, in the Ozark Highlands of Missouri they are found more commonly in woodland habitats with extensive leaf litter and downed logs. In the Ouachita Mountains of Arkansas, they are found in woodland habitats with a closed canopy to reduce predation risk.

Their dens are located belowground in grassy banks, rocky crevices, or along fencerows, as well as aboveground in such places as farm buildings, haystacks, woodpiles, hollow logs or trees, stone piles, or brush heaps. Any locality seems suitable provided the den excludes light and offers protection from the weather and predators. Many of these dens have been built by other animals, such as woodchucks and woodrats, although skunks may dig some homes for themselves. Most dens contain grass or hay as a lining, but some, used only in summer, are often unlined. A den may be used by one or more spotted skunks and seems to be common property of the local spotted skunk population. In the cold winter months a few may even be found together in a single den. However, a female with young usually has a den for her exclusive use.

Habits

During most of the year a spotted skunk remains within approximately 0.6 sq km (¼ sq. mi.), but the exact size of the home range is determined by the quality of food and cover available during all seasons. In spring, males become restless and may range over about 10 sq km (4 sq. mi.). Females tend to maintain their home range size during all seasons. One skunk in Missouri ranged over 4,359 ha (10,771 ac.) in a

Young spotted skunk in defensive pose

13-month period. They apparently do not maintain and defend territories.

When walking, spotted skunks travel up to a rate of 7.2 km (4½ mi.) per hour. When hunting, their usual gait is a bound that may even be increased to a gallop. Spotted skunks are adept at climbing trees and fence posts and commonly do so, particularly as a means of escape from dogs. When threatened, they will do front handstands and stomp their front paws as warning signals, followed by a spray as a last resort (see the striped skunk section for a more detailed account). They will enter shallow water but avoid deep spots.

Spotted skunks are mostly nocturnal and very secretive, although they may come out in daylight on rare occasions. In winter they hunt in fields during good weather but forage in barns or similar protected locations during wet or very cold weather. Their foods are located mostly by their keen senses of smell and hearing. The droppings are usually deposited indiscriminately. Latrines are made only rarely and are then located near or in a den.

Foods

Spotted skunks eat both plant and animal foods, but the latter predominate in the diet. Insects such as grasshoppers, crickets, ground beetles, and scarab beetles are a preferred food, especially in summer. Insects are obtained under logs, in leaf litter, and in other vegetative debris. Mice are regular, important items in the fare, particularly in winter, while rats, cottontails, and various small mammals are taken when other foods are scarce. Large animals are often dragged to the den entrance or some protective shelter before being eaten. Adult, young, and eggs of birds are also consumed. Wild ducks, crippled or killed by hunters but not recovered, are taken when available; chickens are consumed mostly as carrion. When eating an egg, one end or side is bitten off, the contents licked out, and the shells left near the nest. Spotted skunks do not crush the eggshells as much as do striped skunks, and their work can thus be identified. Fruit is relished whenever available, and corn is taken when more esteemed foods are scarce. Miscellaneous food items consist of lizards, snakes, crayfish, salamanders, and mushrooms.

Reproduction

Mating takes place in March and April, and the young are born from late May to July. The gestation period is between 50 and 65 days, and there is probably no delay in development of the embryos as occurs in the western spotted skunk and in the closely related weasel family. Possibly a second litter is produced in late summer. A litter usually contains 5 young with extremes of 2 and 9.

The young are nearly naked at birth, but the black-and-white color pattern is evident on the skin. The

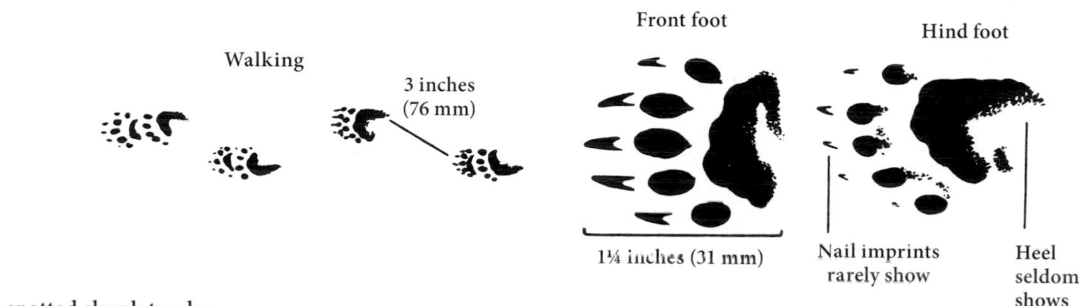

Eastern spotted skunk tracks

eyes and ears are closed, and no teeth are visible. The claws are well developed, and the young are able to crawl about feebly in the nest. They are about 10 cm (4 in.) long and weigh about 9.4 g (⅓ oz.) each. By 21 days of age, dense fur completely covers the body and the coloration is similar to that of adults. When 24 days old, the little spotted skunks raise their tails in warning when frightened. The eyes open about 32 days after birth, and the teeth are visible through the gums. At this stage the young are only able to walk clumsily. They spray for the first time when about 46 days old. Weaning occurs when the little skunks are approximately half-grown at about 54 days of age. Full size is attained between 3 and 4 months, and breeding occurs at 1 year. Males do not participate in the rearing of the young.

Some Adverse Factors

Spotted skunks are preyed upon by great horned owls, coyotes, bobcats, domestic dogs and cats, and humans. Tapeworms, roundworms, mites, ticks, lice, and fleas parasitize them. Spotted skunks sometimes contract rabies and have transmitted it to other animals, including humans. They can also be killed by highway traffic.

Importance

Spotted skunks contribute to the natural control of insects and rodents. They are an asset around farms and should be tolerated and regarded as interesting and valuable members of a farm wildlife community.

The fur of this species has been used to make jackets and trimming for coats. In Missouri, the average annual harvest per year was 18,349 in the 1940s, 758 in the 1950s, 264 in the 1960s, 124 in the 1970s, and 20 in the 1980s. The highest and lowest annual take was 55,440 in 1940 and 0 in 1989 and 1990. The average annual price per pelt was $0.71 in the 1940s, $0.77 in the 1950s, $1.67 in the 1960s, $4.05 in the 1970s, and $3.20 in the 1980s. The highest and lowest average annual prices were $11.00 in 1978 and $0.30 in 1948 and 1949. Beginning in 1991 the harvest of spotted skunks in Missouri was closed due to its rarity and endangerment.

Conservation and Management

To improve habitat for spotted skunks, promote measures for supplying their food sources and homesites, such as preserving small glades and rocky outcrops and leaving down timber and hollow logs in forested areas. A continued understanding of their life history and habitat needs will aid in more detailed conservation efforts.

The eastern (plains) spotted skunk is one of our most endangered mammals and is difficult to survey. If you see a spotted skunk, please report it immediately to your local Missouri Department of Conservation office.

SELECTED REFERENCES

See also discussion of this species in general references, page 23.

Crabb, W. D. 1941. Food habits of the prairie spotted skunk in southeastern Iowa. *Journal of Mammalogy* 22:349–364.

———. 1944. Growth, development, and seasonal weights of spotted skunks. *Journal of Mammalogy* 25:213–221.

———. 1948. The ecology and management of the prairie spotted skunk in Iowa. *Ecological Monographs* 18:201–232.

DeSanty-Combes, J. 2003. Statewide distribution of plains spotted skunks (*Spilogale putorius interrupta*) in Missouri. Final Report to the Missouri Department of Conservation, Jefferson City. 38 pp.

Dragoo, J. W., and R. L. Honeycutt. 1997. Systematics of mustelid-like carnivores. *Journal of Mammalogy* 78:426–443.

Gompper, M. E., and H. M. Hackett. 2005. The long-term, range-wide decline of a once common carnivore: The eastern spotted skunk (*Spilogale putorius*). *Animal Conservation* 8:195–201.

Hackett, H. M. 2008. Occupancy modeling of forest carnivores in Missouri. Ph.D. dissertation, University of Missouri, Columbia. 177 pp.

Kinlaw, A. 1995. *Spilogale putorius. Mammalian Species* 511. 7 pp.

Lesmeister, D. B. 2007. Space use and resource selection by eastern spotted skunks in the Ouachita Mountains, Arkansas. M.S. thesis, University of Missouri, Columbia. 82 pp.

Lesmeister, D. B., M. E. Gompper, and J. J. Millspaugh. 2008. Summer resting and den site selection by eastern spotted skunks (*Spilogale putorius*) in Arkansas. *Journal of Mammalogy* 89:1512–1520.

Lesmeister, D. B., J. J. Millspaugh, M. E. Gompper, and T. W. Mong. 2010. Eastern spotted skunk (*Spilogale putorius*) survival and cause-specific mortality in the Ouachita Mountains, Arkansas. *American Midland Naturalist* 164:52–60.

McCullough, C. 1983. Population status and habitat requirements of the eastern spotted skunk on the Ozark plateau. M.S. thesis, University of Missouri, Columbia. 60 pp.

———. 1986. The spotted skunk's last stronghold in the Ozarks. *Missouri Conservationist* 47:21–22.

McCullough, C. R., and E. K. Fritzell. 1984. Ecological observations of eastern spotted skunks on the Ozark

Plateau. *Transactions of the Missouri Academy of Science* 18:25–32.
Mead, R. A. 1967. Age determination in the spotted skunk. *Journal of Mammalogy* 48:606–616.
———. 1968. Reproduction in eastern forms of the spotted skunk (genus *Spilogale*). *Journal of Zoology* 156:119.
Sasse, D. B., and M. E. Gompper. 2006. Geographic distribution and harvest dynamics of the eastern spotted skunk in Arkansas. *Journal of the Arkansas Academy of Science* 60:119–124.

Striped Skunk (*Mephitis mephitis*)

Name

The scientific name, *Mephitis*, is a Latin word meaning "bad odor"; it refers to the extremely strong scent characteristically given off by this animal upon provocation. The common name, "striped," describes the white markings on the body, while "skunk" is of Algonquian origin. Skunks are sometimes called "polecats," but this name more properly belongs to some of their Old World relatives.

Description (Plate 61)

The striped skunk is a medium-sized, stout-bodied mammal with a small head, short legs, and a prominent, long-haired tail. The eyes are small and the ears short. On each foot there are 5 slightly webbed toes possessing long claws, those on the front feet being much longer than those on the hind. The large foot pad behind the toes on the soles of both feet is undivided. The weight of the body is supported on most of the foot. The short, soft underfur is overlain by long, glossy guard hairs.

Color. The body is predominantly black with a thin white stripe down the center of the face and a broad white one beginning on the back of the head, continuing onto the nape, forking on the shoulders, and extending toward and sometimes onto the tail. In some, the upperparts of the body may have only one pair of white stripes or just a prominent white patch on the back of the head and nape. The tail is mostly black but may have an extension of the white stripes on the sides and a tuft of white hairs at the tip; the black hairs of the tail are white at the base. Striped skunks in Missouri commonly have little white on the tail. Brown, cream-colored, or albino skunks occur rarely. The sexes are colored alike. In winter the pelts of adults are prime, while those of young skunks tend to be unprime.

Measurements

Total length	51–76 cm	20–30 in.
Tail	177–381 mm	7–15 in.
Hind foot	50–82 mm	2–3¼ in.
Ear	19–25 mm	¾–1 in.
Skull length	57–79 mm	2¼–3⅛ in.
Skull width	38–50 mm	1½–2 in.
Weight	1.2–5.3 kg	2⅝–11¹¹⁄₁₆ lb.

Males are heavier than females.

In 2014, a 2.8 kg (6.1 lb.) male became the official record-weight striped skunk confirmed by the Missouri Department of Conservation. The state furbearer record program was started in 2011.

Teeth and skull. The dental formula of the striped skunk is:

$$I\ \frac{3}{3}\ C\ \frac{1}{1}\ P\ \frac{3}{3}\ M\ \frac{1}{2} = 34$$

This is the same as for all members of the weasel and skunk families in Missouri except the river otter. Skulls of striped and eastern spotted skunks can be distinguished from those of weasels in Missouri by the following combination of characteristics: size; number of teeth; length and width of the upper molars nearly equal; and hard palate not extending beyond the upper molars. The skull of the striped skunk is distinguished from that of the eastern spotted skunk as follows: larger size; region above eye sockets well rounded and possessing a definite arch; mastoid region not inflated; auditory bulla surrounding the inner ear not inflated; and an obvious notch on the bottom edge of the lower jaw toward the rear.

Sex criteria. Males are identified by a penis; they have tiny teats, while females have 5 to 7, but usually 6, pairs of relatively prominent teats on the sides of the belly extending from front to hind legs. Pelts of males have a penis scar.

Age criteria and longevity. The age of males can be told by the penis bone, or baculum. The baculum of a

Plate 61

Striped Skunk *(Mephitis mephitis)*

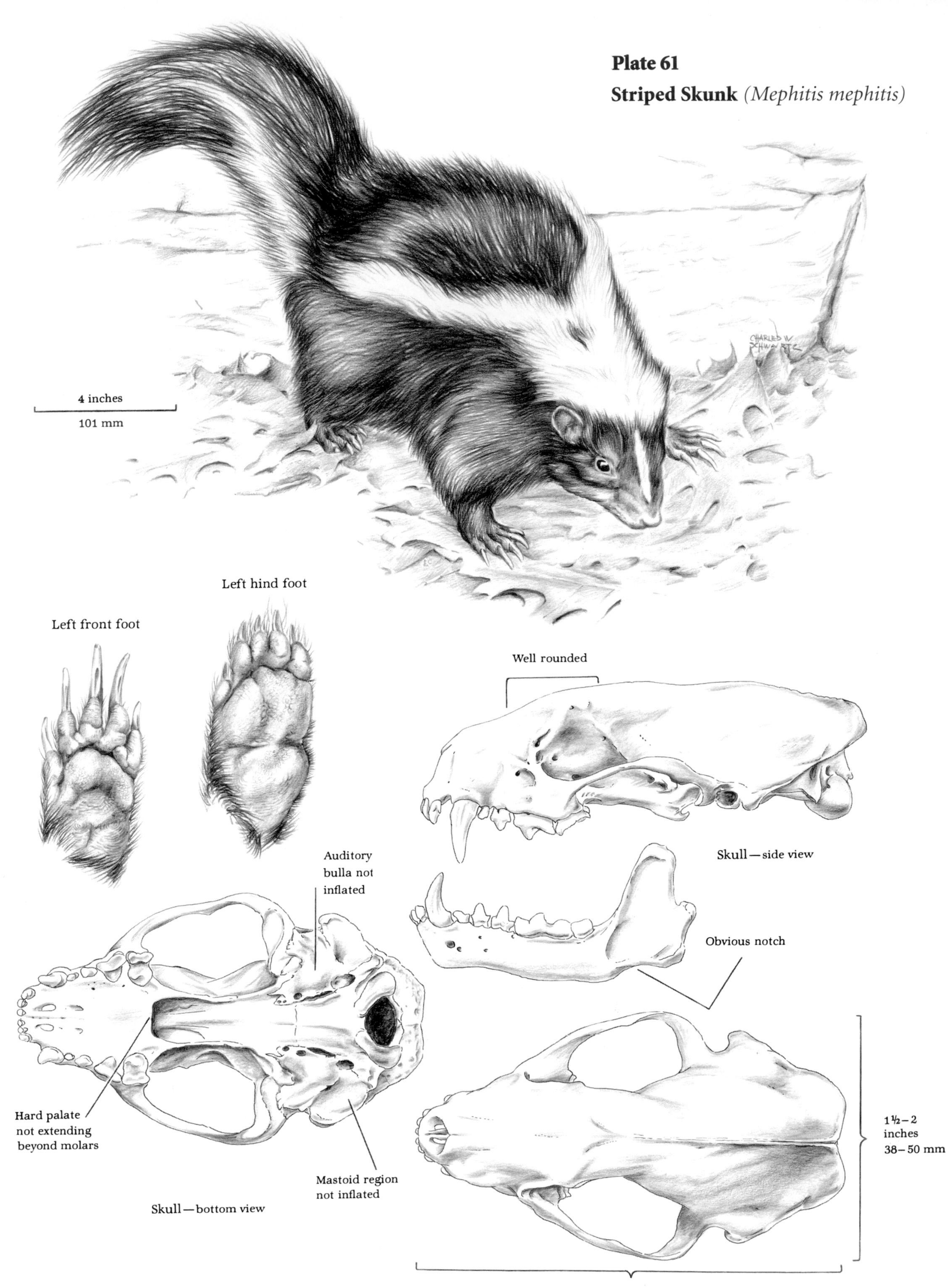

male less than 1 year old is slender, irregularly curved, and without an enlarged basal portion; it measures less than 2 cm (¾ in.) in length. The baculum of an adult male is somewhat stouter and less curved, has an enlarged basal portion, and measures about 2.2 cm (⅞ in.) in length.

In adult females that have bred, the teats are 3 mm (⅛ in.) or more in diameter and quite dark. In young females the teats are smaller, scarcely raised, and flesh colored.

Other methods of age estimation are given under the eastern spotted skunk.

In the wild, striped skunks seldom live over 2–3 years. It is estimated that the population is replaced within a 5-year period. In captivity some may live up to 10 years.

Glands. Skunks are notorious for their ability to discharge a very obnoxious scent upon provocation. Actually the skunk itself is not a foul-smelling creature and has very little unpleasant odor about its body or den. The disagreeable scent or musk is secreted by two internal glands, about 1.3 cm (½ in.) in diameter, located in both sexes at the base of the tail on either side of the anus. These glands are connected by ducts to tiny, paired nipples just within the anus. When the tail is down, the nipples are hidden; when it is raised, the walls of the anus are relaxed and the nipples project outside.

The skunk exercises voluntary control over these scent glands. The glands are sheathed in strong muscles and can be operated separately or together. When discharged in unison, the fluid from the two nipples unites and forms a concentrated stream that gradually disperses into a fine spray. In calm weather this stream can be directed accurately for 1.5 to 3 m (5 to 10 ft.) and somewhat less accurately for much as 6 m (20 ft.). The odor may carry 2.4 km (1½ mi.) downwind. The skunk can aim behind, to either side of, or in front of itself by changing the direction of aim of the nipples, alternating the gland and nipple used, and twisting the body.

The scent glands contain approximately 15 ml (1 tbsp.) of thick, volatile, oily liquid, which is sufficient for 5 or 6 rounds. The scent is produced at a rate of about 9 g (⅓ oz.) per week. This fluid is white, yellow, or greenish yellow in color and is distinctly phosphorescent at night. It is a sulphur compound and has a very pungent odor that is extremely obnoxious to most people. However, the scent of the striped skunk is not as strong or disagreeable as that of the spotted skunk. The musk of skunks is painful to the eyes, causes swelling in the lining of the nose, and may be nauseating. Although interfering temporarily with vision, it does not cause permanent blindness.

Spraying occurs mostly in self-defense. Only rarely is the scent given off when skunks play together, when a skunk engages in a fight with another animal such as a rat, or when females are in heat. Skunks are reluctant to spray and often put up with considerable abuse before doing so. In fact, they usually give one or several warning signals before discharging any scent. As an advance notice of its intentions, a skunk may stamp its front feet very rapidly and loudly enough to be heard several meters (yards) away. It may often walk a short distance on its front feet with the tail high in the air. The skunk may also click its teeth, growl, or hiss. However, the typical warning is to raise the tail, with or without elevating the extreme tip, and erect every hair so that each one is at right angles to the tail's long axis. Just before scenting, the body is usually turned in a U with both head and tail facing the intruder.

Although a skunk generally avoids spraying on its own fur, it may do so if frightened or held in such a position that the tail cannot be raised or twisted to one side. A skunk can be handled without danger of its scenting if it is grasped behind the head with one hand and by the base of the tail with the other. The belly is kept up, the tail pointed away, and the back prevented from humping. Skunks can be carried in a gunnysack with impunity if the sack is not bumped or allowed to touch the ground.

It is very difficult to get rid of the scent if one has been exposed to it. There are some over-the-counter commercial sprays designed to eliminate skunk odor that may help, or one may immediately call a professional company that deals with removing skunk odors. Common household items can be used, including a fresh mixture of 1 quart of 3 percent hydrogen peroxide, ¼ cup baking soda, and 1 to 2 teaspoons liquid dish or laundry soap. For some outdoor items, a dilute bleach with water solution can be effective. Be sure to immediately rinse yourself, your pets, your clothes, and your outdoor items with plenty of water. If you have any concerns about the effectiveness and use of common household items, the best advice is to consult a professional.

Young skunks reared on fur farms are commonly descented when 4 weeks old. The best method is to make an incision through the skin and cut out the scent glands.

Voice and sounds. Striped skunks are ordinarily silent but on occasion make a variety of sounds. They growl, bark, screech, twitter, squeal, hiss, snarl, and purr. They also spit, sniff, and click the teeth. The feet are sometimes stamped rapidly as part of the warning before discharging the scent glands.

Distribution and Abundance

The striped skunk is widespread in North America, ranging from central Canada through most of the United States and into parts of northern Mexico. This range is believed to represent an expansion from presettlement times because new habitat has been created with the opening of forests that accompanied settlement. Striped skunks are most abundant in the central portion of this range and are considerably more numerous than spotted skunks where their ranges coincide. In good habitat, there may be 1 striped skunk per 2 to 4 ha (6 to 11 ac.).

Striped skunks occur throughout the state. Following settlement, the striped skunk population increased in Missouri following the opening of forested areas. Today, the statewide population is stable to slightly increasing. The highest populations occur in the Osage Plains of southwestern Missouri and the adjacent border of the Ozark Highlands. The lowest population is in the Mississippi River Alluvial Basin, where there are a few skunks on the low ridges that extend from north to south, but on the intervening lowlands there are very few. In this area the water table frequently rises to within 0.6 to 0.9 m (2 to 3 ft.) of the surface, which is high enough to flood most den sites.

Habitat and Home

Striped skunks are at home in a variety of habitats but prefer forest borders, brushy field corners, fencerows, cultivated areas, and open grassy fields broken by wooded ravines and rocky outcrops, wherever permanent water is nearby.

During spring and summer, striped skunks may bed aboveground in hay fields, along fencerows, in pastures and croplands, and along grassed waterways, but they usually have a den for their home. The den is customarily belowground but occasionally is in a stump, hollow log, refuse dump, a brush or lumber pile, cave, rock pile, crevice in a cliff, or under a building. Sometimes a skunk digs its own home, particularly for its young, but more often it uses a discarded den of a woodchuck, muskrat, fox, or some other mammal.

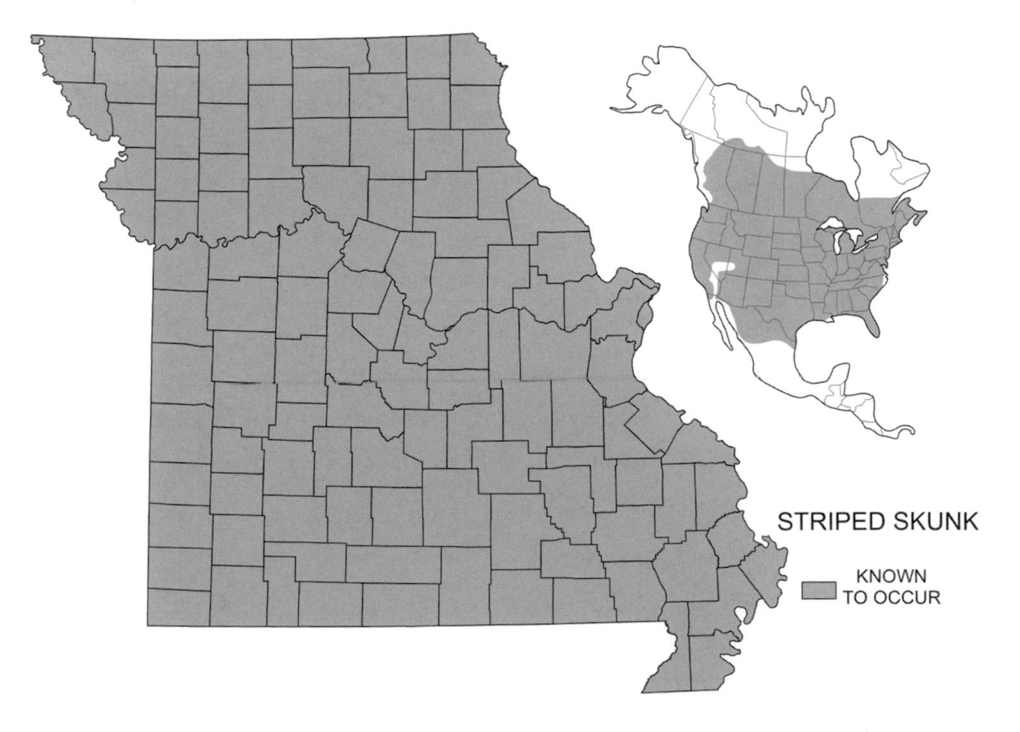

Striped skunk tracks

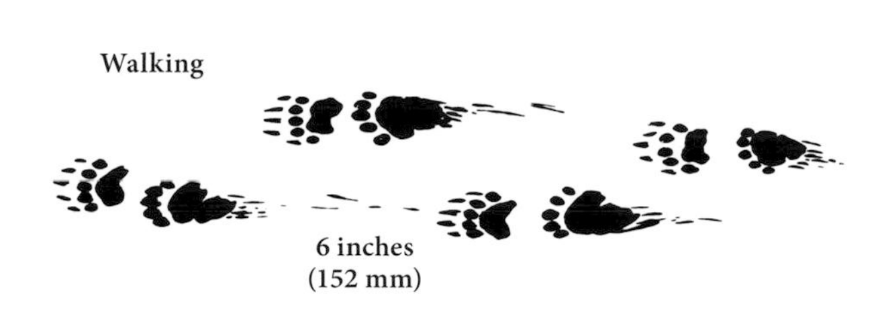

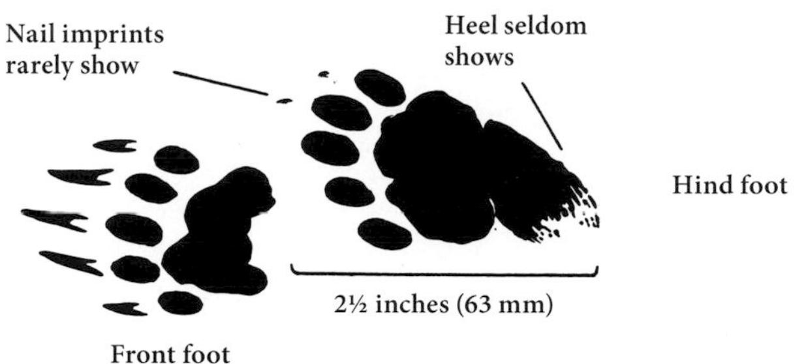

The complexity of the den varies with its originator; one dug by a skunk is very simple, but one taken over from another animal is often quite extensive and elaborate. The location and nature of the terrain also influence the ease with which excavations can be performed. There are from 1 to 5, ordinarily well-hidden, openings about 20.3 cm (8 in.) in diameter that lead to tunnels from 2 to 17 m (6 to 56 ft.) in length opening into 1, 2, or 3 chambers. Winter dens are usually deeper than summer dens and often are well below the frost line.

In one of these subterranean chambers the skunk builds a nest consisting of about 35 l (a bushel) of leaves, which it transports to the den in a peculiar fashion. The skunk gathers a pile of leaves under its body and then shuffles the leaves along. They are either pushed into the den ahead of the animal, or the skunk may back into the tunnel and pull the leaves inside with its mouth. In cold weather the various outside openings are sometimes plugged with leaves.

Habits

While striped skunks may leave their dens at any hour of the day, they usually begin foraging in late afternoon or early evening and are abroad most of the night. Because of these nocturnal habits, they locate their prey more by the senses of smell and hearing than by sight.

The home range of a male is on average 512 ha (1,265 ac.), while that of a female averages 378 ha (934 ac.). In any one night of foraging a skunk may meander over only a small portion of this area, less than 0.8 km (½ mi.). Males always travel more than females, and during the breeding season they may go as far as 6 to 8 km (4 or 5 mi.) a night.

Within this range they have a series of dens that they use from time to time. In a study in northeastern Kansas, the average distance between dens on successive nights was 414 m (1,359 ft.), and the average use of one den was 2 days. Skunks may stay in the same area for at least 2 years.

In autumn, striped skunks acquire a layer of fat; as the weather gets colder, they spend more time in their dens. The amount of activity during winter depends largely upon the temperature. In the northern part of the range skunks may be inactive for many weeks or months, while in the southern part they may be quiescent for only a few days at a time. In general, both sexes are active until the temperature nears freezing, when the females start to become drowsy. Males continue their activities until the temperature reaches about −9°C (15°F), when they too become dormant. During periods of stupor, the body temperature of both sexes does not become lowered as in the case of most truly hibernating mammals, and sleep is usually intermittent. Depending upon the weather conditions of the particular winter, adults seldom remain inactive for more than 1 month, but young skunks may be dormant for as much as 4 months. During these periods males may lose up to 15 percent of their fall weight, while females may lose as much as 40 percent of theirs.

While skunks are not sociable animals, they may den together for warmth. As many as 20 striped skunks have been found in one winter den, although the number is usually much less. Any association of sex and age may occur; even other animals, such as opossums, woodchucks, or cottontails, may be in another part of the same den.

Striped skunks are mainly terrestrial. They normally walk, and their fastest speed for short periods is about 11 to 13 km (7 to 8 mi.) per hour. Striped skunks seldom climb trees. Occasionally they bathe in water and forage in shallow pools. Although capable of swimming, they usually do not take to deep water unless forced.

Their droppings are deposited at random, and only rarely is there a latrine near the den. The droppings, or scats, may have a musty odor but not that of the pungent musk from the anal scent glands.

Foods

Striped skunks eat plant and animal foods in about equal amounts during fall and winter but take considerably more animal matter during spring and summer when insects, their preferred food, are more available. Grasshoppers, beetles, and crickets are the adult insects most frequently taken. Army worms, cutworms, tobacco worms, white grubs, and various other insect larvae injurious to agriculture are included in the diet. Skunks dig and root in the soil and often leave the ground pitted where they forage. Bees, wasps, and their hives, together with larvae and honey, are also eaten. However, these items are taken more often when the ground is dry and difficult to dig and few other insects are available. Skunks strike the bees with their front feet. Apparently they do not mind the stinging, as individuals trapped at hives showed many stings on their bodies and in the mouth and throat.

Striped skunks eat large numbers of mice and rats as well as moles, shrews, ground squirrels, young rabbits, and chipmunks. The larger mammals are usually eaten as carrion, while the smaller ones are caught by the skunk. Birds and their eggs are taken only rarely. When eating an egg, the skunk bites off one end, licks out the contents, and leaves the shells more or less together at the nest. Striped skunks crush the eggshell more than do spotted skunks, and their work can thus be identified. An occasional skunk learns to eat chickens, but this is unusual and not the habit of the species. Miscellaneous animal items at various times of the year are lizards, salamanders, frogs, earthworms, crayfish, clams, minnows, and turtle eggs. Fruits of many kinds are consumed when available, and some grasses, leaves, buds, roots, nuts, grain, and fungi are eaten on occasion. An economic evaluation of the feeding habits of striped skunks shows that about 68 percent of the diet is beneficial to humans, 27 percent neutral, and only 5 percent harmful.

Reproduction

The breeding season of striped skunks begins late in February and extends through March. Females born in the preceding year mate about a month later. The gestation period is 62 to 66 days but may extend to 75 days. In those females with the longest gestation period, the embryos may undergo a period of arrested development similar to that of many members of the related weasel family. Litters of old females are mostly born during the first part of May, while those of young females are born in early June. There is usually only 1 litter annually: it most commonly has 4 to 8 young, but may have from 2 to 10. Younger or smaller females have smaller litters than older or larger females.

At birth the young weigh about 14 g (½ oz.) each and are about 14 cm (5½ in.) long. They are wrinkled and almost naked but possess the adult's characteristic black-and-white color markings. The eyes and ears are closed, and no teeth are visible. However, the claws are well developed and the sexes can be distinguished. About 13 days after birth, the young are fully haired. The eyes open between 17 and 21 days of age, and the young are able to weakly assume a defensive pose at this time. The incisors come through the gums when the young are about 39 days old. The female takes her offspring hunting for the first time when they are about 7 weeks of age. When necessary she carries them in her mouth by the nape of the neck. Weaning is complete at about 2 months of age. The young stay with the female until fall when family ties become loose. The young are capable of mating at 10 months of age. Juvenile mortality is high during summer and the first winter.

Some Adverse Factors

Skunks are not commonly preyed upon because most animals probably fear or respect the skunk's ability to retaliate by using its powerful scent. However, when great horned owls, coyotes, badgers, foxes, and bobcats are pressed by starvation they may prey on young skunks. Domestic dogs may sometimes attack skunks, but most do not become second offenders. Adult male skunks eat the young if the female leaves them unguarded in the nest.

Mites, ticks, lice, fleas, roundworms, flukes, and tapeworms are parasites of this species. A respiratory disease caused by a fungus sometimes occurs. Skunks may carry and transmit rabies if bitten by an infected animal. They are the primary carriers of rabies in the Midwest.

Skunks are frequent victims of highway traffic chiefly because of their slow gait and reluctance to give ground to anything. They are also susceptible to secondary poisoning when scavenging poisoned prey.

Importance

At one time striped skunk pelts were valuable in the fur industry, but they are less valuable in the current market. Striped skunk fur is thick and glossy and is used for trimming, scarves, muffs, jackets, and coats. The pelt is graded according to the amount of white, those with the least being the most valuable. In preparing the fur for use, the white is dyed black, or taken out and the pelt sewed together without it.

In the 1940s the average number of pelts sold annually in Missouri was 57,557. This harvest decreased dramatically in the 1950s to 6,312, then to 2,279 in the 1960s, remained about the same in 1970s at 2,554, and then declined dramatically again to 381 in the 1980s, 184 in the 1990s, and increased slightly to 397 in the first decade of the 2000s. During this seven-decade period, the highest annual take was 156,102 in 1940 and the lowest was 52 in 1990. The average annual price paid per pelt was $1.16 in the 1940s, $0.91 in the 1950s, $0.61 in the 1960s, $1.69 in the 1970s, $1.18 in the 1980s, $1.59 in the 1990s, and $3.73 in the first decade of the 2000s. The highest average annual price per pelt was $5.47 in 2006 and the lowest was $0.40 in 1949. It is hard to measure the relative effects of the several factors that influence harvest numbers and pelt prices. Obviously, population abundance is involved, but to what extent has not been determined. Changes in demand for furs and the fluctuating values of their pelts influence economic reasons for trapping striped skunks. In addition, the numbers of trappers working in Missouri influence the harvest.

Striped skunks are common in rural, suburban, and urban areas. Their activities rarely cause serious economic loss, but they can become a nuisance when their burrowing and feeding habits conflict with humans. They may construct their dens under porches or buildings. They sometimes dig holes in lawns, golf courses, and gardens in search of insect grubs. They may damage beehives when attempting to feed on bees, and they may cause minor damage where they feed on corn. They occasionally kill poultry and eat eggs and are known predators on waterfowl nests.

Striped skunks are efficient mousers and also contribute to the control of insects; they are encouraged around many farms for these reasons.

Conservation and Management

The harvest of this species continues to be regulated. The practice of digging skunks from their dens should be discouraged, because females and young are likely to winter for long periods in dens; this portion of the population is unwisely reduced by den destruction.

As a precaution against depredation by skunks, chicken houses can be properly fenced and beehives raised off the ground. If you are experiencing problems with striped skunks, contact a wildlife professional for advice, assistance, regulations, or special conditions for handling these animals.

SELECTED REFERENCES

See also discussion of this species in general references, page 23.

Allen, D. L., and W. W. Shapton. 1942. An ecological study of winter dens, with special reference to the eastern skunk. *Ecology* 23:59–68.

Bixler, A., and J. L. Gittleman. 2000. Variation in home range and use of habitat in the striped skunk (*Mephitis mephitis*). *Journal of Zoology* 251:525–533.

Dean, F. C. 1965. Winter and spring habits and density of Maine skunks. *Journal of Mammalogy* 46:673–675.

Dragoo, J. W., and R. L. Honeycutt. 1997. Systematics of mustelid-like carnivores. *Journal of Mammalogy* 78:426–443.

Eastland, W. G., and S. L. Beasom. 1986. Potential secondary hazards of Compound 1080 to three mammalian species. *Wildlife Society Bulletin* 14:232–233.

Greenwood, R. J., and A. B. Sargeant. 1994. Age-related reproduction in striped skunks (*Mephitis mephitis*) in the upper Midwest. *Journal of Mammalogy* 75:657–662.

Hamilton, W. J., Jr. 1963. Reproduction of the striped skunk in New York. *Journal of Mammalogy* 44:123–124.

Larivière, S., L. R. Watton, and F. Messier. 1999. Selection by striped skunks (*Mephitis mephitis*) of farmsteads and buildings as denning sites. *American Midland Naturalist* 142:96–101.

Petrides, G. A. 1950. The determination of sex and age ratios in fur animals. *American Midland Naturalist* 43:355–382.

Shirer, H. W., and H. S. Fitch. 1970. Comparison from radiotracking of movements and denning habits of the raccoon, striped skunk, and opossum in northeastern Kansas. *Journal of Mammalogy* 51:491–503.

Storm, G. L. 1972. Daytime retreats and movements of skunks on farmlands in Illinois. *Journal of Wildlife Management* 36:31–45.

Wade-Smith, J., and B. J. Verts. 1982. *Mephitis mephitis. Mammalian Species* 173. 7 pp.

Cats (Family Felidae)

The family name, Felidae, is based on the Latin word for "cat." Thirty-nine species of wild cats are found nearly worldwide, with the exceptions being Antarctica, Australia, New Zealand, and most other islands. Missouri has two wild representatives, but only the bobcat has established populations in our state. No breeding population of mountain lion occurs in Missouri, but because of its status as an extirpated native Missouri mammal, the increase in sightings of wandering individuals, and people's interest in this animal, we have decided to include it in this section. This family also contains the lions, leopards, tigers, ocelots, jaguars, cheetahs, lynx, servals, and other species. In addition, domestic cats have been introduced nearly everywhere humans live.

The members of this family are typically catlike in build. There are 5 toes on each front foot and 4 on each hind foot with long, sharp, curved claws that are usually retractile. The teeth are highly specialized for flesh eating: the canines are very long and sharp, and the carnassial teeth (last upper premolar and first lower molar) are very well developed.

Mountain Lion (*Puma concolor*)

Name

Puma is the Latin word for "cat"; *concolor* is the Latin word for "one color," referring to the uniform coloration. The common name, "mountain," refers to its general habitat, while "lion" indicates its relationship to another member of the cat family. Other common names are cougar, puma, panther, painter, catamount, and mountain cat. This species was formerly referred to as *Felis concolor.*

Description (Plate 62)

The mountain lion is a very large, slender cat with a small head, small rounded ears that are not tufted, very powerful shoulders and hindquarters, and a long, heavy, cylindrical tail. The body fur is short and soft. The front foot has 5 toes, the first of which is high and does not touch the ground and thus leaves no impression in a track. The hind foot has 4 toes. All the toes have long, sharp, strongly curved claws that can be extended and retracted. The mountain lion walks on its toes. The hind feet are smaller than the front and, when walking, the hind foot is often placed in the imprint of the front foot.

The mountain lion is distinguished from the bobcat by its large size, uniform coloration, long tail, and dentition. The track of a mountain lion's foot is 7.6 to 10 cm (3 to 4 in.) long; that of a bobcat's, 5 cm (2 in.).

Color. Upperparts are grizzled gray or dark brown to buff, cinnamon tawny, or rufous. The color is most intense from the top of the head to the base of the tail; the shoulders and flanks are somewhat lighter. Underparts are dull whitish overlaid with buff across the abdomen. The sides of the muzzle are black, while the chin and throat are white. The ears are black externally. The last 5 to 7.6 cm (2 to 3 in.) of the tail are black.

The kittens have black spots on a buffy ground color and rings of brownish black on the tail.

Contrary to popular belief, black panthers do not exist in the wild in North America. Black color phases of leopards and jaguars do occur and are sometimes seen in zoos. There has never been a black mountain lion documented anywhere in their range.

Measurements

Total length	152–259 cm	60–102 in.
Tail 53–89 cm	21–35 in.	
Hind foot	215–279 mm	8½–11 in.
Ear 76–101 mm	3–4 in.	
Skull length	155–234 mm	6⅛–9¼ in.
Skull width	104–139 mm	4⅛–5½ in.
Weight		
Male	36–120 kg	79–265 lb.
Commonly	64–73 kg	140–160 lb.
Female	29–64 kg	64–141 lb.
Commonly	41–50 kg	90–110 lb.

Teeth and skull. The dental formula of the mountain lion is:

$$I\,\frac{3}{3}\;C\,\frac{1}{1}\;P\,\frac{3}{2}\;M\,\frac{1}{1} = 30$$

This is the same as that of the domestic cat, but the cat's skull is much smaller. The mountain lion has one more upper premolar than the bobcat.

Like the bobcat, the skull of the mountain lion is short and broad with a very short facial region, a rounded convex forehead, and a prominent blunt chin.

Sex criteria and sex ratio. Sex is identified as in the bobcat. More males are born than females in the ratio

Plate 62

Mountain Lion *(Puma concolor)*

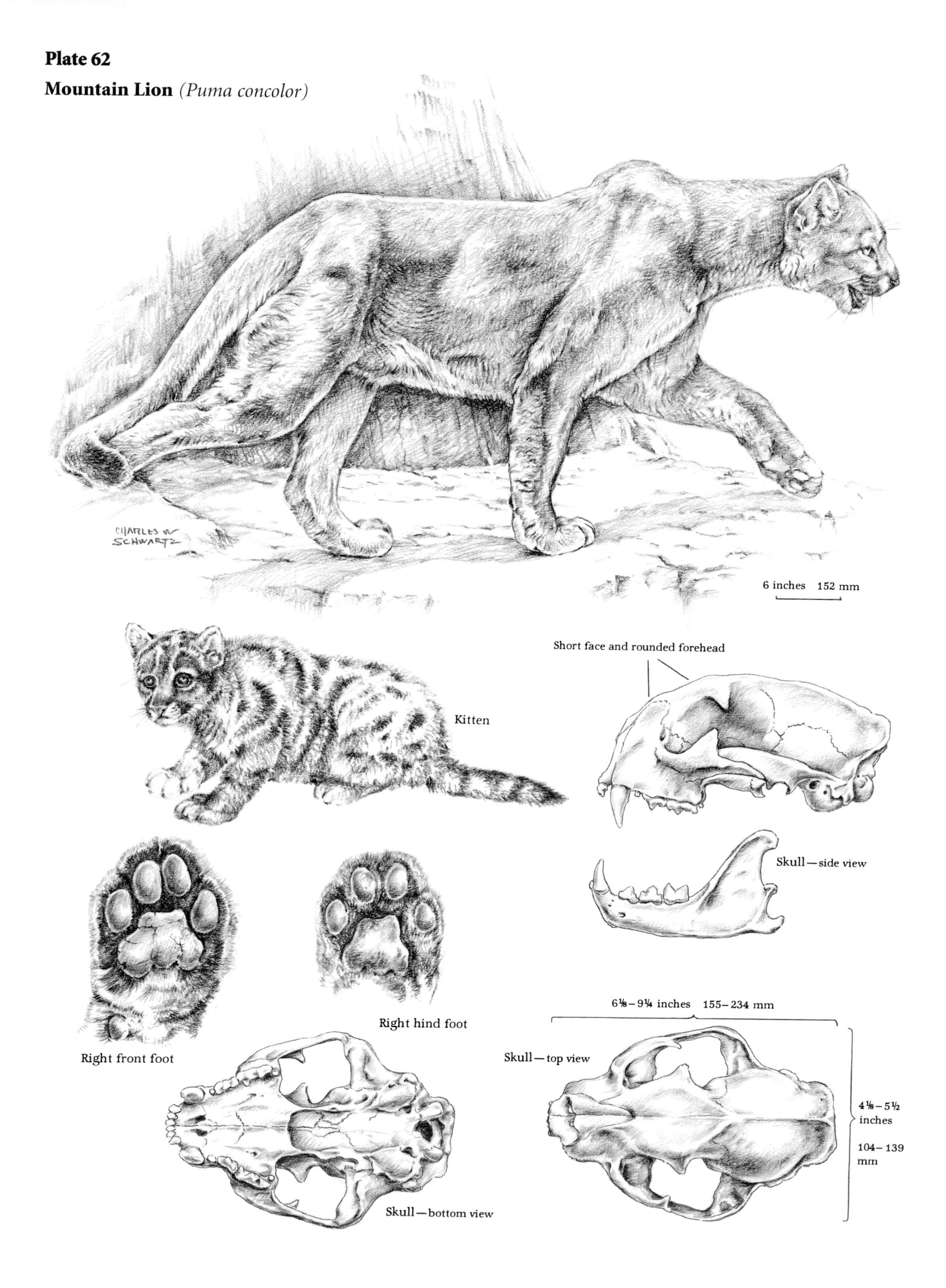

of 120–130 males to 100 females. However, the adult ratio is actually 2 females per male. Females have three pairs of teats.

Age criteria and longevity. Some mountain lions are known to have existed in the wild until 15 or 19 years of age, but most do not reach the age of 12.

Voice and sounds. Mountain lions have a wide range of sounds: cheeps, chirps, whistles, growls, and hisses.

Distribution and Abundance

Mountain lions formerly ranged from northern British Columbia to the southern tip of South America, throughout the swamps and flatlands of the coastal plains to the timbered mountain country. By 1920 in North America, they were restricted to mountainous areas of the West, certain remote swamps in Florida, and scattered areas of Mexico and Central America.

When this book was first written, the mountain lion was listed as a former resident of Missouri because the last definite record was of one killed in 1927 in the Mississippi River Alluvial Basin when it still possessed large-timbered swampland tracts. The demise of mountain lions resulted from the loss of our white-tailed deer herd (their major prey), habitat destruction, and unregulated overhunting. No verifiable evidence exists to indicate that they survived anywhere in the Midwest outside of the Black Hills of South Dakota at that time.

Recently there are increasing reports of scattered, occupied areas within their former range, including the Midwest. There have also been many mountain lion sightings in Missouri, but no conclusive physical evidence existed until 1994. This increase in sightings is due partially to the fact that more cats are dispersing into states east of their current range and partially to new technology, especially trail cameras.

In 1996, the Missouri Department of Conservation formed the Mountain Lion Response Team to investigate sightings, respond to calls, and collect and analyze physical evidence to confirm the presence of mountain lions in Missouri. Most of these sightings turn out to be cases of mistaken identity, but in this search for hard evidence such as mountain lion photos, carcasses, tracks, hair, scat, and videos, 51 instances of mountain lions were confirmed in Missouri between 1994 and 2014. DNA analysis from tissue and hair samples indicates that mountain lions disperse to Missouri from the Black Hills of South Dakota, Colorado, central Montana, and perhaps the Pine Ridge area of Nebraska.

Today, the mountain lion ranges from the Canadian Yukon to the southern Andes of South America with a few small populations in North Dakota, South Dakota, Nebraska, and Florida, and is classified as secure across its range. There is no evidence of a wild, reproducing population of mountain lions in Missouri despite the presence of suitable habitat and abundant prey. The mountain lion is classified as an extirpated species of

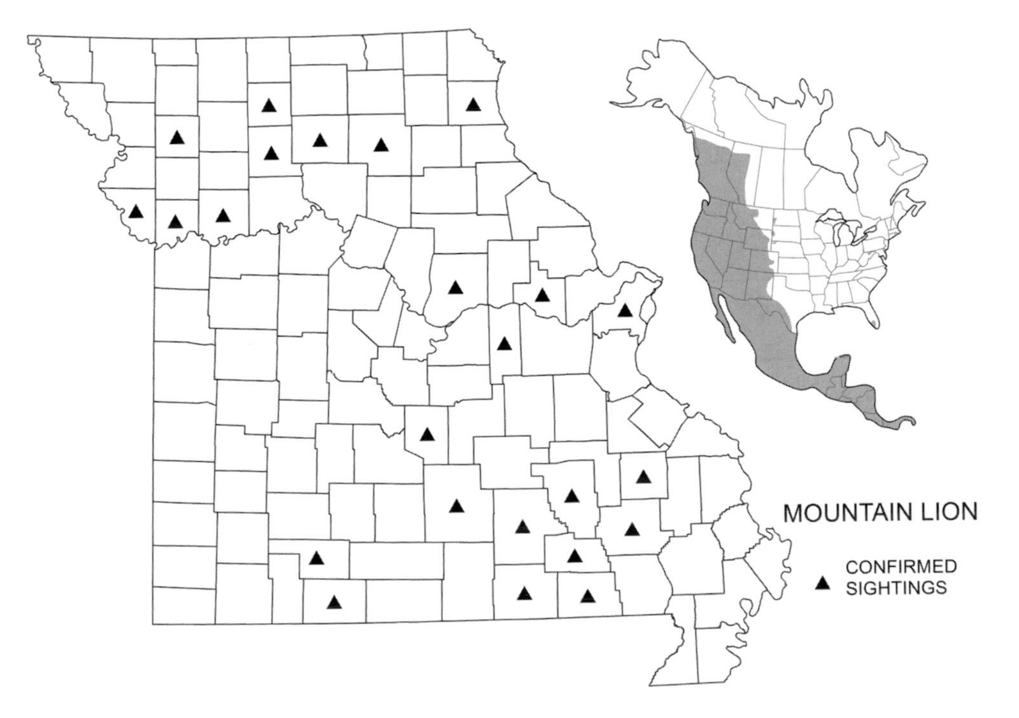

conservation concern in Missouri, and because it was a native resident of the state and sightings continue to occur, we feel that including it here will provide important information for people who want to know more about mountain lions and help in verifying sightings. Whether these animals will eventually establish home ranges and a population in Missouri is unknown.

Habitat and Home

Mountain lions prefer dense, vegetative cover or rocky, rugged terrain, generally in areas of low human habitation, or regions of dense swamps.

These big cats have no special home; they merely seek shelter in rocky crevices, hollow trees or logs, holes in banks, or tall grass or underbrush. No bedding is added for a nest.

Habits

The size of the home range depends upon the food supply, the season, and characteristics of the environment. The usual area of activity is typically 129 to 194 sq km (50 to 75 sq. mi.) for females, and 233 to several hundred sq km (90 to several hundred sq. mi.) for males. Female home ranges overlap extensively with both other females and males, while there is little overlap between ranges of adjacent males. Adult females often share their home range with their female offspring.

The family unit is a female with kittens. Males, and females without kittens, are solitary except during mating time. The kittens may stay together for 2 or 3 months after they leave their mother.

Mountain lions are generally nocturnal and are active near dawn and dusk, but they may be active during the day. The same area is often used by different individuals but seldom at the same time. Various signs, such as scrapes or scratch marks, or spots where individuals have urinated or defecated, serve as markers of use and help other mountain lions identify an occupied area.

Mountain lions avoid areas occupied by other mountain lions yet are attracted to regions where mountain lions live. This paradox may explain why it is difficult for areas that seem optimum for this species' occupancy to be without lions and why it takes so long for this animal to extend its range into new areas away from occupied ranges.

These big cats readily climb trees to obtain food or to escape pursuing dogs. They are capable of swimming but usually avoid doing so because their coats do not resist moisture.

Foods

Deer is a favorite food, but mountain lions also take other mammals including jackrabbits, rabbits, beavers, badgers, opossums, raccoons, skunks, coyotes, minks, small to medium-sized rodents, other mountain lions, and occasionally domestic livestock. The take of deer is not wanton but related to their needs and the deer population. One mountain lion may take about 35 deer a year. They are not particularly destructive to livestock, especially where their natural prey are abundant.

When hunting, a mountain lion stalks its prey to within a distance of about 15 m (50 ft.). Then it rushes and attempts to knock the victim down. Large prey are killed by a bite on the back of the neck or throat. This characteristic method enables researchers to identify mountain lions as the predator when a carcass with the head intact is found. The victim is usually dragged or carried to a hiding place before being fed upon. The stomach and contents are removed and usually dragged away from the carcass. Lions usually first feed on the liver, heart, and lungs, which the mountain lion obtains through an opening in the ribs. If the mountain lion is hungry, the prey may be entirely consumed at a single feeding. But usually, it is not all eaten at once and the remainder of the carcass is covered with leaves or other debris. The mountain lion rests nearby for a day or two, returning to feed when hungry. Mountain lions avoid spoiled or tainted meat. Bobcats and coyotes often feed on mountain lion kills.

Reproduction

It is rare for a female to breed before 2½ and 3 years of age, and thereafter she usually has young at two-year

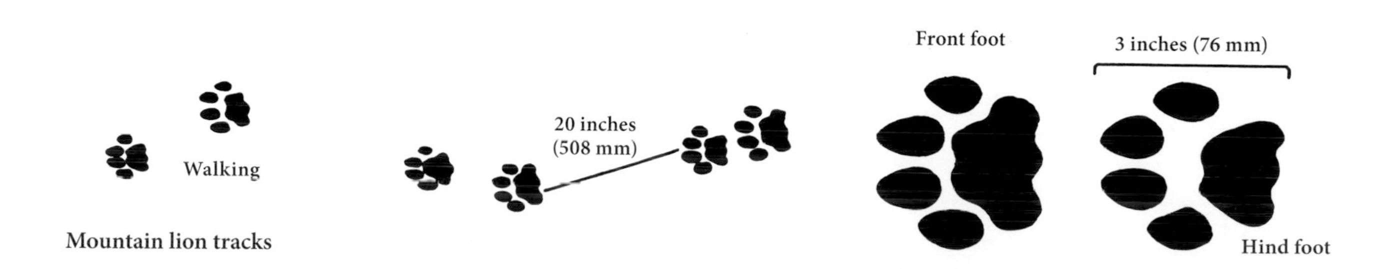

Mountain lion tracks

intervals. Gestation is between 90 and 96 days. The young are born in any month of the year, but there is a peak in July. There may be from 1 to 6 (average 2 to 3) kittens per litter.

At birth the kittens are 30.5 cm (12 in.) long, weigh about 0.5 kg (1 lb.), and are blind. They are buff-colored, spotted with black, and have rings of brownish black on the tail. About 6 or 7 weeks of age, the young are 76 cm (2½ ft.) long, weigh nearly 4½ kg (10 lb.), and begin to chew bones. The female carries food to them until they accompany her hunting, at about 2 months of age. The kittens lose their spots gradually. They often stay with their mother until they are 2 years old.

Some Adverse Factors

Humans are the greatest cause of death, but many other things take their toll, such as highway kills, competition between newborns for milk, male mountain lions attacking young, the physiological drain from parasites such as trichinella, tapeworms, ticks, fleas, and mites and, rarely, diseases such as rabies.

Importance; Conservation and Management

Because of the habit of mountain lions, especially subadult males, to disperse large distances, this species most likely will continue to visit Missouri and will occasionally wander into places densely settled by humans. We have documented cats near St. Louis and Kansas City, but most sightings have been in the Ozarks. Unlike black bears, mountain lions are not interested in human foods. They are occasionally seen by motorists, hunters, or hikers, and their images have been captured on trail cameras. If you see or suspect that you are experiencing problems with a mountain lion, contact your local Missouri Department of Conservation office.

Mountain lions and humans coexist throughout North America. As human populations grow, the likelihood of interactions with these animals will increase. Mountain lion attacks on humans are extremely rare. Across their North America range, roughly 25 fatalities and 95 nonfatal attacks by mountain lions have been reported during the past 100 years.

There is little value to mountain lion fur because the guard hairs are rather stiff and the underfur is short and thin. However, some pelts are used for rugs or wall hangings. The meat is edible.

SELECTED REFERENCES

See also discussion of this species in general references, page 23.

Beier, P., D. Choate, and R. H. Berrett. 1995. Movement patterns of mountain lions during different behaviors. *Journal of Mammalogy* 76:1056–1070.

Beringer, J. 2012. Living with large carnivores. *Missouri Conservationist* 73:22–27.

Clark, D. W., S. C. White, A. K. Bowers, L. D. Lucio, and G. A. Heidt. 2002. A survey of recent accounts of the mountain lion (*Puma concolor*) in Arkansas. *Southeastern Naturalist* 1:269–278.

Currier, M. J. D. 1983. *Felis concolor. Mammalian Species* 200. 7 pp.

Hamilton, D. 2006. Mountain lions in Missouri: Fact or fiction? *Missouri Conservationist* 67:10–15.

Hardin, S. E. 1996. The status of the puma (*Puma concolor*) in Missouri, based on sightings. M.S. thesis, Southwest Missouri State University, Springfield. 63 pp.

Hornocker, M. G. 1969. Winter territoriality in mountain lions. *Journal of Wildlife Management* 33:457–464.

———. 1970. An analysis of mountain lion predation upon mule deer and elk in the Idaho primitive area. *Wildlife Monograph* 21. 39 pp.

LaRue, M. A., and C. K. Nielsen. 2008. Modelling potential dispersal corridors for cougars in Midwestern North America using least-cost path methods. *Ecological Modelling* 212:372–381.

Lewis, J. C. 1969. Evidence of mountain lions in the Ozarks and adjacent areas, 1948–1968. *Journal of Mammalogy* 50:371–372.

McBride, R. T., R. M. McBride, J. L. Cashman, and D. S. Maehr. 1993. Do mountain lions exist in Arkansas? *Proceedings of the Annual Conference of the Southeastern Association of Fish and Wildlife Agencies* 47:394–402.

McKinley, D. 1961. The mountain lion: A history of Missouri's big cat. *Bluebird* 28:6–11.

Pike, J. R., J. H. Shaw, D. M. Leslie Jr., and M. G. Shaw. 1999. A geographic analysis of the status of mountain lions in Oklahoma. *Wildlife Society Bulletin* 27:4–11.

Robb, D. 1955. Cougar in Missouri. *Missouri Conservationist* 16:14.

Robinette, W. L., J. S. Gashwiler, and O. W. Morris. 1961. Notes on cougar productivity and life history. *Journal of Mammalogy* 42:204–217.

Seidensticker, J. C., IV, M. G. Hornocker, W. V. Wiles, and J. P. Messick. 1973. Mountain lion social organization in the Idaho primitive area. *Wildlife Monograph* 35. 60 pp.

Thompson, D. J., D. M. Fecske, J. A. Jenks, and A. R. Jarding. 2009. Food habits of recolonizing cougars in the Dakotas: Prey obtained from prairie and agricultural habitats. *American Midland Naturalist* 161:69–75.

Thompson, D. J., and J. A. Jenks. 2005. Long-distance dispersal by a subadult male cougar from the Black Hills, South Dakota. *Journal of Wildlife Management* 69:818–820.

Young, S. P., and E. A. Goldman. 1946. *The puma: Mysterious American cat.* American Wildlife Institute, Washington, DC. 358 pp.

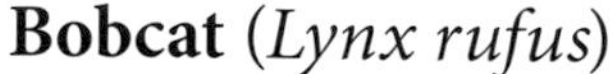

Bobcat (*Lynx rufus*)

Name

The first part of the scientific name, *Lynx*, is a Greek word given to lynxlike cats; it comes from two Greek roots meaning "in lamp" and "to see" and may refer to the bright eyes of the animal as seen in reflected light. The second part, *rufus*, is the Latin word for "reddish," describing the general body color. The common name refers to the short or bobbed tail. Some authorities refer to this species as *Felis rufus*.

Description (Plate 63)

Although the bobcat is only a medium-sized member of the cat family, it is one of the largest wild mammals in Missouri. It has a short, broad face, set off by a slight ruff of fur on the sides of the face extending from the ear down to the lower jaw; prominent, pointed ears, sometimes with tufts that are less than 2.5 cm (1 in.) in height; rather long legs; and a short tail. There are 5 toes on the front foot, of which the first is high and does not touch the ground, and 4 on the hind foot, all with long, sharp, strongly curved claws that can be extended and retracted. The bobcat walks on its toes. The tongue has small, rasping projections on the upper surface. The pupil of the eye is elliptical when contracted in bright light but nearly round when dilated in dim light. The fur is short and sleek, becoming longer and fuller in winter.

The bobcat is distinguished from the domestic cat, which sometimes lives in the wild, by its distinctive color pattern, larger size, proportionately longer legs, much shorter tail, and two fewer upper premolar teeth. It is distinguished from the mountain lion by its smaller size, coloration, shorter tail, and dental formula.

Color. The upperparts and sides are yellowish to reddish brown, streaked and spotted with black. The backs of the ears are black with a central white or gray spot; the tufts, when present, are black. The underparts are white with black spots and several indefinite black bars on the inner sides of the front legs. The upper surface of the tail is yellowish to reddish brown and has 3 or 4 poorly defined brownish black bars toward the end. The bar nearest the end is the broadest and darkest and usually is followed by a small white tip. Below, the tail is white. The sexes are colored alike. There is some variation in color; with age, bobcats lose many of the black spots and streaks, and in winter the fur is paler than in summer. Blackish bobcats occur very rarely. The young have spotted fur.

Measurements

Total length	48–127 cm	18½–50 in.
Tail	90–200 mm	3½–7⅞ in.
Hind foot	127–228 mm	5–9 in.
Ear	63–69 mm	2½–2¾ in.
Skull length	101–139 mm	4–5½ in.
Skull width	69–104 mm	2¾–4⅛ in.
Weight	3.8–22 kg	8–49 lb.

Males are somewhat larger than females.

In 2014, a 16.3 kg (36 lb.) female, and in 2012, a 14.4 kg (31.8 lb.) male, became the official record-weight bobcats confirmed by the Missouri Department of Conservation. The state furbearer record program was started in 2011.

Teeth and skull. The bobcat is the only wild carnivore in Missouri with this dental formula:

$$I\ \frac{3}{3}\ C\ \frac{1}{1}\ P\ \frac{2}{2}\ M\ \frac{1}{1} = 28$$

The skull is broad and relatively short with a very short facial region; a rounded, convex forehead; and a prominent, blunt chin. Mountain lion skulls, which are similar, have one additional upper premolar.

Sex criteria and sex ratio. Males are identified by the penis and scrotum; they do not possess a penis bone. There are three pairs of teats in females. The sex ratio in different parts of the range is from 1 to 3 males per female.

Age criteria and longevity. The young have a spotted coat. The age of bobcats is determined by the various techniques used for other carnivores. The dried weight of the eye lens, time of closure of certain sphenoid bones of the skull, and the replacement of cartilage at the ends of long bones give a general indication of age. The most accurate means of establishing a

Plate 63

Bobcat *(Lynx rufus)*

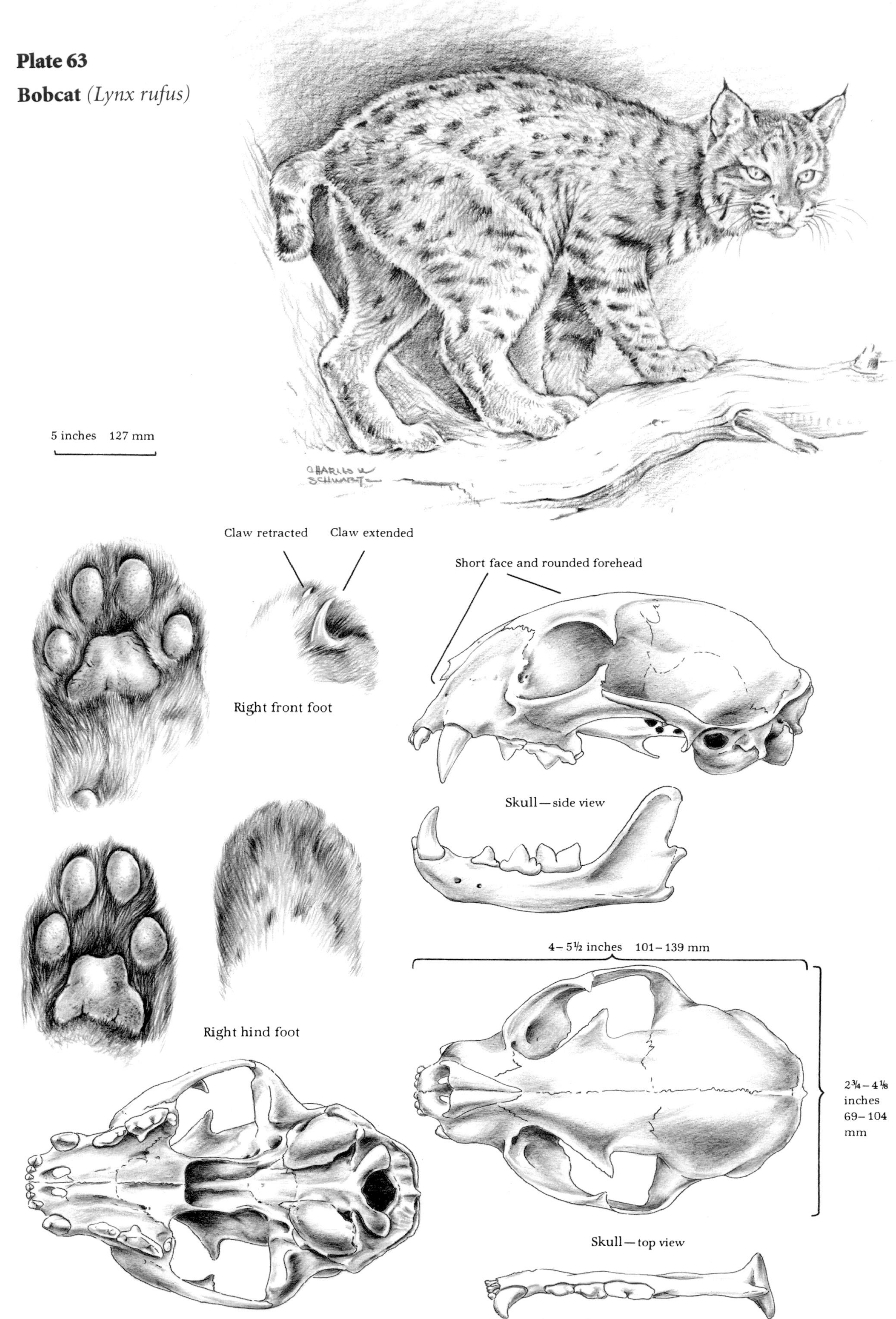

specific year-age appears to be by the number of cementum layers on the canine teeth.

In the wild, bobcats may live to 10 or 12 years of age, with most living 6 to 8 years in the wild, and in captivity up to 25 years.

Glands. Bobcats have a very strong odor, and their dens also develop this characteristic smell. The secretion from anal glands may be used along with urine as a scent marker.

Voice and sounds. Bobcats are generally quiet but may give high-pitched screams or low growls. During the breeding season, when they are more vociferous than at other times, their caterwauling consists of squalls, howls, meows, and yowls. When captured, they growl, hiss, and spit.

Distribution and Abundance

Originally the bobcat range was from southern Canada, throughout the United States and most of Mexico. However, it had been extirpated from much of the central United States by 1900 due to habitat loss resulting from intensive agriculture and exploitation because of real or perceived attacks on livestock. Recent information suggests that bobcats are reoccupying their midwestern landscapes due to changing agricultural and land-use practices, increased habitat and resulting prey species from habitat improvement programs, and improved wildlife management for bobcats. Populations are now stable or increasing throughout most of their range. Today, the bobcat ranges from southern Canada, throughout most of the United States, and into southern Mexico. Marked fluctuations, related to food supply, occur in the population.

In Missouri, bobcats used to live primarily in the Ozark Highlands and Mississippi River Alluvial Basin. From records of bounties paid for bobcat pelts in Missouri, the population was estimated to have been relatively high in the late 1930s, low in the early 1950s, high in the early 1960s, and dwindling into the 1970s. However, bobcat populations have expanded tremendously, beginning in western Missouri in the 1980s, and then moving into the northwestern counties, north-central counties, and now northeastern Missouri. Bobcats now occur statewide and have become established in all suitable habitats. They are most common in the Ozark Highlands, followed by the Osage Plains of western Missouri, and the northwest portion of the Central Dissected Till Plains.

Populations vary from 2 to 7 bobcats per 2.5 sq km (1 sq. mi.) in parts of Texas and California. In the Missouri Ozarks, the estimate was once 1 bobcat per 15 sq km (6 sq. mi.).

Habitat and Home

The bobcat lives in heavy forest cover, preferably second-growth timber with much underbrush, broken with clearings, glades, rocky outcrops, and bluffs, and in timbered swamps. Old fields, clearcuts, and caves are important. In other parts of its range, the bobcat may inhabit very arid and rocky terrain. It occurs in

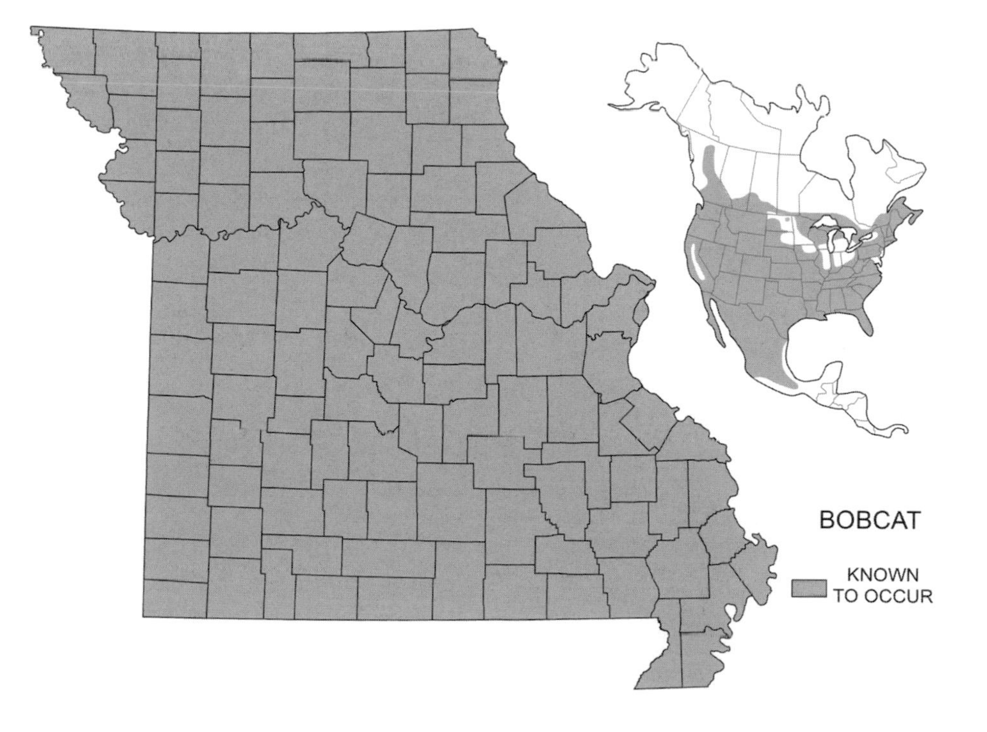

almost every terrestrial habitat type, indicating its role as a habitat generalist.

During most of the year, a fresh rest shelter such as a thicket, standing or fallen hollow tree, or a recess in a rocky cliff is used each day. In the breeding season similar but usually more inaccessible places are chosen for a den. The nest is made of dried leaves and soft moss.

Habits

In Missouri males have an annual home range of 46 to 72 sq km (18 to 28 sq. mi.). They move more in winter and spring than in summer and fall, with the most restricted range occurring during the hot months of July and August. Transient males have larger home ranges and may make extensive movements of up to 64 km (40 mi.).

Females have a home range of 13 to 31 sq km (5 to 12 sq. mi.) from January to March. In spring and summer they live in a more restricted area because they are caring for kittens.

In general, females tend to have almost exclusive home ranges, but males have ranges that overlap those of other males and extensively with females. In addition, a population generally contains transients, which tend to be dispersing juveniles. Evidence suggests that male home ranges increase with time and experience, thereby improving reproductive fitness through increased breeding opportunities. In contrast, with time and experience females improve fitness by decreasing home range size and the associated energy requirements of territory maintenance.

There is little social interaction between individuals. This lack of contact is apparently achieved by marking the home range with scent places containing fecal matter and by squirting urine at various places. These scents serve to prevent encounters of resident individuals and to notify transients that the range is occupied.

Within these home ranges, some individuals travel between 5 and 11 km (3 and 7 mi.) a night and often move distances of 3 to 11 km (2 to 7 mi.) between resting places. Their greatest movement occurs when food is scarce or, for males, during the mating season. Bobcats are very curious and investigate many objects along their way, which accounts for their customary zigzag trail. They usually walk, trot, or take leaps up to 3 m (10 ft.) in length.

In locating prey, a bobcat depends more upon its keen eyesight and hearing than its sense of smell. When stalking, it usually creeps stealthily along, then pounces on its prey; or it may crouch on a game trail or tree limb and await an unwary victim. Bobcats can

kill animals as large as deer by biting the throat at the jugular vein.

Bobcats are both nocturnal and diurnal with most hunting occurring around sunrise and sunset. They are active all year. However, they often remain in a resting place during a storm and avoid deep, soft snow because of the difficulty of walking.

Bobcats are capable of swimming and readily cross streams and small rivers. They are good climbers and take to trees as a refuge from dogs or for resting or observation. They often stretch against some hard snag to sharpen their front claws, much in the manner of domestic cats.

A study of the food habits of 41 bobcats in Missouri showed the following foods and their percentages by volume: rabbits, 67.0; mice, rats, and shrews, 0.7; squirrels, 9.9; white-tailed deer, 8.6 (some of which is probably carrion); opossums, 1.9; domestic cats, 1.7; wild turkeys, 7.9; quail, 1.7; undetermined meat, 0.5; and grasses, 0.1. This agrees with the foods of bobcats in other parts of the range where the diet consists primarily of small mammals supplemented by various birds and reptiles. In Missouri, bobcats kill large white-tailed deer occasionally and fawns often.

Bobcats gorge when food is plentiful, and may not feed again for several days. They seldom return to eat from an old cached kill, or in the case of larger prey such as deer, unless food is scarce. They waste considerable meat and may kill more than they eat. They use their feet to bury any surplus food under snow or leaves.

Reproduction

Breeding begins in December and may extend into June with usually a peak in March. After a 50- to 70-day gestation period, 2 or 3 (with extremes of 1 and 8) young are born. Most litters arrive from mid-May to mid-June, but some are born as late as September or October.

At birth each kitten weighs about 340 g (12 oz.) and measures 25.4 cm (10 in.) in length. They have spotted fur and sharp claws. When about 9 to 11 days old, their eyes open. They soon come outside the den where they gambol and play. Weaning occurs around 2 months of age at which time the young begin to accompany the mother on occasion. The young stay with the female until fall or even later. Females mate when 1 or 2 years old and can produce a single litter each year thereafter, but males do not breed until 2 years of age.

Some Adverse Factors

Humans and domestic dogs are the most important predators and, in the past, excessive persecution for alleged damage to livestock took a heavy toll. Coyotes and mountain lions may prey on adults, and foxes, coyotes, great horned owls, and male bobcats take some kittens. The following parasites are known to occur on or in bobcats: mites, ticks, lice, fleas, roundworms, flukes, tapeworms, spiny-headed worms, and protozoa. The known diseases are distemper (feline panleucopenia) and rabies. The bobcat is also the known host of a fatal disease (Cytauxzoonosis) of domestic cats.

Importance

Bobcat pelts are used for trimming and for making coats, jackets, and hats. Although a heavy demand and take caused a closing of the hunting and trapping season on bobcats in 1977, the season was reopened in 1980. The bobcat was listed in Appendix II of the Convention on International Trade in Endangered Species of Wild Fauna and Flora (CITES) in 1977; it was included not because it is considered threatened with extinction at this time, but because it is similar in appearance to other cat species that are threatened or endangered, and hunting and trading must be closely monitored to avoid the possibility of confusion among bobcat pelts and the pelts of endangered cats.

In the 1940s the average number of pelts sold annually in Missouri was 71. This harvest dwindled in the 1950s to 14, increased slightly to 51 in the 1960s, jumped tenfold in the 1970s to 537, and then increased to 890 in the 1980s, 938 in the 1990s, and 2,976 in the first decade of the 2000s. During this seven-decade period, the highest annual take was 4,453 in 2006 and the lowest was 3 in 1957. The average annual price paid per pelt was $0.66 in the 1940s, $0.55 in the 1950s, $1.77 in the 1960s, $29.10 in the 1970s, $50.25 in the 1980s, $16.49 in the 1990s, and $40.00 in the first decade of the 2000s. The highest average annual price per pelt was $72.92 in 1987 and the lowest was $0.40 in 1948, 1949, and the 1953–1956 seasons. It is hard to measure the relative effects of the several factors that influence harvest numbers and pelt prices. Obviously, population abundance is involved, but to what extent has not been determined. Changes in demand for furs and the fluctuating values of their pelts influence economic reasons for trapping bobcat. In addition, the numbers of trappers working in Missouri influence the harvest.

As predators and scavengers, bobcats play an important role in the wildlife community.

Conservation and Management

Bobcats are opportunistic predators and will feed on poultry, sheep, goats, house cats, small dogs, and exotic birds. Unless they are causing heavy damage to human interests, their presence should be tolerated and regarded as a conservation success story. If you are experiencing problems with bobcats, contact a wildlife professional for advice, assistance, regulations, or special conditions for handling these animals.

SELECTED REFERENCES

See also discussion of this species in general references, page 23.

Bailey, T. N. 1974. Social organization in a bobcat population. *Journal of Wildlife Management* 38:435–446.

Conner, M., B. Plowman, B. D. Leopold, and C. Lovell. 1999. Influence of time-in-residence on home range and habitat use of bobcats. *Journal of Wildlife Management* 63:261–269.

Crowe, D. M. 1972. The presence of annuli in bobcat tooth cementum layers. *Journal of Wildlife Management* 36:1330–1332.

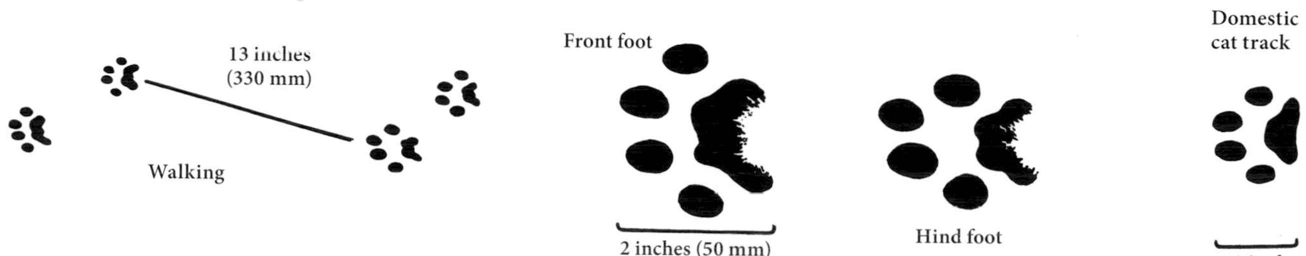

Bobcat tracks

———. 1975. Aspects of ageing, growth, and reproduction of bobcats from Wyoming. *Journal of Mammalogy* 56:177–198.

Erickson, D. W., D. A. Hamilton, and F. G. Sampson. 1981. The status of the bobcat (*Lynx rufus*) in Missouri. *Transaction of the Missouri Academy of Science* 15:49–60.

Foote, L. E. 1945. Sex ratio and weights of Vermont bobcats in autumn and winter. *Journal of Wildlife Management* 9:326–327.

Fritts, S. H., and J. A. Sealander. 1978a. Reproductive biology and population characteristics of bobcats (*Lynx rufus*) in Arkansas. *Journal of Mammalogy* 59:347–353.

———. 1978b. Diets of bobcats in Arkansas with special reference to age and sex differences. *Journal of Wildlife Management* 42:533–539.

Fuller, T. K., S. L. Berendzen, T. A. Decker, and J. E. Cardoza. 1995. Survival and cause-specific mortality rates of adult bobcats (Lynx rufus). *American Midland Naturalist* 134:404–408.

Gashwiler, J. S., W. L. Robinette, and O. W. Morris. 1961. Breeding habits of bobcats in Utah. *Journal of Mammalogy* 42:76–84.

Hamilton, D. A. 1979. *Ecology and status of the bobcat in Missouri.* Semiannual report of the Missouri Cooperative Wildlife Research Unit, July–December 1979: 46–50.

———. 1982. Ecology of the bobcat in Missouri. M.S. thesis, University of Missouri, Columbia. 152 pp.

———. 1998. Internal memo to the Regulation Committee of the Missouri Department of Conservation. 9 pp. + attachments.

Larivière, S., and L. R. Walton. 1997. *Lynx rufus. Mammalian Species* 563:1–8.

Marshall, A. D., and J. H. Jenkins. 1966. Movements and home ranges of bobcats as determined by radio tracking in the upper coastal plain of west-central South Carolina. *Proceedings of the Southeastern Association of Game and Fish Commissioners* 20:206–214.

Marston, M. A. 1942. Winter relations of bobcats to white-tailed deer in Maine. *Journal of Wildlife Management* 6:328–337.

Peterson, R. L., and S. C. Downing. 1952. Notes on the bobcats (*Lynx rufus*) of eastern North America with the description of a new race. *Contributions of the Royal Ontario Museum of Zoology and Paleontology, Toronto, Canada* 33. 23 pp.

Pollack, E. M. 1950. Breeding habits of the bobcat in northeastern United States. *Journal of Mammalogy* 31:327–330.

———. 1951a. Food habits of the bobcat in the New England states. *Journal of Wildlife Management* 15:209–213.

———. 1951b. Observations on New England bobcats. *Journal of Mammalogy* 32:356–358.

Pollack, E. M., and W. G. Sheldon. 1951. *The bobcat in Massachusetts, including analysis of food habits of bobcats from northeastern states, principally New Hampshire.* Massachusetts Division of Fisheries and Game. 24 pp.

Progulske, D. R. 1952. The bobcat and its relation to prey species in Virginia. M.S. thesis, Virginia Polytechnic Institute, Blacksburg. 135 pp.

Roberts, N. M., and Crimmins, S. M. 2010. Bobcat population status and management in North America: Evidence of large-scale population increase. *Journal of Fish and Wildlife Management* 1:169–174.

Rolley, F. E., and W. D. Warde. 1985. Bobcat habitat use in southeastern Oklahoma. *Journal of Wildlife Management* 49:913–920.

Rollings, C. T. 1945. Habits, foods, and parasites of the bobcat in Minnesota. *Journal of Wildlife Management* 9:131–145.

Tucker, S. A., W. R. Clark, and T. E. Gosselink. 2008. Space use and habitat selection by bobcats in the fragmented landscape of south-central Iowa. *Journal of Wildlife Management*, 72:1114–1124.

Tumlison, R., and V. R. McDaniel. 1990. Analysis of the fall and winter diet of the bobcat in eastern Arkansas. *Proceedings of the Arkansas Academy of Science* 44:114–117.

Woolf, A., and G. E. Hubert Jr. 1998. Status and management of bobcats in the United States over three decades: 1970s–1990s. *Wildlife Society Bulletin* 26:287–293.

Young, S. P. 1958. *The bobcat of North America.* Stackpole Books, Harrisburg, PA. 193 pp

———. 1978. *The bobcat of North America: Its history, life habits, economic status and control.* University of Nebraska Press, Lincoln. 193 pp.

9

Even-toed Hoofed Mammals
Order Artiodactyla

The members of this order are hoofed mammals, or ungulates, characterized by 2 elongated, hoofed toes on each foot that support the weight of the body. The other toes are either reduced in size or absent. The name, Artiodactyla, is from two Greek words and means "even-numbered toes." This large order contains the peccaries, camels, giraffes, moose, caribou, and others; plus our domestic cattle, sheep, goats, and pigs. They are found on every continent except Antarctica and were introduced to Australia and New Zealand. Two wild species occur in Missouri: the white-tailed deer and the recently restored elk.

A closely related order, the odd-toed hoofed mammals or order Perissodactyla, has no wild representatives in Missouri, although it contains our domestic horses, asses, and mules.

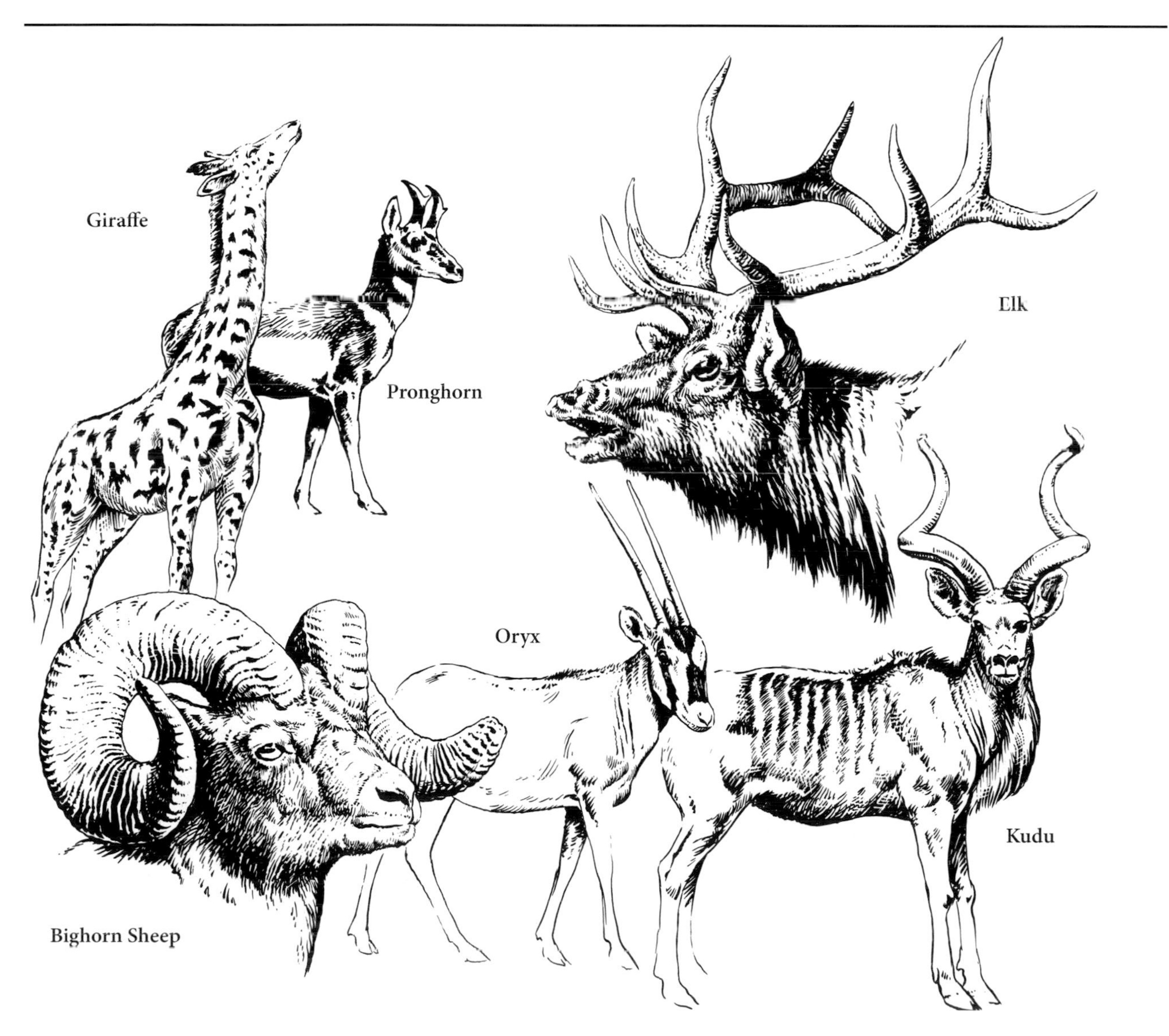

Key to the Species
by Whole Adult Animals

1a. Size large, more than 1.2 m (4 ft.) high at shoulders; tail light tan or straw-colored. **Elk** (*Cervus elaphus*)
1b. Size smaller; less than 1.2 m (4 ft.) high at shoulders; tail brown above, white below. **White-tailed Deer** (*Odocoileus virginianus*)

Key to the Species
by Skulls of Adults[1]

1a. Incisor teeth present in upper jaw. **Go to 2**
1b. Incisor teeth absent in upper jaw. **Go to 3**

2a. (From 1a) Eye socket not enclosed by solid bony ring; upper canine teeth project outward and often upward. **Pig** (*Sus scrofa*)

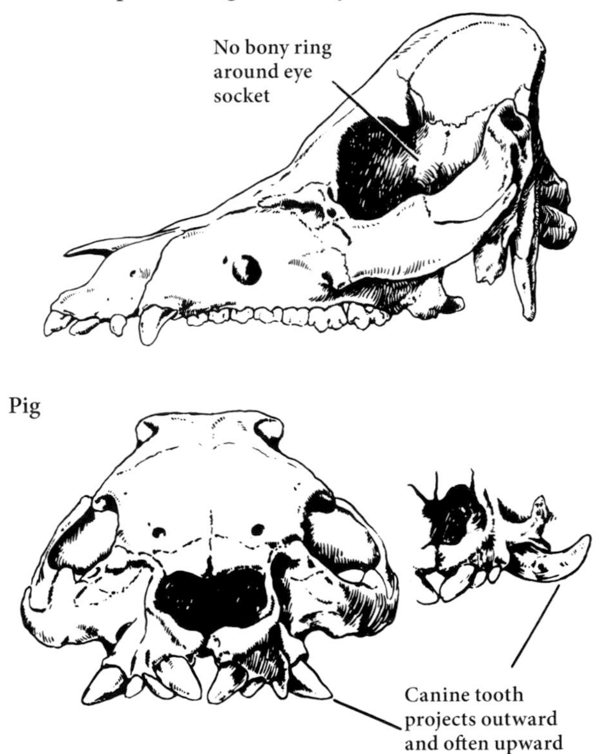

Pig

2b. (From 1a) Eye socket enclosed by solid bony ring; upper canine teeth absent or, when present, project downward. **Horse, Ass,** or **Mule** (*Equus* spp.)

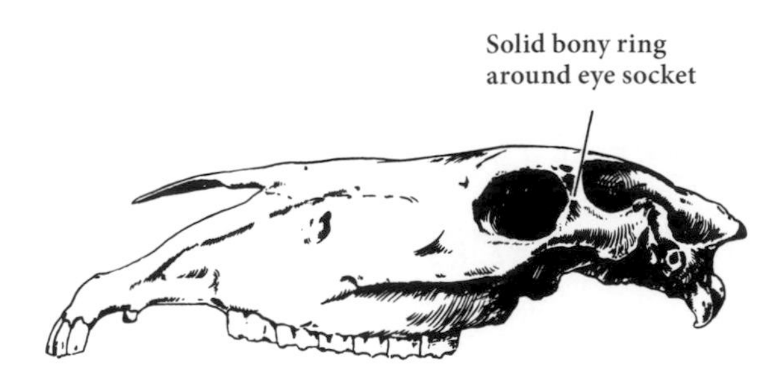

Horse

3a. (From 1b) Large space (15 mm [⅝ in.] or more in width) in front of eye socket between lacrimal and nasal bones; antlers sometimes present. **Go to 4**
3b. (From 1b) No space or small space (less than 15 mm [⅝ in.] in width) in front of eye socket between lacrimal and nasal bones. **Go to 5**

4a. (From 3a) Knob-like canine teeth present on each side of upper jaw; length of upper cheek tooth row more than 110 mm (4⅓ in.). **Elk** (*Cervus elaphus*) p. 410

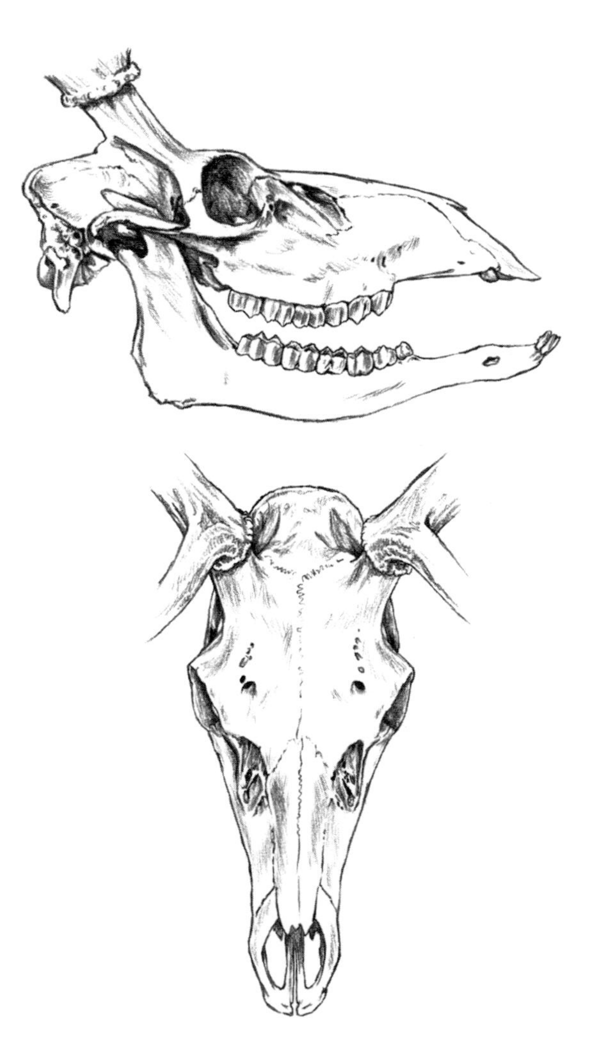

1. This key includes the domestic species belonging to this order and, in addition, the horse, ass, and mule. Because the skeletal remains of these domestic animals may be found in the same general area where white-tailed deer and elk live, their skulls are considered here to avoid any confusion in identification.

4b. (From 3a) Upper canine teeth normally absent; length of upper cheek tooth row less than 110 mm (4⅓ in.). **White-tailed Deer** (*Odocoileus virginianus*) p. 395

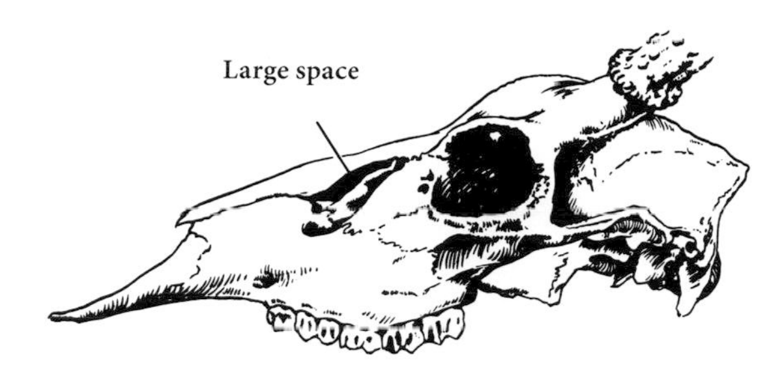

White-tailed deer

5a. (From 3b) Skull 311 mm (12¼ in.) or more in length; cheek tooth row 101 mm (4 in.) or more in length. **Cow** (*Bos taurus*)

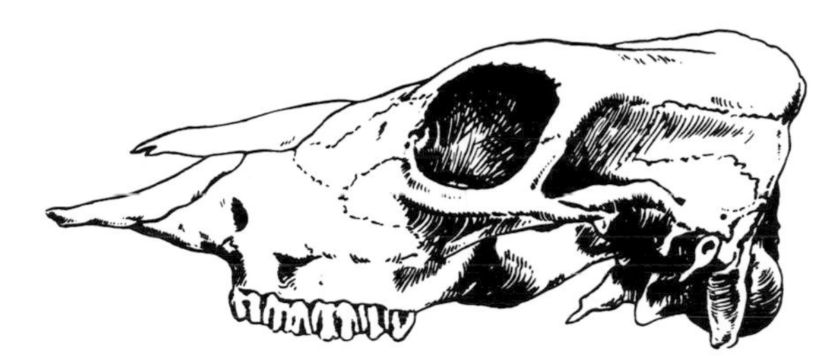

Cow

5b. (From 3b) Skull less than 311 mm (12¼ in.) in length; cheek tooth row less than 101 mm (4 in.) in length. **Go to 6**

6a. (From 4b) No deep depression directly in front of eye socket in lacrimal bone; horns when present nearly parallel and curved backward; groove on top of skull from inner side of base of each horn, or knob, forming a V or U on forehead. **Goat** (*Capra* spp.)

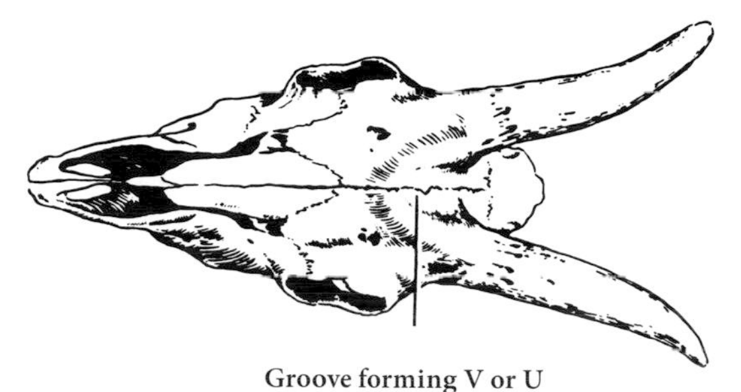

Goat

6b. (From 4b) Deep depression directly in front of eye socket in lacrimal bone; horns when present curved decidedly outward and downward; no groove on top of skull forming a V or U on forehead. **Sheep** (*Ovis aries*)

Sheep

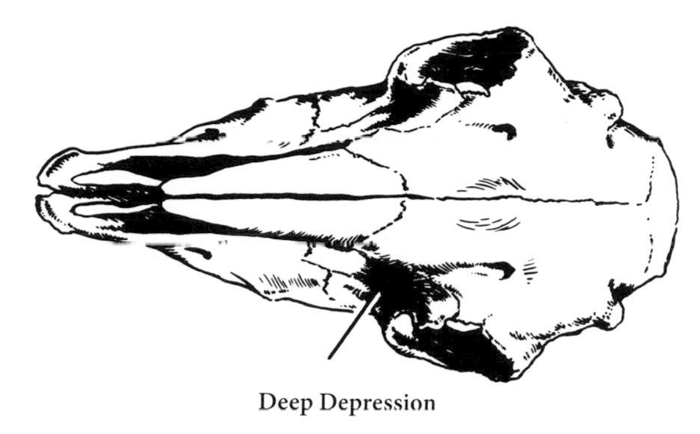

Deer, Elk, and Relatives
(Family Cervidae)

The family name, Cervidae, is based on the Latin word meaning "deer." In this family, the males (and only rarely the females) have bony antlers that are grown and shed each year. Upper incisors are absent and, in most species, upper canines are normally absent too.

Members of this family occur throughout most of the world. The white-tailed deer and elk are the only wild representatives living in Missouri.

White-tailed Deer
(*Odocoileus virginianus*)

Name

The first part of the scientific name, *Odocoileus*, comes from two Greek words and means "hollow tooth" (*odous*, "tooth," and *koilos*, "hollow"), referring to the prominent depressions in the molar teeth. The last part of the scientific name, *virginianus*, is a Latinized word meaning "of Virginia," for the locality cited in the original technical description. The common name, "white-tailed," refers to the white undersurface of the tail, which is very conspicuous when the tail is raised as the deer bounds away. The word "deer" comes from the Anglo-Saxon *deor* or *dior*.

Description (Plate 64)

The white-tailed deer is so well-known that it needs little description. It is readily recognized by its long

Plate 64

White-tailed Deer *(Odocoileus virginianus)*

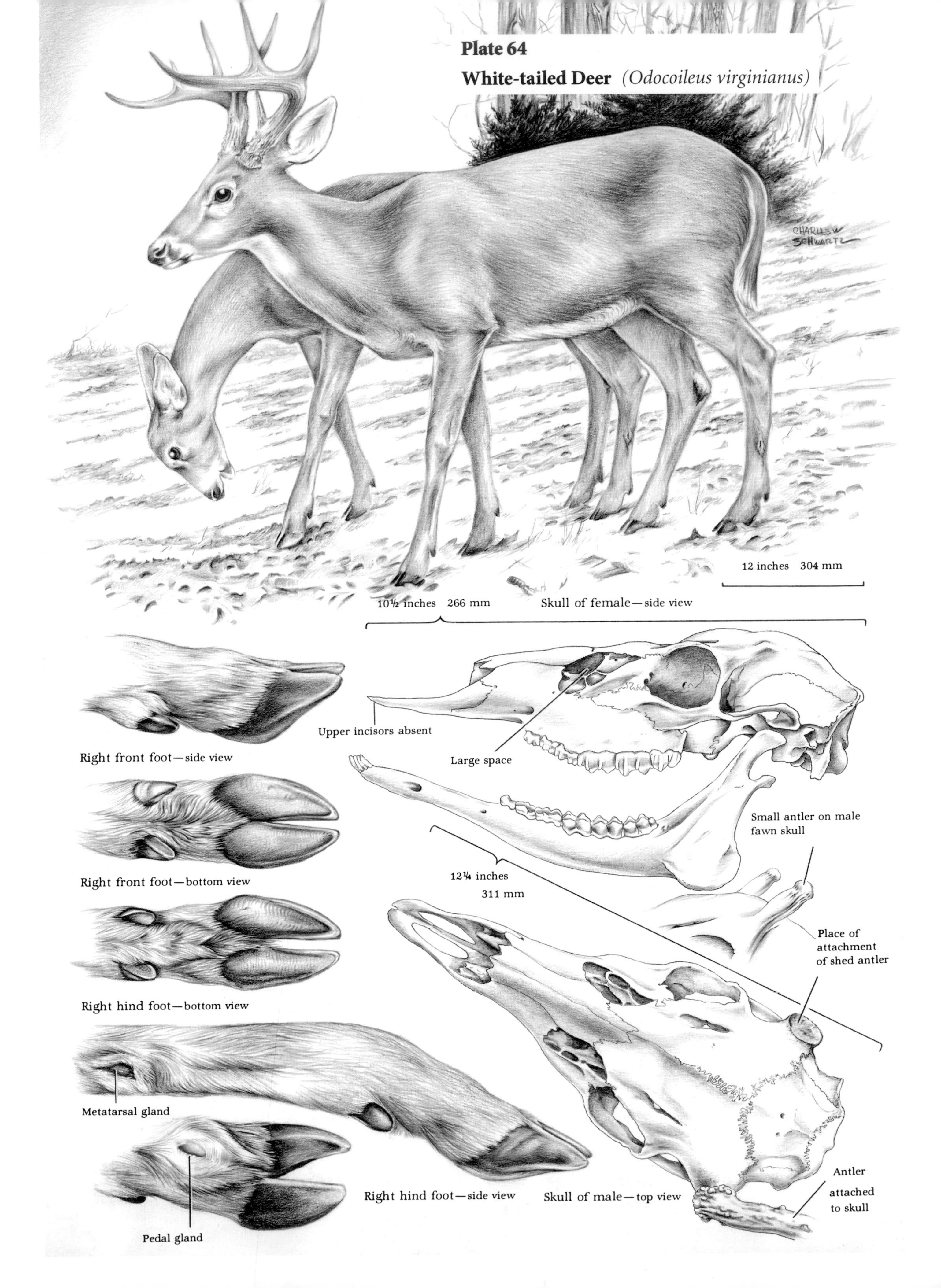

legs and hoofed toes, moderately long and well-haired tail, large size, presence of antlers during part of the year in males, and the presence of spots on fawns during the summer months.

Antlers normally occur only in males and are formed and shed each year. The increasing hours of daylight in spring stimulate the pituitary gland to initiate antler growth. Actual growth starts in April or May when the base of the antler (*pedicle*), located on the frontal bone of the skull, is still covered with soft skin richly supplied with blood vessels. The blood transports the calcium, phosphorus, protein, and other materials from which the antlers are made. During growth, from the time the antlers first appear as "buds" until they assume their final size, the soft skin and the short hair covering each antler have a plush-like quality; deer in this stage are said to be "in velvet." Full size is reached in August or September, shortly before the breeding, or "rutting," season. By this time a male sex hormone, testosterone, is being produced in increasing amounts and initiates calcification of the antler and shedding of the velvet. In this process, the blood supply is cut off at the base of the antler and the velvet skin commences to dry and peel. As the velvet falls off, a buck may rub his antlers against trees and shrubs to help remove the skin. When all the skin has been shed, the lifeless, bony core remains. A buck will continue rubbing, to scent and visually mark his territory; in the process the antlers become polished. The antlers are carried in this condition throughout the rut. Sometime toward the end of the breeding season, usually from the last of December to mid-February, the antlers become loosened around the base by resorption of bone in this region, which is commonly referred to as casting or shedding of the antler. The casting of antlers is the result of a decrease in testosterone production and potentially declining body condition due to rut stress, since bucks in good post-rut condition tend to carry their antlers longer than bucks in poor body condition. After the antlers fall to the ground, they are gnawed and consumed by rodents and rabbits for their minerals and protein.

Each of the paired antlers of adult white-tailed deer typically has 1 main beam, which grows slightly backward at first, and then is directed forward and outward. There is an upright snag near the base on the inside and several unbranched tines, or points, that

arise vertically. All of the points greater than 2.5 cm (1 in.) in length, including the snag on both beams, are counted when determining the total number of points on a set of antlers. Both the size of the antler and the number of points depend upon many factors such as the deer's age, the quality and quantity of food, injury, hormone regulation, and heredity.

In their first fall, fawn bucks have "buttons" that can be felt under the skin or observed as slight swellings. Rarely fawn bucks may have a very short unbranched antler. In yearling bucks (1½ years old), antlers are always visible externally. While some may have only the unbranched main beam or "spike" at this age, many have branched antlers with more than 1 or 2 points. In Missouri there are exceptional cases of well-nourished yearlings having a total of 10 or 11 points. During the succeeding years of the buck's life, the antlers become more massive. In general, the number of points increases to a total of 8 to 10 and occasionally more. Antler size typically increases each successive year until 5½ or 6½ years of age. Following that time period, the size of the antlers generally declines with each year's renewal. Hunting records show that the antlers of bucks in northern Missouri are, as a rule, larger than those of similar-aged bucks from southern Missouri. This is because the more fertile soils of northern Missouri produce foods higher in the calcium and phosphorus from which antlers are made.

Abnormally shaped antlers occur occasionally and sometimes represent injury during growth. Injury probably accounts for extremely large numbers of points, like antlers with 54 and 78 points. An injury to the pedicle, from which the antler regrows each year, is likely the leading cause of abnormal antler growth. Injuries from fighting or during antler casting can affect future antler growth. An upset in the hormone system is one factor that is probably responsible for unusual cases of antlerless bucks or for females having antlers. In some cases, antlerless bucks are the result of hereditary factors.

Color. In summer, both sexes are reddish brown to tan above (often called the "red" coat), being paler on the face and darker along the middle of the back. There is a whitish band across the nose, a white ring around the eyes, and white on the insides of the ears. A blackish spot occurs on each side of the chin. The upper throat, belly, and insides of the legs are whitish. The tail is the same color as the back above, but the edging and undersurface are conspicuously white. The color pattern of the winter coat is similar to the summer coat but is grayish to grayish brown, often called the "blue" coat. The hairs of the back are blacker tipped, giving a darker appearance to the back, but the black chin patch is less sharply defined. Albino deer occur but rarely. Melanistic deer are extremely rare.

Adults have two molts annually. The spring molt begins about the middle of March and continues until around the first of June. The summer coat is lightweight and consists of fine, short hairs that lie close to the skin. The fall molt begins from August to mid-September and is completed by early October. The hairs of this coat are long and heavy; each hair has many air spaces that act as insulators, helping to insure warmth during cold weather. The coat easily repels cold rain and wet snow in winter. Fawns gradually lose their spots between 3 and 5 months of age and acquire a uniform coloration. On rare occasions, late-born fawns will not lose their spots until the following spring.

Measurements

Total length	85–240 cm	33⅛–93⅝ in.
Tail	100–365 mm	3⅞–14¼ in.
Hind foot	362–520 mm	14⅛–20½ in.
Ear	139–228 mm	5½–9 in.
Skull length		
Male	311 mm	12¼ in.
Female	266 mm	10½ in.
Skull width		
Male	120–139 mm	4¾–5½ in.
Female	104 mm	4⅛ in.
Weight	41–141 kg	90–311 lb.

Among similar-aged individuals, females average lighter in weight than males. Deer from the fertile soils of northern Missouri weigh more than those of the same sex and age from southern Missouri. The greatest recorded weight for a buck in Missouri is 167 kg (369 lb.).

Teeth and skull. The dental formula of the white-tailed deer is given in two ways depending upon the name ascribed to the outer, lower incisor-shaped teeth:

$$I\ \frac{0}{4}\ C\ \frac{0}{0}\ P\ \frac{3}{3}\ M\ \frac{3}{3} = 32$$

or

$$I\ \frac{0}{3}\ C\ \frac{0}{1}\ P\ \frac{3}{3}\ M\ \frac{3}{3} = 32$$

Only rarely do upper canine teeth occur.

The absence of upper incisors identifies a deer's skull from the skulls of all native wild mammals, except elk, in Missouri and from those of pigs and horses. The absence of upper canine teeth and length of the upper cheek tooth row less than 110 mm (4.3 in.) distinguishes the deer's skull from the elk's skull. The presence of a space 15 mm (⅝ in.) or more in width in front of the eye socket (between lacrimal and nasal bones) distinguishes a deer's skull from the skulls of cattle, sheep, and goats.

Sex criteria and sex ratio. Females are normally antlerless and possess an udder with four teats. Antlers, when present, usually serve to identify males, but the penis and scrotum are absolute sex identification characteristics at all ages. During the breeding season, the necks of males swell to approximately twice their nonbreeding size, reaching a maximum in mid-November. The factors causing this enlargement are not fully understood.

The sex of a deer cannot be identified by its track because there is no difference in the size and shape between buck and doe hooves.

In white-tailed deer the sex ratios of embryos and of young fawns show only slightly more males than females. Among adults, there are usually more females than males. This slightly higher proportion of females among adults seems to reflect lower natural mortality and lower hunter-harvest when compared to adult males.

Age criteria, age ratio, and longevity. The young, or fawns, are reddish brown or reddish yellow spotted with white. They are easily distinguished from adults until they lose their spotted coats and acquire a uniform coloration, which occurs between 3 and 5 months of age. The size of a buck's antlers or the number of points are not a reliable indicator of age; although, as a buck ages, the antlers do grow larger and a buck will typically have the fewest antler points as a yearling.

The best means of age determination of deer in the field is by replacement and wear on the molariform teeth (see pages 402–403). A more reliable method that is also used is the count of the cementum layers on the root of the first incisor.

There are often discrepancies in the aging of deer by these two techniques. Aging by tooth replacement and wear is most accurate to place deer in the fawn, yearling, and adult categories; and the cementum layer count is useful to obtain specific age estimates from adults.

Deer are in the prime of life between 2½ and 7½ years of age. Some may live for about 15 years in the wild and up to 25 years under protection. There are records of does producing young at 15 and 18 years of age. However, due to hunting mortality few deer survive to older ages in the wild.

Glands. Glands play an important role in scent communications among individual white-tailed deer. The tarsal glands, marked by a tuft of long, coarse hair on the inside of each hind leg at the ankle, or hock, produce an oily secretion with a pronounced ammonia smell. This tuft of hair is raised when the deer is excited by fear or hostility. Fawns a week or so old, as well as adults, stand with their hind legs together, urinate down the insides of the legs so that the fluid saturates the hairs of the tarsal glands, and then rub the legs together. The urine scent, together with the secretion from the tarsal glands, contributes to a body odor that may be recognized by other deer.

Secretion from the metatarsal gland marks the deer's resting spot

Another set of glands, the metatarsals, occurs on the outside of each hind leg between the ankle and hoof (see plate 64). In white-tailed deer, these glands are about 2.5 cm (1 in.) long and are bordered by a conspicuous band of white hair. They have an oily secretion with a pungent, musky odor. Such secretion scents, and possibly serves to identify, the resting spots used by the deer. It also indicates fear.

Pedal glands, lying between the two main toes on each foot (see plate 64), secrete a strong and offensive odor throughout the year. The secretion is conducted by long hairs to the hooves and doubtless scents the tracks of the animal. This would permit a deer to retrace its own steps and to find a fawn or other member of the herd.

"Lip curling" by deer is a threat or a response to a scent from another deer

Small preorbital glands lie just in front of each eye. These probably scent twigs and branches where the deer feeds and less often are rubbed on other deer in a low-level sparring encounter. Marking using the preorbital gland is much more frequent in dominant bucks than subordinate bucks, and much more frequent in bucks than does.

Voice and sounds. Adult deer are usually silent but make certain sounds on occasion. When scared, they sometimes give a loud, hoarse, high-pitched shriek. They frequently snort or blow. This may be a form of communication or may serve to startle motionless animals. Adults of both sexes sometimes bleat. Females call their young by a low murmur or grunt, and the young bleat or *baa* to call their mother. Does and fawns both cry loudly or bawl if in pain. Adult males bawl sometimes when fighting, injured, or restrained. Barking has also been described in deer.

Bucks, does, and fawns may stamp their feet on the ground; this occurs if they are startled or annoyed and is often accompanied by a snort. Deer also communicate with each other by means of their scent glands, by various movements of their tail, ears, head, and face, and by their body posture.

Distribution and Abundance

In primitive times, there was an estimated population of 40 million white-tailed deer ranging from the boreal forest of Canada to the lowlands of Peru. The most populated regions were the Mississippi Valley and what is now the eastern United States. With settlement and conquest of North America by Europeans, the deer population was greatly reduced. Between 1875 and 1915, deer were at their lowest level; in 1908 the white-tailed deer population of the United States was estimated at 500,000.

Restocking and redistribution were begun in the early 1900s and, with added protection, the deer began to increase. By 1920 the deer population of the United States was estimated at 1 million, and by 1948 it had reached an estimated 6 million. In the next 30 years the population doubled, and the current estimate in the United States is around 30 million. Coincident with the original population decline, the range was also altered. Because of the destruction of large sections of the native forests and clearing for agriculture, considerable territory was lost in the east and center of the primitive range; but new areas to the north and northwest were made habitable by the advent of lumbering in the dense forestlands, by the favorable variety of plant growth that often followed clearing, and the creation of extensive edge habitat. In 1948 the nearly 4 million sq km (1.5 million sq. mi.) of occupied range in the United States and Canada generally lay within the boundaries of the primitive range, but consisted of approximately 75 percent of the original area. The heaviest concentrations occurred in the Great Lakes area and New England.

Making allowances for the discrepancies in the estimates of population numbers, the present deer population is still greater in many states than it was in colonial times. Today they again range from northern Canada into South America.

The history of the white-tailed deer in Missouri parallels that for the rest of the range. During historic times, deer were believed to have been abundant everywhere in the state, the largest concentrations occurring on the more fertile soils and in the more varied cover of northern Missouri. At the beginning of the nineteenth century, they occupied about 180,000 sq km (69,000 sq. mi.), and population estimates varied between 345,000 and 690,000. By 1890 they had disappeared from the northern and western counties where extensive cultivation eliminated their habitat, and soon the population was greatly lowered everywhere. In 1925 an estimate showed only 395 deer left in the state, and the season was closed for the first time.

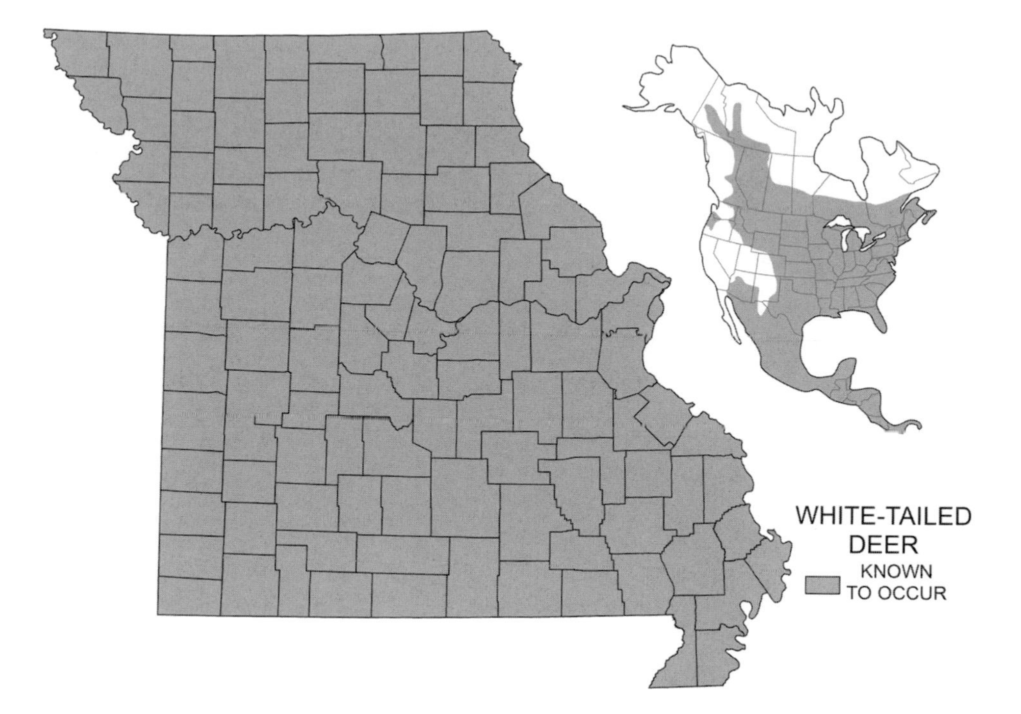

During the next five years, 271 deer were imported from Michigan and, along with 91 Missouri deer purchased from private owners, were used to restock vacant habitat. These introductions and the remnant population formed the nucleus for the present deer herd in Missouri. From 1931 through 1937 the season was reopened, but the harvest was low, the largest season's take being 149 legal bucks. Hunting was again closed from 1938 through 1943. During this period, local deer were trapped in areas of abundance and transplanted to likely habitat in other parts of the state. This transplanting program and the accompanying public support and protection from illegal shooting proved fruitful. By 1944 an increasing population permitted a two-day open season on "bucks only." Limited hunting of "bucks only" continued through 1950, and the legal take increased from 583 in 1944 to 1,622 in 1950. In many parts of Missouri the deer population continued to increase at such a rate that it warranted opening the season on "any" deer (does and fawns as well as bucks); in areas of lower population, the take was restricted to "bucks only" as previously.

The "any deer" season was first inaugurated in 1951. By the late 1980s deer populations across much of the state had been well established and were growing rapidly. This era of rapid population growth was met with increasing liberalization of regulations and expanding hunting opportunities. Prior to 1988, hunters were only allowed to harvest 1 deer during the firearms season, but in 1988, hunters were issued 1–2 additional antlerless deer on bonus permits in several management units throughout the state. Then starting in 2004 hunters were issued unlimited antlerless permits in many management units in the northern, central, and western parts of the state. All of these changes were intended to address issues related to growing deer populations and have been effective at stabilizing deer populations in many parts of rural Missouri.

At present, there are more than 1 million white-tailed deer in Missouri and they occur in every county. Changes in regulations have resulted in stable or reduced deer populations in many parts of northern, western, and central Missouri. Additionally, implementation of the antler-point restriction and shifting hunter preferences for older bucks in the population have resulted in a shift in the composition of harvest, resulting in changing demographics of populations in northern, western, and central Missouri. Across much of southern Missouri, deer populations continue to slowly grow with changing hunter selectivity slowly shifting the sex and age structure of the population, but deer populations in southern Missouri are largely below what the habitat can support and are within levels of human tolerance.

Habitat and Home

White-tailed deer show a preference for timbered areas. However, the pattern and distribution of the timber in a given area influence their presence and abundance to a great extent since they utilize the borders or edges and early successional stages more

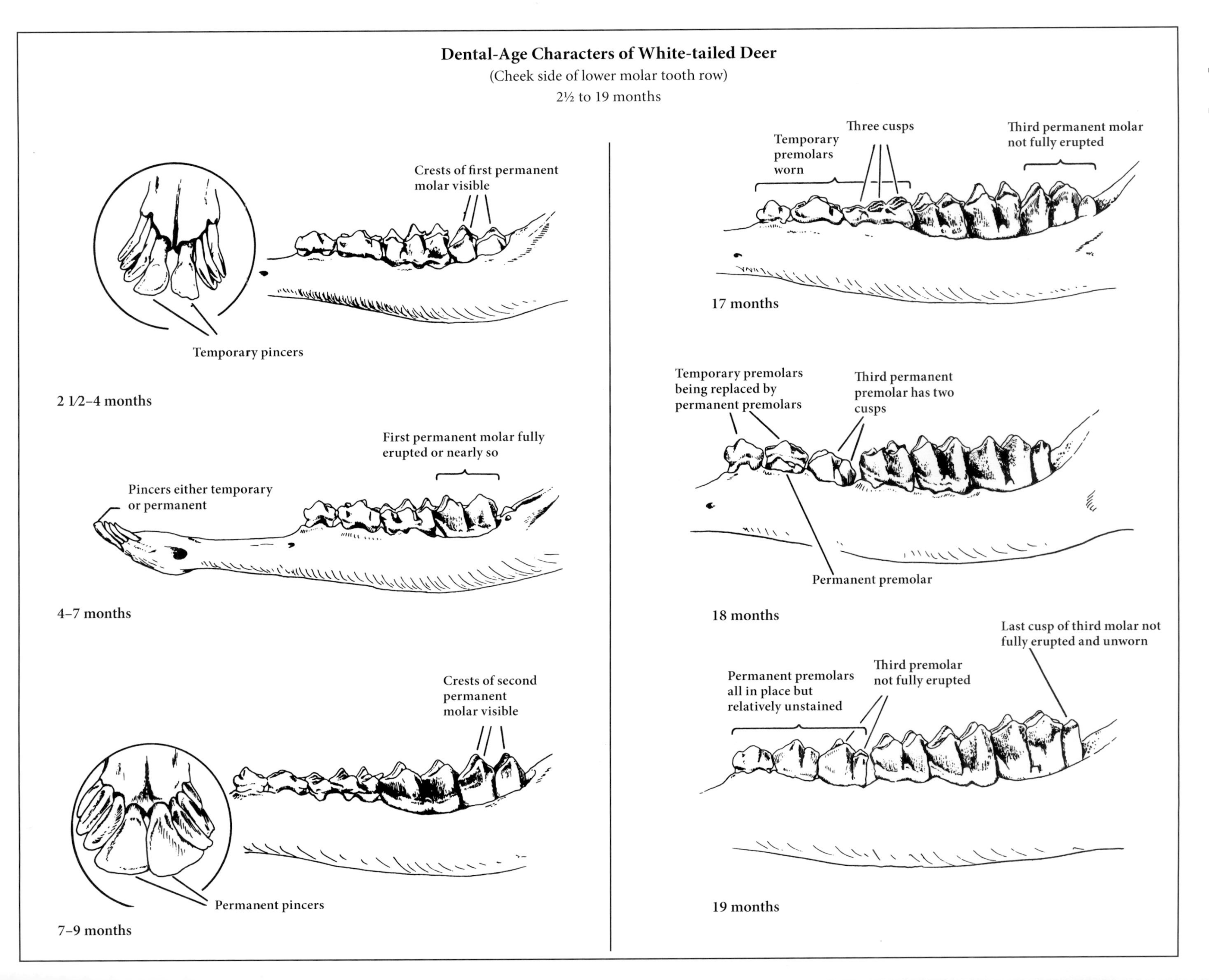
Dental-Age Characters of White-tailed Deer
(Cheek side of lower molar tooth row)
2½ to 19 months
Crests of first permanent molar visible
Temporary pincers
2 1/2–4 months
First permanent molar fully erupted or nearly so
Pincers either temporary or permanent
4–7 months
Crests of second permanent molar visible
Permanent pincers
7–9 months
Temporary premolars worn
Three cusps
Third permanent molar not fully erupted
17 months
Temporary premolars being replaced by permanent premolars
Third permanent premolar has two cusps
Permanent premolar
18 months
Last cusp of third molar not fully erupted and unworn
Permanent premolars all in place but relatively unstained
Third premolar not fully erupted
19 months

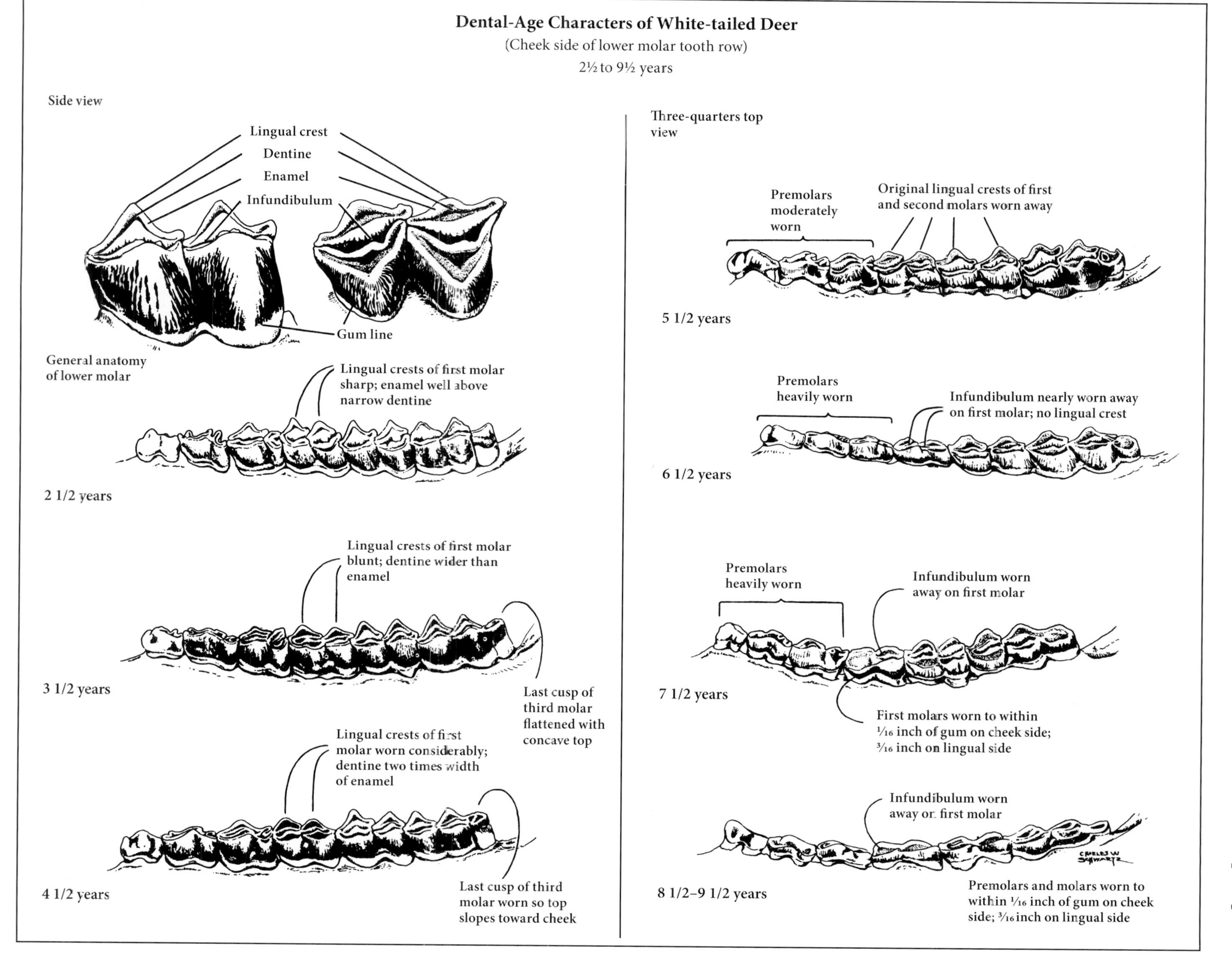
Dental-Age Characters of White-tailed Deer
(Cheek side of lower molar tooth row)
2½ to 9½ years
Side view
Lingual crest
Dentine
Enamel
Infundibulum
Gum line
General anatomy of lower molar
Lingual crests of first molar sharp; enamel well above narrow dentine
2 1/2 years
Lingual crests of first molar blunt; dentine wider than enamel
3 1/2 years
Last cusp of third molar flattened with concave top
Lingual crests of first molar worn considerably; dentine two times width of enamel
4 1/2 years
Last cusp of third molar worn so top slopes toward cheek
Three-quarters top view
Premolars moderately worn
Original lingual crests of first and second molars worn away
5 1/2 years
Premolars heavily worn
Infundibulum nearly worn away on first molar; no lingual crest
6 1/2 years
Premolars heavily worn
Infundibulum worn away on first molar
7 1/2 years
First molars worn to within 1/16 inch of gum on cheek side; 3/16 inch on lingual side
Infundibulum worn away on first molar
8 1/2–9 1/2 years
Premolars and molars worn to within 1/16 inch of gum on cheek side; 3/16 inch on lingual side
CHARLES W. SCHWARTZ

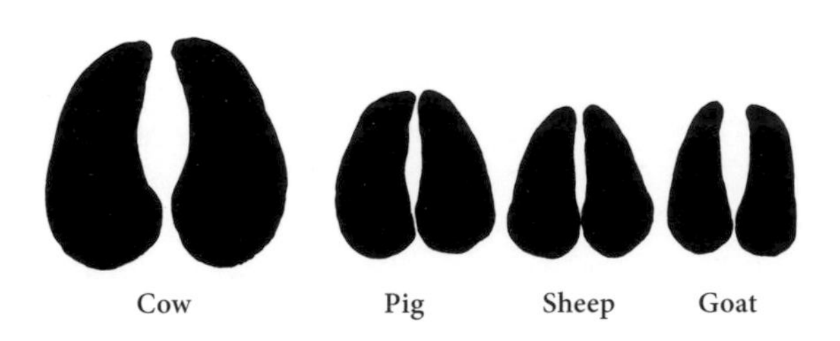

Tracks of hind feet of cow, pig, sheep, and goat, which might be confused with those of white-tailed deer

than dense uniform stands. One of the main reasons for this habitat preference is that the great variety of foods desired by deer grows best along the margins of timbered areas or in clearings in the timber. Another reason, especially true in the more settled sections of the country, is that deer can utilize the forage offered by agricultural crops adjacent to timbered lands and still have the sanctuary and other attractions of the timber itself.

In the Ozark Highlands of Missouri, with its large acreages of timbered land, deer do best where numerous agricultural units in the timber create a maximum amount of edge. In the counties of Missouri that are not primarily timbered, deer live in and about what wooded areas there are, such as small woodlots, timbered bottomlands, and timbered areas bordering the main streams and their tributaries.

Habitat quantity and quality for white-tailed deer has fluctuated over time. Land-use changes such as conversion of forests to cool-season grasses or row crops and back again, and urban sprawl are some of the factors.

Fawns may be born in woody cover or even in open places such as pastures or warm-season grass fields. The fawns' concealing coloration and lack of scent provide protection.

Habits

White-tailed deer tend to live in a very restricted area year after year, having an average annual home range from 1.3 to 3.9 sq km (½ to 1½ sq. mi.). Generations of females in a given line have been found to use the same winter and summer ranges as their ancestors. However, they do not defend territories. Some individuals, particularly bucks during the rut, may cover a larger area. Local movements and range shifts of deer in Missouri are related primarily to seasonal changes in food sources, cover, or mate searching during the rut. For example, where acorns are not abundant in their summer area, some deer may shift to localities where they are available in fall and winter. Also, the harvest of crops in the fall changes cover availability in agricultural habitats.

At the time of fawning, the female leaves her yearlings, or chases them away, and lives alone briefly until the fawns are born. During the summer she is accompanied by the fawns while her yearlings stay together but do not rejoin her. In late summer to early autumn, yearling does join their mother and the fawns, forming the most common family group. Yearling bucks seldom return to the doe. It is at this time that most yearling bucks disperse, sometimes over great distances, and either remain alone or randomly associate with other bucks. In fall, and especially during the rut, a buck spends most of his time in search of females that are nearing estrous. White-tailed deer form a tending bond in which a buck spends a few days with a receptive female before seeking another. After the males shed their antlers in winter, they form small male bands of a few individuals, commonly referred to as bachelor groups, or live separately. Feeding groups of 25 or more deer may occur in local places in late winter, but these are temporary groups and not a herd. The large numbers occur when food is limited and a favorable food supply is available. In Missouri, deer do not "yard," or collect in certain small areas during the winter, as deer do in northern states, because snow is typically not deep enough to restrict their travel or feeding.

Deer normally walk, trot, or bound along in low, smooth jumps interspersed with an occasional high jump for observation. When startled they may run at speeds up to 58 km (36 mi.) per hour for 4.8 to 6.4 km (3 to 4 mi.) but are unable to maintain this pace for

White-tailed deer tracks

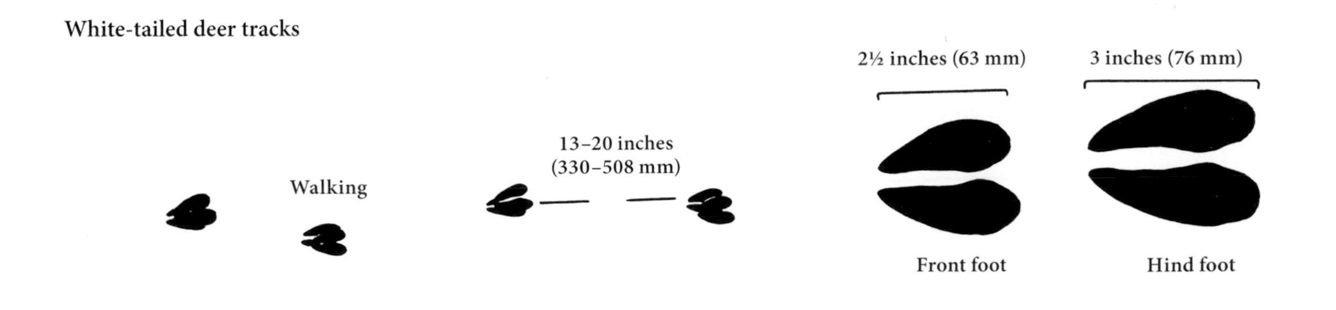

Deer bed in snow

longer distances. Deer are expert jumpers, and when pressed can clear fences 2.6 m (8½ ft.) high. They swim well and attain a speed of 18 to 21 km (11 to 13 mi.) per hour. In summer, deer swim lower in the water than in winter. They are more buoyant in winter because the hairs of their coat have a hollow, air-filled core and because there is an accumulated layer of fat under the skin.

Deer usually spend the day in concealing cover and rarely move about, but toward evening they come out to feed and drink.

The main feeding period is just before dark in the evening. In summer, deer feed around sunset, but in winter they start an hour or so before sunset. They tend to bed down for a few hours and then have another feeding period around dawn. During winter, when food is scarce, they may feed longer hours and even during the day. Does also have more feeding periods when they are nursing fawns as this is a very demanding time.

The location of the bed spot depends largely upon the weather. On sunny warm days, some shady place is selected; on cloudy, windy, or cool days, a sunny spot or one protected from the wind is used. The spot selected for a bed may be either a slight depression or level ground, as long as it permits the deer to lie comfortably. It usually offers a good view of possible intruders.

Deer in an area associate with each other loosely all year and during their contacts establish a relationship of dominance. This is manifested mostly by means of postures and threats. The first signs of aggression are made by a deer laying its ears back against the neck and staring at the other deer while the head is slightly lowered. Usually a deer gives way but, if not, the two deer may move toward each other, turning their sides. If the two deer are bucks, the interaction may be continued by threatening with lowered antlers and, finally, by rushing at the adversary. If they are does, they may threaten by striking with a front foot and, ultimately, by rearing up on the hind feet and attacking with both front feet.

Both sexes show alarm by stamping with their front feet; snorting, if slightly more alarmed; "whistling" (a louder call than a snort) for more intense alarm; and, finally, fleeing with the tail erected.

Bucks mark their home range with rubs and scrapes. When a buck rubs the velvet from his antlers and polishes the hardening core, he vigorously rubs small trees and, perhaps, practices combat by pushing and thrashing them. Other rubs are made when a dominant buck encounters other males, especially prior to the peak of the breeding season. Rubs serve as visual signs of the buck's presence, and a scent left on these rubs gives an olfactory indication as well.

Scrapes, up to 1 m (3 ft.) square, are made in leaves on the ground by pawing with the front feet. The buck urinates in or behind the scraped area, often mixing the urine with scent from the tarsal glands. Scrapes are usually made in spots with an overhanging branch for bucks to rub their preorbital gland. They are often reworked by the buck and even are defended from other males. Females may come to the scrapes and urinate in them to establish contact with the buck.

During the rut, bucks commonly fight each other. They seldom make repeated charges but, once their antlers meet, push and shove without making new contacts. Only rarely do the antlers become entangled permanently. When this happens the bucks are unable to feed properly and die of weakness and starvation. Deer also fight with their front feet, using their sharp-pointed hooves to inflict wounds. Sites of combat are frequently marked by disturbed ground that the deer have torn up with their feet. After establishment of

Buck rubbing antlers on pine tree

Combat between bucks

dominance, fighting and wasted energy are avoided during future encounters by the subordinates.

The rather large, well-haired tail serves many purposes. When the deer is running or bounding, it is held high and the white underhairs are erected as a signal of alarm to other members of the herd. After the deer has inspected the surroundings, the tail is flicked to indicate a lack of danger. In combating flies, the deer switches its tail, flaps its ears, stamps its feet, shakes its head, or twitches its skin.

The droppings consist of from 50 to 100 small pellets that stick together. When fresh, they are blackish, shiny, and soft inside.

Does may aggressively defend their fawns from predators by hitting them with their front feet.

As a signal of alarm, tail is raised and hairs are spread

Tuft of hairs at tarsal gland raised when deer is aroused

Foods

Deer are browsing animals, feeding chiefly on the leaves, twigs, and fruits of trees and shrubs and the foliage of herbaceous plants. They also eat seeds, fungi, mosses, lichens, succulent grasses, farm crops, and sometimes small amounts of animal food like snails and fish. Over 450 different kinds of plants are known to be eaten by white-tailed deer in Missouri, but of these only relatively few are used extensively.

From the time acorns and other mast items first become available in September until they are gone, they are preferred by deer for food. In years of poor acorn production and, as a supplement in years of good acorn crops, deer in fall and winter feed on the fruits of woody plants like sumacs and buckbrush; the cultivated crops of corn, lespedeza, soybeans, wheat, rye, and sorghum; what green leaves are available of grasses, sedges, and perennial plants such as ladies' tobacco; and the twig tips of a few woody plants. In spring they return to the succulent vegetation of perennial forbs, shrubs, and trees; particularly the new growth of white and red elm, fragrant sumac, red clover, Virginia creeper, wild grapes, prickly lettuce, and mushrooms. The summer diet consists largely of the leafy parts and fruits of annual and perennial plants including wild grapes, Korean lespedeza, dwarf sumac, Virginia creeper, red clover, pokeweed, persimmon, Japanese rose, and mushrooms.

Deer show a definite selection of plants and seemingly take first those that are most nutritious and palatable. This selectivity can have serious effects. In ranges having heavy concentrations of deer, overbrowsing occurs. The result is manifested in a lower level of nutrition of the herd and in extreme cases the elimination of these desirable foods. Since many plants preferred by deer grow in openings of the forest, some may also be eliminated by changing conditions caused by too much shade. This natural type of plant succession is as critical to changing the food supply of deer as overbrowsing.

Deer are dainty feeders, nipping a bud here, plucking a leaf there, and moving along in search of another tender morsel. When food is scarce, they rear up on their hind legs and browse twigs and leaves as high as they can reach. When their populations are heavy, deer soon consume all the food they can reach, resulting in a definite "browse line" on the trees.

Deer require water in some form daily. They frequent any mineral licks in the vicinity, especially in spring and early summer.

Deer have a 4-chambered stomach: 3 chambers for storage and breakdown of foodstuff and 1 "true

stomach" for chemical digestion. After being swallowed, the food goes to the paunch for storage. From here most of it returns to the mouth for further chewing in the form of moistened balls known as the cud. It is then swallowed again and goes into the various chambers of the stomach where it is churned and prepared for digestion through the action of microorganisms such as bacteria and protozoa. These microbes actually synthesize nutrients that become critical to survival in certain conditions.

Reproduction

Buck white-tailed deer are capable of mating successfully from September through February and possibly later, but the peak of the mating or rutting season is in November after a period of intense male sparring. The earliest does to come into heat do so in late September; the last, which are young does of the preceding year, come into heat as late as March. Males chase and follow females relentlessly for 5–6 days, including the heat period, or time of receptivity of females. This lasts about 24 hours and in unmated adult does occurs at approximately 28-day intervals throughout the breeding season.

Pregnancy lasts 6½ to 7 months; the young are born most often in late May or early June but occasionally earlier or later. During this period a doe will separate from any grouping and remain apart for about 2 months after the birth. A mature doe usually has twins but sometimes may have a single offspring or triplets. Quadruplets are very rare. Nutrition plays a large role in the number of offspring.

At birth each fawn weighs between 1.8 and 3.2 kg (4 and 7 lb.) and measures 43 to 48 cm (17 to 19 in.) in total length. Its eyes are open and it can stand feebly within an hour. The female leaves the fawns alone in order to feed, but stays within hearing distance of their calls. She nurses them frequently and licks them carefully. The young remain in the close vicinity of their birthplace for several weeks.

When about 3 to 4 weeks old, the fawns begin to follow the doe and start to eat their first solid foods. At this time the mother and fawns may return to the herd, and weaning may begin, although some fawns nurse until they are 6 months old. The young continue to accompany the female until they are old enough to breed but may be chased away by her for a brief time during the birth of a subsequent litter. Many young females in Missouri become sexually mature at 6 to 8 months of age and consequently breed in the year of their birth; they may account for up to 30 percent of the annual increment. The remaining females and young males breed first at 1½ years of age.

Some Adverse Factors

With the extirpation of precolonial predators (wolves and mountain lions), humans are the primary predator for deer. Coyotes and bobcats prey primarily on fawns.

Deer, like other mammals, have their normal complement of parasites; parasitism of deer in Missouri is not, however, severe. The following parasites are reported from white-tailed deer in different parts of their North American range: roundworms, flukes, lungworms, stomachworms, a flatworm, a whipworm,

protozoans, adult and larval tapeworms, mites, ticks, lice, and adult and larval flies.

Hemorrhagic disease is caused by two very closely related groups of viruses, the bluetongue virus and epizootic hemorrhagic disease virus, which are both spread by biting midge flies. Recently, outbreaks have been increasing in frequency across much of the midwestern United States. Outbreaks are most severe in years of extreme summer weather usually accompanied by severe drought.

Missouri has experienced severe hemorrhagic disease outbreaks in 1980, 1988, 1998, 2007, 2012, and 2013. Hemorrhagic disease can cause high mortality rates and significant short-term effects on local populations in Missouri; hemorrhagic disease has never been documented to have long-term population effects, however.

Chronic wasting disease (CWD) is in a family of infectious neurological diseases known as transmissible spongiform encephalopathies. The infectious agent of CWD is an abnormal protein called a prion. CWD prions accumulate in the brain, spinal cord, eyes, spleen, and lymph nodes of infected cervids. The resulting damage causes abnormal behavior, loss of body function leading to emaciation, and eventually death.

CWD is a slowly progressing syndrome that may take in excess of a year for clinical signs to appear. During the prolonged period between infection and clinical signs of CWD, infected cervids excrete infectious prions into the environment via bodily processes (such as defecation, urination, salivation). The shedding of infectious prions results in direct and indirect transmission between cervids. CWD may also be spread directly through the natural movements of infected free-ranging cervids, as well as the interstate movement of infected captive cervids. Indirect transmission may occur through movement of infected carcasses and offal from hunter-harvested cervids and also from contaminated soil and water sources. Once established in a local deer population, CWD will slowly increase in prevalence, causing reduced survival rates leading to long-term population declines.

CWD was first discovered in a captive white-tailed deer in Missouri in 2010. The first CWD-positive free-ranging deer in Missouri was confirmed in 2012.

Some of the diseases known to affect deer in other parts of their range are tuberculosis, tumorous growths, infection of the eyes producing temporary blindness, distemper, black tongue, hoof-and-mouth disease, hoof shedding, tularemia, anthrax, hantavirus, rabies, and pneumonia. Other conditions, specifically reported for Missouri and known to have been fatal, are caused by certain bacteria and the eating of poisonous plants. Adult bucks may get a fatal brain abscess caused by a bacterial infection resulting from rubbing or fighting. A large mass in the jaw may arise when barbed seeds become lodged in the mouth. An associated infection may cause emaciation and death. Very rarely do "madstones" occur in the stomach. These are formed by minerals laid down in concentric layers around some nucleus, like a bullet or nut. The common name comes from folklore, which states that these had some value in curing the bite of a mad or rabid dog.

The most important cause of accidental death is collisions with vehicles. Occasionally deer become entangled in fences or fall and are injured.

Importance

For the Native Americans and the early settlers, deer provided food; hides for clothing, shelter, and bedding; sinews for bowstrings and implements of war, fish lines, and the stitching of bark utensils; brains for bleaching and tanning; and bones and antlers for awls, needles, scrapers, implement-making tools, and ornaments.

Deer still provide us considerable food, sport, and pleasure. Since approximately 57 percent of the live weight of a deer is edible, the venison acquired from legal hunting provides many pounds of meat. The tanned hide, or buckskin, has a limited use for coats, jackets, and gloves. Deer hunting has become a big commercial enterprise and a source of income to many—to manufacturers of arms, ammunition, hunting apparel, and other hunting-related products as well as to persons providing food, lodging, transportation, and a place to hunt.

Where deer populations are heavy, their feeding may damage crops such as corn, soybeans, and sorghum; strawberry patches; vineyards; orchards; nurseries; and the understory of forested lands. However, grazing on crops such as winter wheat and rye seldom affects the ultimate crop production. If you are experiencing problems with white-tailed deer, contact a wildlife professional for advice, assistance, regulations, or special conditions for handling these animals.

Conservation and Management

Deer management of the past focused on growing deer populations. However, as populations grew rapidly over the 1980s and 1990s issues of overabundance occurred, necessitating the need for a shift in management strategy. Efforts to reduce the deer population lead to significant liberalization of regulations, for example increasing season lengths and bag limits, aimed at reducing deer densities. These efforts, along with shifting hunter attitudes toward antlerless harvest, have resulted in stable or reduced deer numbers.

Today, the major consideration of deer management in Missouri is the localized regulation of annual harvest by hunters to maintain stable populations at biologically and socially acceptable levels. Regulating the appropriate number of deer taken by hunters is essential to proper management of the herd and the habitats in which deer live. Uncontrolled herds of high density can alter local plant-deer relationships such that grazing results in the local extirpation of sensitive plants. This effect is exaggerated in fragmented forest communities, which largely dominate landscapes today.

Habitat conditions and hunter density affect population growth and harvest rates, both of which vary greatly across the state. Therefore, effective population management requires the knowledge of sex- and age-specific harvest in given localities each year. To maintain biologically and socially acceptable populations, it is necessary to regulate harvest at the local level. Recently, counties have been used as management units to regulate harvest at the local level. A limit on harvest of antlerless deer (does and young) has also proven to be an important and effective management tool because the harvest of these deer can be more precisely controlled.

SELECTED REFERENCES

See also discussion of this species in general references, page 23.

Augustine, D. J., and L. E. Frelich. 1998. Effects of white-tailed deer on populations of an understory forb in fragmented deciduous forests. *Conservation Biology* 12:995–1004.

Bartlett, I. H. 1949. Whitetail deer: United States and Canada. *Transactions of the Fourteenth North American Wildlife Conference.* Pp. 543–553.

Brohn, A., and D. Robb. 1955. *Age composition, weights, and physical characteristics of Missouri's deer.* Missouri Conservation Commission, Pittman-Robertson Series 13. 28 pp.

Cheatum, E. L., and G. H. Morton. 1946. Breeding season of whitetailed deer in New York. *Journal of Wildlife Management* 10:249–263.

Dalke, P. D. 1941. The use and availability of the more common winter deer browse plants in the Missouri Ozarks. *Transactions of the Sixth North American Wildlife Conference.* Pp. 155–160.

———. 1947. Deer foods in the Missouri Ozarks. *Missouri Conservationist* 8:4–5.

Dimmick, R. W., and M. R. Pelton. 1996. Criteria of sex and age. In *Research and management techniques for wildlife and habitats,* 5th ed., pp. 169–214. The Wildlife Society, Bethesda, MD. 740 pp.

Dunkeson, R. L. 1955. Deer range appraisal for the Missouri Ozarks. *Journal of Wildlife Management* 19:358–364.

Hansen, L. P., J. Beringer, and J. H. Schulz. 1996. Reproductive characteristics of female white-tailed deer in Missouri. *Proceedings of the Southeast Association of Fish and Wildlife Agencies* 50:357–366.

Hawkins, R. E., and W. D. Klimstra. 1970. A preliminary study of the social organization of white-tailed deer. *Journal of Wildlife Management* 34:407–419.

Hewitt, D. G., ed. 2011. *Biology and management of white-tailed deer.* CRC Press, Boca Raton, FL. 674 pp.

Hirth, D. H. 1977. Social behavior of white-tailed deer in relation to habitat. *Wildlife Monographs* 53:1–55.

Kile, T. L., and R. L. Marchington. 1977. White-tailed deer rubs and scrapes: Spatial, temporal, and physical characteristics and social role. *American Midland Naturalist* 97:257–266.

Korschgen, L. J., W. R. Porath, and O. Torgerson. 1980. Spring and summer foods of deer in the Missouri Ozarks. *Journal of Wildlife Management* 44:89–97.

Lockard, G. R. 1972. Further studies of dental annuli for aging white-tailed deer. *Journal of Wildlife Management* 36:46–55.

Marchinton, R. L., and K. V. Miller, eds. 2007. *Quality Whitetails.* Stackpole Books, Harrisburg, PA. 337 pp.

McShea, W. J., H. B. Underwood, and J. H. Rappole, eds. 1997. *The science of overabundance: Deer ecology and population management.* Smithsonian Books, Washington, DC. 402 pp.

Miller, S. G., S. D. Bratton, and J. Hadidien. 1992. Impacts of white-tailed deer on endangered and threatened vascular plants. *Natural Areas Journal* 12:67–74.

Montgomery, G. G. 1963. Nocturnal movements and activity rhythms of white-tailed deer. *Journal of Wildlife Management* 27:422–427.

Mosby, H. S., and C. T. Cushwa. 1969. Deer "madstones" or bezoars. *Journal of Wildlife Management* 33:434–437.

Murphy, D. A., and L. J. Korschgen. 1970. Reproduction, growth, and tissue residues of deer fed dieldrin. *Journal of Wildlife Management* 34:887–903.

Nelson, M. E., and L. D. Mech. 1999. Twenty-year home-range dynamics of a white-tailed deer matriline. *Canadian Journal of Zoology* 77:1128–1135.

Rice, L. A. 1980. Influences of irregular dental cementum layers on aging deer incisors. *Journal of Wildlife Management* 44:266–268.

Robb, D. 1951. *Missouri's deer herd.* Missouri Conservation Commission, Jefferson City. 44 pp.

Rue, L. H., III. 1978. *The deer of North America.* Outdoor Life/Crown Publishers, New York. 463 pp.

Schwede, G., H. Hendrichs, and W. McShea. 1993. Social and spatial organization of female white-tailed deer, *Odocoileus virginianus*, during the fawning season. *Animal Behavior* 45:1007–1017.

Seagle, S. W., and J. D. Close. 1996. Modeling white-tailed deer *Odocoileus virginianus* population control by contraception. *Biological Conservation* 76:87–91.

Severinghaus, C. W. 1949. Tooth development and wear as criteria of age in white-tailed deer. *Journal of Wildlife Management* 13:195–216.

Smith, W. P. 1991. *Odocoileus virginianus. Mammalian Species* 388. 13 pp.

Sumners, J. A. 2009. The evolution of deer management. *Missouri Conservationist* 70:8–13.

———. 2014a. Missouri citizens are key to deer management success. *Missouri Conservationist* 75:25–29

———. 2014b. The tides of change in Missouri deer management. *Missouri Conservationist* 75:28–29.

Taylor, W. P. 1956. *The deer of North America*. Stackpole Books, Harrisburg, PA, and Wildlife Management Institute, Washington, DC. 668 pp.

Thomas, J. W., R. M. Robinson, and R. G. Marburger. 1965. Social behavior in a white-tailed deer herd containing hypogonadal males. *Journal of Mammalogy* 46:314–327.

Toweill. D. E. 1978. Reciprocal forehead rub observed in mule deer. *Murrelet* 59:33–35.

Elk (*Cervus elaphus*)

Name

The first part of the scientific name, *Cervus*, comes from the Latin word *cervos* meaning "deer." The last part of the scientific name, *elaphus*, is from the Greek word *eláfi*, also meaning "deer." The common name is somewhat confusing because in Europe what we consider to be a moose is called an elk. Apparently, early settlers here thought the North American elk looked like the moose that lives in Europe (called an Eurasian elk), and thus named the North American animal an elk. Another common name, *wapiti*, may be more appropriate because it is derived from the Shawnee language meaning "white rump."

There is some disagreement concerning the species designation of elk inhabiting North America. For a long time elk were considered a separate species (*Cervus canadensis*) from the European red deer (*Cervus elaphus*). However, evidence of fertile crossbreeding between red deer and elk and their physical similarity led to the designation of both red deer and elk as subspecies of *Cervus elaphus*. In recent years, newer techniques that allow determination of relatedness by analyzing the molecular characteristics of DNA have led some researchers to conclude that elk and red deer should be considered separate species, and these scientists advocate restoring the elk's earlier name of *Cervus canadensis*.

Description

The elk is the second largest member of the deer family, the largest being the moose. It has many of the general characteristics of white-tailed deer, including having a thick body; long, slender legs; a long head; large ears; and males with antlers that are grown and shed annually. Elk antlers have a different size and conformation than those of white-tailed deer, being much larger and sweeping backward rather than forward. Antlers serve as an indicator of social rank and as weapons during breeding season contests between males. Antler size in elk increases with age up to 7 or 8 years and then plateaus; thereafter the antlers gradually decrease in size. Males cast antlers from late winter to early spring, depending upon age; older males will cast antlers earlier than young males.

Elk also differ from white-tailed deer in that a male elk is called a "bull" rather than "buck," a female elk is a "cow" rather than a "doe," and a young elk is a "calf" rather than a "fawn."

Color. Elk have two coats: a thick winter coat for insulation and a thinner summer coat. The overall coloration is tan, and both males and females have a dark

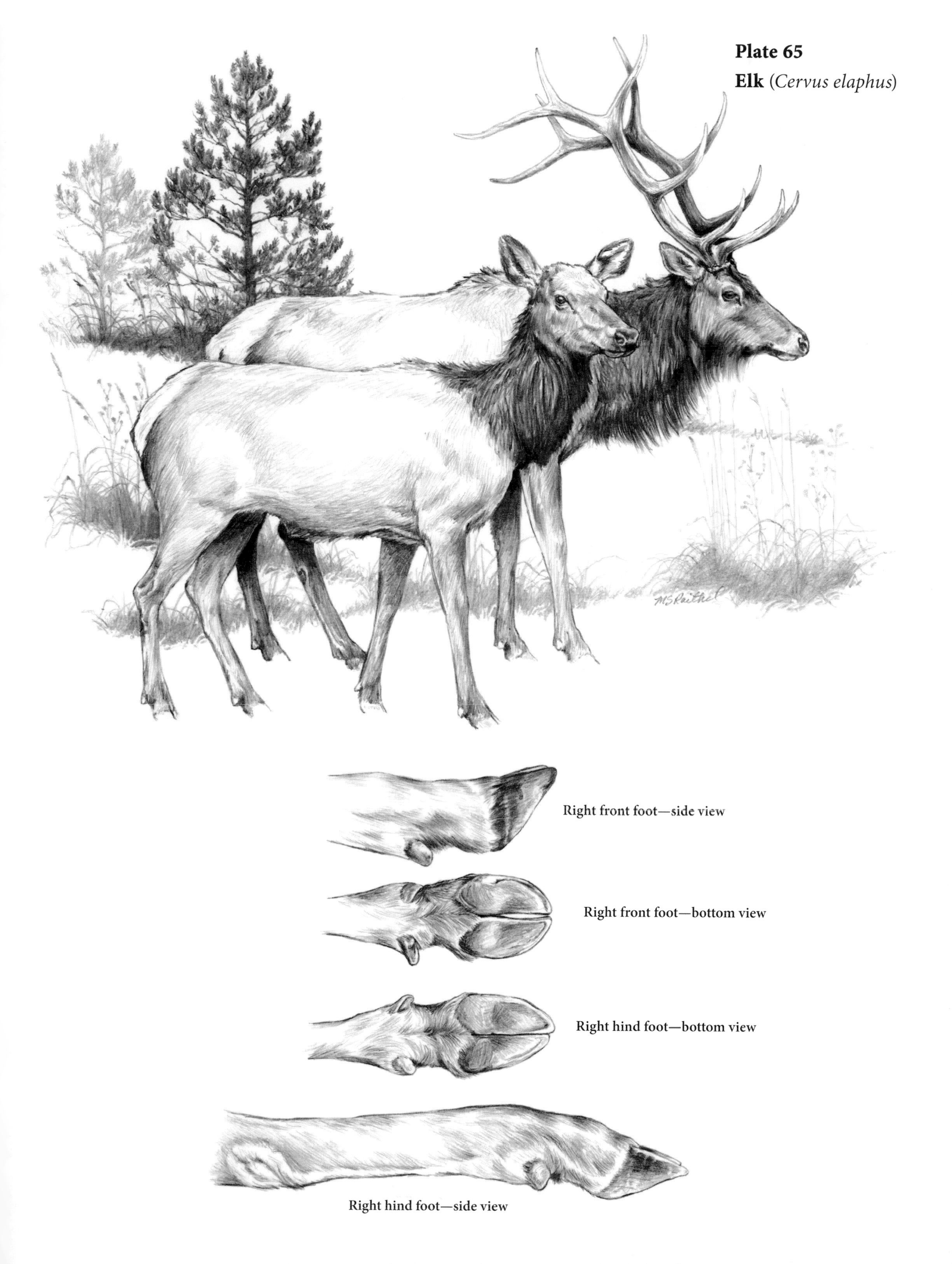

Plate 65
Elk (*Cervus elaphus*)

brown head, neck, legs, and belly. A long, dark, shaggy mane hangs from the neck to the chest and is heaviest during winter. Elk have a light-colored rump and light tan or straw-colored short tail. Newborn elk young are spotted much like white-tailed deer.

Measurements

Total length	214–242 cm	7–9 ft.
Tail	114–155 mm	4½–6¹⁄₁₀ in.
Hind foot	580–710 mm	23–28 in
Ear	184–229 mm	7¼–9 in.
Skull length	410–525 mm	16¹⁄₁₀–20⁷⁄₁₀ in.
Skull width	165–218 mm	6½–8⅗ in.
Weight	227–376 kg	500–830 lb.

Females are generally smaller than males.

Teeth and skull. The dental formula of the elk is:

$$I\,\frac{0}{3}\;C\,\frac{1}{1}\;P\,\frac{3}{3}\;M\,\frac{3}{3} = 34$$

The lack of upper incisors, which are replaced by a dental pad, and the zigzag enamel pattern of the premolars and molars are adaptations to a grazing and browsing diet. Elk of both sexes have two upper canines, called ivories, that may be evolutionary remnants of saberlike tusks used by their ancestors in combat; these rounded, nubby teeth are prized by hunters and are often used in jewelry. These upper canines have no feeding function but in males, which have larger upper canines, they may be used in threat displays. The presence of upper canine teeth, and length of the upper cheek tooth row being greater than 110 mm (4⅓ in.), distinguishes the elk's skull from the deer's skull.

Sex criteria and sex ratio. Male elk have antlers; females generally do not. Males and females can be identified by their sex organs.

In hunted elk populations there are more females than males because a higher proportion of the males are harvested. In unhunted populations the sex ratio may be closer to balanced, but higher natural mortality in males results in more females. Sex ratios at birth may vary annually and geographically but average nearly equal. The condition of the cow may influence fetal sex ratios. Cows in excellent condition with good body fat may produce a preponderance of female calves; when conditions are poor, more males may be produced.

Age criteria, age ratio, and longevity. Elk live 20 years or more in captivity but average 10 to 13 years in the wild. In some subspecies that suffer less predation, they may live 15 years in the wild.

Calf elk are recognizable until 3–5 months of age, when their spots are lost and a normal winter pelage is grown. Teeth provide the best measure of age. The sequence of eruption of permanent incisors and canines provides an indication of age until about 3 years of age. Although there is variability caused by nutrition, in general there is one pair of permanent incisors that erupts between 16 and 23 months, a second pair from 18–24 months, a third pair from 25–29 months, and the incisorform canines from 27–33 months. The upper permanent canines erupt from 15–19 months. Eruption and wear on premolars and molars also can be used to estimate age, with tooth replacement similar to that in white-tailed deer. After all adult teeth have erupted at around 2½ years of age, wear is used to estimate age. Aging from wear patterns, however, is difficult and can produce inaccurate results because the amount of wear depends on the elk's diet. The most accurate technique for aging older elk involves counting cementum layers in the two middle incisors.

Glands. Similar to white-tailed deer, elk have scent glands that are important in communication. The metatarsal gland along the outer part of the hind leg just below the hock and the "belly patch" around the penis are covered by colored hairs. The preorbital gland located in a hollow cavity in front of the eye at times produces large amounts of secretions that change in composition from season to season. Unlike white-tailed deer, elk do not have pedal glands.

Voice and sounds. Elk are vocal mammals issuing a variety of calls and sounds. The most well-known is the "bugle," done mostly by bulls during the breeding season. A bull produces a loud scream followed by a series of grunts, which tells other males to stay away and also attracts cows to his harem. Another common call, the "mew," is variable and may be given as a means of communicating among cows in a herd or as an interaction between cows and calves. The "bark" is an alarm call given when an elk or group of elk is disturbed. Another sound, "knuckle cracking," is produced by the front legs of elk when walking and is a means of maintaining contact when a herd is moving through heavy cover.

Distribution and Abundance

The eastern subspecies of elk once ranged throughout most of Missouri. Most accounts of elk before 1900 were from northern or western Missouri, although there is evidence that elk ranged throughout the Ozark Highlands and the southeastern corner of the state. With the arrival and settlement of Europeans, elk abundance and distribution gradually began to shrink, primarily due to overharvest and habitat

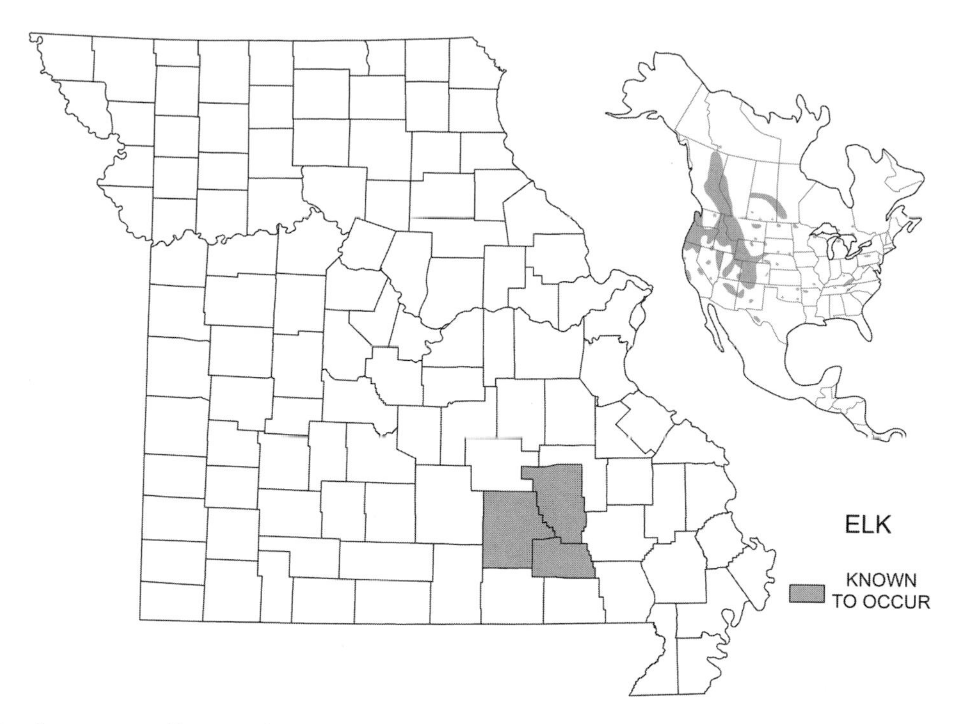

conversion. By the 1830s, elk were becoming scarce throughout Missouri, and the last reported elk was killed in 1886 in Texas County. This decline also occurred throughout the entire range of the eastern subspecies of elk, and by 1900 the subspecies was extinct.

Almost immediately following extinction of the eastern elk, attempts at restoring free-ranging elk populations within their former range began. In most of these attempts, the Rocky Mountain or Manitoban subspecies were used. These two subspecies turned out to be acceptable substitutes for the extinct eastern elk, as genetic variation among the subspecies is relatively low. Some of these early restoration attempts were successful, including Pennsylvania in 1913 and Michigan in 1915, but most failed due to the stresses associated with moving elk from their home area to their restoration area, low initial population sizes, and poaching. Elk restorations in eastern North America primarily after 1980 have been more successful, and there are currently free-ranging elk populations in Arkansas, Kansas, Kentucky, Michigan, Minnesota, Missouri, North Carolina, Oklahoma, Ontario, Pennsylvania, Tennessee, Virginia, and Wisconsin.

Efforts to return elk to Missouri began in the winter of 2010–2011, when 35 Rocky Mountain elk were captured in southeastern Kentucky, transported to the Ozark Highlands, and released at Peck Ranch Conservation Area in Carter and Shannon Counties. This area was identified as suitable for elk restoration due to the large tracts of state and federal lands, low human population density, low road density, and low row crop acreage. An additional 35 and 39 elk were captured in the winters of 2011–2012 and 2012–2013, respectively, and released at the same site. The goal for this restoration program is for 400 to 500 elk to range primarily throughout Shannon, Carter, and Reynolds Counties. The current Missouri elk population is between 110 and 120. In addition, a few wild elk immigrating into Missouri from other states or escaped captive elk have been spotted in different sections of Missouri in recent times.

The information presented below is based mostly on elk natural history across their North American range but also includes information learned about elk in Missouri since their reintroduction.

Habitat and Home

Elk live in a wide range of habitats across North America, from the desert mountain ranges of the Southwest to coniferous rainforests in the Pacific Northwest, and in general, prefer areas of open land for foraging near forested cover. They seek out habitats that provide adequate food and cover; escape from environmental extremes, such as northerly aspects to provide relief from heat during summer and southerly aspects during the winter; and hiding cover to escape from human disturbances and predators. During calving, cow elk typically select areas that provide hiding cover for the calf and nutritious forage to meet the demands of lactation.

Elk tend to avoid areas associated with human disturbance, such as roads and logging activity, although they can adapt to disturbances that occur regularly and are thus predictable, versus those human activities that occur irregularly and are unpredictable. Elk may establish traditional migratory and travel pathways that provide cover between forage patches. Movement paths tend to run parallel to stream drainages and ridgetops presumably to reduce energy expenditures associated with crossing rugged terrain.

The elk restoration zone in southern Missouri is within the Current River Hills subsection of the Ozark Highlands. This area is dominated by a mature mixed oak–hickory–shortleaf pine forest. Here, elk prefer forest and woodland units managed with fire over unburned areas.

Habits

Elk are social mammals that are typically found in groups, an adaptation believed to help avoid predation. During most of the year, adult males are separated from female and juvenile male groups in "bachelor" herds. However, adult males may also be solitary. Female groups may range from 2 to 30 or more individuals, and usually contain related females. A dominant older female will typically guide the groups' movements.

Home range sizes for elk are extremely variable and closely related to the amount of annual precipitation of the area they inhabit, presumably due to the influence of precipitation on forage production. Therefore, elk in eastern North America generally have smaller home range sizes than those in western North America, because eastern North America is generally wetter than western North America. Elk generally have larger winter home ranges relative to summer home ranges due to seasonal differences in food availability. Seasonal home range sizes of elk in Wisconsin range from 21 sq km (8⅕ sq. mi.) in summer to 28 sq km (11 sq. mi.) during winter. Initial home range sizes of elk in Missouri appear to be similar to those reported for Wisconsin. Home ranges for male elk are typically larger than for female elk.

In western North America, it is common for elk to make seasonal migrations between summer and winter ranges, presumably to meet nutritional requirements and to avoid areas of significant snow depths. Elk are typically found at lower elevations during the winter and move up in altitude with the progression of new plant growth associated with snowmelt. In eastern North America, the altitudinal gradient in most areas with elk populations is not as extreme and thus elk in this region rarely make large-scale seasonal migrations.

Foods

Elk consume roughly equal amounts of grasses, forbs (herbaceous nongrass plants), and woody browse such as twigs, bark, seedlings, saplings, and leaves, and they eat acorns in the fall when they are available. They are opportunistic feeders and their diet largely depends on the available flora of the region, but they will seek out areas with their preferred foods. The proportion of grass, forbs, and woody browse in the diet changes by season. In general, the forb component will be highest in the spring and summer and the grass and woody browse components will be highest in the fall and winter. Lichen are consumed in some parts of their range.

In southern Missouri, the newly restored elk rely heavily on deliberately planted food plots year-round.

Reproduction

Shortly before the fall rut, in late September through early October, male elk lose the velvet on their antlers and begin to compete for females. Males are sexually mature at 16 months, although young males do not usually mate until they are a few years old and can compete with more mature males.

In the elk breeding system, a dominant bull advertises his presence to attract a "harem" of cows (he will breed with each cow in his harem) and to discourage other bulls from intruding. Harems are usually made up of 1 bull and up to 6 females with their yearling calves, and are seasonal. Typically prime-age bulls (ages 5 to 10) hold harems, and younger and very old bulls remain on the outskirts of the group.

Fighting between bulls is actually rare during the breeding season because sparring just before the breeding season ensures a dominance hierarchy that is rarely crossed. Sparring allows bulls to learn and gauge the strength of an opponent by sizing up his strength and antlers. Thus, most harem-holding bulls can scare off opponents with threat displays instead of aggressive and costly clashes. Dominance fights occur most often between males with equal-sized antlers; these duels can be intense and may result in injury, exhaustion, or death.

During the breeding season males expend much energy guarding harems and may lose up to 17 percent of their body weight. Other breeding behaviors exhibited by males include rolling and urinating in dirt wallows, thrashing of antlers in vegetation, urinating on their stomachs and necks, and creating scrapes and rubs on trees and shrubs with their antlers.

Elk breed during the fall, and young are born during early summer at a time when food availability is optimal so the nutritional demands on the nursing cow can be met. Females, like males, are sexually mature at 16 months, but cow elk generally have their first young at 3 years of age although in some areas with excellent habitat and mild winters, cows may have young at the age of 2. During spring, pregnant female elk will typically leave groups and seek out calving habitat. After a 240- to 262-day gestation, cows give birth to a single calf; twins are rare. The birthing process may last as long as 2 hours, and the cow may give birth while standing or lying on the ground. At birth, calves weigh around 15 to 16 kg (33 to 35 lb.) and have creamy spots on their back and sides. Their hooves are soft. The calf is typically mobile 1 hour after birth.

The cow and her calf will live alone for several weeks. Newborn calves are hiders, remaining immobile and hidden with the cow visiting only to nurse and remove fecal material that could attract predators. Over a period of days, calves become increasingly mobile and can outrun many potential predators. At around 16 days the cow and calf will join the herd, and weaning is completed within 60 days.

In populations where bulls are heavily hunted, or where populations are new, as is the case early in Missouri's elk restoration program, only young bulls are available to do the breeding. Young bulls may be less efficient breeders, leading to a breeding season that extends over a longer period of time, with calves born throughout the summer rather than during a short period in early summer. This may result in poorer calf survival rates.

Some Adverse Factors

Elk, like other mammals, have a normal complement of parasites. *Parelaphostrongylus tenuis*, commonly known as the meningeal worm or brainworm, can be detrimental to Missouri elk. White-tailed deer are the natural hosts for this nematode parasite and typically harbor them without any signs of disease. However, in most other species within the cervid family, including elk, infection resulting from the parasite can result in neurologic problems and fatal disease. The parasite tends to affect yearling and younger elk more severely than the adults, but elk of any age may become afflicted. Degree of exposure to the parasite, age of the individual elk, and overall population health may all determine the effects meningeal worm parasitism has on elk populations.

Gray wolves, red wolves, and mountain lions historically preyed on elk in Missouri, but populations of these predators no longer occur here. American black bear are known to prey on elk calves, and as Missouri's bear population increases and expands its range, predation on elk may increase.

Missouri winters are mild compared to other areas where elk are adapted to living, and severe winter weather has limited impact on elk populations in Missouri. Elk sometimes can become injured in fences, from falls, and from other accidents, and they can be killed when they collide with vehicles.

In some cases elk become habituated to humans, such as along roadways popular with the elk-viewing public, and can cause injuries or damage vehicles as people try to view them.

Importance

Historically elk provided food, clothing, and shelter for people. Today, recreational elk viewing and tourism are the most important benefits of Missouri's restored population. Elk viewing is popular year-round but is especially so during the mid-September through late October rutting season when elk are bugling. As the herd expands, regulated hunting will become an important activity.

Conservation and Management

Elk restoration in Missouri involves repopulating a small portion of the southern Ozark Highlands, protecting these animals as their population increases and expands, and managing the land so that it meets their habitat needs. There are three general ways to improve the heavily forested Ozark habitat for elk and other wildlife: enhance and manage open ground for high-quality forage; use prescribed fire to restore glade-woodland complexes and stimulate wild grasses and other nutritious plants; and administer properly managed woodland harvest and thinnings that will provide nutritious woody browse and escape cover.

Elk will continue to retain a protected status until the population reaches a level that will support a limited, regulated hunting season. This will help maintain the population size and decrease the chances of elk-human conflicts.

SELECTED REFERENCES

See also discussion of this species in general references, page 23.

Altmann, M. 1952. Social behavior of elk, *Cervus canadensis nelsoni*, in the Jackson Hole area of Wyoming. *Behaviour* 4:116–143.

Anderson, D. P., J. D. Forester, M. G. Turner, J. L. Frair, E. H. Merrill, D. Fortin, J. S. Mao, and M. S. Boyce. 2005. Factors influencing female home range sizes in elk (*Cervus elaphus*) in North American landscapes. *Landscape Ecology* 20:257–271.

Barbknecht, A. E., W. S. Fairbanks, J. D. Rogerson, E. J. Maichak, B. M. Scurlock, and L. L. Meadows. 2011. Elk parturition site selection at local and landscape scales. *Journal of Wildlife Management* 75:646–654.

Beck, J. L., J. M. Peek, and E. K. Strand. 2006. Estimates of elk summer range nutritional carrying capacity constrained by probabilities of habitat selection. *Journal of Wildlife Management* 70:283–294.

Bowyer, R. T. 1981. Activity, movement, and distribution of Roosevelt elk during rut. *Journal of Mammalogy* 62:574–582.

Boyce, M. S. 1991. Migratory behavior and management of elk. *Applied Animal Behaviour Science* 29:239–250.

Cole, E. K., M. D. Pope, and R. G. Anthony. 1997. Effects of road management on movement and survival of Roosevelt elk. *Journal of Wildlife Management* 61:1115–1126.

Craighead, J. J., F. C. Craighead, R. L. Ruff, and B. W. O'Gara. 1973. Home ranges and activity patterns of nonmigratory elk of the Madison Drainage herd as determined by biotelemetry. *Wildlife Monographs* 33:3–50.

Cronin, M. A. 1992. Intraspecific variation in mitochondrial DNA of North American cervids. *Journal of Mammalogy* 73:70–82.

Czech, B. 1991. Elk behavior in response to human disturbance at Mount St. Helens National Volcanic Monument. *Applied Animal Behaviour Science* 29:269–277.

de Vos, A., P. Brokx, and V. Geist. 1967. A review of social behavior of the North American cervids during the reproductive period. *American Midland Naturalist* 77:390–417.

Feighny, J. A., K. E. Williamson, and J. A. Clarke. 2006. North American elk bugle vocalizations: Male and female bugle call structure and context. *Journal of Mammalogy* 87:1072–1077.

Frair, J. L., E. H. Merrill, D. R. Visscher, D. Fortin, H. L. Beyer, and J. M. Morales. 2005. Scales of movement by elk (*Cervus elaphus*) in response to heterogeneity in forage resources and predation risk. *Landscape Ecology* 20:273–287.

Franklin, W. L., A. S. Mossman, and M. Dole. 1975. Social organization and home range of Roosevelt elk. *Journal of Mammalogy* 56:102–118.

Grover, K. E., and M. J. Thompson. 1986. Factors influencing spring feeding site selection by elk in the Elkhorn Mountains, Montana. *Journal of Wildlife Management* 50:446–470.

Hofmann, R. R. 1988. Anatomy of the gastro-intestinal tract. In D. C. Church, ed., *The ruminant animal digestive physiology and nutrition*, pp. 14 43. Prentice Hall, Englewood Cliffs, NJ.

Irwin, L. L., and J. M. Peek. 1983. Elk habitat use relative to forest succession in Idaho. *Journal of Wildlife Management* 47:664–672.

Johnson, D. E. 1951. Biology of the elk calf, *Cervus canadensis nelsoni*. *Journal of Wildlife Management* 15:396–410.

Kie, J. G., A. A. Ager, and R. T. Bowyer. 2005. Landscape-level movements of North American elk (*Cervus elaphus*): Effects of habitat patch structure and topography. *Landscape Ecology* 20:289–300.

Lupardus, J. L. 2005. Seasonal forage availability and diet of reintroduced elk in the Cumberland Mountains, Tennessee. M.S. thesis, University of Tennessee, Knoxville. 98 pp.

McCorquodale, S. M. 2003. Sex-specific movements and habitat use by elk in the Cascade Range of Washington. *Journal of Wildlife Management* 67:729–741.

McKinley, D. 1960. The American elk in pioneer Missouri. *Missouri Historical Review* 54:356–365.

Millspaugh, J. J., K. J. Raedeke, G. C. Brundige, and C. C. Willmott. 1998. Summer bed sites of elk (*Cervus elaphus*) in the Black Hills, South Dakota: Considerations for thermal cover management. *American Midland Naturalist* 139:133–140.

Missouri Conservation Commission. 2010. *Elk restoration in Missouri*. Missouri Department of Conservation, Jefferson City. 24 pp.

Murie, O. L. 1932. Elk calls. *Journal of Mammalogy* 13:331–336.

———. 1951. *The elk of North America*. Stackpole, Harrisburg, PA. 376 pp.

Randi, E., N. Mucci, F. Claro-Hergueta, A. Bonnet, and E. J. Douzery. 2001. A mitochondrial DNA control region phylogeny of the Cervinae: Speciation in *Cervus* and implications for conservation. *Animal Conservation* 4:1–11.

Rearden, S. N., R. G. Anthony, and B. K. Johnson. 2011. Birth-site selection and predation risk of Rocky Mountain elk. *Journal of Mammalogy* 92:1118–1126.

Rowland, M. M., M. J. Wisdom, B. K. Johnson, and J. G. Kie. 2000. Elk distribution and modeling in relation to roads. *Journal of Wildlife Management* 64:672–684.

Sawyer, H., R. M. Nielson, F. G. Lindzey, L. Keith, J. H. Powell, and A. A. Abraham. 2007. Habitat selection of Rocky Mountain elk in a nonforested environment. *Journal of Wildlife Management* 71:868–874.

Toweill, D. E. and J. W. Thomas, eds. 2002. *North American elk: Ecology and management*. Smithsonian Institution Press, Washington, DC. 962 pp.

Witmer, G. W. 1990. Re-introduction of elk in the United States. *Journal of the Pennsylvania Academy of Sciences* 64:131–135.

10

Other Wild Mammals (Extirpated or Rare Visitors)

In addition to the 72 species of wild mammals established in Missouri, there are 8 other species of living wild mammals that will be briefly discussed.

One group includes 5 species (American bison, gray wolf, red wolf, white-tailed jackrabbit, and Brazilian free-tailed bat) that formerly occurred in Missouri within historic times but no longer do so. The second group includes 3 species (big free-tailed bat, Seminole bat, and nutria) that have been recently observed in Missouri but are not permanently established at present.

Sometimes domestic animals, like the hog or cat, successfully take up a wild existence, but such species are not considered here. However, the skulls of certain domestic animals are included in the keys elsewhere to avoid confusion with the skulls of wild species.

American Bison (*Bison bison*)

The American bison, commonly referred to as the American buffalo, once roamed the grasslands of North America in large herds. For the Native Americans, the 363 to 908 kg (800 to 2,000 lb.) bison was

a source of food, shelter, clothing, and utensils, and on the treeless prairie the dried fecal matter—buffalo chips—served as fuel. Bison became nearly extinct across their range mostly because of commercial hunting and slaughter in the nineteenth century. Many were killed for their hides and food, but many more were shot for sport or to deprive Native Americans of their important source for sustenance, forcing them to abandon their traditional homelands. Cattle competed with bison for prairie grass, and the plow further destroyed the prairie sod.

The bison was the largest mammal to occur in Missouri during historic times. Only vague accounts are

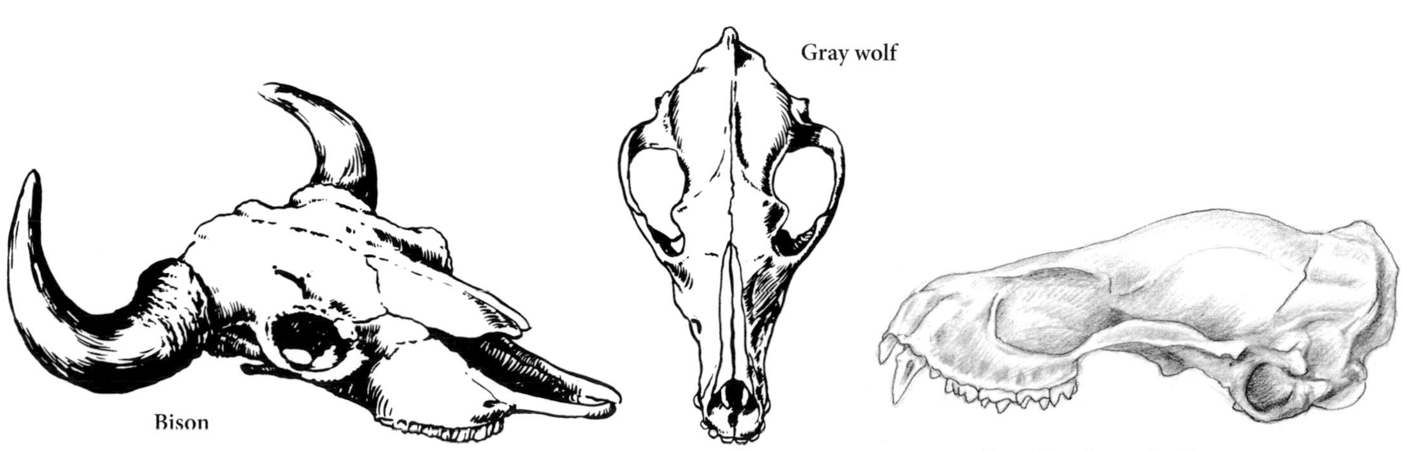

Bison

Gray wolf

Brazilian free-tailed bat

available concerning its former numbers and distribution here, but from these it is concluded that in the period immediately preceding European settlement it occurred sporadically and was never abundant. By 1840 only remnants of this magnificent animal were found in the northwestern and southeastern sections of the state, and these soon disappeared.

Today, there are wild and semiwild herds of bison living on private ranches, in our national parks, and on other state and federal lands in sufficient numbers to allow people to watch them and, in some cases, to permit a harvest. The bison is ranked as apparently secure across its North American range due to the many managed populations on public and private lands.

Gray Wolf (*Canis lupus*)

The gray wolf (also called the timber wolf) once ranged across the wilderness and remote areas of North America, including Missouri, Europe, Asia, and North Africa. The species is now extirpated from most of Western Europe, Africa, southern Asia, the United States, and Mexico. Although it is apparently secure globally, it is a federally endangered species in much of the United States south of Interstate 80.

With the settling of North America, the process of exterminating this 40 to 50 kg (90 to 110 lb.) carnivore proceeded rapidly. The gray wolf was disliked and hunted mostly because of its depredations on livestock and competition for big game. By 1900 this species was gone from most of the eastern and central United States, and by 1915 its population was greatly reduced in the remaining parts of its range in the western United States, Canada, and Alaska.

Today the gray wolf is regarded as an interesting and valuable part of our native wildlife population and has been reintroduced into many areas. Absent from the northern Rocky Mountains since the 1930s, gray wolves occasionally dispersed down from Canada. By 1979, they had reestablished themselves, and reproduction was documented in 1986. In 1995 and 1996, wolves from Canada were released into central Idaho and the Greater Yellowstone area. Research indicated high survival of these wolves during their colonization phase, and they were reproducing; no further releases occurred. Today, wolf mortality due to human threats remains high, and populations need to be monitored and adaptively managed.

The Mexican subspecies has been extinct in the wild for more than three decades, but following the reintroduction of a few wolves into their former range in the southwestern United States and Mexico, some now call Arizona and New Mexico home. The first litter of five Mexican gray wolf pups was discovered in Mexico in 2014.

The gray wolf is listed as extirpated from Missouri and surrounding states except Illinois, where it is listed as critically imperiled. Individuals occasionally wander into Missouri from other states, particularly Minnesota, Wisconsin, or Michigan. Since 2001, three gray wolves, probably from the Great Lakes states, have been confirmed in Missouri. The 2001 wolf was wearing a radio collar and an ear tag linking it to Michigan's Upper Peninsula, more than 600 miles away. Illinois has no self-sustaining populations or packs of gray wolves, but since 2000, it has had eleven confirmed wolves, most likely from Wisconsin.

Red Wolf (*Canis rufus*)

The red wolf is one of the world's most endangered canids, and it once called southern and eastern Missouri home. It is known for the characteristic reddish color of its fur, most apparent behind the ears and along the neck and legs; it is mostly brown and buff colored with some black along the back. It is intermediate in size between the coyote and gray wolf, weighing 20 to 36 kg (45 to 80 lb.). Its taxonomic status has been a controversy for many years; some believe it to be a hybrid of the gray wolf and coyote, while others believe it is a distinct species that has suffered from extensive hybridization with coyotes, and others believe it is a subspecies of the gray wolf. The U.S. Fish and Wildlife Service still considers the red wolf a distinct species.

The red wolf is native to the southeastern United States but was nearly driven to extinction by the mid-1900s due to intensive predator control programs, and degradation and alteration of its habitat. In 1950, a

small female taken in Taney County became the last red wolf on record in Missouri. By the late 1960s, the red wolf occurred only in small numbers in the Gulf Coast of western Louisiana and eastern Texas, and in 1967 it was designated as endangered. The red wolf was declared extinct in the wild in 1980, but a successful captive breeding program has resulted in this canid being successfully reintroduced into a small area in northeastern North Carolina.

White-tailed Jackrabbit
(*Lepus townsendii*)

The white-tailed jackrabbit formerly lived on the prairies and grasslands of northwestern Missouri. In the 1930s and 1940s, the species was reported to be fairly abundant in some areas, but numbers were decreasing mostly due to habitat loss and overharvesting. By 1959, it was reported as rare, and by the late 1960s, biologists reported that there were no known breeding populations of whitetails in Missouri. Both species of Missouri jackrabbits were declared endangered in 1971 and their hunting seasons were closed. The last known white-tailed jackrabbits in Missouri were one hit by a vehicle and killed in 1976 at the Harrison–Mercer county line, and one shot in Gentry County in 1980 or 1981. The species was officially declared as extirpated from Missouri in 1990.

White-tailed jackrabbits are considered globally secure within their current range in western and central North America as they are still fairly abundant with many healthy populations, despite experiencing some loss of habitat in the eastern part of their range.

White-tailed jackrabbits are easily distinguished from black-tailed jackrabbits by the characteristic color of their tails, their white winter pelage, and their shorter ears and smaller hind legs in proportion to the body.

Brazilian Free-tailed Bat
(*Tadarida brasiliensis*)

This bat is identified by its "free tail"; long, narrow wings; and the broad ears that almost meet in the midline but are not joined at the bases across the forehead. Body length of Brazilian free-tailed bats is 88 to 107 mm long (3½ to 4¼ in.) and weight is 7 to 14 g (¼ to ½

oz.). The upperparts are dark brown to grayish brown and the hairs are normally uniformly colored to the bases. The underparts are slightly paler. The dental formula of the Brazilian free-tailed bat is:

$$I\ \frac{1}{3}\ C\ \frac{1}{1}\ P\ \frac{2}{2}\ M\ \frac{3}{3} = 32$$

or

$$I\ \frac{1}{2}\ C\ \frac{1}{1}\ P\ \frac{2}{2}\ M\ \frac{3}{3} = 30$$

The range of the Brazilian free-tailed bat in North America lies primarily in the southern half of the United States and most of Mexico. Those bats in the central portion of this range (Arizona, New Mexico,

Texas, and Oklahoma) live in caves during the summer and then migrate southward to caves in Mexico where they spend the winter. In Missouri, the Brazilian free-tailed bat has been recorded in three different locations: Jackson, Johnson, and Phelps Counties. It is thought these bats probably belong to the migratory population. In such a highly mobile species it is not unlikely that individuals will come into Missouri from time to time, and this species may occur more commonly than is known. It is classified as an accidental species of conservation concern in Missouri and is secure across its North American range.

Big Free-tailed Bat
(*Nyctinomops macrotis*)

The big free-tailed bat is similar to the Brazilian free-tailed bat but is larger, being 107 to 133 mm (4½ to 5½ in.) in total length, and has its large ears joined at their bases across the forehead. The dental formula is:

$$I\,\frac{1}{2}\;C\,\frac{1}{1}\;P\,\frac{2}{2}\;M\,\frac{3}{3} = 30$$

This species has a wide range in the southwestern United States and in Mexico but is generally rare. It has been taken in Kansas, Nebraska, and Iowa and might occur occasionally in Missouri, particularly in the fall after the young have been weaned and the bats are dispersing.

This species was formerly known as *Tadarida macrotis*.

Seminole Bat (*Lasiurus seminolus*)

Seminole bats were long considered to be a subspecies of red bats. Both are of similar size and appearance and are easily confused with each other. The color of the fur is one feature used to distinguish these species: Seminole bats have rich, mahogany brown pelage with whitish tips, as opposed to the brick to rusty red of the red bat. The dental formula is:

$$I\,\frac{1}{3}\;C\,\frac{1}{1}\;P\,\frac{2}{2}\;M\,\frac{3}{3} = 32$$

Seminole bats are distributed widely across the southeastern United States. They prefer lowland wooded areas and roost in trees where Spanish moss

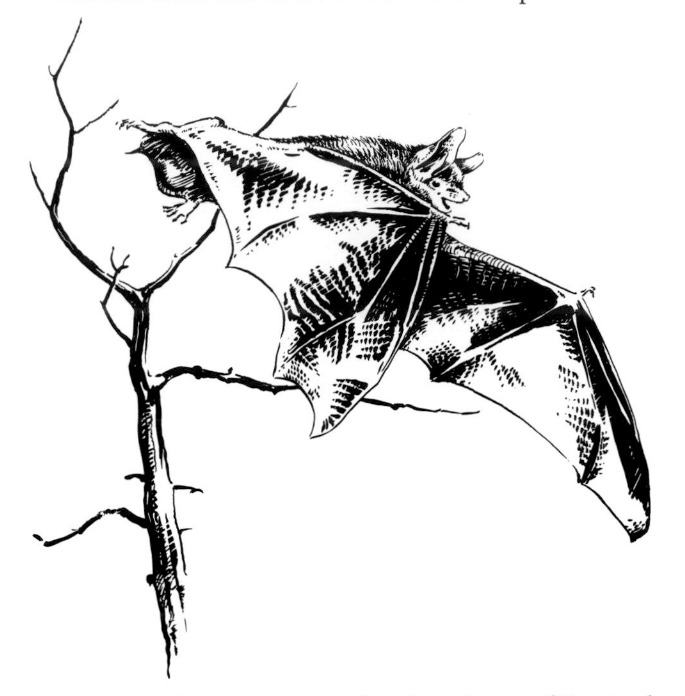

is present. During winter, the abundance of Seminole bats increases in the southern portion of their range and decreases in the northern portion. Bat biologists have captured a few Seminole bats in six southern Missouri counties (Newton, Pulaski, Reynolds, Shannon, Stoddard, and Wayne) beginning in 2006. The Seminole bat is classified as an accidental species of conservation concern in Missouri and is secure across its North American range.

Nutria (*Myocastor coypus*)

The nutria, or coypu, is a robust, overall darkish brown in color, semiaquatic rodent with a large head, small ears, small front legs having feet with unwebbed toes, large hind legs having feet with webbing between the first four toes, and a scaly, scantily haired round tail measuring 30 to 45 cm (12 to 18 in.) in length. Adults typically are 76 to 107 cm (30 and 42 in.) in total length and weigh between 5 and 9 kg (11 and 20 lb.). In 2011, a 4.7 kg (10.4 lb.) male became the official record-weight nutria confirmed by the Missouri Department of Conservation. The state furbearer record program was started in 2011.

The nutria resembles a large muskrat or a small beaver but can easily be distinguished by the round tail (the muskrat's tail is vertically flattened; the beaver's

Feet of nutria

No web between fourth and fifth toes

Left hind foot

Left front foot

tail is horizontally flattened). The tracks of the nutria are distinguished from the beaver's by the imprint of the small thumb on the nutria's front foot and by the lack of webbing between the fourth and fifth toes on the nutria's hind foot. The dental formula is the same as that of the beaver:

$$I\ \frac{1}{1}\ C\ \frac{0}{0}\ P\ \frac{1}{1}\ M\ \frac{3}{3} = 20$$

Nutria can breed year-round and are capable of producing many offspring. Litters average 4–5 young, but they can have up to 13 young per litter and 3 litters per year. Four pairs of mammary glands are located high on the sides of the body, allowing young nutria to nurse while the mother swims.

Nutria are native to South America. They were imported into several states beginning around 1900 for fur ranching. When fur prices collapsed in the early 1940s, animals were released or escaped into the wild where they quickly became established. In addition, nutria were intentionally introduced into the wild in several states to save ponds and lakes choked with vegetation. They are vegetarians and can consume 25 percent of their weight daily. Today, wild nutria populations occur in 18 states (being most numerous in the marshes of Louisiana), some Canadian provinces, and along the Rio Grande in Mexico. Nutria were also introduced and have become established in other countries.

Nutria are detrimental to the wetland ecosystems where they live as they destroy habitat and compete with native fauna for resources, especially muskrats. Their voracious appetite and ability to quickly overpopulate an area can lead to the rapid conversion of a productive marsh into open water. Nutria not only eat the desirable grasses, but also their roots. Once they destroy the root mat—the structure that holds the marsh together—erosion occurs, and more vegetation is lost. These newly created open water areas are called "eat outs." In addition, nutria burrow into banks of ponds, lakes, and levees, compromising these structures. They are also known to forage extensively on crops adjacent to their habitat. Some states are waging war on the nutria in an attempt to recover their wetlands.

The first report of nutria in Missouri was one taken by a trapper in Dunklin County in 1943, and since that time others have been reported, most recently in 2014 in Stoddard County. Missouri has no permanent nutria population as they do not endure Missouri's cold temperatures, but because of their widespread introduction in the United States, great adaptability, and changing climatic conditions, nutria could become permanent members of our mammalian fauna, especially in the southern portions of the state.

Wild nutria are harvested mostly in Louisiana and Texas, but their fur is not very dense and the market for their pelts is poor. Some people eat nutria.

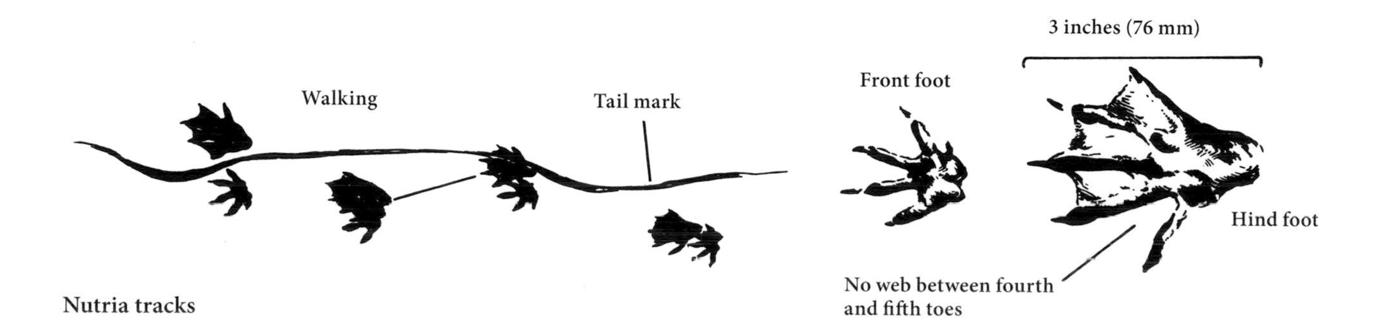

Nutria tracks

INDEX

T

U

V

W

Z

About the Authors

Charles W. Schwartz, who died in 1991, was with the Missouri Department of Conservation for forty years, serving as biologist, author, wildlife photographer, and wildlife artist. His illustrations for Aldo Leopold's *A Sand County Almanac* are particularly well known. Elizabeth R. Schwartz, who died in 2013, was employed with the Department of Conservation for over thirty years as biologist, author, and assistant in wildlife photography. Together, this husband-and-wife team wrote or illustrated thirteen other books and many technical papers for scientific journals and popular articles for magazines. They also produced some twenty-four motion pictures and numerous TV programs, which received both national and international awards. The Schwartzes also received recognition by the North American Wildlife Society for four of their technical publications and motion pictures.